Lecture Notes in Computer Science 2810

Edited by G. Goos, J. Hartmanis, and J. van Leeuwen

Springer
Berlin
Heidelberg
New York
Hong Kong
London
Milan
Paris
Tokyo

Michael R. Berthold Hans-Joachim Lenz
Elizabeth Bradley Rudolf Kruse
Christian Borgelt (Eds.)

Advances in Intelligent Data Analysis V

5th International Symposium on
Intelligent Data Analysis, IDA 2003
Berlin, Germany, August 28-30, 2003
Proceedings

Volume Editors

Michael R. Berthold
Data Analysis Research Lab, Tripos Inc.
South San Francisco, CA, USA
E-mail: berthold@tripos.com

Hans-Joachim Lenz
Freie Universität Berlin
Berlin, Germany
E-mail: ls-lenz@wiwiss.fu-berlin.de

Elizabeth Bradley
University of Colorado
Boulder, CO, USA
E-mail: liz@sogol.cs.colorado.edu

Rudolf Kruse
Christian Borgelt
Otto-von-Guericke-Universität Magdeburg
Magdeburg, Germany
E-mail: {kruse;borgelt}@iws.cs.uni-magdeburg.de

Cataloging-in-Publication Data applied for

A catalog record for this book is available from the Library of Congress.

Bibliographic information published by Die Deutsche Bibliothek
Die Deutsche Bibliothek lists this publication in the Deutsche Nationalbibliografie; detailed bibliographic data is available in the Internet at <http://dnb.ddb.de>.

CR Subject Classification (1998): H.3, I.2, G.3, I.5, I.4.5, J.2, J.3, J.1, I.6

ISSN 0302-9743
ISBN 3-540-40813-4 Springer-Verlag Berlin Heidelberg New York

Springer-Verlag Berlin Heidelberg New York
a member of BertelsmannSpringer Science+Business Media GmbH

http://www.springer.de

Printed in Germany

Typesetting: Camera-ready by author, data conversion by PTP-Berlin GmbH
Printed on acid-free paper SPIN: 10949962 06/3142 5 4 3 2 1 0

Preface

We are glad to present the proceedings of the 5th biennial conference in the Intelligent Data Analysis series. The conference took place in Berlin, Germany, August 28–30, 2003. IDA has by now clearly grown up. Started as a small side-symposium of a larger conference in 1995 in Baden-Baden (Germany) it quickly attracted more interest (both submission- and attendance-wise), and moved from London (1997) to Amsterdam (1999), and two years ago to Lisbon. Submission rates along with the ever improving quality of papers have enabled the organizers to assemble increasingly consistent and high-quality programs. This year we were again overwhelmed by yet another record-breaking submission rate of 180 papers. At the Program Chairs meeting we were – based on roughly 500 reviews – in the lucky position of carefully selecting 17 papers for oral and 42 for poster presentation. Poster presenters were given the opportunity to summarize their papers in 3-minute spotlight presentations. The oral, spotlight and poster presentations were then scheduled in a single-track, 2.5-day conference program, summarized in this book.

In accordance with the goal of IDA, "to bring together researchers from diverse disciplines," we achieved a nice balance of presentations from the more theoretical side (both statistics and computer science) as well as more application-oriented areas that illustrate how these techniques can be used in practice. Work presented in these proceedings ranges from theoretical contributions dealing, for example, with data cleaning and compression all the way to papers addressing practical problems in the areas of text classification and sales-rate predictions. A considerable number of papers also center around the currently so popular applications in bioinformatics.

In addition to the regular conference program we were delighted that Profs. Giacomo della Riccia (Univ. of Udine, Italy) and Lawrence O. Hall (Univ. of South Florida, USA) accepted our invitation to present keynote lectures. Also, for the first time in its history, the conference was preceded by three tutorial lectures: Nada Lavrač on "Rule Learning and Relational Subgroup Discovery," Alfred Inselberg on "Visualization & Data Mining," and Jean-Francois Boulicaut on "Constraint-Based Mining."

Organizing a conference such as this one is not possible without the assistance and support of many individuals. We are particularly grateful to Angelika Wnuk who was in charge of local arrangements. Christian Borgelt was the backbone of the electronic "stuff" (website, electronic review and submission system) all the way to the assembly of the final proceedings, and Andreas Nürnberger was tirelessly active in making sure announcements, reminders and conference calendars were always up to date. Lastly, thanks go to our Program Chairs, Elizabeth Bradley and Rudolf Kruse for all the work they put into assembling an exciting program! But, of course, this was only possible because of the help of the members of the Program Committee, many of whom also submitted papers and attended the conference.

July 2003

Hans-Joachim Lenz, Berlin, Germany
Michael R. Berthold, South San Francisco, USA

Organization

General Chairs	Michael R. Berthold Data Analysis Research Lab, Tripos Inc. South San Francisco, CA, USA `berthold@tripos.com`
	Hans-Joachim Lenz Free University of Berlin Berlin, Germany `ls-lenz@wiwiss.fu-berlin.de`
Program Chairs	Elizabeth Bradley University of Colorado Boulder, CO, USA `lizb@sogol.cs.colorado.edu`
	Rudolf Kruse Otto-von-Guericke-Univerity of Magdeburg Magdeburg, Germany `kruse@iws.cs.uni-magdeburg.de`
Tutorials Chair	Arno Siebes Utrecht University Utrecht, The Netherlands `siebes@cs.uu.nl`
Publication Chair and Webmaster	Christian Borgelt Otto-von-Guericke-Univerity of Magdeburg Magdeburg, Germany `borgelt@iws.cs.uni-magdeburg.de`
Publicity Chair	Andreas Nürnberger University of California Berkeley, CA, USA `anuernb@eecs.berkeley.edu`
Local Arrangements	Angelika Wnuk Free University of Berlin Berlin, Germany `wnuk@wiwiss.fu-berlin.de`

Program Committee

Riccardo Bellazzi (IT)
Michael Berthold (US)
Hans-Georg Beyer (DE)
Hans-Hermann Bock (DE)
Klemens Böhm (DE)
Christian Borgelt (DE)
Pavel Brazdil (PT)
Gerhard Brewka (DE)
Herbert Büning (DE)
Hans-Dieter Burkhard (DE)
Luis M. de Campos (ES)
Akmal Chaudhri (UK)
Robert D. Clark (US)
Paul Cohen (US)
Fabio Crestani (UK)
Francesco d'Ovidio (DK)
Matthew Easley (US)
Fazel Famili (CA)
Giuseppe Di Fatta (IT)
Fridtjof Feldbusch (DE)
Ingrid Fischer (DE)
Douglas Fisher (US)
Peter Flach (UK)
Eibe Frank (NZ)
Karl A. Fröschl (AT)
Alex Gammerman (UK)
Jörg Gebhardt (DE)
Günter Görz (DE)
Adolf Grauel (DE)
Gabriela Guimaraes (PT)
Christopher Habel (DE)
Lawrence O. Hall (US)
Alexander Hinneburg (DE)
Frank Hoffmann (SE)
Adele Howe (US)
Frank Höppner (DE)
Klaus-Peter Huber (DE)
Anthony Hunter (UK)
Alfred Inselberg (IL)
Christian Jacob (CA)
Finn Jensen (DK)
Thorsten Joachims (US)
Daniel A. Keim (DE)
Gabriele Kern-Isberner (DE)
Frank Klawonn (DE)
Aljoscha Klose (DE)
Willi Klösgen (DE)
Alois Knoll (DE)
Joost N. Kok (NL)
Bernd Korn (DE)
Paul J. Krause (UK)
Jens Peter Kreiss (DE)
Mark-A. Krogel (DE)
Rudolf Kruse (DE)
Gerhard Lakemeyer (DE)
Carsten Lanquillon (DE)
Nada Lavrac (SI)
Hans Lenz (DE)
Ramon Lopez de Mantaras (ES)
Bing Liu (US)
Xiaohui Liu (UK)
Rainer Malaka (DE)
Luis Magdalena (ES)
Hans-Peter Mallot (DE)
Bernd Michaelis (DE)
Kai Michels (DE)
Ralf Mikut (DE)
Katharina Morik (DE)
Fernando Moura Pires (PT)
Gholamreza Nakhaeizadeh (DE)
Susana Nascimento (PT)
Olfa Nasraoui (US)
Wolfgang Näther (DE)
Detlef Nauck (UK)
Bernhard Nebel (DE)
Andreas Nürnberger (DE)
Tim Oates (US)
Christian Och (US)
Rick Osborne (US)
Gerhard Paaß (DE)
Simon Parsons (US)
James F. Peters (CA)
Peter Protzel (DE)
Marco Ramoni (US)
Giacomo Della Riccia (IT)
Michael M. Richter (DE)
Martin Riedmiller (DE)
Helge Ritter (DE)

Egmar Rödel (DE)
Raul Rojas (DE)
Dietmar Rösner (DE)
Thomas Runkler (DE)
Gunter Saake (DE)
Tobias Scheffer (DE)
Eckehard Schnieder (DE)
Wolfgang Schröder-Preikschat (DE)
Romano Scozzafava (IT)
Paola Sebastiani (US)
Kaisa Sere (FI)
Roberta Siciliano (IT)
Stefan Siekmann (DE)
Rosaria Silipo (US)
Philippe Smets (BE)
Myra Spiliopoulou (DE)
Martin Spott (UK)
Gerd Stumme (DE)
Thomas Strothotte (DE)
Fay Sudweeks (AU)
Klaus Tönnies (DE)
Shusaku Tsumoto (JP)
Antony Unwin (DE)
Reinhard Viertl (AT)
Jörg Walter (DE)
Matthew O. Ward (US)
Richard Weber (CL)
Gerhard Widmer (AT)
Olaf Wolkenhauer (UK)
Stefan Wrobel (DE)
Fritz Wysotzki (DE)
Andreas Zell (DE)

Table of Contents

Machine Learning

Probability and Topology

Classification and Pattern Recognition

Clustering

Applications

Modeling

(Data) Preprocessing

Pruning for Monotone Classification Trees

Ad Feelders and Martijn Pardoel

Utrecht University, Institute of Information and Computing Sciences,
P.O. Box 80089, 3508TB Utrecht, The Netherlands,
{ad,mapardoe}@cs.uu.nl

Abstract. For classification problems with ordinal attributes very often the class attribute should increase with each or some of the explanatory attributes. These are called classification problems with monotonicity constraints. Standard classification tree algorithms such as CART or C4.5 are not guaranteed to produce monotone trees, even if the data set is completely monotone. We look at pruning based methods to build monotone classification trees from monotone as well as nonmonotone data sets. We develop a number of fixing methods, that make a non-monotone tree monotone by additional pruning steps. These fixing methods can be combined with existing pruning techniques to obtain a sequence of monotone trees. The performance of the new algorithms is evaluated through experimental studies on artificial as well as real life data sets. We conclude that the monotone trees have a slightly better predictive performance and are considerably smaller than trees constructed by the standard algorithm.

1 Introduction

A common form of prior knowledge in data analysis concerns the monotonicity of relations between the dependent and explanatory variables. For example, economic theory would state that, all else equal, people tend to buy less of a product if its price increases. The precise functional form of this relationship is however not specified by theory.

Monotonicity may also be an important model requirement with a view toward explaining and justifying decisions, such as acceptance/rejection decisions. Consider for example a university admission procedure where candidate a scores at least as good on all admission criteria as candidate b, but a is refused whereas b is admitted. Such a nonmonotone admission rule would clearly be unacceptable. Similar considerations apply to selection procedures for applicants for e.g. a job or a loan.

Because the monotonicity constraint is quite common in practice, many data analysis techniques have been adapted to be able to handle such constraints. In this paper we look at pruning based methods to build monotone classification trees from monotone as well as nonmonotone data sets. We develop a number of fixing methods, that make a non-monotone tree monotone by further pruning steps. These fixing methods can be combined with existing pruning techniques

M.R. Berthold et al. (Eds.): IDA 2003, LNCS 2810, pp. 1–12, 2003.

to obtain a sequence of monotone trees. The performance of the new algorithms is evaluated through experimental studies on artificial as well as real life data sets.

This paper is organised as follows. In section 2 we define monotone classification and other important concepts that are used throughout the paper. In section 3 we shortly review previous work in the area of monotone classification trees. Then we discuss a number of methods to make a non-monotone tree monotone by additional pruning steps in section 4. These methods are evaluated experimentally in section 5 on real life and artificial data sets. Finally, in section 6 we end with a summary, and some ideas for further research.

2 Monotone Classification

In this section we define the problem of monotone classification and introduce some notation that will be used throughout the paper. Notation is largely based on [15].

Let $\mathcal{X}$ be a *feature space* $\mathcal{X} = \mathcal{X}_1 \times \mathcal{X}_2 \times \ldots \times \mathcal{X}_p$ consisting of vectors $\mathbf{x} = (x_1, x_2, \ldots, x_p)$ of values on p features or attributes. We assume that each feature takes values x_i in a linearly ordered set $\mathcal{X}_i$. The partial ordering $\leq$ on $\mathcal{X}$ will be the ordering induced by the order relations of its coordinates $\mathcal{X}_i$: $\mathbf{x} = (x_1, x_2, \ldots, x_p) \leq \mathbf{x}' = (x'_1, x'_2, \ldots, x'_p)$ if and only if $x_i \leq x'_i$ for all i. Furthermore, let $\mathcal{C}$ be a finite linearly ordered set of *classes*.

A *monotone* classification rule is a function $f : \mathcal{X} \to \mathcal{C}$ for which

$$\mathbf{x} \leq \mathbf{x}' \Rightarrow f(\mathbf{x}) \leq f(\mathbf{x}')$$

for all instances $\mathbf{x}, \mathbf{x}' \in \mathcal{X}$. It is easy to see that a classification rule on a feature space is monotone if and only if it is non-decreasing in each of its features, when the remaining features are held fixed.

A *data set* is a series $(\mathbf{x}_1, c_1), (\mathbf{x}_2, c_2), \ldots, (\mathbf{x}_n, c_n)$ of n examples $(\mathbf{x}_i, c_i)$ where each $\mathbf{x}_i$ is an element of the instance space $\mathcal{X}$ and c_i is a class label from $\mathcal{C}$. We call a dataset *consistent* if for all i, j we have $\mathbf{x}_i = \mathbf{x}_j \Rightarrow c_i = c_j$. That is, each instance in the data set has a unique associated class. A data set is *monotone* if for all i, j we have $\mathbf{x}_i \leq \mathbf{x}_j \Rightarrow c_i \leq c_j$. It is easy to see that a monotone data set is necessarily consistent.

The classification rules we consider are univariate binary classification trees. For such trees, at each node a split is made using a test of the form $X_i < d$ for some $d \in \mathcal{X}_i, 1 \leq i \leq p$. Thus, for a binary tree, in each node the associated set $T \subset \mathcal{X}$ is split into the two subsets $T_\ell = \{\mathbf{x} \in T : x_i < d\}$ and $T_r = \{\mathbf{x} \in T : x_i \geq d\}$. The classification rule that is induced by a decision tree $\mathcal{T}$ will be denoted by $f_\mathcal{T}$.

For any node or leaf T of $\mathcal{T}$, the subset of the instance space associated with that node can be written

$$T = \{\mathbf{x} \in \mathcal{X} : \mathbf{a} \leq \mathbf{x} < \mathbf{b}\} = [\mathbf{a}, \mathbf{b}) \tag{1}$$

for some $\mathbf{a}, \mathbf{b} \in \overline{\mathcal{X}}$ with $\mathbf{a} \leq \mathbf{b}$. Here $\overline{\mathcal{X}}$ denotes the extension of $\mathcal{X}$ with infinity-elements $-\infty$ and ∞. In some cases we need the infinity elements so we can specify a node as in equation (1).

Below we will call $\min(T) = \mathbf{a}$ the *minimal element* and $\max(T) = \mathbf{b}$ the *maximal element* of T. Together, we call these the *corner* elements of the node T. There is a straightforward manner to test the monotonicity using the maximal and minimal elements of the leaves of the classification tree [15]:

for all pairs of leaves T, T':
  if $\left(f_{\mathcal{T}}(T) > f_{\mathcal{T}}(T') \text{ and } \min(T) < \max(T')\right)$ **or**
    $\left(f_{\mathcal{T}}(T) < f_{\mathcal{T}}(T') \text{ and } \max(T) > \min(T')\right)$
  then stop: $\mathcal{T}$ not monotone

Fig. 1. Algorithm to test the monotonicity of a classification tree

3 Previous Research

In this section we briefly describe previous work on monotone classification trees. Ben-David [1] was the first to incorporate the monotonicity constraint in classification tree construction. He proposed a measure for the quality of a candidate split that takes into account both the impurity reduction that the split achieves, as well as the degree of non-monotonicity of the resulting tree. The degree of non-monotonicity is derived from the number of nonmonotonic leaf pairs of the tree that is obtained after the candidate split is performed. The relative weight of impurity reduction and degree of non-monotonicity in computing the quality of a candidate split can be determined by the user. Ben-David shows that his algorithm results in trees with a lower degree of nonmonotonicity, without a significant deterioration of the inductive accuracy.

Makino et al. [10] were the first to develop an algorithm that guarantees a monotone classification tree. However, their method only works in the two class setting, and requires monotone training data. Potharst and Bioch [11,12] extended this method to the k-class problem, also accommodating continuous attributes. Both Makino as well as Potharst and Bioch enforce monotonicity by adding the corner elements of a node with an appropriate class label to the existing data whenever necessary. While their results are encouraging, applicability of the algorithms is limited by the requirement of monotone training data. In actual practice this is hardly ever the case.

Bioch and Popova [4] extend the work of Potharst to nonmonotone data sets by relabeling data if necessary. While their approach is certainly interesting, a potential disadvantage is that the algorithm may have to add very many data points to the initial training sample. This may result in highly complex trees that have to be pruned back afterwards. Bioch and Popova, in another recent paper [5], perform some experiments with an alternative splitting criterion that was proposed by Cao-Van and De Baets [6]. This *number of conflicts* measure chooses the split with the least number of inconsistencies or conflicts, i.e. the number of non-monotone pairs of points in the resulting branches. An important conclusion of Bioch and Popova [5] is that with the increase of monotonicity noise in the data the number of conflicts criterion generates larger trees with a lower misclassification rate than the entropy splitting criterion.

Feelders [8] argues that the use of a measure of monotonicity in determining the best splits has certain drawbacks, for example that a nonmonotone tree may become monotone after additional splits. He proposes not to enforce monotonicity during tree construction, but to use resampling to generate many different trees and select the ones that are monotone. This allows the use of a slightly modified standard tree algorithm. Feelders concludes that the predictive performance of the monotone trees found with this method is comparable to the performance of the nonmonotone trees. Furthermore, the monotone trees were much simpler and proved to be more stable.

Daniels and Velikova [7] present an algorithm to make the training sample monotone by making minimal adjustments to the class labels. They combine this idea with the resampling approach described above. Their results show that the predictive performance of the trees constructed on the cleaned data is superior to the predictive performance of the trees constructed on the raw data.

Finally, Potharst and Feelders [13] give a short survey of monotone classification trees.

4 Pruning Towards Monotone Trees

The objective of our work is to develop an algorithm that produces a monotone tree with good predictive performance from both monotone and nonmonotone training data. In order to construct a monotone tree, we initially grow a large overfitted tree, and then prune towards monotone subtrees. For growing the initial tree we implemented an algorithm that is very similar to CART [2]; the only major difference is that our implementation records the corner elements of each node during tree construction.

Central to our approach are a number of so-called fixing methods. The basic idea of all our fixing methods is to make minimal adjustments to a nonmonotone tree in order to obtain a monotone subtree. To quantify the degree of nonmonotonicity, we count the number of leaf pairs that are nonmonotone with respect to each other. If a leaf participates in one or more nonmonotone leaf pairs, we call it a nonmonotone leaf. Hence the goal of all fixing methods is to obtain a tree without any nonmonotone leaves. They do so by pruning in a parent node that has at least one nonmonotone leaf for a child; this is repeated until the resulting tree is monotone. The fixing methods differ in the heuristic that is used to select the parent node to be pruned.

In figure 2 we have depicted a nonmonotone tree that we use to illustrate the different heuristics. Each node is numbered (the italicized numbers to the left of each node); the associated split is to the right of each node; and inside each node the classlabel is given. A node is always allocated to the class with the highest relative frequency in that node.

Below each leaf node we have listed the number of observations in that leaf. As the reader can verify, the nonmonotone leaf pairs of this tree are [9,10], [9,12], [9,14], [11,12], [11,14] and [13,14]. For example, the leaf pair [9,10] is nonmonotone because $f_{\mathcal{T}}(9) > f_{\mathcal{T}}(10)$ and $\min(9) < \max(10)$, that is

$$(-\infty, -\infty, 0, -\infty, -\infty) < (0, \infty, \infty, 0, \infty).$$

The leaf pair [8,13] is monotone however, since $f_{\mathcal{T}}(13) > f_{\mathcal{T}}(8)$ but $\min(13) \not< \max(8)$.

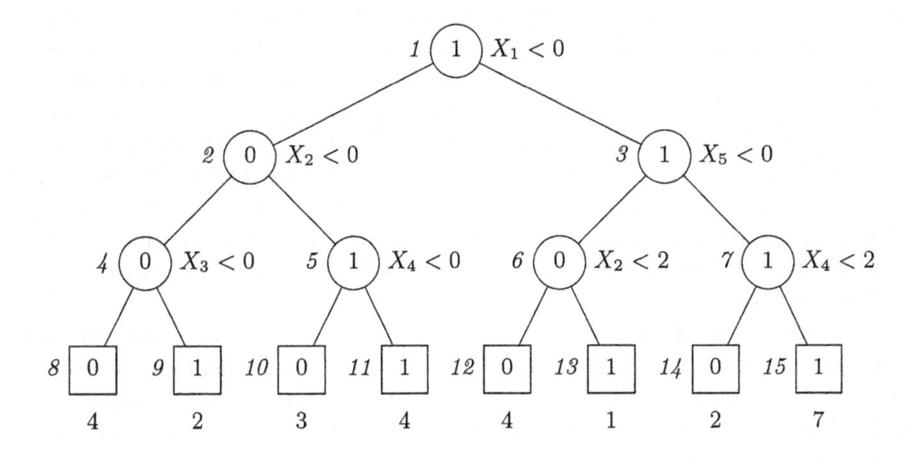

Fig. 2. Example nonmonotone decision tree

The first method prunes in the parent whose children participate in the most nonmonotone leaf pairs. Note that it may be the case that only one of the two children is a leaf node. In that case we only count the number of nonmonotone leaf pairs in which this single child participates. This way of counting aims to discourage pruning in nodes higher up in the tree. We call this heuristic the most nonmonotone parent (MNP) fixmethod. In our example we first count for each parent of a leaf the number of nonmonotonic leaf pairs its children participate in. For node 4 this number is 3 ([9,10],[9,12],[9,14]), node 5 also gives 3 ([9,10], [11,12], [11,14]), and for node 6 and 7 this number is 3 as well. We now face the problem what to do in case of a tie. Apart from choosing one of the equally good fixes at random, we will also consider the alternative of choosing the node with the least number of observations. Henceforth, this rule will be referred to as the LNO tie breaker. The rationale behind this rule is that if the data generating process is indeed monotone, then nonmonotonicity is likely to be due to small erratic leaf nodes. Application of this rule leads to pruning in node 6 since it contains the least observations (5) of the candidates.

A potential disadvantage of the MNP heuristic is that it ignores the fact that after pruning in a node, new nonmonotone leaf pairs, involving the newly created leaf, may appear. Therefore we tested a second method that prunes in the parent of a nonmonotone leaf that gives the biggest reduction in the number of nonmonotone leaf pairs. To prevent pruning too high in the tree too early, we do this levelwise, i.e. we first consider the set of parent nodes with minimal number of descendants for an improvement before we consider parent nodes with higher numbers of descendants. This fixmethod is called the best fix (BF). In our example we first note that all parent nodes have the same number of descendants and consequently will all be considered at the same time. Now, we need to know how many nonmonotone leaf pairs remain after each possible prune action, thus how big the improvement is. To determine this, we need to know with which leaves the candidate leaf nodes 4, 5, 6, and 7 are nonmonotone. These are the pairs [5,12], [5,14], [6,9] and [6,11]; nodes 4 and 7 are monotone with all other leaves, as the reader can verify. We can now compute the number of nonmonotone leaf pairs left after each possible prune. If we prune in node 4 we lose [9,10], [9,12] and [9,14], and are left with [11,12], [11,14] and [13,14], i.e. 3 pairs. If we prune in node 5 we lose [9,10], [11,12] and [11,14], but get [5,12] and [5,14] and still have [9,12], [9,14] and [13,14], i.e. 5 pairs. If we prune in node 6 we are left with 5 pairs and if we prune in node 7 only 3 pairs remain. Thus, we have two best fixes, namely nodes 4 and 7, and node 4 is selected by the LNO tie breaker.

The fixing methods discussed can be combined in different ways with existing pruning techniques. We study how they can be combined with cost-complexity pruning ([2]). An elementary cost-complexity pruning step works as follows. Let $R(\mathcal{T})$ denote the fraction of cases in the training sample that are misclassified by tree $\mathcal{T}$. The total cost $C_\alpha(\mathcal{T})$ of tree $\mathcal{T}$ is defined as

$$C_\alpha(\mathcal{T}) = R(\mathcal{T}) + \alpha|\tilde{\mathcal{T}}|. \quad (2)$$

The total cost of tree $\mathcal{T}$ consists of two components: resubstitution error $R(\mathcal{T})$, and a penalty for the complexity of the tree $\alpha|\tilde{\mathcal{T}}|$. In this expression $\tilde{\mathcal{T}}$ denotes the set of leaf nodes of $\mathcal{T}$, and α is the parameter that determines the complexity penalty. Depending on the value of α (≥ 0) a complex tree that makes no errors may now have a higher total cost than a small tree that makes a number of errors. When considering a tree $\mathcal{T}$, for each node T we can compute the value of α for which the tree obtained by pruning $\mathcal{T}$ in T has equal cost to $\mathcal{T}$ itself. By pruning in the node for which this α-value is minimal, we obtain the tree that is the first to become better than $\mathcal{T}$, when α is increased.

Now, the first way to combine pruning and fixing is to *alternate* between cost-complexity pruning steps and fixing steps. This proceeds as follows. If the initial tree is monotone we add it as the first tree in the sequence; otherwise we first fix it, i.e. make it monotone. Once we have a monotone tree, a cost-complexity pruning step is performed. Subsequently, we check again if the resulting tree is monotone, add it to the sequence if it is, or first fix it otherwise. We continue alternating the pruning and fixing steps until we eventually reach the root node.

A second possibility is to simply take the sequence of trees produced by cost-complexity pruning, and apply the fixing methods to all trees in that sequence. Finally, one could simply take the best tree (in terms of predictive accuracy on a test sample) from the cost-complexity sequence and fix that tree. In the experiments we discuss in the next section we only consider the strategy that alternates between pruning and fixing.

5 Experimental Evaluation

In this section we discuss the experimental results we obtained on real life data sets and artificially generated data. For the real life data (see table 1) we selected a number of data sets from domains where monotonicity is either a plausible assumption or requirement. The *bankruptcy* data set was used by Pompe in [14]. The class label indicates whether or not a company has gone bankrupt within one year of the annual report from which the attributes were taken. These attributes are 11 financial ratios that are indicative for the profitability, activity, liquidity and solvency of the company. The *Windsor housing* data set was used as an example by Koop [9]. It contains data on 546 houses sold in Windsor, Canada. The dependent variable is the sales price of the house in Canadian dollars. This was converted into a class label by taking the quartiles of the sales price.

All other data sets, *adult income, Australian credit approval, Boston housing* and *cpu performance* were taken from the UCI machine learning repository [3]. Whenever necessary, attributes with no monotonicity relation with respect to the class label were removed from the data sets. Also, some negative monotone attributes were inverted to make them positive monotone. For example, in the cpu performance data set the machine cycle time in nanoseconds was converted to clock speed in kilohertz. The class labels of Boston housing and cpu performance were created in the same way as for Windsor housing.

Table 1 gives a number of important characteristics of the data sets. The column *Error* gives the misclassification error of the rule that assigns every observation to the majority class. With regard to this study, the degree of non-monotonicity of the data sets is of special interest. This can be quantified in various ways. In the column *NmDeg*, for degree of nonmonotonicity, we give the ratio of the number of nonmonotone pairs of datapoints to the total number of pairs. However, we require a measure that allows the meaningful comparison of the degree of nonmonotonicity for data sets with highly different characteristics. Important characteristics in this regard are: the number of attributes and class labels, and the fraction of incomparable pairs of points (*FracInc* in table 1). To this end we take the ratio of the observed number of nonmonotone pairs to the number of nonmonotone pairs that would be expected if the class labels were assigned randomly to the observed attribute vectors. This last number is approximated by taking fifty permutations of the observed class labels and computing the average number of nonmonotone pairs of the resulting fifty data sets. The resulting number is given in table 1 in the column *NmRat*. If the ratio is close to zero, then the observed number of nonmonotone pairs is much lower than would be expected by chance.

Table 1. Overview of real life data sets and their characteristics

Description	Classes	Size	Attr.	Error	FracInc	NmDeg	NmRat
Adult Income	2	1000	4	0.500	0.610	0.020	0.21
Aus. Credit Approval	2	690	4	0.445	0.284	0.034	0.19
Boston Housing	4	506	6	0.745	0.678	0.006	0.05
CPU Performance	4	209	6	0.732	0.505	0.007	0.04
Windsor Housing	4	546	11	0.747	0.726	0.009	0.08
Bankruptcy	2	1090	11	0.500	0.877	0.003	0.08

Table 2. Misclassification rate averaged over twenty trials; MNP = Most Nonmonotone Parent, BF = Best Fix, BF-LNO = Best Fix with Least Number of Observations tie breaker

Dataset	Standard	Default	MNP	BF	BF-LNO
AdInc	0.263	0.263	0.260	0.261	0.260
AusCr	0.146	0.146	0.147	0.146	0.146
BosHou	0.331	0.332	0.335	0.336	0.330
CpuPerf	0.394	0.406	0.391	0.388	0.388
DenBHou	0.181	0.191	0.190	0.181	0.181
WindHou	0.531	0.534	0.534	0.533	0.536
Bankr	0.225	0.224	0.223	0.223	0.223
Total	2.071	2.096	2.080	2.068	2.064

For all data sets we follow the same procedure. The data sets are randomly partitioned into a training set (half of the data), a test set (a quarter of the data) and a validation set (also a quarter of the data). This is done twenty times. Each of those times the training set is used to build a tree sequence with the different algorithms considered, and for each algorithm the tree with the lowest misclassification rate on the test set is selected. Subsequently, we test for statistically significant differences in predictive accuracy between the methods using the validation set. The results of these experiments are reported in table 2 and table 3.

The column labeled *Standard* contains the results of a standard CART-like algorithm that we implemented. It uses cost-complexity pruning to create a sequence of trees from which the one with the smallest error on the test set is selected. The resulting trees may be either monotone or nonmonotone. The

Table 3. Tree size averaged over twenty trials; MNP = Most Nonmonotone Parent, BF = Best Fix, BF-LNO = Best Fix with Least Number of Observations tie breaker

Dataset	Standard	Default	MNP	BF	BF-LNO
AdInc	16.40	10.00	13.00	13.95	14.15
AusCr	4.05	3.55	3.95	4.6	4.60
BosHou	13.55	7.65	9.60	10.85	10.65
CpuPerf	15.80	7.40	10.35	10.50	10.50
DenBHou	4.80	2.55	3.75	3.65	3.65
WindHou	44.00	13.15	18.85	19.70	20.10
Bankr	9.15	7.10	8.45	6.65	6.75
Total	107.75	51.40	67.95	69.90	70.40

column *Default* was computed by taking the best monotone tree from the tree sequence produced by the standard algorithm. The last three columns contain the results of the different fixing methods discussed in section 4 combined with the alternating pruning strategy.

From table 2 we conclude that the BF-LNO method performs best, even better than the standard method. However, the differences are not statistically significant: in a total of 140 experiments (20 per dataset) the standard method was significantly better 2 times and significantly worse 1 time, at the 5% level. The default method performs worst, but, again, this is not statistically significant: the standard method is significantly better only once. The slightly worse performance of the default method can be explained by looking at table 3, which contains the tree sizes measured in number of leafs. We see that the trees produced by the default method are somewhat smaller than the trees produced by the MNP, BF and BF-LNO methods: it seems that the default method slightly overprunes. More importantly, we see that the trees produced by the standard method are much larger than the trees produced by the other methods. To take an extreme example: on the *Windsor housing* data set the standard algorithm produces trees with an average of 44 leaf nodes, against an average of 13.15 for the default method, and an average of 18.85 for the MNP heuristic.

Table 4. Average misclassification rate and tree size of all artificial data sets per ratio of nonmonotonicity. The column *Frac. Mon.* gives the fraction of monotone trees generated by the standard algorithm.

NmRat	Standard	Frac. Mon.	Default	MNP	BF	BF-LNO
0	0.096 / 6.0	0.775	0.103 / 5.3	0.102 / 5.6	0.103 / 5.5	0.101 / 5.6
0.05	0.163 / 6.1	0.825	0.164 / 5.5	0.165 / 5.6	0.168 / 5.4	0.167 / 5.4
0.20	0.324 / 5.5	0.850	0.321 / 4.6	0.320 / 4.7	0.324 / 4.6	0.324 / 4.6

Table 4 summarizes the results of our experiments with the artificial data sets. Each artificial data set was generated by making random draws from a pre-specified tree model. The results were averaged along the dimensions *number of class labels* (2 and 4), *number of observations* (100 and 500), and twenty random partitions into training and test data. We generated data sets with differing degrees of nonmonotonicity (0, 0.05 and 0.2 respectively). As with the real life data, the differences between the predictive accuracy of the different fixing heuristics are not significant. Furthermore, we conclude that the predictive accuracy of the standard trees and the monotone trees is comparable.

Also note that the performance of the different methods is comparable for *all* levels of nonmonotonicity considered. More detailed results (not presented here), show that the relative performance is not affected by the other characteristics (the size of the data set and the number of class labels) either. Hence, we conclude that the proposed methods are quite robust.

Another interesting, perhaps somewhat counterintuitive observation is that the fraction of monotone trees produced by the standard algorithm seems to *increase* with the nonmonotonicity ratio. An attempt to explain this in terms of average tree size is not entirely satisfactory, even though the trees at NmRat = 0.20 are somewhat smaller than the trees constructed from completely monotone data. A final observation is that on all real life and artificial data sets, the fixing methods at most had to be applied once, namely to the initial tree. This means that, although we described the algorithm as an alternation of fixing and pruning steps, it turns out that in practice no fixing steps were required once a monotone tree had been constructed.

6 Conclusions and Future Research

We have presented a number of fixing methods, that make a non-monotone tree monotone by pruning towards a monotone subtree. We have shown that these fixing methods can be combined with cost-complexity pruning to obtain a sequence of monotone trees. The performance of the new algorithms has been evaluated through experimental studies on artificial as well as real life data sets. We conclude that in terms of predictive performance the monotone trees are comparable to the trees produced by the standard algorithm; however, they are considerably smaller.

We summarize the strong points of our pruning based methods as we see them. First of all, the methods proposed are *guaranteed* to produce a monotone tree. Secondly, the algorithm works on nonmonotone as well as monotone data sets and therefore has no problems with noisy data. Thirdly, for the problems considered the monotone trees predict slightly, though not significantly, better than the standard trees. Finally, the monotone trees are much smaller than those produced by the standard algorithm.

We see a number of issues for further research. In this study we only applied the the strategy that alternates between pruning and fixing, as presented in section 4. We intend to investigate the other options that we described there.

Preliminary experiments suggest however, that the differences in performance are negligible. Another issue for further research is to extend our work in a different direction. Currently, our algorithm can only be applied to problems for which there is a monotonicity constraint on all variables. For many problems this requirement is too strong: usually there are one or more variables for which the constraint does not apply. We intend to extend the algorithm to handle such cases. Yet another interesting extension would be to consider monotone *regression* trees. In fact, some data sets we used in the experiments are more naturally analysed with regression trees, for example the housing data.

Finally, it seems worthwhile to further investigate the relation between the degree of nonmonotonicity of the data and the fraction of monotone trees generated by a standard algorithm.

Acknowledgements. The authors would like to thank Paul Pompe for the kind donation of the bankruptcy data set, and Eelko Penninkx for helpful suggestions on earlier drafts of this paper.

References

1. Arie Ben-David. Monotonicity maintenance in information-theoretic machine learning algorithms. *Machine Learning*, 19:29–43, 1995.
2. L. Breiman, J.H. Friedman, R.A. Olshen, and C.J. Stone. *Classification and Regression Trees (CART)*. Wadsworth, 1984.
3. C.L. Blake and C.J. Merz. UCI repository of machine learning databases [http://www.ics.uci.edu/~mlearn/mlrepository.html], 1998.
4. Jan C. Bioch and Viara Popova. Monotone decision trees and noisy data. *ERIM Report Series Research in Management*, ERS-2002-53-LIS, 2002.
5. Jan C. Bioch and Viara Popova. Induction of ordinal decision trees: an MCDA approach. *ERIM Report Series Research in Management*, ERS-2003-008-LIS, 2003.
6. Kim Cao-Van and Bernard De Beats. Growing decision trees in an ordinal setting. Submitted to International Journal of Intelligent Systems, 2002.
7. H.A.M Daniels and M. Velikova. Derivation of monotone decision models from non-monotone data. Technical Report 2003-30, Center, Tilburg University, 2003.
8. A.J. Feelders. Prior knowledge in economic applications of data mining. In D.A. Zighed, J. Komorowski, and J. Zytkow, editors, *Principles of Data Mining and Knowledge Discovery*, Lecture Notes in Artificial Intelligence 1910, pages 395–400. Springer, 2000.
9. Gary Koop. *Analysis of Economic Data*. John Wiley and Sons, 2000.
10. Kazuhisa Makino, Takashi Susa, Hirotaka Ono, and Toshihide Ibaraki. Data analysis by positive decision trees. *IEICE Transactions on Information and Systems*, E82-D(1), 1999.
11. R. Potharst and J.C. Bioch. A decision tree algorithm for ordinal classification. In D.J. Hand, J.N. Kok, and M.R. Berthold, editors, *Advances in Intelligent Data Analysis*, Lecture Notes in Computer Science 1642, pages 187–198. Springer, 1999.
12. R. Potharst and J.C. Bioch. Decision trees for ordinal classification. *Intelligent Data Analysis*, 4(2):97–112, 2000.
13. Rob Potharst and Ad Feelders. Classification trees for problems with monotonicity constraints. *SIGKDD Explorations*, 4:1–10, 2002.

Regularized Learning with Flexible Constraints

Eyke Hüllermeier

Informatics Institute, Marburg University, Germany
eyke@informatik.uni-marburg.de

Abstract. By its very nature, inductive inference performed by machine learning methods is mainly data-driven. Still, the consideration of background knowledge – if available – can help to make inductive inference more efficient and to improve the quality of induced models. Fuzzy set-based modeling techniques provide a convenient tool for making expert knowledge accessible to computational methods. In this paper, we exploit such techniques within the context of the regularization (penalization) framework of inductive learning. The basic idea is to express knowledge about an underlying data-generating model in terms of flexible constraints and to penalize those models violating these constraints. Within this framework, an optimal model is one that achieves an optimal trade-off between fitting the data and satisfying the constraints.

1 Introduction

The common framework of learning from data assumes an underlying functional relation $f(\cdot)$ that is a mapping $\mathfrak{D}_X \rightarrow \mathfrak{D}_Y$ from an input space $\mathfrak{D}_X$ to an output space $\mathfrak{D}_Y$. One basic goal of learning from data is to approximate (or even recover) this function from given sample points $(x_\imath, y_\imath) \in \mathfrak{D}_X \times \mathfrak{D}_Y$. Depending on the type of function to be learned, several special performance tasks can be distinguished. For example, in *classification* (pattern recognition) $f(\cdot)$ is a mapping $\mathfrak{D}_X \rightarrow \mathcal{L}$ with $\mathcal{L}$ a finite set of labels (classes). If $f(\cdot)$ is a continuous function, the problem is one of *regression*. Here, the sample outputs are usually assumed to be noisy, e.g. corrupted with an additive error term: $y_\imath = f(x_\imath) + \varepsilon$.

On the basis of the given data, an approximation (estimation) $h_0(\cdot)$ of the function $f(\cdot)$ is chosen from a class $\mathcal{H}$ of candidate functions, called the *hypothesis space* in machine learning [9]. The adequate specification of the hypothesis space is crucial for successful learning, and the complexity of $\mathcal{H}$ is a point of special importance. In fact, if the class of candidate functions is not rich (flexible) enough, it is likely that $f(\cdot)$ cannot be approximated accurately: There might be no $h(\cdot) \in \mathcal{H}$ that is a good approximation of $f(\cdot)$. On the other hand, if $\mathcal{H}$ is too flexible, the learning problem might become ambiguous in the sense that there are several hypotheses that fit the data equally well. Apart from that, too much flexibility involves a considerable risk of *overfitting* the data. Roughly speaking, overfitting means reproducing the data in too exact a manner. This happens if the hypothesis space is flexible enough to provide a good or even exact approximation of any set of sample points. In such cases the induced hypothesis $h_0(\cdot)$

M.R. Berthold et al. (Eds.): IDA 2003, LNCS 2810, pp. 13–24, 2003.

is likely to represent not only the structure of the function $f(\cdot)$ but also the structure of the (noisy) data.

The specification of a hypothesis space allows for the incorporation of background knowledge into the learning process. For instance, if the relation between inputs and outputs is known to follow a special functional form, e.g. linear, one can restrict the hypothesis space to functions of this type. The calibration of such models through estimating free parameters is typical for classical statistical methods. However, if only little is known about the function $f(\cdot)$, restricting oneself to a special class of (simple) functions might be dangerous. In fact, it might then be better to apply *adaptive* methods which proceed from a very rich hypothesis space.

As pointed out above, for adaptive methods it is important to control the complexity of the induced model, that is, to find a good trade-off between the complexity and the accuracy of the hypothesis $h_0(\cdot)$. One way to accomplish this is to "penalize" models which are too complex. Thus, the idea is to quantify both the accuracy of a hypothesis as well as its complexity and to define the task of learning as one of minimizing an objective function which combines these two measures. This approach is also referred to as *regularization.*

Fuzzy set-based (linguistic) modeling techniques provide a convenient way of expressing background knowledge in learning from data [5]. In this paper, we propose fuzzy sets in the form of flexible constraints [4,3,7] as a means for regularization or, more specifically, for expressing background knowledge about a functional relation to be learned.

The rest of the paper is organized as follows: We briefly review the regularization framework in Section 2. The idea of expressing background knowledge about a model in terms of flexible constraints is detailed in Section 3. Section 4 is devoted to "constraint-based regularization", that is the application of flexible constraints in learning from data. We conclude the paper with some remarks in Section 5.

2 The Regularization Framework

Without loss of generality, we can assume $\mathcal{H}$ to be a parameterized class of functions, that is $\mathcal{H} = \{h(\cdot,\omega) \mid \omega \in \Omega\}$. Here, ω is a parameter (perhaps of very high or even infinite dimension) that uniquely identifies the function $h = h(\cdot,\omega)$, and $h(x,\omega)$ is the value of this function for the input x.

Given observed data in the form of a (finite) sample

$$S = \{\, (x_1,y_1),(x_2,y_2),\ldots,(x_n,y_n)\,\} \in (\mathfrak{D}_X \times \mathfrak{D}_Y)^n,$$

the selection of an (apparently) optimal hypothesis $h_0(\cdot)$ is guided by some *inductive principle.* An important and widely used inductive principle is *empirical risk minimization* (ERM) [11]. The risk of a hypothesis $h(\cdot,\omega)$ is defined as

$$R(\omega) \doteq \int_{\mathfrak{D}_X \times \mathfrak{D}_Y} L(y, h(x,\omega))\, p(x,y)\, d(x,y),$$

where $L(\cdot)$ is a loss function and $p(\cdot)$ is a probability (density) function: $p(x, y)$ is the probability (density) of observing the input vector x together with the output y. The empirical risk is an approximation of the true risk:

$$R_{emp}(\omega) \doteq \frac{1}{n} \sum_{i=1}^{n} L(y_i, h(x_i, \omega)) \ . \tag{1}$$

The ERM principle prescribes to choose the parameter ω_0 resp. the associated function $h_0 = h(\cdot, \omega_0)$ that minimizes (1).

Now, let $\phi(\omega) = \phi(h(\cdot, \omega))$ be a measure of the complexity of the function $h(\cdot, \omega)$. The *penalized risk* to be minimized for the *regularization inductive principle* is then given by a weighted sum of the empirical risk and a penalty term:

$$R_{pen}(\omega) \doteq R_{emp}(\omega) + \lambda \cdot \phi(\omega). \tag{2}$$

The parameter λ is called the regularization parameter. It controls the trade-off between accuracy and complexity.

As already indicated above, the function $h(\cdot, \omega_0)$ minimizing (2) is supposed to have a smaller *predictive risk* than the function minimizing the empirical risk. In other words, it is assumed to be a better generalization in the sense that it predicts future outputs more accurately.

Without going into detail let us mention that the practical application of the regularization framework requires, among other things, optimization methods for the minimization of (2) as well as techniques for specifying an optimal regularization parameter λ.

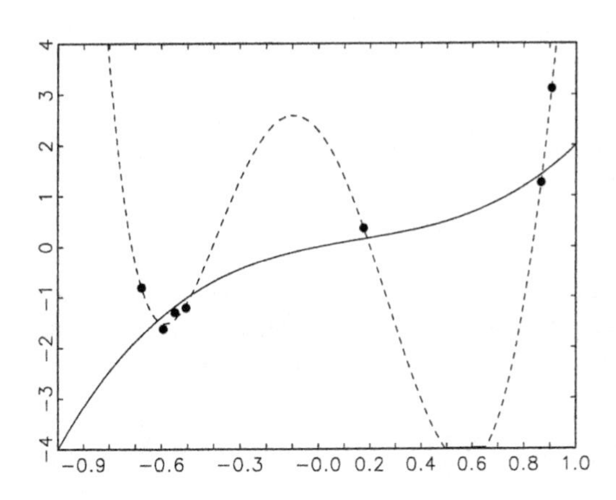

Fig. 1. The function $f(\cdot)$ to be learned (solid line) together with the given sample points and least squares approximation (dashed line).

To illustrate regularized learning, let us consider a very simple example. Suppose that the function to be learned is given by the polynomial $f : [0, 1] \rightarrow \Re$, $x \mapsto 2x^3 - x^2 + x$, and let the hypothesis space consist of all polynomials

$$x \mapsto \omega_0 + \sum_{k=1}^{m} \omega_k \cdot x^k$$

whose degree is at most $m = 5$. We have generated seven sample points at random, assuming a uniform distribution for x and a normal distribution with mean $f(x)$ and standard deviation 1/2 for y. Fig. 1 shows the function $f(\cdot)$ together with the given data. Furthermore, the figure shows the optimal least squares approximation, i.e. the approximation $h_0(\cdot)$ that minimizes the empirical risk with loss function $L(y, h(x, \omega)) = \left(y - h(x, \omega)\right)^2$. As can be seen, $h_0(\cdot)$ does obviously overfit the data. In fact, the approximation of the true function $f(\cdot)$ is very poor.

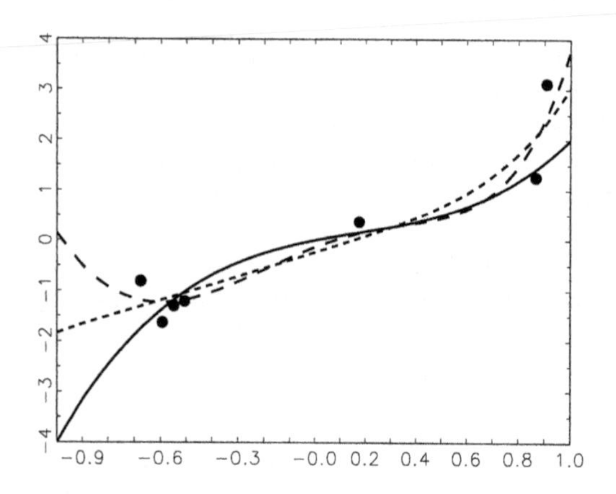

Fig. 2. The function $f(\cdot)$ to be learned (solid line), ridge regression with $\lambda = 0.01$ (dashed line) and $\lambda = 0.1$ (small dashes).

A special type of regularization is realized by so-called (standard) *ridge regression* [8] by using the sum of squared parameters $(\omega_k)^2$ as a penalty term. Thus, ridge regression seeks to minimize the penalized risk

$$R_{pen}(\omega) = \sum_{\imath=1}^{n} \left(y_\imath - h(x_\imath, \omega)\right)^2 + \lambda \cdot \sum_{k=1}^{m} (\omega_k)^2$$

that favors solutions with parameters being small in absolute value.[1] This type of regression problem can still be solved analytically. Fig. 2 shows the approximations for $\lambda = 0.01$ and $\lambda = 0.1$. The regularization effect becomes obvious in both cases. As can be seen, the larger the parameter λ is chosen, the smaller the variation of the approximating function will be.

[1] The intercept ω_0 is usually not considered in the penalty term.

3 Flexible Constraints

This section is meant to provide some background information on fuzzy sets and flexible constraints. Moreover, it will be shown how the framework of flexible constraints can be used for expressing properties of (real-valued) functions.

3.1 Background on Fuzzy Sets

A fuzzy subset of a set $\mathfrak{D}$ is identified by a so-called membership function, which is a generalization of the characteristic function $\mathbb{I}_A$ of an ordinary set $A \subseteq \mathfrak{D}$ [12]. For each element $x \in \mathfrak{D}$, this function specifies the degree of membership of x in the fuzzy set. Usually, membership degrees are taken from the unit interval $[0, 1]$, i.e. a membership function is a mapping $\mathfrak{D} \rightarrow [0, 1]$. We shall use the same notation for ordinary sets and fuzzy sets. Moreover, we shall not distinguish between a fuzzy set and its membership function, that is, $A(x)$ denotes the degree of membership of the element x in the fuzzy set A.

Is it reasonable to say that (in a certain context) 30°C is a high temperature and 29.9°C is not high? In fact, any sharp boundary of the set of high temperatures will appear rather arbitrary. Fuzzy sets formalize the idea of *graded membership* and, hence, allow for "non-sharp" boundaries. Modeling the concept "high temperature" as a fuzzy set A, it becomes possible to express, for example, that a temperature of 30°C is completely in accordance with this concept ($A(30) = 1$), 20°C is "more or less" high ($A(20) = 0.5$, say), and 10°C is clearly not high ($A(10) = 0$). As can be seen, fuzzy sets can provide a reasonable interpretation of linguistic expressions such as "high" or "low". This way, they act as an interface between a quantitative, numerical level and a qualitative level, where knowledge is expressed in terms of natural language.

Membership degrees of a fuzzy set can have different semantic interpretations [6]. For our purposes, the interpretation of fuzzy sets in terms of *preference* is most relevant. Having this semantics in mind, a (fuzzy) specification of the temperature as "high" means that, e.g., a temperature of 30°C is regarded as fully satisfactory, but that other temperatures might also be acceptable to some extent. For example, with the fuzzy set of high temperatures being specified as above, 20°C would be accepted to degree 0.5.

3.2 Flexible Constraints

In connection with the preference semantics, a fuzzy set defines a "flexible constraint" in the sense that an object can satisfy the constraint to a certain degree. Formally, let $C : \mathfrak{D}_X \rightarrow [0, 1]$ be a fuzzy set associated with a constraint on the variable X; $C(x)$ is the degree to which $x \in \mathfrak{D}_X$ satisfies the constraint.

Now, consider constraints $C_1, C_2, \ldots, C_n$ with a common domain $\mathfrak{D}_X$ and let C be the conjunction of these constraints. The degree to which an object $x \in \mathfrak{D}_X$ satisfies C can then be defined as

$$C(x) \doteq C_1(x) \otimes C_2(x) \otimes \ldots \otimes C_n(x), \tag{3}$$

where $\otimes$ is a t-norm. A t-norm is a generalized logical conjunction, that is, a binary operator $[0,1] \times [0,1] \to [0,1]$ which is associative, commutative, non-decreasing in both arguments and such that $\alpha \otimes 1 = \alpha$ for all $0 \leq \alpha \leq 1$. Well-known examples of t-norms include the minimum operator $(\alpha, \beta) \mapsto \min\{\alpha, \beta\}$ and the Lucasiewicz t-norm $(\alpha, \beta) \mapsto \max\{\alpha + \beta - 1, 0\}$. Set-theoretically, C as defined in (3) is the intersection $C_1 \cap C_2 \cap \ldots \cap C_n$ of the fuzzy sets $C_\imath$.

Note that a common domain $\mathfrak{D}_X$ can be assumed without loss of generality. In fact, suppose that C_1' specifies a constraint on a variable Y with domain $\mathfrak{D}_Y$ and that C_2' specifies a constraint on a variable Z with domain $\mathfrak{D}_Z$. The fuzzy sets C_1' and C_2' can then be replaced by their *cylindrical extensions* to the domain $\mathfrak{D}_X \doteq \mathfrak{D}_Y \times \mathfrak{D}_Z$ of the variable $X = (Y, Z)$, that is by the mappings

$$C_1 : (y, z) \mapsto C_1'(y), \qquad C_2 : (y, z) \mapsto C_2'(z).$$

The definition of a cylindrical extension is generalized to higher dimensions in a canonical way.

In connection with the specification of fuzzy constraints, it is useful to dispose of further logical operators:

- The common definition of the negation operator $\neg$ in fuzzy logic is the mapping given by $\neg x = 1 - x$ for all $0 \leq x \leq 1$ (though other negations do exist).
- A t-conorm $\oplus$ is a generalized logical disjunction, that is, a binary operator $[0,1] \times [0,1] \to [0,1]$ which is associative, commutative, non-decreasing in both arguments and such that $\alpha \oplus 0 = \alpha$ for all $0 \leq \alpha \leq 1$. Given a t-norm $\otimes$, an associated t-conorm $\oplus$ can be defined by the mapping $(\alpha, \beta) \mapsto 1 - (1 - \alpha) \otimes (1 - \beta)$ or, more generally, as $(\alpha, \beta) \mapsto \neg(\neg\alpha \otimes \neg\beta)$. For example, the t-conorm associated with the Lucasiewicz t-norm is given by $(\alpha, \beta) \mapsto \min\{\alpha + \beta, 1\}$.
- A (multiple-valued) implication operator $\rightsquigarrow$ is a mapping $[0,1] \times [0,1] \to [0,1]$ which is non-increasing in the first and non-decreasing in the second argument, i.e., $\alpha \rightsquigarrow \beta \leq \alpha' \rightsquigarrow \beta$ for $\alpha' \leq \alpha$ and $\alpha \rightsquigarrow \beta \leq \alpha \rightsquigarrow \beta'$ for $\beta \leq \beta'$. Besides, further properties can be required [2]. An example of an implication is the Lucasiewicz operator $(\alpha, \beta) \mapsto \max\{1 - \alpha + \beta, 0\}$. Implication operators can be used for modeling *fuzzy rules*: Let C_1 and C_2 be two constraints. A new constraint C can then be expressed in terms of a fuzzy rule

$$\text{IF } C_1(x) \text{ THEN } C_2(x).$$

The degree to which an object x satisfies this constraint can be evaluated by means of an implication $\rightsquigarrow$, that is $C(x) = C_1(x) \rightsquigarrow C_2(x)$.

3.3 Flexible Constraints on Functions

Flexible constraints of the above type can be used for expressing knowledge about a functional relation $f : \mathfrak{D}_X \to \mathfrak{D}_Y$. For the sake of simplicity, and without loss of generality, we shall assume that $\mathfrak{D}_Y \subseteq \mathfrak{R}$. If $f(\cdot)$ is a vector-valued function with domain $\mathfrak{D}_Y \subseteq \mathfrak{R}^m$, $m > 1$, then each of its components $f_\imath(\cdot)$ can be considered separately. Subsequently, we shall introduce some basic types of constraints and illustrate them through simple examples.

The most obvious type of constraint is a restriction on the absolute values of the function. Knowledge of this type can be expressed in terms of a fuzzy rule such as [IF x is close to 0 THEN $f(x)$ is approximately 1] or, more formally, as

$$\forall x \in \mathfrak{D}_X : C_1(x) \rightsquigarrow C_2(f(x)),$$

where the fuzzy set C_1 models the constraint "close to 0" and C_2 models the constraint "approximately 1". The degree of satisfaction of this constraint is given by

$$C(f) \doteq \inf_{x \in \mathfrak{D}_X} C_1(x) \rightsquigarrow C_2(f(x)). \tag{4}$$

Note that the infimum operator in (4) generalizes the universal quantifier in classical logic.

Constraints of such type can simply be generalized to the case of $m > 1$ input variables:

$$C(f) \doteq \inf_{x_1,\ldots,x_m \in \mathfrak{D}_X} (C_1(x_1) \otimes \ldots \otimes C_m(x_m)) \rightsquigarrow C_{m+1}(f(x_1,\ldots,x_m)).$$

Note that constraints on the input variables need not necessarily be "non-interactive" as shown by the following example: [IF x_1 is close to x_2 THEN $f(x_1, x_2)$ is approximately 1]. This constraint can be modeled in terms of an implication whose antecedent consists of a constraint such as $C_1 : (x_1, x_2) \mapsto \max\{1 - |x_1 - x_2|, 0\}$.

In a similar way, constraints on first or higher (partial) derivatives of $f(\cdot)$ can be expressed. For instance, let $f'(\cdot)$ denote the (existing) derivative of $f(\cdot)$. Knowing that $f(\cdot)$ is increasing then corresponds to the (non-fuzzy) constraint

$$C(f) \doteq \inf_{x \in \mathfrak{D}_X} f'(x) \geq 0.$$

A flexible version of this constraint could be specified as

$$C(f) \doteq \inf_{x \in \mathfrak{D}_X} C_1(f'(x)),$$

where C_1 is the fuzzy set of derivatives strictly larger than 0, e.g. $C_1(dx) = \min\{1, dx\}$ for $dx \geq 0$ and $C_1(dx) = 0$ otherwise. This version takes into account that $f(\cdot)$ can be increasing to different degrees and actually expresses that $f(\cdot)$ is *strictly* increasing. Of course, a constraint on the derivative of $f(\cdot)$ can also be local in the sense that it is restricted to a certain (fuzzy) subset of $\mathfrak{D}_X$. Thus, a constraint such as "For x close to 0, the derivative of $f(\cdot)$ is close to 1" could be modeled as follows:

$$C(f) \doteq \inf_{x \in \mathfrak{D}_X} C_1(x) \rightsquigarrow C_2(f'(x)),$$

where the fuzzy sets C_1 and C_2 formalize, respectively, the constraints "close to 0" and "close to 1".

Other types of constraints include restrictions on the relative values of the function. For example, the constraint "The values of $f(\cdot)$ for x close to 0 are much smaller than those for x close to 1" can be modeled as follows:

$$C(f) \doteq \inf_{x_1, x_2 \in \mathfrak{D}_X} (C_1(x_1) \otimes C_2(x_2)) \rightsquigarrow C_3(f(x_1), f(x_2)),$$

where the fuzzy relation $C_3 \subseteq \mathfrak{D}_Y \times \mathfrak{D}_Y$ might be defined, e.g., by $C_3(y_1, y_2) = \min\{1, y_2 - y_1\}$ for $y_1 \leq y_2$ and $C_3(y_1, y_2) = 0$ otherwise.

4 Regularization with Flexible Constraints

The regularization framework of Section 2 and the framework of flexible constraints as presented in Section 3 can be combined in order to express background knowledge about a function in learning from data. The basic idea of "constraint-based regularization" is to replace the term $\phi(\omega)$ in the penalized risk functional (2) by a fuzzy constraint $C(\omega) = C(h(\cdot, \omega))$. Thus, one arrives at the following functional:

$$R_{pen}(\omega) \doteq R_{emp}(\omega) - \lambda \cdot C(\omega). \qquad (5)$$

As can be seen, an evaluation $R_{pen}(\omega)$ is a trade-off between the accuracy of the hypothesis $h(\cdot, \omega)$, expressed by $R_{emp}(\omega)$, and the extent to which $h(\cdot, \omega)$ is in accordance with the background knowledge, expressed by $C(\omega)$.

4.1 Example

To illustrate, let us return to our small example presented in Section 2. Again, suppose that the function to be learned is given by the polynomial $f : x \mapsto 2x^3 - x^2 + x$ and let the hypothesis space consist of all polynomials whose degree is at most five. Moreover, the same sample points as before shall be given (cf. Fig. 1). Finally, suppose the prior knowledge about the function $f(\cdot)$ to be reflected by the following flexible constraints:

- If x is almost 0, then $f(x)$ is also close to 0.
- $f'(x)$ is approximately 3 for x around 0.8.
- $f'(x)$ is roughly between 6 and 7 for x close to -0.8.

The fuzzy sets used to model these constraints (i.e. the linguistic terms "almost", "close to", ...) are specified in terms of trapezoidal membership functions (we omit details due to reasons of space). Moreover, as generalized logical operators we employed the Lucasiewicz t-norm, t-conorm, and implication. Fig. 3 shows the function $h(\cdot, \omega_0)$ minimizing the penalized risk functional (5) for $\lambda = 1$. The optimal parameter $\omega_0 = (0.08, 1.17, -0.17, 2.62, -0.53, -0.51)$ has been found by means of a general stochastic search algorithm. As can be seen, the induced function $h(\cdot, \omega_0)$ is already quite close to the true function $f(\cdot)$.

It is worth mentioning that constraint-based regularization naturally supports a kind of "interactive" learning and optimization process. Namely, the

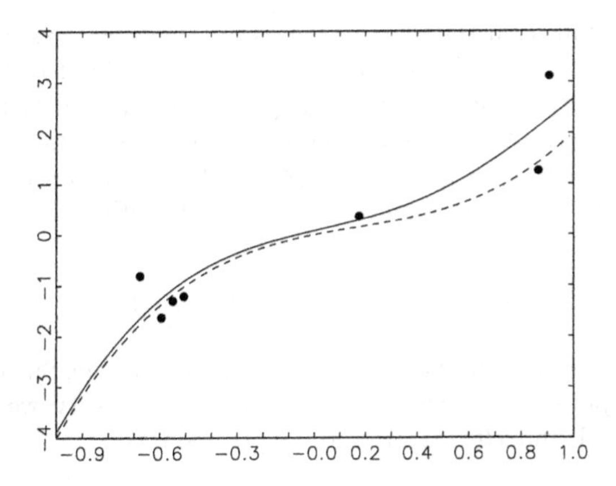

Fig. 3. Optimal approximation (solid line) of the underlying function (dashed line) for $\lambda = 1$.

optimal approximation obtained for a certain set of constraints might give the human expert cause to modify these constraints or to adapt the regularization parameter. That is, the inspection of the apparently optimal function might suggest weakening or strengthening some of the original constraints, or adding further constraints not made explicit so far. This way, learning a function from data is realized in the form of an iterative process.

4.2 Search and Optimization

As already pointed out, the minimization of the penalized risk functional (5) will generally call for numerical optimization techniques, such as stochastic search methods. The search process can often be made more efficient by exploiting the fact that parameters ω with $C(\omega) = 0$ can usually be left out of consideration.

Let ω_1 denote the optimal solution to the unconstrained learning problem, that is, the parameter vector minimizing the empirical risk $R_{emp}(\cdot)$. This vector can often be found by means of analytic methods, at least for common types of loss functions. Moreover, let ω_2 minimize the penalized risk (5) over the subspace

$$\Omega_0 \doteq \{\, \omega \in \Omega \mid C(\omega) > 0 \,\}.$$

For the overall optimal solution ω_0, we obviously have

$$\omega_0 = \begin{cases} \omega_1 & \text{if} \quad \omega_1 \leq \omega_2 \\ \omega_2 & \text{if} \quad \omega_2 < \omega_1 \end{cases}.$$

Consequently, the learning problem can be decomposed into two steps (followed by a simple comparison):

- Find the solution ω_1 to the unconstrained learning task.
- Find the optimal solution ω_2 to the constrained problem in Ω_0.

Essentially, this means that the search process can focus on the subspace $\Omega_0 \subseteq \Omega$, which will often be much smaller than Ω itself. In order to construct Ω_0, or an outer approximation thereof, the constraints on the function $f(\cdot)$ must be translated into constraints on the parameter ω. This translation clearly depends on which type of constraints and parameterized functions are used.

To illustrate the basic principle, consider an absolute value constraint C of the form

$$\text{IF} \quad x \in A \quad \text{THEN} \quad f(x) \in B, \tag{6}$$

where A and B are fuzzy subsets of $\mathfrak{D}_X$ and $\mathfrak{D}_Y$, respectively. Since $1 \rightsquigarrow 0 = 0$ for any implication operator $\rightsquigarrow$ satisfying the neutrality property, we have $C(\omega) = 0$ as soon as $A(x) = 1$ and $B(h(x, \omega)) = 0$, i.e. as soon as

$$\begin{aligned} x \in cr(A) &\doteq \{x \in \mathfrak{D}_X \mid A(x) = 1\}, \\ h(x,\omega) \notin sp(B) &\doteq cl\{x \in \mathfrak{D}_Y \mid B(y) > 0\}. \end{aligned}$$

Here, cr and sp denote, respectively, the so-called core and support of a fuzzy set, and clX is the closure of the set X. It follows that

$$\forall x \in cr(A) : h(x, \omega) \in sp(B) \tag{7}$$

is a necessary (though not sufficient) condition for $\omega \in \Omega_0$.

Now, suppose the functions $h(\cdot, \omega)$ to be expressed in terms of basis functions $\beta_\imath(\cdot)$, that is

$$h(x, \omega) = \sum_{\imath=0}^{m} \alpha_\imath \cdot \beta_\imath(x).$$

Moreover, let $sp(B) = [l, u]$ and

$$l_\imath \doteq \inf_{x \in cr(A)} \beta_\imath(x), \qquad u_\imath \doteq \sup_{x \in cr(A)} \beta_\imath(x)$$

for $0 \leq \imath \leq m$. The following conditions are necessary for the parameters $\alpha_\imath$ to satisfy (7):

$$l \leq \sum_{\imath=0}^{m} l_\imath \alpha_\imath, \qquad u \geq \sum_{\imath=0}^{m} u_\imath \alpha_\imath. \tag{8}$$

Thus, the original absolute value constraint (6) can be translated into constraints on the parameter ω, which are given in the form of linear inequalities.

In our example, we have $cr(T) = [\beta, \gamma]$ and $sp(T) = [\alpha, \delta]$ for a trapezoidal fuzzy set

$$T[\alpha, \beta, \gamma, \delta] : x \mapsto \begin{cases} (x - \alpha)/(\beta - \alpha) & \text{if} \quad \alpha \leq x < \beta \\ 1 & \text{if} \quad \beta \leq x \leq \gamma \\ (\delta - x)/(\delta - \gamma) & \text{if} \quad \gamma < x \leq \delta \\ 0 & \text{if} \quad x \notin [\alpha, \delta] \end{cases}.$$

Moreover, the basis functions are of the form $\beta_\imath(x) = x^\imath$, hence $l_\imath$ and $u_\imath$ are attained at 0 or at the boundary points of the interval $[\beta, \gamma]$.

Finally, note that $0 \otimes t = 0$ for any t-norm $\otimes$ and $0 \leq t \leq 1$. Therefore, conditions on the parameter ω derived from different constraints can be combined in a conjunctive way. Thus, if $\Omega_{0,\imath}^{appr}$ denotes an outer approximation of the set $\Omega_{0,\imath} \doteq \{\omega \mid C_\imath(\omega) > 0\}$, then

$$\Omega_0^{appr} \doteq \bigcap \Omega_{0,\imath}^{appr} \supseteq \Omega_0$$

is an outer approximation of Ω_0. Particularly, this means that linear inequalities (8) stemming from different constraints can simply be lumped together.

4.3 Relation to Bayesian Inference

It is illuminating to compare constraint-based regularization with other types of inductive principles. In this respect, Bayesian inference appears to be particularly interesting. In fact, just like constraint-based regularization, Bayesian inference [1] is concerned with the incorporation of background knowledge in learning from data.[2] (As opposed to this, other frameworks such as *structural risk minimization* [11] or *minimum description length* [10] are more data-driven.) In fact, it can be shown that constraining the class of models as detailed above can be interpreted as specifying a prior probability distribution for Bayesian inference. (Again, we refrain from a detailed discussion due to reasons of space.) This is quite interesting, since the acquisition of a reasonable prior in the Bayesian approach is generally considered a quite difficult problem. Especially, it should be noted that a prior distribution is generally given as a global (parameterized) function on Ω. The specification of such a function is hence difficult for high-dimensional input spaces. For instance, in our example above, one has to define a density over the space of polynomials of degree 5, i.e. over a six-dimensional parameter space. Apart from that, the relation between a parameter vector, such as $(\omega_0, \omega_1, \ldots, \omega_5)$ in our example, and the induced function is often not obvious. In fact, in order to define a distribution over Ω, one first has to "translate" known properties of a function into properties of the parameters. For instance, what does it mean for the prior distribution of $(\omega_0, \omega_1, \ldots, \omega_5)$ that "the derivative of $f(\cdot)$ is small for x close to 1"? In this respect, our constraint-based approach is more flexible and by far less demanding as it allows one to specify constraints that are local and that refer to aspects of the function $f(\cdot)$ directly.

5 Concluding Remarks

We have introduced (fuzzy) constraint-based regularization where the basic idea is to embed fuzzy modeling in regularized learning. This approach provides a simple yet elegant means for considering background knowledge in learning from

[2] The same is true for inductive logic programming and knowledge-based neurocomputing.

data. Using fuzzy set-based (linguistic) modeling techniques, such knowledge can be expressed in terms of flexible constraints on the model to be learned.

Due to limited space, the paper could hardly go beyond presenting the basic idea and its "ingredients" (regularized learning and fuzzy modeling), and the method obviously needs further elaboration. An important aspect of ongoing work concerns its practical realization, namely the development of suitable modeling tools and efficient optimization methods. A first prototype offering a restricted language for expressing constraints already exists. Such restrictions, concerning the type of fuzzy sets and logical operators that can be used as well as the type of constraints that can be specified, are reasonable not only from a modeling point of view, they also enable the development of specialized and hence efficient optimization methods.

There are also several theoretical questions which are worth further investigation. For instance, one such question concerns the sensitivity of constraint-based regularization, that is the dependence of the induced function $h(\cdot, \omega_0)$ on the specification of fuzzy sets and the choice of logical operators.

References

1. J. Bernardo and A. Smith. *Bayesian Theory.* J. Wiley & Sons, Chichester, 1994.
2. B. Bouchon-Meunier, D. Dubois, L. Godo, and H. Prade. Fuzzy sets and possibility theory in approximate reasoning and plausible reasoning. In J.C. Bezdek, D. Dubois, and H. Prade, editors, *Fuzzy Sets in Approximate Reasoning and Information Systems*, pages 15–190. Kluwer, 1999.
3. D. Dubois, H. Fargier, and H. Prade. The calculus of fuzzy restrictions as a basis for flexible constraint satisfaction. In *Second IEEE International Conference on Fuzzy Systems*, pages 1131–1136, 1993.
4. D. Dubois, H. Fargier, and H. Prade. Propagation and satisfaction of flexible constraints. In R.R. Yager and L. Zadeh, editors, *Fuzzy Sets, Neural Networks, and Soft Computing*, pages 166–187. Van Nostrand Reinhold, 1994.
5. D. Dubois, E. Hüllermeier, and H. Prade. Fuzzy set-based methods in instance-based reasoning. IEEE *Transactions on Fuzzy Systems*, 10(3):322–332, 2002.
6. D. Dubois and H. Prade. The three semantics of fuzzy sets. *Fuzzy Sets and Systems*, 90(2):141–150, 1997.
7. H.W. Guesgen. A formal framework for weak constraint satisfaction based on fuzzy sets. Technical Report TR-94-026, ICSI Berkeley, June 1994.
8. A.E. Hoerl and R.W. Kennard. Ridge regression: Biased estimation for nonorthogonal problems. *Technometrics*, 12(3):55–67, 1970.
9. T.M. Mitchell. *Machine Learning.* McGraw-Hill, Boston, Massachusetts, 1997.
10. J. Rissanen. Modeling by shortest data description. *Automatica*, 14:465–471, 1978.
11. V.N. Vapnik. *The Nature of Statistical Learning Theory.* Springer-Verlag, New York, second edition, 2000.
12. L.A. Zadeh. Fuzzy sets. *Information and Control*, 8:338–353, 1965.

Learning to Answer Emails

Michael Kockelkorn[1], Andreas Lüneburg[1], and Tobias Scheffer[2]

[1] Tonxx, Berlin, Germany, {mk, alueneburg}@tonxx.com

[2] University of Magdeburg, FIN/IWS, PO Box 4120, 39016 Magdeburg, Germany
scheffer@iws.cs.uni-magdeburg.de

Abstract. Many individuals, organizations, and companies have to answer large amounts of emails. Often, most of these emails contain variations of relatively few frequently asked questions. We address the problem of predicting which of several frequently used answers a user will choose to respond to an email. Our approach effectively utilizes the data that is typically available in this setting: inbound and outbound emails stored on a server. We take into account that there are no explicit references between inbound and corresponding outbound mails on the server. We map the problem to a semi-supervised classification problem that can be addressed by the transductive Support Vector Machine. We evaluate our approach using emails sent to a corporate customer service department.

1 Introduction

Companies allocate considerable economic resources to communication with their customers. A continuously increasing share of this communication takes place via email; marketing, sales and customer service departments as well as dedicated call centers have to process high volumes of emails, many of them containing repetitive routine questions. It appears overly ambitious to completely automate this process; however, any software support that leads to a significant productivity increase is already greatly beneficial. Our approach to support this process is to predict which answer a user will most likely send in reply to an incoming email, and to propose this answer to the user. The user, however, is free to modify – or to dismiss – the proposed answer.

Our approach is to learn a predictor that decides which of a small set of standard answers a user is most likely to choose. We learn such a predictor from the available data: inbound and outbound email stored on an email server. We transform the email answering problem into a set of semi-supervised text classification problems.

Many approaches are known that learn text classifiers from data. The Support Vector Machine (SVM) [17] is generally considered to be one of the most accurate algorithms; this is supported, for instance, by the TREC filtering challenge [11]. The naive Bayes algorithm (*e.g.,* [12]) is also widely used for text classification problems.

In the naive Bayesian framework, unlabeled training data can be utilized by the EM parameter estimation algorithm. The transductive Support Vector

M.R. Berthold et al. (Eds.): IDA 2003, LNCS 2810, pp. 25–35, 2003.

Machine [8] maximizes the distance between hyperplane and both, labeled and unlabeled data.

The contribution of this paper is threefold. Firstly, we analyze the problem of answering emails, taking all practical aspects into account. Secondly, we present a case study on a practically relevant problem showing how well the naive Bayes algorithm, the Support Vector Machine and the transductive Support Vector Machine can identify instances of particular questions in emails. Thirdly, we describe how we integrated machine learning algorithms into a practical answering assistance system that is easy to use and provides immediate user benefit.

The rest of this paper is organized as follows. In Section 2, we analyze the problem setting. We discuss our general approach and our mapping of the email answering problem to a set of semi-supervised text classification problems in Section 3. In Section 4, we describe the transductive SVM that we used for the case study that is presented in Section 5. In Section 6 we describe how we have integrated our learning approach into the Responsio email management system. Section 7 discusses related approaches.

2 Problem Setting

We consider the problem of predicting which of n (manually identified) standard answers $A_1, \ldots, A_n$ a user will reply to an email. In order to learn a predictor, we are given a repository $\{x_1, \ldots, x_m\}$ of inbound, and $\{y_1, \ldots, y_{m'}\}$ of outbound emails. Typically, these repositories contain at least hundreds, but often (at least) thousands of emails stored on a corporate email server. The problem would become a supervised classification problem if we knew, for each inbound email, which of the standard answers A_i had been used. Unfortunately, for two reasons, this is not the case.

Firstly, although both inbound and outbound emails are stored, it is not trivial to identify which outbound email has been sent in reply to a particular inbound email; neither the emails themselves nor the internal data structures of email clients contain explicit references. Secondly, when an outbound email does not *exactly* match one of the standard answers, this does *not* necessarily mean that none of the standard answers is the correct prediction. The user could have written an answer that is equivalent to one of the answers A_i but uses a few different words.

A characteristic property of the email answering domain is a non-stationarity of the distribution of inbound emails. While the likelihood $P(x|A_i)$ is quite stable over time, the prior probability $P(A_i)$ is not. Consider, for example, a server breakdown which will lead to a sharp increase in the probability of an answer like "we apologize for experiencing technical problems..."; or consider an advertising campaign for a new product which will lead to a high volume of requests for information on that product.

What is the appropriate utility criterion for this problem? Our goal is to assist the user by proposing answers to emails. Whenever we propose the answer that the user accepts, he or she benefits; whereas, when we propose a different answer, the user has to manually select or write an answer. Hence, the optimal predictor proposes the answer A_i which is most likely given x (*i.e.*, maximizes $P(A_i|x)$). and thereby minimizes the probability of the need for the user to write an answer manually. Keeping these characteristics in mind, we can pose the problem which we want to solve as follows.

Problem 1. Given is a repository X of inbound emails and a repository Y of outbound emails in which instances of standard answers $A_1, \ldots, A_n$ occur. There is no explicit mapping between inbound and outbound mails and the prior probabilities $P(A_i)$ are non-stationary. The task is to generate a predictor for the most likely answer A_i to a new inbound email x.

3 Underlying Learning Problem

In this Section, we discuss our general approach that reduces the email answering problem to a semi-supervised text classification problem.

Firstly, we have to deal with the non-stationarity of the prior $P(A_i)$. In order to predict the answer that is most likely given x, we have to choose $\text{argmax}_i P(A_i|x) = \text{argmax}_i P(x|A_i)P(A_i)$ where $P(x|A_i)$ is the likelihood of question x given that it will be answered with A_i and $P(A_i)$ is the prior probability of answer A_i. Assuming that the answer will be exactly one of $A_1, \ldots, A_n$ we have $\sum_i P(A_i) = 1$; when the answer can be any subset of $\{A_1, \ldots, A_n\}$, then $P(A_i) + P(\bar{A}_i) = 1$ for each answer A_i.

We know that the likelihood $P(x|A_i)$ is stationary; only a small number of probabilities $P(A_i)$ has to be estimated dynamically. Equation 1 averages the priors (estimated by counting occurrences of the A_i in the outbound emails) discounted over time.

$$\hat{P}(A_i) = \frac{\sum_{t=0}^{T} e^{-\lambda t} \hat{P}(A_i|t)}{\sum_{t=0}^{T} e^{-\lambda t}} \tag{1}$$

We can now focus on estimating the (stationary) likelihood $P(x|A_i)$ from the data. In order to map the email answering problem to a classification problem, we have to identify positive and negative examples for each answer A_i.

We use the following heuristic to identify cases where an outbound email is a response to a particular inbound mail. The recipient has to match the sender of the inbound mail, and the subject lines have to match up to a prefix ("Re:" for English or "AW:" for German email clients). Furthermore, either the inbound mail has to be quoted in the outbound mail, or the outbound mail has to be sent while the inbound mail was visible in one of the active windows. (We are able to check the latter condition because our email assistance system is integrated into the Outlook email client and monitors user activity.) Using this rule, we are able to identify some inbound emails as positive examples for the answers $A_1, \ldots, A_n$.

Note, however, that for many emails x we will not be able to identify the corresponding answer; these emails are unlabeled examples in the resulting text classification problem. Also note that we are not able to obtain examples of inbound email for which none of the standard answers is appropriate (we would have to decide that the response is *semantically* different from all of the standard answers). Consequently, we cannot estimate $P(\text{no standard answer})$ or $P(\bar{A}_i)$ for any A_i.

We can reasonably assume that no two different standard answers A_i and A_j are semantically equivalent. Hence, when an email has been answered by a standard answer A_i, we can conclude that it is a negative example for A_j. Thus, we have a small set of positive and negative examples for each A_i; additionally, we have a large quantity of emails for which we cannot determine the appropriate answer.

Text classifiers typically return an uncalibrated decision function f_i for each binary classification problem; our decision on the answer to x has to be based on the $f_i(x)$ (Equation 2). Assuming that all information relevant to $P(A_i|x)$ is contained in $f_i(x)$ and that the $f_{j\neq i}(x)$ do not provide additional information, we arrive at Equation 3. Since we have dynamic estimates of the non-stationary $P(A_i)$, Bayes equation (Equation 4) provides us with a mechanism that combines n binary decision functions and the prior estimates optimally.

$$\begin{aligned}
&\text{argmax}_\text{i} P(A_i|x) \\
&= \text{argmax}_\text{i} P(A_i|f_1(x), \ldots, f_n(x)) && (2)\\
&\approx \text{argmax}_\text{i} P(A_i|f_i(x)) && (3)\\
&= \text{argmax}_\text{i} P(f_i(x)|A_i)P(A_i) && (4)
\end{aligned}$$

The decision function values are continuous; in order to estimate $P(f_i(x)|A_i)$ we have to fit a parametric model to the data. Following [3], we assume Gaussian likelihoods $P(f_i(x)|A_i)$ and estimate the μ_i, $\mu_{\bar{i}}$ and σ_i in a cross validation loop as follows. In each cross validation fold, we record the $f_i(x)$ for all held-out positive and negative instances. After that, we estimate μ_i, $\mu_{\bar{i}}$ and σ_i from the recorded decision function values of all examples. Bayes rule applied to a Gaussian likelihood yields a sigmoidal posterior as given in Equation 5.

$$P(A_i|f_i(x)) = \left(1 + e^{\frac{\mu_{\bar{i}} - \mu_i}{\sigma_i^2} f_i(x) + \frac{\mu_{\bar{i}}^2 - \mu_i^2}{2\sigma_i^2} + log \frac{1-P(A_i)}{P(A_i)}}\right)^{-1} \qquad (5)$$

We have now reduced the email answering problem to a semi-supervised text classification problem. We have n binary classification problems for which few labeled positive and negative and many unlabeled examples are available. We need a text classifier that returns a (possibly uncalibrated) decision function $f_i : X \rightarrow real$ for each of the answers A_i.

We considered a Naive Bayes classifier and the Support Vector Machine SVM^{light} [8]. Both classifiers use the bag-of-words representation which considers only the words occurring in a document, but not the word order. As preprocessing operation, we tokenize the documents but do not apply a stemmer. For SVM^{light}, we calculated tf.idf vectors [2]; we used the default values for all SVM parameters.

4 Using Unlabeled Data

In order to calculate the decision function for an instance x, the Support Vector Machine calculates a linear function $f(x) = wx + b$. Model parameters w and b are learned from data $((x_1, y_1), \ldots, (x_m, y_m))$. Note that $\frac{w}{|w|}x_i + b$ is the distance between plain (w, b) and instance x_i; this margin is positive for positive examples $(y_i = +1)$ and negative for negative examples $(y_i = -1)$. Equivalently, $y_i(\frac{w}{|w|}x_i + b)$ is the positive margin for both positive and negative examples.

The optimization problem which the SVM learning procedure solves is to find w and b such that $y_i(wx_i + b)$ is positive for all examples (all instances lie on the "correct" side of the plain) and the smallest margin (over all examples) is maximized. Equivalently to maximizing $y_i(\frac{w}{|w|}x_i + b)$, it is usually demanded that $y_i(wx_i + b) \geq 1$ for all (x_i, y_i) and $|w|$ be minimized.

Optimization Problem 1 *Given data* $((x_1, y_1), \ldots, (x_m, y_m))$*; over all* w, b, *minimize* $|w|^2$, *subject to the constraint* $\forall_{i=1}^m y_i(wx_i + b) \geq 1$.

The SVM^{light} software package [8] implements an efficient optimization algorithm which solves optimization problem 1.

The transductive Support Vector Machine [9] furthermore considers unlabeled data. This unlabeled data can (but need not) be new instances which the SVM is to classify. In transductive support vector learning, the optimization problem is reformulated such that the margin between all (labeled and unlabeled) examples and hyperplain is maximized. However, only for the labeled examples we kwow on which side of the hyperplain the instances have to lie.

Optimization Problem 2 *Given labeled data* $((x_1, y_1), \ldots, (x_m, y_m))$ *and unlabeled data* $(x_1^*, \ldots, x_k^*)$*; over all* w, b, $(y_1^*, \ldots, y_k^*)$, *minimize* $|w|^2$, *subject to the constraints* $\forall_{i=1}^m y_i(wx_i + b) \geq 1$ *and* $\forall_{i=1}^m y_i^*(wx_i^* + b) \geq 1$.

The TSVM algorithm which solves optimization problem 2 is related to the EM algorithm. TSVM starts by learning parameters from the labeled data and labels the unlabeled data using these parameters. It iterates a training step (corresponding to the "M" step of EM) and switches the labels of the unlabeled data such that optimization criterion 2 is maximized (resembling the "E" step). The TSVM algorithm is described in [8].

5 Case Study

The data used in this study was provided by the TELES European Internet Academy, an education provider that offers classes held via the internet. In order evaluate the performance of the predictors, we manually labeled all inbound emails within a certain period with the matching answer. Table 1 provides an overview of the data statistics. Roughly 72% of all emails received can be answered by one of nine standard answers. The most frequent question "product

Table 1. Statistics of the TEIA email data set.

Frequently answered question	emails	percentage
Product inquiries	224	42%
Server down	56	10%
Send access data	22	4%
Degrees offered	21	4%
Free trial period	15	3%
Government stipends	13	2%
Homework late	13	2%
TELES product inquiries	7	1%
Scholarships	7	1%
Individual questions	150	28%
Total	528	100%

inquiries" (requests for the information brochure) covers 40% of all inbound emails.

We briefly summarize the basic principles of ROC analysis which we used to assess the performances of all decision functions in our experiments [14]. The *receiver operating characteristic* (ROC) curve of a decision function plots the number of true positives against the number of false positives for one of the classes. A classifier that compares the decision function against a threshold value achieves some number of true and false positives on the hold-out data; by varying this threshold value, we observe a trajectory of classifiers which is described by the ROC curve of the decision function.

The area under the ROC curve is equal to the probability that, when we draw one positive and one negative example at random, the decision function will assign a higher value to the positive example than to the negative. Hence, the area under the ROC curve (called the *AUC performance*) is a very natural measure of the ability of an uncalibrated decision function to separate positive from negative examples.

In order to estimate the AUC performance and its standard deviation for a decision function, we performed between 7 and 20-fold stratified cross validation and averaged the AUC values measured on the held out data. In order to plot the actual ROC curves, we also performed 10-fold cross validation. In each fold, we filed the decision function values of the held out examples into one global histogram for positives and one histogram for negatives. After 10 folds, We calculated the ROC curves from the resulting two histograms.

First, we studied the performance of a decision function provided by the Naive Bayes algorithm (which is used, for instance, in the commercial Autonomy system) as well as the Support Vector Machine SVM^{light} [8]. Figure 1 shows that the SVM impressively outperforms Naive Bayes in all cases except for one (TELES product inquiries). Remarkably, the SVM is able to identify even very specialized questions with as little as seven positive examples with between 80 and 95% AUC performance. It has earlier been observed that the probability estimates of Naive Bayes approach zero and one, respectively, as the length of analyzed document increases [3]. This implies that Naive Bayes performs poorly when not all documents are equally long, as is the case here.

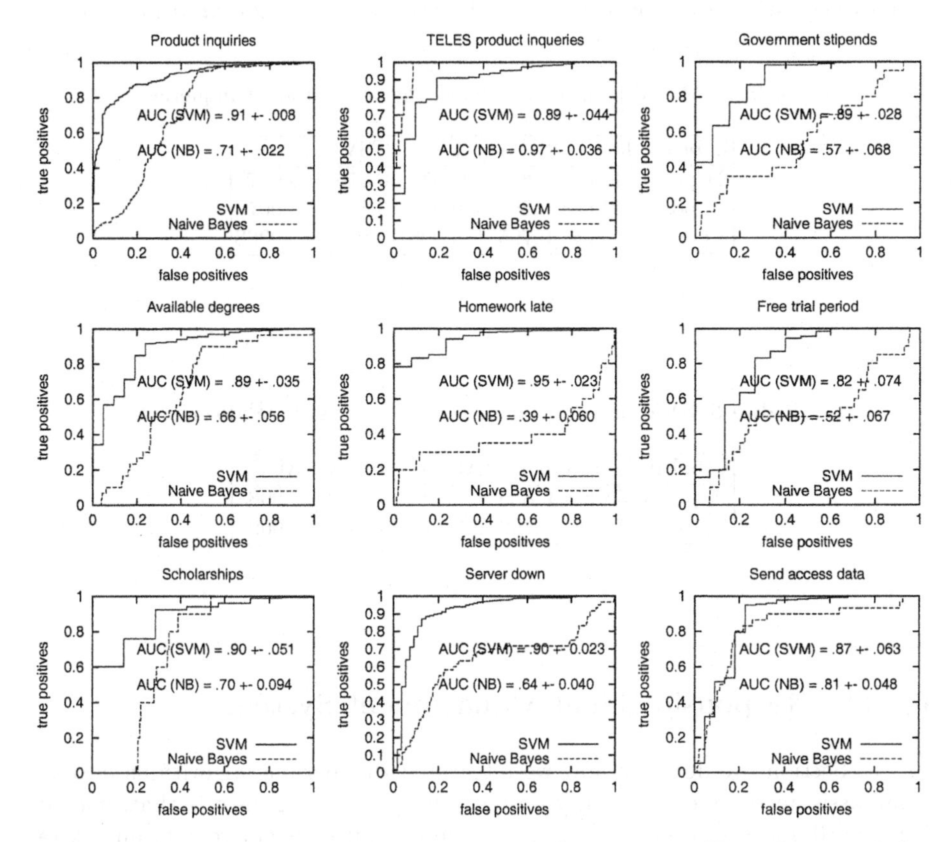

Fig. 1. ROC curves for nine most frequently asked questions of Naive Bayes and the Support Vector Machine.

In the next set of experiments, we observed how the transductive Support Vector Machine improves performance by utilizing the available unlabeled data. We successively reduce the amount of labeled data and use the remaining data (with stripped class labels) as unlabeled and hold-out data. We average five re-sampled iterations with distinct labeled training sets. We compare SVM performance (only the labeled data is used by the SVM) to the performance of the transductive SVM (using both labeled and unlabeled data). Table 2 shows the results for category "general product inqueries"; Table 3 for "server breakdown".

When the labeled sample is of size at least $24 + 33$ for "general product inqueries", or $10 + 30$ for "server breakdown", then SVM and transductive SVM perform equally well. When the labeled sample is smaller, then the transductive

SVM outperforms the regular SVM significantly. We can conclude that utilization of unlabeled data is beneficial and improves recognition significantly if (and only if) only few labeled data are available. Note that the SVM with only 8+11 examples ("general product inqueries") or 5+15 examples ("server breakdown"), respectively, still outperforms the naive Bayes algorithm *with all available data.*

Table 2. SVM and transductive SVM, "general product inqueries".

Labeled data	SVM (AUC)	TSVM (AUC)
24 pos + 33 neg	0.87 ± 0.0072	0.876 ± 0.007
16 pos + 22 neg	0.855 ± 0.007	0.879 ± 0.007
8 pos + 11 neg	0.795 ± 0.0087	0.876 ± 0.0068

Table 3. SVM and transductive SVM, "server breakdown".

Labeled data	SVM (AUC)	TSVM (AUC)
10 pos + 30 neg	0.889 ± 0.0088	0.878 ± 0.0088
5 pos + 15 neg	0.792 ± 0.01	0.859 ± 0.009

6 The Responsio Email Management System

In this Section, we describe how we integrated the learning algorithms into an email assistance system. The key design principle of Responsio is that, once it is installed and the standard answers are entered, it does not require any extra effort from the user. The system observes incoming emails and replies sent, but does not require explicit feedback. Responsio is an add-on to Microsoft Outlook. The control elements (shown in Figure 2) are loaded into Outlook as a COM object.

When an email is selected, the COM add-in sends the email body to a second process which first identifies the language of the email, executes the language specific classifiers and determines the posterior probabilities of the configured answers. The classifier process notifies the COM add-in of the most likely answer which is displayed in the field marked. When the user clicks the "auto answer" button (circled in Figure 2), Responsio extracts first and last name of the sender, identifies the gender by comparing the first names against a list, and formulates a salutation line followed by the proposed standard answer. The system opens a reply window with the proposed answer filled in.

Whenever an email is sent, Responsio tries to identify whether the outbound mail is a reply to an inbound mail by matching recipient and subject line to

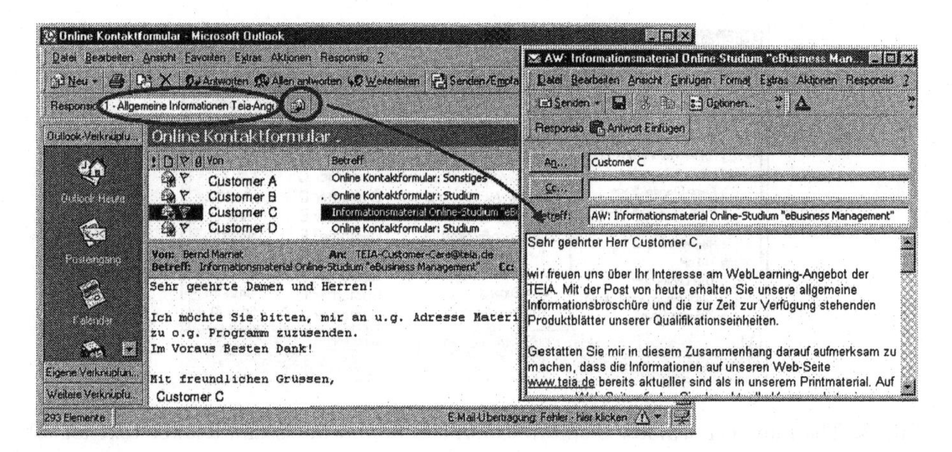

Fig. 2. When an email is read, the most likely answer is displayed in a special field in the Outlook window. Upon clicking the "Auto-Answer" button, a reply window with proposed answer text is created.

sender and subject line of all emails that are visible in one of the Outlook windows. When the sent email includes one of the standard answers, the inbound mail is filed into the list of example mails for that answer. These examples are displayed in the Responsio manager window (Figure 3). It is also possible to manually drag and drop emails into the example folders. Whenever an example list changes, a model update request is posted. The Responsio training unit processes these requests and starts the learning algorithm.

7 Discussion and Related Results

We have discussed the problem of identifying instances of frequently asked questions in emails, using only stored inbound and outbound emails as training data. Our empirical data shows that identifying a relatively small set of standard questions automatically is feasible; we obtained AUC performance of between 80 and 95% using as little as seven labeled positive examples. The transductive Support Vector Machine utilizes the available unlabeled data and improves recognition rate considerably if and only if only few labeled training examples are availeble. The drawback of the transductive SVM is the increase in computation time. Transduction increases the learning time from few seconds to several minutes. For use in a desktop application, the efficiency of the transductive SVM would need to be improved.

A limitation of the data sources that we use is that we cannot determine example emails for which no standard answer is appropriate (we cannot decide whether two syntactically different answers are really semantically different). Hence, we can neither estimate $P(\text{no standard answer})$ nor $P(\bar{A}_i)$ for any A_i.

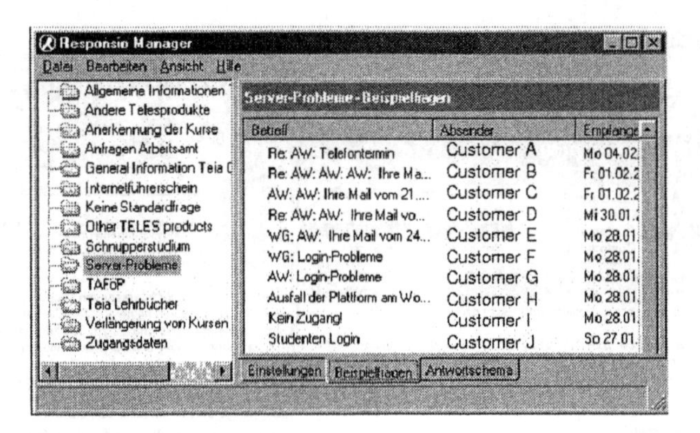

Fig. 3. The manager window displays example inbound mails and answer text for all standard answers.

Information retrieval [2] offers a wide spectrum of techniques to measure the similarity between a question and questions in an FAQ list. While this approach is followed in many FAQ systems, it does not take all the information into account that is available in the particular domain of email answering: emails received in the past. An FAQ list contains only one single instance of each question whereas we typically have many instances of each questions available that we can utilize to recognize further instances of these questions more accurately.

The domain of question answering [18] is loosely related to our email answering problem. In our application domain, a large fraction of incoming questions can be answered by very few answers. These answers can be pre-configured; the difficulty lies in recognizing instances of these frequently asked questions very robustly. Question answering systems solve a problem that is in a way more difficult. For arbitrary questions posed in natural language, these systems select an answer sentence from a large corpus, such as an encyclopedia. Applied to email, any grammar-based approach will face the difficulty of ungrammatical mails.

Several email assistance systems have been presented. [7,4,16,6] use text classifiers in order to predict the correct folder for an email. In contrast to these studies, we study the feasibility of identifying instances of particular questions rather than general subject categories.

Related is the problem of filtering spam email. Keyword based approaches, Naive Bayes [1,13,15,10] and rule-based approaches [5] have been compared. Generating positive and negative examples for spam requires additional user interaction: the user might delete interesting emails just like spam after reading it. By contrast, our approach generates examples for the email answering task without imposing additional effort on the user. Our future efforts concentrate on evaluating the co-training algorithm which also utilizes unlabeled data.

Acknowledgment. We have received support from the German Science Foundation (DFG) under grant SCHE540/10-1, and from Hewlett Packard Consulting, Germany.

References

1. I. Androutsopoulos, J. Koutsias, K. Chandrinos, and C. Spyropoulos. An experimental comparison of naive bayesian and keaword based anti-spam filtering with personal email messsages. In *Proceedings of the International ACM SIGIR Conference*, 2000.
2. R. Baeza-Yates and B. Ribeiro-Neto. *Modern Information Retrieval.* Addison Wesley, 1999.
3. P. Bennett. Assessing the calibration of naive bayes' posterior estimates. Technical report, CMU, 2000.
4. T. Boone. Concept features in Re:Agent, an intelligent email agent. *Autonomous Agents*, 1998.
5. W. Cohen. Learnig rules that classify email. In *Proceedings of the IEEE Spring Symposium on Machine learning for Information Access*, 1996.
6. E. Crawford, J. Kay, and E. McCreath. IEMS – the intelligent email sorter. In *Proceedings of the International Conference on Machine Learning*, 2002.
7. C. Green and P. Edwards. Using machine learning to enhance software tools for internet information management. In *Proceedings of the AAAI Workshop on Internet Information Management*, 1996.
8. T. Joachims. Making large-scale svm learning practical. In B. Schölkopf, C. Burges, and A. Smola, editors, *Advances in Kernel Methods – Support Vector Learning*, 1999.
9. T. Joachims. Transductive inference for text classification using support vector machines. In *Proceedings of the International Conference on Machine Learning*, 1999.
10. A. Kolcz and J. Alspector. Svm-based filtering of e-mail spam with content-specific misclassification costs. In *Proceedings of the ICDM Workshop on Text Mining*, 2001.
11. D. Lewis. The trec-5 filtering track. In *Proceedings of the Fifth Text Retrieval Conference*, 1997.
12. A. McCallum and K. Nigam. Employing em and pool-based active learning for text classification. In *Proceedings of the International Conference on Machine Learning*, 1998.
13. P. Pantel and D. Lin. Spamcop: a spam classification and organization program. In *Proceedings of the AAAI Workshop on Learning for Text Categorization*, 1998.
14. F. Provost, T. Fawcett, and R. Kohavi. The case against accuracy estimation in comparing classifiers. In *Proceedings of the International Conference on Machine Learning*, 1998.
15. M. Sahami, S. Dumais, D. Heckerman, and E. Horvitz. A bayesian approach to filtering junk email. In *Proceedings of the AAAI Workshop on Learning for Text Categorization*, 1998.
16. R. Segal and J. Kephart. Mailcat: An intelligent assistant for organizing mail. In *Autonomous Agents*, 1999.
17. V. Vapnik. *Statistical Learning Theory.* Wiley, 1998.
18. E. Vorhees. The trec-8 question answering track report. In *Proceedings of TREC-8*, 1999.

A Semi-supervised Method for Learning the Structure of Robot Environment Interactions

Axel Großmann[1], Matthias Wendt[1], and Jeremy Wyatt[2]

[1] Department of Computer Science
Technische Universität Dresden
Dresden, Germany
{axg, mw177754}@inf.tu-dresden.de

[2] School of Computer Science
The University of Birmingham
Birmingham, UK, B15 2TT
jlw@cs.bham.ac.uk

Abstract. For a mobile robot to act autonomously, it must be able to construct a model of its interaction with the environment. Oates *et al.* developed an unsupervised learning method that produces clusters of robot experiences based on the dynamics of the interaction, rather than on static features. We present a semi-supervised extension of their technique that uses information about the controller and the task of the robot to (i) segment the stream of experiences, (ii) optimise the final number of clusters and (iii) automatically select the individual sensors to feed to the clustering process. The technique is evaluated on a Pioneer 2 robot navigating obstacles and passing through doors in an office environment. We show that the technique is able to classify high dimensional robot time series several times the length previously handled with an accuracy of 91%.

1 Introduction

We would like our mobile robots to operate successfully in the real world. Independently of whether our aim is truly autonomous behaviour or just reliable and robust operation, this requires the robots to collect information about the interaction with the physical environment. In particular, we want an automatic technique for constructing a model of the world dynamics. Since our particular goal is to use such a model for execution monitoring [4,5] at the level of reactive control, it should support predictions about the qualitative outcome of actions as well as help in explaining situations in which the actions had unintended effects.

As the interaction of a robot with the environment is complex, a description of it will be difficult to obtain. On the one hand, we would favour an unsupervised learning technique, e.g., the work by Oates *et al.* [14] on clustering robot-sensor data using dynamic time warping as similarity measure. On the other hand, learning a world model is fundamentally a supervised learning problem. As the

M.R. Berthold et al. (Eds.): IDA 2003, LNCS 2810, pp. 36–47, 2003.

entire world dynamics will be huge in any realistic application, it will be important to use the robot's goals to focus the model learning so that only useful and important aspects of the interaction are represented [8]. We want to address this dilemma by devising a semi-supervised method.

The approach by Oates *et al.* has a number of deficiencies if it is to be applied to non-trivial robotic tasks. To our knowledge, it has been applied so far only to rather simple settings: a mobile robot moving along a straight line or turning in place, where the actions had durations between 2 and 8 seconds, only 4 sensors were used and these were selected manually. It is not clear how the technique performs for actions lasting up to 30 seconds or so, or whether it can deal with higher dimensional data.

We present a semi-supervised method for clustering robot experiences that can be seen as an extension of the work by Oates *et al.*. Our technique uses information about the reactive controller and the task of the robot to segment the stream of experiences and to optimise the final number of clusters as well as the contribution of the individual sensors to the clustering process. The technique is evaluated on an office-delivery robot passing through doors and navigating dynamic obstacles. Eventually, the robot can reliably distinguish situations in which it was able to pass through a door or avoid an obstacle successfully from situations in which it failed to do so.

The paper is organised as follows. In Section 2, we introduce the main ideas and concepts of learning models of a reactive controller. In Section 3, we describe the individual steps of the learning algorithm. In Section 4, we evaluate the performance of the algorithm in experiments with an office-delivery robot. We conclude in Section 5 and outline future work in Section 6.

2 Obtaining Models of a Reactive Controller

Designing a robot controller for navigation tasks in dynamic environments is hard. In general, we cannot foresee and consider all the possible situations in the interaction of the robot with the environment that might make a difference to success or failure of an action. However, given an existing controller, we can collect a reasonable amount of data and pose this problem as a typical knowledge discovery application [7].

Without any considerable loss of generality, we focus on behaviour-based controllers in the following. This type of controller consists of several interacting, task-specific programs that are referred to as low-level behaviours. Each behaviour program takes the current sensor readings and the state information and computes target values of the robot's actuators. Individual behaviours can overwrite the output of other behaviours. Moreover, behaviours can be temporarily deactivated. The activation context of a behaviour may be specified in terms of the sensory inputs, the state information, or the output of other behaviours. Examples of low-level behaviours for navigation tasks are obstacle avoidance, travelling to a target location, or corridor following.

As we are concerned with learning a model of the world dynamics, we can decide to build such a model for the reactive controller as a whole or for the task-specific low-level behaviours individually. In the first case, we consider the reactive controller to be a black box with a given set of inputs (the current sensor readings and state information obtained by the localisation and object-detection modules) and outputs (the actuator settings) and an evaluation function specifying a task-specific performance measure. In the second case, we model each low-level behaviour as a black box. As each behaviour is supposed to achieve a specific sub-task, the set of inputs and outputs is expected to be smaller than in the first case, and the evaluation function simpler and more specific. In either case, the inputs and the performance values can be represented as multivariate time series.

The overall approach to learning a model of a reactive controller we propose consists of four steps: (1) recording sensory, status, and performance information of the controller during the execution of high-level actions, (2) finding qualitatively different outcomes by clustering the robot's experiences, (3) finding frequently occurring patterns in selected experiences, and (4) building a model the world dynamics for specific purposes such as the detection of exceptional situations and significant events, execution monitoring, or object detection. In our paper, we specifically address steps (1) and (2).

3 Classification of Sensory Data

The clustering algorithm for robot experiences presented in this section is an extension of classical clustering algorithms to handle finite sensor histories. The main challenge thereby is to devise a means for for comparing time series that is suitable for the robotics domain.

Representing signals. At the level of the sensors, the basic objects to be compared are not sequences but are continuous functions of the continuous parameter time. The values of the functions are usually multidimensional, i.e., we obtain readings from the set of sensors. In practise, the continuous functions are converted into discrete sequences by some sampling process. In addition, the time series are segmented by high-level actions of the robot. We assume that the robot can only execute a single high-level action a time.

Let E denote an experience, represented as a multivariate time series containing n measurements from a set of sensors recorded during the robot's operation such that $E = \{\mathbf{e_t} \mid 1 \leq t \leq n\}$. The $\mathbf{e_t}$ are vectors of values containing one element for each sensor. We do not distinguish between raw sensor readings obtained from physical sensors and preprocessed information from virtual sensors.

As the values for each experience E were recorded over the course of engaging in a single high-level action, we store the corresponding high-level action and information about the specific outcome of this action. Suppose an office delivery robot is navigating in an environment. Some delivery actions will be completed successfully while the robot fails at others. We distinguish three possible out-

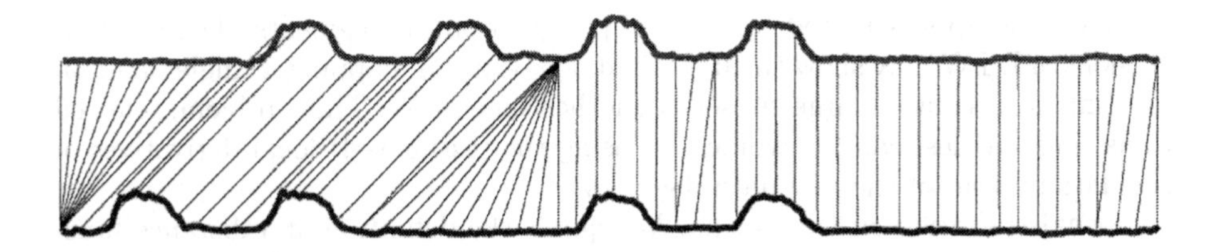

Fig. 1. Given two one-dimensional time series that have an overall similar shape, but are not aligned in the time axis, DTW can efficiently find an alignment between them.

comes: success, failure, and time-out. For each high-level action, we obtain a set of m experiences. For each sensor, the range of the measurements is determined. As a further preprocessing step, the feature values are transformed to the interval $[0, 1]$.

Temporal abstraction. Ideally, we want to find subsets of qualitatively similar experiences. We consider two experiences as qualitatively similar if, in the respective time intervals, the robot's actions had similar effects on the environment as seen from the robot's sensors. Oates *et al.* [14] argued that qualitatively similar experiences may be quantitatively different in at least two ways: (1) they may be of different lengths and (2) the rate at which progress is made can vary non-linearly within a single time series. This is, the similarity measure for time series must support temporal abstraction. The standard approach of embedding the time series in a metric space and using Euclidean distance of similarity measure does not meet this requirement.

Rosenstein and Cohen [15] and Höppner [7] have obtained temporal abstraction by using a pattern alphabet for each type of sensor. That is, the experiences are represented in a unit- and scale-independent manner as sequences of feature vectors, where the feature values are pattern labels. By specifying a time window, the pattern alphabet can be learnt. However, this does not seem possible in our case given the little *a priori* knowledge available about the controller and its interaction with the environment. We need a method that does not impose any constraint on the length of the experiences.

Dynamic time warping. Suppose that two time series have approximately the same overall component shapes, but these shapes do not line up in the time axis. To find the similarity between such sequences, we must 'warp' the time axis of one series to achieve a better alignment. Dynamic time warping (DTW) is a technique for efficiently achieving this warping. It has been used successfully in many domains including speech processing [16], data mining [12], and robotics [14]. See Figure 1 for an example.

DTW appears to be well suited for clustering experiences in a velocity-independent fashion. However, there is also an associated cost. The time complexity of the algorithm is $O(n_1 n_2)$ for comparing two experiences E_1 and E_2, of length n_1 and n_2, respectively. Recently, techniques have been developed that address this problem. For example, it is possible to speed up DTW by one to three orders of magnitude using segmented DTW [12] provided that the time series can be approximated by a set of piecewise linear segments. The classical versions of DTW may show the undesired behaviour of mapping a single point on one time series onto a large subset of another time series. This problem, referred to as singularities, can be avoided by using the (estimated) local derivatives of the data instead of the raw data itself [13].

Local distance measure. DTW requires the definition of a distance function between the values of the sequence at a given time frame. In the following, this is referred to as local distance measure. As in most high-dimensional applications, the choice of the distance metric is not obvious. There is very little literature on providing guidance for choosing the correct distance measure which results in the most meaningful notion of proximity between two records [1]. We have used the following generic local distance function:

$$d(x, y) = \left[\sum_{i=1}^{N} \left(w^i \left| x^i - y^i \right|)^k \right) \right]^{1/k}$$

where x and y are the N-dimensional feature vectors, and w is an N-dimensional weight vector. We obtain Manhattan distance if $k = 1$, Euclidean distance if $k = 2$, and fractional distance if $k < 1$. Aggarwal *et al.* [1] argued that the use of fractional distance may lead to improvements in high-dimensional data-mining problems.

Given two experiences E_1 and E_2, DTW uses dynamic programming to find a warping that minimises the area of the local distance function d in time for E_1 and E_2. The area is used as a measure of similarity between the two experiences.

Sensor weighting. For any given situation, only a small subset of the robot's sensors may contribute to the time-series patterns that characterise the situation. Instead of specifying the contributing sensors *a priori*, a learning algorithm should weigh the importance of each sensor automatically. This may be possible, if the clustering process does not only take into account the sensor readings, but also information about the task to be performed.

The contribution of the individual sensors to the local distance measure is determined by the weight vector w. In order to find suitable parameters for the sensor weights, we propose the following three-step strategy: (1) We try to reduce the dimensionality N by using preprocessed information (virtual sensors) instead of raw sensor readings. Suppose a robot is equipped with five sonar sensors in forward direction. Instead of using the raw readings from all five sensors, we compute estimates of the distances to the closest obstacles for only three areas (front-right, front-centre, and front-left). (2) We exclude individual sensors using knowledge about the task and the implementation of the controller or the low-level behaviour. (3) We optimise the weight parameters automatically using the experiences and information about the outcome of the actions (success or failure). The weights in this paper are binary, and selected by searching through the space of combinations of features. DTW and hierarchical clustering are carried out for each combination of features, and the performance evaluated. If the current feature set being tried outperforms the previous best feature set then the clustering associated with the new best replaces the previous best.

Hierarchical clustering. Given the set of m experiences and an instance of the local distance function, we compute a complete pairwise distance matrix. Using a symmetrised version of the algorithm, this means invoking DTW $m(m-1)/2$ times. Hierarchical clustering is an appealing approach to problems in which there is no single 'obvious' partition of the input space into well-separated clusters. Agglomerative algorithms work by beginning with singleton sets and merging them until the complete input set is achieved. Hence, they produce a hierarchical, binary cluster tree. Given the tree, one can conveniently trade off between the number of clusters and the compactness of each cluster.

Oates *et al.* [14] used the average-linkage method together with DTW. They found that robot experiences with any noise were grouped together and distinguished from those that had none. In fact, most clustering methods are biased toward finding clusters possessing certain characteristics related to size, shape, or dispersion: Ward's method tends to find clusters with roughly the same number of members in each cluster; average linkage is biased toward finding clusters of equal variance; and single linkage performs well for elongated or irregular clusters [3]. Due to the lack of prior knowledge, we selected all three for the experiments.

Search for suitable distance functions. The stopping criterion used by Oates *et al.* [14] is based on distance information. Specifically they do not merge clusters when the mean inter-cluster distance and the mean intra-cluster distance are different and that difference is statistically significant at 5% as measured by a t-test. In this paper we use the same criterion to stop the clustering process. However, we perform multiple clusterings using different sensor weightings, trying to maximise the evaluation function $q = k_1 \frac{1}{C} + k_2 \frac{A}{B}$, where C is the final number of clusters, A is the number of correct classifications, and B is the number of misclassifications. An experience is classified correctly if it has the same action outcome as the majority of elements of the cluster it belongs to. The parameters k_1 and k_2 can be tuned for the specific application.

The complete high-level algorithm is given in Figure 2. After obtaining a suitable partition of the experiences, we have several options on how to proceed. We can select or compute cluster prototypes [14], or we can search for reoccurring patterns within the experiences of each cluster and use them to represent differences between the clusters. In this way, we could predict the outcome of robot actions.

4 Experimental Results

The aim of the experiments is to investigate the usability of DTW as similarity measure and the performance of the proposed learning algorithm in a realistic and sufficiently relevant robotics application.

Application scenario. We have used a system specifically developed for office-delivery tasks [5]. It was used on a Pioneer 2 robot, equipped with a camera, a ring of sonar sensors, and a laser-range-finder. The control software of this system consists of a behaviour-based reactive controller, a position-tracking

Given a robot controller, a selected high-level action a, the set of low-level behaviours $\mathcal{B}$ implementing a, the set of N sensors $\mathcal{S}$ used as inputs by $\mathcal{B}$, a set of m experiences for a, and a method for selecting individual sensors from $\mathcal{S}$.

(0) Let $r = 0$ and $q_0 = 0$. Create an initial instance of the distance function, d_0, with $w_i = 0$ for $i = 1 \dots N$.
(1) Let $r = r + 1$.
(2) Select a single sensor s_k from $\mathcal{S}$ and create a new instance d_r from d_{r-1} by setting $w_k = 1$.
(3) Compute the distance matrix by invoking DTW m(m-1)/2 times using d_r.
(4) Perform hierarchical, agglomerative clustering and obtain a set of clusters $\mathcal{C}$.
(6) Evaluate the current partitions by computing $q_r = q(\mathcal{C})$.
(7) If $q_r > q_{r-1}$ keep $w_k = 1$, set $w_k = 0$ otherwise.
(8) If there are still sensors to select go to (1), terminate otherwise.

Fig. 2. The semi-supervised algorithm for clustering robot experiences.

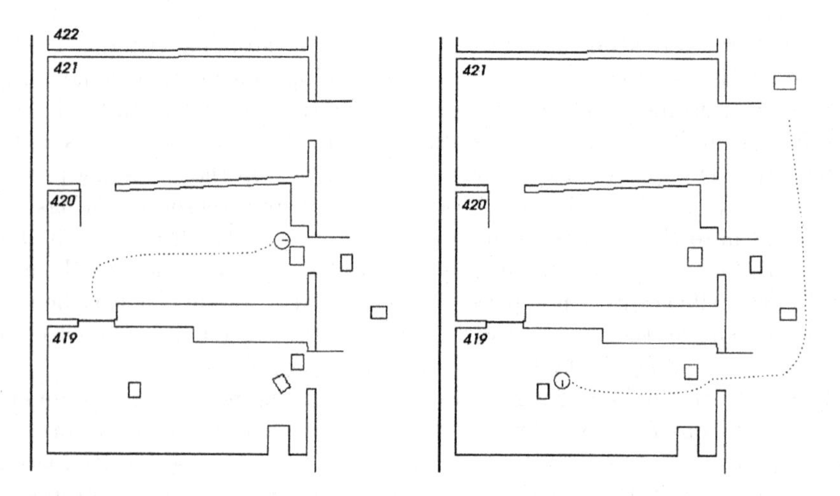

Fig. 3. Typical trajectories in the office environment. *Left*: In Room 420, the low-level behaviour *GoToPos* fails as the way is blocked by an obstacle. *Right*: On the way to Room 419, the robot successfully avoids two obstacles.

system, a path planner, and a high-level planning and reasoning system. We replaced the reactive controller on purpose with a version that had a number of deficiencies. Although the robot was able to pass through open doors, its performance at avoiding dynamic obstacles was rather poor and it did not know how to handle doors that were closed. Typical robot trajectories are shown in Figure 3.

During the execution of a high-level go-to action that takes the robot from its current location to a given goal position, we distinguish the following qualitatively different situations: (A) the robot is operating in open space (no obstacles detected), (B) the robot is avoiding an obstacle successfully, (C) the robot gets stuck at an obstacle, (D) the robot is travelling through a door successfully, and (E) the robot bumps into a door and gets stuck. We want the robot to learn to identify these situations based on its sensory data. In this way, it would be possible to predict the outcome of the robot's actions or to use this information to improve the current implementation of the robot controller.

Robot experiences. We have used the following set of sensors: the distance to the closest obstacle detected in the sonar-sensor buffers front/left (I1), front/right (I2), front (I3), side/right (I4), and side/left (I5), the translational velocity (I6), the rotational velocity (I7), and the angular deviation from the optimal path (I8). These eight data streams correspond to the inputs to the obstacle avoidance behaviour used in the reactive controller (I1-I5) and to state information of the controller (I6-I8), respectively.

As the experiences are recorded during the execution of high-level plans for office-delivery tasks, many different locations are visited during the robot's operation. The durations of the experiences for the go-to action are between 5 and 40 secs. Sequences of about 20 secs are the most frequent. The sampling rate of the sensor values in the given application is 4 Hz.

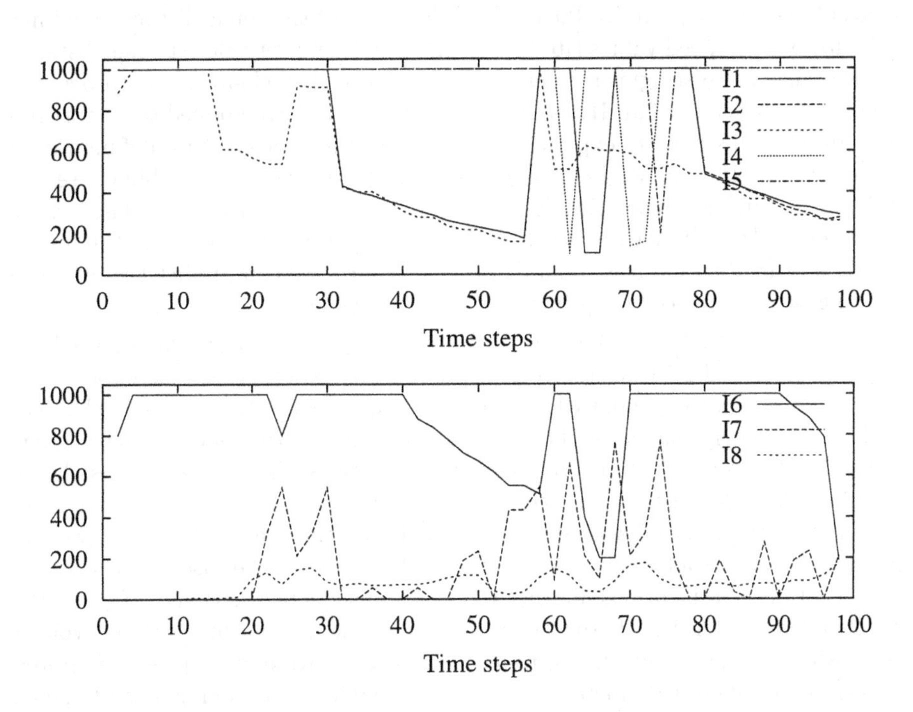

Fig. 4. Robot experience with a duration of 25 secs.

Table 1. Classification results for a set of 314 experiences

Distance function	Selected sensors Ii ($w_i = 1$)	Number of clusters	Number of misclassifications	Percentage of misclassifications
	6	20	61	19.6
Euclidean	6, 8	12	103	33.0
($k = 2$)	6, 7, 8	36	35	11.3
	1, 2, ... , 8	27	48	15.1
Manhattan	6, 8	19	106	34.0
($k = 1$)	6, 7, 8	44	29	9.3
	1, 2, ... , 8	12	110	35.0
Fractional	6, 8	15	74	23.7
($k = 0.3$)	6, 7, 8	23	28	9.0
	1, 2, ... , 8	23	45	14.4

A single robot experience for the go-to action is shown in Figure 4. At the start, the robot does not detect any obstacles in its sonar buffer and moves at maximum speed. At $t = 20$, it performs a 90 deg turn. At $t = 35$, it detects an obstacle situated ahead and begins to slow down. At $t = 55$, it starts navigating around the obstacle. This procedure is finished at $t = 70$. Eventually, the robot stops at the target position. Please note the effect of the sonar distance readings (I1 – I5) on the speed values (I6, I7). As the robot gets very close to the obstacle, it is very hard to distinguish situations in which the obstacle-avoidance procedure is going to succeed or fail. **Iterative clustering.** We have applied the algorithm in Figure 2 to a set of 314 experiences recorded during the execution of high-level go-to actions. In fact, we have used the average-linkage method, binary sensor weights w_i, and an evaluation function q with $k_1 = 0$ and $k_2 = 1$. The results are given in Table 1. We obtained the best classification results using the inputs I6, I7, and I8. Using a fractional distance measure, we can predict the outcome of the go-to actions correctly in about 90 % of the cases.

To investigate the performance of the algorithm in more detail, we have manually selected 97 from the original set of 314 experiences. The experiences in this subset correspond to the situation sequences AA, AB, AC, AD, AE, or DA. The performance of the algorithm for these experiences using average linkage, a fractional distance measure, and several subsets of sensors is shown in Figure 5. The clustering process starts with 97 singleton clusters. For most sensor subsets, it terminates at 11 clusters. Using the sensors I6, I7, I8, we can predict the action outcome, again, in about 90 percent of the cases correctly. The 9 cases of misclassifications all refer to situation sequences AB and AC, i.e., the robot failed to identify whether a dynamic obstacle in front of it can be avoided successfully using the given controller. Please note, we have not yet attempted to use the results of the clustering in connection with execution monitoring, e.g., to stop the robot before it bumps into a closed door or dynamic obstacle.

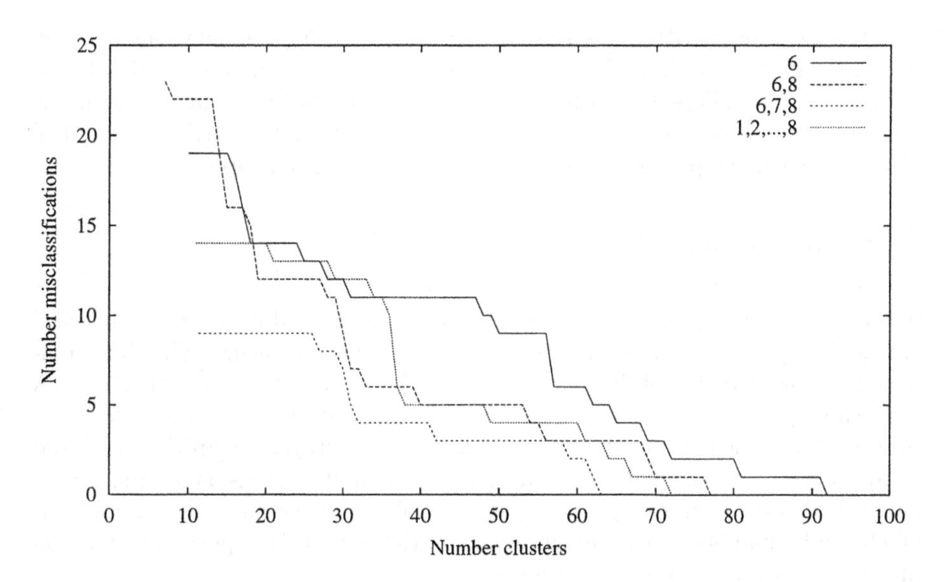

Fig. 5. Hierarchical clustering with a set of 97 experiences using a fractional distance with $k = 0.3$. The clustering process moves from right to left. Clustering stopped when the termination criterion based on the difference between the mean intra-cluster and inter-cluster distances was satisfied.

5 Conclusions

We have presented a semi-supervised method for clustering robot experiences that is applicable to realistic robot tasks. This extends the method of Oates [14] by (i) using the controller to partially segment the time series, (ii) using the labelled sequences to assess the clustering quality, (iii) using the clustering quality to select when to stop clustering and which sensors to use in the DTW process. The experimental results show that this extends the usefulness of Oates' approach so that significantly longer time series of double the dimensionality can be handled; and that accurate classifications can be generated.

Finding a suitable clustering of robot experiences is certainly an important first step in building a model of the world dynamics. Given the clusters of experiences, it would be desirable to be able to identify short sequences of sensor data which can be used for the prediction of success or failure of high-level actions. Thus we face the problem of segmenting the time series at a finer level than currently, and detecting dependencies between the subsequences and the high-level action's outcomes.

The original purpose of the DTW algorithm was to compare one-dimensional time series. In the current approach, we use the rather ad-hoc solution of using the local distance function on the multi-dimensional data items to project the multi-dimensional time series to a one-dimensional one. It is clear that this causes a loss of information, and we have reasons to believe that the most interesting patterns, i.e., those with a good prediction rate, are interdependencies between the different sensory data. The relationship between the distance estimates to obstacles in the different directions yield possibly useful information about the situation the robot is in. Thus, we should also look for alternative ways to deal effectively with the problem of multi-dimensionality of our input data.

6 Future Work

We will investigate the problem of segmenting a multi-dimensional numerical time series, incorporating somehow the DTW distance measure. The data mining algorithms capable of dealing with all of these properties are rather rare. A promising approach however seems to be the combination of algorithms for finding frequent sequences in categorical data, like SPADE [17] or FreeSpan [6] and an appropriate categorisation of the numerical multi-dimensional time series. This would provide an easy solution to the problem of the multi-dimensionality, which can be handled in the categorical case. However, this imposes a restriction on the categorisation of the time series used.

The discretised representation of the time series should support a distance measure lower bounding the DTW distance, otherwise there is no possibility of combining the categorical algorithm and the DTW distance measure. There have been some efficient discretisation techniques proposed, among them piecewise constant and piecewise linear approximations [9,10,11] which despite their simplicity have some remarkable properties in lower bounding several distance measures.

Another approach could be a combination of the discretisation technique used by Das *et al.* [2], sliding a window over the time series and clustering the subsequences using DTW. This yields a discretised version of the time series, which then can be searched for frequent patterns.

It will have to be subject for future research which of the discretisation techniques proposed in the literature can be used in combination with categorical algorithms for frequent sequence detection to correctly classify the recorded sensor data.

References

1. C. C. Aggarwal, A. Hinneburg, and D. A. Keim. On the surprising behaviour of distance metrics in high dimensional space. In *Proc. of 8th Int. Conf. on Database Theory (ICDT-2001)*, pages 4200–434, 2001.
2. G. Das, K.-I. Lin, H. Mannila, et al. Rule discovery from time series. In *Proc. of the 4th Int. Conf. on Knowledge Discovery and Data Mining (KDD-98)*, pages 16–22, 1998.
3. B. S. Everitt. *Cluster analysis*. John Wiley, 1993.
4. G. D. Giacomo, R. Reiter, and M. Soutchanski. Execution monitoring of high-level robot programs. In *Principles of Knowledge Representation and Reasoning*, pages 453–465, 1998.

5. A. Großmann, A. Henschel, and M. Thielscher. A robot control system integrating reactive control, reasoning, and execution monitoring. In *Proceedings of the First International Workshop on Knowledge Representation and Approximate Reasoning (KRAR-2003)*, Olsztyn, Poland,, May 2003.
6. J. Han, J. Pei, B. Mortazavi-Asl, et al. FreeSpan: Frequent pattern-projected sequential pattern mining. In *Proc. of the 6th Int. Conf. on Knowledge Discovery and Data Mining (KDD-2000)*, pages 20–23, 2000.
7. F. Höppner. Discovery of temporal patterns. Learning rules about the qualitative behaviour of time series. In *Proc. of the 5th European Conf. on Principles of Data Mining and Knowledge Discovery (PKDD-2001)*, pages 192–203, 2001.
8. L. P. Kaelbling, T. Oates, N. H. Gardiol, and S. Finney. Learning in worlds with objects. In *Working Notes of the AAAI Stanford Spring Symposium on Learning Grounded Representations*. AAAI, 2001.
9. E. J. Keogh. A fast and robust method for pattern matching in time series databases. In *Proc. of the 9th Int. Conf. on Tools with Artificial Intelligence (ICTAI-97)*, pages 578–584, 1997.
10. E. J. Keogh, S. Chu, D. Hart, and M. J. Pazzani. An online algorithm for segmenting time series. In *Proc. of the IEEE Int. Conf. on Data Mining*, pages 289–296, 2001.
11. E. J. Keogh and M. J. Pazzani. An enhanced representation of time series which allows fast and accurate classification clustering and relevance feedback. In *Proc. of the 4th Int. Conf. on Knowledge Discovery and Data Mining (KDD-98)*, pages 239–243, 1998.
12. E. J. Keogh and M. J. Pazzani. Scaling up dynamic time warping to massive datasets. In *Proc. of the 3rd European Conf. on Principles of Data Mining and Knowledge Discovery (PKDD-1999)*, pages 1–11, 1999.
13. E. J. Keogh and M. J. Pazzani. Derivative dynamic time warping. In *Proc. of the 1st SIAM Int. Conf. on Data Mining (SDM-2001)*, Chicago, IL, USA,, 2001.
14. T. Oates, M. D. Schmill, and P. R. Cohen. A method for clustering the experiences of a mobile robot that accords with human judgements. In *Proc. of the 17th National Conf. on Artificial Intelligence (AAAI-2000)*, pages 846–851, 2000.
15. M. T. Rosenstein and P. R. Cohen. Continuous categories for a mobile robot. In *Proc. of the 16th National Conf. on Artificial Intelligence (AAAI-99)*, pages 634–641, 1999.
16. D. Sankoff and J. B. Kruskal. *Time warps, string edits, and macromolecules: The theory and practice of sequence comparison.* CLSI Publications, 3rd edition, 1999.
17. M. J. Zaki. SPADE: An efficient algorithm for mining frequent sequences. *Machine Learning*, 42(1/2):31–60, 2001.

Using Domain Specific Knowledge for Automated Modeling

Ljupčo Todorovski and Sašo Džeroski

Jožef Stefan Institute, Jamova 39, 1000 Ljubljana, Slovenia
{Ljupco.Todorovski,Saso.Dzeroski}@ijs.si

Abstract. In this paper, we present a knowledge based approach to automated modeling of dynamic systems based on equation discovery. The approach allows for integration of domain specific modeling knowledge in the process of modeling. A formalism for encoding knowledge is proposed that is accessible to mathematical modelers from the domain of use. Given a specification of an observed system, the encoded knowledge can be easily transformed into an operational form of grammar that specifies the space of candidate models of the observed system. Then, equation discovery method LAGRAMGE is used to search the space of candidate models and find the one that fits measured data best. The use of the automated modeling framework is illustrated on the example domain of population dynamics. The performance and robustness of the framework is evaluated on the task of reconstructing known models of a simple aquatic ecosystem.

1 Introduction

Scientists and engineers build mathematical models in order to analyze and better understand the behavior of real world systems. Mathematical models have the ability to integrate potentially vast collections of observations into a single entity. They can be used for simulation and prediction of the behavior of the system under varying conditions. Another important aspect of mathematical models is their ability to reveal the basic processes that govern the behavior of the observed system.

Establishing an acceptable model of an observed system is a very difficult task that occupies a major portion of the work of the mathematical modeler. It involves observations and measurements of the system behavior under various conditions, selecting a set of system variables that are important for modeling, and formulating the model itself. This paper deals with automated modeling task, i.e., the task of formulating a model on the basis of observed behavior of the selected system variables. A framework for automated modeling of real world systems is based on equation discovery methods [6,10,12].

Most of the work in equation discovery is concerned with assisting the empirical (data driven) approach to modeling physical systems. Following this approach, the observed system is modeled on a trial-and-error basis to fit observed data. The scientist first chooses a structure of the model equations from some general class of structures (such as linear or polynomial) that is believed to be

M.R. Berthold et al. (Eds.): IDA 2003, LNCS 2810, pp. 48–59, 2003.

adequate, fits the constant parameters, and checks whether the simulation match the observed data. If not, the procedure is repeated until an adequate model is found. None of the available domain knowledge about the observed system (or a very limited portion thereof) is used in the modeling process. The empirical approach is in contrast to the theoretical (knowledge driven) approach to modeling, where the basic physical processes involved in the observed system are first identified. A human expert then uses domain knowledge about the identified processes to write down a proper structure of the model equations. The values of the constant parameters of these equations are fitted against the observed data using standard system identification methods [7]. If the fit of the model to the observed data is not adequate, human expert can decide to go through a feedback loop and refine the proposed structure of the model equations.

While most of the equation discovery methods are data driven, here we present a knowledge driven method capable of integrating domain specific modeling knowledge in the process of equation discovery. The presented method allows for integration of theoretical and empirical approaches to automated modeling of real world systems. To achieve that goal, we first present a formalism for encoding modeling knowledge from a domain of interest in a form of taxonomy of process classes. Given a specific modeling task, the encoded knowledge can be easily transformed in an operational form of grammar that specifies the space of candidate models for the given task. This grammar representation allows us to use the existing equation discovery method LAGRAMGE [10] in order to search through the space of candidate models and find the one that fits the measurement data best.

The paper is organized as follows. The formalism for encoding domain specific modeling knowledge is presented in Section 2. The automated modeling framework based on the equation discovery method LAGRAMGE is presented in Section 3. The results of experiments on several tasks of reconstructing several models of a simple aquatic ecosystem are presented in Section 4. Finally, Section 5 concludes the paper with a summary, discusses the related work and proposes directions for further work.

2 Encoding Domain Specific Modeling Knowledge

The formalism for encoding domain specific modeling knowledge allows for organization of the knowledge in three parts. The first part contains knowledge about what types of variables are measured and observed in systems from the domain of interest. The second part contains models of typical basic processes that govern the behavior of systems in the observed domain. The third part encodes knowledge about how to combine models of individual basic processes into a single model of the whole system.

The use of the formalism will be illustrated on population dynamics domain. Note that focusing on a single domain is due to the lack of space and does not correspond to the generality of the formalism. The presented formalism allows for encoding modeling knowledge from variety of domains, including chemical kinetics and spring mechanics. The formalism is also general enough to allow representing domain independent knowledge about measurement scales used to measure the system variables. The use of the formalism in these domains is further elaborated in [9].

Population ecology studies the structure and dynamics of populations, where each population is a group of individuals of the same species sharing a common environment. In this paper, we consider modeling the dynamics of populations, especially the dynamics of change of their concentrations (or densities) in the observed environment. The models take the form of systems of differential equations [8].

2.1 Taxonomy of Variable Types

The first part of the domain specific modeling knowledge is the taxonomy of types of the variables that can be used in the models. An example of a taxonomy of types for the population dynamics domain is presented in Table 1.

Table 1. A taxonomy of variable types that are can be used in population dynamics models.

```
type Concentration is nonnegative_real_number
type Inorganic is Concentration
type Population is Concentration
```

The generic variable type in the population dynamics domain is `Concentration`, since we are interested in the change of concentrations of different populations. The definition of the concentration type specifies that concentrations are nonnegative real numbers. The concentration type has two sub-types. The first is `Population`, used to denote a concentration of organic species. The second type `Inorganic` is used to denote a concentration of an inorganic nutrient that can be consumed by organic species. Note that the inheritance rule applies to the taxonomy of types, resulting in the fact that the population and inorganic types are also nonnegative real numbers.

2.2 Taxonomy of Basic Process Classes

The most important part of the modeling knowledge is the taxonomy of process classes. Each of them represents an important class of basic processes that govern or influence the behavior of dynamic systems in the domain of interest. The complete taxonomy of process classes in the population dynamics domain is presented in Figure 1.

On the highest level the taxonomy distinguishes between processes that involve single species or inorganic nutrient and processes that represent interactions between two or more species and nutrients. Down the taxonomy tree, the process classes become more specific. Each leaf node represents a particular modeling alternative usually used by the experts in the domain. The formalization of several process classes from the taxonomy is presented in Table 2.

The first process class `Growth` represents the processes of a single species growth in the absence of any interaction with other populations in the observed environment. The `Growth` process class has two sub-classes, specifying two

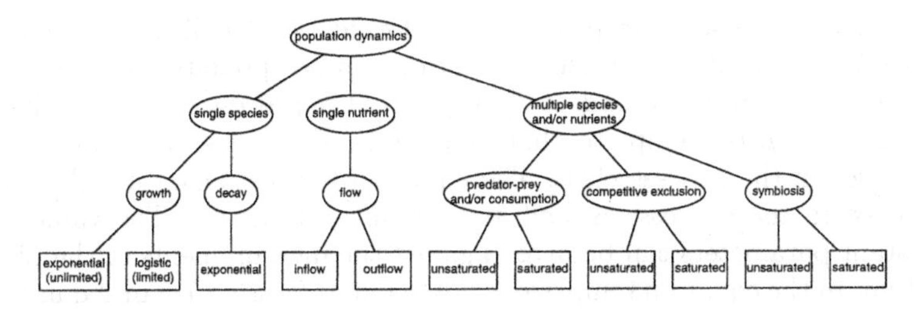

Fig. 1. A taxonomy of classes of processes used for modeling population dynamics systems.

Table 2. A formalization of several process classes from the population dynamics taxonomy presented in Figure 1.

```
process class Growth(Population p)
process class Exponential_growth is Population_growth
  expression const(growth_rate,0,1,Inf) * p
process class Logistic_growth is Population_growth
  expression const(growth_rate,0,1,Inf) * p *
    (1 - p / const(capacity,0,1,Inf))

process class Feeds_on(Population p, set of Concentration cs)
  condition p ∉ cs
  expression p * ∏_{c∈cs} Saturation(c)
```

modeling alternatives for the growth processes. The `Exponential_growth` class specifies unlimited exponential growth of the observed population. For many real-world systems the unlimited growth model is appropriate, since real-world environments typically have limited carrying capacity for the population. In such cases, alternative `Logistic_growth` model is used.

The definition of each process class consist of three parts, each of them specifying one aspect of the processes in the class and/or the models thereof. Each part is presented in more detail below.

Types of the variables involved. This part of the process class definition is used to specify what are the types of the variables that are involved in the processes from the class. For example, each process from the `Growth` process class involves a single population `p`. Furthermore, the processes in the `Feeds_on` class involve a population `p` that depends on several food sources specified as a set of concentrations `cs`. The declarations of the types of variables are inherited through the taxonomy, i.e., processes in the `Exponential_growth` class inherit from the parent class `Growth` the fact that they involve a single variable `p` of `Population` type.

Conditions on the variables involved. This part of the process class definition specifies constraints on the variables involved in the processes. The condition p $\notin$ cs in the Feeds_on process class forbids cannibalism within a single species, i.e., specifies that the population can not predate on itself.

Declaration of the process models. This part of the process class definition specifies the equation template that is used by domain experts for modeling processes in the process class. The equation template can include variables involved in the process, as well as constant parameters. The values of the constant parameters can be fitted against the measurements of the observed system. In the equation template, symbol const(name, lower_bound, default, upper_bound) is used to specify a constant parameter. The symbol specifies the name of the constant parameter, along with the lower and upper bound on its value, as well as its initial or default value.

Table 3. A taxonomy of function classes specifying different expressions for modeling the saturation of predation rate in the population dynamics domain.

```
function class Saturation(Concentration c)
function class No_saturation is Saturation
  expression c
function class Saturation_type_1 is Saturation
  expression c / (c + const(saturation_rate(0,1,Inf)))
```

Note that the expression used to declare the model of the Feeds_on processes involves Saturation(c) expression. It is used to denote whether the predation (or grazing) rate is saturated or not. The latter means that the predation is not unlimited, i.e., when the prey (or nutrient) population becomes abundant, the predation capacity saturates to some limit value. The last function class in Table 3 specifies one of different alternative expressions used for modeling the saturation [8]. The second function class specifies an expression used for modeling unsaturated (unlimited) predation. The declarations of functions have the same structure as declarations of process classes. The difference is that function classes are not used to define processes, but rather specify expression templates that are commonly used in the model templates.

2.3 Scheme for Combining Models of Individual Processes

The modeling knowledge formalism also specifies the scheme that is used to combine the models of individual processes into a model of the whole system. Two combining schemes used to combine models of basic processes in the population dynamics domain are presented in Table 4.

The first combining scheme specifies how the equation that models the time change of a concentration of an inorganic nutrient i is composed from the individual process models. The change of i (assessed by the time derivative di$/d$t of i) decreases with all the Feeds_on interactions, in which a population p consumes the inorganic nutrient i.

The second scheme is a bit more complicated. The change of the concentration of the population p (again assessed by the time derivative of p) increases with the population growth and decreases with the decay. On the other hand,

Table 4. Combining schemes that specify how to combine the models of individual population dynamics processes into a model of the whole system.

```
combining scheme Population_dynamics(Inorganic i)
  di/dt = - Σ_{food,i∈food} const(_,0,1,Inf) * Feeds_on(p, food)

combining scheme Population_dynamics(Population p)
  dp/dt = +Growth(p) - Decay(p)
    + Σ_{food} const(_,0,1,Inf) * Feeds_on(p, food)
    - Σ_{pred,food,p∈food} const(_,0,1,Inf) * Feeds_on(pred, food)
```

Feeds_on processes can positively or negatively influence the change, depending on the role of p in the interaction. The processes where p is involved as a consumer or predator, positively influence the change of p, while the processes where p is involved as a prey (or nutrient), negatively influence the change of p (see the last two lines in Table 4).

3 An Automated Modeling Framework Based on Equation Discovery

The integration of domain specific modeling knowledge in the process of equation discovery is schematized in Figure 2. The integration is supported using the transformation principle. First, the domain specific knowledge is transformed into a grammar based on modeling task specification. The obtained grammar specifies the space of candidate models for the observed system. The equation discovery method LAGRAMGE [10] can be then used to search through the space of candidate models and find the one that fits the measured data best.

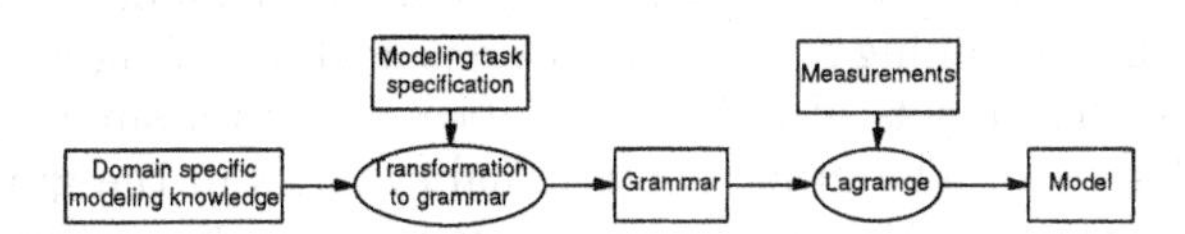

Fig. 2. An automated modeling framework based on the integration of domain specific modeling knowledge in the process of equation discovery.

3.1 Modeling Task Specification

The encoded modeling knowledge is general, in the sense that it allows for modeling of an arbitrary system from the population dynamics domain. In order to use the knowledge for automated modeling of a particular system, a specification of the system has to be provided first. An example of modeling task specification is presented in Table 5.

A specification consists of two parts. The first part specifies the list of observed system variables. Each variable is assigned a type from the taxonomy of

Table 5. A task specification used for modeling a simple aquatic ecosystem consisting of two consumption interactions between three populations of inorganic nutrient, phytoplankton, and zooplankton.

```
variable Inorganic nut
variables Population phyto, zoo

processes Decay(phyto), Decay(zoo)
processes Feeds_on(phyto, nut), Feeds_on(zoo, phyto)
```

types. The example of a simple aquatic ecosystem involves inorganic nutrient, phytoplankton and zooplankton. The second part of the modeling task specification consist of a list of processes that govern the dynamics of the observed system. Each process in the list should belong to a process class from the taxonomy of process classes: the process class and the system variables involved in the process are specified. The two `Decay` processes specify that the populations of phytoplankton and zooplankton tend to decrease in absence of any interactions with the environment and other species. The two `Feeds_on` processes specify that phytoplankton consumes inorganic nutrient and zooplankton consumes phytoplankton.

3.2 Transforming the Modeling Knowledge into a Grammar

The algorithm for transforming the encoded modeling knowledge into grammar takes two arguments as input. The first is the library of modeling knowledge for the domain at hand, and the second is the modeling task specification, specifying the types of the variables and the processes of the observed system.

The transformation of the modeling knowledge proceeds in a top-down manner. It starts with the starting symbol and assigns productions to it, then proceeds with other nonterminal symbols. The starting symbol of the grammar corresponds to the combining schemes in the encoded modeling knowledge. It is used to combine the models of individual processes into a single model of the whole observed system. The other nonterminal symbols in the grammar correspond to process classes. Alternative productions for each nonterminal symbol specify alternative models of the corresponding process class.

For example, consider the aquatic ecosystem example from Table 5. The start symbol will use the combining schemes from Table 4 to compose a model of the whole ecosystem as illustrated in Table 6. The right hand side of the production for the starting symbol builds three equations, one for each variable of the observed ecosystem. Each equation reflects the corresponding combining scheme from Table 4. For example, the second equation is built by summing up the growth processes for the `phyto` population (the leading `0` means that no such process is specified), the `phyto` decay processes (i.e., `Decay(phyto)`, the consumption processes where `phyto` is involved as consumer (i.e., `Feeds_on(phyto, {nut})`), and the predator-prey processes where `phyto` has the role of prey (i.e., `Feeds_on(zoo, {phyto})`). The productions for the other nonterminal symbols are built according to the taxonomies of process and function classes from Tables 2 and 3.

Table 6. A part of the grammar specifying the candidate models for modeling the simple aquatic ecosystem presented in Table 5.

```
Start ->
  time_deriv(nut) = - const[_:0:1:] * Feeds_on_phyto_nut;
  time_deriv(phyto) = 0 - Decay_phyto + const[_:0:1:] * Feeds_on_phyto_nut
    - const[_:0:1:] * Feeds_on_zoo_phyto;
  time_deriv(zoo) = 0 - Decay_zoo + const[_:0:1:] * Feeds_on_zoo_phyto

Feeds_on_phyto_nut -> phyto * Saturation_nut;
Feeds_on_zoo_phyto -> zoo * Saturation_phyto;

Decay_phyto -> Exponential_decay_phyto;
Exponential_decay_phyto -> const[decay_rate:0:1:] * phyto;

Saturation_nut -> No_saturation_nut | Saturation_type_1_nut;
No_saturation_nut -> nut;
Saturation_type_1_nut -> nut / (nut + const[saturation_rate:0:1:]);
```

The grammar from Table 6 specifies 16 possible model structures of the observed system. In the final step from Figure 2, LAGRAMGE is used to fit the parameters of each model structure against measured data and find the model that fits the data best. Strictly speaking, the grammar in Table 6 is not context free. The production for the starting symbol generates two Feeds_on_phyto_nut symbols, one in the first and another one in the second equation. In a context free grammar, these two nonterminal symbols can generate two different expressions. In population dynamics models, however, these two expressions have to be the same. Adding context dependent constraints to the grammar can overcome this limitation of context free grammars used in previous versions of LAGRAMGE [10].

4 Experimental Evaluation

The experiments are performed that evaluate the performance of the proposed automated modeling framework on 16 different tasks of reconstructing 16 models of the simple aquatic ecosystem, specified in Table 5, from simulated data. This experimental setup is useful since it allows a comparison of the discovered model with the original one and an evaluation of the success of reconstruction. This is the first evaluation criterion used in the experiments.

The synthetic experimental data is obtained by simulating the 16 chosen models. Each of the models is simulated from ten randomly chosen initial states to obtain ten simulation traces.[1] These traces are then used for a cross-validation estimate of the model error on test data, unseen during the discovery process. Root mean squared error (RMSE) was used to measure the quality of the reconstructed models. The RMSE measures the discrepancy between the provided

[1] The initial values of the system variables were randomly chosen using a random variable uniformly distributed over the $[0.5, 1.5]$ interval. Each simulation trace included 100 time steps of 0.1 time units.

values of the system variables and the values obtained by simulating the discovered model. The cross-validated RMSE of the model is the second evaluation criterion used in the experiments.

In addition to reconstructing models from noise free simulation data, we also tested the noise robustness of our automated modeling approach. To each simulation trace, artificially generated random Gaussian noise is added. The noise at five relative noise levels is added: 1%, 2%, 5%, 10%, and 20%. The relative noise level of l% means that we replaced the original value x with the noisy value of $x \cdot (1 + l \cdot r/100)$, where r is a normally distributed random variable with mean zero and standard deviation one.

We compare the performance of our automated modeling framework with the performance of previous version of equation discovery method LAGRAMGE that use context free grammars and can not use the lower and upper bounds on the values of the constant parameters [2]. The results of the comparison are summarized in Figure 3 and Table 7.

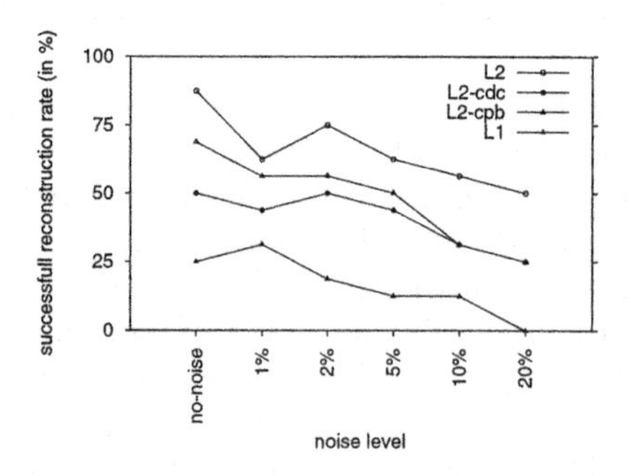

Fig. 3. The percentage of successfully reconstructed model structures (out of 16) with different versions of the grammar. Legend: `L2` = using completely constrained context dependent grammar, `L2-cdc` = using context free grammar, `L2-cpb` = using context dependent grammar without bounds on the values of the constant parameters, and `L1` = using completely unconstrained grammar (as in previous version of LAGRAMGE).

The graph in Figure 3 shows that the use of both context dependent grammars and constraints (bounds) on the values of the constant parameters contribute to the noise robustness of our automated modeling approach. The `L2` line in the graph shows that LAGRAMGE with completely constrained grammar was able to reconstruct almost 90% of the original models (14 out of 16) from noise free data and 50% of the original models from noisy data with 20% relative noise. The comparison of `L2` line with each of the `L2-cdc` and `L2-cpb` lines shows that the use of both kinds of constraints in the grammar are almost equally important. When context free grammar is used (`L2-cdc`), only 50% of the original models are successfully reconstructed from noise free data. Note that this reconstruction rate is equal to the one gained using completely constrained grammar on the data with 20% relative noise. Similarly, the successful reconstruction rate significantly drops if no bounds on the values of the constant parameters are specified (line `L2-cpb`). Note also that none of the 16 models can be success-

Table 7. The average of root squared mean errors (RMSE) of the 16 models, discovered using different constraints in the grammar from noise free data and data with added noise at five different noise levels. The RMSE of each model was estimated using 10-fold cross validation. For legend, see Figure 3.

noise level (in %)	L2	L2-cdc	L2-cpb		L1
no-noise	0.00643	0.00364	0.00404		0.00379
1	0.00880	0.00756	0.00775		0.0119
2	0.0140	0.0131	0.0132		0.0131
5	0.0319	0.0310	0.0310		0.0309
10	0.0293	0.0278	0.0279	(*1)	0.118
20	0.0549	0.0629	0.0458	(*5)	0.0647

fully reconstructed from the data with 20% relative noise using the completely unconstrained grammar (L1).

Furthermore, the comparison of all four lines on the graph in Figure 3 shows that the improvements due to each of the two different kinds of constraints almost sum up. That means that each kind of grammar constraints improves different aspect of reconstruction capabilities of LAGRAMGE. Note that context dependent grammars also reduce the complexity of the space of possible models, sought by LAGRAMGE 2.0. While the context dependent grammar specifies 16 distinct model structures, the context free grammar (i.e., the one without context dependent constraints) specifies the space of 256 distinct model structures. Thus, the use of context dependent constraints reduces the complexity of the search space by factor of sixteen.

The figures in Table 7 show the average RMSE of of the 16 rediscovered models using different grammars. Note that the results are unexpected. Note that the models discovered with unconstrained versions of the grammar (i.e., L2-cdc, L2-cpb, and L1) are on average more accurate than the models discovered with completely constrained grammar. Why? Context free grammars allow for models with inappropriate structure where different expressions are used to model the same process in different equations. Grammars without bounds on the values of the constant parameters allow for models with inappropriate (unrealistic) values of the constant parameters (e.g., negative growth or saturation rate). By analyzing the reconstructed models, we can see that most of these inappropriate models can better fit the data than the original models with proper structure and parameter values. Thus, using RMSE only is not an appropriate criterion for validating the quality of the reconstructed models.

Finally, note that inappropriate structure or inappropriate values of the constant parameters can lead to problematic models that can not be simulated.[2] The (*N) symbols beside the last two figures from Table 7 denote that N out of 16 discovered models could not be simulated properly. On the other hand, all the models discovered using the fully constrained grammar can be properly simulated.

[2] In population dynamics domain the usual reason for inability to simulate a model is "division by zero" singularities encountered due to a negative value of the saturation rate constant parameter.

5 Summary and Discussion

In this paper, we presented a knowledge based approach to automated modeling of dynamic systems based on equation discovery. The approach allows for representation and integration of domain specific modeling knowledge in the process of equation discovery. The preliminary experimental evaluation shows the importance of the modeling knowledge for the performance of the approach. Using domain knowledge to constrain the space of possible model structures and possible values of the constant parameters can greatly improve the noise robustness of the equation discovery methods. Due to the lack of space, the use of the framework was illustrated on a single experimental series in the population dynamics domain only. Further experiments show that the approach is capable of reconstructing models of more complex aquatic systems including different kinds of interactions between species [9]. Note however, that the presented encoding formalism is general enough to allow for encoding and integration of different kinds of modeling knowledge from a variety of domains [9].

The presented approach follows the compositional modeling (CM) paradigm [3], proposed within the area of qualitative modeling [4]. CM is automated modeling paradigm used for automated building qualitative models from observations in presence of domain specific modeling knowledge. In the CM approaches, modeling knowledge is organized as a library of model fragments (equivalent to the models of individual processes in our formalism). Given a modeling task specification (or scenario, in the terminology of CM), CM composes a set of appropriate model fragments into a model that is suitable for modeling the observed system. The obtained model is evaluated by qualitative simulation. Although the concepts introduced within the CM area are relevant also for automated building of quantitative models of real world systems, this venue of work has not been explored much. A notable exception is the PRET reasoning system for automated modeling of dynamic systems [1]. PRET employs other kinds of knowledge in the process of automated modeling in form of "conservation rules". The example of such rule in electrical circuits domain is Kirchoff's law specifying that the sum of input and output currents at any observed point in the circuit is zero. Similarly, the force balance rule in mechanics domain specifies that the sum of forces at any observed coordinate of the mechanical system is zero. Thus, the knowledge used in PRET is much more general and cross-domain, as opposed to domain specific knowledge used in our approach.

The approach presented here follows our previous work on building process based models from measured data, presented in [2] and [5]. Note however, that the taxonomy of process classes introduced here allows for better organization of knowledge. Also, here we made the combining scheme explicit (i.e., declarative), as opposed to the implicit combining scheme used in [2] and assumption that models of individual processes are always additively combined in [5].

The usability of the approach should be further confirmed by applying it to the tasks of reconstructing dynamic system models from other domains. The ultimate goal of further and ongoing empirical evaluation is to apply the approach to different tasks of modeling real-world systems from measured data.

Although the scope of this paper is modeling dynamic change of the observed system through time, the approach can be extended to incorporate knowledge about spatial processes also. In the population dynamics domain, these processes would reflect the spatial diffusion of populations through the environment. The extended approach would allow modeling of spatio-temporal dynamic systems with partial differential equations (PDEs). The PADLES method [11] can be used for discovery of PDEs.

Acknowledgements. We acknowledge the support of the cInQ (Consortium on discovering knowledge with Inductive Queries) project, funded by the European Commission under contract IST-2000-26469.

References

1. E. Bradley, M. Easley, and R. Stolle. Reasoning about nonlinear system identification. *Artificial Intelligence*, 133:139–188, 2001.
2. S. Džeroski and L. Todorovski. Encoding and using domain knowledge on population dynamics for equation discovery. In L. Magnani and N. J. Nersessian, editors, *Logical and Computational Aspects of Model-Based Reasoning*, pages 227–248. Kluwer Academic Publishers, Boston, MA, 2002.
3. B. Falkenheiner and K. D. Forbus. Compositional modeling: Finding the right model for the job. *Artificial Intelligence*, 51:95–143, 1991.
4. B. Kuipers. *Qualitative reasoning: modeling and simulation with incomplete knowledge.* MIT Press, Cambridge, MA, 1994.
5. P. Langley, J. Sanchez, L. Todorovski, and S. Džeroski. Inducing process models from continuous data. In *Proceedings of the Fourteenth International Conference on Machine Learning*, pages 347–354, San Mateo, CA, 2002. Morgan Kaufmann.
6. P. Langley, H. A. Simon, G. L. Bradshaw, and J. M. Żythow. *Scientific Discovery.* MIT Press, Cambridge, MA, 1987.
7. L. Ljung. *System Identification - Theory For the User.* PTR Prentice Hall, Upper Saddle River, NJ, 2000. Second Edition.
8. J. D. Murray. *Mathematical Biology.* Springer, Berlin, 1993. Second, Corrected Edition.
9. L. Todorovski. *Using domain knowledge for automated modeling of dynamic systems with equation discovery.* PhD thesis, Faculty of computer and information science, University of Ljubljana, Ljubljana, Slovenia, 2003.
10. L. Todorovski and S. Džeroski. Declarative bias in equation discovery. In *Proceedings of the Fourteenth International Conference on Machine Learning*, pages 376–384, San Mateo, CA, 1997. Morgan Kaufmann.
11. L. Todorovski, S. Džeroski, A. Srinivasan, J. Whiteley, and D. Gavaghan. Discovering the structure of partial differential equations from example behavior. In *Proceedings of the Eighteenth International Conference on Machine Learning*, pages 991–998, San Mateo, CA, 2000. Morgan Kaufmann.
12. T. Washio and H. Motoda. Discovering admissible models of complex systems based on scale-types and identity constraints. In *Proceedings of the Fifteenth International Joint Conference on Artificial Intelligence*, volume 2, pages 810–817, San Mateo, CA, 1997. Morgan Kaufmann.

Resolving Rule Conflicts with Double Induction

Tony Lindgren and Henrik Boström

Department of Computer and Systems Sciences,
Stockholm University and Royal Institute of Technology,
Forum 100, 164 40 Kista, Sweden
{tony, henke}@dsv.su.se

Abstract. When applying an unordered set of classification rules, the rules may assign more than one class to a particular example. Previous methods of resolving such conflicts between rules include using the most frequent class in the conflicting rules (as done in CN2) and using naïve Bayes to calculate the most probable class. An alternative way of solving this problem is presented in this paper: by generating new rules from the examples covered by the conflicting rules. These newly induced rules are then used for classification. Experiments on a number of domains show that this method significantly outperforms both the CN2 approach and naïve Bayes.

1 Introduction

Two major induction strategies are used in top-down rule induction: Divide-and-Conquer [9] and Separate-and-Conquer (SAC) [5]. The former strategy does not generate overlapping rules and hence no conflict between rules can occur, while the latter strategy can give rise to overlapping rules. The rules can be ordered to avoid conflicts (which results in so called decision lists) [10] or they may be used without any ordering [3,2]. In the latter case one has to deal with conflicts that can occur when two or more rules cover the same example but assign different classes to it.

This work addresses the problem of handling conflicting rules and how to solve the conflicts in a more effective way than using existing methods. Previous methods that deals with conflicts among rules either returns the most frequent class covered by the conflicting rules (as done in CN2 [2]) here referred to as *frequency-based classification*, or uses naïve Bayes [8] to calculate the most probable class, as done in the rule induction system RDS [1].

We propose another way of resolving such conflicts: by applying rule induction on the examples covered by the conflicting rules, in order to generate a new set of rules that can be used instead of the conflicting rules. The motivation for this method is that by focusing on the examples within the region of interest (i.e. where the example to be classified resides), one is likely to obtain rules that better separate classes in this region compared to having to consider all examples, since examples outside the region may dominate the rule induction process, making the separation of classes within the region of marginal importance. This novel method is called *Double Induction.*

M.R. Berthold et al. (Eds.): IDA 2003, LNCS 2810, pp. 60–67, 2003.

The paper is organized as follows. In section 2, we first review frequency-based classification and naïve Bayes to resolve rule conflicts, and then describe Double Induction. In section 3, the three different methods are compared empirically and the results are presented. Finally, in section 4, we discuss the results and give pointers to future work.

2 Ways of Resolving Classification Conflicts

In this section, we first recall two previous methods for resolving classification conflicts among overlapping rules and then introduce the novel method, Double Induction.

2.1 Frequency-Based Classification

The system CN2 [2] resolves classification conflicts between rules in the following way. Given the examples in Fig. 1, the class frequencies of the rules that covers the example to be classified (marked with '?') are calculated:

$$C(+) = covers(R_1, +) + covers(R_2, +) + covers(R_3, +) = 32$$

and

$$C(-) = covers(R_1, -) + covers(R_2, -) + covers(R_3, -) = 33$$

where $covers(R, C)$ gives the number of examples of class C that are covered by rule R. This means that CN2 would classify the example as belonging to the negative class (-). More generally:

$$FreqBasedClassification = argmax_{Class_i \in Classes} \sum_{j=1}^{|CovRules|} covers(R_j, C_i)$$

where $CovRules$ is the set of rules that cover the example to be classified.

2.2 Naïve Bayes Classification

Bayes theorem is as follows:

$$P(C|R_1 \wedge \ldots \wedge R_n) = P(C)\frac{P(R_1 \wedge \ldots \wedge R_n|C)}{P(R_1 \wedge \ldots \wedge R_n)}$$

where C is a class label for the example to be classified and $R_1 \ldots R_n$ are the rules that cover the example. As usual, since $P(R_1 \wedge \ldots \wedge R_n)$ does not affect the relative order of different hypotheses according to probability, it is ignored. Assuming (naively) that $P(R_1 \wedge \ldots \wedge R_n|C) = P(R_1|C) \ldots P(R_n|C)$, the maximum a posteriori probable hypothesis (MAP) is:

$$h_{MAP} = argmax_{Class_i \in Classes} P(Class_i) \prod_{Rj \in Rules}^{|Rules|} P(R_j|Class_i)$$

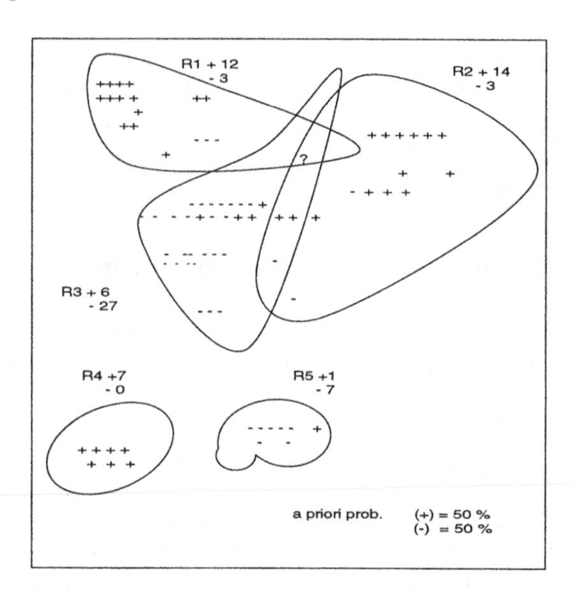

Fig. 1. Three rules covering an example to be classified (marked with ?). The training examples are labeled with their respective classes (+ and -).

where *Rules* is the set of rules that cover the example to be classified.

If we again consider the example shown in Fig. 1, we get:

$$P(+|R_1 \wedge R_2 \wedge R_3) = P(+) * P(R_1|+) * P(R_2|+) * P(R_3|+) =$$

$$40/80 * 12/40 * 14/40 * 6/40 = 0.0079$$

$$P(-|R_1 \wedge R_2 \wedge R_3) = P(-) * P(R_1|-) * P(R_2|-) * P(R_3|-) =$$

$$40/80 * 3/40 * 3/40 * 27/40 = 0.0019$$

Hence naïve Bayes assigns the positive (+) class to the example. Note that if a rule involved in a conflict does not cover any examples of a particular class, this would eliminate the chances for that class to be selected, even if there are several other rules that cover the example with a high probability for that class. To overcome this problem, we use Laplace-m correction (described in [6]) in the experiments.

2.3 Double Induction

The idea of Double Induction is to induce new rules based on the examples that are covered by the rules in conflict. By doing this we obtain a completely fresh set of rules that are tailor made to separate the examples in this subset of the whole domain. By concentrating on a small subspace of the example space there is a higher chance to find rules that separate the classes better. The training set is randomly divided into two halves in order to create a grow set (to induce rules) and a prune set (to prune rules). This division is likely to further reduce the probability that we will find the same rules as before. The Double Induction algorithm is given in Fig. 2.

```
Input: R1 = rules from the first induction round, e = example to be
classified,
E1 = training examples
Output: C = a class assigned to e

collect all rules R1,e ⊆ R1 that cover e
if conflictingRules(R1,e) then
   collect all training examples E2 ⊆ E1 covered by R1,e
   induce new rules R2 from E2
   collect all rules R2,e ⊆ R2 that cover e
   if conflictingRules(R2,e) then
      let C = naiveBayes(R2,e,E2)
   else let C = majorityClass(R2,e)
else let C = majorityClass(R1,e)
```

Fig. 2. The Double Induction algorithm.

It is worth noting that when solving conflicts with naïve Bayes on the newly induced rules, the a priori probability is computed from the examples covered by the previously conflicting rules. This means that the a priori probability reflects the probability distribution of this subspace (contained by the rules in conflict), rather than the a priori probability of the whole domain.

Consider again the scenario shown in Fig. 1. Given the examples covered by R_1, R_2 and R_3, Separate-and-Conquer may come up with the three new rules shown in Fig. 3. The unlabeled example is then classified using these newly induced rules. In our hypothetical scenario, the example is covered by R_6 resulting in that the positive class is assigned to the example.

In our experiments we use two different sets of examples as input in the second rule induction round: one which has no duplicates of examples and one with the concatenation of all the covered examples of every rule, in which some examples may be present more than once (i.e., a multi-set). The latter is in a sense a weighting scheme of the examples in the example set. One reason for using such a weighting scheme is that it has been empirically demonstrated that the examples in the intersection are more important (and hence should have more weight) than other examples when doing the classification, see [7]. Note that this type of weighting is implicitly used in both frequency-based classification and naïve Bayes.

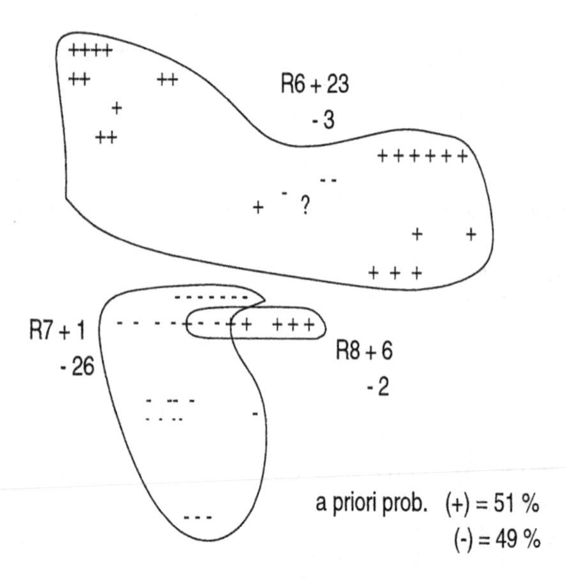

Fig. 3. Three new rules from a second induction round.

3 Empirical Evaluation

Double Induction has been implemented as a component to the Rule Discovery System (RDS) [1], which is a rule based machine learning system that supports various rule induction techniques, e.g., separate-and-conquer, divide-and-conquer, and ensemble methods like bagging and boosting.

Double Induction has been implemented both with and without weighting of the examples according to the coverage of previous learned rules. Both of these variants are compared to frequency-based classification and naïve Bayes.

3.1 Experimental Setting

The employed search strategy was Separate-and-Conquer together with information gain to greedily choose what condition to add when refining a rule. One half of the training set was used as grow set and the second half was used as pruning set. The rules were pruned using incremental reduced error pruning (IREP) [4]. These settings were used in both induction rounds, i.e. both when inducing the initial set of rules and when resolving rule conflicts. The experiments were performed using 10 fold cross-validation.

All datasets used were taken from the UCI Machine Learning Repository except the King-Rook-King-Illegal (KRKI) database which comes from the Machine Learning group at the University of York. In Table 1, the domains used in the experiment are shown, as well as their main characteristics.

Table 1. The domains used in the experiment

Domain	Classes	Class distribution	Examples
The Glass	2	24.1, 75.9	112
Balance Scale	3	8, 46, 46	625
Breast Cancer	2	29.7, 70.3	286
Liver-disorders	2	42, 58	345
Car Evaluation	4	4, 4, 22, 70	1728
Dermatology	6	5.5, 13.3, 14.2, 16.7, 19.7, 30.6	366
Congressional Voting	2	45.2, 54.8	435
Ionosphere	2	36, 64	351
Lymphography	4	1.4, 2.7, 41.2, 54.7	148
New thyroid	3	69.8, 16.3, 13.9	215
Primary tumor	22	24.8, 5.9, 2.7, 4.1, 11.5, 0.3, 4.1, 1.8, 0, 0.6, 8.3, 4.7, 2.1, 7.1, 0.6, 0.3, 2.9, 8.6, 1.8, 0.6, 0.3, 7.1	339
Sonar	2	53, 47	208
Nursery	5	33.3, 0.0, 2.5, 32.9, 31.2	12960
Shuttle Landing Control	2	47.8, 52.2	278
KRKI	2	34, 66	1000

3.2 Experimental Results

The rule conflict resolution methods were tested on fifteen domains. In four of these domains, the rules learned were without conflict. These domains were: Glass, Liver-disorders, Ionosphere and Sonar.

The results from the domains with rule conflicts are shown in Table 2, where the result for each domain has been obtained by ten-fold cross-validation. Exactly the same folds and generated rules are used by the four classification methods.

In Table 2, the first column gives the name of the domain. The second column shows the accuracy of Double Induction with weighted examples, while the third column gives the accuracy of Double Induction with no weighting. The fourth column shows the accuracy of frequency-based classification and the fifth column shows the accuracy of naïve Bayes. The sixth column shows the percentage of all classifications that involve conflicting rules. This gives an upper-bound of how much the accuracy can be improved.

Table 3 gives numbers that are used for the significance test. Inside each parenthesis, two numbers are given: the number of domains in which the method in the leftmost column has a higher respectively lower accuracy compared to the method in the upper row (wins, losses). The value after each parenthesis shows the p-value according to an exact version of McNemar's test. An asterisk (*) is used to signal that the result is statistically significant, using the threshold $p < 0.05$.

It is worth noting that Double Induction with weighting is significantly more accurate than both naïve Bayes and frequency-based classification. It is not significantly more accurate than Double Induction without weighting, but is still better (6 wins and 2 losses). Double Induction without weighting performs significantly better than frequency-based classification, but not significantly better than naïve Bayes. There is no significant difference in accuracy between naïve Bayes and frequency-based classification.

Table 2. Accuracy of Double Induction (with and without weights), frequency based classification and naïve Bayes.

Domain	D. Ind. w weight	D. Ind.	frequency-b.	naïve B.	conflicts
Balance scale	83.30	82.43	81.20	82.78	50.8
Breast cancer	75.00	74.60	73.81	75.00	21.4
Car	86.82	87.21	78.04	77.59	34.1
Dermatology	91.69	91.69	91.69	91.08	5.2
C. Votes	95.20	94.95	94.44	93.94	6.6
Lymphography	80.00	80.00	79.26	80.00	26
New thyroid	85.71	87.24	85.71	85.71	5.6
Primary tumor	38.33	38.33	37.50	37.50	49.2
Nursery	85.63	85.42	79.91	84.32	26.0
Shuttle	94.86	94.07	93.68	94.86	7.5
KRKI	98.35	98.24	92.20	98.02	11.2

Table 3. Result of McNemar's test.

	D. Ind. w. weight	D. Induction	frequency-b.	naïve Bayes
D. Ind. w. weight	(0, 0), 1	(6, 2), 0.289	(9, 0), 3.91e-3*	(7, 0), 1.56e-2*
D. Induction	(2, 6), 0.289	(0, 0), 1	(10, 0), 1.953e-3*	(7, 3), 0.344
frequency-b.	(0, 9), 3.91e-3*	(0, 10), 1.953e-3*	(0, 0), 1	(3, 7), 0.344
naïve Bayes	(0, 7), 1.56e-2*	(3, 7), 0.344	(7, 3), 0.344	(0, 0), 1

To minimize the amount of work within Double Induction, the conflicts already seen (for a particular fold) are saved to allow re-use of the generated rules in the second round whenever the same conflict is observed again (within the same fold). This can greatly speed up the classification process. For the current domains, this reuse of classifications varies from 2 percent to 60 percent of all classifications with conflicts.

4 Discussion

Instead of using the information conveyed in a set of conflicting rules (and the class distributions of the whole training set, if using naïve Bayes) and making a qualified guess of which class is most probable given the rules, we suggest doing a second round of rule induction in hope of finding a better separation of the examples. This is often possible due to that the induction algorithm only has to consider the examples that are relevant to the region in which the example to be classified resides. This idea has been empirically demonstrated to significantly outperform two previous methods for resolving rule conflicts: frequency-based classification and naïve Bayes.

Furthermore, as seen by the experiments, it is worthwhile to use the weighting of the examples that is given by the first set of rules, because it helps in the second round to produce a better set of rules. This confirms the earlier observation in [7], that examples in the intersection of conflicting rules are more important than other examples.

One of the major drawbacks of using Double Induction is of course the computational cost. However, if accuracy is of uttermost importance and quick response time is not, then it is a useful technique.

One issue that needs further investigation is the use of other weighting schemes than the one used in our experiments which is quite conservative. Another issue is to resolve new conflicts that occur between rules obtained in the second round, by continuing recursively instead of using naïve Bayes after the second round. This recursive step could be done until no further conflicts are present in the resulting rules or until no new rules can be found.

References

1. Henrik Boström. Rule discovery system user manual. *Compumine AB*, 2003.
2. P. Clark and R. Boswell. Rule induction with CN2: Some recent improvements. In *Proceedings Fifth European Working Session on Learning*, pages 151–163, Berlin, 1991. Springer.
3. P. Clark and T. Niblett. The CN2 induction algorithm. *Machine Learning, 3, 261–283*, 1989.
4. J. Fürnkranz and G. Widmer. Incremental reduced error pruning. In W.W. Cohen and H. Hirsh, editors, *Proceedings of the 11th International Conference on Machine Learning*, pages 70–77. Morgan Kaufmann, 1994.
5. Johannes Fürnkranz. Separate-and-conquer rule learning. *Artificial Intelligence Review*, 1999.
6. R. Kohavi, B. Becker, and D. Sommerfield. Improving simple bayes. In *Proceedings of the European Conference on Machine Learning*, 1997.
7. Tony Lindgren and Henrik Boström. Classification with intersecting rules. In *Proceedings of the 13th International Conference on Algorithmic Learning Theory (ALT'02)*, pages 395–402. Springer-Verlag, 2002.
8. Duda R. O. and Hart P. E. *Pattern Classification and Scene Analysis*. Wiley, 1973.
9. J. R. Quinlan. Induction of decision trees. *Machine Learning*, 1:81–106, 1986.
10. R. Rivest. Learning decision lists. *Machine Learning, 2(3), 229–246*, 1987.

A Novel Partial-Memory Learning Algorithm Based on Grey Relational Structure

Chi-Chun Huang[1] and Hahn-Ming Lee[2]

[1] Department of Electronic Engineering,
National Taiwan University of Science and Technology
Taipei 106, Taiwan
cjh84@ms26.hinet.net

[2] Department of Computer Science and Information Engineering,
National Taiwan University of Science and Technology
Taipei 106, Taiwan
hmlee@mail.ntust.edu.tw

Abstract. In instance-based learning, the storage of instances must increase along with the number of training instances. In addition, it usually takes too much time to classify an unseen instance because all training instances have to be considered in determining the 'nearness' between instances. This paper proposes a novel partial-memory learning method based on the grey relational structure. That is, only some of the training instances are adopted for classification. The relationships among instances are first determined according to the grey relational structure. In this relational structure, the inward edges of each training instance, indicating how many times each instance is used as the nearest neighbor or neighbors in determining the class labels of other instances, can be found. This approach excludes the training instances with no or few inward edges for learning. By using the proposed approach, new instances can be classified with a few training instances. Five datasets are used for demonstrating the performance of the proposed approach. Experimental results indicate that the classification accuracy can be maintained when most of the training instances are pruned prior to learning. Meanwhile, the number of remained training instances is comparable to that of other existing pruning techniques.

1 Introduction

Over the past decade, instance-based learning (IBL) [1,2,12] has gained great popularity in machine learning. Also known as memory-based learning [16], an IBL system generally collects a set of training instances, where each instance is represented by n attributes with an output label (or value). To classify a new, unseen instance, some similarity functions are first used to determine the relationships (i.e., the 'nearness') among the new instance and all training instances. Accordingly, the output label of the new instance is assigned as that of its "nearest" training instance or instances. Within the nearest neighbor rule

M.R. Berthold et al. (Eds.): IDA 2003, LNCS 2810, pp. 68–75, 2003.

[4,8], instance-based learning can yield excellent performance, compared with others classification algorithms.

In general, since all training instances are stored prior to learning in the IBL system, the storage will increase along with the number of training instances. Furthermore, in such case, it usually takes much time to classify a new, unseen instance because all training instances have to be considered in determining the 'nearness' between instances. To overcome these problems, various partial-memory learning or instance pruning techniques have been investigated [9,15,18]. That is, only some of the training instances are adopted for classifying new, unseen instances. In [10], the Condensed Nearest Neighbor rule was proposed to select representative instances from the original set of training instances for learning. In this method, a set S of q instances corresponding to q classes in a specific domain is first randomly selected from the original training set. If a training instance i is misclassified by using S, then this instance will be added to S (i.e., S will be updated). This procedure is repeated until all training instances are checked. As a result, the final training set S is used for learning. In this method, however, the final training set S changes along with the initial set of selected instances. In another approach, namely the Reduced Nearest Neighbor rule [9], a training instance j will be removed if the classification accuracy of other training instances is not decreased without considering instance j.

Generally, the classification accuracy may be decreased when some of the training instances in IBL are not adopted for learning, implying that, the above pruning tasks should be done carefully. In other words, reducing the number of training instances in classification problems becomes increasingly difficult. Thus, in this paper, we propose a novel partial- memory learning method for instance-based learning based on the grey relational structure, a directed graph used to determine the relationships among instances. For classifying a new, unseen instance, the relationships among instances are first determined according to the grey relational structure. In the grey relational structure, the inward edges of each training instance, indicating how many times each instance is used as the nearest neighbor or neighbors in determining the class labels of other instances, can be found. This approach excludes the training instances with no or few inward edges for solving classification problems (That is, these instances are rarely used in determining the class label of other instances and thus are considered to be eliminated). By using the proposed approach, new instances can be classified with a few training instances. Five datasets are used for demonstrating the performance of the proposed learning approach. Experimental results indicate that the classification accuracy can be maintained when most of the training instances are pruned prior to learning. Meanwhile, the number of remained training instances in the proposal presented here is comparable to that of other existing pruning techniques.

The rest of this paper is structured as follows. The concept of grey relational structure is reviewed in Sections 2. In Section 3, a novel partial-memory learning algorithm based on grey relational structure is presented. In Section 4, experiments performed on five datasets are reported. Finally, conclusions are made in Section 5.

2 Grey Relational Structure

In [11], an instance-based learning approach based on *grey relational structure* is proposed. In this relational structure, grey relational analysis [5,6,7,14] is employed, with calculating the *grey relational coefficient* (GRC) and the *grey relational grade* (GRG), to determine the relationships among instances. Assume that a set of m+1 instances $\{x_0, x_1, x_2, \ldots, x_m\}$ is given, where x_0 is the referential instance and $x_1, x_2, \ldots, x_m$ are the compared instances. Each instance x_e includes n attributes and is represented as $x_e = (x_e(1), x_e(2), \ldots, x_e(\mathrm{n}))$. The grey relational coefficient can be expressed as,

$$GRC\,(x_0(p), x_i(p)) = \frac{\Delta\mathrm{min} + \zeta\Delta\mathrm{max}}{|x_0(p) - x_i(p)| + \zeta\Delta\mathrm{max}}, \tag{1}$$

where $\Delta\mathrm{min} = \min_{\forall j} \min_{\forall k} |x_0(k) - x_j(k)|$, $\Delta\max = \max_{\forall j} \max_{\forall k} |x_0(k) - x_j(k)|$, $\zeta \in [0,1]$ (Generally, $\zeta = 0.5$), $i = j =$ 1, 2, ..., m, and $k = p =$ 1, 2, ..., n.

Here, $GRC\,(x_0(p), x_i(p))$ is viewed as the similarity between $x_0(p)$ and $x_i(p)$. If $GRC\,(x_0(p), x_1(p))$exceeds $GRC\,(x_0(p), x_2(p))$ then the similarity between $x_0(p)$ and $x_1(p)$ is larger than that between $x_0(p)$ and $x_2(p)$; otherwise the former is smaller than the latter. In addition, if x_0 and x_i have the same value for attribute p, $GRC\,(x_0(p), x_i(p))$ (i.e., the similarity between $x_0(p)$ and $x_i(p)$) will be one. By contrast, if x_0 and x_i are different to a great deal for attribute p, $GRC\,(x_0(p), x_i(p))$ will be close to zero. Notoably, if attribute p of each instance x_e is numeric, the value of $GRC\,(x_0(p), x_i(p))$ is calculated by Eq. (1). If attribute p of each instance x_e is symbolic, the value of $GRC\,(x_0(p), x_i(p))$ is calculated as: $GRC\,(x_0(p), x_i(p)) = 1$, if $x_0(p)$ and $x_i(p)$ are the same; $GRC\,(x_0(p), x_i(p)) = 0$, if $x_0(p)$ and $x_i(p)$ are different.

Accordingly, the grey relational grade between instances x_0 and x_i is calculated as,

$$GRG\,(x_0, x_i) = \frac{1}{\mathrm{n}} \sum_{k=1}^{n} GRC(x_0(k), x_i(k)) \tag{2}$$

where $i =$ 1, 2, ..., m.

The main characteristic of grey relational analysis is stated as follows.

If $GRG\,(x_0, x_1)$*is larger than* $GRG\,(x_0, x_2)$ *then the difference between* x_0 *and* x_1 *is smaller than that between* x_0 *and* x_2*; otherwise the former is larger than the latter.*

According to the degree of *GRG*, the grey relational order (GRO) of observation x_0 can be stated as follows.

$$GRO\,(x_0) = (y_{01}, y_{02}, \ldots, y_{0m}) \tag{3}$$

where $GRG\,(x_0, y_{01}) \geq GRG\,(x_0, y_{02}) \geq \ldots \geq GRG\,(x_0, y_{0\mathrm{m}})$, $y_{0r} \in \{x_1, x_2, x_3, \ldots, x_m\}$, r=1, 2, ..., m, and $y_{0a} \neq y_{0b}$ if a≠b.

The grey relational orders of observations $x_1, x_2, x_3, \ldots, x_m$can be similarly obtained as follows.

$$GRO\,(x_q) = (y_{q1}, y_{q2}, \ldots, y_{qm}) \tag{4}$$

where q=1, 2, ..., m, $GRG\,(x_q, y_{q1}) \geq GRG\,(x_q, y_{q2}) \geq \ldots \geq GRG\,(x_q, y_{q\mathrm{m}})$, $y_{qr} \in \{x_0, x_1, x_2, \ldots, x_m\}$, $y_{qr} \neq x_q$, r=1, 2, ..., m, and $y_{qa} \neq y_{qb}$ if a≠b.

In Eq. (4), the *GRG* between observations x_q and y_{q1} exceeds those between x_q and other observations (y_{q2}, y_{q3}, ..., y_{qm}). That is, the difference between x_q and y_{q1} is the smallest.

Here, a graphical structure, namely *h-level grey relational structure* (h≦m), can be defined as follows to describe the relationships among referential instance $x_q$and all other instances, where the total number of nearest instances of referential instance x_q(q=0, 1, ..., m) is restricted to h.

$$GRO^*\,(x_q, \mathrm{h}) = \left(y_{q1}, y_{q2}, \ldots, y_{q\mathrm{h}}\right) \tag{5}$$

where $GRG\,(x_q, y_{q1}) \geq GRG\,(x_q, y_{q2}) \geq \ldots \geq GRG\left(x_q, y_{q\mathrm{h}}\right)$, $y_{qr} \in \{x_0, x_1, x_2, \ldots, x_m\}$, $y_{qr} \neq x_q$, r=1, 2, ..., h, and $y_{qa} \neq y_{qb}$ if a≠b.

That is, a directed graph, shown in Fig. 1, is used to express the relational space, where each instance x_q (q=0, 1, ..., m) and its "nearest" instances (i.e., y_{qr}, r=1, 2, ..., h) are represented by vertices and each expression $GRO^*\,(x_q, \mathrm{h})$ is represented by h directed edges (i.e., $x_q$to y_{q1}, $x_q$to y_{q2}, ..., $x_q$to y_{qh}).

Here, the characteristics of the above h-level grey relational structure are detailed. First, for each instance x_q, h instances (vertices) are connected by the inward edges from instance x_q. That is, these instances are the nearest neighbors (with small difference) of instance x_q, implying that, they evidence the class label of instance x_q according to the nearest neighbor rule [8,10]. Also, in the one-level grey relational structure, instance x_{q1}, with the largest similarity, is the nearest neighbor of instance x_q. Therefore, a new, unseen instance can be classified according to its nearest instance in the one-level grey relational structure or its nearest instances in the h-level grey relational structure. Obviously, for classifying a new, unseen instance i, only h inward edges connected with instance i in the above h-level grey relational structure are needed.

3 A Partial-Memory Learning Algorithm

In this section, a novel partial-memory learning algorithm based on the above grey relational structure is proposed. The main idea is stated as follows. In the above h-level grey relational structure, for example, an instance may not be connected by any inward edges from other instances. In other words, this instance is rarely used in determining the class labels of other instances, implying that, it is probably a good choice for instance pruning in a learning system. In the grey relational structure, the inward edges of each training instance, indicating how many times each instance is used as the nearest neighbor or neighbors in determining the class labels of other instances, can be found. This approach excludes

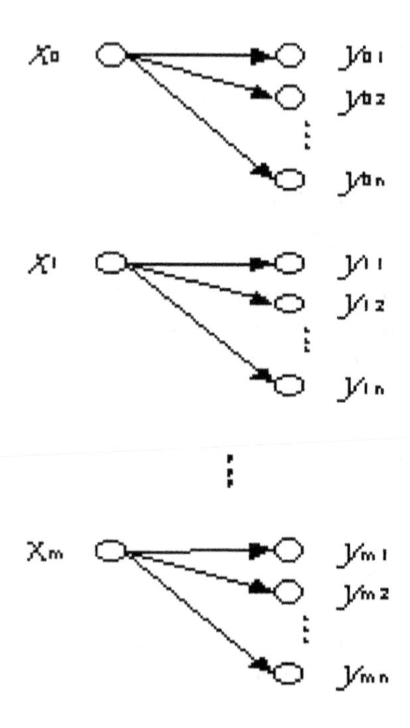

Fig. 1. h-level grey relational structure

the training instances with no or few inward edges for solving classification problems (That is, these instances are rarely used in determining the class label of other instances and thus are considered here to be eliminated).

Consider that a training set T of m labeled instances is given, denoted by $T = \{x_1, x_2, \ldots, x_m\}$, where each instance x_e has n attributes and is denoted as $x_e = (x_e(1), x_e(2), \ldots, x_e(\mathrm{n}))$. Let S_{lk} be the set of training instances having no less than l inward edges in the above k-level grey relational structure, where $S_{lk} \subseteq T$, $1 \leqq l \leqq W_k$ and W_k is the largest number of inward edges that an instance in the k-level grey relational structure may have. For classifying a new, unseen instancex_0, the proposed partial-memory learning procedure is performed as follows.

1. Let $R_{lk} = T - S_{lk}$, where $1 \leqq l \leqq W_k$, $1 \leqq k \leqq h$ and W_k is the largest number of inward edges that an instance in the k-level grey relational structure may have (here, h is set to 10, i.e., at most 10 nearest neighbors of each instance are considered here). Consider that S_{lk} and R_{lk} are used as training and test sets, respectively. Further, calculate the classification accuracy ACC_{lk} corresponding to each S_{lk} and R_{lk} by using the instance-based learning ap-

proach presented in [11]. (That is, each R_{lk} is used as a test set to validate the corresponding training set S_{lk}).

2. Let CS be a set containing each training set S_{lk} with ACC_{lk} greater than α (Here, α is set to the average accuracy of classification obtained by using T as the validating data set in ten-fold cross-validation). Select a training set V with the smallest number of instances from CS as the final training set for classifying the new instancex_0. That is, the above instance-based learning approach is used again with the final training set V for the classification task.

As stated earlier, S_{lk}, in each k-level grey relational structure, is the set of training instances having no less than l inward edges. The training setS_{lk} means that k nearest neighbors of an instance will be considered to evidence the class label of that instance. It is easily seen that $S_{0k} = T$ for every k; meanwhile, $||S_{ik}||$ $\leq||S_{jk}||$ for every $i > j$, where $||S_{ik}||$ is the total number of training instances in S_{ik}. By considering different value of $||S_{ik}||$, the size of training set used for partial-memory learning here can be varied (i.e., using different S_{ik}). In addition, the number of nearest neighbors (i.e., k) used to evidence the class label of an instance is also considered. A training set V will be selected, from all S_{ik}, for partial-memory learning if acceptable classification accuracy (as stated in Step 1, $T - V$ is used as the test set for validation) with small training size can be obtained. This selection process is done in Step 2.

4 Experimental Results

In this section, experiments performed on five application domains (derived from [3]) are reported to demonstrate the performance of the proposed learning approach.

In the experiments, ten-fold cross validation [17] was used and applied ten times for each application domain. That is, the entire data set of each application domain was equally divided into ten parts in each trial; each part was used once for testing and the remaining parts were used for training. The proposed partial-memory learning approach was applied for selecting a suitable subset from each original training set. Accordingly, the average accuracy of classification is obtained. Table 1 gives the performances (average accuracy of classification and data retention rate [13]) of the proposed approach, the ICPL and RT3 [13,18]. The data retention rate is stated as follows,

$$\text{data retention rate} = \frac{\text{No. of remained instances}}{\text{No. of training instances}}, \tag{6}$$

The classification accuracy and data retention rate of the ICPL and RT3 on the above five datasets are derived from [13]. It can be easily seen that the proposed approach yields high performance with small size of training set, compared with other existing pruning techniques.

Table 1. Comparison with existing techniques

Dataset	RT3		ICPL		Proposed approach	
	Average accuracy	Data retention rate	Average accuracy	Data retention rate	Average accuracy	Data retention rate
Iris	0.947	0.080	0.947	0.043	0.9513	0.047
Glass	0.672	0.211	0.677	0.154	0.682	0.149
Shuttle	0.998	0.007	0.998	0.003	0.998	0.003
Vowel	0.956	0.293	0.923	0.151	0.957	0.151
Wine	0.938	0.114	0.960	0.040	0.960	0.042

5 Conclusions

In this paper, a novel partial-memory learning approach based on grey relational structure is proposed. Grey relational analysis is used to precisely describe the entire relational structure of all instances in a specific application domain. By using the above-mentioned grey relational structure, instances having no or few inward edges can be eliminated prior to learning (i.e., these instances are rarely used in determining the class labels of other instances and thus can be pruned in a learning system). Accordingly, a suitable training subset with small size can be obtained for classifying new instances. Experiments performed on five application domains are reported to demonstrate the performance of the proposed approach. It can be easily seen that the proposed approach yields high performance with small size of training set, compared with other existing pruning techniques.

References

1. Aha, D. W., Kibler, D., Albert, M. K.: Instance-based learning algorithms. Machine Learning, Vol. 6, (1991) 37–66
2. Aha, D. W.: Tolerating noisy, irrelevant and novel attributes in instance-based learning algorithms. Int. J. Man-Machine Studies, vol. 36, no. 2, (1992), 267–287
3. Blake, C. L., Merz, C. J.: UCI Repository of machine learning databases [http://www.ics.uci.edu/~mlearn/MLRepository.html]. Irvine, CA: University of California, Department of Information and Computer Science, (1998)
4. Cover, T. M., Hart, P. E.: Nearest neighbor pattern classification. IEEE Transactions on Information Theory, vol. 13, no. 1, (1967), 21–27
5. Deng, J.: The theory and method of socioeconomic grey systems. Social Sciences in China, vol. 6, (in Chinese), (1984), 47–60
6. Deng, J.: Introduction to grey system theory. The Journal of Grey System, vol. 1, (1989), 1–24
7. Deng, J.: Grey information space. The Journal of Grey System, vol. 1, (1989), 103–117
8. Fix, E., Hodges, J. L.: Discriminatory analysis: nonparametric discrimination: consistency properties. Technical Report Project 21-49-004, Report Number 4, USAF School of Aviation Medicine, Randolph Field, Texas, (1951)

9. Gates, G.W.: The Reduced Nearest Neighbor Rule. IEEE Trans. Information Theory, vol. 18, no. 3, (1972), 431–433
10. Hart, P.E.: The Condenced Nearest Neighbor Rule. IEEE Trans. Information Theory, vol. 14, no. 3, (1968), 515–516
11. Huang, C. C., Lee, H. M.: An Instance-based Learning Approach based on Grey Relational Structure. Proc. of 2002 UK Workshop on Computational Intelligence (UKCI-02), Birmingham, U.K, (2002)
12. Kibler, D., Aha, D. W.: Learning Representative Exemplars of Concepts: An Initial Case Study. Proc. of the Fourth International Workshop on Machine Learning, Irvine, CA: Morgan Kaufmann, (1987), 24–30
13. Lam, W., Keung, C.K., Liu, D.: Discovering Useful Concept Prototypes for Classification Based on Filtering and Abstraction. IEEE Trans. Pattern Analysis and Machine Intelligence, vol. 24, no. 8, (2002), 1075–1090
14. Lin, C. T., Yang, S. Y.: Selection of home mortgage loans using grey relational analysis. The Journal of Grey System, vol. 4, (1999), 359–368
15. Maloof, M.A., Michalski R.S.: Selecting Examples for Partial Memory Learning. Machine Learning, vol. 41, no. 1, (2000), 27–52
16. Stanfill, C., Waltz, D.: Towards memory-based reasoning. Communications of the ACM, vol. 29, no. 12, (1986), 1213–1228
17. Stone, M.: Cross-validatory choice and assessment of statistical predictions. Journal of the Royal Statistical Society, B vol. 36, (1974), 111–147
18. Wilson, D.R., Martinez, T.R.: Reduction Techniques for Instances-based Learning Algorithm. Machine Learning, vol. 38, (2000), 257–286

Constructing Hierarchical Rule Systems

Thomas R. Gabriel and Michael R. Berthold

Data Analysis Research Lab
Tripos, Inc., 601 Gateway Blvd., Suite 720
South San Francisco, CA 94080, USA.
{tgabriel,berthold}@tripos.com

Abstract. Rule systems have failed to attract much interest in large data analysis problems because they tend to be too simplistic to be useful or consist of too many rules for human interpretation. We present a method that constructs a hierarchical rule system, with only a small number of rules at each stage of the hierarchy. Lower levels in this hierarchy focus on outliers or areas of the feature space where only weak evidence for a rule was found in the data. Rules further up, at higher levels of the hierarchy, describe increasingly general and strongly supported aspects of the data. We demonstrate the proposed method's usefulness on several classification benchmark data sets using a fuzzy rule induction process as the underlying learning algorithm. The results demonstrate how the rule hierarchy allows to build much smaller rule systems and how the model—especially at higher levels of the hierarchy—remains interpretable. The presented method can be applied to a variety of local learning systems in a similar fashion.

1 Introduction

Data sets obtained from real world systems often contain missing, noisy, or simply wrong records. If one attempts to build a rule model for such data sources, the result is either an overfitted and inherently complex model, which is impossible to interpret, or the model ends up being too simplistic, ignoring most of the interesting aspects of the underlying system as well as the outliers. But even this effect of outlier ignorance is often also not desirable since the very example that was excluded from the model may be caused by a rare but still extremely interesting phenomena. The real challenge is therefore to build models that describe all interesting properties of the data while still being interpretable by the human expert.

Extracting rule models from data is not a new area of research. In [9] and [13] algorithms were described that construct hyperrectangles in feature space. The resulting set of rules encapsulates regions in feature space that contain patterns of the same class. Other approaches, which construct fuzzy rules instead of crisp rules, were presented in [1,6,11] and [12]. All of these approaches have in common that they tend to build very complex rule systems for large data sets originating from a complicated underlying system. In addition, high-dimensional feature

M.R. Berthold et al. (Eds.): IDA 2003, LNCS 2810, pp. 76–87, 2003.

spaces result in complex rules relying on many attributes and increase the number of required rules to cover the solution space even further. An approach that aims to reduce the number of constraints on each rule individually was recently presented in [4]. The generated fuzzy rules only constrain few of the available attributes and hence remain readable even in case of high-dimensional spaces. However, this algorithm also tends to produce many rules for large, complicated data sets.

In this paper we describe a method that attempts to tackle this inherent problem of interpretability in large rule models. We achieve this by constructing a hierarchy of rules with varying degrees of complexity. Early attempts to build model hierarchies have been published previously. In [5] an approach to build unsupervised hierarchical cluster models was described, however, the resulting system of clusters does not offer great interpretability. A method to build hierarchies of rule systems was described in [8]. Here, an ensemble of rule sets with different granulation is built at the beginning. Starting with a coarse granulation (usually having only two membership functions for each attribute), the remaining rule sets exhibit increasingly finer granulation. In the resulting multi-rule table the grade of certainty for each rule is then used for pruning, that is, rules with a low grade of certainty will be removed from the rule set. However, this method still exhibits the complexity problem described above—in case of high-dimensional feature spaces, the resulting number of rules will increase exponentially with the dimensionality of the space. As a result, each rule relies on all attributes, making it very hard to interpret.

The approach presented here builds a rule hierarchy for a given data set. The rules are arranged in a hierarchy of different levels of precision and each rule only depends on few, relevant attributes thus making this approach also feasible for high-dimensional feature spaces. Lower levels of the hierarchy describe regions in input space with low evidence in the given data, whereas rules at higher levels describe more strongly supported concepts of the underlying data. The method is based on the fuzzy rule learning algorithm mentioned above [4], which builds a single layer of rules autonomously. We recursively use the resulting rule system to determine rules of low relevance, which are then used as a filter for the next training phase. The result is a hierarchy of rule systems with the desired properties of simplicity and interpretability on each level of the resulting rule hierarchy. We evaluate the classification performance of the resulting classifier-hierarchy using benchmark data sets from the European StatLog–Project [7]. Experimental results demonstrate that fuzzy models at higher hierarchical levels indeed show a dramatic decrease in number of rules while still achieving better or similar generalization performance than the fuzzy rule system generated by the original, non-hierarchical algorithm.

2 The Underlying Fuzzy Rule Algorithm

The underlying, non-hierarchical fuzzy rule learning algorithm is described in [4]. The algorithm constructs a set of fuzzy rules from given training data. The resulting set of rules can then be used to classify new data. We briefly summarize the used type of fuzzy rules before explaining the main structure of the training algorithm.

2.1 Fuzzy Rule Systems

The underlying fuzzy rule systems are based on a local granulation of the input space for each rule, so that rules can be independently interpreted:

$$\begin{array}{llll}
\mathcal{R}_1^1 : \texttt{IF } x_1 \texttt{ IS } \mu_{1,1}^1 & \texttt{AND} \cdots \texttt{AND } x_n & \texttt{IS } \mu_{n,1}^1 & \texttt{THEN class } 1 \\
\vdots & & \vdots & \vdots \\
\mathcal{R}_{r_1}^1 : \texttt{IF } x_1 \texttt{ IS } \mu_{1,r_1}^1 & \texttt{AND} \cdots \texttt{AND } x_n & \texttt{IS } \mu_{n,r_1}^1 & \texttt{THEN class } 1 \\
\vdots & & \vdots & \vdots \\
\mathcal{R}_{r_k}^k : \texttt{IF } x_1 \texttt{ IS } \mu_{1,r_k}^k & \texttt{AND} \cdots \texttt{AND } x_n & \texttt{IS } \mu_{n,r_k}^k & \texttt{THEN class } c
\end{array}$$

where $\mathcal{R}_j^k$ represents rule j for class k. The rule base contains rules for c classes and r_k indicates the number of rules for class k ($1 \leq j \leq r_k$ and $1 \leq k \leq c$).

The fuzzy sets $\mu_{i,j}^k : \mathbb{R} \mapsto [0,1]$ are defined for every feature i ($1 \leq i \leq n$) and the overall degree of fulfillment of a specific rule for an input pattern $\boldsymbol{x} = (x_1, \ldots, x_n)$ can be computed using the minimum-operator as fuzzy-AND:

$$\mu_j^k(\boldsymbol{x}) = \min_{i=1,\cdots,n} \left\{ \mu_{i,j}^k(x_i) \right\}.$$

The combined degree of membership for all rules of class k can be computed using the maximum-operator as fuzzy-OR:

$$\mu^k(\boldsymbol{x}) = \max_{j=1,\cdots,r_k} \left\{ \mu_j^k(\boldsymbol{x}) \right\}.$$

From these membership values we then derive the predicted class k_{best} for an input pattern $\boldsymbol{x}$ as:

$$k_{\text{best}}(\boldsymbol{x}) = \arg\max_k \left\{ \mu^k(\boldsymbol{x}) \mid 1 \leq k \leq c \right\}.$$

The used fuzzy membership functions are trapezoidal and can be described using four parameters $<a_i, b_i, c_i, d_i>$, where a_i and d_i indicate the fuzzy rule's support-, and b_i and c_i its core-region for each attribute i of the input dimension. The training algorithm shown in the next section usually only constrains few attributes, that is, most support-regions will remain infinite, leaving the rules interpretable even in case of high-dimensional input spaces.

2.2 Construction of Fuzzy Rules

The original algorithm [4] will be abbreviated as FRL (**F**uzzy **R**ule **L**earner) in the following. During the training process all existing example patterns are subsequentially presented to the growing model. The algorithm then introduces new fuzzy rules when necessary and adjusts the coverage (support- and core-regions) of already existing rules in case of conflicts. The complete training data is presented once during each epoch. For each training pattern, three different steps are executed:

- **Cover**: If a new training pattern lies inside the support-region of an already existing fuzzy rule of the correct class, its core-region is extended to cover the new pattern. In addition, the weight of this rule is incremented.
- **Commit**: If the new pattern is not yet covered, a new fuzzy rule belonging to the corresponding class will be created. The new example is assigned to its core-region, whereas the overall rule's support-region is initialized "infinite", that is, the new fuzzy rule is unconstrained and covers the entire domain.
- **Shrink**: If a new pattern is incorrectly covered by an existing fuzzy rule of conflicting class, this fuzzy rule's support-region will be reduced, so that the conflict with the new pattern is avoided. The underlying heuristic of this step aims to minimize the loss in volume (see [4] for details).

The algorithm usually terminates after only few iterations over the training data. The final set of fuzzy rules can be used to compute a degree of class membership for new input patterns as described in Section 2.1.

3 Learning Hierarchies of Rule Systems

The fuzzy rule learner described in the previous section has the usual problems when encountering noisy data or outliers. In such cases an excessive number of rules is being introduced simply to model noise and outliers. This is due to the fact that the algorithm aims to generate conflict free rules, that is, examples encountered during training will result in a non-zero degree of membership only for rules of the correct class. Unfortunately, in case of noisy data, which distorts class boundaries, such an approach will result in many superfluous rules introduced simply to model these artifacts.

Using an already existing model we can, however, easily determine these parts of a rule model since they tend to have very low relevance. The main idea behind the hierarchical training algorithm is to use those parts of the rule model with low relevance as a filter for the input data, which in turn is used to create the next layer. This recursive process of data cleansing results in higher levels of the hierarchy, which are built only for examples with strong support in the original data set. For this to work, we obviously need to define the notion of rule relevance or importance more precisely.

3.1 Measuring Rule Relevance and Filtering Outliers

To measure a rule's relevance often the weight parameter $\omega^{(\mathcal{R})}$ is used, which represents the number of training patterns covered by rule $\mathcal{R}$. A measure for the importance or relevance of each rule can be derived by simply using the percentage of patterns covered by this rule:

$$\Phi(\mathcal{R}) = \frac{\omega^{(\mathcal{R})}}{|\mathbf{T}|},$$

where $\mathbf{T}$ indicates the set of all training examples. Other measures which are also used determine the loss of information if rule $\mathcal{R}$ is omitted from all rules $\mathbf{R}$:

$$\Phi(\mathcal{R}) = I(\mathbf{R}) - I(\mathbf{R}\backslash\{\mathcal{R}\})$$

where $I(\cdot)$ indicates a function measuring the information content of a rule set. Most commonly used are the Gini-index and the fuzzy entropy function (see [10] for details):

- Gini-index:

$$I_{\text{Gini}}(\mathbf{R}) = 1 - \sum_{k=1}^{c} V(\mathbf{R}^k)^2,$$

- Fuzzy entropy:

$$I_{\text{Entropy}}(\mathbf{R}) = -\sum_{k=1}^{c} \left(V(\mathcal{R}^k) \log_2 V(\mathbf{R}^k)\right)$$

where $\mathbf{R} = \bigcup_{k=1}^{c} \mathbf{R}^k$ and $V(\mathbf{R}^k)$ indicate the volume for all rules of class k ($\mathbf{R}^k = \{\mathcal{R}_1^k, \ldots, \mathcal{R}_{r_k}^k\}$):

$$V(\mathbf{R}^k) = \int_{\boldsymbol{x}} \max_{\mathcal{R}^k \in \mathbf{R}^k} \{\mu^k(\boldsymbol{x})\} \, d\boldsymbol{x}.$$

In [10] it is shown how this volume can be computed efficiently based on a system of fuzzy rules. More extensive overviews can be found in [2].

The choice of relevance metric is made depending on the nature of the underlying rule generation algorithm, as well as the focus of analysis, i. e. the interpretation of important vs. unimportant or irrelevant data points.

Using any of the above-mentioned measures of (notably subjective) relevance, together with a relevance threshold θ_{outlier}, we can now extract rules with low relevance from the model:

$$\mathbf{R}^{\text{outlier}} = \{\mathcal{R} \in \mathbf{R} \mid \Phi(\mathcal{R}) \leq \theta_{\text{outlier}}\}.$$

Using this outlier model as a filter for a second training phase will then generate a new fuzzy model, which has fewer rules with higher significance. In effect, the original training data is filtered and only data points, which are not covered by the outlier model, are passed onto the next level for training[1]. For this a filter-threshold θ_{filter} ($0 \leq \theta_{\text{filter}} \leq 1$) is used:

$$\mathbf{T}^{\text{relevant}} = \mathbf{T} \setminus \left\{(\boldsymbol{x}, k) \in \mathbf{T} \mid \exists \mathcal{R} \in \mathbf{R}^{\text{outlier}} : \mu_{\mathcal{R}}(\boldsymbol{x}) \geq \theta_{\text{filter}}\right\}.$$

This procedure was used in [3] to remove potential outliers from the training data. There, the initial model is used to extract an outlier model, which is subsequentially used to filter the data and generate the final rule model. Figure 1 illustrates the flow of this procedure.

[1] Note that this procedure is similar to well-known pruning strategies. In this case, however, the pruned model parts are not discarded but instead used as explicit outlier model.

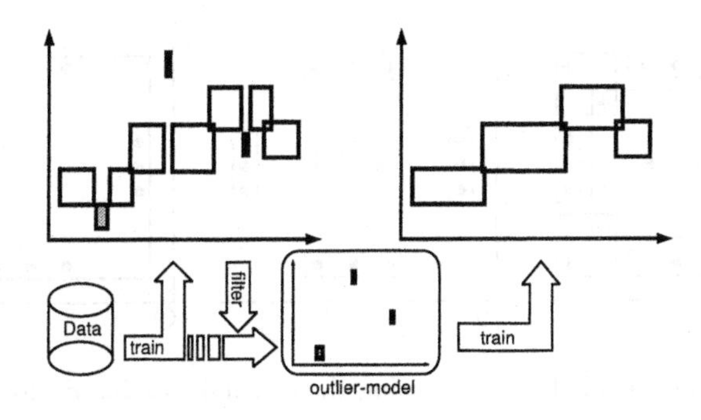

Fig. 1. Two-stage outlier filtering as described in [3].

3.2 Constructing Rule Hierarchies

An entire model hierarchy can be obtained by a multi-stage fuzzy rule induction process based on the two-stage procedure described above. This hierarchical training process recursively continues to apply the filtering procedure until a final rule model is generated, where all rules lie above the relevance threshold. The resulting **h**ierarchical **F**uzzy **R**ule **L**earner (or hFRL) can then formally be described as shown in Table 1.

After some initializations, line (4) shows the generation of the fuzzy rule model using the original FRL for the input data separately for each level of the hierarchy. From the resulting rule set, we extract rules with low relevance using the outlier-threshold θ_{outlier} at line (5). The next step extracts all examples from the rule set using the filter parameter θ_{filter}. This procedure is repeated until all rules are above the outlier-threshold and the outlier model remains empty. The model hierarchy consists of all outlier models $\mathbf{R}_i$ and the rule model of the last iteration, which we denote by $\mathbf{R}_x$.

Figure 2 shows a two-dimensional example with patterns of two classes (*cross* and *circle*). The rectangles depict the rule's core-regions only. In this example the three-level hierarchy is built using an outlier-threshold $\theta_{\text{outlier}} = 2$ and a filter-threshold $\theta_{\text{filter}} = 1.0$. That is, rules covering two or less patterns are moved to the outlier model, and only patterns falling inside the core-region of an outlier rule (hence resulting in a degree of membership of at least one) are removed.

Table 1. ALGORITHM hFRL

(1)	$\mathbf{T}_0 = \mathbf{T}$
(2)	$i \leftarrow 0$
(3)	`REPEAT`
(4)	$\mathbf{R}_i = \text{FRL}(\mathbf{T}_i)$
(5)	$\mathbf{R}_i^{\text{outlier}} = \{\mathcal{R} \in \mathbf{R}_i \mid \Phi(\mathcal{R}) \leq \theta_{\text{outlier}}\}$
(6)	$\mathbf{T}_{i+1} = \mathbf{T}_i \setminus \{(\boldsymbol{x}, k) \in \mathbf{T}_i \mid \exists \mathcal{R} \in \mathbf{R}_i^{\text{outlier}} : \mu_{\mathcal{R}}(\boldsymbol{x}) \geq \theta_{\text{filter}}\}$
(7)	$i \leftarrow i + 1$
(8)	`WHILE` $\lvert\mathbf{R}_{i-1}^{\text{outlier}}\rvert > 0$

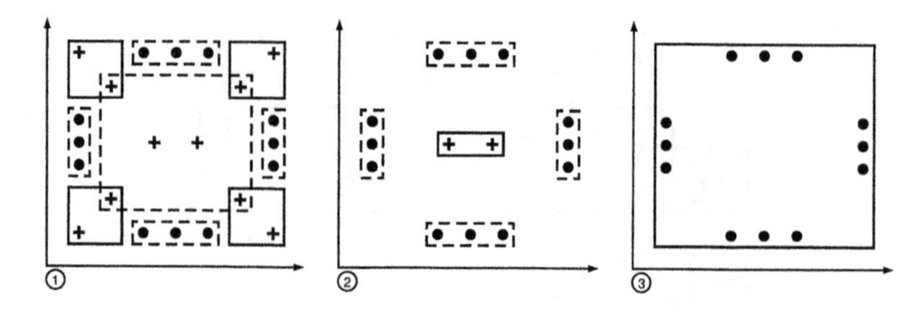

Fig. 2. An example rule hierarchy, using the original *rule-based* filtering strategy.

Figure 2(1) illustrates the original rule model consisting of nine rules generated at the beginning of the process. The four rules in the corners cover only two patterns each and are therefore added to the first outlier model. The remaining five rules (dashed lines) are not used at this level. The algorithm then removes the corresponding patterns from the set of training data. Figure 2(2) displays the next model, containing five rules. Here, only the rule in the center does not cover more than two examples and composes the second outlier model. The last step generates the final model with one rule covering all remaining patterns of class *circle* (Figure 2(3)).

The example in Figure 2 shows how this rule-based filtering mechanism can potentially remove large portions of the training data during subsequent stages. This is due to the fact that only individual rule's coverage of patterns is considered. More reasonable results can be achieved using a global, model-based approach.

Our filter criterion can easily be extended accordingly. In addition to the test in line (6), if an outlier rule exists that covers a certain pattern, we now also ensure that no other non-outlier rule covers this pattern:

$$\mathbf{T}_{i+1} = \mathbf{T}_i \setminus \{ (\boldsymbol{x}, k) \in \mathbf{T}_i \mid \ldots \wedge \nexists \mathcal{R} \in \mathbf{R}_i \setminus \mathbf{R}_i^{outlier} : \mu_{\mathcal{R}}(\boldsymbol{x}) \geq \theta_{\text{filter}} \}.$$

Figure 3 demonstrates the impact of this change using the same data as before. Figure 3(1) again shows the first model in the hierarchy. The rules in the corners are again added to the first outlier model. But this time, the inner data

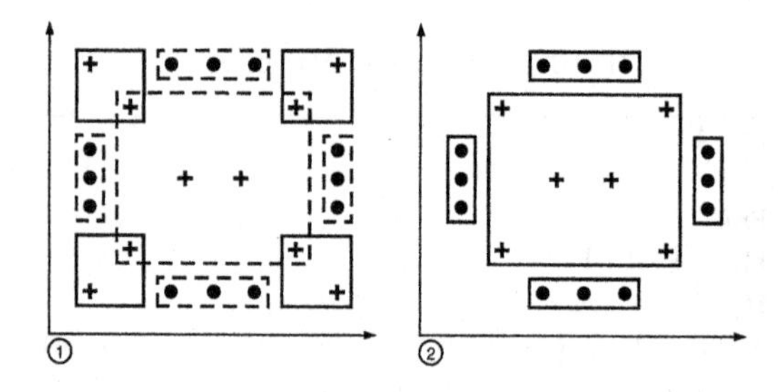

Fig. 3. An example of a *model-based* rule hierarchy, using the same data as in Figure 2.

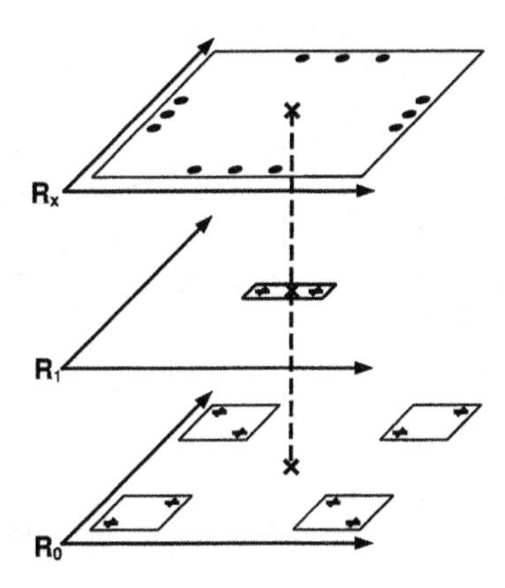

Fig. 4. Classification of a new pattern using a three-level rule hierarchy.

points of these rules are also covered by the rule in the center, which is not part of the outlier model. Therefore these points remain in the training data for the next iteration. The final model in Figure 3(2) contains five rules, all of which are above the outlier-threshold.

This model-based approach enables us to only remove patterns from the next stage that are solely covered by rules of the outlier model and avoids the overly pessimistic problem demonstrated above.

3.3 Classification Using Hierarchies

In Section 2.1 we showed how a non-hierarchical rule system can be used for classification of new input patterns, resulting in a set of fuzzy membership values for each class. In case of a hierarchical rule system, the outputs of the rule systems at different levels need to be combined.

Figure 4 shows an example using the hierarchical fuzzy rule model of Figure 2. Each rule model, $\mathbf{R}_0$, $\mathbf{R}_1$, and $\mathbf{R}_x$, provides its own output fuzzy membership degrees for a given input pattern. For the displayed rule system, $\mathbf{R}_0$ does not make a prediction for the given example, $\mathbf{R}_1$ outputs class *cross*, and $\mathbf{R}_x$ predicts class *circle*. For our tests we use two classification strategies. One accumulates the fuzzy membership degrees of each level by summing them (FuzzySum $\mathbf{R}_x \to \mathbf{R}_i$), the other one (FirstHit $\mathbf{R}_i[\to \mathbf{R}_x]$), determines the output based on the model-level that covers the pattern first using a bottom–up strategy.

In the following section we report experimental results using varying depths of the hierarchy. That is, we use the top level model along with an increasing number of outlier models for classification.

4 Experimental Results

Eight benchmark data sets from the StatLog–Project [7] are used to demonstrate the performance of the proposed methodology. We follow the division in

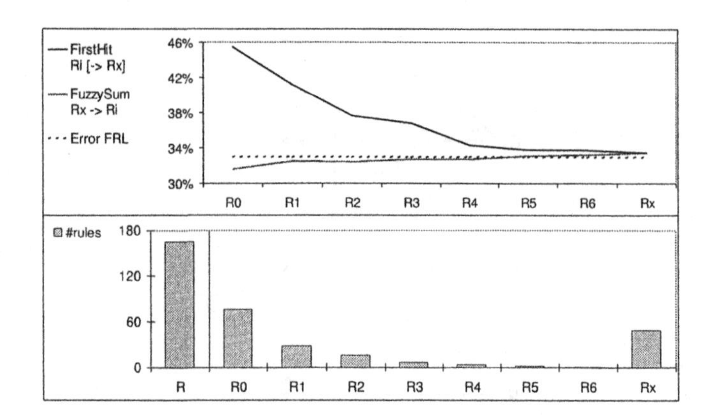

Fig. 5. Vehicle Silhouette data (752 patterns, 8-level hierarchy).

training/test set resp. number of cross validation folds as specified in [7]. For the tests we choose an outlier-threshold $\theta_{\text{outlier}} = 5$, that is, rules which cover less than five pattern are moved to the outlier model. We ran preliminary experiments with various choices for this threshold, and this choice seems to be an appropriate value to generate an acceptable depth of the hierarchy for all data sets.

The following paragraphs discuss the results of three of the eight data sets in more detail. In each of the corresponding diagrams, the graph on top shows the error of the non-hierarchical fuzzy rule learner (dashed line), and the performance curves of both classification strategies introduced above. For those two cases, the number of outlier models used for classification was varied (left: using all models down to $\mathbf{R}_0$; right: using only the top level model $\mathbf{R}_x$). In addition, the graph on the bottom shows the number of rules generated by the original fuzzy rule learner ($\mathbf{R}$), as well as the number of rules in each level $\mathbf{R}_i$ to $\mathbf{R}_x$ of the hierarchy. At the end of this section we summarize the results on these and the remaining four data sets.

4.1 Vehicle Silhouette Data Set

Figure 5 illustrates the results on the Vehicle Silhouette data set. The proposed method generates 8 hierarchy levels using the training data. The classical fuzzy model consists of 165 rules producing an error of 33.0% on the test data. Using only the rule base at the top level ($\mathbf{R}_x$) a similar performance (33.5%) can be achieved with only 49 rules. Slightly better performance (32.7%) results from using the FuzzySum-strategy and models $\mathbf{R}_x$ to $\mathbf{R}_4$ with 57 rules in total. Subsequentially adding models decreases this error further to 31.6% with altogether 186 rules. On the other hand, using the FirstHit-strategy and all levels of the generated hierarchy produces substantially worse classification results. However, this error declines quickly, when the lower few levels of hierarchy are ignored. This indicates that these parts of the model hierarchy represent artifacts in the training data that are not representative for the underlying concept.

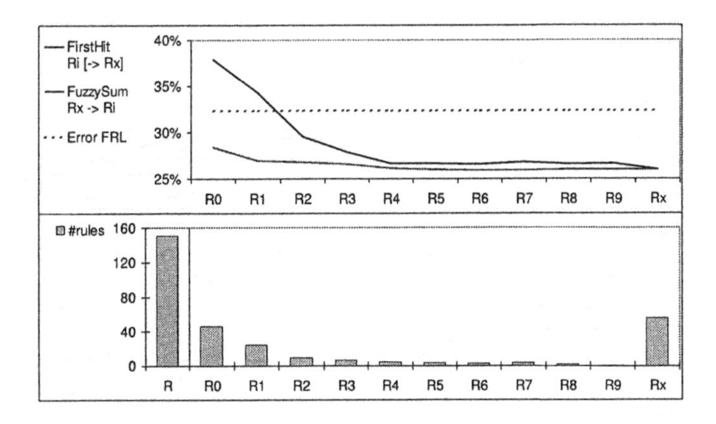

Fig. 6. Pima Indians Diabetes data base (704 patterns, 11-level hierarchy).

4.2 Pima Indians Diabetes Data Base

The second test is conducted using the Pima Indians Diabetes data base. Figure 6 shows the results. The non-hierarchical fuzzy model (151 rules) achieves a generalization performance of 32.3%. The hierarchical rule learner generates 11 levels. It is interesting to note that in this case, even model $\mathbf{R}_x$ at the top level reaches a substantially better classification performance (26.0%) using only 56 rules. The entire hierarchy consists of 162 rules. In this example adding models from lower levels does not increase the performance. But instead, similar to before, including the bottom level harms the performance.

4.3 Australian Credit Approval

Experiments on the the Australian Credit Approval with 690 patterns are shown in Figure 7. The non-hierarchical fuzzy rule learning algorithm generates 125 rules with an error of 18.8% on the test data. The hierarchical rule learner with 5 levels achieves a noticeably better performance of 15.4% using only 31 rules at the top level. The entire rule hierarchy consists of 91 rules with an minimum error rate of 15.2%. Again, note how when using the FirstHit-strategy the addition of the first outlier model $\mathbf{R}_0$ significantly increases the generalization error, indicating a substantial amount of artifacts in the original training data, which were filtered out during the first phase of training.

4.4 Summary Results

Table 2 summarizes the results of all eight benchmark data sets from the StatLog-Project used for our experiments. The table shows the results in comparison to the non-hierarchical fuzzy rule learner.

The table is separated into three parts. The first column shows the name of the data set, followed by the results of the FRL algorithm ($\mathbf{R}$). Column (2) and (3) list the corresponding number of rules and error rate in percent. The

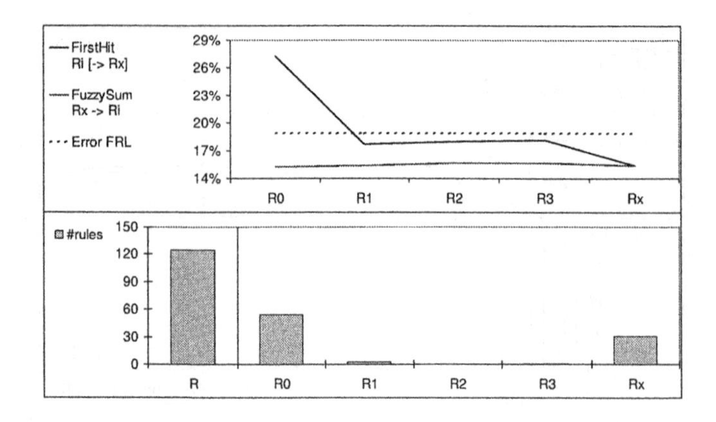

Fig. 7. Australian Credit Approval (690 patterns, 5-level hierarchy).

next column shows the number of generated rule levels, followed by the number of rules and the performance of the top model in the hierarchy ($\mathbf{R}_x$). Finally, the last two columns show the number of rules and the classification error ($\mathbf{R}_x \to \mathbf{R}_0$) of the entire hierarchy using the FuzzySum-strategy.

Note how for all eight data sets the number of rules at the top level of the hierarchy is substantially smaller than in the original model. For the Pima Indians Diabetes data base and Australia Credit Approval an even better performance can be achieved using just the top level model. Often even the entire hierarchy is smaller than the classical non-hierarchical rule system (e. g. Australia Credit Approval and Shuttle data set). This clearly indicates that the hierarchical rule learning strategy does indeed extract rules—and subsequentially patterns—with low relevance throughout the training process. The results on the DNA data are not representative since the feature space consists of 240 binary variables. The underlying fuzzy rule learner is highly unstable in this case.

Table 2. Summary of the result for all data sets from the StatLog–Project.

Data set	$\mathbf{R}$			$\mathbf{R}_x$		FuzzySum $\mathbf{R}_x \to \mathbf{R}_0$	
	#rules	error	#level	#rules	error	#rules	error
Diabetes	151	32.3	11	**56**	**26.0**	162	**28.4**
Aust. Cred.	125	18.8	5	**31**	**15.4**	**91**	**15.2**
Vehicle	165	33.0	8	**49**	33.5	186	**31.6**
Segment	96	3.9	6	**70**	4.9	96	4.5
Shuttle	60	0.06	3	**30**	0.15	**57**	0.09
SatImage	416	13.8	14	**132**	14.0	490	**13.2**
DNA	742	32.7	9	**158**	44.9	850	39.6
Letter	2369	14.4	13	**714**	23.0	3252	14.8

5 Conclusion

We have described a method that allows to build a hierarchy of rules based on an underlying, non-hierarchical rule learning algorithm. Using several benchmark data sets we showed how the classification accuracy was comparable and sometimes even better than the underlying, non-hierarchical algorithm and how the resulting rule systems are of substantially reduced complexity. Due to the general nature of the method proposed here, it is possible to apply it to other local learning methods. As long as it is possible to extract parts of the model with low relevance, those pieces can then be used as a filter for subsequent levels in the hierarchy [4].

References

1. S. Abe and M.-S. Lan. A method for fuzzy rules extraction directly from numerical data and its application to pattern classifiction. *IEEE Transactions on Fuzzy Systems*, 3(1):18–28, 1995.
2. C. Apte, S. Hong, J. Hosking, J. Lepre, E. Pednault, and B. K. Rosen. Decomposition of heterogeneous classification problems. *Intelligent Data Analysis*, 2(2):17–28, 1998.
3. M. R. Berthold. Learning fuzzy models and potential outliers. In *Computational Intelligence in Data Mining*, pages 111–126. Springer-Verlag, 2000.
4. M. R. Berthold. Mixed fuzzy rule formation. *International Journal of Approximate Reasoning (IJAR)*, 32:67–84, 2003.
5. A. B. Geva. Hierarchical unsupervised fuzzy clustering. *IEEE Transactions on Fuzzy Systems*, 7(6):723–733, Dec. 1999.
6. C. M. Higgins and R. M. Goodman. Learning fuzzy rule-based neural networks for control. In *Advances in Neural Information Processing Systems*, 5, pages 350–357, California, 1993. Morgan Kaufmann.
7. D. Michie, D. J. Spiegelhalter, and C. C. Taylor, editors. *Machine Learning, Neural and Statistical Classification.* Ellis Horwood Limited, 1994.
8. K. Nozaki, H. Ishibuchi, and H. Tanaka. Adaptive fuzzy rule-based classification systems. *IEEE Transactions on Fuzzy Systems*, 4(3):238–250, 1996.
9. S. Salzberg. A nearest hyperrectangle learning method. In *Machine Learning*, 6, pages 251–276, 1991.
10. R. Silipo and M. R. Berthold. Input features impact on fuzzy decision processes. *IEEE Transcation on Systems, Man, and Cybernetics, Part B: Cybernetics*, 30(6):821–834, 2000.
11. P. K. Simpson. Fuzzy min-max neural networks – part 1: Classification. *IEEE Transactions on Neural Networks*, 3(5):776–786, Sept. 1992.
12. L.-X. Wang and J. M. Mendel. Generating fuzzy rules by learning from examples. *IEEE Transactions on Systems, Man, and Cybernetics*, 22(6):1313–1427, 1992.
13. D. Wettschereck. A hybrid nearest-neighbour and nearest-hyperrectangle learning algorithm. In *Proceedings of the European Conference on Machine Learning*, pages 323–335, 1994.

Text Categorization Using Hybrid Multiple Model Schemes

In-Cheol Kim[1] and Soon-Hee Myoung[2]

[1] Dept. of Computer Science, Kyonggi University
San94-6, Yiui-dong, Paldal-gu, Suwon-si, Kyonggi-do, Korea,
kic@kyonggi.ac.kr

[2] Dept. of IMIS, Yong-In Songdam College
571-1 Mapyong-dong, Yongin-si, Kyonggi-do, Korea,
shmyoung@ysc.ac.kr

Abstract. Automatic text categorization techniques using inductive machine learning methods have been employed in various applications. In this paper, we review the characteristics of the existing multiple model schemes which include bagging, boosting, and stacking. Multiple model schemes try to optimize the predictive accuracy by combining predictions of diverse models derived from different versions of training examples or learning algorithms. In this study, we develop hybrid schemes which combine the techniques of existing multiple model schemes to improve the accuracy of text categorization, and conduct experiments to evaluate the performances of the proposed schemes on MEDLINE, Usenet news, and Web document collections. The experiments demonstrate the effectiveness of the hybrid multiple model schemes. Boosted stacking algorithms, that are a kind of the extended stacking algorithms proposed in this study, yield higher accuracies relative to the conventional multiple model schemes and single model schemes.

1 Introduction

The increasing amount of textual data available in digital form with the advancement of WWW, compounded by the lack of standardized structure, presents a formidable challenge for text management. Among various techniques to find useful information from these huge resources, automatic text categorization has been the focus of many research efforts in the areas of information retrieval and data mining. In the 1980s, expensive expert systems were widely used to perform the task. In recent years, inductive machine learning algorithms that are capable of generating automatic categorization rules have been adopted to text categorization with success. They offer less expensive, but more effective alternatives[1][2]. Naive Bayesian, k-nearest neighbors, and decision tree algorithms are the subset of machine learning algorithms employed in our study.

More recently, several multiple model schemes have been explored in an effort to achieve increased generalization accuracy by means of model combination.

M.R. Berthold et al. (Eds.): IDA 2003, LNCS 2810, pp. 88–99, 2003.

Multiple models are derived from different versions of training examples or different types of learning algorithms. In various practical applications, such as pattern recognition and data mining, methods of combining multiple classifiers have been widely used. Results demonstrate the effectiveness of this new approach. However, a great variation in error reduction has been reported from domain to domain or from dataset to dataset[5].

In this paper, we briefly review the characteristics of basic multiple model approaches and propose new hybrid approaches that include stacked bagging and boosting as well as bagged and boosted stacking. We also evaluate the performances of the hybrid multiple model schemes using real world document collection from the MEDLINE database, Usenet news articles, and web document collection.

2 Standard Multiple Model Schemes

The current standard multiple model schemes, namely bagging[6], boosting[7], and stacking[8] have evolved mainly from the need to address the problems inherent in models derived from single learning algorithms based on limited training examples. One of the problems is the instability observed in learned models. For example, while decision trees generated with the C4.5 algorithm is easy to comprehend, small changes in data can lead to large changes in the resulting tree[4]. The effect of combining multiple models can be evaluated with the concept of bias-variance decomposition. High variance in predictors is inherent in models derived from single algorithms based on a finite number of training examples. Inductive learning methods employ different representations for the learned model and different methods for searching the space of hypotheses. Representation and search methods make up the source of persistent learning biases for different algorithms.

Recent approaches deal with the challenge by using multi-strategy techniques. These multiple model approaches can intervene in all phases of the conventional machine learning process by modifying input, learning, and/or output phases. In the input phase, multiple different subsets of data can be selected as an input data set for the learning phase by applying different sampling methods into a given dataset. In the learning phase, multiple models can be induced by applying different learning algorithms on the same single set of training data or by applying the same learning algorithm on multiple different sets of training data. In the former case, homogeneous classifiers sharing the identical representation are derived. On the other hand, in the latter case, heterogeneous ones are generated. In the output phase, for the final decision, the predictions by multiple classifiers can be combined according to different combination methods, such as the weighted vote in boosting or the additional meta-learning in stacking. All in all, the approaches using multiple models can be characterized by many aspects: the data sampling methods, the model representing methods, the model learning methods, and the prediction combining methods. Whereas each model has its own bias and variance portions of errors, homogeneous models mainly

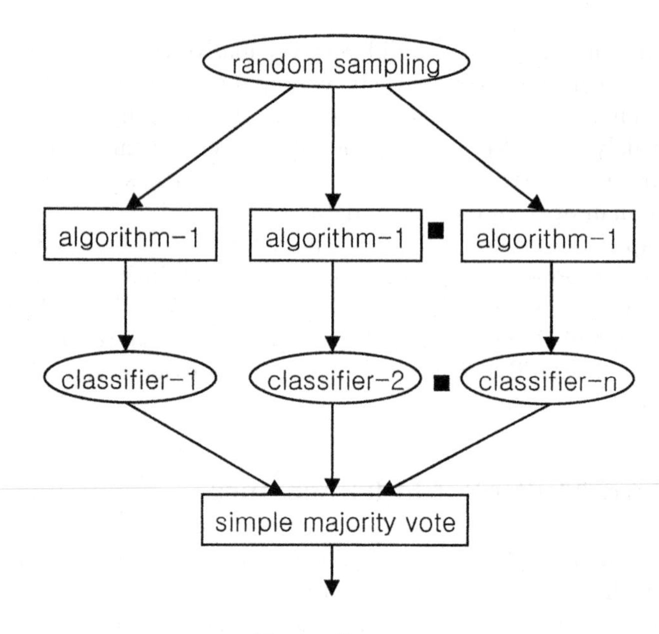

Fig. 1. Bagging

address the variance effects of the error, using perturbation of training data. Heterogeneous models mitigate the bias resulting from participating algorithms by combining the models induced from different learning algorithms by means of meta-learning. Bagging, the simplest of voting or committee schemes, uses random sampling with replacement (also called bootstrap sampling) in order to obtain different versions of a given dataset. The size of each sampled dataset equals the size of the original dataset. On each of these versions of the dataset the same learning algorithm is applied. Classifiers obtained in this manner are then combined with majority voting. Fig. 1 illustrates the concept of bagging. Boosting, such as the popular AdaBoost, is drastically different from bagging in practice in its method for sampling and drawing decisions. Fig. 2 represents the concept of boosting. Boosting first builds a classifier with some learning algorithm from the original dataset. The weights of the misclassified examples are then increased and another classifier is built using the same learning algorithm. The procedure is repeated several times. Classifiers derived in this manner are then combined using a weighted vote. In boosting, the weights of training data reflect how often the instances have been misclassified by the classifiers produced so far. Hence the weights are altered depending on the current classifier's overall error. More specifically, if e denotes the classifiers's error on the weighted data, then weights are updated by the expression (1).

$$weight \leftarrow weight \cdot \frac{e}{1-e} \tag{1}$$

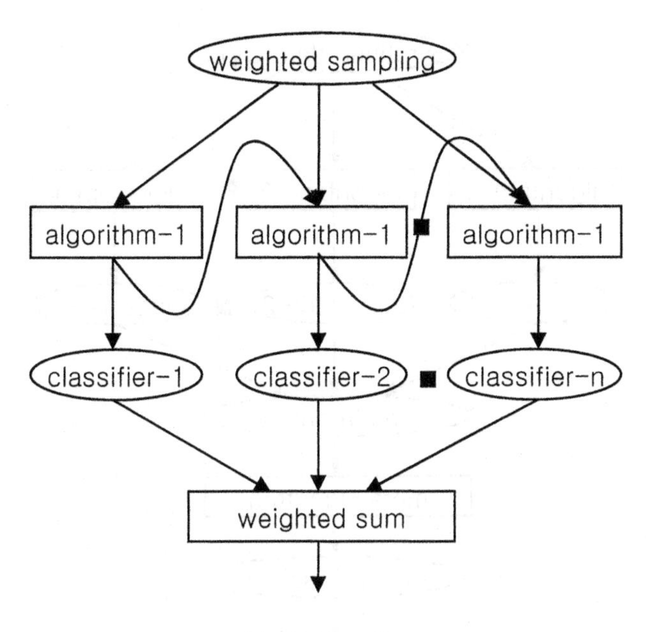

Fig. 2. Boosting

In order to make a prediction, the weights of all classifiers that vote for a particular class are summed, and the class with the greatest total is chosen. To determine the weights of classifiers, note that a classifier that performs well on the weighted training data from which it was built should receive a high weight, and a classifier that performs badly should receive a low one. More specifically, we use the expression (2) to determine the weight of each classifier.

$$weight = -\log \frac{e}{1-e} \tag{2}$$

Better performance is observed in boosting compared with bagging, but the former varies more widely than the latter. In addition, bagging is amenable to parallel or distributed processing. One practical issue in machine learning is how to deal with multiplicity problem [9]. A critical problem is to select a learning algorithm that is expected to yield optimal result for a given problem domain. The strategy of stacked generalization or stacking is to combine the learning algorithms rather than to choose one amongst them. Stacking achieves a generalization accuracy using two phases of processing: one by reducing biases employing a mixture of algorithms, and the other by learning from meta-data the regularities inherent in base-level classifiers. Stacking introduces the concept of a meta-learner, which replaces the voting procedure. Stacking tries to learn which base-level classifiers are the reliable ones, using another learner to discover how best to combine the output of the base learners. Figure 3 depicts the way stacking is conceptualized. The input to the meta model (also called the level-1 model) are the predictions of the base models, or level-0 models. A level-1

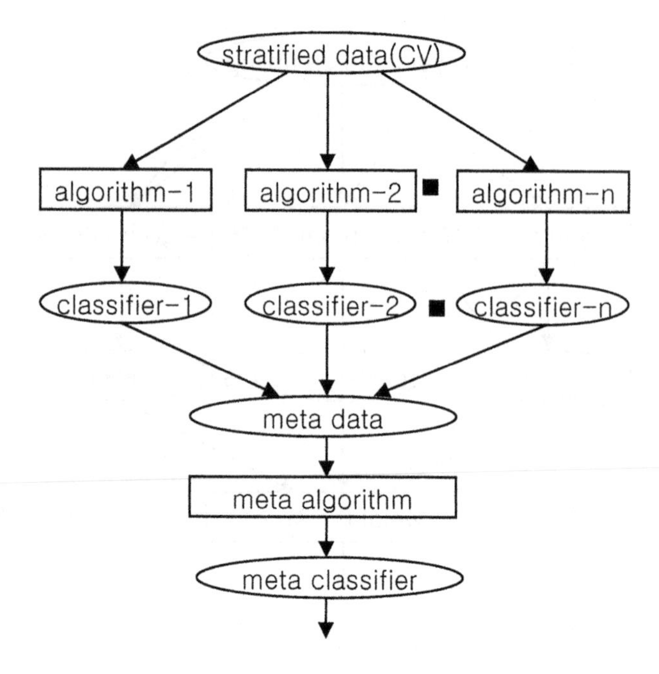

Fig. 3. Stacking

instance has as many attributes as there are level-0 learners, and the attribute values give the predictions of these learners on the corresponding level-0 instance. When the stacked learner is used for classification, an instance is first fed into the level-0 models, and each one guesses a class value. These guesses are fed into the level-1 model, which combines them into the final prediction. In stacking, cross-validation is used as a means for error estimation. It performs a cross-validation for every level-0 learner. Each instance in the training data occurs in exactly one of the test folds of the cross-validation, and the predictions of the level-0 inducers built from the corresponding training fold are used to build a level-1 training instance. Because a level-0 classifier has to be trained for each fold of the cross-validation, the level-1 classifier can make full use of the training data.

3 Hybrid Multiple Model Schemes

In this study, we develop two classes of hybrid schemes: one is the extended vote algorithms employing a meta-level learning phase instead of a simple majority vote to combine homogenous classifiers' predictions, and the other is the extended stacking algorithms, also augmented with different versions of dataset to train multiple heterogeneous base-level classifiers. Our implementation incorporates all the concepts and techniques of existing multiple model schemes.

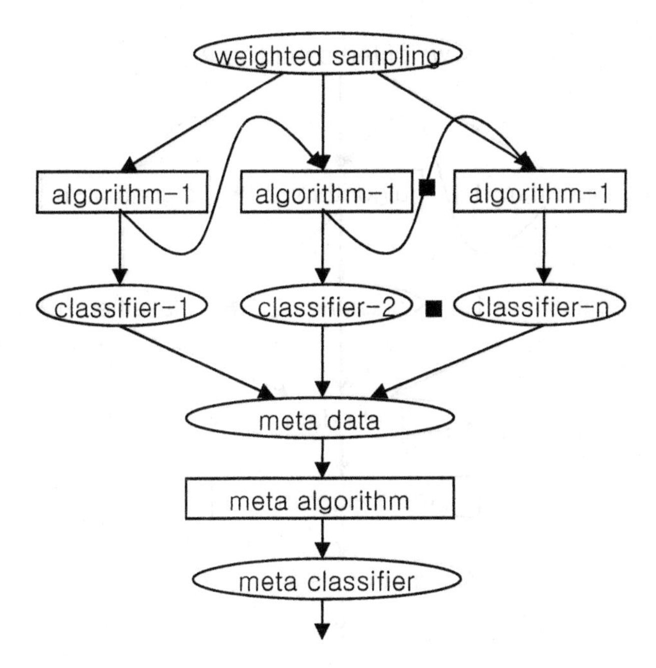

Fig. 4. Stacked boosting

Fig. 4 presents the concept of stacked boosting as an extended vote algorithm. Stacked bagging differs from stacked boosting only in that it samples training data for base model generation with uniform probability. The outputs generated by base-level classifiers are turned into a sequence of class probabilities attached to the original class to form meta-data. Instead of taking majority vote for the final decision as seen in simple bagging or boosting, the hybrid vote algorithm applies a meta-learning algorithm to meta-data to induce a meta-classifier which produces the final decision.

As for the extended stacking, we implement the stacking algorithms combined with model generation modules adapted from bagging and boosting. We call these extended stacking algorithms as bagged stacking and boosted stacking algorithms, respectively. The meta-data are produced in the same manner as in the extended vote algorithms, except that multiple models are generated from different algorithms independently and the output of multiple classifiers from each algorithm are linearly combined and turned into meta-data. Meta-level algorithm is applied to learn the regularities inherent in base models again, and the meta-classifier induced works as an arbiter of these heterogeneous base models. Fig. 5 shows the concept formulated for the implementation of the boosted stacking.

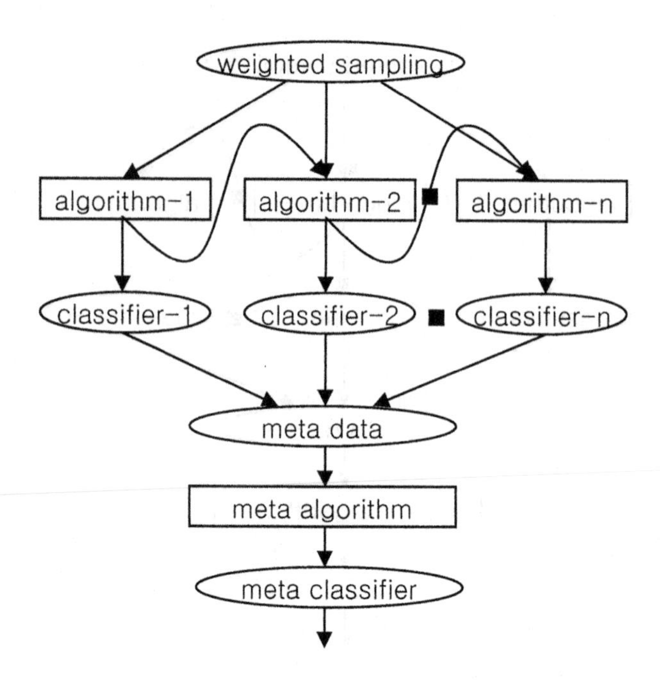

Fig. 5. Boosted stacking

4 Experiments

In this section, we compare the results of the empirical test of the existing basic multiple model algorithms and those of our proposed schemes, all based on Naive Bayesian, k-NN, and decision tree algorithms. We first describe the conditions under which the experiments were carried out, the data used, and how the experiments were executed. Next, we then evaluate the performances of various classifiers.

At the practical level, text categorization techniques can be regarded as a mixture of machine learning methods and advanced document retrieval techniques for document modelling. Documents need to be transformed into vectors of attributes prior to training to construct models or classifiers. A typical solution uses text preprocessing techniques such as stemming and indexing, the processes that turn documents into *Bag of Words* [10]. Feature selection should be conducted so as to reduce computation substantially without damaging the categorization performance. To achieve this, mainly filtering and wrapper approaches are taken into account. The filtering approach determines the features with some measures, and the wrapper approach searches for a subset of attributes more effective to a specific learning algorithm. Examples of the filtering approach are TFIDF, Information Gain(IG) and LSI techniques. For our experimentation, Information Gain(IG) is used as filtering measure in favor of its high accuracy based on entropy reported in various research outcomes [11].

Table 1. Classification accuracy(%): MEDLINE collection

Classification schemes	Generalizers	Document models			
		50 bin	50 wt	100 bin	100 wt
Single classifiers	k-NN	63.38	63.43	63.14	61.62
	C4.5	68.87	70.58	68.24	67.89
	NB	76.91	74.11	76.76	74.12
Basic multiple models(vote)	BagC4.5	72.51	69.80	70.49	70.59
	BoostC4.5	68.48	67.76	67.99	68.09
Basic multiple models(stacking)	StackDS	33.18	30.78	31.27	30.25
	StackC4.5	78.48	77.94	79.07	79.07
	StackNB	76.78	76.42	73.92	73.48
Hybrid multiple models(vote)	StackedBag	75.58	75.65	75.95	75.67
	StackedBoost	75.50	75.03	75.48	75.38
Hybrid multiple models(stacking)	BaggedStack	75.98	74.20	76.48	75.32
	BoostedStack	76.80	77.17	77.68	78.35

Table 2. Classification accuracy(%): USENET news collection

Classification schemes	Generalizers	Document models			
		50 bin	50 wt	100 bin	100 wt
Single classifiers	k-NN	96.94	91.94	97.47	89.47
	C4.5	97.29	96.59	96.88	86.70
	NB	97.65	66.65	97.06	70.00
Basic multiple models(vote)	BagC4.5	97.35	97.18	97.48	97.00
	BoostC4.5	97.41	97.18	97.76	97.41
Basic multiple models(stacking)	StackDS	38.94	38.88	39.00	38.88
	StackC4.5	97.65	96.65	97.52	96.71
	StackNB	97.82	96.71	98.00	96.79
Hybrid multiple models(vote)	StackedBag	97.38	96.94	97.38	97.00
	StackedBoost	97.42	97.32	97.42	96.92
Hybrid multiple models(stacking)	BaggedStack	97.40	97.08	97.50	97.50
	BoostedStack	97.48	97.26	97.40	97.08

Table 3. Classification accuracy(%): Web document collection

Classification schemes	Generalizers	Document models			
		50 bin	50 wt	100 bin	100 wt
Single classifiers	k-NN	74.30	75.80	74.62	75.21
	C4.5	77.03	77.55	77.88	79.05
	NB	76.97	69.88	81.33	69.49
Basic multiple models(vote)	BagC4.5	79.90	79.77	82.56	83.41
	BoostC4.5	77.29	80.09	81.78	83.80
Basic multiple models(stacking)	StackDS	52.37	51.01	53.03	51.59
	StackC4.5	78.59	77.90	82.50	79.12
	StackNB	79.38	78.66	82.30	75.63
Hybrid multiple models(vote)	StackedBag	78.80	78.71	78.51	78.71
	StackedBoost	78.49	77.78	77.82	78.38
Hybrid multiple models(stacking)	BaggedStack	80.15	77.29	83.98	80.06
	BoostedStack	80.61	78.84	82.63	80.13

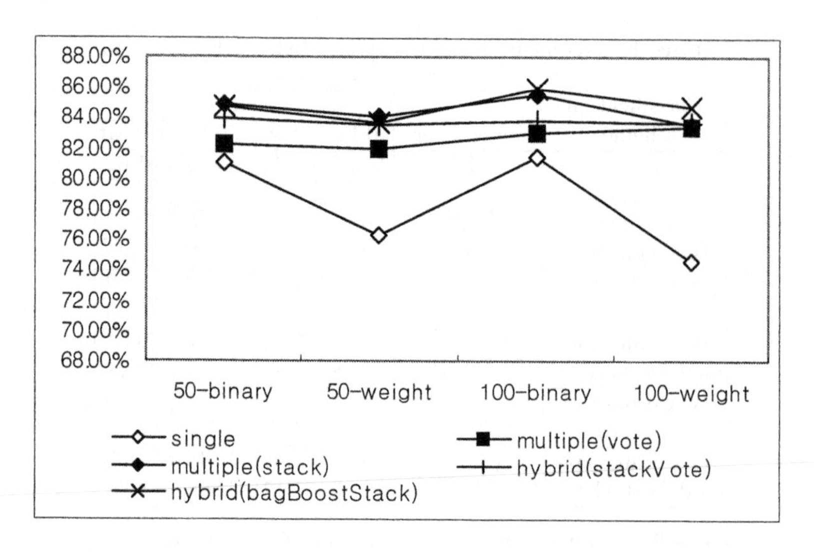

Fig. 6. Comparison among different schemes: classification accuracy (%)

The experiments were performed on a computer system equipped with Linux operating system, Intel Pentium IV processor, and 256 MB of memory. Rainbow[4] was used as the tool for text preprocessing and, WEKA[9] for categorization. These programs were modified with Java and C programming languages for the purpose of our research. Since we intend to compare the performance of each of these methods, two standard datasets and a real world data from MEDLINE database were used to evaluate each method. The MEDLINE data collection was obtained as the result of queries posed to MEDLINE database. The size of MEDLINE dataset is 6000 documents with 1000 documents assigned to each of six classes. Usenet news articles collection consists of five classes, each of which also holds 1000 documents. Web collection has 4,518 documents with six classes, and the class distribution varies. Two thirds of each collection was used as training examples and the rest of the data was set aside as test examples.

In order to examine the effect of suitable meta-level learner, the decision stump algorithm known to be an inefficient method is tested as meta algorithm. The category denoted as StackDS contains the results. Table 1, 2, and 3 summarize the results of the experiments to measure and compare the predictive accuracy of multiple classifiers with baseline performance by single classifiers. All of the error estimates were obtained using ten-fold cross validation.

Table 1, 2, and 3 present the performance data of various classifiers and Fig. 6 illustrates the performance of classifiers represented with average cross-collection accuracy. The accuracy rates of StackDS were excluded since this algorithm gives unrealistically pessimistic statistics. It is concluded from the results of decision stump algorithm that the meta algorithm as combining function itself substantially affects the overall performance.

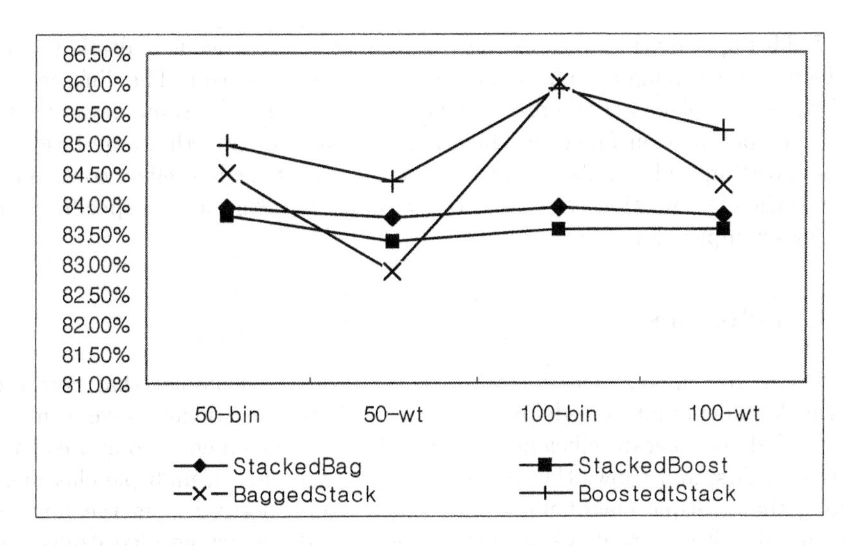

Fig. 7. Comparison among hybrid multiple model schemes: classification accuracy (%)

According to the results of the experiments, the methods of document representation developed in IR community has a trifling effect in performance with binary document models showing slightly better performance than weighted models. The size of the feature subset, however, appears to have no correlation with the distribution of accuracy rates. Among the document collection, the Usenet news articles records the highest classification accuracy, apparently owing to discriminating properties of keywords contained in computer newsgroup articles.

The utility of the predictive models derived is in its performance assessed in terms of generalization accuracy. The hybrid schemes show the highest accuracy rate, and the lowest standard deviation in their performance which indicates the stability and reliability of the predictions produced by the models. Single classifiers records 78.36 % of accuracy rate in cross-collection average while standard multiple models show 83.55 %, and the hybrid multiple models 84.30 percent of accuracy rate throughout the experiments. The standard deviation of single classifiers is 0.066 while those for basic multiple classifiers and hybrid classifiers are 0.023 and 0.01 respectively. Our research results in text categorization using multiple classifiers clearly demonstrate that the model combination approaches are more effective than single classifiers.

The experiments prove the utility of meta-learner as a combination function. The only difference between the standard vote algorithms and stacking algorithms is in model combining method. The extended vote algorithms implemented with meta-learning technique demonstrate higher accuracy than the standard methods by 1.09 %. Among the experiments dealing with hybrid algorithms, it is noted that the extended stacking methods like boosted and bagged stacking yield higher accuracy than the extended vote approach as shown in

Fig. 7. This apparently indicates that combining heterogeneous multiple models is effective in reducing the bias component of prediction error. The difference in performance is observed whereas all the algorithms use the same dataset with the same combination function which is meta-learning, and the same base and meta-algorithms. This boils down to a conclusion that the bias effect contributes more to the classification error than the variance resulting from a specific set of training examples does.

5 Conclusions

A primary goal in text categorization is to develop a classifier that correctly assigns documents into pre-defined categories. This paper reviews several methods applied to generate efficient classifiers for text documents, compares their relative merits, and shows the utility of the combination of multiple classifiers. Among the multiple classifiers experimented, those derived from the boosted stacking algorithms, that are a kind of the extended stacking algorithms proposed in this study, produce the best results. Our study demonstrates that the variants of the stacking algorithm contribute most to error reduction, mitigating the bias effect from learning algorithms. And meta-learning is proved to be a powerful tool for combining models. We also have found that the perturbation of training data reduces the error component caused by the variance due to the training dataset used for induction.

Our research findings confirm that hybrid model combination approaches improve the existing standard multiple model schemes. All of these findings can be a basis for formulating an inexpensive and efficient text classification system. Even though computation cost increases substantially with the implementation of hybrid multiple schemes, the cost may be justified by the enhanced reliability, stability and improved classification accuracy of the new approaches. We conclude that the techniques to address the problems of variance of training data and bias of learning algorithms can be implemented within a single classification system.

References

1. Kjersti A. and Eikvil L.: Text categorization: a survey. (1999)
2. Apte et al.: Automated learning of decision rules for text categorization. ACM Transactions on Information Systems, Vol.12, No.3, (1994) 233–251
3. Mladeni'c, D. and Grobelnik, M.: Efficient text categorization. In Text Mining workshop on the 10th European Conference on Machine Learning ECML-98 (1998)
4. Mitchell, Tom: Machine learning. New York: McGraw-Hill (1997)
5. Bauer, Eric and Ron Kohavi.: An empirical comparison of voting classification algorithms: bagging boosting and variants. Machine Learning Vol.36, (1999) 105–142
6. Breiman, Leo: Bagging predictors. Machine Learning Vol.24 (1996) 49–64
7. Schaphire, Robert E.: Theoretical views of boosting. Proceedings of the European Conference on Computational Learning Theory EuroCOLT-99 (1999)

8. Wolpert, D. and Macready, W.: Combining stacking with bagging to improve a learning algorithm. Technical report. Santa Fe Institute (1996)
9. Witten, Ian H. and Eibe, Frank: Data Mining: practical machine learning tools and techniques with Java implementations. Morgan Kaufman (2000)
10. Hong, Se June and Sholom M. Weiss: Advances in predictive model generation for data mining. IBM Research Report RC-21570 (1999)
11. Salton, Gerard: Introduction to information retrieval. McGraw-Hill (1983)
12. Yang, Yiming and Jan O. Pedersen: A comparative study on feature selection in text categorization. Proceedings of the 4th International Conference on Machine Learning (1997)

Learning Dynamic Bayesian Networks from Multivariate Time Series with Changing Dependencies

Allan Tucker and Xiaohui Liu

Department of Information Systems and Computing,
Brunel University, Uxbridge, Middlesex, UB8 3PH, UK
Allan.Tucker@brunel.ac.uk

Abstract. Many examples exist of multivariate time series where dependencies between variables change over time. If these changing dependencies are not taken into account, any model that is learnt from the data will average over the different dependency structures. Paradigms that try to explain underlying processes and observed events in multivariate time series must explicitly model these changes in order to allow non-experts to analyse and understand such data. In this paper we have developed a method for generating explanations in multivariate time series that takes into account changing dependency structure. We make use of a dynamic Bayesian network model with hidden nodes. We introduce a representation and search technique for learning such models from data and test it on synthetic time series and real-world data from an oil refinery, both of which contain changing underlying structure. We compare our method to an existing EM-based method for learning structure. Results are very promising for our method and we include sample explanations, generated from models learnt from the refinery dataset.

1 Introduction

There are many examples of Multivariate Time Series (MTS) where dependencies between variables change over time. For example, variables in an oil refinery can be affected by the way operators control the processes and by the different products being refined at a particular time. If these changing dependencies are not taken into account, any model that is learnt from the data will average over the different dependency structures. This will not only apply to chemical processes but also to many other dynamic systems in general. There has previously been work in the modelling of time series with changing dependencies. Methods have been explored to model or cluster the hidden states of a system, which change over time [6,11]. However, our focus is on making the underlying processes understood to non-statisticians through the automatic generation of explanations. Previously, we have developed methods for learning Dynamic Bayesian Networks (DBNs) from Multivariate Time Series (MTS) [12]. In this paper we extend this work to handle changing dependencies and at the same

M.R. Berthold et al. (Eds.): IDA 2003, LNCS 2810, pp. 100–110, 2003.

time remain transparent so that the resulting models and explanations account for these changes.

In the next section, the dynamic cross correlation function is introduced which is used to analyse the changes in dependency structure between two variables within MTS. We then outline a method to learn DBNs from MTS with changing dependency structures, followed by details of the experiments and their results when applied to synthetic and real world MTS data from an oil refinery process. Finally, explanations are generated from the resulting structures and conclusions are drawn.

2 Methods

During the analysis of MTS we have found it useful to explore how the Cross Correlation Function (CCF) [2] between two variables changes over time. The CCF is used to measure the correlation between two time series variables over varying time lags by time shifting the data before calculating the correlation coefficient. Our analysis of MTS with changing dependencies has involved developing a Dynamic CCF (DCCF), whereby the CCF is calculated for a window of data over lags, l, between two variables, a_i and a_j, in a MTS, A, and moved over the MTS by incrementing the window position by set amounts. The DCCF is generated by calculating the CCF for all lags and window increments [13], which can be visualised as a surface plot such as the example in Figure 3a. Unlike the CCF, the DCCF will not only measure correlation between two variables over various time lags but it will also measure the *change* in correlations over time lags. Whilst this DCCF will not be able to tell us about the more complex relationships within the MTS that involve more than two variables, it will assist us in making preliminary judgments about where likely dependency changes occur.

Bayesian Networks (BNs) are probabilistic models that can be used to combine expert knowledge and data [10]. They also facilitate the discovery of complex relationships in large datasets. A BN consists of a directed acyclic graph consisting of links between nodes that represent variables in the domain. The links are directed from a parent node to a child node, and with each node there is an associated set of conditional probability distributions. The Dynamic Bayesian Network (DBN) is an extension of the BN that models time series [4]. There has been much research into learning BNs from data, such as [3]. We propose the use of a hidden node to control the dependency structures for each variable in a DBN. Previously, hidden nodes have been used in BNs for modelling missing data and unobserved variables [7], and Markov chains have been used to control dependencies in DBNs [1]. The challenge of learning models with changing dependencies is that not only do we have to discover the network structure, but we also need to discover the parameters related to the link between each variable and its hidden controller (as the actual states of the controller will be unknown).

In [5] Structural Expectation Maximisation (SEM) was developed that searches for hidden variables whilst learning structure. We apply this method to our datasets and compare the results to our proposed algorithm.

2.1 Hidden Controller Hill Climb: Representation

We introduce a hill climb to search for DBNs with a specifically designed representation. By including a hidden controller node as a parent for each node at

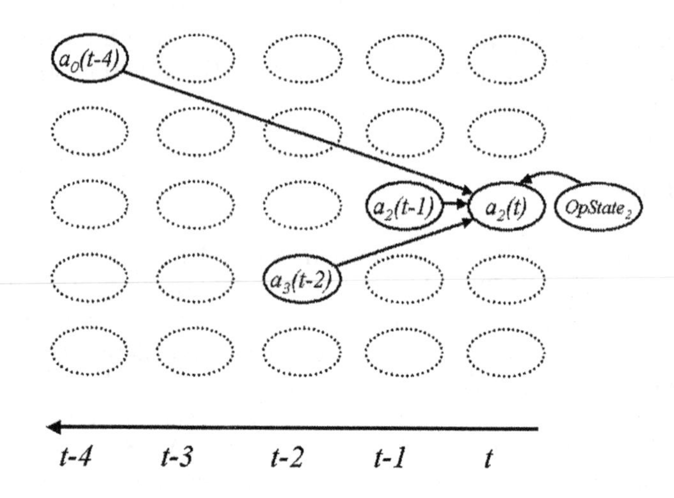

Fig. 1. Using a Hidden Variable, $OpState_2$, to Control Variable a_2 at Time t

time t representing variable a_i, which we will refer to from now on as $Opstate_i$, the dependencies associated with that variable can be controlled (see Figure 1).

OpState nodes will determine how the variables in the MTS behave based upon their current state. The structure is represented using a list of triples of the form (a_i,a_j,l), which represents a link from a_i to a_j with a lag of l. We will call the list of triples $DBNList$. For learning models with changing dependencies, the *OpState* node is automatically inserted before evaluation. The changing of state of *OpState* is not likely to happen very often in process data and so the relative stability of these variables can be used to speed convergence of a hill climb. For this reason, the hill climb will use a list, *SegmentList*, of pairs (*state*, *position*) to represent the switches in each *OpState* variable, where *state* represents its new state and *position* represents the position of change in the MTS.

A heuristic-based hill climb procedure is employed by making small changes in $DBNList$ and the *SegmentList* for each *OpState* variable. In order to search over the DBN structure and the parameters of *OpState* for each variable, the Hidden Controller Hill Climb (HCHC) algorithm uses the representation described above. HCHC takes as input the initial random $DBNList$ and a $SegmentList$ for each *OpState* node, along with the MTS, A, the maximum time lag, $MaxT$, and the number of iterations for the structure search, $DBNIterations$. $DBNList$ and the N *SegmentLists* are output and used to generate the final DBN structure and *OpState* parameters. The *OpState* variables can then be reconstructed

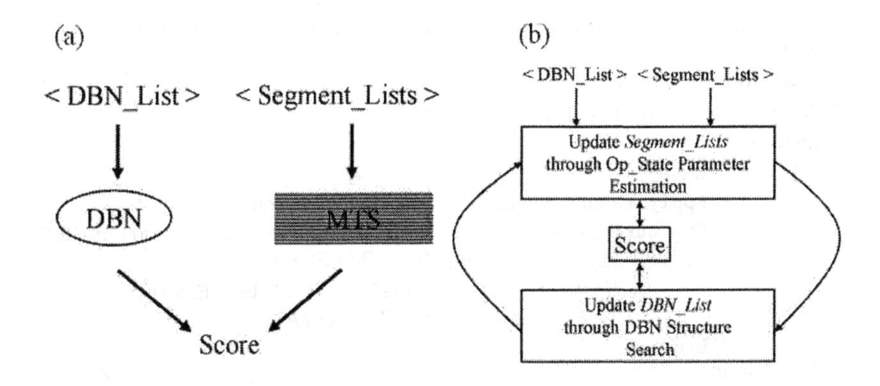

Fig. 2. (a) Building a DBN from *SegmentLists* and *DBNList*. (b) The HCHC procedure

using the *SegmentLists*. Once all *OpState* variables have been constructed, DBN parameters can be learnt from the MTS.

2.2 Hidden Controller Hill Climb: Search

HCHC applies a standard hill climb search [9] to the *OpState* parameters by making small changes to each *SegmentList* and keeping any changes that result in an improved log likelihood score. The Log likelihood function requires the DBN constructed from the *DBNList* and the MTS with *OpState* nodes constructed from the *SegmentLists* (see Figure 2a). It is calculated as follows:

$$logp(D|bn_D) = \prod_{i=1}^{N}\prod_{j=1}^{q_i}\frac{(r_i-1)!}{(F_{ij}+r_i-1)!}\prod_{k=1}^{r_i}F_{ijk}! \tag{1}$$

where N is the number of variables in the domain, r_i denotes the number of states that a node x_i can take, q_i denotes the number of unique instantiations of the parents of node x_i, F_{ijk} is the number of cases in D, where x_i takes on its kth unique instantiation and, the parent set of i takes on its jth unique instantiation. $F_{ij} = \sum_{k=1}^{r} F_{ijk}$.

Structure search is then carried out by repeatedly making small changes to *DBNList* and keeping any changes that improve log likelihood. These changes involve adding and removing links or applying a lag mutation operator [13]. Note that the $Opstate_i$ nodes are fixed so that they cannot have parents and their only child is their respective variable, i. After structure search is repeated *DBNIterations* times, the entire process is repeated, returning to the *OpState* parameter search. The search ends when the function calls, FC, reach some pre-defined value, *Iterations*. HCHC is illustrated in Figure 2b and defined formally below.

```
Input A, MaxT, DBNIterations, Random DBNList, NSegmentLists
1       FC = 0
2       Use current DBNList and SegmentLists to construct DBN
3       BestScore = log likelihood of DBN
4       Repeat
5           For i=1 to N
6                   Apply random change to the ith SegmentList
7                   Use current DBNList and SegmentLists to construct DBN
8                   If log likelihood of DBN > BestScore Then
9                           BestScore = log likelihood of DBN Else
10                          Undo change to ith SegmentList
11                  End If
12          End For
13          For j=1 to DBNIterations
14                  Apply random change to DBNList
15                  Use current DBNList and SegmentLists to construct DBN
16                  If log likelihood of DBN > BestScore Then
17                          BestScore = log likelihood of DBN Else
18                          Undo change to DBNList
19                  End If
20          End For
21          FC = FC + 1
22      Until Convergence or FC>Iterations
OutputDBNList and N SegmentLists used to construct DBN
```

The HCHC Segmentation Algorithm

3 Results

3.1 Synthetic Data

Synthetic data has been generated from hand-coded DBNs using stochastic simulation [8]. The MTS datasets consisted of between five and ten variables with time lags of up to 60 time slices. In order to incorporate changing dependencies within the datasets, different DBN structures were used to generate different sections of the MTS. Essentially, several MTS were appended together, having been generated from DBNs with varying structures and parameters. For example, MTS 1 consists of three MTS with five variables appended together, each of length 1000. Characteristics of the datasets are described in Table 1. The experiments involved applying the HCHC search procedure to learn the dependency structure and segment the data for each variable according to its *OpState*. The resultant structures and segmentations at convergence were then compared to the original structures that generated each segment of the MTS by calculating the Structural Differences (SD) [15]. We also apply Friedman's SEM and compare results. Having applied both HCHC and the SEM algorithm to the synthetic datasets, we now document the results. The average log likelihood at convergence is shown in Table 2. It is apparent that the log likelihoods of the DBNs constructed with the SEM method are much higher than those resulting from HCHC. This could be due to HCHC being limited in its number of segmentations whereas the SEM is not. However, as the true number of segmentations

Table 1. Details of the Synthetic Data with Changing Dependencies

-	Num of Variables	MTS Length	Segment Length	Maximum Lag	Num of Segments
MTS1	5	3000	1000	5	3
MTS2	10	3000	1000	60	3
MTS3	5	2500	500	25	5

for each dataset was within the limits set for HCHC, this implies either an inefficient HCHC or overfitting in the SEM. We now investigate these hypotheses. Table 3 shows the SD between the original and the discovered structures and the

Table 2. Resulting Log Likelihoods on Synthetic Data

-	HCHC	SEM
MTS1	-3115.665	-220.738
MTS2	-15302.017	-2032.371
MTS3	-4494.706	-239.346

percentage of correct links discovered for both SEM and HCHC. For all datasets the HCHC generates networks with SDs that are relatively small and the percentage of correct links discovered is high. The SEM appears to have performed less well with consistently higher SD and smaller percentages of correct links found. This implies that the high scores in Table 2 are due to spurious correlations causing high SD. Upon investigating the segmentation from SEM, it was

Table 3. Structural Difference Results on Synthetic Data

-	-	HCHC	-	-	SEM	-
-	MTS1	MTS2	MTS3	MTS1	MTS2	MTS3
Total SD	5.6	6.7	5.8	9.6	16.8	10.8
Correct	0.916	0.892	0.913	0.725	0.681	0.488

found that the expected states of the *OpState* variables varied from segment to segment. For example in one segment they may remain steadily in one state whilst in another they may fluctuate rapidly between 2 or more. However, they do not neatly allocate a different state to different MTS segments and so statistics could not easily be used to measure the quality, suffice to say that there would have been hundreds of segmentations in each dataset. This could well be due to the model overfitting identified in the SD analysis. On the other hand, segmentation for the HCHC results was very good. We now discuss this with respect to MTS 3. We make use of the DCCF in order to further analyse the results of the HCHC on MTS 3. Whilst this is not really necessary for synthetic data where the original network structures are available, it will be useful when

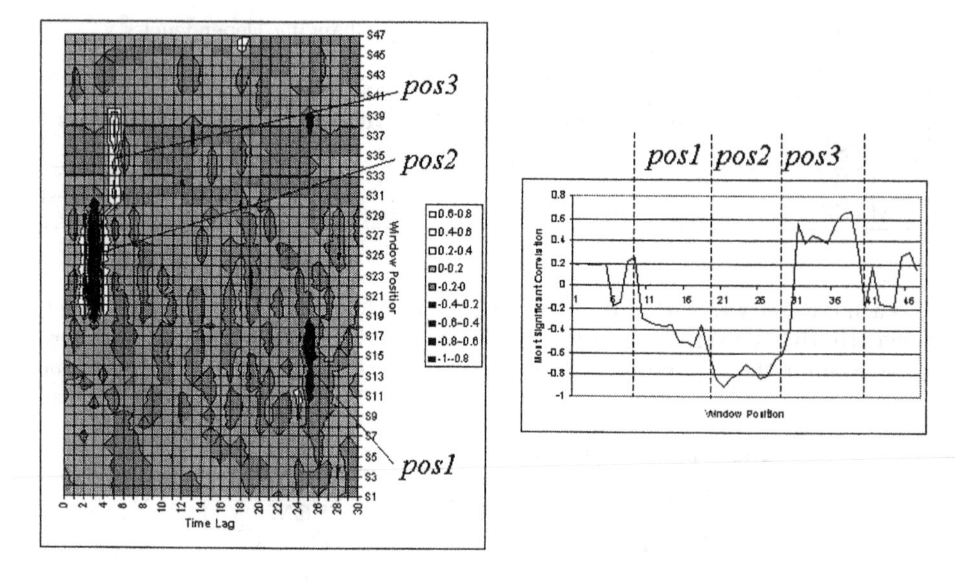

Fig. 3. (a) The DCCF for Variable a_3 and a_4 in MTS 3. (b) The Most Significant Correlation for Each Window Position in (a)

analysing real world data where these structures are unknown. Figure 3a shows the DCCF for variable a_3 and a_4 in MTS 3. The maximum time lag, $MaxT$, calculated for each CCF was 30. Note the varying peaks (in white) and troughs (in black), which signify stronger positive and negative correlations, respectively. Below are the discovered $DBNList$ and $SegmentList$ for $OpState_4$.

DBNList: ((1,0,8), (0,1,5), (0,1,9), (2,1,3), (2,1,6), (3,2,2), (1,2,7), (3,2,20), (4,3,3), (2,3,5), **(3,4,3), (3,4,25), (3,4,5)***)*

SegmentList for $OpState_4$: *((0,0),(498,1),(981,2),(1502,3),(1997,4))*

Figure 3b shows a graph of the most significant correlation (the maximum absolute correlation) for each window position in the DCCF. Firstly there is no correlation from a_3 to a_4 for window positions 1-10 (MTS position 1-500). Then an inverse correlation occurs (black trough denoted by $pos1$) with a lag of 25 in window positions 10-20 (MTS position 500-1000). This lag then switches to 3 for MTS positions 1000-1500 ($pos2$). Then a positive correlation can be seen (white area denoted by $pos3$) with a lag of 5 in MTS positions 1500-2000, and in the final sections of data, no significant correlation from a_3 to a_4 can be found. These correlations and lags correspond well to the links in $DBNList$ from a_3 to a_4 (in bold) and the $SegmentList$ for $OpState_4$, discovered for MTS 3.

3.2 Oil Refinery Data

The real world MTS in this paper is from an oil refinery process which contains 300 variables and 40 000 time points. We now look at an example of applying HCHC to an oil refinery variable and comparing the discovered segmentations with the DCCF between certain variables. Due to the lack of space as well as the poor results of SEM on both the synthetic and the oil refinery MTS, we focus on segmentations and explanations generated using HCHC. In Figure 4,

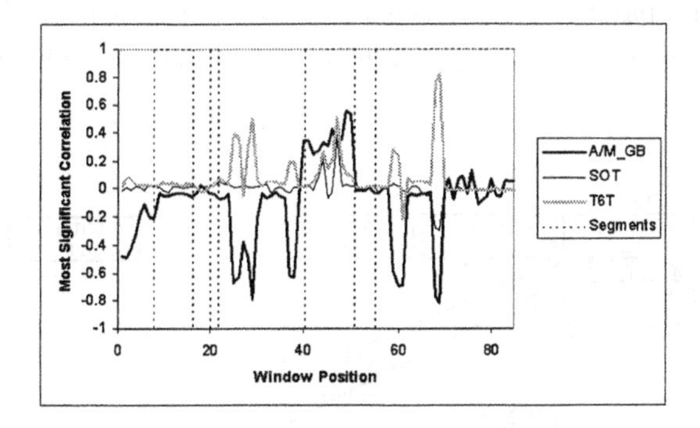

Fig. 4. Most Significant Correlations for DCCFs from Oil Refinery Variables

the most significant correlation graphs for each window position of the DCCF are shown for variable TGF with each of its three discovered parents. There is generally a fluctuation between no and strong negative correlations between TGF and A/MGB throughout the length of the MTS. However, at about window position 40, a positive correlation occurs. This continues until position 50 where the fluctuating returns until the end of the series. This closely follows the segmentation found using HCHC, which are marked in Figure 4 by dotted lines. The same applies for the other two parent variables with SOT also showing positive and negative correlations and T6T showing varying amounts of positive correlation. The segmentation appears to successfully separate out each of these regions. Similar results were discovered for all variables in the oil refinery data with most pair-wise relationships being successfully segmented.

Note that most of the segmentations that have been discovered occur where there are switches from positive to negative correlation rather than between regions of strong and weak correlation. Also some of the results appear to find a number of segmentations that do not correspond with the correlation changes. This could be because the relationships that are changing are more complex than the pair-wise relationships identified in a DCCF. We explore this in the next section.

3.3 Explanations Incorporating Hidden Controllers

We generated a DBN structure for a subset of 21 oil refinery variables in order to explore some generated explanations. In general, the results from the process data have been very encouraging with many of the relationships identified using the DCCF being discovered. It must be noted that other more complex relationships may also have been discovered and we now look at some sample explanations that have been generated from the DBNs with *OpState* nodes included to identify these relationships. Given the discovered DBN structure and parameters for each node in the network including the set of *OpStates*, inference can be applied to generate explanations. This involves a process whereby certain observations are made about variables in the DBN and inference is used to generate posterior probability distributions over the unobserved variables. *OpStates* can also be included as part of the explanation. Figure 5 shows some explana-

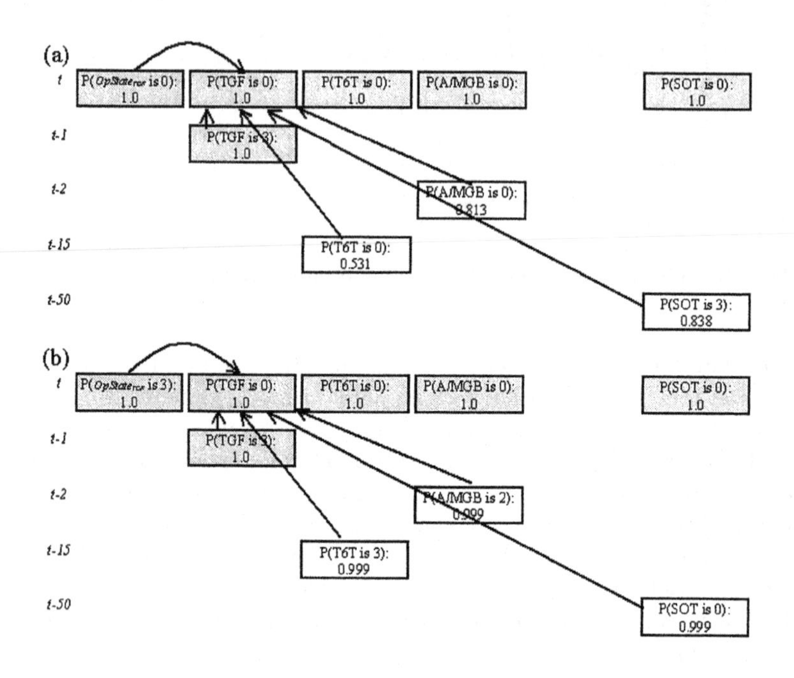

Fig. 5. Sample of Generated Explanations from the Oil Refinery DBN

tions that have been generated using the DBN discovered from the oil refinery data. Shaded nodes represent observations that have been entered into the DBN (hence the probability is 1.0). It can be seen in Figure 5a that SOT has a strong likelihood of being in state 3, 50 time points previously, given the instantiations and $OpState_{TGF}$ being 0. When $OpState_{TGF}$ changes to 3 as shown in Figure 5b the most probable state for each parent variable alters (SOT now most likely to be in state 0 and A/MGB in state 2). This shows how *OpStates* can affect explanation.

In Figure 6a the effect of adding new evidence to Figure 5b is shown, which changes all of the variable states again. In Figure 6b TGF and its set of parents are instantiated resulting in the controller variable $OpState_{TGF}$ being most likely in state 0, showing how an *OpState* node can be included as part of an explanation. These sample explanations indicate how complex relationships involving 3 or more variables can be modelled, illustrating how the segmentation

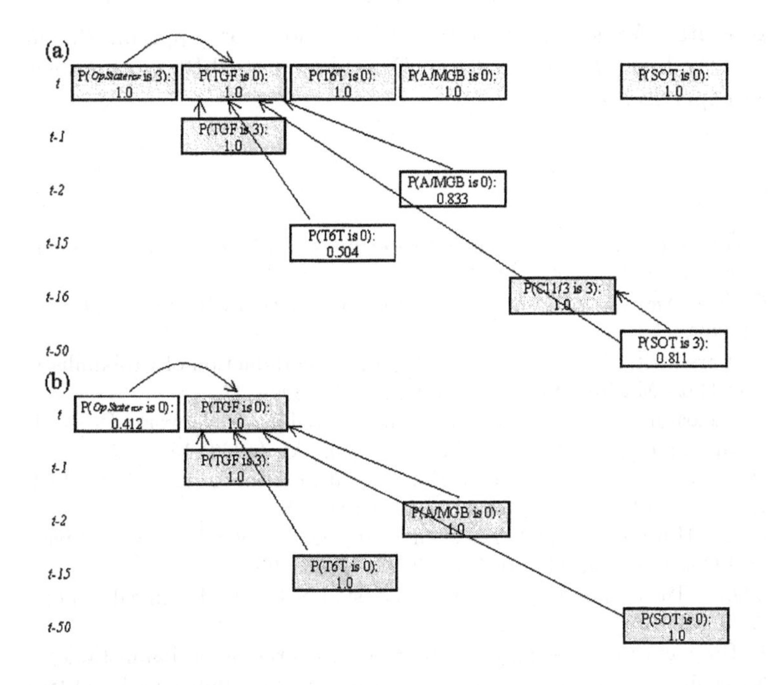

Fig. 6. Sample of Generated Explanations from the Oil Refinery DBN

of the MTS in Figure 4 may represent more complex relationships than those in its DCCF.

4 Conclusions

In this paper, we have used hidden nodes in a dynamic Bayesian network to model changing dependencies within a Multivariate Time Series (MTS). These hidden nodes can be interpreted as 'the current state of the system' and, therefore, be used within explanations of events. We introduce an algorithm, HCHC, which reduces the search space drastically through the use of a specific representation. It performs far better than EM methods for learning hidden nodes with BN structure. HCHC has allowed good models to be learnt from oil refinery MTS, reflecting expected dependency structures and changes in the data. Future work will involve trying to improve the EM results using deterministic annealing [14] or some method of *smoothing* the control transition states by fixing their transition probabilities. We also intend to look at other datasets such as gene expression and visual field data where dependency structure is based upon experimental conditions. We will use HCHC to classify these MTS into such operating states.

Acknowledgements. We would like to thank BP-Amoco for supplying the oil refinery data and Andrew Ogden-Swift and Donald Campbell-Brown for their control engineer expertise.

References

1. Bilnes, J.A.: Dynamic Bayesian Multinets Proceedings of the 16th Annual Conference on Uncertainty in AI, (2000) 38–45
2. Chatfield, C.: The Analysis of Time Series – An Introduction Chapman and Hall, 4th edition, (1989)
3. Cooper, G.F. Herskovitz, E: A Bayesian Method for the Induction of Probabilistic Networks from Data Machine Learning, (1992) Vol. 9, 309–347.
4. Friedman, N.: Learning the Structure of Dynamic Probabilistic Networks Proceedings of the 14th Annual Conference on Uncertainty in AI, (1998) 139–147
5. Friedman, N.: The Bayesian Structural EM Algorithm Proceedings of the 14th Annual Conference on Uncertainty in AI, (1998) 129–138
6. Ghahramani, Z., Hinton, G. E.: Variational Learning for Switching State-Space Models Neural Computation, (1999), Vol. 12, No. 4, 963–996
7. Heckerman, D.: A Tutorial on Learning with Bayesian Networks Technical Report, MSR-TR-95-06, Microsoft Research, (1996)
8. Henrion, M.: Propagating uncertainty in Bayesian networks by probabilistic logic sampling Proceedings of the 2nd Annual Conference on Uncertainty in AI, (1988) 149–163
9. Michalewicz, Z., Fogel, D.B.: How To Solve It: Modern Heuristics,Springer, 1998
10. Pearl, J.: Probabilistic Reasoning in Intelligent Systems: Networks of Plausible Inference Morgan Kaufmann, (1988)
11. Ramoni, M, Sebstiani, P., Cohen, P.: Bayesian Clustering by Dynamics, Machine Learning, (2002), Vol. 47, No. 1 91–121
12. Tucker, A., Liu, X., Ogden-Swift, A.: Evolutionary Learning of Dynamic Probabilistic Models with Large Time Lags. International Journal of Intelligent Systems **16**, Wiley, (2001) 621–645
13. Tucker, A.: The Automatic Explanation of Multivariate Time Series PhD Thesis, Birkbeck College, University of London, (2001)
14. Ueda, N., Nakano, R.: Deterministic Annealing Variant of the SEM algorithm Advances in Neural Information Processing Systems 7, (1995) 545–552
15. Wong, M., Lam, W., Leung, S.: Using Evolutionary Programming and Minimum Description Length Principle for Data Mining of Bayesian Networks IEEE Transactions on Pattern Analysis and Machine Intelligence, (1999), Vol. 21, No.2, 174–178

Topology and Intelligent Data Analysis*

V. Robins[1], J. Abernethy[2], N. Rooney[2], and E. Bradley[2]

[1] Department of Applied Mathematics
Research School of Physical Sciences and Engineering
The Australian National University
ACT 0200 Australia

[2] University of Colorado
Department of Computer Science
Boulder, CO 80309-0430

Abstract. A broad range of mathematical techniques, ranging from statistics to fuzzy logic, have been used to great advantage in intelligent data analysis. Topology—the fundamental mathematics of shape—has to date been conspicuously absent from this repertoire. This paper shows how topology, properly reformulated for a finite-precision world, can be useful in intelligent data analysis tasks.

1 Introduction

Topology is the fundamental descriptive machinery for shape. Putting its ideas into real-world practice, however, is somewhat problematic, as traditional topology is an infinite-precision notion, and real data are both limited in extent and quantized in space and time. The field of *computational topology* grew out of this challenge[5,9]. Among the formalisms in this field is the notion of *variable-resolution topology*, where one analyzes the properties of the data—*e.g.*, the number of components and holes, and their sizes—at a variety of different precisions, and then deduces the topology from the limiting behavior of those curves. This framework, which was developed by one of the authors of this paper (Robins)[20,21,22], turns out to be an ideal tool for intelligent data analysis.

Our approach to assessing connectedness and components in a data set has its roots in the work of Cantor. We define two points as *epsilon (ϵ) connected* if there is an ϵ-chain joining them; all points in an *ϵ-connected set* can be linked by an ϵ-chain. For the purposes of this work, we use several of the fundamental quantities introduced in [21]: the number $C(\epsilon)$ and maximum diameter $D(\epsilon)$ of the ϵ-connected components in a set, as well as the number $I(\epsilon)$ of ϵ-isolated points—that is, ϵ-components that consist of a single point. As demonstrated in [20,22], one can compute all three quantities for a *range* of ϵ values and deduce the topological properties of the underlying set from their limiting behavior. If the underlying set is connected, the behavior of C and D is easy to understand.

* Supported by NSF #ACI-0083004 and a Grant in Aid from the University of Colorado Council on Research and Creative Work.

M.R. Berthold et al. (Eds.): IDA 2003, LNCS 2810, pp. 111–122, 2003.

When ϵ is large, all points in the set are ϵ-connected and thus it has one ϵ-component ($C(\epsilon) = 1$) whose diameter $D(\epsilon)$ is the maximum diameter of the set. This situation persists until ϵ shrinks to the largest interpoint spacing, at which point $C(\epsilon)$ jumps to two and $D(\epsilon)$ shrinks to the larger of the diameters of the two subsets, and so on.

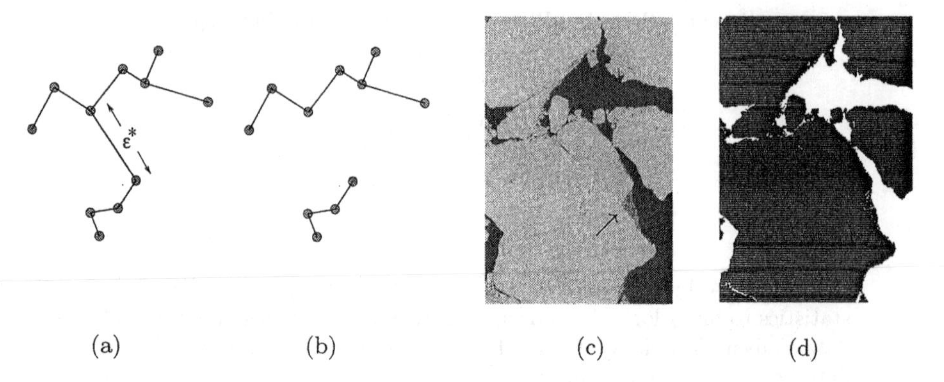

Fig. 1. Computing connectedness: (a) The *minimal spanning tree* (MST) whose edges connect nearest neighbors in a data set (b) This set contains two ϵ-connected components if ϵ is slightly less than ϵ^* (c) An aerial image of a patch of the arctic ocean; the arrow indicates a bay full of small ice floes (d) The MST of the ice pixels in (c)

When ϵ reaches the *smallest* interpoint spacing, every point is an ϵ-connected component, $C(\epsilon) = I(\epsilon)$ is the number of points in the data set, and $D(\epsilon)$ is zero. If the underlying set is a disconnected fractal, the behavior is similar, except that C and D exhibit a stair-step behavior with changing ϵ because of the scaling of the gaps in the data. When ϵ reaches the largest gap size in the middle-third Cantor set, for instance, $C(\epsilon)$ will double and $D(\epsilon)$ will shrink by 1/3; this scaling will repeat when ϵ reaches the next-smallest gap size, and so on. [21] derives, explains, and demonstrates these results in detail, and discusses the obvious link to fractal dimension. Pixellation can also cause stair-step effects because it quantizes inter-point distances; this can be confusing, as it mistakenly suggests that the underlying data set has a disconnected fractal structure.

Our computer implementation of these connectedness calculations relies on constructs from discrete geometry: the minimal spanning tree (MST) and the nearest neighbor graph (NNG). The former is the tree of minimum total branch length that spans the data; see Figure 1(a) for an example. To construct the MST, one starts with any point in the set and its nearest neighbor, adds the closest point, and repeats until all points are in the tree[1]. The NNG is a directed graph that has an edge from x_A to x_B if x_B is the nearest neighbor of x_A. To

[1] Our implementation actually employs Prim's algorithm[7], which is a bit more subtle. It begins with any vertex as the root and grows the MST in stages, adding at each stage an edge (x, y) and vertex y to the tree if (x, y) is minimal among all edges where x is in the tree and y is not.

construct it, one starts with the MST and keeps the shortest edge emanating from each point. Both algorithms may be easily implemented in R^d; the computational complexity of the MST is $O(N^2)$ in general and $O(N \log N)$ in the plane, where N is the number of data points. To compute C and I from these graphs, one simply counts edges. $C(\epsilon)$, for example, is one more than the number of MST edges that are longer than ϵ, and $I(\epsilon)$ is the number of NNG edges that are longer than ϵ. Note that one must count NNG edges with multiplicity, since x_A being x_B's nearest neighbor does not imply that x_B is x_A's nearest neighbor (*i.e.*, if a third point x_C is even closer to x_A). Note, too, that the MST and NNG need only be constructed once; all of the C and I information for different ϵ values is captured in their edge lengths. To identify the individual ϵ-components, one simply removes edges that are longer than ϵ. Diameters $D(\epsilon)$ of ϵ-connected components are then found using standard computational geometry techniques.

These trees and the information encoded in their edges are extremely useful in intelligent data analysis. Vertical jumps in $C(\epsilon)$, for instance, take place at ϵ values that correspond to sizes of gaps in the dataset. $D(\epsilon)$ is of obvious utility in describing the sizes of objects, and $I(\epsilon)$ can be used to filter out noise. The MST itself can bring out structure. Figure 1(c), for example, shows a black-and-white of a patch of the arctic ocean. The small, right-facing bay just below and to the left of center (arrow), which faces into the prevailing wind, is full of small ice floes. The floes are so small that this region simply appears as grey in the raw image, but a MST of the white (ice) pixels brings out the small pockets of ice and water, as shown in part (d). All of these quantities mesh well with experimental reality: the MST of a sea-ice image, for instance, captures exactly how many floes will be resolved if the instrument has a precision of 10m, 5m, 1m, etc. The examples in Section 2 expand upon some of these ideas. MSTs can also be used to aggregate points into structures, and so they have been widely used in the kinds of clustering tasks that arise in pattern recognition[10] and computer vision[4]. Their branching structure can also be exploited—*e.g.*, to identify orbit types in dynamical systems[27] or to find discontinuities in bubble-chamber tracks[28]. Clustering and coherent-structure extraction are not the only applications of the MST; additive noise creates small transverse 'hairs' on these trees, and so filtering out those points is simply a matter of pruning the associated edges. Section 3 covers this idea in more depth.

Another fundamental topological property is the number, size, and shape of the *holes* in an object. One way to characterize this mathematically is via homology. This branch of topology describes structure using algebraic groups that reflect the connectivity of an object in each dimension. The rank of each homology group is called the *Betti number*, b_k. The zeroth-order Betti number, b_0, counts the number of connected components. Geometric interpretations of the other Betti numbers depend on the space the object lives in: in 2D, b_1 is the number of holes, while in 3D b_1 is (roughly speaking) the number of open-ended tunnels and b_2 is the number of enclosed voids. See our website[1] for some images that make these definitions more concrete. The definition of the homology groups requires a discrete geometric representation of the object, *e.g.*, a triangulation of subsets of 2D, and simplicial complexes in three or more dimensions. See Munkres[16] for further details.

In order to put these hole-related ideas into computational practice, we use the α-shape algorithm developed by Edelsbrunner[12], which computes the Betti numbers from triangulations that capture the topology of the coarse-grained data at different resolutions. The full algorithm involves some fairly complex computational geometry techniques, but the basic idea is to "fatten" the data by forming its α-neighborhood: that is, to take the union of balls of radius α centered at each data point. When α is large (on the order of the diameter of the data) this α-neighborhood is a single connected blob. As α decreases, the α-neighborhood shrinks and more shape detail is resolved in the data. When α is just small enough that a ball of radius α can fit inside the data without enclosing any data points, a hole is created in the α-neighborhood. For very small α—less than half the minimum spacing of the data—the individual data points are resolved. The associated calculations are non-trivial, but fast algorithms have been developed for α-shapes of 2D and 3D data[8], and associated software is available on the world-wide web[2].

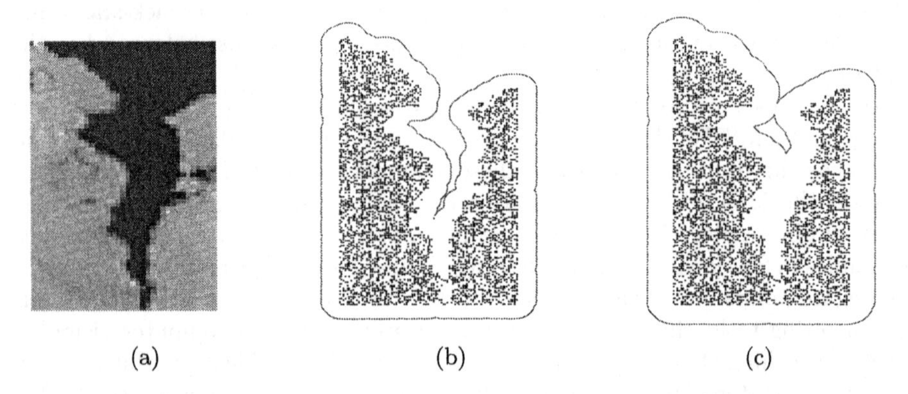

Fig. 2. Computing holes: α-neighborhoods can be used to detect whether the image in part (a) contains a bay, and, if so, how wide its mouth is. Images (b) and (c) show α-neighborhoods of the set of white (ice) pixels in (a) for a few interesting α values.

As in the connectedness discussion above, one can compute the number of holes in an α-neighborhood of a data set while varying α, and then use that information to deduce the topological properties of the underlying set[19,20]. There is one important difference, however. The *geometry* of the set can create holes in the α-neighborhoods, even if the set contains no holes. This effect is demonstrated in Figure 2. Mathematically, this problem can be resolved by incorporating information about how the set maps inside its α-neighborhood. This leads to the definition of a *persistent Betti number*, which was introduced in [19]: for $\epsilon < \alpha$, $b_k(\epsilon, \alpha)$ is the number of holes in the α-neighborhood for which there are corresponding holes in the ϵ-neighborhood (equivalently, the number of holes in the ϵ-neighborhood that do not get filled in by forming the fatter α-

neighborhood). These persistent Betti numbers are well defined for sequences of complexes that provide successively better approximations to a manifold[19] and are computable using linear algebra techniques. Recently, Edelsbrunner and collaborators have made a similar definition of persistent Betti number specifically for α-shapes, and devised an incremental algorithm for their calculation[11].

While the non-persistent holes effect makes it difficult to make a correct diagnosis of the underlying topology, it has important implications for intelligent data analysis because it gives us geometric information about the embedding of the set in the plane. This can be very useful in the context of coherent structure extraction. Consider a narrow bay in an icepack, as shown in Figure 2. In this case, $b_1(\alpha)$ computed for the set of white (ice) pixels would be zero for α smaller than half the width of the mouth of the bay, zero for α larger than the largest radius of its interior, and one in between—that is, where an α ball fits inside the bay, but not through its mouth. Note that the α value where the spurious hole first appears is exactly the half-width of the entrance to the bay. If there were a *hole* in the ice, rather than a bay, $b_1(\alpha)$ would be a step function: zero when α is greater than the largest radius of the hole and one when it is less. This technique for finding and characterizing coherent structures whose defining properties involve holes and gaps of various shapes—ponds, isthmuses, channels, tunnels, etc.—is potentially quite useful in intelligent data analysis. More examples and discussion follow in Section 2.

2 Coherent Structure Extraction

In this section, we present three examples that demonstrate how to use our variable-resolution topology techniques to find coherent structures in aerial images of sea ice[2]. Scientists look for several things in these kinds of images: open water or "leads," ice floes, and melt ponds. all of which are visible in Figure 3. Climatology studies that track the seasonal evolution of these coherent structures are a major current thrust of Arctic science, but they require an automated way of analyzing large numbers of images. While it is certainly possible for someone to sit down and outline the floes in an image with a mouse, modern scientific instruments can easily generate gigabytes of data in an hour[13], and any human analyst will rapidly be overwhelmed (and bored) by the task. Traditional image-processing and machine-learning tools can help with tasks like this, to a point—*e.g.*, edge-finding, contrast enhancement, etc.—but topology-based methods are an even better solution. The following paragraphs treat three examples in detail: finding a lead through an icepack, distinguishing regions of different ice/water concentration, and studying how the number and size of melt ponds are distributed in sea ice.

The size, shape, and position of floes and leads is of obvious interest to travelers in the polar regions. Finding a path for a ship through a complicated region of the ice pack, in particular, is a common practical problem, and the

[2] courtesy of D. Perovich from CRREL.

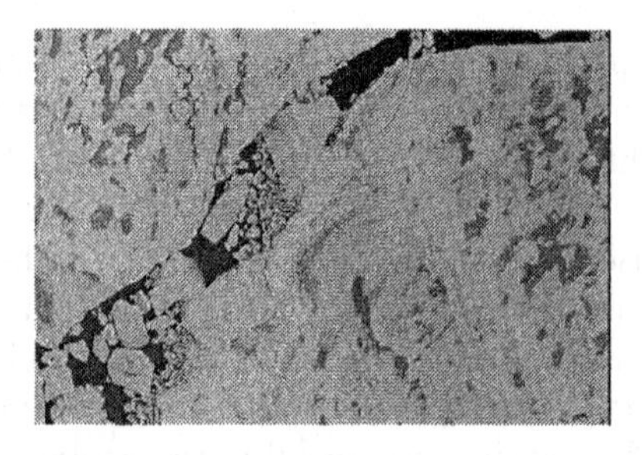

Fig. 3. The arctic ice pack is made up of open water, ice, and melt ponds, which image as black, white, and grey, respectively. Studying the climatology of this system—the seasonal evolution of the different kinds of *coherent structures*—is a major current thrust of Arctic science, and doing so requires an automated way of analyzing large numbers of images.

variable-resolution computational topology techniques described in Section 1 are quite useful in doing so. Consider an nm-wide ship that must traverse the patch of ocean depicted in Figure 4(a), from left to right. Solving this problem

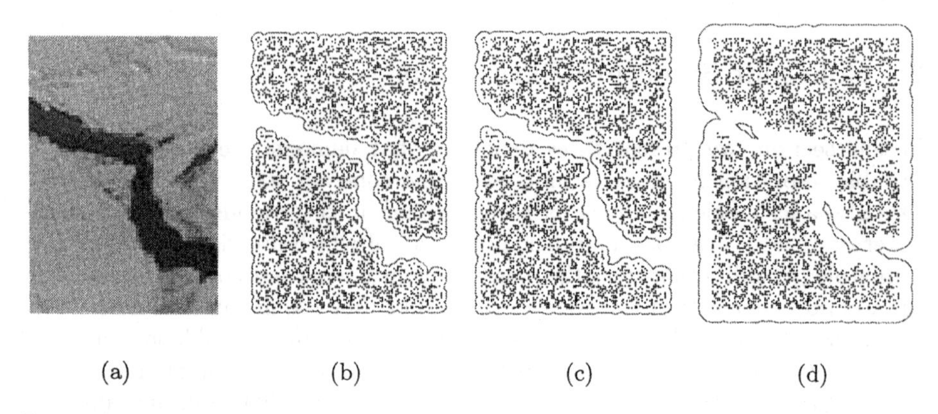

(a) (b) (c) (d)

Fig. 4. Using α-shapes to find a path through an icepack: (b), (c), and (d) show three different α-neighborhoods of the set of ice pixels in (a).

requires assessing the holes in the ice, so we threshold the data and discard the dark water pixels, then apply the α-shape algorithm with a radius $\alpha = n/2$. The results, shown in Figure 4(b), identify all holes that are at least nm wide. There are actually two holes in this α-shape, one corresponding to the kinked, diagonal channel and a much smaller one about halfway up the right-hand edge of Figure 4(b). The next step is to use simple computational geometry techniques on these holes to ascertain which ones touch both sides of the image and then determine which is the shortest[3]. Parts (c) and (d) of the Figure show the effects of raising α beyond $n/2$m, demonstrating the relationship between the channel shape and the holes in the α-neighborhood—*e.g.*, successively resolving the various tight spots, or identifying regions of water that are at least α meters from ice.

More generally, α shapes can be used to assess the mix of water and ice in a region—not just the relative area fractions, but also the distribution of sizes of the regions involved—or to find and characterize regions of different ice concentration. Depending on the task at hand, one can analyze either the

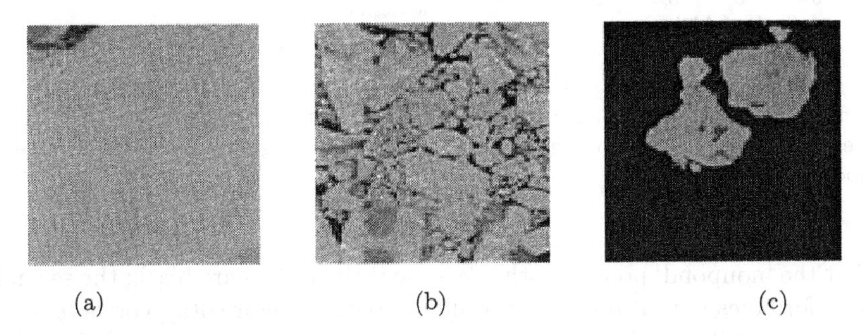

Fig. 5. Using α-shapes to distinguish different ice/water distributions: (a) almost-solid ice (b) a lead filled with various-size floes (c) open water with a few big floes. An α-shapes analysis of these images yields a direct measure of the morphology of the ice and water.

water pixels (where α-holes are α-size ice floes) or the ice pixels, where holes are leads. When the ice is close to solid, as in Figure 5(a), the narrow water channel is resolved as a small α-hole in the set of ice pixels, similar to the lead in the previous paragraph. When the image contains large areas of open water and a few floes, as in part (c), an α-shape analysis of the water pixels resolves a few large holes over a wide range of α values—after a band of smaller αs where the α-neighborhoods are filling in the various gaps and bumps in the floes and creating spurious holes, as discussed in conjunction with Figures 2 and 4. The middle image is more interesting; the wide distribution of floe sizes translates to a wide α range where a small change in that parameter resolves a few new holes in the set of water pixels (*i.e.*, floes). All of these results accurately reflect important properties of the data.

The temporal evolution of the albedo of Arctic sea ice is of great importance to climate modelers because of its role in heat transfer. A key factor governing the sea-ice albedo in summer is the size distribution of the melt ponds, which appears to take the form of a power law[18]. We can easily automate these distribution calculations using α-shapes. The first step is to threshold the data

[3] Note that this development assumes a circular ship; for oblong ships, one should use an α value that is the largest chord of the hull.

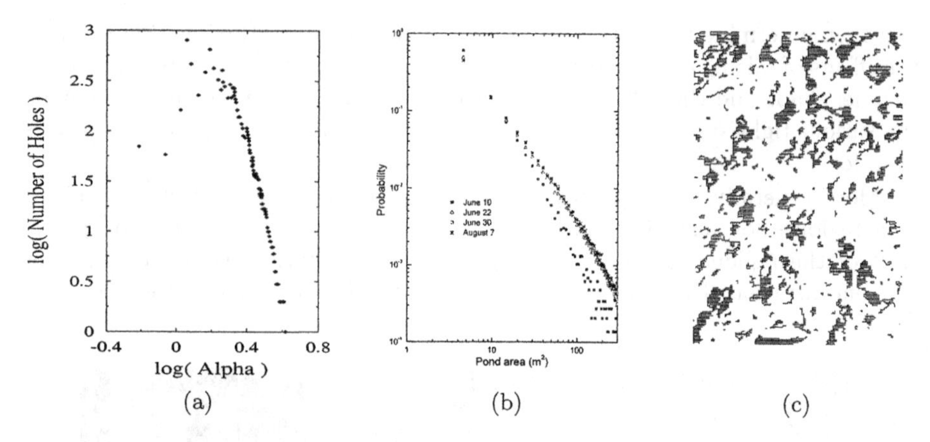

Fig. 6. Computational topology and melt ponds: (a) An α-shapes analysis of the ponds in the lower-right quadrant of Figure 3 (b) Corresponding figure from [18] (c) Connected components in pond data

so that the 'nonpond' points (both the ice and the water) are black; the second is to look for holes in that data. The results, shown in Figure 6(a), corroborate the power-law distribution hypothesis quite nicely; after an initial region below about $\log \alpha = 0.3$, where changes in this parameter are resolving promontories and bays in the melt ponds, the log-log plot is fairly linear. Part (b) reproduces the corresponding figure from [18] for comparison. The power-law slopes in (a) and (b) are different because the former plot has pond *radius* on the horizontal axis, while the latter has pond *area*. If the extra power of two is taken into account, the slopes of (a) and (b) actually match to within 8%[4]. We can obtain similar curves using the connectedness tools described in Section 1: *e.g.*, computing a minimal spanning tree of the 'pond' points in the image, as in Figure 6(c), and then extracting the connected components and computing their areas and/or diameters.

Our next step will be to combine these notions of multiple-resolution topology with some ideas from spatial statistics and artificial intelligence in order to handle the ill-defined nature of real coherent structures and to recognize the spatial and temporal patterns in which those structures move, evolve, and interact. Spatial statistics are useful in this application because computational topology produces aggregate measures over a whole dataset. Adding a sliding window to the process, and manipulating its geometry intelligently—an approach that we term *topospatial analysis*—allows one to distinguish sub-regions where those measures are different. Artificial intelligence (AI) techniques are useful because coherent structures are not crisp. The Gulf Stream, for instance, is a hot-water jet extending from the eastern seaboard of the US into the Atlantic. It does not have a natural, well-defined boundary. Rather, the temperature at its edges falls off smoothly, and so any threshold-based definition is necessarily somewhat

[4] The residual discrepancy is due to the fact that not all ponds are perfectly circular.

arbitrary. Moreover, coherent structures are much easier to recognize than to describe, let alone define in any formal way. The AI community has developed a variety of representations and techniques for problems like these, and these solutions mesh well with topospatial analysis.

3 Filtering

Real-world data invariably contain noise of one form or another. Traditional linear filters are useful in removing some kinds of noise, but not all. Chaotic behavior, for instance, is both broad band and sensitively dependent on system state, so linear or Fourier-based filters—which simply remove all signal in some band of the power spectrum—can alter important features of the dynamics, and in a significant manner[26]. In cases like this, topology-based filtering can be a much better solution. As mentioned in Section 1, noise often manifests as isolated points (*e.g.*, from an intermittent glitch in a sensor), and variable-resolution topology is an effective way to identify these points. Figure 7 demonstrates the basic ideas. The spanning tree clearly brings out the displacement of the noisy

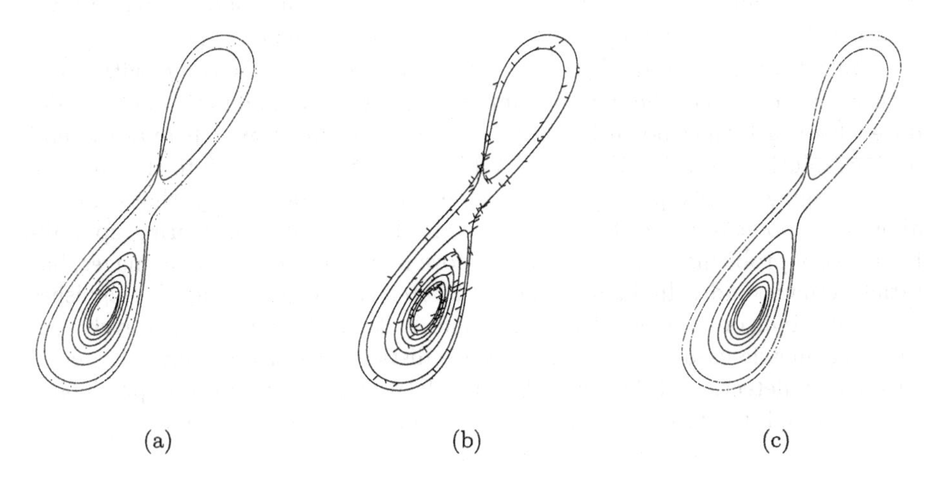

Fig. 7. The effects of noise upon data topology. The dataset in part (a) is a canonical example in dynamical systems—the Lorenz attractor—each point of which was perturbed, with probability 0.01, by a small amount of additive noise. The MST in part (b) clearly shows these noisy points. Part (c) shows the same data set after processing with a topology-based filter that prunes the 'hairs' off the MST.

points from the rest of the orbit, both in direction and in edge length. Specifically, if the noise magnitude is large compared to the inter-point spacing, the edges joining the noisy points to the rest of the tree are longer than the original edges. (Of course, if the magnitude of the noise is *small* compared to that spacing, the associated MST edges will not be unusually long, and the noisy points are not so obvious.) This kind of *separation of scale* often arises when two processes are at work in the data—*e.g.*, signal and noise. Because variable-resolution topology is good at identifying scale separation, we can use it to identify and remove the noisy points. The first step is to look for a second peak in the edge-length distribution of the MST (equivalently, a shoulder in the plot $I(\epsilon)$ of the number of isolated points as a function of ϵ). The maximum edge length of the MST of the *non*-noisy data occurs in the interval between the two peaks (or at the location of the $I(\epsilon)$ breakpoint). At that value, most of the noisy points—and few of the regular points—are ϵ-isolated. The next step is to prune all MST edges that exceed that length, removing the points at their termini[5]. Part (c) of the Figure shows the results; in this case, the topology-based filter removed 534 of the 545 noisy points and 150 of the 7856 non-noisy points. This translates to 98.0% success with a 1.9% false positive rate. These are promising results—better than any linear filter, and comparable to or better than the noise-reduction techniques used by the dynamical systems community[3]. Increasing the pruning length, as one would predict, decreases the false-positive rate; somewhat less intuitively, though, larger pruning lengths do *not* appear to significantly affect the success rate—until they become comparable to the length scales of the noise. As described in [23], these rates vary slightly for different types and amounts of noise, but remain close to 100% and 0%; even better, the topology-based filter does not disturb the dynamical invariants like the Lyapunov exponent.

While topology-based filtering is very effective, it does come with some caveats. For example, the method simply *removes* noisy points; it does not deduce where each one 'should be' and move it in that direction. There are several obvious solutions to this, all of which add a bit more geometry into the recipe—*e.g.*, averaging the two points on either side of the base of the edge that connects an isolated point to the rest of the trajectory. Rasterized images bring up a different set of problems, as noise does not move points around in these images, but rather reshades individual pixels. The computer vision community distinguishes these two kinds of noise as "distortion" and "salt and pepper," respectively. Because the metric used in the MST captures distances between points, it is more effective at detecting the former than the latter. Space limitations preclude a more-thorough treatment of this issue here; for more details, see[6].

4 Conclusion

The mathematical framework of variable-resolution topology has tremendous potential for intelligent data analysis. It is, among other things, an efficient way

[5] The exact details of the pruning algorithm are somewhat more subtle than is implied above because not all noisy points are terminal nodes of the spanning tree. Thus, an algorithm that simply deletes *all* points whose connections to the rest of the tree are longer than the pruning length can sever connections to other points, or clusters of points. This is an issue if one noisy point creates a 'bridge' to another noisy point and only one of the associated MST edges is longer than the pruning length.

to automate the process of finding and characterizing coherent structures. Figure 6(b), for instance, was constructed by hand-processing roughly 2000 images, which required roughly 50 working days[17]; Figure 6(a) required only a few minutes of CPU time. Because it reveals separation of scale, variable-resolution topology can also be useful in noise removal. In a series of numerical and laboratory experiments, a topology-based filter identified close to 100% of the noisy points from dynamical system data sets, with a false-positive rate of only a few percent. Separation of scale is fundamental to many other processes whose results one might be interested in untangling, so this method is by no means limited to filtering applications—or to dynamical systems.

There have been a few other topology-based approaches to filtering. Mischaikow *et al.*[15], for example, use algebraic topology to construct a coarse-grained representation of the data. This effectively finesses the noise issue, and thus constitutes a form of filtering. Rather than use algebraic topology to construct a useful coarse-grained representation of the dynamics, our approach uses *geometric* topology to remove noisy points while working in the original space, which allows us to obtain much finer-grained results.

A tremendous volume of papers on techniques for reasoning about the structure of objects has appeared in various subfields of the computer science literature. Very few of these papers focus on distilling out the topology of the coherent structures in the data. Those that do either work at the pixel level (*e.g.*, the work of Rosenfeld and collaborators [24]) or are hand-crafted for a particular application (*e.g.*, using wavelets to find vortices in images of sea surface temperature[14]), and all are constrained to two or three dimensions. Our framework is algorithmically much more general, and it works in arbitrary dimension. Many geographic information systems[25] and computer vision[4] tools incorporate simple topological operations like adjacency or connectedness; these, too, generally only handle 2D gridded data, and none take varying resolution into account—let alone exploit it.

References

1. `http://www.cs.colorado.edu/~lizb/topology.html`.
2. `http://www.alphashapes.org`.
3. H. Abarbanel. *Analysis of Observed Chaotic Data.* Springer, 1995.
4. D. Ballard and C. Brown. *Computer Vision.* Prentice-Hall, 1982.
5. M. Bern and D. Eppstein. Emerging challenges in computational topology, 1999.
6. E. Bradley, V. Robins, and N. Rooney. Topology and pattern recognition. Preprint; see `www.cs.colorado.edu/~lizb/publications.html`.
7. T. Cormen, C. Leiserson, R. Rivest, and C. Stein. *Introduction to Algorithms.* The MIT Press, 2001. pp 570–573.
8. C. Delfinado and H. Edelsbrunner. An incremental algorithm for Betti numbers of simplicial complexes on the 3-sphere. *Computer Aided Geometric Design*, 12:771–784, 1995.
9. T. Dey, H. Edelsbrunner, and S. Guha. Computational topology. In B. Chazelle, J. Goodman, and R. Pollack, editors, *Advances in Discrete and Computational Geometry.* American Math. Society, Princeton, NJ, 1999.

10. R. Duda and P. Hart. *Pattern Classification.* Wiley, New York, 1973.
11. H. Edelsbrunner, D. Letscher, and A. Zomorodian. Topological persistence and simplification. In *IEEE Symposium on Foundations of Computer Science*, pages 454–463, 2000.
12. H. Edelsbrunner and E. Mücke. Three-dimensional alpha shapes. *ACM Transactions on Graphics*, 13:43–72, 1994.
13. U. Fayyad, D. Haussler, and P. Stolorz. KDD for science data analysis: Issues and examples. In *Proceedings of the ACM SIGKDD International Conference on Knowledge Discovery and Data Mining (KDD)*, 1996.
14. H. Li. Identification of coherent structure in turbulent shear flow with wavelet correlation analysis. *Transactions of the ASME*, 120:778–785, 1998.
15. K. Mischaikow, M. Mrozek, J. Reiss, and A. Szymczak. Construction of symbolic dynamics from experimental time series. *Physical Review Letters*, 82:1144–1147, 1999.
16. J. Munkres. *Elements of Algebraic Topology.* Benjamin Cummings, 1984.
17. D. Perovich. Personal communication.
18. D. Perovich, W. Tucker, and K. Ligett. Aerial observations of the evolution of ice surface conditions during summer. *Journal of Geophysical Research: Oceans*, 10.1029/2000JC000449, 2002.
19. V. Robins. Towards computing homology from finite approximations. *Topology Proceedings*, 24, 1999.
20. V. Robins. *Computational Topology at Multiple Resolutions.* PhD thesis, University of Colorado, June 2000.
21. V. Robins, J. Meiss, and E. Bradley. Computing connectedness: An exercise in computational topology. *Nonlinearity*, 11:913–922, 1998.
22. V. Robins, J. Meiss, and E. Bradley. Computing connectedness: Disconnectedness and discreteness. *Physica D*, 139:276–300, 2000.
23. V. Robins, N. Rooney, and E. Bradley. Topology-based signal separation. *Chaos.* In review; see `www.cs.colorado.edu/~lizb/publications.html`.
24. P. Saha and A. Rosenfeld. The digital topology of sets of convex voxels. *Graphical Models*, 62:343–352, 2000.
25. S. Shekhar, M. Coyle, B. Goyal, D. Liu, and S. Sarkar. Data models in geographic information systems. *Communications of the ACM*, 40:103–111, 1997.
26. J. Theiler and S. Eubank. Don't bleach chaotic data. *Chaos*, 3:771–782, 1993.
27. K. Yip. *KAM: A System for Intelligently Guiding Numerical Experimentation by Computer.* Artificial Intelligence Series. MIT Press, 1991.
28. C. Zahn. Graph-theoretical methods for detecting and describing Gestalt clusters. *IEEE Transactions on Computers*, C-20:68–86, 1971.

Coherent Conditional Probability as a Measure of Information of the Relevant Conditioning Events

Giulianella Coletti[1] and Romano Scozzafava[2]

[1] Dipartimento di Matematica, Università di Perugia,
Via Vanvitelli, 1 – 06100 Perugia (Italy)
coletti@dipmat.unipg.it

[2] Dipartimento Metodi e Modelli Matematici, Università "La Sapienza",
Via Scarpa, 16 – 00161 Roma (Italy)
romscozz@dmmm.uniroma1.it

Abstract. We go on with our study of a coherent conditional probability looked on as a general non-additive "uncertainty" measure $\varphi(\cdot) = P(E|\cdot)$ of the conditioning events. In particular, we proved in [6] (see also the book [5]) that φ can be interpreted as a *possibility* measure. In a previous paper [7] we gave a relevant characterization, showing that φ is a *capacity* if and only if it is a *possibility*. In this paper we give a characterization of φ as an (antimonotone) *information measure* in the sense of Kampé de Feriet and Forte.

1 Introduction

By resorting to the most effective concept of conditional probability (for a complete account of the relevant theory, see [5]), we have been able not only to define fuzzy subsets, but also to introduce in a very natural way the basic continuous T-norms and the relevant dual T-conorms, bound to the former by *coherence*. Moreover, we have given, as an interesting and fundamental by-product of our approach, a natural interpretation of *possibility functions*, both from a semantic and a syntactic point of view. In [7] we deepened the study of the properties of a coherent conditional probability looked on as a general *non-additive uncertainty measure* of the conditioning events, and we proved that this measure is a *capacity* if and only if it is a *possibility*.

Some of the addressed results (e.g., concerning possibility measures and the justification of t-norms in possibility theory) have been studied by various authors in the last decade - with quite different formal settings, but with quite similar intentions. As an example, see the work [9] on the "context model" for a uniform view of uncertainty in knowledge based-systems.

On the other hand, we recall that Kampé de Feriet and Forte ([11], [10]) have put forward a notion of *information* looked on as a suitable *antimonotone* function. In fact, in the Shannon-Wiener theory of information, the information furnished by the realization of an event A of probability $P(A)$ is defined as

M.R. Berthold et al. (Eds.): IDA 2003, LNCS 2810, pp. 123–133, 2003.

$J(A) = -c \log P(A)$, while the mean information, obtained by noting which of the events $\{A_1, \cdots, A_n\}$ constituting a partition π of the sample space occurs, is defined as $H(\pi) = -\sum_{j=1}^{n} P(A_j) \log P(A_j)$. Arguing that the notion of information can be regarded as more primitive than that of probability, Kampé de Feriet and Forte considered the notion of information furnished by the realization of an event to be "naturally" simpler than that of a mean information such as $H(\pi)$, and gave axioms for a measure of the information from this point of view.

Starting from this idea, in this paper we take the concept of information measure (of events) as a function defined on an algebra of events, satisfying a system of axioms and *without any reference to the probability of these events*. In this context the classical information theory introduced by Wiener and Shannon can be regarded just as a particular way to *compute* the information.

The concepts of conditional event and conditional probability (as dealt with in this paper: we recall the main points in the next Section) play a central role for the probabilistic reasoning. In particular, here we study the properties of a coherent conditional probability looked on as a suitable antimonotone *information measure*.

2 Preliminaries

Our approach is based on the general notion of *coherent conditional probability*: the starting point is a synthesis of the available information, expressing it by one or more *events*: to this purpose, the concept of event must be given its more general meaning, *i.e.* it must not looked on just as a possible outcome (a subset of the so–called "sample space"), but expressed by a *proposition*. Moreover, events play a two–fold role, since we must consider not only those events which are the direct object of study, but also those which represent the relevant "state of information": in fact *conditional* events and *conditional* probability are the tools that allow to manage specific (conditional) statements and to update degrees of belief on the basis of the evidence.

On the other hand, what is usually emphasized in the literature – when a conditional probability $P(E|H)$ is taken into account – is only the fact that $P(\cdot|H)$ *is a probability for any given* H: this is a very restrictive (and misleading) view of conditional probability, corresponding trivially to just a modification of the so-called "sample space" Ω. It is instead essential – for a correct handling of the subtle and delicate problems concerning the use of conditional probability – to regard the conditioning event H as a "variable", *i.e.* the "status" of H in $E|H$ is not just that of something representing a given *fact*, but that of an (uncertain) *event* (like E) for which the knowledge of its truth value is not required.

2.1 Conditional Events and Coherent Conditional Probability

In a series of paper (see, e.g., [3] and [4], and the recent book [5]) we showed that, if we do not assign the same "third value" $t(E|H) = u$ (undetermined) to *all* conditional events, but make it suitably depend on $E|H$, it turns out that this function $t(E|H)$ can be taken as a general conditional *uncertainty measure* (for example, conditional *probability* and conditional *possibility* correspond to particular choices of the relevant operations between conditional events, looked on as particular random variables, as shown, respectively, in [3] and [1]).

By taking as *partial* operations the ordinary sum and product in a given family $\mathcal{C}$ of conditional events, with the only requirement that the resulting value of $t(E|H)$ be a *function* of (E, H), we proved in [3] that this function $t(\cdot|\cdot)$ satisfies "familiar" rules. In particular, if the set $\mathcal{C} = \mathcal{G} \times \mathcal{B}$ of conditional events $E|H$ is such that $\mathcal{G}$ is a Boolean algebra and $\mathcal{B} \subseteq \mathcal{G}$ is closed with respect to (finite) logical sums, then, putting $t(\cdot|\cdot) = P(\cdot|\cdot)$, these rules *coincide with the usual axioms* for a *conditional probability* on $\mathcal{G} \times \mathcal{B}^o$, where $\mathcal{B}^o = \mathcal{B} \setminus \{\emptyset\}$, due to de Finetti [8].

And what about an assessment P on an *arbitrary* set $\mathcal{C}$ of conditional events?

Definition 1 - We say that the assessment $P(\cdot|\cdot)$ on $\mathcal{C}$ is *coherent* if there exists $\mathcal{C}' \supset \mathcal{C}$, with $\mathcal{C}' = \mathcal{G} \times \mathcal{B}^o$ ($\mathcal{G}$ a Boolean algebra, $\mathcal{B}$ an additive set), such that $P(\cdot|\cdot)$ can be extended from $\mathcal{C}$ to $\mathcal{C}'$ as a *conditional probability*.

Some of the peculiarities (which entail a large flexibility in the management of any kind of uncertainty) of this concept of *coherent* conditional probability - versus the usual one - are the following: due to the *direct* assignment of $P(E|H)$ as a whole, the knowledge (or the assessment) of the "joint" and "marginal" unconditional probabilities $P(E \wedge H)$ and $P(H)$ is not required; the *conditioning* event H (which *must* be a *possible* one) may have *zero probability*; a suitable interpretation of the extreme values 0 and 1 of $P(E|H)$ for situations which are different, respectively, from the trivial ones $E \wedge H = \emptyset$ and $H \subseteq E$, leads to a "natural" treatment of the default reasoning; it is possible to represent "vague" statements as those of fuzzy theory (as done in [6], see also [5]).

The following characterization of coherence has been discussed in many previous papers: we adopt the formulation given in [5] (see also [2]).

Theorem 1 - Let $\mathcal{C}$ be an *arbitrary* family of conditional events, and consider, for every $n \in \mathbb{N}$, any finite subfamily $\mathcal{F} = \{E_1|H_1, \ldots, E_n|H_n\} \subseteq \mathcal{C}$; denote by $\mathcal{A}_o$ the set of atoms A_r generated by the (unconditional) events $E_1, H_1, \ldots, E_n, H_n$. For a real function P on $\mathcal{C}$ the following three statements are equivalent:

(a) P is a *coherent* conditional probability on $\mathcal{C}$;

(b) for every finite subset $\mathcal{F} \subseteq \mathcal{C}$, there exists a class of coherent (unconditional) probabilities $\{P_\alpha\}$, each probability P_α being defined on a suitable subset $\mathcal{A}_\alpha \subseteq \mathcal{A}_o$, and for any $E_i|H_i \in \mathcal{C}$ there is a unique P_α satisfying

$$(1) \qquad \sum_{A_r \subseteq H_i}^{r} P_\alpha(A_r) > 0 \quad , \quad P(E_i|H_i) = \frac{\sum_{A_r \subseteq E_i \wedge H_i}^{r} P_\alpha(A_r)}{\sum_{A_r \subseteq H_i}^{r} P_\alpha(A_r)} \; ;$$

moreover $\mathcal{A}_{\alpha'} \subset \mathcal{A}_{\alpha''}$ for $\alpha' > \alpha''$ and $P_{\alpha''}(A_r) = 0$ if $A_r \in \mathcal{A}_{\alpha'}$;

(c) for every finite subset $\mathcal{F} \subseteq \mathcal{C}$, all systems of the following sequence, with unknowns $x_r^\alpha = P_\alpha(A_r) \geq 0$, $A_r \in \mathcal{A}_\alpha$, $\alpha = 0, 1, 2, \ldots, k \leq n$, are compatible:

$$(S_\alpha) \quad \begin{cases} \sum_{A_r \subseteq E_i \wedge H_i}^{r} x_r^\alpha = P(E_i|H_i) \sum_{A_r \subseteq H_i}^{r} x_r^\alpha , \\ \left[\text{for all } E_i|H_1 \in \mathcal{F} \text{ such that } \sum_{A_r \subseteq H_i}^{r} x_r^{\alpha-1} = 0 \, , \; \alpha \geq 1, \right] \\ \sum_{A_r \subseteq H_o^\alpha}^{r} x_r^\alpha = 1 \end{cases}$$

where $H_o^o = H_o = H_1 \vee \ldots \vee H_n$ and $\sum_{A_r \subseteq H_i}^{r} x_r^{-1} = 0$ for **all** H_i's, while H_o^α denotes, for $\alpha \geq 1$, the union of the H_i's such that $\sum_{A_r \subseteq H_i}^{r} x_r^{\alpha-1} = 0$.

Another fundamental result is the following, essentially due (for unconditional events, and referring to an *equivalent* form of coherence in terms of betting scheme) to de Finetti [8] .

Theorem 2 - Let $\mathcal{C}$ be a family of conditional events and P a corresponding assessment; then there exists a (possibly not unique) coherent extension of P to an *arbitrary* family $\mathcal{K} \supseteq \mathcal{C}$, *if and only if* P is coherent on $\mathcal{C}$.

The following theorem (see [5]) shows that a coherent assignment of $P(\cdot|\cdot)$ to a family of conditional events whose conditioning ones are a *partition* of Ω is essentially unbound.

Theorem 3 - Let $\mathcal{C}$ be a family of conditional events $\{E_i|H_i\}_{i \in I}$, where $card(I)$ is arbitrary and the events H_i's are a *partition* of Ω. Then *any* function $f : \mathcal{C} \to [0,1]$ such that

$$(2) \quad f(E_i|H_i) = 0 \quad \text{if} \quad E_i \wedge H_i = \emptyset \, , \qquad \text{and} \quad f(E_i|H_i) = 1 \quad \text{if} \quad H_i \subseteq E_i$$

is a coherent conditional probability.

Corollary 1 - Under the same conditions of Theorem 3, let $P(\cdot|\cdot)$ be a coherent conditional probability such that $P(E|H_i) \in \{0, 1\}$. Then the following two statements are equivalent

(i) $P(\cdot|\cdot)$ is the *only* coherent assessment on $\mathcal{C}$;

(ii) $H_i \wedge E = \emptyset$ for every $H_i \in \mathcal{H}_o$ and $H_i \subseteq E$ for every $H_i \in \mathcal{H}_1$, where $\mathcal{H}_r = \{H_i : P(E|H_i) = r\}$, $r = 0, 1$.

The latter two theorems constitute the main basis for the aforementioned interpretation of a fuzzy subset (through the membership function) and of a possibility distribution in terms of coherent conditional probability.

2.2 Fuzzy Sets and Possibility

We recall from [6] the following two definitions (Definitions 2 and 3 below).

If X is a (not necessarily numerical) random quantity with range C_X, let A_x, for any $x \in C_X$, be the event $\{X = x\}$. The family $\{A_x\}_{x \in C_x}$ is obviously a *partition* of the certain event Ω. Let φ be any *property* related to the random quantity X : notice that a *property*, even if expressed by a proposition, does not single–out an *event*, since the latter needs to be expressed by a *nonambiguous* statement that can be either *true* or *false*. Consider now the **event**

$$E_\varphi = \text{"You claim } \varphi \text{"}$$

and a coherent conditional probability $P(E_\varphi|A_x)$, looked on as a real function

$$\mu_{E_\varphi}(x) = P(E_\varphi|A_x)$$

defined on C_X. Since the events A_x are incompatible, then (by Theorem 3) every $\mu_{E_\varphi}(x)$ with values in $[0,1]$ is a coherent conditional probability.

Definition 2 - Given a random quantity X with range C_X and a related property φ, a *fuzzy subset* E^*_φ of C_X is the pair

$$E^*_\varphi = \{E_\varphi \,, \, \mu_{E_\varphi}\},$$

with $\mu_{E_\varphi}(x) = P(E_\varphi|A_x)$ for every $x \in C_X$.

So a coherent conditional probability $P(E_\varphi|A_x)$ is a measure of how much You, given the event $A_x = \{X = x\}$, are willing to *claim* the property φ, and it plays the role of the membership function of the fuzzy subset E^*_φ.

Moreover, we have been able not only to define fuzzy subsets, but also to introduce in a very natural way the basic continuous T-norms and the relevant dual T-conorms, bound to the former by *coherence*.

Definition 3 - Let E be an arbitrary event and P any coherent conditional probability on the family $\mathcal{G} = \{E\} \times \{A_x\}_{x \in C_X}$, admitting $P(E|\Omega) = 1$ as (coherent) extension. A *distribution of possibility* on C_X is the real function π defined by $\pi(x) = P(E|A_x)$.

2.3 Conditional Probability as a Non Additive Uncertainty Measure

Actually, the previous definition can be used as well to introduce any general distribution φ, to be called just *uncertainty measure*.

Definition 4 - Under the same conditions of Definition 3, a *distribution of uncertainty measure* on C_X is the real function φ defined by $\varphi(x) = P(E|A_x)$.

Remark - When C_X is finite, since every extension of $P(E|\cdot)$ must satisfy axioms *(i), (ii)* and *(iii)* of a conditional probability, condition $P(E|\Omega) = 1$ gives

$$P(E|\Omega) = \sum_{x \in C_x} P(A_x|\Omega)P(E|A_x) \qquad \text{and} \qquad \sum_{x \in C_x} P(A_x|\Omega) = 1\,.$$

Then

$$1 = P(E|\Omega) \leq \max_{x \in C_x} P(E|A_x)\,;$$

therefore $P(E|A_x) = 1$ for at least one event A_x.

On the other hand, we notice that in our framework (where *null probabilities for possible conditioning events are allowed*) it does not necessarily follow that

$P(E|A_x) = 1$ for every x; in fact we may will have $P(E|A_y) = 0$ (or else equal to any other number between 0 and 1) for some $y \in C_X$. Obviously, the constraint $P(E|A_x) = 1$ for some x is not necessary when the cardinality of C_X is infinite.

We list now (without proof) some of the properties (established in [7]) of coherent extensions of the function φ, seen as coherent conditional probability $P(E|\cdot)$, to the algebra spanned by the events A_x. First of all we note that if $\mathcal{A}$ is an algebra, E an arbitrary event and $P(E|\cdot)$ a coherent conditional probability on $\{E\} \times \mathcal{A}^o$, then function $f(\cdot)$ defined on $\mathcal{A}$ by putting $f(\emptyset) = 0$ and $f(A) = P(E|A)$ *is not necessarily additive*. On the contrary (as shown in the following Corollary 2) f is necessarily sub-additive (0-alternating).

Lemma 1 - Let $\mathcal{C}$ be a family of conditional events $\{E|H_i\}_{i \in I}$, where $card(I)$ is arbitrary and the events H_i's are a *partition* of Ω, and let $P(E|\cdot)$ an arbitrary (coherent) conditional probability on $\mathcal{C}$. Then any coherent extension of P to $\mathcal{C}' = \{E|H : H \in \mathcal{H}^o\}$, where $\mathcal{H}$ is the algebra spanned by H_i, is such that, for every $H, K \in \mathcal{H}$, with $H \wedge K = \emptyset$,

$$\min\{P(E|H), P(E|K)\} \leq P(E|H \vee K) \leq \max\{P(E|H), P(E|K)\}. \tag{3}$$

Corollary 2 - Under the same conditions of Lemma 1, any coherent extension of P to $\mathcal{C}' = \{E|H : H \in \mathcal{H}^o\}$, where $\mathcal{H}$ is the algebra spanned by H_i, is such that, for every $H, K \in \mathcal{H}$, with $H \wedge K = \emptyset$,

$$P(E|H \vee K) \leq P(E|H) + P(E|K).$$

The function $f(H) = P(E|H)$, with P a coherent conditional probability, in general *is not a capacity*.

The next theorem focuses the condition assuring the monotonicity of $P(E|\cdot)$.

Theorem 4 - Under the same conditions of Lemma 1, a coherent extension of P to $\mathcal{C}' = \{E|H : H \in \mathcal{H}^o\}$, where $\mathcal{H}$ is the algebra spanned by H_i, is monotone with respect to $\subseteq$ if and only if for every $H, K \in \mathcal{H}$

$$P(E|H \vee K) = \max\{P(E|H), P(E|K)\}.$$

The question now is: are there coherent conditional probabilities $P(E|\cdot)$ monotone with respect to $\subseteq$? We reached a positive answer by means of the following theorem (given in [6]), which represents the main tool to introduce *possibility functions* in our context referring to coherent conditional probabilities.

Theorem 5 - Let E be an arbitrary event and $\mathcal{C}$ be the family of conditional events $\{E|H_i\}_{i \in I}$, where $card(I)$ is arbitrary and the events H_i's are a *partition* of Ω. Denote by $\mathcal{H}$ the algebra spanned by the H_i's and let $f : \mathcal{C} \rightarrow [0,1]$ be *any* function such that (2) holds (with $E_i = E$ for every $i \in I$). Then any P extending f on $\mathcal{K} = \{E\} \times \mathcal{H}^o$ and such that

$$P(E|H \vee K) = \max\{P(E|H), P(E|K)\}, \quad \text{for every } H, K \in \mathcal{H}^o \tag{4}$$

is a coherent conditional probability.

Definition 5 - Let $\mathcal{H}$ be an algebra of subsets of C_X (the range of a random quantity X) and E an arbitrary event. If P is any coherent conditional probability on $\mathcal{K} = \{E\} \times \mathcal{H}^o$, with $P(E|\Omega) = 1$ and such that

$$P(E|H \vee K) = \max\{P(E|H), P(E|K)\}\,, \quad \text{for every}\, H, K \in \mathcal{H}^o\,,$$

then a *possibility measure* on $\mathcal{H}$ is a real function Π defined by $\Pi(H) = P(E|H)$ for $H \in \mathcal{H}^o$ and $\Pi(\emptyset) = 0$.

Remark - Theorem 5 assures (in our context) that any possibility *measure* can be obtained as coherent extension (unique, in the finite case) of a possibility *distribution*. Vice versa, given any possibility measure Π on an algebra $\mathcal{H}$, there exists an event E and a coherent conditional probability P on $\mathcal{K} = \{E\} \times \mathcal{H}^o$ agreeing with Π, *i.e.* whose extension to $\{E\} \times \mathcal{H}$ (putting $P(E|\emptyset) = 0$) coincides with Π.

The following Theorem 6 is an immediate consequence of Theorems 4 and 5.

Theorem 6 - Under the same conditions of Theorem 5, any coherent P extending f on $\mathcal{K} = \{E\} \times \mathcal{H}^o$ **is a capacity if and only if it is a possibility.**

3 Conditional Probability as Information Measure

First of all, we recall the definition of information measure as given by Kampé de Feriet and Forte.

Definition 6 - A function I from an arbitrary set of events $\mathcal{E}$, and with values on $R^* = [0, +\infty]$, is an *information measure* if it is *antimonotone*, that is the following condition holds: for every $A, B \in \mathcal{E}$

$$A \subseteq B \quad \Longrightarrow \quad I(B) \leq I(A). \tag{5}$$

So, if both $\emptyset$ and Ω belong to $\mathcal{E}$, it follows that

$$0 \leq I(\Omega) = \inf_{A \in \mathcal{E}} I(A) \leq \sup_{A \in \mathcal{E}} I(A) = I(\emptyset).$$

Kampé de Feriet [11] claims that the above inequality is necessary and sufficient to build up an information theory; nevertheless, only to attribute an universal value to $I(\Omega)$ and $I(\emptyset)$ the further conditions $I(\emptyset) = +\infty$ and $I(\Omega) = 0$ are given. The choice of these two values is obviously aimed at reconciling with the Wiener-Shannon theory. In general, axiom (5) implies only that

$$I(A \vee B) \leq \min\{I(A), I(B)\}\,;$$

we can specify the rule of composition by introducing a binary operation $\odot$ to compute $I(A \vee B)$ by means of $I(A)$ and $I(B)$.

Definition 7 - An information measure defined on an additive set of events $\mathcal{A}$ is $\odot$- *decomposable* if there exists a binary operation $\odot$ from $[0, +\infty] \times [0, +\infty]$ such that, for every $A, B \in \mathcal{A}$ with $A \wedge B = \emptyset$, we have

$$I(A \vee B) = I(A) \odot I(B).$$

So "min" is only one of the possible choices of $\odot$.

Obviously, every function defined on an algebra of events with values on $[0, a]$ and with $I(\Omega) = 0$, satisfying $I(A \vee B) = \min\{I(A), I(B)\}$, can be regarded as an increasing transformation g of an information measure. Therefore in this framework it is natural the give the following definition:

Definition 8 - Let F be an arbitrary event and P any coherent conditional probability on the family $\mathcal{G} = \{F\} \times \{A_x\}_{x \in C_X}$, admitting $P(F|\Omega) = 0$ as (coherent) extension. We define *pointwise information measure* on C_X the real function ψ defined by $\psi(x) = P(F|A_x)$.

Remark - When C_X is finite, since every extension of $P(F|\cdot)$ must satisfy the well-known axioms of a conditional probability [8], considering the condition $P(F|\Omega) = 0$, we necessarily have

$$P(F|\Omega) = \sum_{x \in C_x} P(A_x|\Omega) P(F|A_x) \qquad \text{and} \qquad \sum_{x \in C_x} P(A_x|\Omega) = 1\,.$$

Then

$$0 = P(F|\Omega) \geq \min_{x \in C_x} P(F|A_x),$$

so $P(F|A_x) = 0$ for at least one event A_x. On the other hand, we notice that in our framework it does not necessarily follow that $P(F|A_x) = 0$ for every x; in fact we may well have $P(F|A_y) = 1$ (or to any other number between 0 and 1) for some $y \in C_X$. Obviously, the constraint $P(F|A_x) = 0$ for some x is not necessary when the cardinality of C_X is infinite.

We study now the properties of the coherent extensions of the function ψ to the algebra spanned by the events A_x, as coherent conditional probability $P(F|\cdot)$. An immediate consequence of Lemma 1 is the following

Theorem 7- Let $\mathcal{C}$ be a family of conditional events $\{F|H_i\}_{i \in I}$, where $card(I)$ is arbitrary and the events H_i's are a *partition* of Ω, and let $P(F|\cdot)$ be an arbitrary (coherent) conditional probability on $\mathcal{C}$. A coherent extension of P to $\mathcal{C}' = \{F|H\,, H \in \mathcal{H}\}$, where $\mathcal{H}$ is the algebra spanned by H_i, is *antimonotone* with respect to $\subseteq$ if and only if for every $H, K \in \mathcal{H}$

$$P(F|H \vee K) = \min\{P(F|H), P(F|K)\}\,.$$

The following theorem proves the existence of coherent conditional probabilities $P(F|\cdot)$ antimonotone with respect to $\subseteq$. It represents also the main tool to introduce *information measures* in our context referring to coherent conditional probabilities.

Theorem 8 - Let F be an arbitrary event and $\mathcal{C}$ be the family of conditional events $\{F|H_i\}_{i \in I}$, where $card(I)$ is arbitrary and the events H_i's are a *partition* of Ω. Denote by $\mathcal{H}$ the algebra spanned by the H_i's and let $f : \mathcal{C} \to [0,1]$ be *any* function such that (2) holds (with $E_i = F$ for every $i \in I$). Then any P extending f on $\mathcal{K} = \{F\} \times \mathcal{H}^o$ and such that

$$(6) \qquad P(F|H \vee K) = \min\{P(F|H), P(F|K)\}\,, \quad \text{for every } H, K \in \mathcal{H}^o\,.$$

is a coherent conditional probability.

Proof - The proof goes essentially along the same lines of that (given in [5]) of Theorem 5. We start by assuming $card(I) < \infty$. If $H_1, ..., H_n$ are the (exhaustive and mutually exclusive) conditioning events, then any A_r belonging to the set $\mathcal{A}_o$ of atoms generated by the events $\{F, H_i : F|H_i \in \mathcal{C}\}$ is of the kind either $A'_r = H_r \wedge F$ or $A''_r = H_r \wedge F^c$. To prove the theorem in the finite case, it is sufficient to prove that the function P obtained by putting, for every element $K_j \neq \emptyset$ of the algebra $\mathcal{H}$ generated by $\mathcal{A}_o$,

$$P(F|K) = \min_{H_i \subseteq K} f(F|H_i)$$

is a coherent conditional probability.

So, taking into account Theorem 1, we need to prove that there exists a class $\{P_\alpha\}$ (agreeing with P) satisfying the sequence of systems $(\mathcal{S}_\alpha)$: the first system $(\mathcal{S}_o)$, with unknowns $P_o(A_r) \geq 0$, is the following, with $K_j \in \mathcal{H}^o$:

$$\begin{cases} \sum\limits_{A'_r \subseteq F \wedge K_j}^{r} P_o(A'_r) = \min\limits_{H_i \subseteq K_j} P(F|H_i) \left[\sum\limits_{A'_r \subseteq F \wedge K_j}^{r} P_o(A'_r) + \sum\limits_{A''_r \subseteq F^c \wedge K_j}^{r} P_o(A''_r) \right], \\ \sum\limits_{A'_r \subseteq F \wedge H^o_o}^{r} P_o(A'_r) + \sum\limits_{A''_r \subseteq F^c \wedge H^o_o}^{r} P_o(A''_r) = 1 \end{cases}$$

where $H^o_o = H_1 \vee \ldots \vee H_n = \Omega$.

Put $m_o = \min\limits_{H_i \subseteq H^o_o} P(F|H_i)$. Notice that among the above equations we have (choosing respectively either $K_j = H_r$, with $r = 1, 2, \ldots, n$, or $K_j = H^o_o$):

$$P_o(A'_r) = P(F|H_r)[P_o(A'_r) + P_o(A''_r)], \quad r = 1, ..., n$$

$$\sum\limits_{A'_r \subseteq F \wedge H^o_o}^{r} P_o(A'_r) = \min\limits_{H_i \subseteq H^o_o} P(F|H_i) \left[\sum\limits_{A'_r \subseteq F \wedge H^o_o}^{r} P_o(A'_r) + \sum\limits_{A''_r \subseteq F^c \wedge H^o_o}^{r} P_o(A''_r) \right] = m_o.$$

Moreover, choosing $K_j = H_{m_o} = \bigvee\limits_r \{H_r : P(F|H_r) = m_o\}$, the corresponding equation is

$$\sum\limits_{A'_r \subseteq F \wedge H_{M_o}}^{r} P_o(A'_r) = m_o \left[\sum\limits_{A'_r \subseteq F \wedge H_{M_o}}^{r} P_o(A'_r) + \sum\limits_{A''_r \subseteq F^c \wedge H_{m_o}}^{r} P_o(A''_r) \right].$$

Consider, among the first n equations of this subsystem of $n + 2$ equations, those with $P(F|H_r) = m_o$: clearly, the last equation is linearly dependent on them. We give the atoms corresponding to these equations nonnegative values summing up to 1 with the only constrains $P_o(A''_r) = P_o(A'_r)[(1 - m_o)/m_o]$. Then the other equations (those with $P(F|H_r) > m_o$) can be trivially satisfied (as $0 = 0$) giving zero to all the relevant atoms A_r. Finally, it is now easy to check (by suitably grouping the values obtained in the left and right-hand side of the first n equations of the subsystem) that this solution is also a solution of the system $(\mathcal{S}_o)$.

We need now to consider the next system $(\mathcal{S}_1)$, with unknowns $P_1(A_r)$, relative to the atoms A_r contained on those K_j such that $P_o(K_j) = 0$ (for example, those which are disjunction of H_i's with $P(F|H_i) > m_o$):

$$\begin{cases} \sum_{A'_r \subseteq F \wedge K_j} P_1(A'_r) = \min_{H_i \subseteq K_j} P(F|H_i) \left[\sum_{A'_r \subseteq F \wedge K_j} P_1(A'_r) + \sum_{A''_r \subseteq F^c \wedge K_j} P_1(A''_r) \right] \\ \sum_{A'_r \subseteq F \wedge H_o^1} P_1(A'_r) + \sum_{A''_r \subseteq F^c \wedge H_o^1} P_1(A''_r) = 1 \end{cases}$$

where H_o^1 denotes the disjunction of the H_i's such that $P_o(H_i) = 0$.

Putting $m_1 = \min_{H_i \subseteq H_o^1} P(F|H_i)$, we can proceed along the same line as for system $(\mathcal{S}_o)$, so proving the solvability of system $(\mathcal{S}_1)$; then we possibly need to consider the system $(\mathcal{S}_2)$, and so on. In a finite number of steps we obtain a class of $\{P_\alpha\}$ agreeing with P, and so, by Theorem 1, P is coherent.

Consider now the case $card(I) = \infty$. Let $\mathcal{F} = \{F|K_1, ...F|K_n\}$ be any finite subset of $\mathcal{K}$. If $\mathcal{A}_o$ is the (finite) set of atoms F_r spanned by the event $\{F, K_i : F|K_i \in \mathcal{F}\}$, denote by F'_r and F''_r, respectively, the atoms contained in F and in F^c. To prove the coherence of P, it is enough to follow a procedure similar to that used for the finite case, where now the role of the $H_i's$ is played by the events $F'_r \vee F''_r$.

In the case that the assessment $P(F|H_i)$ admits $P(F|\Omega) = 0$ as coherent extension, we obtain a coherent extension as well by requiring both $P(F|\Omega) = 0$ and that the extension of P is made by using "min" as combination rule.

Theorem 9 - Under the same conditions of Theorem 8, any P extending f on $\mathcal{K} = \{F\} \times \mathcal{H}^o$, such that $P(F|\Omega) = 0$ and

$$(7) \qquad P(F|H \vee K) = \min\{P(F|H), P(F|K)\}, \quad \text{for every } H, K \in \mathcal{H}^o .$$

is a coherent conditional probability.

Proof - The proof goes along the lines of that of Theorem 8, taking into account that the first system must be replaced by the following

$$\begin{cases} \sum_{A'_r \subseteq F \wedge K_j} P_o(A'_r) = \min_{H_i \subseteq K_j} P(F|H_i) \left[\sum_{A'_r \subseteq F \wedge K_j} P_o(A'_r) + \sum_{A''_r \subseteq F^c \wedge K_j} P_o(A''_r) \right], \\ \sum_{A''_r \subseteq F^c \wedge H_o^o} P_o(A''_r) = 1 \end{cases}$$

where $H_o^o = H_1 \vee \ldots \vee H_n = \Omega$.

The above results make us able to introduce (from our point of view) a "convincing" definition of information measure.

Definition 9 - Let $\mathcal{H}$ be an algebra of subsets of C_X (the range of a random quantity X) and F an arbitrary event. If P is any coherent conditional probability on $\mathcal{K} = \{F\} \times \mathcal{H}^o$, with $P(F|\Omega) = 0$ and such that

$$(8) \qquad P(F|H) \leq P(F|K) \quad \text{for every } H, K \in \mathcal{H}^o , \quad \text{with} \quad K \subseteq H ,$$

then an *information measure* on $\mathcal{H}$ is a real function I defined, for $H \in \mathcal{H}^o$, by $I(H) = g(P(F|H))$, with g any increasing function from $[0,1]$ to $[0,+\infty]$ and $I(\emptyset) = +\infty$.

Definition 10 - Under the same conditions of Definition 9, if P is any coherent conditional probability on $\mathcal{K} = \{F\} \times \mathcal{H}^o$, with $P(F|\Omega) = 0$ and such that

$$P(F|H \vee K) = \min\{P(F|H), P(F|K)\}, \quad \text{for every } H, K \in \mathcal{H}^o, \tag{9}$$

then a *min-decomposable information measure* on $\mathcal{H}$ is a real function I defined, for $H \in \mathcal{H}^o$, by $I(H) = g(P(F|H))$, with g any increasing function from $[0,1]$ to $[0,+\infty]$ and $I(\emptyset) = +\infty$.

In conclusion, we have proved the following

Theorem 10 - Under the same conditions of Theorem 8, any P extending f on $\mathcal{K} = \{F\} \times \mathcal{H}^o$ is an increasing transformation of an information measure *if and only if* it is an increasing transformation of a *min-decomposable* information measure.

References

1. Bouchon-Meunier, B., Coletti, G., Marsala, C.: Conditional Possibility and Necessity. In: Bouchon-Meunier, B., Gutiérrez-Rios, J., Magdalena, L., Yager, R.R. (eds.): Technologies for Constructing Intelligent Systems, Vol.2, Springer, Berlin (2001) 59–71.
2. Coletti, G., Scozzafava, R.: Characterization of Coherent Conditional Probabilities as a Tool for their Assessment and Extension. International Journal of Uncertainty, Fuzziness and Knowledge-Based System **4** (1996) 103–127.
3. Coletti, G., Scozzafava, R.: Conditioning and Inference in Intelligent Systems. Soft Computing **3** (1999) 118–130.
4. Coletti, G., Scozzafava, R.: From conditional events to conditional measures: a new axiomatic approach. Annals of Mathematics and Artificial Intelligence **32** (2001) 373–392.
5. Coletti, G., Scozzafava, R.: Probabilistic logic in a coherent setting. Trends in Logic, n.15, Kluwer, Dordrecht / Boston / London (2002).
6. Coletti, G., Scozzafava, R.: Conditional probability, fuzzy sets and possibility: a unifying view. Fuzzy Sets and Systems (2003), to appear.
7. Coletti, G., Scozzafava, R.: Coherent conditional probability as a measure of uncertainty of the relevant conditioning events. In: Lecture Notes in Computers Science/LNAI (ECSQARU-2003, Aalborg), 2003, in press.
8. de Finetti, B.: Sull'impostazione assiomatica del calcolo delle probabilità. Annali Univ. Trieste **19** (1949) 3–55. (Engl. transl.: Ch.5 in: Probability, Induction, Statistics, Wiley, London, 1972)
9. Gebhardt, J., Kruse, R.: Parallel combination of information sources. In: Gabbay, D., Smets, P., Handbook of Defeasible Reasoning and Uncertainty Management Systems, Vol. 3, Kluwer Academic, Dordrecht (1998) 329–375.
10. Kampé de Feriet, J., Forte, B.: Information et Probabilité. *Comptes Rendus Acad. Sci. Paris* **265** **A**(1967) 110–114, 142–146, 350–353.
11. Kampé de Feriet, J: Measure de l'information fournie par un événement. In: *Colloques Internationaux C.N.R.S.* **186**(1969) 191–221.

Very Predictive Ngrams for Space-Limited Probabilistic Models

Paul R. Cohen and Charles A. Sutton

Department of Computer Science,
University of Massachusetts, Amherst, MA 01002
{cohen, casutton}@cs.umass.edu

Abstract. In sequential prediction tasks, one repeatedly tries to predict the next element in a sequence. A classical way to solve these problems is to fit an order-n Markov model to the data, but fixed-order models are often bigger than they need to be. In a fixed-order model, all predictors are of length n, even if a shorter predictor would work just as well. We present a greedy algorithm, VPR, for finding variable-length predictive rules. Although VPR is not optimal, we show that on English text, it performs similarly to fixed-order models but uses fewer parameters.

1 Introduction

In *sequential prediction* tasks, one repeatedly tries to predict the next element in a sequence. More precisely, given the first $i \geq 0$ elements in a sequence, one tries to predict the next element x_{i+1}. After making a prediction, one is shown x_{i+1} and challenged to predict x_{i+2}. In general, it is very difficult to predict the first element x_0 in a sequence (imagine trying to predict the first letter in this paper). Often it is easier to predict x_1 having seen x_0, easier still to predict x_2, and so on. For instance, given the sequence $w\,a$ most English speakers will predict the next letter is s; indeed, in English text s is about seven times more likely to follow $w\,a$ than y, the next most likely letter. One might form a *prediction rule* $[wa \rightarrow s]$ that says, "When I am trying to predict the next letter in a corpus of English text and I have just observed $w\,a$, I should predict s." However, if there are N_{wa} occurrences of $w\,a$ in a corpus, then one must expect at least $N_{wa}/8$ incorrect predictions, because roughly one-eighth of the occurrences of $w\,a$ are followed by y, not s. In English, the sequence $l\,w\,a$ is almost always followed by y, so one can add the prediction rule $[lwa \rightarrow y]$ to "mop up" most of the prediction errors committed by the rule $[wa \rightarrow s]$.

This paper is concerned with finding the best k rules, the rules that together reduce prediction errors the most. We present a greedy algorithm, VPR, for finding sets of k very predictive rules. Although VPR finds suboptimal rules in some cases, it performs well on English-language text. If our algorithm always found the most predictive rules we would call it MPR; instead, we call it VPR, for Very Predictive Rules.

One motivation for VPR is from previous work on ngram models and Markov chains in which ngrams are of fixed length. ngrams are sequences $x_0 x_1 \ldots x_n$

M.R. Berthold et al. (Eds.): IDA 2003, LNCS 2810, pp. 134–142, 2003.

that may be used as prediction rules of the form $x_0x_1 \ldots x_{n-1} \to x_n$, whereas Markov chains are conditional probability distributions which have the form $\Pr(X_n = x|X_0 ... X_{n-1})$. Typically, n is fixed, but fixed-order models are often bigger than they need to be. Consider the case of the letter q in English. q is invariably followed by u, so the prediction rule, $[q \to u]$ has a very low error rate. If we insist that all prediction rules be of length three (say), then we would have to write additional rules — $[aq \to u]$, $[bq \to u]$, $[cq \to u]$ ... — none of which reduces the errors made by the original, shorter rule. Similarly, an order-3 Markov chain requires conditional probabilities $\Pr(X_2 = u|X_0, X_1 = q)$ though none of these is different from $\Pr(X_2 = u|X_1 = q)$.

Reducing the number of predictors is good both for minimizing space and increasing accuracy. For a given amount of space, one can increase accuracy by using variable-length predictors, because one short rule, like $[q \to u]$, can do the work of many longer ones. And also, for a desired level of accuracy, having few predictors makes the parameter learning problem easier. A lot of data is needed to train a full 7-gram model; considerably less is needed to train a model that has mostly 3- and 4-grams, but a few highly predictive 7-grams.

The VPR algorithm finds k very predictive rules which are no longer than they should be. The rule $[aq \to u]$ will not be among these k because it makes identical predictions with, so cannot reduce more errors than, the shorter rule $[q \to u]$. In contrast, the rule $[lwa \to y]$ was shown to reduce errors that would be committed by $[wa \to s]$, but it might not reduce enough errors to warrant including it in the k rules found by VPR.

2 Reducing Errors by Extending Rules

It is best to introduce the algorithm with a series of examples. Suppose someone opens an English text to a random location and says to you, "I am looking at the letter n, predict the letter that follows it." If the text is the first 50 000 letters in George Orwell's *1984* (the source of all examples and statistics in this article), then your best guess is t, because $n\,t$ appears 548 times in the text, more than any other subsequence of length two that begins with n. A prediction rule $[X \to Y]$ is called *maximal* if XY is more frequent in a corpus than $XZ, (Z \neq Y)$. The rule $[n \to t]$ is maximal because the sequence $n\,t$ occurs 548 times whereas $n\,d$, the next most frequent subsequence that begins with n, occurs only 517 times. Maximal prediction rules incur fewer errors; for example, $[n \to t]$ incurs $548 - 517 = 31$ fewer errors than $[n \to d]$.

Although it is maximal, $[n \to t]$ incurs *many* errors: because n occurs 3471 times, and $n\,t$ only 548 times, the rule is wrong 2923 times. To cover some of these errors, we can *extend* the maximal rule, by adding an additional letter to the left-hand side, producing a new rule of the form $[\alpha n \to \beta]$. Of course, then both $[n \to t]$ and $[\alpha n \to \beta]$ will match the sequence $a\,n$, so we need a *precedence rule* to decide which to use. Since the longer rule is designed to be an exception to the shorter one, the precedence rule we use is: Always use the rule with the longest left-hand side. It turns out that the best extension to $[n \rightarrow t]$ is $[in \rightarrow g]$. This rule eliminates 341 errors, more than any other extension of $[n \rightarrow t]$.

The reduction comes from two sources, one obvious, the other less so, as shown in Table 1. The first two rows of the table show what we already know, that there are 548 instances of $n\,t$ and 3471 instances of n, so predicting t following n incurs $3471 - 548 = 2923$ errors. The next two lines show the number of errors incurred by the maximal rule $[in \rightarrow g]$. There are 1095 occurrences of $i\,n$ in the corpus and 426 occurrences of $i\,n\,g$, so $1095 - 426 = 669$ errors are incurred by this rule. The surprise comes in the final two lines of Table 1. The rule $[in \rightarrow g]$ applies whenever one sees the sequence $i\,n$, but what if one sees, instead, $o\,n$? Should we continue to use the rule $[n \rightarrow t]$ in this case? No. Once we have a rule $[in \rightarrow g]$, we can distinguish two kinds of sequence: those that include n and start with i and those that include n and start with something other than i, denoted $\neg i$ in Table 1. To treat these cases as identical, to use $[n \rightarrow t]$ for both, is to ignore the information that we're looking at n and the previous letter was *not* i. This information can reduce errors. Table 1 shows 2376 instances of $\neg in$ and in these cases the maximal rule predicts not t but d. Whereas $[n \rightarrow t]$ incurs 2923 errors, the pair of rules $[in \rightarrow g]$ and $[\neg in \rightarrow d]$ incur $669 + 1913 = 2582$ errors. The reduction in errors by substituting these rules for the original one, $[n \rightarrow t]$, is 341 errors. As it happens, this substitution is the best available for the rule $[n \rightarrow t]$, the one that reduces errors the most.

Table 1. How error rates change when maximal rules are extended

LHS x	Prediction α	Frequency $N(x)$	Frequency $N(x\alpha)$	Error
n	t	3471	548	2923
in	g	1095	426	669
¬ i n	d	2376	463	1913

Although $[in \rightarrow g]$ is the best extension of $[n \rightarrow t]$, it is not the only one, and others may reduce errors, too. In Figure 1 we show how a second extension of $[n \rightarrow t]$ is added to a tree of prediction rules. Panel A shows the rule $[n \rightarrow t]$. Panel B shows the substitution of two rules, $[in \rightarrow g]$ and $[\neg(i)n \rightarrow d]$ for the original one. The original 3471 instances of n are broken into 1095 occurrences of in and 2376 instances of $\neg(i)n$. Correspondingly, the best prediction for $\neg(i)n$ switches from t to d and the errors for both rules sum to 2582, a reduction of 341 over the single rule in panel A. A further substitution is shown in panel C. This time $[\neg(i)n \rightarrow d]$ is replaced by two rules, $[\neg(i,a)n \rightarrow t]$ and $[an \rightarrow d]$. The 2376 cases of $\neg(i)n$ are split into 1704 occurrences of $\neg(i,a)n$ and 672 occurrences of an. By pulling the an cases out of the $\neg(i)n$ cases, we gain information that helps to reduce errors. If we observe an and predict d we make 328 errors ($[an \rightarrow d]$). If we observe n preceded by *neither* i nor a and predict t we make 1359 errors ($[\neg(i,a)n \rightarrow t]$). The two new rules together incur 228 fewer errors than $[\neg(i)n \rightarrow d]$.

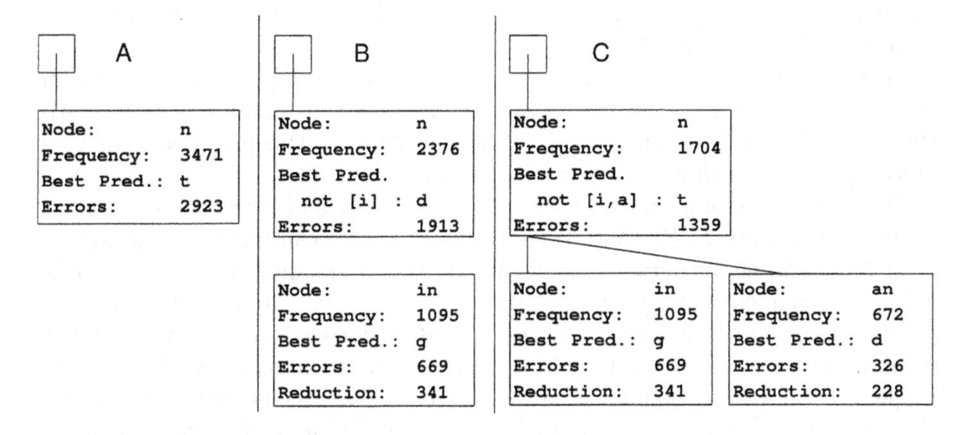

Fig. 1. The evolution of a tree of prediction rules

3 The SPACE-LIMITED-NGRAMS Problem

Now we formally define the problem. Let X be a string of length n with alphabet Σ. We define a *prediction set* P for X as a set of rules $[r \to \alpha]$ for ngrams r and characters α. We require that P contain a rule for the empty string ϵ; this is the rule of last resort. To add an ngram r to P is to add the rule $[r \to \gamma]$, where $r\gamma$ is maximal. An extension of the rule $[r \to \gamma]$ is any rule $[\alpha r \to \beta]$.

We can organize a prediction set P as a tree, as shown in Figure 1, where each ngram has its extensions as descendants. For example, the tree in Figure 1(C) corresponds to the prediction set $\{[\neg(i,a)n \to t], [in \to g], [an \to d]\}$. Using this tree it is easy to find the longest rule that matches a given sequence. One simply "pushes" the sequence into the tree. Suppose the tree is as shown in Figure 1(C) and the sequence is $o\,n$: This sequence "stops" at the rule $[\neg(i,a)n \to t]$ because on matches $\neg(i,a)n$. Alternatively, if the sequence were $i\,n$ it would stop at the rule $[in \to g]$.

The error $E(P)$ of a prediction set is the number of mistakes it makes in predicting all the characters of X. More formally, for a string Y, let $f_P(Y)$ be the prediction made by the longest rule in P that matches Y. Let $X(i) = x_0 x_1 \ldots x_i$. Let $\overline{\delta}(a,b)$ be the function that is 0 if its arguments are equal, and 1 otherwise. Then we can write

$$E(P) = \sum_i \overline{\delta}(f_P(X(i)), x_i). \tag{1}$$

For trees like those in Figure 1, we can compute $E(P)$ by summing the `Errors:` field for all the nodes.

This error metric is an interesting one for judging rules because it combines precision and recall. Rules with longer antecedents are correct more often when they apply, but they apply less frequently. So a rule that is very precise but is uncommon, such as [*declaratio* $\rightarrow$ *n*], will not reduce the error of P as much as a more fallible rule like [*th* $\rightarrow$ *e*].

The SPACE-LIMITED-NGRAMS problem is, given an input string X and an integer k, to return the prediction set P^* of size k that minimizes $E(P^*)$. Although we do not have a polynomial time algorithm for this problem, we have a greedy polynomial-time approximation, which we call VPR.

The VPR algorithm, shown in Figure 2, repeatedly adds the ngram to P that would most reduce $E(P)$. With some bookkeeping we can quickly compute the error after adding an ngram (line 4). We do this by maintaining for each node $q \in P$ its frequency $N(q)$ (that is, the number of times it is needed to make predictions), and its follower table $\boldsymbol{v}(q)$, that is, for each $\alpha \in \Sigma$, the number of times α follows q. When we add an ngram r that extends q, we modify the frequency and follower count so that they do not include the portions of the string that are handled by the exception. More specifically, note that every point in the string X that r predicts was previously predicted by q, for q is a suffix of r. So the new frequency $N'(q)$ is simply $N(q) - N(r)$. And the new follower table $\boldsymbol{v}'(q)$ is $\boldsymbol{v}(q) - \boldsymbol{v}(r)$.

The running time of VPR depends on the loop in line 4 in Figure 2, that is, how large the set Q can be. For each ngram we add to P, we add at most Σ new ngrams to Q, so the size of Q is at most $k\Sigma$. We can cache the frequencies and follower counts of the ngrams using a linear preprocessing step over the string. So the running time of VPR is $O(n + k^2\Sigma)$.

Note that VPR does not necessarily find the optimal prediction set. This can happen if, for example, none of the bigrams appear very predictive, but some trigrams are highly predictive. An example is the string "abccbbacaabccaacbaabccbbbac," where the bigrams are equally distributed, but the trigrams are not.

An alternate greedy algorithm, VPR*, is given in Figure 3. Although slower, VPR* finds better predictive sets. Whereas VPR chooses greedily among one-step extensions of existing rules, VPR* chooses greedily among all the ngrams that occur in the corpus. Clearly, VPR* can perform no worse than VPR, but this comes at a price: in the worst case, VPR* will iterate over all the ngrams that appear in the corpus to select each of the k prediction elements. At most $\binom{n}{2}$ distinct ngrams occur in the corpus—n of length 1, $n-1$ of length 2, and so on. So VPR* has a worst-case running time of $O(kn^2)$.

4 Experiments

Although VPR does not always find the optimal predictor set, we believe that it performs well on naturally occurring sequences. To demonstrate this, we compared VPR to fixed-order classifiers on the first 50 000 characters of George Orwell's *1984*. Each algorithm was trained on the entire corpus, and then the resulting classifier was used to predict each character in the corpus, given the preceding characters. This task is challenging only because k is small; if k were 50 000, for example, it would be trivial to find an optimal predictor set.

```
VPR(seq, k)
1  P ← an ngram-set predictor, initially {ε}
2  Q ← a list of all 1-grams in seq
3  repeat k times
4      r ← arg min_{r∈Q} E(P ∪ {r})
5      Add r to P
6      Remove r from Q
7      Add all one-character extensions of r (that is, αr for all α ∈ Σ) to Q
8  return P
```

Fig. 2. The VPR algorithm

```
VPR*(seq, k)
1  P ← a rule set, initially {ε}
2  Q ← the list of all ngrams in seq
3  repeat k times
4      r ← arg max_{r∈Q} E(P ∪ {r})
5      Add r to P
6      Remove r from Q
7  return P
```

Fig. 3. The VPR* algorithm. This differs from VPR* in that it considers all possible rules, not just one-step extensions of current rules

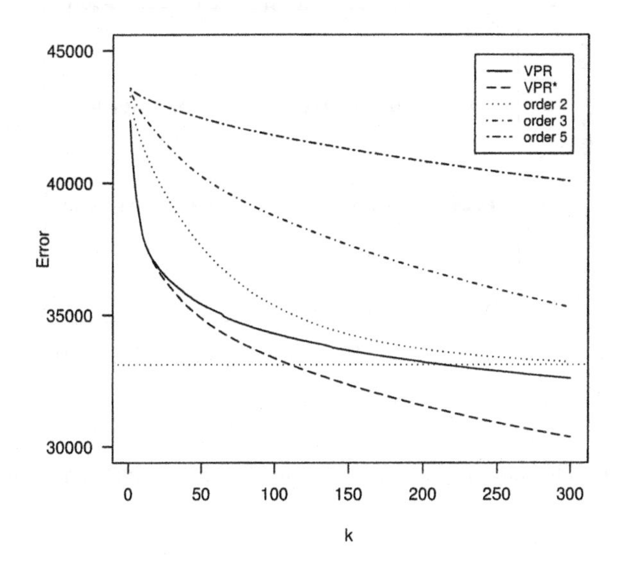

Fig. 4. Comparison of VPR, VPR*, and space-limited fixed-order models on the Orwell data set. The horizontal line is the number of errors made by a full bigram model

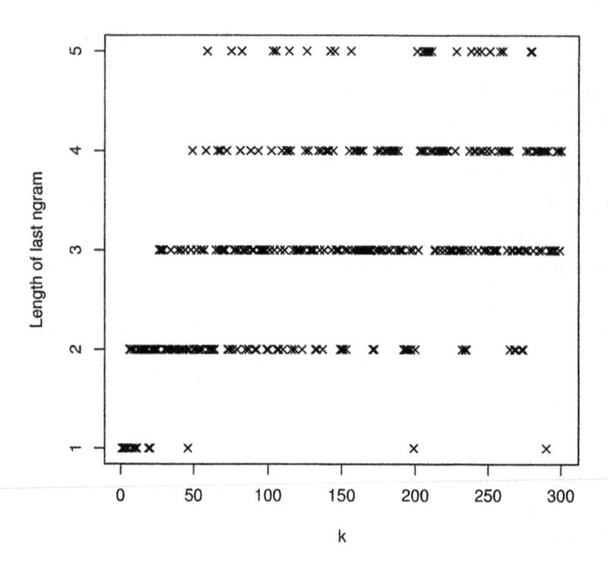

Fig. 5. The length of the k-th ngram chosen by VPR, for each k

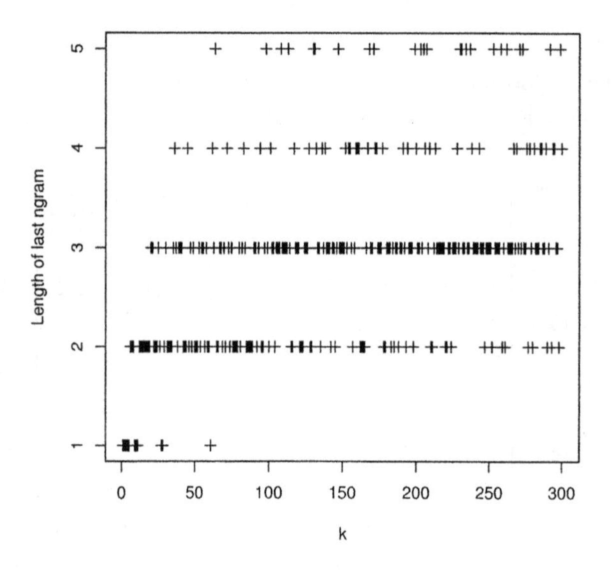

Fig. 6. The length of the k-th ngram chosen by VPR*, for each k

We ran VPR and VPR* with k ranging from 1 to 300. For the fixed-order classifiers, we trained an order-n Markov chain on the text, for n ranging from 1 to 7. But using a full fixed-order model would not be fair to VPR, because a full bigram model, for example, contains 676 predictors, while the largest variable-length model we built has only 300. We made the fixed-order models space-limited by selecting the k fixed-length ngrams that most reduced error. Figure 4 shows the number of errors out of 50 000 made by VPR, VPR*, and the space-limited fixed-length models for different values of k.

With 300 ngrams, VPR makes 32 600 errors on the Orwell corpus out of 50 000 characters. VPR* makes 30 364 errors, about 7% better. By comparison, a full bigram model makes 33 121 errors but uses 676 ngrams. VPR gets that level of performance with 214 ngrams.

To see how useful variable-length ngrams are, we plotted for VPR and VPR*, the length of the k-th ngram added for each k, as shown in Figures 5 and 6. After an initial period of adding 1-grams and 2-grams, the algorithm choose 3-, 4-, and 5-grams alternately. In fact, many 5-grams are chosen before the last few 1-grams.

These results suggest that using variable-length predictors can generate classifiers with the same predictive power but fewer parameters. In addition, the fact that VPR* performs only marginally better than VPR despite having more flexibility suggests that VPR performs well for natural inputs.

5 Related Work

Several authors have used suffix trees to represent variable-memory Markov probabilistic models: Ron et al. [6] call them Probabilistic Suffix Trees (PSTs), and Laird and Saul [4] call them Transition Directed Acyclic Graphs (TDAGs). The rule sets we generate are essentially deterministic versions of these, although we can use the follower tables to build a probabilistic classifier. Ron et al. apply PSTs to finding errors in corrupted text and identifying coding sequences in DNA. PSTs have also been applied to protein family modeling [2]. Laird and Saul describe applications to text compression, dynamic optimization of Prolog programs, and prefetching in mass-storage systems.

Given a corpus generated by a PST T,[1] Ron et al. [6] give a polynomial-time algorithm that with probability $1-\delta$, returns a PST $\hat{T}$ such that the KL-distance between T and $\hat{T}$ is less than ϵ, for any choices of δ and ϵ, assuming that the sample size is sufficiently large. They give an upper bound on the number of states in $\hat{T}$, but in terms of the size of the unknown generator T. A later algorithm for PST-learning has linear time and space complexity [1].

Laird and Saul [4] construct a full suffix tree up to a fixed depth, but omit states whose relative frequency are below a user-specified threshold.

Both of these algorithms use variable-memory models to reduce the space used by the model while retaining predictive power, as does ours. Our approach differs from these in that VPR takes the number of predictors k as a parameter, and guarantees that the model it returns will not exceed this. Furthermore, the question of what is the best accuracy that can be attained with a fixed predictor size has not been explored in the previous work.

[1] There is actually a minor restriction on the source PST.

Our error metric $E(P)$, the number of errors P would make in predicting the input, is also novel. Ron et al. choose to add an ngram if its follower distribution is sufficiently different from its parent's. Laird and Saul include an ngram in a model if it occurs a minimum number of times and its probability of being reached, no matter the confidence in its prediction, is above some threshold. A comparison of these methods is left for future work.

Many methods exist for fixed-memory sequence prediction. Commonly-used techniques include Markov chains, HMMs [5], and recurrent neural networks [3].

6 Conclusion

We have given a greedy algorithm for sequential prediction tasks that uses variable-sized predictors, rather than the fixed-size predictors used in classical Markov models. By doing this, we can build predictors that are equally good as Markov models, but have a smaller number of predictors.

Although both VPR and VPR* require time polynomial in the length n of the corpus, neither of them is optimal. We do not have a polynomial-time optimal algorithm for the SPACE-LIMITED-NGRAMS problem. Future work includes either giving a polynomial time algorithm or showing that the associated decision problem is **NP**-complete. If the problem is **NP**-complete, then it would be good to derive an approximation ratio for VPR.

It is also natural to extend the SPACE-LIMITED-NGRAMS problem so that the input is a sample drawn from a language. Then the problem would be to learn a predictor set from the sample that has low expected error on the language. We suspect, but have not demonstrated, that similar algorithms would be useful in this case.

References

1. Alberto Apostolico and Gill Bejerano. Optimal amnesic probabilistic automata or how to learn and classify proteins in linear time and space. In *RECOMB*, pages 25–32, 2000.
2. G. Bejerano and G. Yona. Variations on probabilistic suffix trees: statistical modeling and the prediction of protein families. *Bioinformatics*, 17(1):23–43, 2001.
3. Richard O. Duda, Peter E. Hart, and David G. Stork. *Pattern Classification.* Wiley, 2nd edition, 2001.
4. Philip Laird and Ronald Saul. Discrete sequence prediction and its applications. *Machine Learning*, 15:43–68, 1994.
5. Lawrence Rabiner. A tutorial on hidden markov models and selected applications in speech recognition. *Proceedings of the IEEE*, 77(2):257–285, 1989.
6. Dana Ron, Yoram Singer, and Naftali Tishby. The power of amnesia: Learning probabilistic automata with variable memory length. *Machine Learning*, 25(2–3):117–149, 1996.

Interval Estimation Naïve Bayes

V. Robles[1], P. Larrañaga[2], J.M. Peña[1], E. Menasalvas[1], and M.S. Pérez[1]

[1] Department of Computer Architecture and Technology, Technical University of Madrid, Madrid, Spain, {vrobles, jmpena, emenasalvas, mperez}@fi.upm.es

[2] Department of Computer Science and Artificial Intelligence, University of the Basque Country, San Sebastián, Spain, ccplamup@si.ehu.es

Abstract. Recent work in supervised learning has shown that a surprisingly simple Bayesian classifier called naïve Bayes is competitive with state of the art classifiers. This simple approach stands from assumptions of conditional independence among features given the class. In this paper a new naïve Bayes classifier called *Interval Estimation naïve Bayes* is proposed. Interval Estimation naïve Bayes is performed in two phases. First, an interval estimation of each probability necessary to specify the naïve Bayes is calculated. On the second phase the best combination of values inside these intervals is calculated using a heuristic search that is guided by the accuracy of the classifiers. The founded values in the search are the new parameters for the naïve Bayes classifier. Our new approach has shown to be quite competitive related to simple naïve Bayes. Experimental tests have been done with 21 data sets from the UCI repository.

1 Introduction

The naïve Bayes classifier [3,9] is a probabilistic method for classification. It can be used to determine the probability that an example belongs to a class given the values of the predictor variables. The naïve Bayes classifier guarantees optimal induction given a set of explicit assumptions [1]. However, it is known that some of these assumptions are not compliant in many induction scenarios, for instance, the condition of variable independence respecting to the class variable. Improvements of accuracy have been demonstrated by a number of approaches, collectively named semi naïve Bayes classifiers, which try to adjust the naïve Bayes to deal with a-priori unattended assumptions.

Previous semi naïve Bayes classifiers can be divided into three groups, depending on different pre/post-processing issues: (i) to manipulate the variables to be employed prior to application of naïve Bayes induction [13,15,19], (ii) to select subsets of the training examples prior to the application of naïve Bayes classification [11,14] and (iii) to correct the probabilities produced by the standard naïve Bayes [5,22].

We propose a new semi naïve Bayes approach named *Interval Estimation naïve Bayes (IENB)*, that tries to relieve the assumption of independence of the

M.R. Berthold et al. (Eds.): IDA 2003, LNCS 2810, pp. 143–154, 2003.

variables given the class by searching for the best combination of naïve Bayes probabilities inside theirs confidence intervals. There is some related work with this approach. In [7] it is described an algorithm named *Iterative Bayes*, that tries to improve the conditional probabilities of naïve Bayes in an iterative way with a hill-climbing strategy. In [20] is presented an algorithm named *Robust Bayes Classifier (RBC)*, a naïve Bayes classifier designed for the case in which learning instances are incomplete. RBC takes into account all the possibilities for the missing values, calculating an interval for each naïve Bayes probability. In [23] a naïve Bayes extension named *naïve Credal Classifier (NCC)* is presented. Credal sets represent probability distributions as points that belong to a close and fenced geometrical regions. In [24] the authors propose to find the parameters that maximize the conditional likehood in a Bayesian network in spite of the sample joint likehood.

Interval Estimation naïve Bayes has been implemented in Visual C++ and the experimental evaluation has been done with 21 problems from the UCI database [18].

The outline of this paper is as follows: Section 2 presents the naïve Bayes classifier. Section 3 is a brief introduction to statistical inference. Section 4 presents the new algorithm Interval Estimation naïve Bayes. Section 5 illustrates the results with the UCI experiments. Section 6 gives the conclusions and suggests further future work.

2 Naïve Bayes

The naïve Bayes classifier [3,9] is a probabilistic method for classification. It performs an approximate calculation of the probability that an example belongs to a class given the values of predictor variables. The simple naïve Bayes classifier is one of the most successful algorithms on many classification domains. In spite of its simplicity, it is shown to be competitive with other more complex approaches in several specific domains.

This classifier learns from training data the conditional probability of each variable X_k given the class label c. Classification is then done by applying Bayes rule to compute the probability of C given the particular instance of $X_1, \ldots, X_n$,

$$P(C = c | X_1 = x_1, \ldots, X_n = x_n)$$

Naïve Bayes is founded on the assumption that variables are conditionally independent given the class. Therefore the posterior probability of the class variable is formulated as follows,

$$P(C = c | X_1 = x_1, \ldots, X_n = x_n) \propto P(C = c) \prod_{k=1}^{n} P(X_k = x_k | C = c) \quad (1)$$

This equation is highly appropriate for learning from data, since the probabilities $p_i = P(C = c_i)$ and $p^i_{k,r} = P(X_k = x^r_k | C = c_i)$ may be estimated from training data. The result of the classification is the class with highest posterior probability.

In naïve Bayes these parameters are estimated using a point estimation (see section 3.1). The first step in the new algorithm we propose is based on an interval estimation for the parameters (see section 3.2).

3 Parameter Estimation

Statistical inference studies a collection of data based on a sample of these ones. This sample represents the part of population considered in the analysis. Amongst other things, statistical inference studies the problem known as "estimation problem".

There are two ways of accomplishing this task:

- *Point Estimation:* Point estimation uses a sample with the aim of assigning a single value to a parameter. The maximum likelihood method is used in this context.
- *Interval Estimation:* This technique calculates for each sample an interval that probably contains the parameter. This interval is called confidence interval. The probability of a parameter to be included in an interval is known as confidence level.

3.1 Point Estimation of Parameters in Naïve Bayes

Considering the instances of the database $\mathcal{D} = \{(\mathbf{x}^{(1)}, c^{(1)}), \ldots, (\mathbf{x}^{(N)}, c^{(N)})\}$ as a random sample on size N where the predictor variables $X_1, \ldots, X_n$ follow Bernoulli distributions and using the maximum likelihood method, the next intuitive results are reached:

$$\hat{p}_i = \frac{N_i}{N} \tag{2}$$

where,
N is the size of the database
N_i is the number of instances where $C = c_i$

$$\hat{p}^i_{k,r} = \frac{N_{kri}}{N_i} \tag{3}$$

where,
N_{kri} denotes the number of instances where $X_k = x^r_k$ and $C = c_i$
N_i denotes the number of instances where $C = c_i$

3.2 Interval Estimation of Parameters in IENB

In the case of IENB the calculation of confidence intervals of the parameters is required. This estimation is achieved by the calculation of the sum of the variables in the sample, which generate a binomial distribution that can be approximated by a normal distribution, given the next result, obtaining that the interval that contains the parameter value, p, with a confidence level of $1-\alpha$ is given by:

$$\left(\hat{p} - z_\alpha\sqrt{\frac{\hat{p}(1-\hat{p})}{M}}; \hat{p} + z_\alpha\sqrt{\frac{\hat{p}(1-\hat{p})}{M}}\right) \tag{4}$$

where,
$\hat{p}$ is the point estimation of the probability
z_α is the $(1-\frac{\alpha}{2})$ percentil in the $\mathcal{N}(0,1)$ distribution
M is the corresponding sample size (N_i or N)

4 Interval Estimation Naïve Bayes – IENB

We propose a new semi naïve Bayes approach named *Interval Estimation naïve Bayes (IENB)*. In this approach, instead of calculating the point estimation of the conditional probabilities from data, as simple naïve Bayes makes, confidence intervals are calculated. According to this, by searching for the best combination of values into these intervals, we aim to relieve the assumption of independence among variables the simple naïve Bayes makes –because the effect of the combination of probabilities is evaluated with the overall classifier accuracy–. This search is carry out by a heuristic search algorithm and is guided by the accuracy of the classifiers.

As it is represented in figure 3 at the end of the paper, there are three main important aspects in IENB algorithm:

- **Calculation of confidence intervals**
  Given the dataset, the first step is to calculate the confidence intervals for each conditional probability and for each class probability. For the calculation of the intervals first the point estimations of these parameters (see section 3.1) must be computed.
  In this way, each conditional probability $p_{k,r}^i = P(X_k = x_k^r | C = c_i)$, that has to be estimated from the dataset must be computed with the next confidence interval, as it is introduced in section 3.1. The contribution of this paper is addressed towards this approach.

For $k = 1, \ldots, n; i = 1, \ldots, r_0; r = 1, \ldots, r_k$

$$\left(\hat{p}_{k,r}^i - z_\alpha\sqrt{\frac{\hat{p}_{k,r}^i(1-\hat{p}_{k,r}^i)}{N_i}}; \hat{p}_{k,r}^i + z_\alpha\sqrt{\frac{\hat{p}_{k,r}^i(1-\hat{p}_{k,r}^i)}{N_i}}\right) \tag{5}$$

denotes the interval estimation for the conditional probabilities $p_{k,r}^{i}$, where,
r_k is the possible values of variable X_k
r_0 is the possible values of the class
$\hat{p}_{k,r}^{i}$ is the point estimation of the conditional probability $P(X_k = x_k^r | C = c_i)$
z_α is the $(1 - \frac{\alpha}{2})$ percentil in the $\mathcal{N}(0,1)$ distribution.
Also, in a similar way, the probabilities for the class values $p_i = P(C = c_i)$ are estimated with the next confidence interval,

$$\left(\hat{p}_i - z_\alpha \sqrt{\frac{\hat{p}_i(1-\hat{p}_i)}{N}}; \hat{p}_i + z_\alpha \sqrt{\frac{\hat{p}_i(1-\hat{p}_i)}{N}}\right) \tag{6}$$

where,
$\hat{p}_i^i$ is the point estimation of the probability $P(C = c_i)$
z_α is the $(1 - \frac{\alpha}{2})$ percentil in the $\mathcal{N}(0,1)$ distribution
N is the number of cases in dataset

- **Search space definition**
  Once the confidence intervals are estimated from the dataset, it is possible to generate as many naïve Bayes classifiers as we want. The parameters of these naïve Bayes classifiers must only be taken inside theirs corresponding confidence intervals.
  In this way, each naïve Bayes classifier is going to be represented with the next tupla of dimension $r_0(1 + \sum_{i=1}^{n} r_i)$

$$(p_1^*, \ldots, p_{r_0}^*, p_{1,1}^{*1}, \ldots, p_{1,1}^{*r_0}, \ldots, p_{1,r_1}^{*r_0}, \ldots, p_{n,r_n}^{*r_0}) \tag{7}$$

  where each component in the tupla p^* is the selected value inside its corresponding confidence interval.
  Thus, the search space for the heuristic optimization algorithm is composed of all the valid tuplas. A tupla is valid when it represents a valid naïve Bayes classifier. Formally,

$$\sum_{i=1}^{r_0} p_i^* = 1; \forall k \forall i \sum_{r=1}^{r_k} p_{k,r}^{*i} = 1 \tag{8}$$

  Finally, each generated individual must be evaluated with a fitness function. This fitness function is based on the percentage of successful predictions on each dataset, which means that we are carrying out one wrapper approach.

- **Heuristic search for the best individual**
  Once the individuals and the search space are defined, one heuristic optimization algorithm is ran in order to find the best individual.
  To deal with the heuristic search, the continuous variant of EDAs –estimation of distribution algorithms– algorithms have been selected. EDAs [16,17] are

non-deterministic, stochastic and heuristic search strategies that belong to the evolutionary computation approaches. In EDAs, a number of solutions or individuals is created every generation, evolving once and again until a satisfactory solution is achieved. In brief, the characteristic that most differentiates EDAs from other evolutionary search strategies, such as GAs, is that the evolution from a generation to the next one is done by estimating the probability distribution of the fittest individuals, and afterwards by sampling the induced model. This avoids the use of crossing or mutation operators, and, therefore, the number of parameters that EDAs require is reduced considerably.

EDA

$D_0 \leftarrow$ Generate M individuals (the initial population) randomly

Repeat for $l = 1, 2, \ldots$ until a stopping criterion is met

$D^S_{l-1} \leftarrow$ Select $S \leq M$ individuals from D_{l-1} according to a selection method

$\rho_l(\boldsymbol{x}) = \rho(\boldsymbol{x}|D^S_{l-1}) \leftarrow$ Estimate the probability distribution of an individual being among the selected individuals

$D_l \leftarrow$ Sample M individuals (the new population) from $\rho_l(\boldsymbol{x})$

Fig. 1. Pseudocode for the EDA approach

In EDAs, the variables belonging to an individual are analyzed looking for dependencies. Also, while in other heuristics from evolutionary computation the interrelations among the different variables representing the individuals are kept in mind implicitly (e.g. building block hypothesis), in EDAs the interrelations are expressed explicitly through the joint probability distribution associated with the individuals selected at each iteration. The task of estimating the joint probability distribution associated with the database of the selected individuals from the previous generation constitutes the hardest work to perform, as this requires the adaptation of methods to learn models from data developed in the domain of probabilistic graphical models.
Figure 1 shows the pseudocode of EDA, in which the main steps of this approach are sketched:

1. At the beginning, the first population D_0 of M individuals is generated, usually by assuming a uniform distribution (either discrete or continuous) on each variable, and evaluating each of the individuals.
2. Secondly, a number S ($S \leq M$) of individuals are selected, usually the fittest.
3. Thirdly, the n–dimensional probabilistic model expressed better the interdependencies between the n variables is induced.

4. Next, the new population of M new individuals is obtained by simulating the probability distribution learnt in the previous step.

Steps 2, 3 and 4 are repeated until a stopping condition is verified. The most important step of this new paradigm is to find the interdependencies between the variables (step 3). This task will be done using techniques from the field of probabilistic graphical models.

In the particular case where every variable in the individuals are continuous and follows a gaussian distribution, the probabilistic graphical model is called *Gaussian network.*

Figure 2 shows the result of the execution of the IENB algorithm in the *pima* dataset with a z_α value of 1.96. In the figure can be observed all the confidence intervals calculated for this dataset and the final values selected by the algorithm.

In this execution IENB gets a result of 79.84% of successful classified while naïve Bayes gets 77.73%.

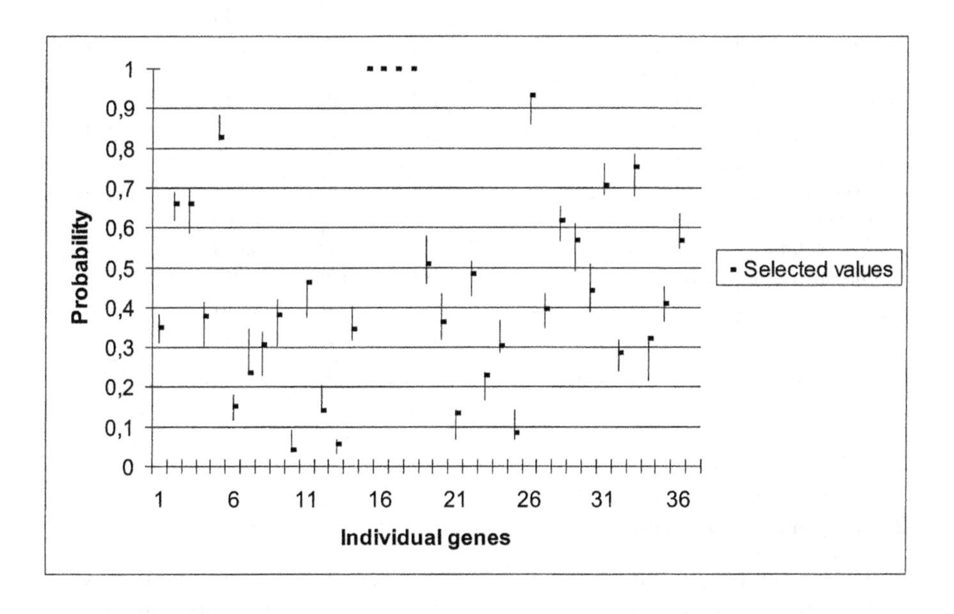

Fig. 2. Confidence intervals and final values obtained by IENB for the *pima* dataset

5 Experimentation

5.1 Datasets

Results are compared for 21 classical datasets –see table 1–, also used by other authors [6]. All the datasets belong to the UCI repository [18], with the exception of *m-of-n-3-7-10* and *corral.* These two artificial datasets, with irrelevant and correlated attributes, were designed to evaluate methods for feature subset selection [10].

Table 1. Description of the data sets used in the experiments

Name	Attributes			Classes	Instances	
	Total	Continuous	Nominal		Learning	Validation
breast	10	10	-	2	699	-
chess	36	-	36	2	3196	-
cleve	13	6	7	2	303	-
corral	6	-	6	2	128	-
crx	15	6	9	2	692	-
flare	10	2	8	2	1066	-
german	20	7	13	2	1000	-
glass	9	9	-	7	214	-
glass2	9	9	-	2	163	-
hepatitis	19	6	13	2	155	-
iris	4	4	-	3	150	-
lymphography	18	3	15	4	148	-
m-of-n-3-7-10	10	-	10	2	300	1024
pima	8	8	-	2	768	-
satimage	36	36	-	6	6435	-
segment	19	19	-	7	2310	-
shuttle-small	9	9	-	7	5800	-
soybean-large	35	-	35	19	683	-
vehicle	18	18	-	4	846	-
vote	16	-	16	2	435	-
waveform-21	21	21	-	3	300	4700

5.2 Experimental Methodology

To estimate the prediction accuracy for each classifier our own implementation of a naïve Bayes classifier has been programmed. This implementation uses the Laplace correction for the point estimation of the conditional probabilities [8,10] and deals with missing values as recommended by [1].

However, our new algorithm does not handle continuous attributes. Thus, a discretization step with the method suggested by [2] has been performed usign MLC++ tools [12]. This discretization method is described by Ting in [21] that is a global variant of the method of Fayyad and Irani [4].

Nineteen of the datasets has no division between training and testing sets. On these datasets the results are obtained by a *leave-one-out* method inside of the heuristic optimization loop.

On the other hand, two out of these twenty one datasets include separated training and testing sets. For these cases, the heuristic optimization algorithm uses only the training set to tune the classifier. A *leave-one-out* validation is perfomed internally inside of the optimization loop, in order to find the best classifier. Once the best candidate is selected, it is validated using the testing set.

5.3 Results

Experiments were ran in a Athlon 1700+ with 256MB of RAM memory. The parameters used to run EDAs were: population size 500 individuals, selected individuals for learning 500, new individuals on each generation 1000, learning type UMDA (*Univariate Marginal Distribution Algorithm*) [17] and elitism. Experiments were ran 10 times with the percentile 0.95 ($z_\alpha = 1.96$).

Table 2. Experiment results for Interval Estimation naïve Bayes

Dataset	*Naïve Bayes*	*IENB*	*Time(min)*	*Improv.*	*Iter. Bayes impr*
breast	97.14	97.71 ± 0.00 †	1	0.57	-0.16
chess	87.92	93.31 ± 0.10 †	221	5.39	
cleve	83.82	86.29 ± 0.17 †	2	2.47	0.1
corral	84.37	93.70 ± 0.00 †	1	9.33	
crx	86.23	89.64 ± 0.10 †	14	3.41	
flare	80.86	82.69 ± 0.13 †	3	1.83	
german	75.40	81.57 ± 0.10 †	81	6.17	-0.05
glass	74.77	83.00 ± 0.20 †	2	8.23	-1.19
glass2	82.21	88.27 ± 0.00 †	0	6.06	
hepatitis	85.16	92.60 ± 0.34 †	1	7.44	1.41
iris	94.67	95.97 ± 0.00 †	0	1.30	1.4 †
lymphography	85.14	94.56 ± 0.00 †	5	9.42	
monf-3-7-10	86.33	95.31 ± 0.00 †	64	8.98	
pima	77.73	79.84 ± 0.09 †	2	2.11	
satimage	82.46	83.88 ± 0.32 †	496	1.42	3.59 †
segment	91.95	96.38 ± 0.08 †	692	4.43	1.5 †
shuttle-small	99.36	99.90 ± 0.00 †	32	0.54	
soybean-large	92.83	95.67 ± 0.10 †	585	2.84	
vehicle	61.47	71.16 ± 0.25 †	64	9.69	4.39 †
vote	90.11	95.07 ± 0.19 †	4	4.96	1.45 †
waveform-21	78.85	79.81 ± 0.10 †	4	0.96	

Results for the UCI problems are shown in the table 2. Columns in the table are: first, the value obtained by the naïve Bayes algorithm, second, the mean value ± standard deviation from IENB, third, the average time –in minutes– for the executions, fourth, the improvement (*IENB-Naïve Bayes*) and finally the improvement respect to naïve Bayes obtained by *Iterative Bayes*, the most similar semi naïve Bayes approach.

Results are really interesting. Respect to naïve Bayes, using the 0.95 percentile, we obtained an average improvement of 4.30%. Besides, although this

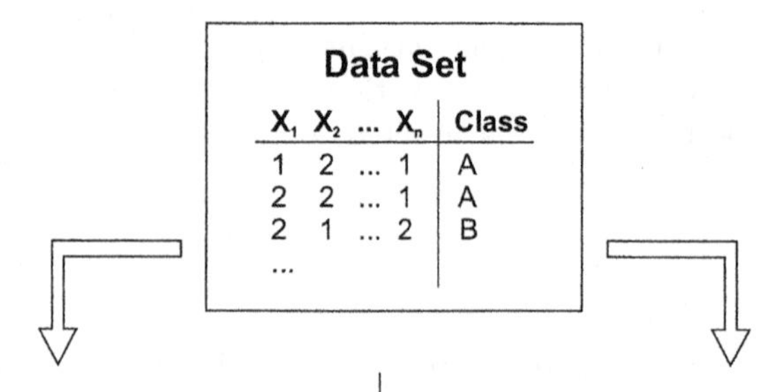

Naïve Bayes

Learning

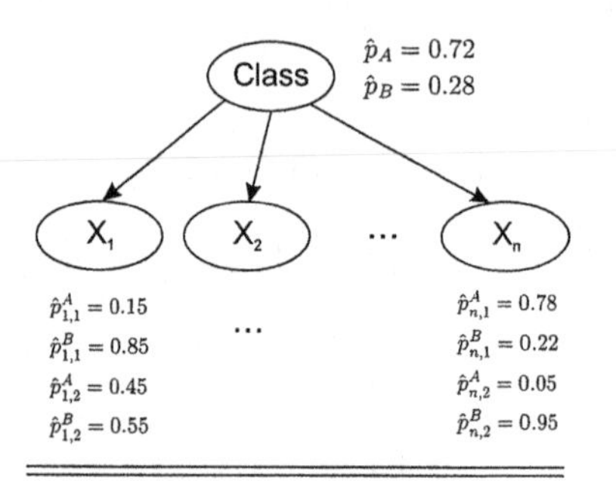

Abbrevations

$\hat{p}^* \equiv Value\ inside\ interval$

$p^{*i}_{k,r} \in (\hat{p}^i_{k,r} - z_\alpha \sqrt{\frac{\hat{p}^i_{k,r}(1-\hat{p}^i_{k,r})}{N_i}}; \hat{p}^i_{k,r} + z_\alpha \sqrt{\frac{\hat{p}^i_{k,r}(1-\hat{p}^i_{k,r})}{N_i}})$

$p^*_i \in (\hat{p}_i - z_\alpha \sqrt{\frac{\hat{p}_i(1-\hat{p}_i)}{N}}; \hat{p}_i + z_\alpha \sqrt{\frac{\hat{p}_i(1-\hat{p}_i)}{N}})$

$\hat{p}^i_{k,r} = \hat{P}(X_k = x^r_k | C = c_i)$

$\hat{p}_i = \hat{P}(C = c_i)$

$N \equiv number\ of\ instances$

$N_i \equiv number\ of\ instances\ where\ C = c_i$

Interval Estimation Naïve Bayes

Learning

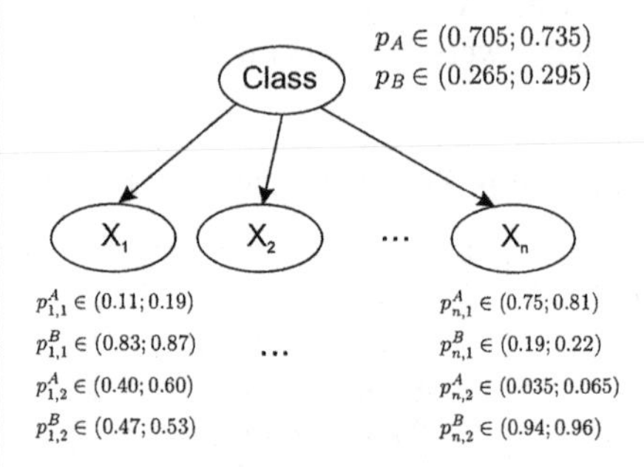

Generation of the valid combinations of probabilities (individuals for the heuristic search)

Individual = $(p^*_A, p^*_B, p^{*A}_{1,1}, p^{*B}_{1,1}, \ldots, p^{*B}_{n,2})$

With $\sum_i p^*_i = 1; \forall k \forall i \sum_r p^{*i}_{k,r} = 1$

Search Space

$(0.712, 0.288, 0.13, 0.87, \ldots, 0.95)$

$(0.734, 0.266, 0.17, 0.83, \ldots, 0.942)$

...

Heuristic Optimization

Best individual

Fig. 3. Illustration of the differences between naïve Bayes and Interval Estimation naïve Bayes

is not always true, better improvements are obtained in the datasets with less number of cases, as the complexity of the problem is lower.

The non-parametric tests of Kruskal-Wallis and Mann-Whitney were used to test the null hypothesis of the same distribution densities for all of them. This task was done with the statistical package S.P.S.S. release 11.50. The results for the tests applied to all the algorithms are shown below.

- Between naïve Bayes algorithm and IENB:
  - Naïve Bayes vs. IENB. Fitness value: $p < 0.001$

These results show that the differences between naïve Bayes and Interval Estimation naïve Bayes are significant in all the datasets, meaning that the behavior of selecting naïve Bayes or IENB is very different.

6 Conclusion and Further Work

In this work a new semi naïve Bayes approach has been presented. In our experiments this approach has an average improvement of 4.30% respect to the simple naïve Bayes.

As this is the first time we use this approach, many issues remain for future research. For instance, Interval Estimation naïve Bayes can also be used with continuous variables. It is possible to estimate intervals for the parameters μ and σ of the Gaussian distribution.

IENB can also be executed with different percentiles, in order to enlarge the confidence intervals. This also means an increment in the search space of the heuristic algorithm and therefore the execution time also increases.

The accuracy increment for the cases with correlated and redundant attributes (*corral* and *m-of-n-3-7-10*) is much better than the other cases. This can be interpreted as an improvement in terms of the assumptions problems of the original naïve Bayes.

Also, it is possible to change the objective of the heuristic search. We can try to maximize the area under the ROC curve instead of the percentage of successful predictions. Another factible idea is to combine Interval Estimation naïve Bayes with a feature subset selection. On a first phase it is possible to make a subset selection, and on a second phase to apply interval estimation to the previous results.

References

1. P. Domingos and M. Pazzani. Beyond independence: conditions for the optimality of the simple Bayesian classifier. In *Proceedings of the 13th International Conference on Machine Learning*, pages 105–112, 1996.
2. J. Dougherty, R. Kohavi, and M. Sahami. Supervised and unsupervised discretization of continuous features. In *Proceedings of the 12th International Conference on Machine Learning*, pages 194–202, 1995.
3. R. Duda and P. Hart. *Pattern Classification and Scene Analysis.* John Wiley and Sons, 1973.
4. U. Fayyad and K. Irani. Multi-interval discretization of continuous-valued attributes for classification learning. In *Proceedings of the 13th International Conference on Artificial Intelligence*, pages 1022–1027, 1993.

5. J.T.A.S. Ferreira, D.G.T. Denison, and D.J. Hand. Weighted naïve Bayes modelling for data mining. Technical report, Deparment of Mathematics, Imperial College, May 2001.
6. N. Friedman, D. Geiger, and D.M. Goldszmidt. Bayesian network classifiers. *Machine Learning*, 29(2-3):131–163, 1997.
7. J. Gama. Iterative Bayes. *Intelligent Data Analysis*, 4:475–488, 2000.
8. I.J. Good. *The Estimation of Probabilities: An Essay on Modern Bayesian Methods.* MIT Press, 1965.
9. D.J. Hand and K. Yu. Idiot's Bayes - not so stupid after all? *International Statistical Review*, 69(3):385–398, 2001.
10. J. Kohavi, B. Becker, and D. Sommerfield. Improving simple Bayes. Technical report, Data Mining and Visualization Group, Silicon Graphics, 1997.
11. R. Kohavi. Scaling up the accuracy of naïve-Bayes classifiers: a decision-tree hybrid. In *Proceedings of the Second International Conference on Knowledge Discovery and Data Mining*, pages 202–207, 1996.
12. R. Kohavi, G. John, R. Long, D. Manley, and K.Pfleger. MLC++: A machine learning library in C++. *Tools with Artificial Intelligence. IEEE Computer Society Press*, pages 740–743, 1994.
13. I. Kononenko. Semi-naïve Bayesian classifier. In *Sixth European Working Session on Learning*, pages 206–219, 1991.
14. P. Langley. Induction of recursive Bayesian classifiers. In *European Conference on Machine Learning. Berlin: Springer-Verlag*, pages 153–164, 1993.
15. P. Langley and S. Sage. Induction of selective Bayesian classifiers. In Morgan Kaufmann, editor, *Proceedings of the Tenth Conference on Uncertainty in Artificial Intelligence*, pages 399–406, Seattle, WA, 1994.
16. P. Larrañaga and J.A. Lozano. *Estimation of Distribution Algorithms. A New Tool for Evolutionary Computation.* Kluwer Academic Publisher, 2001.
17. H. Mühlenbein. The equation for response to selection and its use for prediction. *Evolutionary Computation*, 5:303–346, 1998.
18. P. M. Murphy and D. W. Aha. UCI repository of machine learning databases. http://www.ics.uci.edu/~mlearn/, 1995.
19. M. Pazzani. Searching for dependencies in Bayesian classifiers. In *Proceedings of the Fifth International Workshop on Artificial Intelligence and Statistics*, pages 239–248, 1996.
20. M. Ramoni and P. Sebastiani. Robust learning with missing data. *Machine Learning*, 45(2):147–170, 2001.
21. K.M. Ting. Discretization of continuous-valued attributes and instance-based learning. Technical Report 491, University of Sydney, 1994.
22. G.I. Webb and M.J. Pazzani. Adjusted probability naïve Bayesian induction. In *Australian Joint Conference on Artificial Intelligence*, pages 285–295, 1998.
23. M. Zaffalon, K. Wesnes, and O. Petrini. Credal classification for dementia screening. *Artificial Intelligence in Medicine, LNAI 2101*, pages 67–76, 2001.
24. W. Zhou and R. Greiner. Learning accurate belief nets. Technical report, University of Alberta, 1999.

Mining Networks and Central Entities in Digital Libraries. A Graph Theoretic Approach Applied to Co-author Networks

Peter Mutschke

Social Science Information Centre, Lennéstr. 30, D–53113 Bonn, Germany,
mutschke@bonn.iz-soz.de

Abstract. Based on graph theoretic concepts, the paper introduces the notion of a social network for mining central entities in bibliographic Digital Libraries. Due to this issue, several concepts of centrality in social networks have been employed that evaluate the strategic position of actors in a social network. Moreover, the paper discusses a structural technique (MPA*) that optimizes network propagation and the evaluation of actor centrality without imposing low depth threshold. The models proposed have been applied to co–author networks for finding human experts in scientific domains. The expressiveness of our network model is demonstrated by testing the precision of document sets that have been ranked by author centrality.

1 Network Structures in Digital Information

It is a known fact that standard search services do not meet the wealth of information material supplied by Digital Libraries (DLs). Traditional retrieval systems are strictly document oriented such that a user is not able to exploit the full complexity of information stored. Bibliographic data, for instance, offer a rich information structure that is usually "hidden" in traditional information systems. This particularly regards link structures among authors (given by co–author- or (co)citation relationships) and, more importantly, the strategic position of entities (e.g. authors) in a given scientific collaboration and communication structure. On the other hand, the continuous growth of accessible information implies that the user is more concerned with potentially irrelevant information. Moreover, relevant information is more and more distributed over several heterogeneous information sources and services (bibliographic reference services, citation indices, full–text services).

DLs are therefore only meaningfully usable if they provide both high–level search services that fully exhaust the information structures stored as well as services that reduce the complexity of information to highly relevant items. Data mining, on the other hand, addresses the "identification of interesting structure" [1] and, based on the structures identified, the extraction of "high–level knowledge (patterns) from low–level data" [2]: A user who searches for literature on a particular field might be just as much interesting in finding human experts

M.R. Berthold et al. (Eds.): IDA 2003, LNCS 2810, pp. 155–166, 2003.

on the field studied as well as some meta information on the scholarly structure of the field, e.g. cooperation and citation networks. This strongly suggests the development of data mining techniques that overcome the strict document orientation of standard indexing and retrieval methods by providing a deeper analysis of link structures and the centrality of entities in a given network structure.

In order to make a contribution to this issue, this paper focuses on graph theoretic concepts for mining social networks and central actors in bibliographic data. The network models proposed are applied to co–author networks in scientific domains. The main goal of our approach is to take fully advantage of the potentials of graph theoretic concepts and network analysis methods to enhance information retrieval in bibliographic DLs. As to this issue, we can fall back to outcomes of the AKCESS project where network analysis methods on the basis of co–author networks have been used for information retrieval purposes for the first time [3,4]. Currently, some of the models proposed in this paper are used in the DAFFODIL[1] system [5] which provides high–level search services on Federated DLs in the field computer science. This regards particularly searching for central authors and ranking documents by author centrality (see [6] for details). It is planned to provide a version of the techniques proposed by the INFOCONNEX[2] systems that supplies a portal to several social science, education, and psychology data sources.

The relevance of central actors in network structures has been confirmed by several empirical studies. Two studies using AKCESS and cognitive mapping methods have shown that central actors are actually most likely working on central subjects of the research field under study [7,8]. A recent study analyzing the carrier of scientists in the social sciences under consideration of their social status in collaboration networks indicates that well–connectedness strongly conditions the placement of actors within a scientific community [9]. The expressiveness of co–author networks has been moreover demonstrated in a number of scientometric studies (see e.g. [10,11]).

Apart from the systems mentioned above, graph theoretic concepts and particularly social networks have been used in some referral systems, such as REFERRALWEB [12], where referral paths from a user to a human expert are evaluated. However, these systems are restricted to the evaluation of relationships between particular individuals and do not take under consideration the whole network structure. A further notion of centrality used in search engines is given by Kleinberg [13] and in Google[3] where the structural prominence of Web pages is determined by analyzing link structures. However, Googles' PageRank as well as Kleinberg's Hubs and Authorities do not exhaust the full complexity of network information, and centrality is reduced to just a local attribute of the entities studied (see also [14]).

In contrast, the basic approach of our network model is to reason about the whole network structure using graph theoretic approaches that have been par-

[1] www.daffodil.de

[2] www.infoconnex.de

[3] www.google.com

ticularly further developed in *social network analysis* that describe relationships among social actors [15]. The special focus of this paper is on co–author networks that appear hidden in bibliographic data. Founded on social network theory, the basic philosophy of mining scholarly collaboration networks is to take advantage of knowledge about the (social) interaction among the actors of a scientific community and the development of super–individual structures (networks, clusters, cliques) within a research field.

In the following we describe the concept of a social network and especially the concept of centrality of actors in a social network (Sec. 2.1). Sect. 2.2 discusses an optimization strategy for network propagation and centrality evaluation. In Sect. 3 the results of a retrieval pre–test on the basis of co–author networks are presented. Sect. 4, finally, outlines current and future research issues.

2 Social Networks and Centrality

2.1 Basic Concepts and Methods

According to graph theory[4], a *social network* in our model is described as a graph $G=(V,E)$, where V is the set of vertices (or *actors*), and E the set of edges. A *walk* in a graph from $s \in V$ to $t \in V$ is a sequence of linked pairs of vertices, starting from s and ending in t, such that s and t are connected. A *path* is a walk where all vertices and edges are distinct. The *length* of a path is given by its number of edges. A *shortest path* (or *geodesic*) is the path with the smallest length among all possible paths connecting two vertices in G. The graph is *connected* if any two vertices are connected by a walk. The maximal connected subgraph is called a *component*. Note that if the graph is disconnected walks, and as a consequence shortest paths calculations, can only be performed for each connected component.

In our application model, vertices are represented by authors, and edges are represented by co–authorships. Co–author networks are propagated by recursively pursuing the transitive closure of co–author relations found in bibliographic DLs. Two scenarios are realized for propagating a co–author network: (1) A co–author network is propagated by pursuing all co–author relations that originate from a given start author. This is performed by a breadth–first search on the basis of the entire document collection until the propagation process stops at a designated depth cut–off point. A depth cut-off of 2, for instance, yields a connected graph where all vertices are one or two links away from the start actor. This scenario might be useful when a user intends to browse the co–author network of a particular author. (2) A co–author network is propagated on the basis of *all* co–authorships that appear in a given document set, for instance the result set of a query. This might be a suitable scenario when a user intends to evaluate the entire collaboration structures of a particular field. The major difference between these scenarios is that scenario (2) might yield a number of disconnected components since it is not sure that all actors in a given document

[4] For a comprehensive work see [16]

set are (transitively) connected via co–authorships, whereas scenario (1) yield always a connected graph. Fig. 1 shows an example of a co-author graph (here: a number of Information Retrieval (IR) authors supplied by a DAFFODIL search).

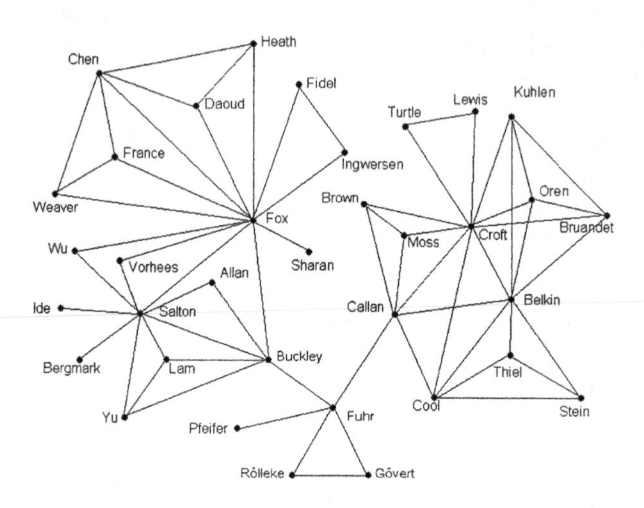

Fig. 1. A co–author network of the field Information Retrieval

On social networks a number of calculations can be performed. Apart from basic statistics that address general network properties, such as the size and density of a particular network, or the number of components in a network (see scenario (2)), more valuable calculations are measures that emphasize properties of the vertices, particularly their strategic position within a social network under study. Here, an important structural attribute of an actor is its *centrality* [17] within a given network. Centrality measures the contribution of a network position to the actor's prominence, influence or importance within a social structure. A basic calculation, *degree* centrality, counts the number of incoming or outgoing edges incident to a node. Degree centrality just measures an actors local "prominence", in terms of the number of choices an actor receives from others, but does not reflect its strategic position within the whole structure. More effective centrality indices are based on shortest paths calculations, such as *closeness* and *betweenness* centrality [17].

Closeness centrality $C_C(v)$ of a vertex v computes the sum of the length of the shortest paths between v and all $t \in V$ in G. The actor with the highest closeness is therefore considered as the actor having the shortest distances to all other vertices in the graph.[5] In contrast, betweenness centrality $C_B(v)$, $v \in V$,

[5] To handle distance in disconnected graphs, where the distance between two disconnected vertices is infinity, closeness is multiplied with a weight of size n_v/n, n_v being the size of the v's component, n being the size of the graph, such that vertices in larger components is given more importance [18].

focuses on the ratio of shortest paths an actor lies on, i.e. the number of shortest paths from all $s \in V$ to all $t \in V$ having v as one of the succeeding or preceding vertices on the shortest path connecting s and t in G. The actor having the highest betweenness is therefore a vertex connecting most of the actors in the network. Whereas closeness stresses the level of an actor's efficiency within a given network structure, betweenness indicates an actor's degree of control or influence of interaction processes that construct a network structure.[6]

The normalized centrality scores for the actors from Fig. 1 indicate evidently the structural prominence particularly of Fox, Salton, Buckley, Callan, Croft, and Belkin within this network (see Table 1). As mentioned above, several studies on

Table 1. Normalized centrality scores of the actors from Fig. 1 having more than four co–authorships, expressed as a number on a [0,1] scale

Author	Degree	Closeness	Betweenness
Fox	0.35	0.37	0.44
Croft	0.29	0.31	0.20
Salton	0.26	0.36	0.19
Belkin	0.21	0.31	0.12
Buckley	0.18	0.42	0.54
Callan	0.18	0.37	0.48
Chen	0.15	0.28	0.004
Fuhr	0.15	0.41	0.59
Cool	0.15	0.30	0.05

the basis of social science research fields have indicated that central authors are relevant for the topic under study [7,8]. Accordingly, an index of centrality in a scientific collaboration and communication structure might indicate the degree of relevance of an actor for the domain in question because of his key position in the network. In our application model information on the centrality of authors is used to find human experts for a particular field: As to scenario (2), *central author search* is performed on the basis of a given (large enough) result set retrieved from a subject search. The initial result set is re–ranked by descending author centrality such that the documents of central authors are displayed first, to be used as starting points for further search activities (e.g. citation search, similarity search). This is particular interesting in a situation where a user is confronted with a (too) large result set, and he/she do not want or do not know how to narrow the search. A retrieval pre–test actually indicates that by centrality based rankings the precision of a subject based search for relevant documents can be improved significantly (see Sect. 3). Note that author centrality based rankings provided by an integrating service might be particularly useful in the

[6] As an efficient mechanism to compute centrality in fairly large networks with thousands of actors an adaptation of the algorithm of Brandes [19] that also computes closeness is used.

case of Federated DLs since mixing term based rankings supplied by different DL services is still an unsolved problem. The basic idea of scenario (1) is to locate central actors in the network of a known author. In our model the paths between all actors in the network propagated are evaluated such that the user can explore the relationship between any two authors in the network. See [6] for further scenarios of using author networks in information systems.

2.2 Optimizing Central Author Search by Extended Main Path Analysis

Propagating a social network by pursuing all link structures between vertices in the graph starting with a given actor (see scenario (1)) is done by performing a breadth–first search that considers all successor vertices of a given vertex first before expanding further. In propagating a network this way we are concerned with the principle problem to define when searching should terminate. A possible solution, that is provided by our model, is to let the user specify a depth cut–off, i.e. network propagation stops when a specified depth of the network is achieved. However, the major conceptual problem of setting a cut–off is that more central vertices might be located outside of the cut–off point. Here, specifying a maximum depth of the network is a completely arbitrary cut–off that might induce a misrepresentation of the real link structure of the domain studied. The latter issue is closely connected with the technical problem that breadth–first search within a large and distributed search space might by very time consuming, so that the user is forced to choose a low threshold. Note that even a cut–off of two may yield hundreds of vertices.

In order to make the propagation and analysis of social networks more efficient, we extended a structural technique used in citation graph analysis called *main path analysis* (MPA) [20,21] by hill-climbing strategies, referred to as *MPA** in the following. In citation analysis this technique is used to evaluate the main stream literature by pursuing citation links that over time connect the most heavily cited publications, starting with a given paper. MPA* can be therefore seen as a best–first search that reduces the number of paths to be pursued to the *best* ones by applying a particular evaluation function. In our model the evaluation function is given by *degree* centrality, i.e. by MPA* using degree centrality as evaluation criterion a path is traced through the valued network from node to node, always choosing as the next tie(s) that (those) edge(s) that connect(s) to (a) node(s) with the highest degree centrality score. But unlike MPA in citation graph analysis the propagation process in our model terminates when a local maximum is achieved, i.e. the path terminates in a node that is locally the vertex that has a higher degree centrality score than all its neighbors. This yields a sequence of edge choices tracing a *main path* through the network that starts with a given vertex and terminates in a central one. This is either a central node in a subgraph or the global optimum, i.e. the most central node in the graph. This way, MPA* yields last–in–first–out lists of vertices found on main paths, having the terminal nodes on their top.

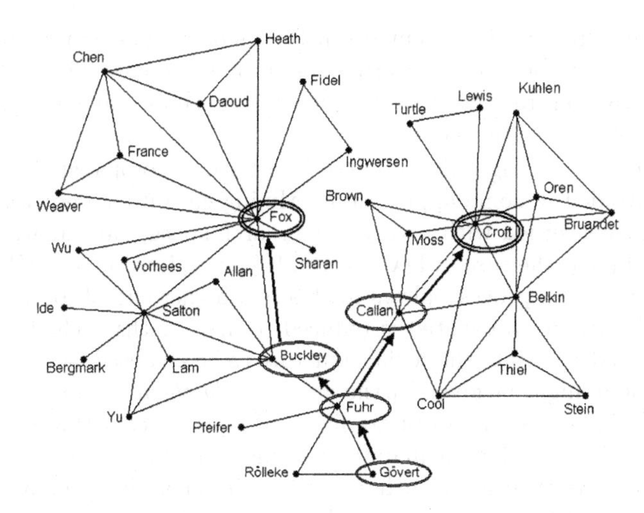

Fig. 2. Main paths in a co–author network of IR authors starting with Gövert

Given the co–author network from Fig. 1, a main path starting with Gövert, for instance (see Fig. 2), will terminate in Fox (via Fuhr and Buckley) and Croft (via Fuhr and Callan), who are the most central actors in the network by degree centrality (see Table 1) — and, indisputably, important authors in IR. Note that by applying breadth–first search with a depth cut–off of two, for instance, Fox and Croft would not have been found. The main paths generated include moreover Fuhr, Buckley and Callan, the most central actors by betweenness centrality (see Table 1), likewise important IR authors.

However, a conceptual problem of MPA* is that it, according to the link structure of a network, possibly cannot find *all* local maxima. This is the case when a successor of a node would lead to a local maximum vertex but has been rejected due to the evaluation function. To deal with this problem, and unlike pure hill–climbing, MPA* uses a node list as in breadth–first search to allow backtracking, such that the method is able to tread alternative ("secondary") main paths through the graph. By using backtracking, MPA* stepwise includes successors having the second-highest (third–highest etc.) degree centrality score and proceeds in propagating "primary" main paths emanating from these successors. Those alternatives walks through the graph are moreover necessary to avoid the "plateau" effect known in hill–climbing where a whole set of neighbouring nodes have the same value such that it is not possible to determine the best direction in which to move. Note that secondary main paths emanating from Buckley and Callan would also include Salton and Belkin (see Fig. 2), two further important IR authors.

The notion of a "*main* path" emphasizes the idea that those paths lead to the "centers of excellence" in a particular domain, e.g. the area where the main research in scientific domains takes place (keeping in mind that all connecting actors on a main path might be particularly relevant as well). MPA* is therefore a good method to drastically cut down the amount of search by reducing the number of paths to be followed to the most promising ones. Thus, MPA* allows a user to locate central actors in the wider network of a given author without imposing a low depth threshold.

None the less, setting a depth cut–off for network propagation cannot completely be avoided in some applications. Mining co–author networks from DLs is potentially concerned with a very large search space, particularly when they are federated (as it the case in DAFFODIL). However, introducing MPA* into the network propagation process allows to choose a higher cut–off because the complexity of the graph propagated is reduced to main paths such that the search process can go deeper into the social structure to be analysed. A clue for a meaningful threshold might be given by the *small–world phenomenon* [22] saying that any two individuals in a network are likely to be connected through a sequence of at most six intermediate acquaintances, known as the "six degrees of separation" principle. As regards co–author networks in scientific communities it is our intuition that a depth cut–off of six will accordingly produce networks that suitably represent the collaboration structure of a scientific community, such that any main path (given large enough networks) will converge on actors who are actually relevant in the field[7].

3 Evaluation. A Retrieval Pre-test Using Author Centrality

The expressiveness of our model has been evaluated by a retrieval pre–test using centrality in co–author networks for document ranking. The retrieval test measures the precision of a given result set that has been ranked by the authors' centrality, i.e. the degree in which a centrality–based ranking provides only relevant documents among the first n documents. The precision test was conducted for ten Boolean queries on the basis of the German social science database SOLIS[8]. All queries have been constructed by two concepts that are combined by Boolean AND, whereby a single concept represents a series of search terms combined by Boolean OR. This yields queries that looks like (a_1 or a_2 ... or a_n) and (b_1 or b_2 ... or b_n), where a and b are search terms. The queries were carried out as controlled term searches using thesaurus terms [9]. Each of the ten result sets was than ranked by publication year in descending order (the standard output of SOLIS), and by author closeness (ACL) in co–author networks that were propagated on the basis of the documents retrieved, according to scenario (2).

To allow also a comparison with standard information retrieval techniques based on term frequencies, such as inverse document frequency, each query was

[7] This is consistent with an analysis of connectivity in a citation network using MPA (see [21]).

[8] http://www.social-science-gesis.de/en/information/SOLIS/index.htm

[9] http://www.social-science-gesis.de/en/information/support/index.htm#Thesaurus

executed by invoking the inverse frequency algorithm of the ORACLE TEXT[10] `contains` operator using its *accumulate* scoring variant (IDF). For this, the initial queries were converted to a text vector and carried out as free text searches on the basis of titles and abstracts. The ORACLE accumulate operator searches for documents that contain at least one occurrence of any of the query terms, and ranks the documents retrieved according to the total term weight of a document.

The precision of each ranked result set was determined by evaluating the number of highly relevant documents among the first 20 documents in the ranking list (the *top set*). Divided by 20, this yields a score on a [0,1] scale. The larger the score the better the precision. As an evaluation criterion for the relevance of a document for the query in question the occurrence of search terms, synonyms respectively, in the title of a document was taken.

Table 2. Precision of ten result sets from SOLIS, ordered by publication year (PY), ranked by inverse document frequency (IDF), and by author closeness (ACL). The *Surplus Value of* ACL indicates the ratio of the number of relevant documents of the ACL top set that do not appear in the IDF top set, related to the number of relevant documents of the latter set, expressed as a percentage

Query	Result set ranked by			Surplus Value
	PY	*IDF*	*ACL*	of *ACL*
Youth – Violence	0.25	0.60	0.55	92
Right-Wing Extremism – East Germ.	0.35	0.45	0.60	122
Woman – Personnel Policy	0.35	0.60	0.65	100
Resistance – Third Reich	0.40	0.65	0.95	138
Forced Labour – II. World War	0.55	0.65	0.70	92
Elites – FRG	0.40	0.70	0.85	107
Poverty – Urban Life	0.30	0.35	0.55	157
Labour Movement – 19./20. Century	0.55	0.55	0.90	164
Change of Attitudes – Youth	0.40	0.50	0.30	50
Terrorism – Democracy	0.20	0.35	0.60	129
Average	**0.38**	**0.54**	**0.67**	**115**

As indicated in Table 2, in all cases, except for one, the result sets ranked by author centrality achieved a much higher precision than the initial result sets ordered by publication year: The ACL sets yield a precision of 0.67 on average[11], whereas the PY sets only achieved a precision of 0.38 on average. Thus, by ranking using author centrality the precision has increased by 76 per cent. Moreover, it turned out that author centrality ranked result sets (at least on the basis of controlled term searches) seem to yield a better precision than standard statistical rankings that are based on term frequencies, although the relevance criterion used for this pre–test, search term occurrence in document

[10] http://otn.oracle.com/products/text/content.html

[11] Result sets ranked by author betweenness yield nearly the same results.

titles, rather prove advantageous to the IDF ranking. However, the difference between the precisions of the ACL and IDF result sets of this pre–test is not significant enough to allow a final conclusion. More research is needed here[12].

It is more important to stress that author centrality based rankings obviously tend to provide a very different set of relevant documents than IDF result sets: The test turned out that a number of relevant documents of the ACL top sets did not appear in the IDF top sets, i.e. ACL rankings favour other documents than IDF rankings do. As indicated in the latter column of Table 2, the average informational surplus value of the ACL ranking lists in comparison with the IDF rankings, given by the number of additional relevant documents yield by the ACL top sets, is more than 100 per cent. Thus, the retrieval quality can be enhanced immensely by combining IDF and ACL result sets. This result strongly suggests the use of author centrality as a particular ranking strategy, in addition to standard techniques such as the vector space retrieval model.

As a result, the retrieval test clearly indicates that central authors in a scientific collaboration structure provide relevant documents for the field under study because of their key position in the network. This significantly confirms the strengths of mining networks and central actors in scientific domains, and the usefulness of document rankings based on author centrality in a given scholarly collaboration network.

4 Conclusion and Future Work

Based on graph theoretic concepts, we introduced the notion of a *social network* for data mining purposes, here the detection of central actors in co–author networks in order to find human experts for a particular field. Moreover, we presented a structural technique (MPA*) that locate central actors in the wider network of a given actor more efficiently than a breadth–first search with a (possibly too low) depth cut-off. The models proposed have been applied to co–author networks in bibliographic DLs. To provide this, two scenarios for propagating co-author networks are described (network propagation on the basis of a given document set, and network propagation per breadth-first search, MPA* respectively, starting with a given author). We intend to apply the method also to citation networks[13], keyword networks, and shared interest networks.

The retrieval pre–test turned out that information on the strategic position of actors in a social network (centrality) might enrich the information seeking process immensely because it strongly emphasizes the cooperation and communication structure of a particular community. Author centrality seems to be not only a powerful concept to improve the precision of a given result set, but also a useful strategy to provide a user with a number of relevant document that

[12] An interesting issue to be investigated is whether or not a retrieval test that also include the relevance of authors in relevance judgements will increase the precision of author centrality based rankings.

[13] Citation networks are particularly valuable for discovering important authors that are publishing mainly as a single author.

are not absolutely favoured by traditional retrieval techniques. This suggests the use of author centrality based rankings as a particular strategy to enhance the overall retrieval quality, in addition to standard search term related techniques such as vector space retrieval.

The models proposed have been implemented as Java services. These services will be enhanced by the computation of author cliques, further measurements of structural attributes of actors (e.g. "bridges" that connect cliques), and heuristics for evaluating author identity. An ongoing main activity stresses the refinement of the networks propagated by a deeper edge analysis. A possible approach, which is currently investigated [23], is Pathfinder network scaling [24] that yield minimum cost networks. A further ongoing activity focuses on the visualization of author networks, in order to provide the user with more insights into how a particular research field is socially structured. A currently investigated approach is to restrict the complexity of a graph to be drawn to main paths provided by MPA*.

A further important research issue concerns the adequacy of the models proposed for particular purposes, e.g. the use of centralities for document ranking. We are aware that central author search and rankings based on author centrality are only useful if the networks propagated provide an appropriate representation of the real communication structures of the particular community under study. Due this issue, a future research work will address the development of appropriate criteria of adequacy of the networks detected, taking account of the size of a network, its internal coherence, and, more importantly, its *community* character.

References

1. Fayyad, U., Uthurusamy, R.: Evolving data mining into solutions for insights. Communications of the ACM **45** (2002) 28–31
2. Fayyad, U., Piatetsky-Shapiro, G., Smyth, P.: From data mining to knowledge discovery in databases. AI Magazine **17** (1996) 37–54
3. Mutschke, P.: Processing scientific networks in bibliographic databases. In Bock, H., Lenski, W., Richter, M., eds.: Information Systems and Data Analysis. Prospects – Foundations – Applications, Heidelberg – Berlin, Springer–Verlag (1994) 127–133
4. Mutschke, P.: Uncertainty and actor-oriented information retrieval in μ–AKCESS. an approach based on fuzzy set theory. In Bock, H.H., Polasek, W., eds.: Data Analysis and Information Systems. Statistical and Conceptual Approaches, Berlin – Heidelberg, Springer–Verlag (1996) 126–138
5. Fuhr, N., Schaefer, A., Klas, C.P., Mutschke, P.: Daffodil: An integrated desktop for supporting high-level search activities in federated digital libraries. In Agosti, M., Thanos, C., eds.: Research and Advanced Technology for Digital Libraries: 6th European Conference, ECDL 2002, Proceedings. Volume 2458 of Lecture Notes in Computer Science., Berlin – Heidelberg – New York, Springer–Verlag (2002) 597–612

6. Mutschke, P.: Enhancing information retrieval in federated bibliographic data sources using author network based stratagems. In Constantopoulos, P., Sölvberg, I., eds.: Research and Advanced Technology for Digital Libraries: 5th European Conference, ECDL 2001, Proceedings. Volume 2163 of Lecture Notes in Computer Science., Berlin – Heidelberg – New York, Springer–Verlag (2001) 287–299
7. Mutschke, P., Renner, I.: Akteure und Themen im Gewaltdiskurs. Eine Strukturanalyse der Forschungslandschaft. In Mochmann, E., Gerhardt, U., eds.: Gewalt in Deutschland. Soziale Befunde und Deutungslinien, München, Oldenbourg Verlag (1995) 147–192
8. Mutschke, P., Quan Haase, A.: Collaboration and cognitive structures in social science research fields. Towards socio-cognitive analysis in information systems. Scientometrics **52** (2001) 487–502
9. Güdler, J.: Kooperation in der Soziologie: Langfristige Entwicklungen, strukturbildende Wirkungen, individuelle Plazierungseffekte. Volume 5 of Forschungsberichte. Informationszentrum Sozialwissenschaften, Bonn (2003)
10. Newman, M.: Who is the best connected scientist? A study of scientific coauthorship networks. Phys. Rev. **E64** (2001)
11. Newman, M.: The structure of scientific collaboration networks. Proc. Natl. Acad. Sci. USA **98** (2001) 404–409
12. Kautz, H., Selman, B., Shah, M.: The hidden web. AI Magazine **18** (1997) 27–36
13. Kleinberg, J.: Authoritative sources in a hyperlinked environment. Journal of the Association for Computing Machinery (1999)
14. Brandes, U., Cornelsen, S.: Visual ranking of link structures. Journal of Graph Algorithms and Applications (2003) (to appear).
15. Wasserman, S., Faust, K.: Social Network Analysis: Methods and Applications. Cambridge University Press, New York (1994)
16. Palmer, E.: Graphical Evolution. Wiley, New York (1985)
17. Freeman, L.: Centrality in social networks: Conceptual clarification. Social Networks **1** (1979) 215–239
18. Tallberg, C.: Centrality and random graphs. Technical Report 7, Stockholm University, Department of Statistics (2000)
19. Brandes, U.: A faster algorithm for betweenness centrality. Journal of Mathematical Sociology **25** (2001) 163–177
20. Carley, K., Hummon, N., Harty, M.: Scientific influence. An analysis of the main path structure in the journal of conflict resolution. Knowledge: Creation, Diffusion, Utilization **14** (1993) 417–447
21. Hummon, N., Doreian, P.: Connectivity in a citation network: The development of DNA theory. Social Networks **11** (1989) 39–63
22. Kleinberg, J.: The small-world phenomenon: An algorithmic perspective. Technical report, Cornell Computer Science Technical Report, 99-1776 (1999)
23. Mutschke, P.: Socio–cognitive analysis and pathfinder network scaling. International Sunbelt Social Network Conference XXII, New Orleans, LA, USA. (2002) (http://www.spcomm.uiuc.edu/Conf/SunbeltXXII/index2.htm).
24. Schvaneveldt, R.: Pathfinder Associative Networks: Studies in Knowledge Organization. Ablex Publishing, Norwood (1990)

Learning Linear Classifiers Sensitive to Example Dependent and Noisy Costs

Peter Geibel, Ulf Brefeld, and Fritz Wysotzki

TU Berlin, Fak. IV, ISTI, AI Group, Sekr. FR5-8
Franklinstr. 28/29, D-10587 Berlin, Germany
{geibel, brefeld, wysotzki}@cs.tu-berlin.de

Abstract. Learning algorithms from the fields of artificial neural networks and machine learning, typically, do not take any costs into account or allow only costs depending on the classes of the examples that are used for learning. As an extension of class dependent costs, we consider costs that are example, i.e. feature and class dependent. We derive a cost-sensitive perceptron learning rule for non-separable classes, that can be extended to multi-modal classes (DIPOL) and present a natural cost-sensitive extension of the support vector machine (SVM).

1 Introduction

The consideration of cost-sensitive learning has received growing attention in the past years ([9,4,6,8]). The aim of the inductive construction of classifiers from training sets is to find a hypothesis that minimizes the mean predictive error. If costs are considered, each example not correctly classified by the learned hypothesis may contribute differently to this error. One way to incorporate such costs is the use of a cost matrix, which specifies the misclassification costs in a class dependent manner (e.g. [9,4]). Using a cost matrix implies that the misclassification costs are the same for each example of the respective class.

The idea we discuss in this paper is to let the cost depend on the single example and not only on the class of the example. This leads to the notion of example dependent costs (e.g. [7]). Besides costs for misclassification, we consider costs for correct classification (gains are expressed as negative costs). Because the individual cost values are obtained together with the training sample, we allow the costs to be corrupted by noise.

One application for example dependent costs is the classification of credit applicants to a bank as either being a "good customer" (the person will pay back the credit) or a "bad customer" (the person will not pay back parts of the credit loan).

The gain or the loss in a single case forms the (mis-) classification cost for that example in a natural way. For a good customer the cost for correct classification corresponds to the gain of the bank expressed as negative costs. This cost will in general depend on the credit loan, the interest rate etc. Generally there are no costs to be expected (or a small loss related to the handling expenses) if the

M.R. Berthold et al. (Eds.): IDA 2003, LNCS 2810, pp. 167–178, 2003.

customer is rejected, for he is incorrectly classified as a bad customer[1]. For a bad customer the misclassification cost is simply the actual loss that has occured. The costs of correct classification is zero (or small positive if one considers handling expenses of the bank).

As opposed to the construction of a cost matrix, we claim that using the example costs directly is more natural and will lead to more accurate classifiers.

In this paper we consider single neuron perceptron learning and the algorithm DIPOL introduced in [10,12,15] that brings together the high classification accuracy of neural networks and the interpretability gained from using simple neural models (threshold units).

Another way of dealing with non-linearly separable, non-separable or multimodal data is the Support Vector Machine (SVM, [14]). We will demonstrate how to extend the SVM with example dependent costs, and compare its performance to the results obtained using DIPOL.

This article is structured as follows. In section 2 the Bayes rule in the case of example dependent costs is discussed. In section 3, the learning rule is derived for a cost-sensitive extension of a perceptron algorithm for non-separable classes. In section 4 the extension of the learning algorithm DIPOL for example dependent costs is described, and in section 5 the extension of the SVM is presented. Experiments on two artificial data sets, and on a credit data set can be found in section 6. The conclusion is presented in section 7.

2 Example Dependent Costs

In the following we consider binary classification problems with classes –1 (negative class) and +1 (positive class). For an example $\mathbf{x} \in \mathbf{R}^d$ of class $y \in \{+1, -1\}$, let

- $c_y(\mathbf{x})$ denote the cost of misclassifying $\mathbf{x}$ belonging to class y
- and $g_y(\mathbf{x})$ the cost of classifying $\mathbf{x}$ correctly.

In our framework, gains are expressed as negative costs. I.e. $g_y(\mathbf{x}) < 0$ if there is a gain for classifying $\mathbf{x}$ correctly into class y. $\mathbf{R}$ denotes the set of real numbers. d is the dimension of the input vector.

Let $r : \mathbf{R}^d \longrightarrow \{+1, -1\}$ be a classifier (decision rule) that assigns $\mathbf{x}$ to a class. Let $X_y = \{\mathbf{x} \mid r(\mathbf{x}) = y\}$ the region where class y is decided. According to [14] the risk of r with respect to the probability density function p of $(\mathbf{x}, y)$ is given with $p(\mathbf{x}, y) = p(\mathbf{x}|y)P(y)$ as

$$R(r) = \sum_{\substack{y_1, y_2 \in \{+1,-1\} \\ y_1 \neq y_2}} \left[\int_{X_{y_1}} g_{y_1}(\mathbf{x}) p(\mathbf{x}|y_1) P(y_1) d\mathbf{x} + \int_{X_{y_1}} c_{y_2}(\mathbf{x}) p(\mathbf{x}|y_2) P(y_2) d\mathbf{x} \right] \quad (1)$$

[1] Where it is assumed that the money can be given to another customer without loss of time.

$P(y)$ is the prior probability of class y, and $p(\mathbf{x}|y)$ is the class conditional density of class y. The first integral expresses the cost for correct classification, whereas the second integral expresses the cost for misclassification. We assume that the integrals defining R exist. This is the case if the cost functions are integrable and bounded.

The risk $R(r)$ is minimized if $\mathbf{x}$ is assigned to class +1, if

$$0 \leq (c_{+1}(\mathbf{x}) - g_{+1}(\mathbf{x}))p(\mathbf{x}|+1)P(+1) - (c_{-1}(\mathbf{x}) - g_{-1}(\mathbf{x}))p(\mathbf{x}|-1)P(-1) \quad (2)$$

holds. This rule is called the Bayes classifier (see e.g. [3]). We assume $c_y(\mathbf{x}) - g_y(\mathbf{x}) > 0$ for every example $\mathbf{x}$, i.e. there is a real benefit for classifying $\mathbf{x}$ correctly.

From (2) it follows that the classification of examples depends on the *differences* of the costs for misclassification and correct classification, not on their actual values. Therefore we will assume $g_y(\mathbf{x}) = 0$ and $c_y(\mathbf{x}) > 0$ without loss of generality. E.g. in the credit risk problem, for good customers the cost of correct classification is set to zero. The misclassification cost of good customers is defined as the gain (that is lost).

2.1 Noisy Costs

If the cost values are obtained together with the training sample, they may be corrupted due to measurement errors. I.e. the cost values are prone to noise. A probabilistic noise model for the costs can be included into the definition of the risk (1) by considering a common distribution of $(\mathbf{x}, y, c)$ where c is the cost. (1) can be reformulated (with $g_y = 0$) to

$$R(r) = \sum_{y_1 \neq y_2} \int_{X_{y_1}} \left[\int_{\mathbf{R}} c\, p(c|\mathbf{x}, y_2) p(\mathbf{x}|y_2) P(y_2) dc \right] d\mathbf{x},$$

where $p(c|\mathbf{x}, y)$ is the probability density function of the cost given $\mathbf{x}$ and y.

It's easy to see that the cost functions c_y can be obtained as the expected value of the costs, i.e.

$$c_y(\mathbf{x}) := E_c[c\, p(c|\mathbf{x}, y)] \quad (3)$$

where we assume that the expected value exists. In the learning algorithms presented in the next sections, it's not necessary to compute (3) or estimate it before learning starts.

3 Perceptrons

Now we assume, that a training sample $(\mathbf{x}^{(1)}, y^{(1)}, c^{(1)}), \ldots, (\mathbf{x}^{(l)}, y^{(l)}, c^{(l)})$ is given with example dependent cost values $c^{(i)}$. We allow the cost values to be noisy, but for the moment, we will require them to be positive. In the following we derive a cost-sensitive perceptron learning rule for linearly non-separable classes, that is based on a non-differentiable error function. A perceptron (e.g. [3]) can be seen as representing a parameterized function defined by a vector $\mathbf{w} = (w_1, \ldots, w_n)^T$ of *weights* and a threshold θ. The vector $\bar{\mathbf{w}} = (w_1, \ldots, w_n, -\theta)^T$ is called the extended weight vector, whereas $\bar{\mathbf{x}} = (x_1, \ldots, x_n, 1)^T$ is called the extended input vector. We denote their scalar product as $\bar{\mathbf{w}} \cdot \bar{\mathbf{x}}$. The output function $y : R^d \longrightarrow \{-1, 1\}$ of the perceptron is defined by $y(\mathbf{x}) = \text{sign}(\bar{\mathbf{w}} \cdot \bar{\mathbf{x}})$. We define $\text{sign}(0) = 1$.

A weight vector having *zero* costs can be found in the *linearly separable case*, where a class separating hyperplane exists, by choosing an initial weight vector, and adding or subtracting examples that are not correctly classified (for details see e.g. [3]).

Because in many practical cases as the credit risk problem the classes are *not* linearly separable, we are interested in the behaviour of the algorithm for linearly non-separable classes. If the classes are linearly non-separable, they can either be non-separable at all (i.e. overlapping), or they are separable but not linearly separable.

3.1 The Criterion Function

In the following we will present the approach of Unger and Wysotzki for the linearly non-separable case ([13]) extended to the usage of individual costs. Other perceptron algorithms for the linearly non-separable case are discussed in [16,3]. Let the step function σ be defined by $\sigma(u) = 1$ for $u \geq 0$, and $\sigma(u) = 0$ if $u < 0$. In the following, σ will be used as a function that indicates a classification error.

Let S_{+1} contain all examples from class $+1$ together with their cost value. S_{-1} is defined accordingly. For the derivation of the learning algorithm, we consider the **criterion function**

$$I_\epsilon(\bar{\mathbf{w}}) = \frac{1}{l} \left[\sum_{(\mathbf{x},c) \in S_{+1}} c(-\bar{\mathbf{w}} \cdot \bar{\mathbf{x}} + \epsilon)\sigma(-\bar{\mathbf{w}} \cdot \bar{\mathbf{x}} + \epsilon) + \sum_{(\mathbf{x},c) \in S_{-1}} c(\bar{\mathbf{w}} \cdot \bar{\mathbf{x}} + \epsilon)\sigma(\bar{\mathbf{w}} \cdot \bar{\mathbf{x}} + \epsilon) \right] \quad (4)$$

The parameter $\epsilon > 0$ denotes a *margin* for classification. Each correctly classified example must have a geometrical distance of at least $\frac{\epsilon}{|\mathbf{w}|}$ to the hyperplane. The margin is introduced in order to exclude the zero weight vector as a minimizer of (4), see [13,3].

The situation of the criterion function is depicted in fig. 1. In addition to the original hyperplane $H : \bar{\mathbf{w}} \cdot \bar{\mathbf{x}} = 0$, there exist two margin hyperplanes $H_{+1} : \bar{\mathbf{w}} \cdot \bar{\mathbf{x}} - \epsilon = 0$ and $H_{-1} : -\bar{\mathbf{w}} \cdot \bar{\mathbf{x}} - \epsilon = 0$. The hyperplane H_{+1} is now responsible for the classification of the class $+1$ examples, whereas H_{-1} is responsible for class -1 ones. Because H_{+1} is shifted into the class $+1$ region, it causes at least as much errors for class $+1$ as H does. For class -1 the corresponding holds.

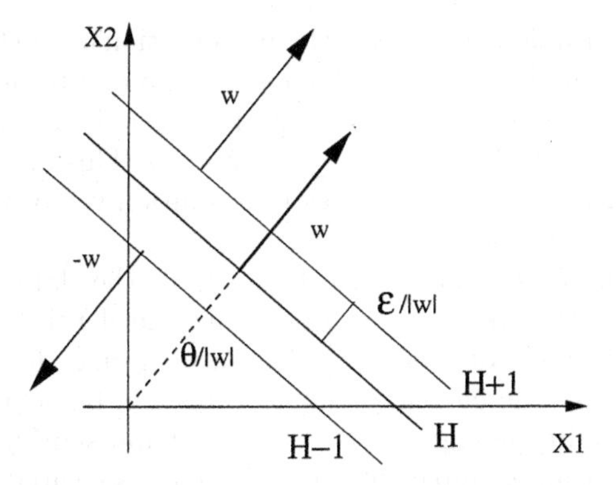

Fig. 1. Geometrical interpretation of the margins, 2-dimensional case

It's relatively easy to see that I_ϵ is a convex function by considering the convex function $h(z) := k\,z\sigma(z)$ (where k is some constant), and the sum and composition of convex functions. From the convexity of I_ϵ it follows that there exists a unique minimum value.

It can be shown that the choice of an $\epsilon > 0$ is not critical, because the hyperplanes minimizing the criterion function are identical for every $\epsilon > 0$, see also [5].

3.2 The Learning Rule

By differentiating the criterion function I_ϵ, we derive the learning rule. The gradient of I_ϵ is given by

$$\nabla_{\bar{w}} I_\epsilon(\bar{\mathbf{w}}) = \frac{1}{l}\left[\sum_{(\mathbf{x},c)\in S_{+1}} -c\,\bar{\mathbf{x}}\sigma(-\bar{\mathbf{w}}\cdot\bar{\mathbf{x}}+\epsilon) + \sum_{(\mathbf{x},c)\in S_{-1}} c\,\bar{\mathbf{x}}\sigma(\bar{\mathbf{w}}\cdot\bar{\mathbf{x}}+\epsilon)\right] \quad (5)$$

To handle the points, in which I_ϵ cannot be differentiated, in [13] the gradient in (5) is considered to be a *subgradient.* For a subgradient a in a point $\bar{\mathbf{w}}$, the condition $I_\epsilon(\bar{\mathbf{w}}') \geq I_\epsilon(\bar{\mathbf{w}}) + a\cdot(\bar{\mathbf{w}}' - \bar{\mathbf{w}})$ for all $\bar{\mathbf{w}}'$ is required. The subgradient is defined for convex functions, and can be used for incremental learning and stochastic approximation (see [13,1,11]).

Considering the gradient for a single example, the following *incremental* rule can be designed. For learning, we start with an arbitrary initialisation $\bar{\mathbf{w}}(0)$. The following weight update rule is used when encountering an example $(\mathbf{x}, y)$ with cost c at time (learning step) t:

$$\bar{\mathbf{w}}(t+1) = \begin{cases} \bar{\mathbf{w}}(t) + \gamma_t c\,\bar{\mathbf{x}} & \text{if } y = +1 \text{ and} \\ & \bar{\mathbf{w}}(t)\cdot\bar{\mathbf{x}} - \epsilon \leq 0 \\ \bar{\mathbf{w}}(t) - \gamma_t c\,\bar{\mathbf{x}} & \text{if } y = -1 \text{ and} \\ & \bar{\mathbf{w}}(t)\cdot\bar{\mathbf{x}} + \epsilon \geq 0 \\ \bar{\mathbf{w}}(t) & \text{else} \end{cases} \quad (6)$$

We assume either a randomized or a cyclic presentation of the training examples.

In order to guarantee convergence to a minimum and to prevent oscillations, for the factors γ_t the following conditions for stochastic approximation are imposed: $\lim_{t\to\infty} \gamma_t = 0$, $\sum_{t=0}^{\infty} \gamma_t = \infty$, $\sum_{t=0}^{\infty} \gamma_t^2 < \infty$. This algorithm for linearly non-separable distributions will converge to a solution vector in the linearly separable case (for a discussion see [5]).

If the cost value c is negative due to noise in the data, the example can just be ignored. This corresponds to modifying the density $p(\mathbf{x}, y, c)$ which is in general *not* desirable. Alternatively, the learning rule (6) must be modified in order to *mis*classify the current example. This can be achieved by using the modified update conditions $\text{sign}(c)\bar{\mathbf{w}}(t) \cdot \bar{\mathbf{x}} - \epsilon \leq 0$ and $\text{sign}(c)\bar{\mathbf{w}}(t) \cdot \bar{\mathbf{x}} + \epsilon \geq 0$ in (6). This means that an example with negative cost is treated as if it belongs to the other class.

4 Multiple and Disjunctive Classes

In order to deal with multi-class/multimodal problems (e.g. XOR), we have extended the learning system DIPOL ([10,12,15]) in order to handle example dependent costs. DIPOL can be seen as an extension of the perceptron approach to multiple classes and multi-modal distributions. If a class possesses a multi-modal distribution (disjunctive classes), the clusters are determined by DIPOL in a preprocessing step using a minimum-variance clustering algorithm (see [12, 3]) for every class.

After the (optional) clustering of the classes, a separating hyperplane is constructed for each pair of classes or clusters if they belong to different classes. When creating a hyperplane for a pair of classes or clusters, respectively, all examples belonging to other classes/clusters are *not* taken into account. Of course, for clusters originating from the same class in the training set, no hyperplane has to be constructed.

After the construction of the hyperplanes, the whole feature space is divided into decision regions each belonging to a single class, or cluster respectively. For classification of a new example $\mathbf{x}$, it is determined in which region of the feature space it lies, i.e. a region belonging to a cluster of a class y. The class y of the respective region defined by a subset of the hyperplanes is the classification result for $\mathbf{x}$.

In contrast to the version of DIPOL described in [10,12,15], that uses a quadratic criterion function and a modified gradient descent algorithm, we used the criterion function I_ϵ and the incremental learning rule in sect. 3.2 for the experiments.

5 Support Vector Machines

DIPOL constructs a classifier by dividing the given input space into regions belonging to different classes. The classes are separated by hyperplanes computed with the algorithm in sect. 3. In the SVM approach, hyperplanes are not computed by gradient descent but by directly solving an optimization problem, see below. More complex classifiers are formed by an implicit transformation of the given input space into a so called feature space by using kernel functions.

Given a training sample $(\mathbf{x}^{(1)}, y^{(1)}, c^{(1)}), \ldots, (\mathbf{x}^{(l)}, y^{(l)}, c^{(l)})$, the optimization problem of a standard soft margin support vector machine (SVM) [14,2] can be stated as

$$\begin{aligned} \min_{\mathbf{w},b,\boldsymbol{\xi}} \quad & \frac{1}{2}|\mathbf{w}|^2 + C\sum_{i=1}^{l} \xi_i^k \\ \text{s.t.} \quad & y^{(i)}\left(\mathbf{w}\cdot\mathbf{x}^{(i)} + b\right) \geq 1 - \xi_i \\ & \xi_i \geq 0, \end{aligned} \tag{7}$$

where the regularization constant $C > 0$ determines the trade-off between the complexity term $\frac{1}{2}|\mathbf{w}|^2$ and the sum. It holds $b = -\theta$. The sum takes all examples into account for which the corresponding pattern $\mathbf{x}^{(i)}$ has a geometrical margin of less than $\frac{1}{|\mathbf{w}|}$, and a functional margin of less than 1. For such an example, the slack variable $\xi_i > 0$ denotes the difference to the required functional margin. Different values of k lead to different versions of the soft margin SVM, see e.g. [2].

For k=1, the sum of the ξ_i can be seen as an upper bound of the empirical risk. Hence we can extend the optimization problem (7) in a natural way by weighting the slack variables ξ_i with the corresponding costs $c^{(i)}$. This leads for $k = 1$ to the cost-sensitive optimization problem

$$\begin{aligned} \min_{\mathbf{w},b,\boldsymbol{\xi}} \quad & \frac{1}{2}|\mathbf{w}|^2 + C\sum_{i=1}^{l} c^{(i)}\,\xi_i \\ \text{s.t.} \quad & y^{(i)}\left(\mathbf{w}\cdot\mathbf{x}^{(i)} + b\right) \geq 1 - \xi_i \\ & \xi_i \geq 0. \end{aligned} \tag{8}$$

Introducing non-negative Lagrangian multipliers α_i, $\mu_i \geq 0$, $i = 1, \ldots, l$, we can rewrite the optimization problem (8), and obtain the following primal Lagrangian

$$\begin{aligned} L_P(\mathbf{w}, b, \boldsymbol{\xi}, \boldsymbol{\alpha}, \boldsymbol{\mu}) = & \frac{1}{2}|\mathbf{w}|^2 + C\sum_{i=1}^{l} c^{(i)}\,\xi_i \\ & -\sum_{i=1}^{l}\alpha_i\left[y^{(i)}\left(\mathbf{w}\cdot\mathbf{x}^{(i)} + b\right) - 1 + \xi_i\right] - \sum_{i=1}^{l}\mu_i\,\xi_i. \end{aligned}$$

Substituting the derivatives with respect to $\mathbf{w}$, b and $\boldsymbol{\xi}$ into the primal, we obtain the dual Langragian that has to be maximized with respect to the α_i,

$$L_D(\boldsymbol{\alpha}) = \sum_{i=1}^{l}\alpha_i - \frac{1}{2}\sum_{i,j=1}^{l}\alpha_i\,\alpha_j\,y^{(i)}y^{(j)}\mathbf{x}^{(i)}\cdot\mathbf{x}^{(j)}. \tag{9}$$

Equation (9) defines the 1-norm soft margin SVM. Note that the example dependent costs do not occur in L_D, but restrict the α_i by the so called box constraints

$$0 \leq \alpha_i \leq c^{(i)} C, \quad \forall i,$$

that depend on the cost value for the respective example and therefore limit its possible influence. The box constraints can be derived from the optimization problem, see e.g. [2].

If the optimal decision function is not a linear function of the data, we map the input data to some other Euclidean Space $\mathcal{H}$ (possibly of infinite dimension), the feature space, by means of a mapping $\boldsymbol{\phi} : \mathbf{R}^d \rightarrow \mathcal{H}$. Substituting the mapped data into the optimization problem leads to the dual Lagrangian

$$L_D(\boldsymbol{\alpha}) = \sum_{i=1}^{l} \alpha_i - \frac{1}{2} \sum_{i,j=1}^{l} \alpha_i \alpha_j y^{(i)} y^{(j)} \boldsymbol{\phi}(\mathbf{x}^{(i)}) \cdot \boldsymbol{\phi}(\mathbf{x}^{(j)}). \tag{10}$$

By means of kernel functions $K : \mathbf{R}^d \times \mathbf{R}^d \rightarrow \mathbf{R}$, with the property $K(\mathbf{x}, \mathbf{x}') = \boldsymbol{\phi}(\mathbf{x}) \cdot \boldsymbol{\phi}(\mathbf{x}')$, we are able to evaluate the inner product in $\mathcal{H}$ without explicitly knowing $\boldsymbol{\phi}$. Furthermore it is always possible to achieve linear separability in the feature space if we use i.e. radial basis function kernels.

For $k = 2$ analogous results can be obtained for the 2-norm soft margin SVM, where the dual Lagrangian depends directly on the individual costs.

In order to show the relationship between perceptron and support vector learning, note that for $k = 1$ the ξ_i in (8) correspond to $-y^{(i)} \bar{\mathbf{w}} \cdot \bar{\mathbf{x}}^{(i)} + \epsilon$ of the criterion function I_ϵ in (4) for patterns misclassified by $H_{y^{(i)}}$. Thus, in the limit $C \rightarrow \infty$, both methods are equivalent for $\epsilon = 1$.

6 Experiments

6.1 The Uni-Modal Case

If the classes are linearly separable, each separating hyperplane also minimizes the cost-sensitive criterion function I_ϵ. We therefore do not present results for the linearly separable case here. In our first experiment, we used the perceptron algorithm for the linearly non-separable case (sect. 3.2), that is part of DIPOL, and the extended SVM with a radial basis function kernel.

We have constructed an artificial data set with two attributes x_1 and x_2. For each class, 1000 randomly chosen examples were generated using a modified Gaussian distribution with mean $(0.0, \pm 1.0)^T$. The covariance matrix for both classes is the unit matrix.

The individual costs of class +1 are defined using the function $c_{+1}(x_1, x_2) = 2\frac{1}{1+e^{-x_1}}$. The costs of the class –1 examples were defined in a similar way by the function $c_{-1}(x_1, x_2) = 2\frac{1}{1+e^{x_1}}$. I.e. for $x_1 > 0$ the +1-examples have larger misclassification costs, whereas for $x_1 < 0$ the -1-examples have larger costs.

The dataset together with the resulting hyperplane for $\epsilon = 0.1$ is depicted in fig. 2 (left, bold line). Other ϵ-values produced similar results. Without costs, a line close to the x_1-axis was produced (fig. 2, left, dashed line). With class dependent misclassification costs, lines are produced that are almost parallel to

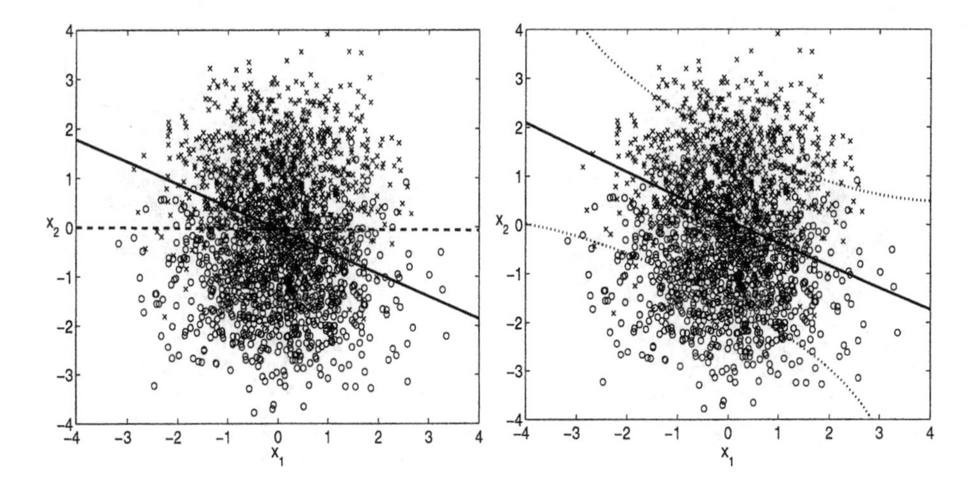

Fig. 2. Results for the non-seperable case. The two classes +1 and −1 are visualized by crosses and circles respectively. Hyperplanes (bold) generated by DIPOL (left), and the class boundary for the extended SVM (right). The DIPOL solution for the cost-free case (dashed) and the margin of the SVM (dotted) are shown additionally.

the x_1 axis and that are shifted into the class region of the less dangerous class (not displayed in fig. 2). Analogous results are achieved by the extended SVM (fig. 2, right).

Our selection of the individual cost functions caused a *rotation* of the class boundary, see fig. 2. This effect cannot be reached using cost matrices alone. I.e. our approach is a genuine extension of previous approaches for including costs, which rely on class dependent costs or cost matrices.

6.2 The Multi-modal Case

To test an augmented version of DIPOL that is capable of using individual costs for learning, we have created the artificial dataset that is shown in fig. 3. Each class consists of two modes, each defined by a Gaussian distribution.

For class +1, we have chosen a constant cost $c_{+1}(x_1, x_2) = 1.0$. For class −1 we have chosen a variable cost, that depends only on the x_1-value, namely $c_{-1}(x_1, x_2) = 2\frac{1}{1+e^{-x_1}}$. This means, that the examples of the left cluster of class −1 (with $x_1 < 0$) have smaller costs compared to the class +1 examples, and the examples of the right cluster (with $x_1 > 0$) have larger costs.

For learning, the augmented version of DIPOL was provided with the 2000 training examples together with their individual costs. The result of the learning algorithm is displayed in fig. 3. It is clear that, for reasons of symmetry, the separating hyperplanes that would be generated *without* individual costs must coincide with one of the bisecting lines. It is obvious in fig. 3, that this is not the case for the hyperplanes that DIPOL has produced for the dataset *with* the individual costs: The left region of class −1 is a little bit smaller, the right

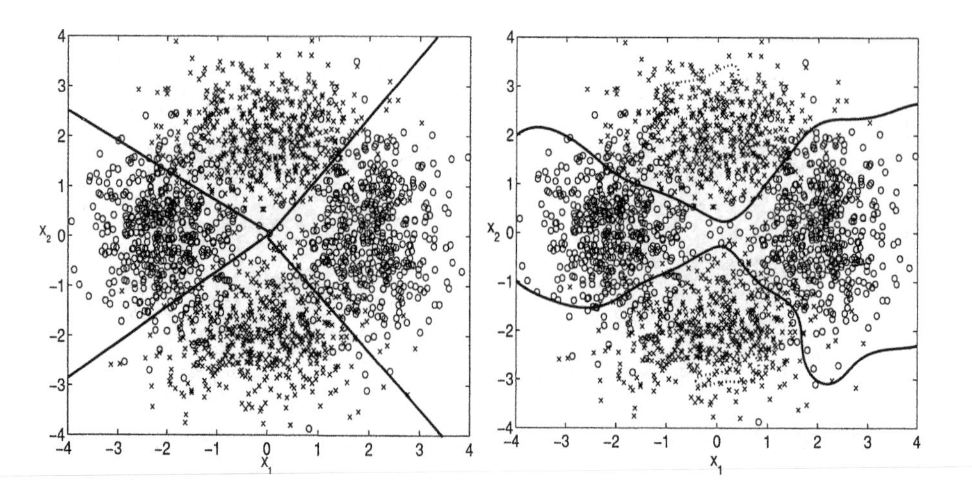

Fig. 3. Results for the multi-modal non-seperable case. Bold lines visualize the learned hyperplanes by DIPOL (left) and the extended SVM (right).

region is a little bit larger compared to learning without costs. Both results are according to the intuition.

The solution of the extended SVM with a radial basis function kernel results in the same shift of the class regions. Due to a higher sensitivity to outliers the decision boundary is curved in contrast to the piecewise linear hyperplanes generated by DIPOL.

6.3 German Credit Data Set

In order to apply our approach to a real world domain, we also conducted experiments on the German Credit Data Set ([10], chapter 9) from the STATLOG project (the dataset can be downloaded from the UCI repository). The data set has 700 examples of class "good customer" (class +1) and 300 examples of class "bad customer" (class −1). Each example is described by 24 attributes. Because the data set does not come with example dependent costs, we assumed the following cost model: If a good customer is incorrectly classified as a bad customer, we assumed the cost of $0.1\frac{duration}{12} \cdot amount$, where *duration* is the duration of the credit in months, and *amount* is the credit amount. We assumed an effective yearly interest rate of $0.1 = 10\%$ for every credit, because the actual interest rates are not given in the data set. If a bad customer is incorrectly classified as a good customer, we assumed that 75% of the whole credit amount is lost (normally a customer will pay back at least part of the money). In the following, we will consider these costs as the real costs of the single cases.

In our experiments we wanted to compare the results using example dependent costs with the results when a cost matrix is used. We constructed the cost matrix $\begin{pmatrix} 0 & 6.27 \\ 29.51 & 0 \end{pmatrix}$, where 6.27 is the average cost for the class +1 examples, and 29.51 is the average cost for the class -1 examples (the credit amounts were normalized to lie in the interval [0,100]).

In our experiment we used cross validation to find the optimal parameter settings (cluster numbers) for DIPOL, i.e. the optimal cluster numbers, and to estimate the mean *predictive* cost T using the 10%-test sets. When using the individual costs, the estimated mean predictive cost was 3.67.

In a second cross validation experiment, we determined the optimal cluster numbers when using the cost matrix for learning and for evaluation. For these optimal cluster numbers, we performed a second cross validation run, where the classifier is constructed using the cost matrix for the respective training set, but evaluated on the respective test set using the example dependent costs. Remember, that we assumed the example dependent costs as described above to be the real costs for each case. This second experiment leads to an estimated mean predictive cost of 3.98.

I.e. in the case of the German Credit Dataset, we achieved a 7.8% reduction in cost using example dependent costs instead of a cost matrix. The classifiers constructed using the cost matrix alone performed worse than the classifiers constructed using the example dependent costs.

The extended SVM generated similar results for the usage of the cost matrix and the example dependent costs respectively, i.e. we found no substantially increased performance. The reason is presumably that the results for DIPOL, the SVM, and other learning algorithms are not much better than the default rule, see [10], though DIPOL and SVM perform comparably well.

7 Conclusion

In this article we discussed a natural cost-sensitive extension of perceptron learning with example dependent costs for correct classification and misclassification. We stated an appropriate criterion function and derived a costs-sensitive learning rule for linearly non-separable classes from it, that is a natural extension of the cost-insensitive perceptron learning rule for separable classes.

We showed that the Bayes rule only depends on differences between costs for correct classification and for misclassification. This allows us to define a simplified learning problem where the costs for correct classification are assumed to be zero. In addition to costs for correct and incorrect classification, it would be possible to consider example dependent costs for rejection, too.

The usage of example dependent costs instead of class dependent costs leads to a decreased misclassification cost in practical applications, e.g. credit risk assignment.

Experiments with the extended SVM approach verified the results of perceptron learning. Its main advantage lies in a lower generalisation error at the expense of non-interpretable decision boundaries. The piecewise linear classifier of DIPOL can easily be transformed to disjunctive rules with linear inequalities. Compared to the SVM, the gradient descent described in sect. 3.2 is very slow in general. The original version of DIPOL described in [10,12,15] incorporates a much faster modified learning procedure (e.g. [5]) that leads to results similar to those presented in this article.

References

1. F. H. Clarke. *Optimization and Nonsmooth Analysis.* Canadian Math. Soc. Series of Monographs and Advanced Texts. John Wiley & Sons, 1983.
2. N. Cristianini and J. Shawe-Taylor. *An Introduction to Support Vector Machines (and Other Kernel-Based Learning Methods).* Cambridge University Press, 2000.
3. R. O. Duda and P. E. Hart. *Pattern Classification and Scene Analysis.* John Wiley & Sons, New York, 1973.
4. Charles Elkan. The foundations of Cost-Sensitive learning. In Bernhard Nebel, editor, *Proceedings of the seventeenth International Conference on Artificial Intelligence (IJCAI-01)*, pages 973–978, San Francisco, CA, August 4–10 2001. Morgan Kaufmann Publishers, Inc.
5. P. Geibel and F. Wysotzki. Using costs varying from object to object to construct linear and piecewise linear classifiers. Technical Report 2002-5, TU Berlin, Fak. IV (WWW http://ki.cs.tu-berlin.de/~geibel/publications.html), 2002.
6. M. Kukar and I. Kononenko. Cost-sensitive learning with neural networks. In Henri Prade, editor, *Proceedings of the 13th European Conference on Artificial Intelligence (ECAI-98)*, pages 445–449, Chichester, 1998. John Wiley & Sons.
7. A. Lenarcik and Z. Piasta. Rough classifiers sensitive to costs varying from object to object. In Lech Polkowski and Andrzej Skowron, editors, *Proceedings of the 1st International Conference on Rough Sets and Current Trends in Computing (RSCTC-98)*, volume 1424 of *LNAI*, pages 222–230, Berlin, June 22–26 1998. Springer.
8. Yi Lin, Yoonkyung Lee, and Grace Wahba. Support vector machines for classification in nonstandard situations. *Machine Learning*, 46(1-3):191–202, 2002.
9. Dragos D. Margineantu and Thomas G. Dietterich. Bootstrap methods for the cost-sensitive evaluation of classifiers. In *Proc. 17th International Conf. on Machine Learning*, pages 583–590. Morgan Kaufmann, San Francisco, CA, 2000.
10. D. Michie, D. H. Spiegelhalter, and C. C. Taylor. *Machine Learning, Neural and Statistical Classification.* Series in Artificial Intelligence. Ellis Horwood, 1994.
11. A. Nedic and D.P. Bertsekas. Incremental subgradient methods for nondifferentiable optimization. *SIAM Journal on Optimization*, pages 109–138, 2001.
12. B. Schulmeister and F. Wysotzki. Dipol - a hybrid piecewise linear classifier. In R. Nakeiazadeh and C. C. Taylor, editors, *Machine Learning and Statistics: The Interface*, pages 133–151. Wiley, 1997.
13. S. Unger and F. Wysotzki. *Lernfähige Klassifizierungssysteme (Classifier Systems that are able to Learn).* Akademie-Verlag, Berlin, 1981.
14. V. N. Vapnik. *The Nature of Statistical Learning Theory.* Springer, New York, 1995.
15. F. Wysotzki, W. Müller, and B. Schulmeister. Automatic construction of decision trees and neural nets for classification using statistical considerations. In G. DellaRiccia, H.-J. Lenz, and R. Kruse, editors, *Learning, Networks and Statistics*, number 382 in CISM Courses and Lectures. Springer, 1997.
16. J. Yang, R. Parekh, and V. Honavar. Comparison of performance of variants of single-layer perceptron algorithms on non-separable data. *Neural, Parallel and Scientific Computation*, 8:415–438, 2000.

An Effective Associative Memory for Pattern Recognition

Boris Kryzhanovsky[1], Leonid Litinskii[1], and Anatoly Fonarev[2]

[1] Institute of Optical Neural Technologies Russian Academy of Sciences,
Vavilov str., 44/2, 119333, Moscow, Russia
{kryzhanov, litin}@mail.ru
[2] Department of Chemistry and Physics, Kean University,
Union, NJ 07083, USA

Abstract. Neuron models of associative memory provide a new and prospective technology for reliable date storage and patterns recognition. However, even when the patterns are uncorrelated, the efficiency of most known models of associative memory is low. We developed a new version of associative memory with record characteristics of its storage capacity and noise immunity, which, in addition, is effective when recognizing correlated patterns.

1 Introduction

Conventional neural networks can not efficiently recognize highly correlated patterns. Moreover, they have small memory capacity. Particularly, the Hopfield neural network [1] can store only $p_0 \approx N/2\ln N$ randomized N-dimensional binary patterns. When the patterns are correlated this number falls abruptly. Few algorithms that can be used in this case (e.g., the projection matrix method [2]) are rather cumbersome and do not allow us to introduce a simple training principle [3]. We offer a simple and effective algorithm that allows us to recognize a great number of highly correlated binary patterns even under heavy noise conditions. We use *a parametrical neural network* [4] that is a fully connected vector neural network similar to the Potts-glass network [5] or optical network [6],[7]. The point of the approach is given below.

Our goal is to recognize p distorted N-dimensional *binary* patterns $\{Y^\mu\}$, $\mu \in \overline{1,p}$. The associative memory that meets this aim works in the following way. Each pattern Y^μ from the space R^N is put in one-to- one correspondence with an image X^μ from a space $\Re$ of greater dimensionality (Sect. 3). Then, the set of images $\{X^\mu\}$ is used to build a parametrical neural network (Sect. 2). The recognition algorithm is as follows. An input binary vector $Y \in \mathrm{R}^\mathrm{N}$ is replaced by corresponding image $X \in \Re$, which is recognized by the parametrical neural network. Then the result of recognition is mapped back into the original N-dimensional space. Thus, recognition of many correlated binary patterns is reduced to recognition of their images in the space $\Re$. Since parametrical neural network has an extremely large memory capacity (Sect. 2) and the mapping process $Y \to X$ allows us to eliminate the correlations between patterns almost completely (Sect. 3), the problem is simplified significantly.

M.R. Berthold et al. (Eds.): IDA 2003, LNCS 2810, pp. 179–186, 2003.

2 Parametrical Neural Network and Their Recognition Efficiency

Let us described the parametrical neural network (PNN) [4]. We consider fully connected neural network of n vector-neurons (*spins*). Each neuron is described by unit vectors $\boldsymbol{x}_i = x_i \boldsymbol{e}_{l_i}$, where an amplitude x_i is equal ± 1 and $\boldsymbol{e}_{l_i}$ is the l_ith unit vector-column of Q-dimensional space: $i = 1, ..., n;\ 1 \le l_i \le Q$. The state of the whole network is determined by the set of vector-columns $X = (\boldsymbol{x}_1, \boldsymbol{x}_2, ..., \boldsymbol{x}_n)$. According [4], we define the Hamiltonian of the network the same as in the Hopfield model

$$H = -\frac{1}{2} \sum_{i,j=1}^{n} \boldsymbol{x}_i^{+} \hat{\mathbf{T}}_{ij} \boldsymbol{x}_j, \tag{1}$$

where $\boldsymbol{x}_i^{+}$ is the Q-dimensional vector-row. Unlike conventional Hopfield model, where the interconnections are scalars, in Eq.(1) an interconnection $\hat{\mathbf{T}}_{ij}$ is a $(Q \times Q)$-matrix constructed according to the generalized Hebb rule [5],

$$\hat{\mathbf{T}}_{ij} = (1 - \delta_{ij}) \sum_{\mu=1}^{p} \boldsymbol{x}_i^{\mu} \boldsymbol{x}_j^{\mu+}, \tag{2}$$

from p initial patterns:

$$X^{\mu} = (\boldsymbol{x}_1^{\mu}, \boldsymbol{x}_2^{\mu}, ..., \boldsymbol{x}_n^{\mu}),\ \mu = 1, 2, ..., p. \tag{3}$$

It is convenient to interpret the network with Hamiltonian (1) as a system of interacting Q-dimensional spins and use the relevant terminology. In view of (1)-(3), the input signal arriving at the ith neuron (i.e., *the local field* that acts on ith spin) can be written as

$$\boldsymbol{h}_i = \sum_{j=1}^{n} \hat{\mathbf{T}}_{ij} \boldsymbol{x}_j = \sum_{l=1}^{Q} A_l^{(i)} \boldsymbol{e}_l, \tag{4}$$

where

$$A_l^{(i)} = \sum_{j(\ne i)}^{n} \sum_{\mu=1}^{p} (\boldsymbol{e}_l \boldsymbol{x}_i^{\mu})(\boldsymbol{x}_j^{\mu} \boldsymbol{x}_j),\ l = 1, ..., Q. \tag{5}$$

The behavior of the system is defined in the natural way: under action of the local field $\boldsymbol{h}_i$ the ith spin gets orientation that is as close to the local field direction as possible. In other words, the state of the ith neuron in the next time step, $t + 1$, is determined by the rule:

$$\boldsymbol{x}_i(t+1) = x_{\max} \boldsymbol{e}_{\max},\ x_{\max} = \operatorname{sgn}\left(A_{\max}^{(i)}\right), \tag{6}$$

where *max* denotes the greatest in modulus amplitude $A_l^{(i)}$ in (5).

The evolution of the network is a sequence of changes of neurons states according to (6) with energy (1) going down. Sooner or later the network finds itself at a fixed point.

Let us see how efficiently PNN recognizes noisy patterns. Let the distorted mth pattern X^m come to the system input, i.e. the neurons are in the initial states described as $\boldsymbol{x}_i = \hat{a}_i \hat{b}_i \boldsymbol{x}_i^m$. Here $\hat{a}_i$ is the multiplicative noise operator, which changes the sign of the amplitude x_i^m of the vector-neuron $\boldsymbol{x}_i^m = x_i^m \boldsymbol{e}_{l_i^m}$ with probability a and keeps it the same with probability $1 - a$; the operator $\hat{b}_i$ changes the basis vector $\boldsymbol{e}_{l_i^m} \in \{\boldsymbol{e}_l\}_1^Q$ by any other from $\{\boldsymbol{e}_l\}_1^Q$ with probability b and retains it unchanged with probability $1 - b$. In other words, a is the probability of an error in the sign of the neuron ("-" in place "+" and vice versa), b is the probability of an error in the vector state of the neuron. The network recognizes the reference pattern X^m correctly, if the output of the ith neuron defined by Eq.(6) is equal to $\boldsymbol{x}_i^m$, that is $\boldsymbol{x}_i = \boldsymbol{x}_i^m$. Otherwise, PNN fails to recognize the pattern X^m. According to the Chebyshev-Chernov method [8] (for such problems it is described comprehensively in [7], [4]) the probability of recognition failure is

$$\mathrm{P}_{err} \leq n \exp\left[-\frac{nQ^2}{2p}(1-2a)^2(1-b)^2\right]. \tag{7}$$

The inequality sets the upper limit for the probability of recognition failure for PNN. The memory capacity of PNN (i.e., the greatest number of patterns that can be recognized) is found from (7):

$$p_c = nQ^2 \frac{(1-2a)^2(1-b)^2}{2\ln n}. \tag{8}$$

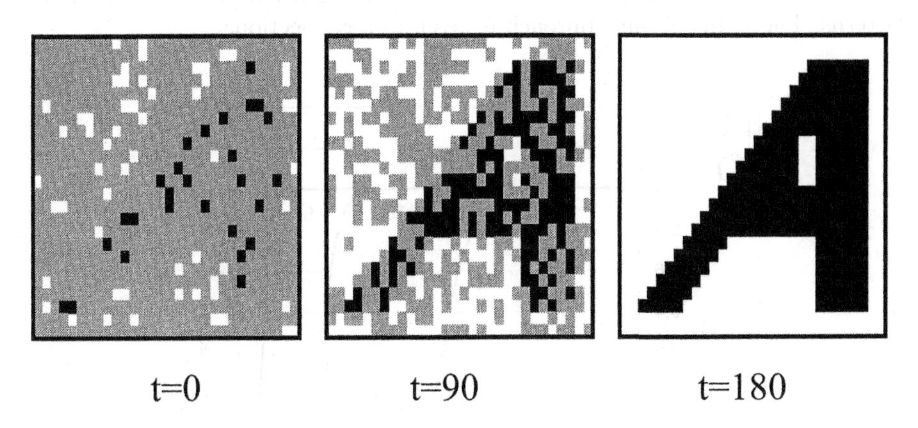

Fig. 1. Process of recognition of letter "A" by PNN at $p/n = 2, Q = 32, b = 0.85$. Noise-distorted pixels are marked in gray.

Comparison of (8) with similar expressions for the Potts-glass neural network [5] and the optical network [6],[7] shows that the memory capacity of PNN is approximately twice as large as the memories of both aforementioned models. That means that under other conditions being equal its recognition power is 20-30% higher.

It is seen from (7) that with growing Q the noise immunity of PNN increases exponentially. The memory capacity also grows: it is Q^2 times as large as that of the Hopfield network. For example, if $Q = 32$, the 180-neuron PNN can recognize 360 patterns ($p/n = 2$) with 85% noise-distorted components (Fig. 1). With smaller distortions, $b = 0.65$, the same network can recognize as many as 1800 patterns ($p/n = 10$), *etc.* Let us note, that some time ago the restoration of 30% noisy pattern for $p/n = 0.1$ was a demonstration of the best recognition ability of the Hopfield model [9].

Of course, with regard to calculations, PNN is more complex than the Hopfield model. On the other hand the computer code can be done rather economical with the aid of extracting bits only. It is not necessary to keep a great number of $(Q \times Q)$-matrices $\hat{\mathbf{T}}_{ij}$ (2) in your computer memory. Because of the complex structure of neurons, PNN works Q-times slower than the Hopfield model, but it makes possible to store Q^2-times greater number of patterns. In addition, the Potts-glass network operates Q-times slower than PNN.

Fig. 2 shows the recognition reliability $\bar{P} = 1 - \mathrm{P}_{err}$ as a function of the noise intensity b when the number of patterns is twice the number of neurons ($p = 2n$) for $Q = 8, 16, 32$. We see that when the noise intensity is less than a threshold value

$$b_c = 1 - \frac{2}{Q}\sqrt{\frac{p}{n}},$$

the network demonstrates reliable recognition of noisy patterns. We would like to use these outstanding properties of PNN for recognition of correlated binary patterns. The point of our approach is given in next Sections.

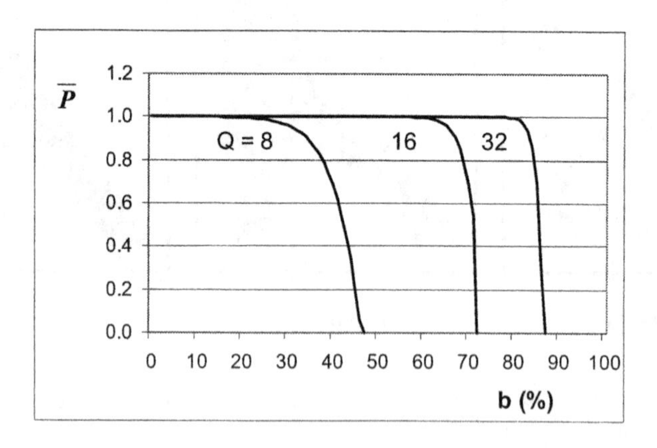

Fig. 2. Recognition reliability of PNN $\bar{P} = 1 - \mathrm{P}_{err}$ as a function of noise intensity b.

3 Mapping Algorithm

Let $Y = (y_1, y_2, ..., y_N)$ be an N-dimensional binary vector, $y_i = \{\pm 1\}$. We divide it mentally into n fragments of $k+1$ elements each, $N = n(k+1)$. With each fragment we associate an integer number $\pm l$ according the rule: the first element of the fragment defines the sign of the number, and the other k elements determine the absolute value of l:

$$l = 1 + \sum_{i=1}^{k} (y_i + 1) \cdot 2^{i-2}; \quad 1 \leq l \leq 2^k.$$

Now we associate each fragment with a vector $\boldsymbol{x} = \pm \boldsymbol{e}_l$, where $\boldsymbol{e}_l$ is the lth unit vector of the real space R^{Q} and $Q = 2^k$. We see that any binary vector $Y \in \mathrm{R}^{\mathrm{N}}$ one-to-one corresponds to a set of n Q-dimensional unit vectors, $X = (\boldsymbol{x}_1, \boldsymbol{x}_2, ..., \boldsymbol{x}_n)$, which we call *the internal image* of the binary vector Y. (In the next Section we use the internal images X for PNN construction.) The number k is called *a mapping parameter.*

For example, the binary vector $Y = (-1, 1, -1, -1, 1, -1, -1, 1)$ can be split into two fragments of four elements: (-11-1-1) and (1-1-11); the mapping parameter k is equal 3, $k = 3$. The first fragment ("-2" in our notification) corresponds to the vector $-\boldsymbol{e}_2$ from the space of dimensionality $Q = 2^3 = 8$, and the second fragment ("+5" in our notification) corresponds to the vector $+\boldsymbol{e}_5 \in \mathrm{R}^8$. The relevant mapping can be written as $Y \to X = (-\boldsymbol{e}_2, +\boldsymbol{e}_5)$.

It is important that the mapping is biunique, i.e., the binary vector Y can be restored uniquely from its internal image X. It is even more important that the mapping eliminates correlations between the original binary vectors. For example, suppose we have two 75% overlapping binary vectors

$$\begin{aligned} Y_1 &= (\ 1, -1, -1, -1, -1, -1, -1, 1\ ) \\ Y_2 &= (\ 1, -1, -1, \ \ 1, \ \ 1, -1, -1, 1\ ) \end{aligned}.$$

Let us divide each vector into four fragments of two elements. In other words, we map these vectors with the mapping parameter $k = 1$. As a result we obtain two internal images $X_1 = (+\boldsymbol{e}_1, -\boldsymbol{e}_1, -\boldsymbol{e}_1, -\boldsymbol{e}_2)$ and $X_2 = (+\boldsymbol{e}_1, -\boldsymbol{e}_2, +\boldsymbol{e}_1, -\boldsymbol{e}_2)$ with $\boldsymbol{e}_l \in \mathrm{R}^2$. The overlapping of these images is 50%. If the mapping parameter $k = 3$ is used, the relevant images $X_1 = (+\boldsymbol{e}_1, -\boldsymbol{e}_5)$ and $X_2 = (+\boldsymbol{e}_5, +\boldsymbol{e}_5)$ with $\boldsymbol{e}_l \in \mathrm{R}^8$ do not overlap at all.

4 Recognition of the Correlated Binary Patterns

In this Section we describe the work of our model as a whole, i.e. the mapping of original binary patterns into internal images and recognition of these images with the aid of PNN. For a given mapping parameter k we apply the procedure from Sect. 3 to a set of binary patterns $\{Y^\mu\} \in \mathrm{R}^{\mathrm{N}}$, $\mu \in \overline{1,p}$. As a result we obtain a set of internal images $\{X^\mu\} \in \Re$, allowing us to build PNN with $n = N/(k+1)$ Q-dimensional vector-neurons, where $Q = 2^k$. Note, the dimension

Q of vector-neurons increases exponentially with k increasing, and this improves the properties of PNN.

In further analysis we use so-called *biased* patterns Y^μ whose components y_i^μ are either $+1$ or -1 with probabilities $(1+\alpha)/2$ and $(1-\alpha)/2$ respectively $(-1 \leq \alpha \leq 1)$. The patterns will have a mean activation of $\overline{Y^\mu} = \alpha$ and a mean correlation $\overline{(Y^\mu, Y^{\mu\prime})} = \alpha^2$, $\mu \neq \mu\prime$. Our aim is to recognize the mth noise-distorted binary pattern $\tilde{Y}^m = (s_1 y_1^m, s_2 y_2^m, ..., s_N y_N^m)$, where a random quantity s_i changes the variable y_i^m with probability s, and the probability for y_i^m to remain the same is $1-s$. In other words, s is the distortion level of the mth binary pattern. Mapped into space $\Re$, this binary vector turns into the mth noise-distorted internal image $\tilde{X}^m$, which has to be recognized by PNN. Expressing multiplicative noises a and b as functions of s and substituting the result in (7), we find that the probability of misrecognition of the internal image X^m is

$$\mathrm{P}_{err} = n\left(\mathrm{ch}(\nu\alpha^3) - \alpha \mathrm{sh}(\nu\alpha^3)\right) \cdot \exp\left[-\frac{\nu}{2\beta}(1+\beta^2\alpha^6)\right], \quad (9)$$

where

$$\nu = n(1-2s)^2(1-s)^k,\ \beta = p[A/(1-s)]^k,\ A = (1+\alpha^2)[1+\alpha^2(1-2s)]/4.$$

Expression (9) describes the Hopfield model when $k=0$. In this case, even without a bias ($\alpha = 0$) the memory capacity does not exceed a small value $p_0 \approx N/2\ln N$. However, even if small correlations ($\alpha > N^{-1/3}$) are present, the number of recognizable patterns is $(N\alpha^3)$-times reduced. In other words, the network almost fails to work as associative memory. The situation changes noticeably when parameter k grows: the memory capacity increases and the influence of correlations decreases. In particular, when correlations are small ($\nu\alpha^3 < 1$), we estimate memory capacity from (9) as

$$p = p_0 \frac{[(1-s)^2/A]^k}{k+1}.$$

Let us return to the example from [9] with $\alpha = 0$. When $k=5$, in the framework of our approach one can use 5-times greater number of randomized binary patterns, and when $k=10$, the number of the patterns is 80-times greater.

When the degree of correlations is high ($\nu\alpha^3 > 1$), the memory capacity is somewhat smaller:

$$p = \alpha^{-3}[(1-s)/A]^k.$$

Nevertheless, in either case increasing k gives us *the exponential* growth of the number of recognizable patterns and a rise of recognition reliability.

Fig. 3 shows the growth of the number of recognizable patterns when k increases for distortions s of 10% to 50% ($\alpha = 0$).

Fig. 4 demonstrates how the destructive effect of correlations diminishes when k increases: the recognition power of the network is zero when k is small; however, when the mapping parameter k reaches a certain critical value, the misrecognition probability falls sharply (the curves are drawn for $\alpha = 0, 0.2, 0.5, 0.6$, $p/N = 2$ and $s = 30\%$).

Our computer simulations confirm these results.

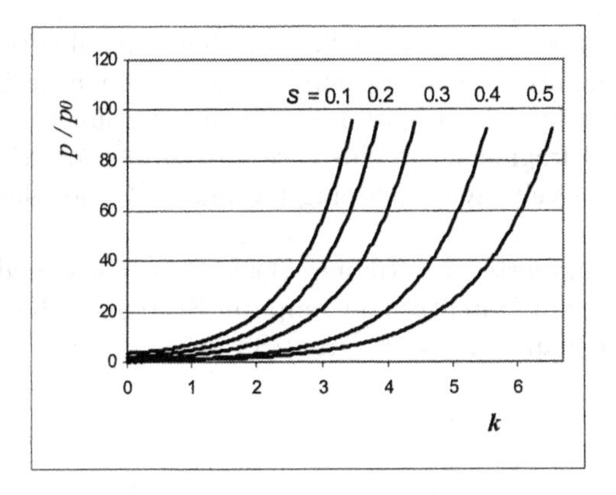

Fig. 3. Growth of memory capacity with the mapping parameter k ($s = 0.1 - 0.5$).

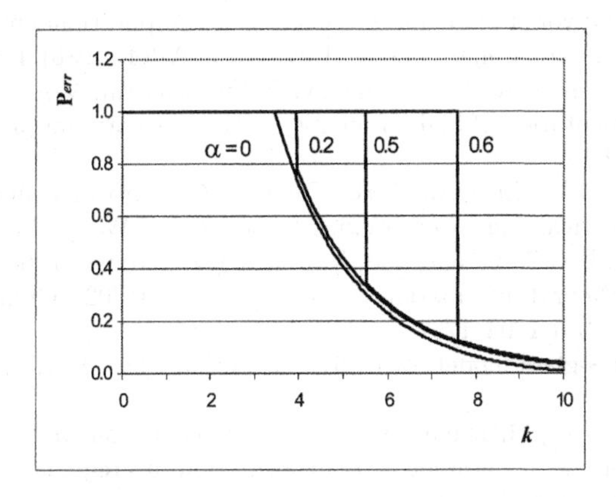

Fig. 4. Fall of the misrecognition probability with the mapping parameter k.

5 Concluding Remarks

Our algorithm using the large memory capacity of PNN and its high noise immunity, allows us to recognize many correlated patterns reliably. The method also works when all patterns have the same fragment or fragments. For instance, for $p/N = 1$ and $k = 10$, this kind of neural network permits reliable recognition of patterns with 40% coincidence of binary components.

When initially there is a set of correlated vector-neuron patterns which must be stored and recognized, the following algorithm can be used to suppress correlations. At first, the patterns with vector coordinates of original dimensionality q_1 are transformed into binary patterns. Then they are mapped into patterns with vector neurons of dimensionality $q_2 (q_2 > q_1)$. A PNN is built using patterns with q_2-dimensional vector neurons. This double mapping eliminates correlations and provides a reliable pattern recognition.

In conclusion we would like to point out that a reliable recognition of correlated patterns requires a random numbering of their coordinates (naturally, in the same fashion for all the patterns). Then the constituent vectors do not have large identical fragments. In this case a decorrelation of the patterns can be done with a relatively small mapping parameter k and, therefore, takes less computation time.

The work was supported by Russian Basic Research Foundation (grant 01-01-00090), the program "Intellectual Computer Systems" (the project 2.45) and by President of Russian Federation (grant SS-1152-2003-1).

References

1. Hopfield, J.J.: Neural networks and physical systems with emergent collective computational abilities. Proc. Natl. Acad. Sci. *USA* **79** (1982) 2554–2558.
2. Personnaz, L., Guyon, I., Dreyfus, G.: Collective computational properties of neural networks: New learning mechanisms. Phys. Rev. A **34** (1986) 4217–4228.
3. Mikaelian, A.L., Kiselev, B.S., Kulakov, N.Yu., Shkitin, V.A. and Ivanov, V.A.: Optical implementation of high-order associative memory. Int. J. of Opt. Computing **1** (1990) 89–92.
4. Kryzhanovsky, B.V., Litinskii, L.B., Fonarev A.: Optical neural network based on the parametrical four-wave mixing process. In: Wang, L., Rajapakse, J.C, Fukushima, K., Lee, S.-Y., and Yao, X. (eds.): Proceedings of the 9th International Conference on Neural Information Processing (ICONIP'02), Orchid Country Club, Singapore, (2002) **4** 1704–1707.
5. Kanter, I.: Potts-glass models of neural networks. Phys. Rev. A **37** (1988) 2739–2742.
6. Kryzhanovsky, B.V., Mikaelian, A.L.: On the Recognition Ability of a Neural Network on Neurons with Parametric Transformation of Frequencies. Doklady Mathematics, **65(2)** (2002) 286–288.
7. Kryzhanovsky, B.V., Kryzhanovsky, V.M., Mikaelian, A.L., Fonarev, A.: Parametric dynamic neural network recognition power. Optical Memoryu & Neural Networks **10** (2001) 211–218.
8. Chernov, N.: A measure of asymptotic efficiency for tests of hypothesis based on the sum of observations. Ann. Math. Statistics **23** (1952) 493–507.
9. Kinzel, W.: Spin Glasses and Memory. Physica Scripta **35** (1987) 398–401.

Similarity Based Classification

Axel E. Bernal[1], Karen Hospevian[2], Tayfun Karadeniz[3], and Jean-Louis Lassez[3]

[1] University of Pennsylvania. CIS Department
Phildelphia, PA 19104. USA
axel@snowball.pcbi.upenn.edu

[2] New Mexico Institute of Mining and Technology. Computer Science Dept.
Socorro, New Mexico. USA
karen@nmt.edu

[3] Coastal Carolina University. Department of Computer Science
Conway, SC 29528. USA
{jlassez, tkarade}@coastal.edu

Abstract. We describe general conditions for data classification which can serve as a unifying framework in the study of kernel based Machine Learning Algorithms. From these conditions we derive a new algorithm called SBC (for Similarity Based Classification), which has attractive theoretical properties regarding underfitting, overfitting, power of generalization, computational complexity and robustness. Compared to classical algorithms, such as Parzen windows and non-linear Perceptrons, SBC can be seen as an optimized version of them. Finally it is a conceptually simpler and a more efficient alternative to Support Vector Machines for an arbitrary number of classes. Its practical significance is illustrated through a number of benchmark classification problems.

1 Introduction

Research in Machine Learning has recently received considerable attention due to massive proliferation of data in Bioinformatics and the Internet. In particular, Support Vector Machines (SVM) have been shown to play a major role in microarray data analysis [1], and web data classification [7]. We refer the reader to [5] for a thorough treatment and further references regarding the wide range of applications of SVM. The main significance of Support Vector Machines is their theoretical underpinning and their handling of non-linear decision functions. The key to this significance is the use of kernels, as was done with Parzen windows and the non-linear Perceptron [6]. A thorough study on kernel methods can be found in [13]

In this paper we initially introduce a unifying framework to describe these "classical" and show that they are different stages of refinement of the same decision function ; the only difference across is the way the training set itself is classified: Parzen windows do not require that the decision function separate correctly the

M.R. Berthold et al. (Eds.): IDA 2003, LNCS 2810, pp. 187–197, 2003.

training sets, the non linear Perceptron will be seen as a version of Parzen windows that requires that the training set be properly classified, while SVM will be seen as a version of the non linear Perceptron that gives an optimal solution to that classification. From this unifying framework we gain two major insights: the first is that all these algorithms, Parzen windows, SVM, non linear Perceptron, SBC, and arguably other algorithms based on the concept of maximizing similarity, assuming appropriate normalization of the data, converge to the same solution, as the "tightness" of the similarity measure increases. This property is established at the mathematical level and verified empirically. The other main insight we derive is the SBC algorithm, which in a sense subsumes these three algorithms and is an answer we propose to the open problem of finding a suitable "direct" multi-class SVM [2]. This is because we replace the concept of maximal margin, which is essentially a binary concept, by a concept of robustness of the decision function which is independent of the number of classes (but equivalent to the maximal margin in the binary case).
This paper discusses the issues of underfitting, overfitting, power of generalization, computational complexity and robustness within a simple framework using basic definitions of similarity matrix and tightness of similarity measures. We also compare at the mathematical level, the asymptotic behavior of these various algorithms and show that they converge with increasing tightness, assuming some normalization of the data, towards the same solution: the Maxsim formula. The SBC algorithm, derived from this study, uses naturally introduced similarity functions (not requiring mappings in high-dimensional Hilbert spaces), and optimizes "a la" SVM but without the need to solve the problem in a dual space with a complex quadratic objective function and more importantly with the ability to handle large multi-class problems. All this comes with a small penalty in the binary case, as the SVM decision function is smaller than the corresponding SBC function. In a concluding section we report briefly on an empirical validation of our claims, a full study is the object of a forthcoming paper.

2 Similarity Matrix

As in [13] we introduce the problem of classifying an object X by comparing its similarity to sets of previously classified training objects; X will be assigned to the class whose objects are most similar to X.
Given a set of I class-labeled training objects $\{X_i, \xi(X_i)\}$, $i = 1..I$, where $\xi(X_i)$ is the class of X_i, and for an unclassified object X, we define the *class similarity* of X with respect to a class C as

$$S_C(X) = \sum_{X_k \in C} \alpha_k s(X_k, X) \tag{1}$$

Where s is the similarity function and $\alpha_k \geq 0$ reflects the relative importance given to each X_k with respect to the classification.
We can therefore predict the class of X using the following decision function:

$$\xi(X) = arg_C\{max(S_C(X))\} \tag{2}$$

Or the strong version, which requires that not only X be more similar to class C than it is to any other class, but also be more similar to class C than it is to the union of any other collection of classes.

$$\xi(X) = arg_C\{max(S_C(X) > \sum_{D \neq C} S_D(X))\} \tag{3}$$

In order to compare our approach with classical Machine Learning algorithms that deal with binary classification, we will also consider the problem of supervised classification of objects from only two different classes A and B. In this case, (1) can be rewritten as:

$$\begin{gathered} S_A(X) - S_B(X) > 0 \rightarrow \xi(X) = A \\ S_A(X) - S_B(X) < 0 \rightarrow \xi(X) = B \\ S_A(X) - S_B(X) = 0 \rightarrow \xi(X) \text{ is } \textit{not defined} \end{gathered}$$

Let α be the vector of coefficients α_i. The problem of a classification algorithm is to choose the criteria to compute these coefficients. In that respect some algorithms will be satisfied with finding a simply empirically acceptable solution, as Parzen windows do, while others will want a solution that satisfies predefined criteria, as the Perceptron does with a requirement of linear separability, and finally others, such as Support Vector Machines will find an optimal solution to these criteria. But all share a decision function with the structure of (1).
Among the many possible similarity measures, we will use mainly Radial Basis functions and polynomial kernels. This is because they are among the most popular and efficient, and also because they are the main ones used in SVM and the non-linear Perceptron, making comparisons more meaningful.
The Radial Basis Function (RBF in short): $s(x, y) = e^{\|X-Y\|^2/2\sigma^2}$ can be viewed as a similarity measure between two points in $\Re^n$, as it is a measure of the inverse of the distance between two points, its range is the interval (0,1) instead of an infinite range.
Depending on the applications, the data is represented by points in $\Re^n$, or normalized vectors. In this last case we consider bs; that is similarities based on a measure of the angle between two vectors of length 1. Those are quite suitable for many applications, such as text processing [7]
In any case it is easy conceptually, and trivial algebraically, to approximate a hyperplane by a spherical area, given a radius of sufficient magnitude. We just need to add a value corresponding to the radius of the sphere to the list of coefficients of each point, and normalize. So we can still address general classification problems of points in $\Re^n$, where we use vectors of uniform length to approximate the points.
Representatives of similarity measures for vectors of uniform length are the polynomial functions: $s(X, Y) = \|X, Y\|^n$ and of course the RBF.
We will say that similarity measure s is tighter than similarity measure t iff $\forall\ i, j \in I$, $s(X_i, X_j) \leq t(X_i, Y_j)$. The *similarity matrix* is defined as:

$$S = [\delta s(X_i, X_j)] \tag{4}$$

Where $\delta = 1$ if $\xi(X_i) = \xi(X_j)$, and $\delta = -1$ otherwise.

3 Fitting Conditions

We would like our decision function to satisfy two "fitting conditions", the first one simply expresses that the training vectors are properly classified. The second deals with parameterized measures, like RBF's σ parameter and the polynomial kernel's degree, which allow the user to control tightness. The condition essentially reflects the fact that extreme values of these parameters correspond to extreme tightness and overfitting, or no tightness at all and underfitting.

3.1 No Underfitting Condition

$$S\alpha > 0, \alpha \geq 0 \textit{ is solvable} \tag{5}$$

As each row of the system expresses that a training vector's similarity to its own class is superior to its similarity to the union of all other classes, we have:

Proposition 1: *The strong decision function correctly classifies the training vectors if and only if the set of constraints in (5) is solvable.*
We will say that a solution α (fits) the training data if and only if it satisfies the (no underfitting condition).

3.2 Absolute Over/Under Fitting Condition

We assume that the similarity functions are parameterized using a *tightness parameter* q, which allows the user to control the tightness of the similarity measure, and have the following properties:

$$\exists\, q,\ \forall\, i, j \in I / S_q(X_i, X_j) = 1$$
$$\exists\, q, \forall i, j \in I / S_q(X_i, X_j) = 1 \iff i = j \wedge S_q(X_i, X_j) = 0 \text{ for } i \neq j$$

At one extreme setting the parameter q will make all objects equally similar. This corresponds to a situation where we have no means of comparing the objects, and leads to a decision function that assigns an object systematically to the class with the largest number of objects. This case we call absolute underfitting. At the other extreme q will make the similarity function become tighter and tighter until it becomes the identity step function. As the similarity function converges towards the identity step function, the decision function converges towards absolute overfitting; that is it will classify correctly only the training objects.
Both conditions are met by the RBF, or by polynomial kernels when the data has been normalized to vectors of length one.
The motivation for this condition, which is not fundamentally restrictive in nature, is that we find that the similarity matrix, which is a square symmetric matrix, has interesting theoretical and computationally advantageous properties as the tightness of the similarity measure increases:

Proposition 2: *As the tightness of the similarity measure increases, the similarity matrix becomes positive definite and consequently invertible.*
Proof (sketch):
We use Rayleigh's conditions and assume for sake of simplicity that the vectors are not duplicated. As the tightness increases the matrix S converges to the identity matrix and therefore min $\frac{\|SX,X\|}{(\|X\|\|Y\|)}$, the smallest eigenvalue, is strictly positive as the vector SX will be close to the vector X.

3.3 The Tight Fitting Zone

As the tightness increases the matrix S converges to the identity matrix and we have absolute overfitting. Then, of course the no underfitting condition becomes satisfied, as the unit vector (all coefficients equal to one) is a trivial solution to the system $I\alpha > 0,\ \alpha \geq 0$.
Now $\exists\alpha, S\alpha \geq 0$ is equivalent to stating that the vectors which are the rows of the matrix S are all on the same side of the hyperplane Ψ whose normal vector is α. As the tightness increases these vectors tend towards the vectors that form a basis of the vector space. Clearly they will pass, in general, on the same side of the hyperplane Ψ, defined by the unit vector, well before they reach the basis vectors. Furthermore, we also see that the unit vector provides a solution which is more and more b the basis. As a consequence we will say that a similarity measure is *tight fitting* when the unit vector is a solution to the no underfitting condition.
This brief analysis suggests that a similarity measure will be tight fitting, and consequently will admit a trivially computed decision function that satisfies the no underfitting condition, well before it becomes overfitting. This analysis also suggests that as we have data of increasingly complex shape/distribution, we have to use similarity measures of increased tightness, and as a consequence move closer to overfitting, but then the unit solution becomes closer to an optimal solution. This claim will be justified in later sections. It is an important claim as it essentially states that as the structure of the data becomes increasingly complex, a straightforward solution becomes more acceptable.
This solution is represented by the MaxSim (for Maximum Similarity) formula, which is the decision function 1 with the unit vector:

$$\xi(X) = arg_C\{max(\sum_{k \in C} S(X_k, X))\} \tag{6}$$

4 Asymptotic Behavior and Maximum Similarity

We proceed to formally establish how the MaxSim formula is related to some classification algorithms that we have used here as motivating examples. Specifically, we will show that the behavior of these algorithms converge asymptotically to equation (6)

4.1 Parzen Windows

The connection with Parzen windows is immediate, as the Maxsim formula (6) is a clear generalization of Parzen windows, and, as a corollary to proposition 2 we can see that Parzen windows become closer to SBC as the tightness of the similarity measure increases:

Proposition 3: *As the tightness of the similarity measure increases, Parzen windows will satisfy the no underfitting condition.*

4.2 Non-linear Perceptron

The Perceptron algorithm dates from 1956, designed by Rosenblatt at Cornell University. It provided the algorithmic foundation for the area of Neural Nets. As opposed to the notion of maximum similarity, the Perceptron uses the notion of linear separability, which we briefly describe: The issue is to find a hyperplane Ψ separating two clouds of points in Rn, assuming that they are separable. Ψ will then be used as a decision function; a new point is assigned to the cloud which is on the same side of the hyperplane as the point.
There is a very elegant proof due to Novikov showing that when the two sets are separable the algorithm is guaranteed to find a solution, see [5] where the authors stress the significance of a dual view of the Perceptron. We use here this duality to show that the Perceptron satisfies the framework that we introduced in the previous section.
We know that the solution to the Perceptron algorithm can be expressed as $M^t\alpha$, which is as a linear combination of input vectors; from this remark and assuming a normalization of the input data, one can show:

Proposition 4: (Solving the system $MX > 0$ is equivalent to solving its dual $MM^t\alpha > 0$. The notion of linear separability is equivalent to the notion of no underfitting).
MM^t is a similarity matrix with the dot product as similarity measure. This means that the Perceptron algorithm classifies a test point according to a weighted sum of its similarities with the training points.

Corollary Replacing the dot product by an appropriate similarity measure we generalize the dual form of the Perceptron algorithm to a non linear multi-class version whose decision function converges to the Maxsim formula as the tightness of the similarity measure increases.

4.3 Support Vector Machines

MAXSIM and *SVM* are classifiers of a different nature. In the former all vectors play an equal role, in the latter as many as possible are eliminated. It does not seem evident that they could have similar behavior. However [6] have shown that the non-linear Perceptron can give results that are similar in accuracy to the SVM. Our version of the non-linear Perceptron might therefore also give

similar results. Furthermore [7] in his study of SVM for large scale applications indicates that a number of support vectors have their coefficients α_i equal to their limit C. It is also clear mathematically, and verified in practice, that when one increases the dimension (by using kernels of greater non linearity), the number of support vectors increases. It seems that with increasing complexity of the data, and therefore increasing non linearity, the number of coefficients α_i that take non-zero values increases, which seems to indicate that the behavior of the SVM in such cases tends towards the behavior of the Maxsim formula.
We will give a formal proof of this, under the following conditions:

- The training vectors are all different (no duplicates)
- The training vectors have been normalized to length one.
- The kernels are polynomial and we consider increasing powers, or the kernels are radial Basis Functions and we consider decreasing O in which case the preceding normalization is not compulsory.
- The two subsets of the training set that have to be separated should have the same size.

These restrictions do not affect the generality of our argument, as we know that the existence of duplicate vectors can affect the execution of the SVM, making the matrix singular, the normalization is done in many applications, and can be done easily in all cases, the kernels mentioned are used in most applications, and the property we will use may well be shared with other kernels. We require that the two subsets have the same size to make the presentation clearer, it is not needed otherwise.
Under these conditions we will establish the following proposition:

Proposition 5: *In the limit of non-linearity, the decision function of Support Vector Machines converges to the Maxsim formula.*

Proof (sketch):
There are a number of equivalent mathematical formulations of the SVM, we will choose one from [3] and [4].

$$\begin{gathered} min(\alpha^t Q \alpha) \\ 0 \leq \alpha_i < 1 \\ e^t \alpha = \nu l \\ y^t \alpha = 0 \end{gathered}$$

And the decision function is:

$$sgn(\sum y_i \frac{\alpha_i}{\rho}(k(X_i, X) + b)) \tag{7}$$

Under our hypothesis, the matrix Q is equal to a similarity matrix $[\delta k(X_i, X_j)]$. For polynomial kernels with degrees tending to infinity, or for a RBF whose parameter σ tends to 0, the matrix Q converges to the *identity matrix.*

The solution of the SVM is then the point on the plane $e^t\alpha = \nu l$, whose distance to the origin is minimum, that is a point such that all coordinates α_i are equal, and as the two subsets have the same number of vectors, the last equality is trivially verified.

5 The SBC Algorithm

As a consequence of Proposition 2, we know that by choosing a tight enough similarity measure, the No Underfitting condition will be satisfied. We also know that the unit vector may or may not provide a solution, but even if it does, there will be infinitely many weight vectors to choose from. As we have seen in the last section, Parzen Windows and Non-linear perceptrons are only two possible choices for these vectors; on the other hand, SVM provides an optimal solution by finding a maximal margin hyperplane. In our case, we also would like to find a solution which is optimal in some sense. We will justify our proposed optimization in two ways:

- Robustness: we want the decision function to give similar values for two vectors of similarities that are very close to each other (similar points should be similarly classified). So its values should not vary too steeply.
- Power of generalization: the less it varies (and provided the no underfitting conditions are met), the more points can be safely assigned to any given class.

As a consequence we want to minimize the norm of the gradient, which controls the variations of the values of the decision function. We give now a more geometric justification, which also shows a link with the SVM's choice of maximizing the margin.
First, we replace the condition $S\alpha > 0$ by the equivalent $S\alpha > 1$.
We also assume, for sake of simplicity, that the rows of S have been normalized to length 1.
A row of the system now reads as:

$$\cos(\theta_{i\alpha}) > \frac{1}{\|\alpha\|} \tag{8}$$

Where $\theta_{i\alpha}$ is the angle between the ith row vector and the weight vector α. A robust solution is one for which small changes in the weight vector will still give a weight vector solving the system. Therefore a more robust solution is achieved when one minimizes $\|\alpha\|$. These considerations lead us to the following quadratic program, which determines the values of the weight vectors of the decision function, which we call SBC:

$$\begin{gathered} min(\|\alpha\|^2) \\ S\alpha > 1 \\ \alpha \geq 0 \end{gathered}$$

5.1 Implementation Issues

This quadratic program has many attractive features from a computational point of view. The objective function is a very simple convex function, the underlying quadratic form uses the identity matrix, and according to proposition 2, as the

tightness increases, the system of linear constraints admits a trivial solution (the unit vector). So it will run well with standard quadratic programming software such as MatLab, but there is much scope for more efficient implementations. For instance, because the objective function is convex, one can use the Karush-Kuhn-Tucker conditions to compute the optimal solution via a simple modification to the simplex algorithm [10]. Also, we can remark that the program computes the distance of a polyhedral set to the origin, so one could derive an algorithm based on such geometrical consideration, in a way inspired by [8], which computes the maximal margin as the distance between two polytopes.
Finally, as with "pure" SVM, this "pure" version of SBC is sensitive to noisy data, even though it performs quite well on a number of simple KDD benchmark data (see next section). This problem can be taken care of by adopting SVM's solution(s) to this problem [2] [5], for instance introducing a parameter C to control the amount of allowed training data misclassifications.

6 Empirical Validation

We have chosen to compare experimentally the Maxsim formula and our more refined SBC algorithm with Chang and Lin's implementation of Support Vector Machines, LIBSVM (version 2.33) [3]. It is based on the formulation of the quadratic programming optimization that was our basis for a theoretical comparison, but also it is an excellent implementation of SVM which won supervised learning competitions, it also provides software for multi-class, density function and other extensions of the SVM technology. We have chosen data sets which are benchmarks in machine learning for classification and were found on the Silicon Graphics site http://www.sgi.com/tech/mlcdb/index.html, which contains a wealth of information about the origin, format, uses of these sets, and also accuracy comparisons of various machine learning algorithms. It is worth mentioning that SVM edged out the other methods in most of the tests sets we used, which is the reason we only considered SVM for comparison. Also, the reported results in each case are the best we could obtain by trying different sets of parameters.
We provide here only a summary, as a complete and detailed treatment including implementation issues and an analysis of the statistical significance of the differences in accuracy is to be found in a forthcoming paper.
Surprisingly, the best fit between the Maxsim formula and the SVM was found for small data sets: they achieved the same accuracy, up to two digits: 96.67% for the wine recognition data; for the Pima Indians diabetes data, they both achieved 75%. Slight differences were found for the liver disorder data: for polynomial kernels the SVM achieved 73.91% accuracy, while the Maxsim formula achieved 77.39%, while with RBF kernels, the SVM achieved 79% and the Maxsim formula 77%. Similarly, for the satellite image recognition data, the SVM achieved 92% to 91% for Maxsim. For the letter recognition problem, Maxsim achieved 96.4% and SVM achieved 94.8% with polynomial kernels and 95.6% and SVM 96.7%, with RBF.

The SBC gave either exactly the same results as the SVM, or slightly better.
We also have experimented with the classification of biological signals [11]. It is known that present day data bases of genomic and proteomic data contain a large number of errors (sometimes up to 25%), or hypothetic information solely verified by visual inspection of patterns. In order to avoid interference of noisy or unreliable data with our comparison, we have used data whose quality has been experimentally confirmed. We performed two experiments: splice site detection with the data set obtained from [14] and E. Coli start sites prediction with the dataset obtained from [9] and stored in ECOGENE [12]. We found out by cross validation that Maxsim, SVM and SBC give excellent results in the range of 92% to 100% depending on the sets, Maxsim was consistently slightly lower or equal to SVM, while the SVM was consistently slightly lower or equal to SBC. However for more noisy problems, the SBC has to be implemented with the same considerations as the SVM; that is with a C parameter which controls noise and avoids overfitting.
Finally, we also compared the SVM and Maxsim when given a large multi-class problem. For this we derived a problem to recognize two-letter words. That is, we have now 676 classes instead of 26 (as in the original letter-recognition problem). SVM took 17'01" to run, while Maxsim took 42" ; this difference is due to the quadratic complexity of both training and testing phases of the SVM on the number of classes.

7 Conclusion

We have described a general framework for the study of kernel-based machine learning classification algorithms in a simple and systematic way. We also have shown that these algorithms converge to the Maxsim formula (4.3), as the tightness increases. In practice, the Maxsim formula seems to give another solution 'in the middle' with the added capability of handling multi-class problems in a natural way. The SBC algorithm, described in section 5, also handles multi-class problems naturally and more importantly it gives a solution which is optimal. The experiments confirmed this, as the performance of SBC was at least as good as the SVM in all cases. We believe our main contribution is that we accomplished SVM-like accuracy numbers while maintaining a transparent and efficient handling of multi-class problems.

References

1. MPS Brown WN Grundy, D Lin, N Cristianini, CW Sugnet, TS Furey, M Ares, D Haussler.: Knowledge-based analysis of microarray gene expression data by using support vector machines. Proceedings of the National Academy of Sciences 97, 262–267, 2000.
2. Burges, C. J. C.: A tutorial on support vector machines for pattern recognition. Data Mining and Knowledge Discovery, 2(2):121–167. 1998
3. Chang C. C., and C. J. Lin.: Training ν-support vector classifiers, Theory and Algorithms. Neural Computation (13) 9 2119–2147, 2001.

4. Crisp D.J., and C.J.C. Burges 2000. In: S. Solla, T. Leen and K.R. Muller (Eds). A Geometric Interpretation of N=bsvm Classifiers. Advances in Neural Information Processing Systems Vol 12 Cambridge MA MIT Press, 2000.
5. N. Cristianini and J. Shawe-Taylor.: An Introduction to Support Vector Machines. Cambridge University Press, 2000.
6. Freund Y., and R.E. Schapire.: Large Margin Classification using the Perceptron Algorithm. Machine Learning (37) 3 277–296, 1999
7. T. Joachims.: Learning to Classify Text using Support Vector Machines.: Kluwer, 2002
8. S.S. Keerthi, S.K. Shevade, C. Bhattacharyya and K.R.K. Murthy.: A fast iterative nearest point algorithm for support vector machine classifier design. IEEE Transactions on Neural Networks, Vol. 11, pp.124–136, Jan 2000.
9. Link, A.J., Robinson, K. and Church, G.M.: Comparing the predicted and observed properties of proteins encoded in the genome of Escherichia Coli. Electrophoresis., 18, 1259–1313, 1997
10. K. G. Murty.: Linear Complementarity, Linear and Nonlinear Programming.: Helderman-Verlag, 1988.
11. Martin G. Reese, Frank H. Eeckman, David Kulp, David Haussler.: Improved Splice Site Detection in Genie. RECOMB. Santa Fe, ed. M. Waterman, 1997.
12. Rudd, K. E.: Ecogene: a genome sequence database for Escherichia Coli K-12. Nucleic Acid Research, 28, 60–64, 2000
13. B Scholkopf, A Smola.: Learning with Kernels. MIT Press, 2001
14. T.A. Thanaraj.: A clean data set of EST-confirmed splice sites from Homo sapiens amnd standards for clean-up procedures. Nucleic Acids Research. Vol 27, No. 13 2627–2637, 1999

Numerical Attributes in Decision Trees: A Hierarchical Approach

Fernando Berzal, Juan-Carlos Cubero, Nicolás Marín, and Daniel Sánchez

IDBIS Research Group – Dept. of Computer Science and A.I.,
E.T.S.I.I. – University of Granada, 18071, Granada, Spain
{fberzal, jc.cubero, nicm, daniel}@decsai.ugr.es

Abstract. Decision trees are probably the most popular and commonly-used classification model. They are recursively built following a top-down approach (from general concepts to particular examples) by repeated splits of the training dataset. When this dataset contains numerical attributes, binary splits are usually performed by choosing the threshold value which minimizes the impurity measure used as splitting criterion (e.g. C4.5 gain ratio criterion or CART Gini's index). In this paper we propose the use of multi-way splits for continuous attributes in order to reduce the tree complexity without decreasing classification accuracy. This can be done by intertwining a hierarchical clustering algorithm with the usual greedy decision tree learning.

1 Introduction

It is well-known [6] [17] that decision trees are probably the most popular classification model. Decision trees are usually built following a top-down approach, from general concepts to particular examples. That is the reason why the acronym TDIDT, which stands for Top-Down Induction of Decision Trees, is used to refer to this kind of algorithms.

The TDIDT algorithm family includes classical algorithms, such as CLS (Concept Learning System), ID3 [16], C4.5 [18] and CART (Classification And Regression Trees) [2], as well as more recent ones, such as SLIQ [15], SPRINT [22], QUEST [12], PUBLIC [21], RainForest [7], BOAT [5], and even ART [1].

Some of the above algorithms build binary trees, while others induce multi-way decision trees. However, when working with numerical attributes, most TDIDT algorithms choose a threshold value in order to perform binary tests, unless the numerical attributes are discretized beforehand.

Our paper is organized as follows. Section 2 introduces the binary splits which are commonly employed to branch decision trees when continuous attributes are present. Section 3 describes how to build multi-way decision trees using a hierarchical clustering algorithm. In Section 4, we present some empirical results we have obtained by applying our alternative approach to build decision trees. Finally, some conclusions and pointers to future work are given in Section 5.

M.R. Berthold et al. (Eds.): IDA 2003, LNCS 2810, pp. 198–207, 2003.

2 Binary Splits for Numerical Attributes

The splitting criteria used in TDIDT algorithms, such as C4.5 gain ratio or CART Gini's index, provide a mechanism for ranking alternative partitions of the training set when building decision trees. Given a splitting criterion we will use to rank alternative partitions, we just enumerate some possible partitions of the training set.

Most TDIDT algorithms define a template of possible tests so that it is possible to examine all the tests which match with the template. Those tests usually involve a single attribute because it makes the tree nodes easier to understand and sidesteps the combinatorial explosion that results if multiple attributes can appear in a single test [18].

When an attribute A is continuous, a binary test is usually performed. This test compares the value of A against a threshold t: $A \leq t$. Sorting the values of A we obtain $\{v_1, v_2..v_n\}$. Any threshold between v_i and v_{i+1} will have the same effect when dividing the training set, so that you just have to check $n-1$ possible thresholds for each numerical attribute A.

Once you can determine that the best possible threshold is between v_i and v_{i+1}, you must choose an accurate value for the threshold to be used in the decision tree. Most algorithms choose the midpoint of the interval $[v_i, v_{i+1}]$ as the actual threshold, although C4.5, for instance, chooses the largest value of A in the entire training set that does not exceed the above interval midpoint. C4.5 approach ensures that any threshold value used in the decision tree actually appears in the data, although it could be misleading if the value sample in the training dataset is not representative.

We prefer a slightly different approach which consists of choosing a threshold between v_i and v_{i+1} depending on the number of training instances below and above the threshold. If there are L instances with $v \leq v_i$, and R examples $v \geq v_{i+1}$, then the threshold t_i is

$$t_i = \frac{R * v_i + L * v_{i+1}}{L + R}$$

This way, the threshold will be nearer to v_i than to v_{i+1} if there are less training instances falling in the left branch of the tree, which is the branch corresponding to the examples with $v \leq v_i$. Similarly, the threshold will be nearer to v_{i+1} if the majority of the training examples have values of A below or equal to v_i.

Although this slight modification to the usual algorithm will usually not yield any classifier accuracy improvement, we find that the decision trees it produces are more appealing, specially when the resulting trees are quite unbalanced (which, on the other hand, is a common situation when continuous attributes are present).

3 Multi-way Splits for Numerical Attributes

If we had only categorical attributes, we could use any C4.5-like algorithm in order to obtain a multi-way decision tree, although we would usually obtain a binary tree if our dataset included continuous attributes.

Using binary splits on numerical attributes implies that the attributes involved could appear several times in the paths from the root of the tree to its leaves. Although these repetitions can be simplified when converting the decision tree into a set of rules, they make the constructed tree more leafy, unnecessarily deeper, and harder to understand for human experts. As stated in [19], "non-binary splits on continuous attributes make the trees easier to understand and also seem to lead to more accurate trees in some domains".

We could group all numeric attribute values before building the decision tree (a.k.a. global discretization). Numerical attributes would then be treated as if they were categorical. However, it would also be desirable to find some way to cluster the numeric values of continuous attributes depending on the particular situation (i.e. local discretization), since the most adequate intervals could vary among the different nodes in the decision tree.

3.1 Classical Value Clustering

Clustering techniques can be used to group numerical values into suitable intervals. Given a set of points in a high-dimensional space, clustering algorithms try to group those points into a small number of clusters, each cluster containing similar points in some sense. Good introductions to this topic can be found in [8] and [23].

As TDIDT algorithms, most clustering methods are heuristic, in the sense that local decisions are made which may or may not lead to optimal solutions (i.e. good solutions are usually achieved although no optimum is guaranteed). A great number of clustering algorithms have been proposed in the literature, most of them from the Pattern Recognition research field. In general, we can classify clustering algorithms into centroid-based methods and hierarchical algorithms.

- Centroid-based clustering algorithms characterize clusters by central points (a.k.a. centroids) and assign patterns to the cluster of their nearest centroid. The k-Means algorithm is the most popular algorithm of this class, and also one of the simplest.
- Hierarchical clustering methods "produce a nested data set in which pairs of items or clusters are successively linked until every item in the data set is connected" [20]. Agglomerative (a.k.a. merging or bottom-up) hierarchical clustering methods begin with a cluster for each pattern and merge nearby clusters until some stopping criterion is met. Alternatively, divisive (a.k.a. splitting or top-down) hierarchical clustering methods begin with a unique cluster which is split until the stopping criterion is met. In this paper we propose a hierarchical algorithm in order to build multi-way decision trees using numeric attributes, although we use a novel approach to decide how to cluster data.

3.2 Discretization Techniques

Splitting a continuous attribute into a set of adjacent intervals, also known as discretization, is a well-known enabling technique to process numerical attributes in Machine Learning algorithms [11]. It not only provides more compact and simpler models which are easier to understand, compare, use, and explain; but also makes learning more accurate and faster [3]. In other words, discretization extends the borders of many learning algorithms [11].

Discretization methods used as an aid to solve classification tasks are usually univariate (for the sake of efficiency) and can be categorized according to several dimensions:

- **Supervised vs. unsupervised** (depending on their use of class information to cluster data). Supervised methods employ the class information to evaluate and choose the cut-points, while unsupervised methods do not.
- **Local (dynamic) vs. global (static).** In a global method, discretization is performed beforehand. Local methods discretize continuous attributes in a localized region of the instance space (as in C4.5) and their discretization of a given attribute may not be unique for the entire instance space.

The simplest discretization methods create a specified number of bins (e.g. equiwidth and equidepth). Supervised variants of those algorithms include Holte's One Rule Discretizer [10].

Information Theory also provides some methods to discretize data. Fayyad and Irani's multi-interval discretization based on Rissanen's Minimum Description Length Principle (MDLP) is a prominent example [4]. Mantaras Distance can be used as a variation of the original Fayyad and Irani's proposal [13] and Van de Merckt's contrast measure provides another alternative [24].

Other approaches include, for example, Zeta [9], which employs a measure of strength of association between nominal values: the maximum achievable accuracy when each value of a feature predicts a different class value. Good surveys and categorizations of discretization techniques can be found in [3] and [11].

3.3 A General-Purpose Supervised Discretization Technique Based on Hierarchical Clustering Algorithms

We can also formulate a clustering algorithm which will help us to group numeric values into intervals, as discretization techniques do. In a supervised setting, those intervals will then be used to build multi-way decision trees.

Supervised discretization techniques (see Section 3.2) do take class information into account, although they usually focus on some kind of purity measure associated to each interval resulting from the discretization process. We prefer to measure the similarity between adjacent intervals in order to choose the appropriate cut-points which delimit them.

Clustering algorithms usually compute the distance between patterns using their feature space. Using a hierarchical approach, we need to introduce a criterion to decide which pair of adjacent values or intervals to combine (if we follow a bottom-up approach) or where to split a given interval (if we decide to employ a top-down algorithm). Here we will use the distance between class distributions corresponding to adjacent values.

Given a set of adjacent intervals $I_1, I_2 .. I_n$ for attribute A, we will characterize each one of them with the class distribution of the training examples it covers. If there are J different problem classes, each interval I will have an associated characteristic vector $V_I = (p_1, p_2, .., p_J)$, where p_j is the probability of the jth class in the training examples which have their A value included in the interval I.

Let us consider, for example, a three-class classification problem. Figure 1 shows how the most similar pair of adjacent class distributions are merged to discretize a numerical attribute into a set of intervals.

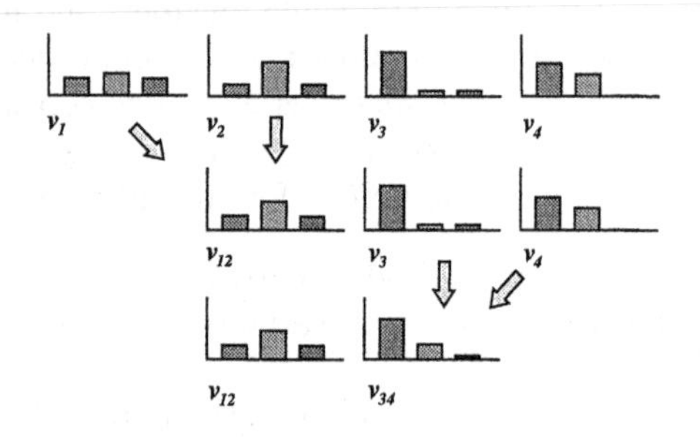

Fig. 1. Iterative grouping of adjacent intervals.

It should be noted that only two distances must be recomputed in each iteration of our algorithm. Once v_i is merged with v_{i+1}, the resulting class distribution $V_{i,i+1}$ is computed. We just need to evaluate the similarity between $V_{i,i+1}$ and its neighbours (V_{i-1} and V_{i+2}) because the other class distributions and similarity measurements remain unchanged.

Interleaving the hierarchical clustering algorithm with the TDIDT evaluation process. The hierarchical clustering algorithm described in the previous paragraphs can be intertwined with the evaluation of alternative training set partitions during decision tree construction in order to build a multi-way decision tree where the decision tree branching factor is not established beforehand.

Each time we merge two adjacent intervals, we can evaluate the decision tree we would build using the current set of intervals to branch the tree. The resulting algorithm can be expressed as shown in Figure 2

Although traditional hierarchical agglomerative clustering methods (HACM) are usually $O(N^2)$ in time and $O(N)$ in space, our special-purpose HACM is only $O(N \log N)$ in time.

1. Create one interval per attribute value in the training dataset.
2. Identify the two closest adjacent intervals and combine them (according to their class distributions).
3. Evaluate the decision tree which would result from using the current set of intervals to branch the tree. If it is the best split known so far, remember it.
4. If more than two interval remain, return to step 2.

Fig. 2. Context discretization interleaved within the TDIDT process.

Alternatively, we could have used a divisive top-down hierarchical clustering algorithm starting from a unique interval covering the whole instance space and splitting it in each iteration. A top-down approach would be less sensitive to noise in the training set than the agglomerative method, although the noise effect could also be reduced using a standard binning discretization algorithm beforehand (such as equidepth).

4 Experimental Results

Table 1 summarizes the results we have obtained building decision trees using different local discretization methods for 16 standard small and medium-sized datasets, most of them from the UCI Machine Learning Repository[1]. All the results reported in this section were obtained using 10-CV (ten-fold cross-validation), Quinlan's gain ratio criterion, and pessimistic pruning (with $CF = 0.25$), as described in [18].

The standard binary-split C4.5 algorithm is compared with several multi-way techniques: our contextual discretizer, three previous supervised discretizers (Holte's One Rule, Fayyad and Irani's MDLP, and Ho and Scott's Zeta), and three standard unsupervised discretization methods (the ubiquitous K-Means and two binning algorithms: Equiwidth and Equidepth). In order to improve our algorithm robustness in the presence of noise, we used an agglomerative clustering algorithm preceded by Equidepth binning (using 25 bins as a starting point for our hierarchical algorithm).

Table 1 shows that our method is competitive in terms of classification accuracy and it tends to build small decision trees. Regarding to the tree size, our method obtains trees which contain 84% of the nodes required by the corresponding binary tree on average and it is only outperformed by Fayyad and Irani's MDLP discretizer. As previous studies have suggested and Table 1 confirms, global discretization improves TDIDT efficiency, reduces tree complexity, and maintains classification accuracy.

[1] `http://www.ics.uci.edu/~mlearn/MLRepository.html`

Table 1. Discretization experiments summary

	C4.5	Context	OneRule	MDLP	Zeta	KMeans	Equiwidth	Equidepth
LOCAL DISCRETIZATION								
CV Accuracy	82.00%	82.04%	76.90%	81.44%	74.87%	81.53%	81.06%	81.01%
Training time (seconds)	6.93	12.17	7.30	7.92	15.53	20.82	8.61	8.26
Tree size	110.9	83.8	96.1	42.5	116.8	84.3	93.5	89.0
- Internal nodes	49.9	32.1	21.8	15.8	15.3	24.0	29.0	27.0
- Leaves	60.9	51.7	74.3	26.7	101.6	60.3	64.5	62.0
Average tree depth	3.91	3.40	2.37	3.02	1.84	2.80	4.43	2.70
GLOBAL DISCRETIZATION								
CV Accuracy	-	81.92%	77.24%	82.40%	74.91%	81.25%	79.24%	81.16%
Training time (seconds)	-	1.62	6.30	1.02	10.54	1.03	1.05	1.02
Tree size	-	70.7	77.6	62.7	92.1	85.5	75.4	82.4
- Internal nodes	-	21.8	10.1	20.8	9.0	24.4	23.2	24.1
- Leaves	-	48.9	67.4	41.8	83.0	61.1	52.2	58.2
Average tree depth	-	2.76	1.70	3.03	1.55	2.80	4.09	2.60

Table 2. Classifier accuracy (local discretization)

Dataset	C4.5	Context	OneRule	MDLP	Zeta	KMeans	Equiwidth	Equidepth
ADULT	82.60%	82.58%	77.10%	85.27%	80.56%	84.73%	82.92%	82.42%
AUSTRALIAN	85.65%	84.93%	85.80%	85.07%	84.78%	85.51%	84.64%	85.65%
BREAST	93.99%	93.84%	94.85%	94.28%	95.57%	95.28%	95.28%	94.27%
BUPA	66.10%	67.83%	59.76%	57.71%	57.71%	64.71%	62.96%	66.72%
CAR	92.88%	92.88%	92.88%	92.88%	92.88%	92.88%	92.88%	92.88%
GLASS	70.06%	68.77%	59.70%	66.34%	49.46%	70.58%	67.38%	63.46%
HAYESROTH	73.75%	73.75%	73.75%	73.75%	73.75%	73.75%	73.75%	73.75%
HEART	77.04%	76.67%	74.07%	76.67%	73.33%	78.89%	78.52%	79.63%
IONOSPHERE	90.60%	90.32%	87.21%	91.16%	79.76%	88.31%	89.17%	88.60%
IRIS	94.00%	93.33%	94.00%	93.33%	93.33%	95.33%	94.67%	94.67%
PIMA	74.85%	76.81%	68.74%	74.72%	58.19%	74.20%	76.03%	72.76%
SPAMBASE	90.52%	90.24%	87.09%	92.00%	87.83%	91.68%	82.22%	92.11%
THYROID	96.18%	95.61%	96.46%	97.29%	96.04%	95.32%	96.04%	94.54%
WAVEFORM	76.80%	76.50%	58.86%	76.80%	55.36%	74.38%	75.74%	74.58%
WINE	92.71%	91.57%	75.07%	90.92%	74.51%	87.09%	92.68%	88.20%
YEAST	54.25%	56.94%	45.15%	54.79%	44.81%	51.76%	52.09%	51.89%
Average	82.00%	82.04%	76.90%	81.44%	74.87%	81.53%	81.06%	81.01%
Win-Loss-Tie (1%)		3-2-11	0-10-2	3-3-10	1-10-5	5-5-6	3-7-6	2-7-7

10-CV accuracy results are displayed in Table 2. Multi-way decision trees for numerical attributes improved TDIDT accuracy in some experiments. The accuracy results are more relevant when we take into account that the resulting tree complexity diminishes when multi-way splits are used. The tree size (internal nodes plus leaves) and the average tree depth are reduced. Although it is commonly assumed that binary decision trees have less leaves but are deeper than multi-way trees [14], we have found that this fact does not hold (at least after pruning) since the pruned binary tree can have more leaves than its multi-way counterpart as Table 1 shows for our discretizer and MDLP.

5 Conclusions

In this paper we have proposed an alternative way of handling numerical attributes when building multi-way decision trees so that the resulting trees tend to be smaller while their accuracy is preserved. Since our technique is ortoghonal to the underlying TDIDT implementation, the overall TDIDT asymptotic complexity is preserved and, therefore, our approach is applicable to data mining problems where decision trees play an important role [6].

Our algorithm, like Fayyad and Irani's MDLP, is remarkable because it dramatically reduces tree complexity without hurting classifier accuracy at a reasonable cost. Our method can also be used as an alternative discretization technique to be added to the computer scientist toolkit, since it has proven to be efficient and suitable for both local and global discretization.

References

1. Berzal, F., Cubero, J.C., Sánchez, D., and Serrano, J.M. (2003). *ART: A hybrid classification model.* Machine Learning, to be published.
2. Breiman, L., Friedman, J. H., Olshen, R. A., and Stone, C. J. (1984). *Classification and Regression Trees.* Wadsworth, California, USA, 1984. ISBN 0-534-98054-6.
3. Dougherty, J., Kohavi, R., and Sahami, M. (1995). *Supervised and unsupervised discretization of continuous features.* Proceedings of the 12th International Conference on Machine Learning, Los Altos, CA, Morgan Kaufmann, 1995, pp. 194–202.
4. Fayyad, U.M., and Irani, K.B. (1993). *Multi-interval discretization of continuous-valued attributes for classification learning.* Proceedings of the 13th International Joint Conference on Artificial Intelligence, pp. 1022–1027.
5. Gehrke, J., Ganti, V., Ramakrishnan, R., and Loh, W.-Y. (1999a). *BOAT - Optimistic Decision Tree Construction.* Proceedings of the 1999 ACM SIGMOD international conference on Management of Data, May 31 – June 3, 1999, Philadelphia, PA USA, pp. 169–180
6. Gehrke, J., Loh, W.-Y., and Ramakrishnan, R. (1999b). *Classification and regression: money can grow on trees.* Tutorial notes, KDD'99, August 15–18, 1999, San Diego, California, USA, pp. 1–73
7. Gehrke, J, Ramakrishnan, R., and Ganti, V. (2000). *RainForest – A Framework for Fast Decision Tree Construction of Large Datasets.* Data Mining and Knowledge Discovery, Volume 4, Numbers 2/3, July 2000, pp. 127–162

8. Han, J., and Kamber, M. (2001). *Data Mining. Concepts and Techniques.* Morgan Kaufmann, USA, 2001. ISBN 1-55860-489-8.
9. Ho, K.M., and Scott, P.D. (1997). *Zeta: A global method for discretization of continuous variables.* 3rd International Conferemce on Knowledge Discovery and Data Mining (KDD'97), Newport Beach, CA, AAAI Press, pp. 191–194.
10. Holte, R.C. (1993). *Very simple classification rules perform well on most commonly used datasets.* Machine Learning, 11:63–90.
11. Hussain, F., Liu, H., Tan, C.L., and Dash, M. (1999). *Discretization: An enabling technique.* The National University of Singapore, School of Computing, TRC6/99, June 1999.
12. Loh, W.-Y., and Shih, Y.-S. (1997). *Split Selection Methods for Classification Trees.* Statistica Sinica, Vol.7, 1997, pp. 815–840
13. Lopez de Mantaras, R. (1991). *A Distance-Based Attribute Selection Measure for Decision Tree Induction.* Machine Learning, 6, pp. 81–92.
14. Martin, J. K. (1997). *An Exact Probability Metric for Decision Tree Splitting and Stopping.* Machine Learning, 28, pp. 257–291.
15. Mehta, M., Agrawal, R., and Rissanen, J. (1996). *SLIQ: A Fast Scalable Classifier for Data Mining.* EDBT'96, Avignon, France, March 25–29, 1996, pp. 18–32
16. Quinlan, J.R. (1986a). *Induction on Decision Trees.* Machine Learning, 1, 1986, pp. 81–106
17. Quinlan, J.R. (1986b). *Learning Decision Tree Classifiers.* ACM Computing Surveys, 28:1, March 1986, pp. 71–72
18. Quinlan, J.R. (1993). *C4.5: Programs for Machine Learning.* Morgan Kaufmann, 1993. ISBN 1-55860-238-0.
19. Quinlan, J.R. (1996). *Improved use of continuous attributes in C4.5.* Journal of Artificial Intelligence Research, 4:77–90.
20. Rasmussen, E. (1992). *Clustering algorithms.* In W. Frakes and E. Baeza-Yates (eds.): *Information Retrieval: Data Structures and Algorithms*, Chapter 16.
21. Rastogi, R., and Shim, K. (2000). *PUBLIC: A Decision Tree Classifier that integrates building and pruning.* Data Mining and Knwoledge Discovery, Volume 4, Number 4, October 2000, pp. 315–344
22. Shafer, J.C., Agrawal, R., and Mehta, M. (1996). *SPRINT: A Scalable Parallel Classifier for Data Mining.* VLDB'96, September 3–6, 1996, Mumbai (Bombay), India, pp. 544–555
23. Ullman, J. (2000). *Data Mining Lecture Notes.* Stanford University CS345 Course on Data Mining, Spring 2000.
http://www-db.stanford.edu/~ullman/mining/mining.html
24. Van de Merckt, T. (1993). *Decision trees in numerical attribute spaces.* Proceedings of the 13th International Joint Conference on Artificial Intelligence, pp. 1016–1021.

Similarity-Based Neural Networks for Applications in Computational Molecular Biology

Igor Fischer

Wilhelm-Schickard-Institut für Informatik, Universität Tübingen,
Sand 1, 72076 Tübingen, Germany
fischer@informatik.uni-tuebingen.de

Abstract. This paper presents an alternative to distance-based neural networks. A distance measure is the underlying property on which many neural models rely, for example self-organizing maps or neural gas. However, a distance measure implies some requirements on the data which are not always easy to satisfy in practice. This paper shows that a weaker measure, the similarity measure, is sufficient in many cases. As an example, similarity-based networks for strings are presented. Although a metric can also be defined on strings, similarity is the established measure in string-intensive research, like computational molecular biology. Similarity-based neural networks process data based on the same criteria as other tools for analyzing DNA or amino-acid sequences.

1 Introduction

In respect to underlying mathematical properties, most artificial neural networks used today can be classified as scalar product-based or distance-based. Of these, multi-layer perceptrons and LVQ [1] are typical representatives.

In distance-based models, each neuron is assigned a pattern to which it is sensitive. Appearance of the same or a similar pattern on the input results in a high activation of that neuron — similarity being here understood as the opposite of distance.

For numerical data, distance-based neural networks can easily be defined, for there is a wide choice of distance measures, the Euclidean distance being certainly the best known. In some applications, however, the data cannot be represented as numbers or vectors. Although it may sometimes still be possible to define a distance on them, such a measure is not always natural, in the sense that it well represents relationships between data.

One such example are symbol strings, like DNA or amino-acid sequences which are often subject to research in computational molecular biology. There, a different measure – similarity – is usually used. It takes into account mutability of symbols, which is determined through complex observations on many biologically close sequences. To process such sequences with neural networks, it is preferable to use a measure which is well empirically founded.

M.R. Berthold et al. (Eds.): IDA 2003, LNCS 2810, pp. 208–218, 2003.

This paper discusses the possibility of defining neural networks to rely on similarity instead of distance and shows examples of such networks for symbol strings. Some results of processing artificial and natural data are presented below.

2 Distance and Similarity

Distance is usually, but not necessarily, defined on a vector space. For $\boldsymbol{x}, \boldsymbol{y}, \boldsymbol{z} \in \mathcal{X}$, any function $d : \mathcal{X}^2 \rightarrow R$ fulfilling the following properties is a distance measure on $\mathcal{X}$:

1. $d(\boldsymbol{x}, \boldsymbol{y}) \geq 0$
2. $d(\boldsymbol{x}, \boldsymbol{y}) = 0 \Leftrightarrow x = y$
3. $d(\boldsymbol{x}, \boldsymbol{y}) = d(\boldsymbol{y}, \boldsymbol{x})$
4. $d(\boldsymbol{x}, \boldsymbol{y}) + d(\boldsymbol{y}, \boldsymbol{z}) \geq d(\boldsymbol{x}, \boldsymbol{z})$

For strings, one such measure is the Levenshtein distance [2], also known as edit distance, which is the minimum number of basic edit operations - insertions, deletions and replacements of a symbol - needed to transform one string into another. Edit operations can be given different costs, depending on the operation and the symbols involved. Such weighted Levenshtein distance can, depending on the chosen weighting, cease to be distance in the above sense of the word.

Another measure for quantifying how much two strings differ is "feature distance" [3]. Each string is assigned a collection of its substrings of a fixed length. The substrings – the features – are typically two or three symbols long. The feature distance is then the number of features in which two strings differ. It should be noted that this measure is not really a distance, for different strings can have a zero distance: consider, for example, strings `AABA` and `ABAA`. Nevertheless, feature distance has a practical advantage over the Levenshtein by being much easier to compute.

A similarity measure is simpler than distance. Any function $s : \mathcal{X}^2 \rightarrow R$ can be declared similarity – the question is only if it reflects the natural relationship between data. In practice, such functions are often symmetrical and assign a higher value to two identical elements than to distinct ones, but this is not required.

For strings, similarity is closely related to alignment.

> A (global) alignment of two strings S1 and S2 is obtained by first inserting chosen spaces (or dashes), either into or at the ends of S1 and S2, and then placing the two resulting strings one above the other so that every character or space in either string is opposite a unique character or a unique space in the other string. [4]

The spaces (or dashes) are special symbols (not from the alphabet over which the strings are defined) used to mark positions in a string where the symbol from the other string is not aligned with any symbol. For example, for strings `AABCE` and `ABCD`,

```
AABC-E
-ABCD-
```

is an alignment, not necessarily optimal. Each alignment can be assigned a score according to certain rules. In most simple cases, a similarity score is assigned to each pair of symbols in the alphabet, as well as for pairs of a symbol and a space. The score for two aligned strings is computed as the sum of similarity scores of their aligned symbols. Such similarity measure is called "additive". There are also more complex schemes, for example charging contiguous spaces less then isolated. The similarity of the strings is defined as the score of their highest-scoring alignment.

In computational molecular biology, similarity is most often computed for DNA or amino-acid sequences (sequence and string are used as synonyms here), where similarity between symbols is established empirically to reflect observed mutability/stability of symbols. Because each pair of symbols can have a different similarity and no obvious regularity exists, similarities are stored in look-up tables, which have the form of a quadratic matrix. Among scoring matrices, the PAM (point accepted mutations) [5] and BLOSUM (block substitution matrix) [6] families are the most often used.

It is intuitively clear that distance and similarity are somehow related, but quantifying the relationship is not always straightforward. For additive measures, if the similarity of each symbol with itself is the same, it can be established by a simple equation [7]:

$$\langle \mathbf{s}_1 | \mathbf{s}_2 \rangle + d(\mathbf{s}_1, \mathbf{s}_2) = \frac{M}{2} \cdot (|\mathbf{s}_1| + |\mathbf{s}_2|). \tag{1}$$

The notation $\langle \mathbf{s}_1 | \mathbf{s}_2 \rangle$ is used to symbolize the optimal alignment score of strings $\mathbf{s}_1$ and $\mathbf{s}_2$, $|\mathbf{s}|$ the string length, and M is an arbitrary constant. The distance $d(\mathbf{s}_1, \mathbf{s}_2)$ is defined as the minimal cost for transforming one string into the other. The cost is the sum of the symbol replacement costs $c(\alpha_i, \beta_i)$, α_i and β_i being the symbols comprising $\mathbf{s}_1$ and $\mathbf{s}_2$, and the insertion/deletion costs, each of the operations costing some $h > 0$ per symbol. These values are related to the values comprising the string similarity:

$$p(\alpha, \beta) = M - c(\alpha, \beta) \quad \text{and}$$
$$g = \frac{M}{2} - h. \tag{2}$$

where $p(\alpha, \beta)$ is the similarity score for the symbols α and β and g is the value of space in the alignment (usually a negative one). For the condition (2) on the distance to be satisfied, $c(\alpha, \alpha)$ must be always zero, that is, $p(\alpha, \alpha)$ must be constant for all α. Unfortunately, the above mentioned PAM and BLOSUM scoring matrices do not satisfy this condition.

Especially for searching large databases of protein and DNA sequences, sophisticated, fast heuristics like FASTA [8] and BLAST [9] have been developed. They have been fine-tuned not only for speed, but also for finding biologically meaningful results. For example, from an evolutionary point of view, a compact gap of 10 symbols in an aligned string is not twice as probable as a gap of 20 symbols at the same position. Thus, non-additive scoring schemes are applied. This is another reason why the above described method for computing the distance from similarity is not applicable.

A simple method for computing "distance" from similarity score for proteins was applied in [10]. For computing the score normalized scoring matrices with values scaled to $[0, 1]$ were used, and for spaces a fixed value of 0.1 were applied. The scores for all pairs of proteins from the data set were computed and ordered into a new matrix $\boldsymbol{S}$. The element $\boldsymbol{S}[i][j]$ was the similarity score for the i-th and j-th protein from the data set. This matrix was subsequently also scaled to $[0, 1]$. The distance between i-th and j-th protein was then computed as

$$\boldsymbol{D}[i][j] = 1 - \boldsymbol{S}[i][j]. \tag{3}$$

This approach has several disadvantages: First, the computational and storage overheads are obvious. In most applications pair-wise similarity scores of all data are not needed. Also, this method is not applicable for on-line algorithms, with data sets of unknown and maybe even infinite sizes. But more than that, it is not clear, if the above function is a distance at all. It is easy to see that simple scaling of the $\boldsymbol{S}$ matrix can lead to a situation where the requirement (2) for distance is violated. Such a case appears when the diagonal elements – standing for self-similarities of strings – are not all equal. A workaround, like attempt to scale the matrix row- or column-wise, so that the diagonal elements are all ones, would cause a violation of the symmetry relationship (3). Element $\boldsymbol{S}[i][j]$ would generally be scaled differently than $\boldsymbol{S}[j][i]$ so the two would not be equal any more. And finally, the triangle inequality – requirement (4) – is not guaranteed to be satisfied.

3 Distance-Based Neural Networks

As presented in the introduction, distance-based neural networks rely on distance for choosing the nearest neuron. However, in the training phase the distance is also involved, at least implicitly. All distance based neural networks known to the author try to place or adapt neurons in such a way that they serve as good prototypes of the input data for which they are sensitive. If the Euclidean distance is used, the natural prototype is the arithmetic mean. It can be expressed in terms of distance as the element having the smallest sum of squared distances over the data set. This property is helpful if operations like addition and division are not defined for the data, but a distance measure is provided.

Another kind of prototype is the median. For scalar data, the median is the middle element of the ordered set members. Like mean, it can be expressed through distance. It is easy to see that the median has the smallest sum of distances over the data set. Thus the concept of median can be generalized on arbitrary data for which a distance measure is defined.

Based on distance measures for strings and relying on the above mentioned properties of mean and median, neural networks for strings have been defined. Self organizing map and learning vector quantization – have been defined both in the batch [11] and on-line [12] form. As is often the case, the on-line version has been shown to be significantly faster than the batch form.

Finding the mean or the median of a set of strings is not quite easy. The first algorithms [13] performed an extensive search through many artificially generated strings in order to find the one with the smallest sum of (squared) distances. A later and much faster algorithm [12] used a by-product of computing the distance – the edit transcript – for finding the prototype string. The edit transcript is a list of operations needed to transform one string into another. It is not actually needed for computing the distance, but it can be directly deduced by backtracing through the so-called dynamic programming table, which is an intermediate result when computing Levenshtein or related distance. Starting from an approximation of the prototype, the edit operations needed for transforming it into every of the set strings were computed. Then, at each position, the edit operation appearing most frequently in all transcripts was applied, as long as it minimized the sum of (squared) distances. This is only a best effort approach and it can get stuck in a local optimum. Moreover, since strings are discrete structures, the mean or the median are not always unambiguous. For example, for the set of two strings, { `A`, `B` }, both `A` and `B` are equally good mean and median.

4 Similarity-Based Neural Networks

As mentioned in section 2, for some applications similarity is a more natural measure than distance. Taking the above discussed inconveniences that can appear when using distance, one could attempt to circumvent it altogether. If one were able to define a prototype based on similarity, a neural network could be described by an algorithm almost identical to the distance-based networks: one only needs to search for the neuron with the highest similarity instead of the one with the minimum distance. Of course, if the distance is defined in a way that properly reflects the reality, similarity can be defined as the negative distance and the two paradigms converge.

Computing the prototype based on similarity can indeed be performed, at least for strings. Recall that the similarity between two strings is computed from their optimal alignment by summing the similarity scores of aligned symbols. Three or more strings can also be aligned, in which case one speaks of a multiple alignment. Having computed the multiple alignment (let the question how to compute it efficiently be put aside for the moment), the prototype string $\bar{s}$ can be defined as the string maximizing the sum of similarities with all strings from the set:

$$\bar{s} : \sum_i \langle \bar{s} | s_i \rangle \geq \sum_i \langle s_k | s_i \rangle \quad \forall s_k \neq \bar{s}. \tag{4}$$

Not quite identical, but in practice also often used prototype is the "consensus string", defined as the string having at each position the most common symbol appearing at that position in all aligned strings.

5 Implementation Considerations

Many of the ideas presented here rely on approximate string matching algorithms. Efficient implementation of a neural network based on string similarity should take care of the following issues: computing the similarity, finding the most similar string in the given set and computing multiple alignment.

5.1 Computing the Similarity

Similarity is computed from aligned strings, and alignment can be found using dynamic programming. The simplest algorithms have quadratic time and space complexity. Both can be reduced under certain circumstances, but also increased if more complex scoring schemes are used (for literature cf. [4,7,14]).

An adaptation of the original algorithm, which includes a divide-and-conquer strategy [15], computes the alignment in linear space for the price of roughly doubling the computation time. If the alignment itself is not needed, but only the similarity score, the computation time can remain the same.

5.2 Finding the Most Similar String

Such a case appears in the search for the neuron most similar to the input string. A simple approach would be to go through all neurons and compute their similarity to the string, retaining the most similar. Alignments between neurons and the string are never needed. This algorithm can further be refined: not only alignment, but even exact similarity need not be computed for all neurons.

The idea is roughly the following: suppose one expects the most similar neuron to have the similarity score above a certain threshold. This expectation can be based on experience from previous iterations or on knowledge about the strings. Then one can go through neurons, start computing the similarity for each and break off the computation as soon as it becomes obvious that the score would be below the threshold. If the threshold was correct, the algorithm finds the most similar neuron by computing the similarity only on a fraction of them. Else, the threshold is reduced and the whole process repeated.

The above idea is implemented by computing only the similarity matrix elements in a limited band around its diagonal, thus reducing the complexity to $O(kn)$ [16], k being the band width and n the string length. The observed speed-up is about an order of magnitude.

5.3 Computing Multiple Alignments

Multiple alignment is considered the "holy grail" in computational molecular biology [4]. Exact computation can be performed by dynamic programming, but with time complexity exponential in the number of sequences. Therefore a number of approximate algorithms have been developed. In this work, the "star alignment" [17] is used, not only for its speed and simplicity, but also because it's close relatedness to the problem of finding a prototype string.

In star alignment, the multiple alignment is produced by starting from one string – the "star center" – and aligning it successively with other strings. Spaces inserted into the center in course of an alignment are retained for further alignments and simultaneously inserted into all previously aligned strings. That way, the previous alignments with the center are preserved.

The choice of the star center is not critical. At the beginning, one can use the string from the set that maximizes equation (4), and the current prototype in all consecutive iterations. Measurements presented below show that iterative search for the prototype is not very sensitive to the initialization.

6 Experimental Results

To show that computing the string prototype from a set of similar strings is quite robust to initialization, a series of experiments has been performed. In one, the prototype string was computed from a set of strings, all derived from an original random string by adding noise (insertions, deletions and replacements of random symbols), and starting from one of these strings. In repeated runs, keeping the noise level at 50%, the computed prototype converged to the original string in 97% of cases. Even increasing the noise level to 75% still resulted in 80% of prototypes converging to the original string.

In another type of experiment, everything was left the same, except that the algorithm started from a completely random string. Even then, a 89% convergence was observed in case of 50% noise and 62% in case of 75% noise.

Small-scale experiments on similarity-based self-organizing maps were performed on a set of 1750 words generated artificially by adding noise to 7 English words (Figure 1). Care was taken that no original words appear in the set. Even at the noise level of 75% and with random initialization, the resulting 8×12 map converged to a state with the original 7 words placed at distinct closed regions.

In a real-world test, a self-organizing map was tested on a set of 68 sequences from seven protein families. The 12×10 map was initialized by random sequences ordered along their Sammon projection [18]. The map was computed using BLOSUM62 similarity matrix and additive scoring. The map (Figure 2) showed all families distributed on it, although one of them (transducin) much weaker. On the class boundaries, unidentified "mutations" appeared.

Another real-world example is shown in figure 3. The mapping is produced from 320 hemoglobine alpha and beta chain sequences of different species, as used in [19].

vhmgzschebmopje usodrqvse ufralsd wcgolf misabnnhc macoiotics pxvfrjosplher macrohiotics
gsce ditance nlf dostqknrzi ide phiosvwhve dxstftwhdcke wosf
pmiloomsa udepasdg uicshance hcaka wpoqcrce cpqilwztpueu icie dstane
maobioyics distdanc ndtancw dzvnseeaqe hce mvwxctobqitjtcp viddselaged jiccte
wof ix ueawmkhobvfaik raixlewfa udrgcd rlwo of miwvny
mcg dosane lailwahy railea raiwdpw zalsay ij macrxojotycs
maccirobiopgos racka imstvonce wobf orfw ze dibisticef vpsltopfhwk
wxistancsj unerrgbv jhdivkhe isne gcg demoraytehd gollr tbdy
negkgezx tistnce pmhilzosynhh ucdfqaged phildsonphhwer yjilay mawhrowwictcs uhvilopjgscppe
fgilwaevn rwflwy uxmdaeeragedm rfaimuj vlwpay roln uxpneraed dqwsatcm
lilme pdioiopmheu qstauck railwnyz undvergck rkailwaazy zoavjfkyictvh bcimoht
lia zaauyv qe rilnwat woqw iyge underagned hlsy

wolf wolf wolf wlf mwc macobioics macrobiotics macrobiotics
wolf wolf wol wol moi maotis macrotics macrobiotics
wolf wolf wol wl ice ici maic mcrobics
wolf wlf ile ile ice ice ic ic
dnf woie ilce ice ice ic ic ic
distanc distac dine ice ice ie ie ie
distance distance disce dnce ice ie ie ie
distance distance ditace dice raly rai ri ut
distance distnce dtance railay raiay raiy raay undra
philophe philer paile raily railway railway unlrad undaged
philopher philosopher philher railwy railway railway underagd underaged
philopher philopher philose rilnway railway railway underage underaged

Fig. 1. A simple example of similarity string SOM. The mapping is produced from the set of English words corrupted by 75% noise. **Above:** Initial map with corrupted words randomly distributed over it. **Below:** The converged map after 300 iterations. The algorithm has extracted the original (prototypic) words from the noisy set and ordered them into distinct regions on the map. Artificial "transition" words appear on regions borders.

Table 1. LVQ classification of seven protein family samples, using 15 prototypes, on average two per class. $\boldsymbol{\mu}_{ij}$ denotes the j-th prototype of the i-th class and N_i is the number of elements from class i assigned to the prototype in the corresponding column. In this set, six sequences are incorrectly classified.

	μ_{11}	μ_{12}	μ_{13}	μ_{21}	μ_{22}	μ_{31}	μ_{41}	μ_{42}	μ_{51}	μ_{52}	μ_{61}	μ_{62}	μ_{63}	μ_{71}	μ_{72}
N_1	3	3	4												
N_2				9	2										
N_3					2	6									
N_4							4	5		1					
N_5									5	5					
N_6						1					7	1	1		
N_7				1				1						4	4

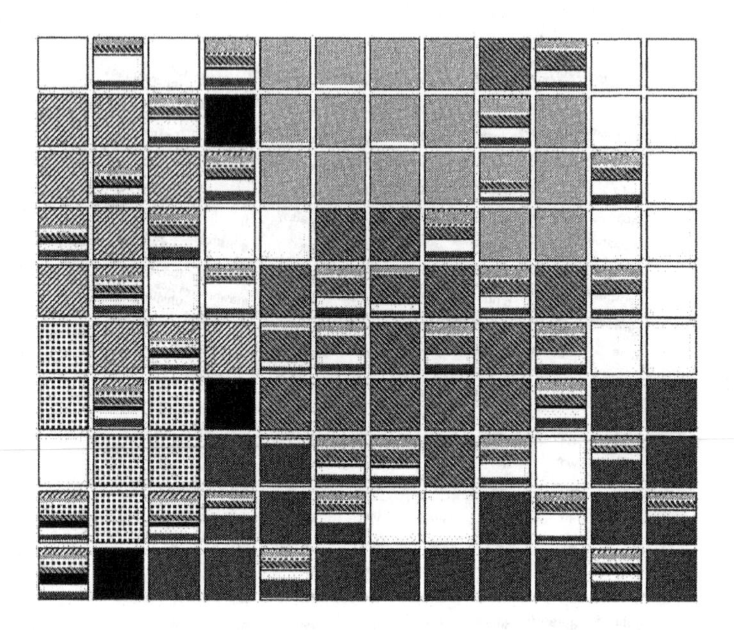

Fig. 2. Similarity string SOM of seven protein families. Each square represents a node in the map , and different families are represented by different fill patterns. The area filled by a pattern corresponds to the node similarity to the protein. All families are well expressed except transducin (the black squares).

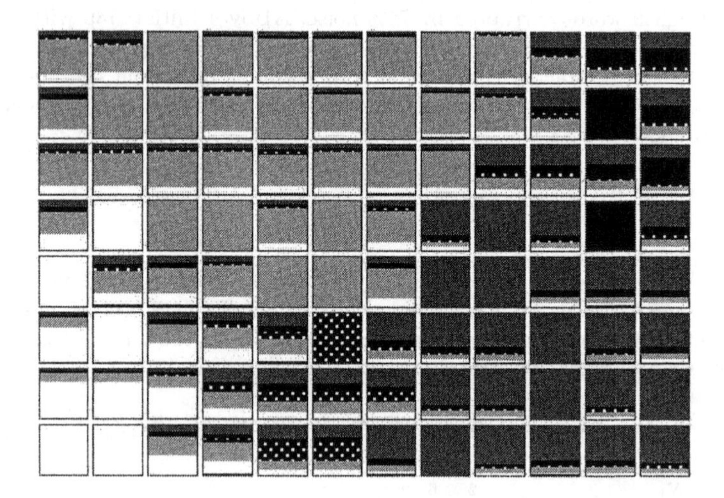

Fig. 3. Similarity string SOM of hemoglobine sequences. Two different chains, α (dark shades) and β (light shades), occupy distinct map areas. Moreover, weakly expresse subclusters can be recognized in each of the chains (white-dotted and black for α, ar white for β).

Attempting to classify the seven protein families by LVQ, using 15 prototype strings led to the results presented in Table 1. Six sequences (about 9% of the data set) were incorrectly classified. On the simpler hemoglobine data, perfect classification is obtained by already two prototypes.

7 Conclusion and Outlook

This paper shows that already a simple similarity measure, combined with an algorithm for its local maximization, is sufficient to define a large class of neural networks. A distance measure is not necessary. Even if it is defined, like, for example, for symbol strings, it is not always a 'good' measure. In some applications, like molecular biology, a similarity measure is more natural than distance and is preferred in comparing protein sequences. It is shown that such data can be successfully processed by similarity-based neural networks. It can therefore be concluded that similarity-based neural networks are a promising tool for processing and analyzing non-metric data.

In the presented work, simple, additive similarity measures for strings were used. Further experiments, for example embedding BLAST search engine, are in preparation.

References

1. Kohonen, T.: Learning vector quantization. Neural Networks **1** (1988) 303
2. Levenshtein, L.I.: Binary codes capable of correcting deletions, insertions, and reversals. Soviet Physics–Doklady **10** (1966) 707–710
3. Kohonen, T.: Self-Organization and Associative Memory. Springer, Berlin Heidelberg (1988)
4. Gusfield, D.: Algorithms on Strings, Trees, and Sequences. Cambridge University Press (1997)
5. Dayhoff, M., Schwartz, R., Orcutt, B.: A model of evolutionary change in proteins. In Dayhoff, M., ed.: Atlas of Protein Sequence and Structure. Volume 5., Washington, DC., Natl. Biomed. Res. Found. (1978) 345–352
6. Henikoff, S., Henikoff, J.G.: Amino acid substitution matrices from protein blocks. In: Proceedings of the National Academy of Sciences. Volume 89., Washington, DC (1992) 10915–10919
7. Setubal, J.C., Meidanis, J.: Intorduction to Computational Molecular Biology. PWS Publishing Company, Boston (1997)
8. Pearson, W.R., Lipman, D.J.: Improved tools for biological sequence comparison. In: Proceedings of the National Academy of Sciences of the U.S.A. Volume 85., Washington, DC, National Academy of Sciences of the U.S.A (1988) 2444–2448
9. Altschul, S.F., Gish, W., Miller, W., Meyers, E.W., Lipman, D.J.: Basic local alignment search tool. Journal of Molecular Biology **215** (1990) 403–410
10. Agrafiotis, D.K.: A new method for analyzing protein sequence relationships based on Sammon maps. Protein Science **6** (1997) 287–293
11. Kohonen, T., Somervuo, P.: Self-organizing maps of symbol strings. Neurocomputing **21** (1998) 19–30

12. Fischer, I., Zell, A.: String averages and self-organizing maps for strings. In Bothe, H., Rojas, R., eds.: Proceedings of the Neural Computation 2000, Canada/Switzerland, ICSC Academic Press (2000) 208–215
13. Kohonen, T.: Median strings. Pattern Recognition Letters **3** (1985) 309–313
14. Sankoff, D., Kruskal, J.B.: Time Warps, String Edits, and Macromolecules: the Theory and Practice of Sequence Comparison. Addison-Wesley, Reading, MA (1983)
15. Hirshberg, D.: A linear space algorith for computing maximal common subsequences. Communications of the ACM **18** (1975) 341–343
16. Landau, G., Vishkin, U.: Introducing efficient parallelism into approximate string matxhing and a new serial algorithm. In: Proceedings of the ACM Symposium on the Theory of Computing. (1986) 220–230
17. Altschul, S.F., Lipman, D.J.: Trees, stars, and multiple biological sequence alignment. SIAM Journal of Applied Mathematics **49** (1989) 197–209
18. Sammon, Jr., J.: A nonlinear mapping for data structure analysis. IEEE Transactions on Computers **18** (1969) 401–409
19. Apostol, I., Szpankowski, W.: Indexing and mapping of proteins using a modified nonlinear Sammon projection. Journal of Computational Chemistry (1999)

Combining Pairwise Classifiers with Stacking

Petr Savicky[1] and Johannes Fürnkranz[2]

[1] Institute of Computer Science
Academy of Sciences of the Czech Republic
Pod Vodarenskou Vezi 2, 182 07 Praha 8, Czech Republic
savicky@cs.cas.cz

[2] Austrian Research Institute for Artificial Intelligence
Schottengasse 3, A-1010 Wien, Austria
juffi@oefai.at

Abstract. Pairwise classification is the technique that deals with multi-class problems by converting them into a series of binary problems, one for each pair of classes. The predictions of the binary classifiers are typically combined into an overall prediction by voting and predicting the class that received the largest number of votes. In this paper we try to generalize the voting procedure by replacing it with a trainable classifier, i.e., we propose the use of a meta-level classifier that is trained to arbiter among the conflicting predictions of the binary classifiers. In our experiments, this yielded substantial gains on a few datasets, but no gain on others. These performance differences do not seem to depend on quantitative parameters of the datasets, like the number of classes.

1 Introduction

A recent paper [6] showed that pairwise classification (aka *round robin* learning) is superior to the commonly used one-against-all technique for handling multi-class problems in rule learning. Its basic idea is to convert a c-class problem into a series of two-class problems by learning one classifier for each pair of classes, using only training examples of these two classes and ignoring all others. A new example is classified by submitting it to each of the $c(c-1)/2$ binary classifiers, and combining their predictions. The most important finding of [6,7] was that this procedure not only increases predictive accuracy, but it is also no more expensive than the more commonly used one-against-all approach.

Previous work in this area only assumed voting techniques for combining the predictions of the binary classifiers, i.e., predictions were combined by counting the number of predictions for each individual class and assigning the class with the maximum count to the example (ties were broken in favor of larger classes). In this paper, we evaluate the idea of replacing the voting procedure with a trainable classifier.

We will start with a brief recapitulation of previous results on round robin learning (Section 2). Section 3 presents our algorithm based on stacking. The experimental setup is sketched in Section 4. Section 5 shows our results for using stacking for combining the predictions of the individual classifiers.

M.R. Berthold et al. (Eds.): IDA 2003, LNCS 2810, pp. 219–229, 2003.

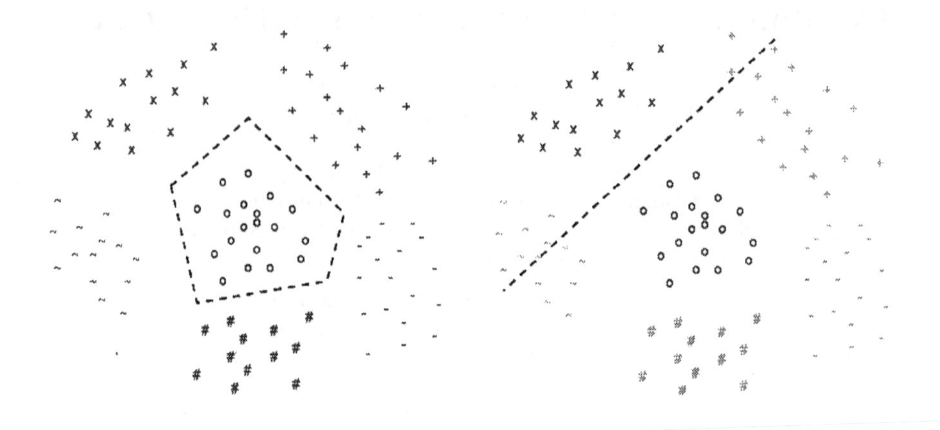

Fig. 1. *One-against-all class binarization* (left) transforms each c-class problem into c binary problems, one for each class, where each of these problems uses the examples of its class as the positive examples (here o), and all other examples as negatives. *Round robin class binarization* (right) transforms each c-class problem into $c(c-1)/2$ binary problems, one for each pair of classes (here o and x) ignoring the examples of all other classes.

2 Round Robin Classification

In this section, we will briefly review round robin learning (aka pairwise classification) in the context of our previous work in rule learning [6,7]. Separate-and-conquer rule learning algorithms [5] are typically formulated in a concept learning framework. The goal is to find an explicit definition for an unknown concept, which is implicitly defined via a set of positive and negative examples. Within this framework, multi-class problems, i.e., problems in which the examples may belong to (exactly) one of several categories, are usually addressed by defining a separate concept learning problem for each class. Thus the original learning problem is transformed into a set of binary concept learning problems, one for each class, where the positive training examples are those belonging to the corresponding class and the negative training examples are those belonging to all other classes.

On the other hand, the basic idea of round robin classification is to transform a c-class problem into $c(c-1)/2$ binary problems, one for each pair of classes. Note that in this case, the binary decision problems not only contain fewer training examples (because all examples that do not belong to the pair of classes are ignored), but that the decision boundaries of each binary problem may also be considerably simpler than in the case of one-against-all binarization. In fact, in the example shown in Fig. 1, each pair of classes can be separated with a linear decision boundary, while more complex functions are required to separate each class from all other classes. Evidence that the decision boundaries of the binary problems are in fact simpler can also be found in practical applications:

[12] observed that the classes of a digit recognition task were pairwise linearly separable, while the corresponding one-against-all task was not amenable to single-layer networks. Similarly, [10] obtained better results using one-against-one technique than one-against-all for SVM with both linear and non-linear kernels.

The main contributions of previous work [6,7] were on the one hand to empirically evaluate the technique for rule learning algorithms and to show that it is preferable to the one-against-all technique that is used in most rule learning algorithms. More importantly, however, we analyzed the computational complexity of the approach, and demonstrated that despite the fact that its complexity is quadratic in the number of classes, the algorithm is no slower than the conventional one-against-all technique. It is easy to see this, if we consider that in the one-against-all case, each training example is used c times (namely in each of the c binary problems), while in the round robin approach, each examples is only used $c - 1$ times, namely only in those binary problems, where its own class is paired against one of the other $c-1$ classes. Furthermore, the advantage of pairwise classification increases for computationally expensive (super-linear) learning algorithms. The reason is that expensive learning algorithms learn many small problems much faster than a few large problems. For more information we refer to [7] and [8].

3 Algorithm

The general framework of combining the predictions of multiple classifiers via a separate, trainable classifier is commonly referred to as *stacking* [16]. Its basic idea may be derived as a generalization of voting as follows. Let us consider the voting step as a separate classification problem, whose input is the vector of the responses of the base classifiers. Simple voting uses a predetermined algorithm for this, namely to count the number of predictions for each class in the input and to predict the most frequently predicted class. Stacking replaces this with a trainable classifier. This is possible, since for the training set, we have both the predictions of the base learners and the true class. The matrix containing the predictions of the base learners as predictors and the true class for each training case will be called the *meta-data set.* The classifier trained on this matrix will be called the *meta-classifier* or the classifier at the *meta-level.*

In our case, the round robin procedure learns a number of theories, one for each pair of classes. At the classification time, a new example is tested on each of the binary classifiers. Those, which are trained to distinguish the true class of the new example from some other class, will be called *relevant* predictors. The others, which distinguish a pair of classes, none of which is the correct one for this example, will be called *irrelevant.*

Simple voting relies on the assumption that the relevant classifiers frequently predict the correct class and, hence, give more votes to the true class than any other class can get from the irrelevant ones. If all relevant classifiers predict the correct class, then the overall prediction is correct as well, independently of the

Classifier			*original*
c_1/c_2	c_1/c_3	c_2/c_3	*class*
c_1	c_1	c_2	c_1
c_2	c_3	c_2	c_2
c_1	c_3	c_3	c_1
...	...	...	...
c_2	c_3	c_3	c_3

Fig. 2. Combining stacking with pairwise classification: The original problem was a three-class problem with classes (c_1,c_2,c_3). Each training example consists of the predictions of *all* pairwise classifiers c_i/c_j for this example and the original example's class label.

predictions of irrelevant classifiers. However, if some of the relevant classifiers give a wrong prediction and several of the irrelevant predictors predict the same wrong class, the overall prediction may be wrong. Occurrence of such an event depends on the behavior of both relevant and irrelevant predictors. Although the training process tries to optimize the behavior of the relevant predictors, it provides no guarantee concerning the behavior of the irrelevant ones.

The idea of our approach is as follows. Construct the pairwise classifiers as usual in round robin. When they are finished, each of them is tested on all the training data. This gives us information about the possible misleading votes caused by irrelevant predictors, which is then used to minimize their influence to the classification of new cases. A natural framework for such an approach is the stacking idea described above.

The fact that we have to test each training example on *each* of the binary classifiers, adds a quadratic complexity term (in the number of classes) to the otherwise linear complexity of pairwise classification [7]. However, testing is usually cheap in comparison to training, so we would expect that at least for moderate numbers of classes these additional costs are not noticeable.

Fig. 2 shows the meta-data training set that results from stacking the pairwise classifiers of a hypothetical 3-class problem. Note that the last example of the original training set (which has class c_3) was not part of the training set of the classifier c_1/c_2, but that the meta-data nevertheless contain the response of this classifier.

To see the potential gain of this procedure, consider the third example of Fig. 2. Straight-forward voting would predict class c_3 because the classifier that should discriminate class c_1 from class c_3 erroneously classified this example as c_3, and by chance the classifier c_2/c_3 also considered this example of class c_1 more likely to belong to class c_3. Thus, class c_3 receives 2 votes, whereas the correct class c_1 receives only one vote. However, if cases with this property occur already in the training set, there is a chance that the meta-classifier learns that in such cases, c_1 is actually the more likely prediction. In our experiments, we indeed found situations, when stacking with a nearest neighbor classifier was able to correct some of the mistakes of simple voting. For example, in the letter dataset, we found that in 65% (163 out of 250) cases, for which the true class was not among the classes with the largest number of votes, nearest neighbor classifier still provided a correct classification (see Table 3 in Section 5).

4 Experimental Setup

To evaluate the combination of stacking and pairwise classification, we used the rule learner ripper [3] as the base learning algorithm in a round robin procedure. The round robin learning procedure was implemented in Perl as a wrapper around the base learning algorithm. The wrapper reads in a dataset, provides the binary training and test sets for each pair of the classes to the base learner (by writing them to the disk), and calls the base learner on these sets. The predictions of the learned theories are then used to form the features of the meta dataset for a meta-level classifier. Because ripper is not symmetric in the classes (it produces a rule set for one of the classes and classifies all examples that are not covered by this rule set as belonging to the other class), we performed a double round robin procedure, i.e., we transformed a c-class problem into $c(c-1)$ pairwise problems, where discriminating c_i/c_j and discriminating c_j/c_i are different problems. For details on this we refer to [6].

We implemented the combination of stacking and pairwise classification as discussed in the previous section. At the meta-level, we experimented with three different classifiers:

- the decision-tree learner c4.5 [14]
- ripper in its default settings using ordered binarization to handle the multi-class meta-problem
- a simple nearest neighbor classifier that computes the distance between examples by counting the number of attributes in which they differ. For a given test example, the algorithm predicts the class of the nearest training example. Should there be more than one example at the same minimum distance, the class that appears most frequently in these cases is predicted.

The meta-level problem can become quite large. For example, for the 26-class letter dataset we had to store $26 \times 25 = 650$ predictions for each of the 16,000 training examples, and the nearest neighbor classifier had to compare each of them to each of the 4000 test examples. We handled this quite efficiently by encoding the binary attributes in the form of bit vectors, which also reduced the computation of the similarity measure to counting the number of on-bits in the result of an XOR of two example representations. Nevertheless, these problems may have an unusually high dimension compared to other stacking problems, where the number of classifiers (and hence the number of features at the meta-level) is typically independent of the number of classes in the problem.[1]

[1] An exception is the proposal of [15], who encode the meta-level using the classifiers' probability estimates for each class. But even there, the encoding is only linear in the number of classes (for a fixed number of classifiers), while we have a quadratic size at the meta level.

Table 1. *Data sets used.* The first column shows the total number of examples in this domain. The second column shows the evaluation methodology (an independent test set for *letter*, 10-fold cross-validation for all others). The next three columns show the number of symbolic and numeric attributes as well as the number of classes. The last column shows the default error, i.e., the error one would get by always predicting the majority class of the dataset.

name	train	test	sym	num	classes	def. error
letter	16,000	4000	0	16	26	95.9
abalone	4177	10-x-val	1	7	29	83.5
car	1728	10-x-val	6	0	4	30.0
glass	214	10-x-val	0	9	7	64.5
image	2310	10-x-val	0	19	7	85.7
lr spectrometer	531	10-x-val	1	101	48	89.6
optical	5620	10-x-val	0	64	10	89.8
page-blocks	5473	10-x-val	0	10	5	10.2
sat	6435	10-x-val	0	36	6	76.2
solar flares (c)	1389	10-x-val	10	0	8	15.7
solar flares (m)	1389	10-x-val	10	0	6	4.9
soybean	683	10-x-val	35	0	19	94.1
thyroid (hyper)	3772	10-x-val	21	6	5	2.7
thyroid (hypo)	3772	10-x-val	21	6	5	7.7
thyroid (repl.)	3772	10-x-val	21	6	4	3.3
vehicle	846	10-x-val	0	18	4	74.2
vowel	990	10-x-val	0	10	11	90.9
yeast	1484	10-x-val	0	8	10	68.8

Because of this, one may expect that the training examples of stacked round robin problems are less dense than in conventional stacking problems, with larger differences for a larger number of classes.

The experiments were performed on the 18 multi-class datasets shown in Table 1. These are the same datasets as used by [7] except for the largest sets *vowel* and *covertype*. The datasets were chosen arbitrarily among datasets with ≥ 4 classes available at the UCI repository [2]. The implementation of the algorithm was developed independently and not tuned on these datasets. On all datasets but one we performed a 10-fold paired stratified cross-validation using Johann Petrak's Machine Learning Experimentation Environment (MLEE) [2]. The results on the *letter* datasets used the same training/test split that is commonly used in the literature. In addition to the error rates, we also report improvement ratios over round robin learning with unweighted voting. These improvement rates are the ratio of the error rate of the classifier over the error rate of the base-line classifier. For averaging these ratios, we use geometric averages so that performance ratios r and $1/r$ average to 1.

[2] `http://www.ai.univie.ac.at/~johann`

Table 2. Error rates of pairwise classification (using unweighted voting) and three versions of pairwise stacking, using nearest-neighbor, c4.5 and ripper at the meta level (and ripper at the base level). For comparison, we also show previous results (taken from [8]) using weighted voting and the results of unmodified ripper. Shown are the error rates and (where applicable) the performance ratio (i.e., the error rate divided by the error rate of unweighted voting). The averages in the last line exclude the letter dataset, which was evaluated with the designated test set. All other results are from 10-fold cross-validations.

dataset	unweighted voting	pairwise stacking			weighted voting	default ripper
		nearest	c4.5	ripper		
letter	7.85	4.23 *0.539*	14.15 *1.803*	12.90 *1.643*	7.90 *1.006*	15.75
abalone	74.34	79.03 *1.063*	76.51 *1.029*	81.25 *1.093*	73.69 *0.991*	81.18
car	2.26	0.64 *0.282*	0.87 *0.385*	0.93 *0.410*	2.14 *0.949*	12.15
glass	25.70	24.77 *0.964*	28.50 *1.109*	28.50 *1.109*	25.70 *1.000*	34.58
image	3.42	2.34 *0.684*	3.42 *1.000*	3.16 *0.924*	3.16 *0.924*	4.29
lr spectrometer	53.11	59.89 *1.128*	52.17 *0.982*	56.69 *1.067*	54.05 *1.018*	61.39
optical	3.74	3.08 *0.824*	7.15 *1.914*	5.93 *1.586*	3.58 *0.957*	9.48
page-blocks	2.76	2.81 *1.020*	2.76 *1.000*	2.96 *1.073*	2.72 *0.987*	3.38
sat	10.35	11.08 *1.071*	11.58 *1.119*	11.84 *1.144*	10.55 *1.020*	13.04
solar flares (c)	15.77	16.05 *1.018*	16.41 *1.041*	15.77 *1.000*	15.69 *0.995*	15.91
solar flares (m)	5.04	5.47 *1.086*	5.18 *1.029*	5.33 *1.057*	4.90 *0.971*	5.47
soybean	5.86	7.03 *1.200*	6.15 *1.050*	6.44 *1.100*	5.86 *1.000*	8.79
thyroid (hyper)	1.11	1.11 *1.000*	1.22 *1.095*	1.17 *1.048*	1.19 *1.071*	1.49
thyroid (hypo)	0.53	0.34 *0.650*	0.48 *0.900*	0.34 *0.650*	0.45 *0.850*	0.56
thyroid (repl.)	1.01	1.03 *1.026*	1.06 *1.053*	1.11 *1.105*	0.98 *0.974*	0.98
vehicle	29.08	28.49 *0.980*	28.96 *0.996*	29.67 *1.020*	28.61 *0.984*	30.38
vowel	19.29	6.46 *0.335*	20.81 *1.079*	21.21 *1.099*	18.18 *0.942*	27.07
yeast	41.78	47.17 *1.129*	43.94 *1.052*	42.32 *1.013*	41.44 *0.992*	42.39
average	17.36	17.46 *0.852*	18.07 *1.012*	18.51 *0.997*	17.23 *0.977*	20.74

5 Results

The results of the experiments are shown in Table 2. They are very mixed. To some extent, the first line (the letter dataset) illustrates the main finding: Stacking may in some cases lead to a considerable reduction in error, in particular when the nearest-neighbor algorithm is used at the meta-level. In this case, the error can be reduced to about 85% of the error of unweighted pairwise classification, whereas there is no average gain when c4.5 or ripper are used as meta-learners. In fact, their performance (only 3–4 wins on 18 datasets) is significantly worse than unweighted voting at the 5% level according to a sign test.

A more detailed analysis of the performance of stacking with nearest neighbor classifier at the meta level is presented in Table 3. The table is calculated on the letter data set with 26 classes. The test cases are splitted into several groups according to the properties of the set of classes, which received the largest number of votes. Namely, we distinguish the cases according to whether the true class belongs to this set and according to the size of this set. The last two lines of the table show that nearest neighbor remains relatively succesfull even on cases,

Table 3. Results of nearest neighbor meta-classifier on the test sample (4000 cases) splitted into several subsets of cases according to whether the true class received the maximum number of votes and the number of classes, which received the maximum number of votes.

true class among max. votes	no. classes max. votes	n.n. predictions wrong	n.n. predictions correct
yes	1	68	3574
yes	2	21	82
yes	3	0	5
no	1	77	159
no	2	10	4

Table 4. Error reduction ratio vs. number of classes, with a linear least-squares fit.

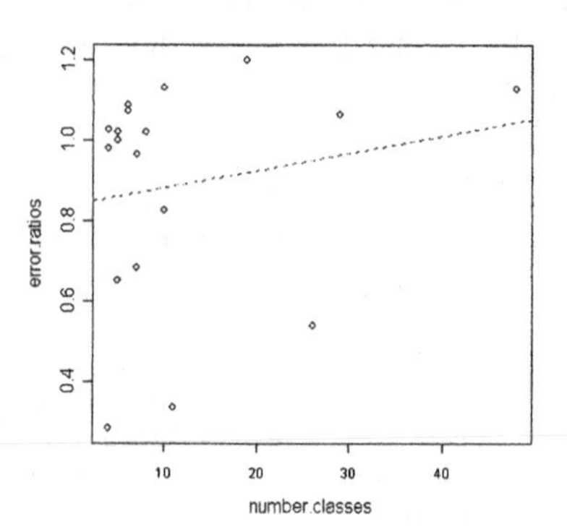

where the true class received strictly smaller number of votes than some other classes. Simple voting, of course, cannot predict the correct class for any of these cases.

For all meta classifiers, in particular for the nearest-neighbor classifier, there are a few selected datasets, where stacking can lead to a substantial gain (*letter*, *car*, and *vowel* for the nearest-neighbor case). The obvious question at this point is in which cases we can expect a gain from the use of stacking for combining the predictions of pairwise classifiers. This does not seem to depend on simple quantitative parameters of the dataset, like the number of classes in the original problem. There is no correlation ($r^2 = 0.0328$) between the error reduction ratio and the number of classes of the problem (see Fig. 4). We also looked at the correlation of the error reduction ratio with several other data characteristics, including default error, sample size, and the ratio of sample size over the number of classes, with similarly negative results. The performance gain of stacking depends on the structure of the problem and may be tested e.g. by cross-validation.

For comparison, we also show the results of regular ripper (without the use of pairwise classification) and previous results using weighted voting (taken from [8]) where the vote of each binary classifier is weighted with its confidence into its prediction, estimated by the precision of the rule that was used to make the classification. Weighted voting leads to more consistent gains, but never to such an extent as stacking in those cases where it works.

The question when stacking can outperform unweighted voting remains open. However, our experiments demonstrate that it is an alternative that is worth to be tried. For example, in the *letter* dataset, the combination of stacking and pairwise classification has produced the best result that we have seen with ripper as a base learner.

6 Related Work

The basic idea of pairwise classification is fairly well-known from the literature. It can be found in the areas of statistics, neural networks, support vector machines, and others. We refer to [7, Section 8] for a brief survey of the literature on this topic.

Unweighted voting is regularly used for combining the predictions of multiple classifiers, but there is some work on alternative procedures. [9] give several examples where votes that depend on class probability estimates may result in considerably different predictions than unweighted voting. Whether this is desirable or not, it can even happen that a class that wins all pairwise comparisons against all other classes, is not predicted because the sum of the probabilities of the predictions are lower than the weights for all other classes. An empirical evaluation of this issue in a similar setup as in this paper confirmed that weighted voting leads to small but consistent performance improvements over unweighted voting [8].

Allwein et al. [1] treat both one-against-all and pairwise classification (and other class binarization techniques) as special cases of a generalization of error-correcting output codes [4]. They also use a nearest neighbor classifier to combine the predictions of the base learners, however, the distance is measured to prototypes given by a predetermined learning scheme, which does not depend on the training data. In particular, with an appropriate matrix this is equivalent to the simple voting in the one-against-all or pairwise setting. This technique is further analyzed in [11], where experimental evidence showed that unweighted voting performed at least as well (for support-vector machines as the base learner). In [13] another approach for combining binary classifiers is suggested, where separate classifiers are trained for deciding whether a binary classifier is relevant or irrelevant for a given example.

The proposed combination of round robin learning and stacking may also be viewed as analogous to a similar proposal for integrating bagging with stacking [17]. In this sense, this work complements previous investigation of combining round robin learning with other ensemble techniques, namely bagging and boosting [8].

7 Conclusions

In this paper, we evaluated a technique for combining pairwise classification with stacking. The basic idea is to replace the voting procedure that combines the predictions of the binary classification by a trainable classifier. The training set of this meta-classifier consists of all predictions of the binary classifiers for each training example. We evaluated the use of a rule learner, a decision tree learner and a nearest-neighbor learning algorithm. Stacking with the nearest-neighbor classifier at the meta level achieved a substantial improvement in some cases. The performance gain does not seem to depend on the number of classes of the problem and similar characteristics. However, it may be estimated e.g. by crossvalidation in concrete situations. The results show that stacking should be kept in mind as a potential candidate for optimizing the accuracy of the final prediction.

Acknowledgments. We would like to thank Johann Petrak for his experimentation scripts and help using them, and the maintainers of and contributors to the UCI collection of machine learning databases. Petr Savicky was supported by the Ministry of Education of the Czech Republic under Research and Development project LN00A056, Institute of Theoretical Computer Science - ITI. Johannes Fürnkranz is supported by an APART stipend (no. 10814) of the *Austrian Academy of Sciences*. The Austrian Research Institute for Artificial Intelligence is supported by the Austrian Federal Ministry of Education, Science and Culture.

References

1. E. L. Allwein, R. E. Schapire, and Y. Singer. Reducing multiclass to binary: A unifying approach for margin classifiers. *Journal of Machine Learning Research*, 1:113–141, 2000.
2. C. L. Blake and C. J. Merz. UCI repository of machine learning databases. `http://www.ics.uci.edu/ mlearn/MLRepository.html`, 1998. Department of Information and Computer Science, University of California at Irvine, Irvine CA.
3. W. W. Cohen. Fast effective rule induction. In A. Prieditis and S. Russell, editors, *Proceedings of the 12th International Conference on Machine Learning (ML-95)*, pages 115–123, Lake Tahoe, CA, 1995. Morgan Kaufmann.
4. T. G. Dietterich and G. Bakiri. Solving multiclass learning problems via error-correcting output codes. *Journal of Artificial Intelligence Research*, 2:263–286, 1995.
5. J. Fürnkranz. Separate-and-conquer rule learning. *Artificial Intelligence Review*, 13(1):3–54, February 1999.
6. J. Fürnkranz. Round robin rule learning. In C. E. Brodley and A. P. Danyluk, editors, *Proceedings of the 18th International Conference on Machine Learning (ICML-01)*, pages 146–153, Williamstown, MA, 2001. Morgan Kaufmann.
7. J. Fürnkranz. Round robin classification. *Journal of Machine Learning Research*, 2:721–747, 2002.
8. J. Fürnkranz. Round robin ensembles. *Intelligent Data Analysis*, 7(5), 2003. In press.
9. T. Hastie and R. Tibshirani. Classification by pairwise coupling. In M. Jordan, M. Kearns, and S. Solla, editors, *Advances in Neural Information Processing Systems 10 (NIPS-97)*, pages 507–513. MIT Press, 1998.
10. C.-W. Hsu and C.-J. Lin. A comparison of methods for multi-class support vector machines. *IEEE Transactions on Neural Networks*, 13(2):415–425, March 2002.
11. A. Klautau, N. Jevtić, and A. Orlitsky. On nearest-neighbor ECOC with application to all-pairs multiclass SVM. *Journal of Machine Learning Research*, 4:1–15, 2003.
12. S. Knerr, L. Personnaz, and G. Dreyfus. Handwritten digit recognition by neural networks with single-layer training. *IEEE Transactions on Neural Networks*, 3(6): 962–968, 1992.

13. M. Moreira and E. Mayoraz. Improved pairwise coupling classification with correcting classifiers. In C. Nédellec and C. Rouveirol, editors, *Proceedings of the 10th European Conference on Machine Learning (ECML-98)*, Chemnitz, Germany, 1998. Springer-Verlag.
14. J. R. Quinlan. *C4.5: Programs for Machine Learning.* Morgan Kaufmann, San Mateo, CA, 1993.
15. K. M. Ting and I. H. Witten. Issues in stacked generalization. *Journal of Artificial Intelligence Research*, 10:271–289, 1999.
16. D. H. Wolpert. Stacked generalization. *Neural Networks*, 5(2):241–260, 1992.
17. D. H. Wolpert and W. G. Macready. Combining stacking with bagging to improve a learning algorithm. Technical Report SFI-TR-96-03-123, Santa Fe Institute, Santa Fe, New Mexico, 1996.

APRIORI-SD: Adapting Association Rule Learning to Subgroup Discovery

Branko Kavšek, Nada Lavrač, and Viktor Jovanoski

Institute Jožef Stefan, Jamova 39, 1000 Ljubljana, Slovenia
{branko.kavsek,nada.lavrac,viktor.jovanoski@ijs.si}

Abstract. This paper presents a subgroup discovery algorithm APRIORI-SD, developed by adapting association rule learning to subgroup discovery. This was achieved by building a classification rule learner APRIORI-C, enhanced with a novel post-processing mechanism, a new quality measure for induced rules (weighted relative accuracy) and using probabilistic classification of instances. Results of APRIORI-SD are similar to the subgroup discovery algorithm CN2-SD while experimental comparisons with CN2, RIPPER and APRIORI-C demonstrate that the subgroup discovery algorithm APRIORI-SD produces substantially smaller rule sets, where individual rules have higher coverage and significance.

1 Introduction

Classical rule learning algorithms are designed to construct classification and prediction rules [12,3,4,7]. In addition to this area of machine learning, referred to as *supervised learning* or *predictive induction*, developments in *descriptive induction* have recently gained much attention, in particular *association rule learning* [1] (e.g., the APRIORI association rule learning algorithm), *subgroup discovery* (e.g., the MIDOS subgroup discovery algorithm [18,5]), and other approaches to non-classificatory induction.

As in the MIDOS approach, a subgroup discovery task can be defined as follows: given a population of individuals and a property of those individuals we are interested in, find population subgroups that are statistically 'most interesting', e.g., are as large as possible and have the most unusual statistical (distributional) characteristics with respect to the property of interest [18,5].

Some of the questions on how to adapt classical classification rule learning approaches to subgroup discovery have already been addressed in [10] and a well-known rule learning algorithm CN2 was adapted to subgroup discovery. In this paper we take a rule learner APRIORI-C instead of CN2 and adapt it to subgroup discovery, following the guidelines from [10].

We have implemented the new subgroup discovery algorithm APRIORI-SD in C++ by modifying the APRIORI-C algorithm. The proposed approach performs subgroup discovery through the following modifications of the rule learning algorithm APRIORI-C: (a) using a weighting scheme in rule post-processing, (b) using weighted relative accuracy as a new measure of the quality of the rules in

M.R. Berthold et al. (Eds.): IDA 2003, LNCS 2810, pp. 230–241, 2003.

the post-processing step when the best rules are selected, (c) probabilistic classification based on the class distribution of covered examples by individual rules, and (d) area under the ROC curve rule set evaluation. The latter evaluation criterion was used in addition to the standard evaluation criteria: rule coverage, rule significance, rule set size and rule set accuracy.

This paper presents the APRIORI-SD subgroup discovery algorithm, together with its experimental evaluation in selected domains of the UCI Repository of Machine Learning Databases [13]. Results of APRIORI-SD are similar to the subgroup discovery algorithm CN2-SD while experimental comparisons with CN2, RIPPER and APRIORI-C demonstrate that the subgroup discovery algorithm APRIORI-SD produces substantially smaller rule sets, where individual rules have higher coverage and significance. These three factors are important for subgroup discovery: smaller size enables better understanding, higher coverage means larger subgroups, and higher significance means that rules describe subgroups whose class distribution is significantly different from the entire population by no loss in terms of the area under the ROC curve and accuracy.

This paper is organized as follows. In Section 2 the background for this work is explained: the APRIORI-C rule induction algorithm, including the post-processing step of selecting the best rules. Section 3 presents the modified APRIORI-C algorithm, called APRIORI-SD, adapting APRIORI-C for subgroup discovery together with the weighted relative accuracy measure, probabilistic classification and rule evaluation in the ROC space. Section 4 presents the experimental evaluation on selected UCI domains. Section 5 concludes by summarizing the results and presenting plans for further work. While in Section 4 we present the summary results of the experiments, the complete results on all the UCI data sets are presented in the Appendix.

2 Background: The APRIORI-C Algorithm

The idea of using association rules for classification is not new [11]. The main advantage of APRIORI-C over its predecessors is lower memory consumption, decreased time complexity and improved understandability of results. The reader can find a detailed description of the APRIORI-C algorithm in [7]. We describe here just the parts of the APRIORI-C that are essential for the reader to understand the derived APRIORI-SD algorithm.

The APRIORI-C algorithm is derived from the well known association rule learning algorithm APRIORI [1,2] which was extensively studied, adopted to other areas of machine learning and data mining, and successfully applied in many problem domains.

An association rule has the following form:

$$X \rightarrow Y, \tag{1}$$

where $X, Y \subset I$, X and Y are itemsets, and I is the set of all items.

The quality of each association rule is defined by its *confidence* and *support*. *Confidence* of a rule is an estimate of the conditional probability of Y given X:

$p(Y|X)$. *Support* of a rule is an estimate of the probability of itemset $X \cup Y$: $p(XY)$. Confidence and support are computed as follows:

$$Confidence = \frac{n(XY)}{n(X)} = \frac{p(XY)}{p(X)} = p(Y|X), Support = \frac{n(XY)}{N} = p(XY) \quad (2)$$

where $n(X)$ is the number of transactions that are supersets of itemset X and N is the number of all the transactions.

The association rule learning algorithm APRIORI is then adopted for classification purposes (APRIORI-C) by implementing the following steps: (1) Discretize continuous attributes. (2) Binarize all (discrete) attributes. (3) Run the APRIORI algorithm by taking in consideration only rules whose right-hand side consists of a single item, representing the value of the class attribute (while running APRIORI). (4) Post-process this set of rules, selecting the best among them and use this rules to classify unclassified examples.

These steps of the APRIORI-C algorithm, as well as the approaches to feature subset selection, are described in detail in [7]. Here we describe just the last step, the post-processing of rules and classification of unclassified examples, which are the ones we changed to obtain APRIORI-SD.

Post-processing by rule subset selection. The APRIORI-C algorithm induces rules according to the parameters *minimal confidence* and *minimal support* of a rule [7]. The setting of these two parameters is often such that the algorithm induces a large number of rules, which hinders the understandability and usability of induced rules. Moreover, there are the problems of rule redundancy, incapability of classifying examples and poor accuracy in domains with unbalanced class distribution. A way to avoid these problems is to select just some best rules among all the induced rules. APRIORI-C has three ways of selecting such best rules:

Use N best rules: The algorithm first selects the best rule (rule having the highest support), then eliminates all the covered examples, sorts the remaining rules according to support and repeats the procedure. This procedure is repeated until N rules are selected or there are no more rules to select or there are no more examples to cover. The algorithm then stops and returns the classifier in the form of an IF-THEN-ELSE rule list.

Use N best rules for each class: The algorithm behaves in a similar way as the 'use N best rules' case, selecting N best rules for each class (if so many rules exist for each class). This way the rules for the minority class will also find their way into the classifier.

Use a weighting scheme to select the best rules: The algorithm again behaves in a similar way as 'use N best rules'. The difference is that covered examples are not eliminated immediately, but instead their weight is decreased. They are then eliminated when the weight falls below a certain threshold (e.g., when an example has been covered more than K times). The details of the weighting scheme together with the threshold used are given in Section 3, describing APRIORI-SD.

3 APRIORI-SD

The main modifications of the APRIORI-C algorithm, making it appropriate for subgroup discovery, involve the implementation of a new weighting scheme in post-processing, a different rule quality function (the weighted relative accuracy), the probabilistic classification of unclassified examples and the area under the ROC curve rule set evaluation.

3.1 Post-processing Procedure

The post-processing procedure is performed as follows:

```
repeat
- sort rules from best to worst in terms of
  the weighted relative accuracy quality measure (see Section 3.3)
- decrease the weights of covered examples (see Section 3.2)
until
  all the examples have been covered or
  there are no more rules
```

3.2 The Weighting Scheme Used in Best Rule Selection

In the 'use a weighting scheme to select best rules' post-processing method of APRIORI-C described in Section 2, the examples covered by the 'currently' best rule are not eliminated but instead re-weighted. This approach is more suitable for the subgroup discovery process which is, in general, aimed at discovering interesting properties of subgroups of the entire population. The weighting scheme allows this.

The weighting scheme treats examples in such a way that covered positive examples are not deleted when the currently 'best' rule is selected in the post-processing step of the algorithm. Instead, each time a rule is selected, the algorithm stores with each example a count that shows how many times (with how many rules selected so far) the example has been covered so far. Initial weights of all positive examples e_j equal 1, $w(e_j, 0) = 1$, which denotes that the example has not been covered by any rule, meaning 'among the available rules select a rule which covers this example, as this example has not been covered by other rules', while lower weights mean 'do not try too hard on this example'.

Weights of positive examples covered by the selected rule decrease according to the formula $w(e_j, i) = \frac{1}{i+1}$. In the first iteration all target class examples contribute the same weight $w(e_j, 0) = 1$, while in the following iterations the contributions of examples are inverse proportional to their coverage by previously selected rules. In this way the examples already covered by one or more selected rules decrease their weights while rules covering many yet uncovered target class examples whose weights have not been decreased will have a greater chance to be covered in the following iterations.

3.3 The Weighted Relative Accuracy Measure

Weighted relative accuracy is used in subgroup discovery to evaluate the quality of induced rules. We use it instead of support when selecting the 'best' rules in the post-processing step.

We use the following notation. Let $n(X)$ stand for the number of instances covered by a rule $X \rightarrow Y$, $n(Y)$ stand for the number of examples of class Y, and $n(YX)$ stand for the number of correctly classified examples (true positives). We use $p(YX)$ etc. for the corresponding probabilities. Rule accuracy, or rule confidence in the terminology of association rule learning, is defined as $Acc(X \rightarrow Y) = p(Y|X) = \frac{p(YX)}{p(X)}$. Weighted relative accuracy [9,16] is defined as follows.

$$WRAcc(X \rightarrow Y) = p(X).(p(Y|X) - p(Y)). \tag{3}$$

Weighted relative accuracy consists of two components: generality $p(X)$, and relative accuracy $p(Y|X) - p(Y)$. The second term, relative accuracy, is the accuracy gain relative to the fixed rule $true \rightarrow Y$. The latter rule predicts all instances to satisfy Y; a rule is only interesting if it improves upon this 'default' accuracy. Another way of viewing relative accuracy is that it measures the utility of connecting rule body X with a given rule head Y. However, it is easy to obtain high relative accuracy with highly specific rules, i.e., rules with low generality $p(X)$. To this end, generality is used as a 'weight', so that weighted relative accuracy trades off generality of the rule ($p(X)$, i.e., rule coverage) and relative accuracy ($p(Y|X) - p(Y)$). All the probabilities in Equation 3 are estimated with relative frequencies e.g., $p(X) = \frac{n(X)}{N}$, where N is the number of all instances.

Modified WRAcc function with Example Weights. WRAcc - the rule quality function used in APRIORI-SD was further modified to enable handling example weights, which provide the means to consider different parts of the instance space with each application of a selected rule (as described in Section 3.2).

The modified *WRAcc* measure is defined as follows:

$$WRAcc(X \rightarrow Y) = \frac{n'(X)}{N'}\left(\frac{n'(YX)}{n'(X)} - \frac{n'(Y)}{N'}\right). \tag{4}$$

where N' is the sum of the weights of all examples, $n'(X)$ is the sum of the weights of all covered examples, and $n'(YX)$ is the sum of the weights of all correctly covered examples.

3.4 Probabilistic Classification

In general the induced rules can be treated as ordered or unordered. Ordered rules are interpreted as an IF-THEN-ELSE decision list [15] in a straight-forward manner: when classifying a new example, the rules are sequentially tried and the first rule that covers the example is used for prediction. This interpretation of rules is also used by APRIORI-C, when classifying unclassified examples.

In the case of unordered rule sets, the distribution of covered training examples among classes is attached to each rule. Rules of the form:

$$X \rightarrow Y \ [ClassDistribution]$$

are induced, where numbers in the *ClassDistribution* list denote, for each individual class, the percentage of training examples of this class covered by the rule. When classifying a new example, all rules are tried and those covering the example are collected. If a clash occurs (several rules with different class predictions cover the example), a voting mechanism is used to obtain the final prediction: the class distributions attached to the rules are considered to determine the most probable class. If no rule fires, a default rule is invoked which predicts the majority class of uncovered training instances. This type of interpretation of induced rules for classification, also called probabilistic classification, is used by the APRIORI-SD algorithm.

3.5 Area under ROC Curve Evaluation

A point on the *ROC curve* (ROC: Receiver Operating Characteristic) [14] shows classifier performance in terms of false alarm or *false positive rate* $FPr = \frac{FP}{TN+FP} = \frac{FP}{Neg}$ (plotted on the X-axis; 'Neg' standing for the number of all negative examples) that needs to be minimized, and sensitivity or *true positive rate* $TPr = \frac{TP}{TP+FN} = \frac{FP}{Pos}$ (plotted on the Y-axis; 'Pos' standing for the number of all positive examples) that needs to be maximized. The confusion matrix shown in Table 1 defines the notions of TP (true positives), FP (false positives), TN (true negatives) and FN (false negatives).

Table 1. Confusion matrix.

	predicted positive	predicted negative
actual positive	TP	FN
actual negative	FP	TN

Applying the notation used to define *confidence* and *support* (see Equation 2) FPr and TPr can be expressed as: $FPr = \frac{n(X\overline{Y})}{Neg}$, $TPr = \frac{n(XY)}{Pos}$. In the ROC space, an appropriate tradeoff, determined by the expert, can be achieved by applying different algorithms, as well as by different parameter settings of a selected data mining algorithm or by taking into the account different misclassification costs. The ROC space is appropriate for measuring the success of subgroup discovery, since subgroups whose TPr/FPr tradeoff is close to the diagonal can be discarded as insignificant. The reason is that the rules with TPr/FPr on the diagonal have the same distribution of covered positives and negatives as the distribution in the entire data set.

The area under the ROC curve (AUC) can be used as a quality measure for comparing the success of different learners. AUC can be computed by employing combined probabilistic classifications of all subgroups (rules) [10], as indicated below. If we always choose the most likely predicted class, this corresponds to setting a fixed threshold 0.5 on the positive probability: if the positive probability is larger than this threshold we predict positive, else negative. A ROC curve can be constructed by varying this threshold from 1 (all predictions negative, corresponding to (0,0) in the ROC space) to 0 (all predictions positive, corresponding to (1,1) in the ROC space). This results in $n + 1$ points in the ROC space, where n is the total number of classified examples. Equivalently, we can order all the testing examples by decreasing predicted probability of being positive, and tracing the ROC curve by starting in (0,0), stepping up when the tested example is actually positive and stepping to the right when it is negative, until we reach (1,1); in the case of ties, we make the appropriate number of steps up and to the right at once, drawing a diagonal line segment. The area under this ROC curve indicates the combined quality of all subgroups (i.e., the quality of the entire rules set). This method can be used with a test set or in cross-validation. A detailed description of this method can be found in [6].

This method of computing the AUC was used in the experimental evaluation in Section 4.

4 Experimental Evaluation

For subgroup discovery, expert's evaluation of results is of ultimate interest. Nevertheless, before applying the proposed approach to a particular problem of interest, we wanted to verify our claims that the mechanisms implemented in the APRIORI-SD algorithm are indeed appropriate for subgroup discovery.

We experimentally evaluated our approach on 23 data sets from the UCI Repository of Machine Learning Databases [13]. In Table 2, the selected data sets are summarized in terms of the number of attributes, number of classes, number of examples, and the percentage of examples of the majority class. All continuous attributes were discretized with a discretization method described in [8] using the WEKA tool [17].

The comparison of APRIORI-SD with APRIORI-C, and RIPPER was performed in terms of coverage of rules, size of the rule sets, significance of rules, area under the ROC curve and predictive accuracy (despite the fact that optimizing predictive accuracy is not the intended goal of subgroup discovery algorithms). The area under the ROC curve evaluation was computed only on two-class problems (first 16 data sets in Table 2). The method we used for evaluation was 10-fold stratified cross validation. The parameters used to run the algorithms APRIORI-SD and APRIORI-C were: *minimal confidence* 0.8, *minimal support* 0.03 and *'use a weighting scheme to select best rules'* as the post-processing scheme. We used the version of RIPPER implemented in WEKA [17] with default parameters.

Furthermore we present also the results showing the comparison of APRIORI-SD with the algorithms CN2 and CN2-SD. Note that these results were obtained from the original (non-discretized) UCI data sets and are included to enable an additional comparison with a standard classification rule learner (CN2) and a state-of-the-art subgroup discovery algorithm (CN2-SD).

Table 3 presents summary results of the comparisons on UCI data sets, while details can be found in Tables 4, 5, 6 and 7 in the Appendix. For each performance measure, the summary table shows the average value over all the data sets, the significance of the results compared to APRIORI-SD (p-value), and the

Table 2. Data set characteristics.

	Data set	#Attr.	#Class.	#Ex.	Maj. Class (%)
1	australian	14	2	690	56
2	breast-w	9	2	699	66
3	bridges-td	7	2	102	85
4	chess	36	2	3196	52
5	diabetes	8	2	768	65
6	echo	6	2	131	67
7	german	20	2	1000	70
8	heart	13	2	270	56
9	hepatitis	19	2	155	79
10	hypothyroid	25	2	3163	95
11	ionosphere	34	2	351	64
12	iris	4	3	150	33
13	mutagen	59	2	188	66
14	mutagen-f	57	2	188	66
15	tic-tac-toe	9	2	958	65
16	vote	16	2	435	61
17	balance	4	3	625	46
18	car	6	4	1728	70
19	glass	9	6	214	36
20	image	19	7	2310	14
21	soya	35	19	683	13
22	waveform	21	3	5000	34
23	wine	13	3	178	40

Table 3. Summary of the experimental results comparing APRIORI-SD with different algorithms using 10-fold stratified cross-validation.

Performance Measure	Data Sets	**APRIORI-SD**	**APRIORI-C**	**RIPPER**	**CN2**	**CN2-SD**
Coverage	23	**0.53** ± 0.26	0.36 ± 0.19	0.19 ± 0.19	0.13 ± 0.14	0.46 ± 0.25
• sig. (p value)			0.000	0.000	0.000	0.012
• loss/win/draw			22/1/0	22/1/0	22/1/0	19/4/0
Size	23	**3.6** ± 1.96	5.6 ± 2.84	16.1 ± 27.47	18.2 ± 21.77	6.4 ± 4.58
• sig. (p value)			0.000	0.035	0.003	0.000
• loss/win/draw			21/2/0	20/3/0	22/1/0	17/4/2
Likelihood ratio	23	**12.37** ± 7.26	2.60 ± 0.55	2.36 ± 0.55	2.11 ± 0.46	18.47 ± 9.00
• sig. (p value)			0.000	0.000	0.000	0.000
• loss/win/draw			22/1/0	22/1/0	22/1/0	8/15/0
AUC	16	**82.80** ± 8.70	80.92 ± 9.95	80.11 ± 10.23	82.16 ± 16.81	86.33 ± 8.60
• sig. (p value)			0.190	0.027	0.871	0.011
• loss/win/draw			10/6/0	12/4/0	5/11/0	3/13/0
Accuracy	23	**79.98** ± 16.67	81.02 ± 16.50	83.46 ± 10.24	81.61 ± 11.66	79.36 ± 16.24
• sig. (p value)			0.039	0.282	0.489	0.414
• loss/win/draw			10/13/0	13/10/0	15/8/0	12/11/0

LOSS/WIN/DRAW in terms of the number of domains in which the results are worse/better/equal compared to APRIORI-SD. The analysis shows the following:

- In terms of the average coverage per rule APRIORI-SD produces rules with significantly higher coverage (higher the coverage better the rule) than both APRIORI-C and RIPPER. Moreover it produces rules with significantly higher coverage than both CN2 and CN2-SD.
- APRIORI-SD induces rule sets that are significantly smaller than those induced by APRIORI-C and RIPPER (smaller rule sets are better). It induces also significantly smaller rule sets than both CN2 and CN2-SD.

- APRIORI-SD induces significantly better rules in terms of significance (rules with higher significance are better) - computed by the average likelihood ratio - than APRIORI-C, RIPPER and CN2, while being significantly worse than CN2-SD. Note: APRIORI-SD, CN2 and CN2-SD rules are already significant at the 99% level.
- As the comparisons in terms of the area under the ROC curve (AUC) are restricted to binary class data sets, only the 16 binary data sets were used in this comparison. Notice that while being better than APRIORI-C and RIPPER, APRIORI-SD is comparable to CN2, but worse than CN2-SD.

Despite the fact that subgroup discovery is not intended at maximizing accuracy, it is worthwhile noticing that in terms of predictive accuracy APRIORI-SD is just insignificantly worse than the other three rule learners (APRIORI-C, RIPPER and CN2) and insignificantly better than CN2-SD.

5 Conclusions

Following the ideas presented in [10] we have adapted the APRIORI-C algorithm to subgroup discovery (APRIORI-SD). Experimental results on 23 UCI data sets demonstrate that APRIORI-SD produces smaller rule sets, where individual rules have higher coverage and significance compared to other rule learners (APRIORI-C, RIPPER and CN2). These three factors are important for subgroup discovery: smaller size enables better understanding, higher coverage means larger support, and higher significance means that rules describe discovered subgroups that are significantly different from the entire population.

When comparing APRIORI-SD to another subgroup discovery algorithm (CN2-SD) it turns out that APRIORI-SD still produces smaller rule sets, where individual rules have higher coverage, but they are less significant than the CN2-SD rules. Note that these results need further analysis as CN2-SD rules were induced from non-discretized data while APRIORI-SD used discretized data sets.

We have evaluated the results of APRIORI-SD also in terms of classification accuracy and AUC and shown a small increase in terms of the area under the ROC curve compared to APRIORI-C and RIPPER. A decrease in AUC compared to (CN2 and) CN2-SD could again be attributed to the use of non-discretized attributes in CN2(-SD).

In further work we plan to study the theoretical properties of the weighted covering approach. Our plans are also to compare our algorithm with the subgroup discovery algorithm MIDOS. Finally, we plan to use the APRIORI-SD subgroup discovery algorithm for solving practical problems, in which expert evaluations of induced subgroup descriptions is of ultimate interest.

Acknowledgements. The work reported in this paper was supported by the Slovenian Ministry of Education, Science and Sport.

References

1. R. Agrawal, T. Imielinski and R. Shrikant. Mining Association Rules between Sets of Items in Large Databases. In *Proceedings of ACM SIGMOD Conference on Management of Data*, Washington, D.C., 1993. pp. 207–216.
2. R. Agrawal and R. Srikant. Fast Algorithms for Mining Association Rules. In *Proceedings of the 20th International Conference on Very Large Databases*, Santiago, Chile, 1994. pp. 207–216.
3. P. Clark and T. Niblett. The CN2 induction algorithm. *Machine Learning*, 3(4): 261–283, 1989.
4. W.W. Cohen. Fast effective rule induction. In *Proceedings of the 12th International Conference on Machine Learning*, Morgan Kaufmann, 1995. pp. 115–123.
5. S. Džeroski, N. Lavrač. (Eds). *Relational Data Mining*, Springer, 2001. pp. 74–99.
6. C. Ferri-Ramírez, P.A. Flach, and J. Hernandez-Orallo. Learning decision trees using the area under the ROC curve. In *Proceedings of the 19th International Conference on Machine Learning*, Morgan Kaufmann, 2002. pp. 139–146.
7. V. Jovanoski, N. Lavrač. Classification Rule Learning with APRIORI-C. In P. Brazdil, A. Jorge (Eds.): *10th Portuguese Conference on Artificial Intelligence*, EPIA 2001, Porto, Portugal, 2001. pp. 44–51.
8. I. Kononenko. On Biases in Estimating Multi-Valued Attributes. In *Proceedings of the 14th Int. Joint Conf. on Artificial Intelligence*, 1995. pp. 1034–1040.
9. N. Lavrač, P. Flach, and B. Zupan. Rule evaluation measures: A unifying view. In *Proceedings of the 9th International Workshop on Inductive Logic Programming*, Springer, 1999. pp. 74–185.
10. N. Lavrač, P. Flach, B. Kavšek, L. Todorovski. Adapting classification rule induction to subgroup discovery. In *Proceedings of the IEEE International Conference on Data Mining (ICDM'02)*, Maebashi City, Japan, 2002. pp. 266–273.
11. B. Liu, W. Hsu and Y. Ma. Integrating Classification and Association Rule Mining. In *Proceedings of the Fourth International Conference on Knowledge Discovery and Data Mining KDD'98*, New York, USA, 1998. pp. 80–86.
12. R.S. Michalski, I. Mozetič, J. Hong, and N. Lavrač. The multi-purpose incremental learning system AQ15 and its testing application on three medical domains. In *Proc. 5th National Conference on Artificial Intelligence*, Morgan Kaufmann, 1986. pp. 1041–1045.
13. P.M. Murphy and D.W. Aha. *UCI repository of machine learning databases* [http://www.ics.uci.edu/~mlearn/MLRepository.html]. Irvine, CA: University of California, Department of Information and Computer Science, 1994.
14. F. Provost and T. Fawcett. Robust classification for imprecise environments. *Machine Learning*, 42(3): 203–231, 2001.
15. R.L. Rivest. Learning decision lists. *Machine Learning*, 2(3): 229–246, 1987.
16. L. Todorovski, P. Flach, and N. Lavrač. Predictive performance of weighted relative accuracy. In D.A. Zighed, J. Komorowski, and J. Zytkow, editors, *Proceedings of the 4th European Conference on Principles of Data Mining and Knowledge Discovery*, Springer, 2000. pp. 255–264.
17. I.H. Witten and E. Frank. *Data Mining: Practical Machine Learning Tools and Techniques with Java Implementations*. Morgan Kaufmann, 1999.
18. S. Wrobel. An algorithm for multi-relational discovery of subgroups. In *Proceedings of the 1st European Symposium on Principles of Data Mining and Knowledge Discovery*, Springer, 1997. pp. 78–87.

Appendix

Table 4. Average size (*S*), coverage (*CVG*) and likelihood ratio (*LHR*) of rules with standard deviations using 10-fold stratified cross-validation - algorithms: APRIORI-SD, APRIORI-C and RIPPER.

	APRIORI-SD			APRIORI-C			RIPPER		
#	$S \pm sd$	$CVG \pm sd$	$LHR \pm sd$	$S \pm sd$	$CVG \pm sd$	$LHR \pm sd$	$S \pm sd$	$CVG \pm sd$	$LHR \pm sd$
1	3.5 ± 0.15	0.55 ± 0.06	8.40 ± 0.04	2.6 ± 0.51	0.43 ± 0.04	2.29 ± 0.03	11.6 ± 1.01	0.09 ± 0.01	2.77 ± 0.18
2	4.2 ± 0.43	0.30 ± 0.03	14.24 ± 0.02	8.0 ± 0.26	0.19 ± 0.02	3.08 ± 0.14	10.7 ± 0.12	0.10 ± 0.01	1.51 ± 0.55
3	2.4 ± 0.51	0.54 ± 0.05	8.16 ± 0.02	2.7 ± 0.04	0.60 ± 0.06	2.77 ± 0.05	1.4 ± 0.15	0.80 ± 0.08	3.30 ± 0.12
4	1.4 ± 0.20	0.53 ± 0.05	9.78 ± 0.15	3.2 ± 0.38	0.50 ± 0.05	2.87 ± 0.06	17.5 ± 0.83	0.05 ± 0.00	1.81 ± 0.05
5	4.4 ± 0.30	0.30 ± 0.03	16.40 ± 0.06	3.9 ± 0.35	0.28 ± 0.03	2.52 ± 0.07	10.2 ± 0.32	0.09 ± 0.01	2.31 ± 0.19
6	1.0 ± 0.00	1.00 ± 0.10	10.35 ± 0.03	3.5 ± 0.00	0.71 ± 0.07	2.37 ± 0.04	2.9 ± 0.12	0.39 ± 0.04	1.80 ± 0.75
7	6.2 ± 0.05	0.30 ± 0.03	10.98 ± 0.05	9.7 ± 0.83	0.11 ± 0.01	2.53 ± 0.02	11.5 ± 1.13	0.07 ± 0.01	2.57 ± 0.02
8	1.4 ± 0.14	0.67 ± 0.07	5.22 ± 0.06	4.4 ± 0.10	0.28 ± 0.03	2.69 ± 0.03	5.2 ± 0.04	0.16 ± 0.02	1.91 ± 0.08
9	2.8 ± 0.72	0.85 ± 0.08	10.53 ± 0.12	4.4 ± 0.28	0.52 ± 0.05	2.75 ± 0.06	2.5 ± 0.69	0.40 ± 0.04	2.76 ± 0.60
10	1.4 ± 0.71	0.52 ± 0.05	4.15 ± 0.04	3.0 ± 0.08	0.50 ± 0.05	1.83 ± 0.08	9.3 ± 0.14	0.10 ± 0.01	2.31 ± 0.65
11	3.5 ± 0.63	0.24 ± 0.02	1.86 ± 0.02	5.8 ± 0.13	0.22 ± 0.02	2.09 ± 0.03	6.6 ± 0.02	0.16 ± 0.02	1.81 ± 0.04
12	2.1 ± 0.49	0.84 ± 0.08	7.45 ± 0.03	2.5 ± 0.61	0.52 ± 0.05	2.86 ± 0.06	1.8 ± 0.75	0.52 ± 0.05	2.41 ± 0.04
13	2.8 ± 0.14	0.91 ± 0.09	15.25 ± 0.05	4.1 ± 0.07	0.48 ± 0.05	2.25 ± 0.18	3.3 ± 0.16	0.28 ± 0.03	2.48 ± 0.64
14	2.3 ± 0.27	0.88 ± 0.09	15.26 ± 0.03	4.0 ± 0.23	0.47 ± 0.05	3.43 ± 0.02	2.4 ± 0.05	0.23 ± 0.02	2.12 ± 0.13
15	7.1 ± 0.22	0.29 ± 0.03	15.18 ± 0.17	10.3 ± 1.00	0.13 ± 0.01	3.24 ± 0.04	25.9 ± 1.94	0.04 ± 0.00	3.33 ± 0.48
16	2.0 ± 0.44	0.71 ± 0.07	12.00 ± 0.03	4.2 ± 0.41	0.68 ± 0.07	3.33 ± 0.06	5.7 ± 0.55	0.19 ± 0.02	1.54 ± 0.02
17	4.2 ± 0.18	0.38 ± 0.04	5.61 ± 0.06	6.2 ± 0.55	0.23 ± 0.02	3.00 ± 0.04	24.0 ± 1.60	0.04 ± 0.00	3.05 ± 0.06
18	5.8 ± 0.56	0.26 ± 0.03	25.02 ± 0.07	6.8 ± 0.17	0.16 ± 0.02	2.31 ± 0.05	34.5 ± 3.01	0.03 ± 0.00	2.05 ± 0.33
19	2.8 ± 0.09	0.84 ± 0.08	2.53 ± 0.08	5.3 ± 0.45	0.37 ± 0.04	1.17 ± 0.06	5.9 ± 0.17	0.15 ± 0.02	2.97 ± 0.60
20	5.3 ± 0.20	0.24 ± 0.03	29.47 ± 0.19	9.7 ± 0.94	0.14 ± 0.01	2.22 ± 0.12	21.7 ± 1.34	0.04 ± 0.00	2.66 ± 0.66
21	8.2 ± 0.24	0.28 ± 0.03	16.78 ± 0.05	12.9 ± 1.12	0.10 ± 0.01	2.63 ± 0.04	17.2 ± 1.17	0.06 ± 0.01	1.54 ± 0.33
22	5.1 ± 0.15	0.24 ± 0.03	26.36 ± 0.18	7.3 ± 0.20	0.19 ± 0.02	3.52 ± 0.07	135.3 ± 12.7	0.01 ± 0.00	2.75 ± 0.07
23	2.4 ± 0.21	0.62 ± 0.06	13.52 ± 0.04	4.5 ± 0.36	0.55 ± 0.06	2.06 ± 0.02	3.4 ± 0.20	0.38 ± 0.04	2.52 ± 0.05
∅	3.6 ± 1.96	0.53 ± 0.26	12.37 ± 7.26	5.6 ± 2.83	0.36 ± 0.19	2.60 ± 0.55	16.1 ± 27.46	0.19 ± 0.19	2.36 ± 0.55

Table 5. Accuracy (*Acc*) and area under the ROC curve (*AUC*) of rules with standard deviations using 10-fold stratified cross-validation - algorithms: APRIORI-SD, APRIORI-C and RIPPER.

	APRIORI-SD		APRIORI-C		RIPPER	
#	$Acc \pm sd$	$AUC \pm sd$	$Acc \pm sd$	$AUC \pm sd$	$Acc \pm sd$	$AUC \pm sd$
1	87.26 ± 7.80	84.14 ± 2.00	89.99 ± 8.29	82.11 ± 2.01	82.03 ± 7.40	83.22 ± 7.80
2	97.48 ± 8.90	88.99 ± 3.05	95.85 ± 9.36	91.50 ± 3.11	94.76 ± 8.60	90.07 ± 8.02
3	86.02 ± 7.97	81.15 ± 2.03	87.17 ± 8.08	85.96 ± 2.03	86.17 ± 7.80	84.14 ± 7.59
4	96.16 ± 8.62	90.79 ± 3.02	94.52 ± 9.13	90.97 ± 3.00	98.90 ± 8.95	88.94 ± 7.91
5	75.00 ± 7.41	76.94 ± 4.06	74.95 ± 7.32	76.25 ± 4.09	71.29 ± 6.53	76.34 ± 7.41
6	67.90 ± 5.84	66.48 ± 1.05	71.66 ± 6.51	67.18 ± 1.08	67.85 ± 6.31	63.27 ± 5.69
7	69.52 ± 5.98	74.25 ± 4.25	71.19 ± 6.22	70.98 ± 4.08	72.52 ± 6.62	66.95 ± 6.66
8	79.83 ± 7.08	85.13 ± 2.07	79.57 ± 7.22	75.47 ± 2.09	69.88 ± 6.02	72.70 ± 6.84
9	82.30 ± 7.46	84.08 ± 3.06	82.44 ± 7.73	78.86 ± 3.03	81.36 ± 7.82	79.58 ± 7.82
10	99.91 ± 9.11	93.16 ± 4.00	99.20 ± 9.82	97.29 ± 4.00	99.16 ± 8.99	96.36 ± 9.48
11	88.97 ± 8.70	90.09 ± 2.08	92.44 ± 8.71	75.58 ± 2.04	86.34 ± 8.50	88.52 ± 8.28
12	95.19 ± 9.33	90.82 ± 2.00	95.59 ± 8.65	89.83 ± 2.01	96.01 ± 9.60	90.20 ± 8.99
13	79.40 ± 7.30	78.84 ± 3.10	81.72 ± 7.69	77.50 ± 3.12	76.51 ± 6.79	75.31 ± 7.25
14	79.17 ± 7.01	72.32 ± 3.08	81.08 ± 7.89	77.98 ± 3.00	74.21 ± 7.05	74.30 ± 6.98
15	75.21 ± 7.21	71.69 ± 3.68	80.15 ± 7.23	62.90 ± 4.02	85.79 ± 8.44	63.42 ± 6.29
16	96.47 ± 8.79	96.00 ± 1.06	94.63 ± 9.07	94.38 ± 1.05	93.47 ± 9.04	88.42 ± 8.43
17	75.13 ± 7.21		77.93 ± 6.86		79.48 ± 7.87	
18	85.21 ± 7.11		84.47 ± 7.64		92.73 ± 8.50	
19	66.49 ± 6.23		66.09 ± 6.05		64.84 ± 5.70	
20	19.98 ± 1.89		18.25 ± 1.37		86.42 ± 7.92	
21	68.21 ± 6.54		71.89 ± 7.17		89.39 ± 7.96	
22	75.58 ± 6.32		81.01 ± 7.47		78.85 ± 6.90	
23	93.23 ± 9.18		91.64 ± 8.29		91.51 ± 8.51	
Avg	79.98 ± 16.67	82.80 ± 8.70	81.02 ± 16.50	80.92 ± 9.95	83.46 ± 10.24	80.11 ± 10.23

Table 6. Average size (S), coverage (CVG), likelihood ratio (LHR), accuracy (Acc) and area under the ROC curve (AUC) of rules with standard deviations using 10-fold stratified cross-validation for the algorithm CN2.

	CN2				
#	$S \pm sd$	$CVG \pm sd$	$LHR \pm sd$	$Acc \pm sd$	$AUC \pm sd$
1	12.4 ± 0.01	0.07 ± 0.01	2.0 ± 0.01	81.62 ± 0.01	33.39 ± 0.01
2	12.6 ± 0.10	0.08 ± 0.10	2.7 ± 0.10	92.28 ± 0.10	90.74 ± 0.10
3	1.8 ± 0.03	0.63 ± 0.03	2.1 ± 0.03	82.45 ± 0.03	84.51 ± 0.03
4	14.6 ± 0.01	0.05 ± 0.01	2.4 ± 0.01	94.18 ± 0.01	96.22 ± 0.01
5	12.8 ± 0.08	0.06 ± 0.08	2.0 ± 0.08	72.77 ± 0.08	71.33 ± 0.08
6	3.7 ± 0.06	0.31 ± 0.06	1.9 ± 0.06	68.71 ± 0.06	70.53 ± 0.06
7	15.1 ± 0.08	0.05 ± 0.08	2.0 ± 0.08	72.40 ± 0.08	71.99 ± 0.08
8	6.4 ± 0.09	0.11 ± 0.09	1.9 ± 0.09	74.10 ± 0.09	74.17 ± 0.09
9	3.0 ± 0.04	0.21 ± 0.04	2.7 ± 0.04	80.74 ± 0.04	78.81 ± 0.04
10	10.1 ± 0.00	0.09 ± 0.00	1.4 ± 0.00	98.58 ± 0.00	96.22 ± 0.00
11	7.6 ± 0.05	0.10 ± 0.05	2.0 ± 0.05	91.44 ± 0.05	94.46 ± 0.05
12	3.8 ± 0.01	0.38 ± 0.01	1.9 ± 0.01	91.33 ± 0.01	99.17 ± 0.01
13	4.7 ± 0.11	0.16 ± 0.11	2.1 ± 0.11	80.87 ± 0.11	83.20 ± 0.11
14	5.2 ± 0.01	0.14 ± 0.01	2.5 ± 0.01	72.28 ± 0.01	75.06 ± 0.01
15	21.2 ± 0.01	0.03 ± 0.01	2.5 ± 0.01	98.01 ± 0.01	97.90 ± 0.01
16	7.1 ± 0.01	0.13 ± 0.01	2.6 ± 0.01	94.24 ± 0.01	96.88 ± 0.01
17	28.7 ± 0.00	0.02 ± 0.00	2.7 ± 0.00	74.71 ± 0.00	
18	83.8 ± 0.05	0.02 ± 0.05	1.5 ± 0.05	89.82 ± 0.05	
19	12.9 ± 0.01	0.07 ± 0.01	1.0 ± 0.01	60.60 ± 0.01	
20	32.8 ± 0.11	0.04 ± 0.11	1.5 ± 0.11	58.88 ± 0.11	
21	35.1 ± 0.01	0.04 ± 0.01	2.4 ± 0.01	88.73 ± 0.01	
22	77.3 ± 0.01	0.00 ± 0.01	2.6 ± 0.01	69.18 ± 0.01	
23	5.5 ± 0.01	0.23 ± 0.01	2.0 ± 0.01	89.16 ± 0.01	
Avg	18.2 ± 21.77	0.13 ± 0.14	2.11 ± 0.46	81.61 ± 11.66	82.16 ± 16.81

Table 7. Average size (S), coverage (CVG), likelihood ratio (LHR), accuracy (Acc) and area under the ROC curve (AUC) of rules with standard deviations using 10-fold stratified cross-validation for the algorithm CN2-SD.

	CN2-SD (additive)				
#	$S \pm sd$	$CVG \pm sd$	$LHR \pm sd$	$Acc \pm sd$	$AUC \pm sd$
1	3.5 ± 0.54	0.42 ± 0.03	4.6 ± 0.03	88.35 ± 0.03	85.12 ± 0.03
2	9.2 ± 0.25	0.26 ± 0.04	26.6 ± 0.04	92.60 ± 0.04	94.52 ± 0.04
3	1.8 ± 0.06	0.33 ± 0.02	22.9 ± 0.02	81.46 ± 0.02	83.03 ± 0.02
4	8.5 ± 0.32	0.51 ± 0.04	30.2 ± 0.04	96.08 ± 0.04	92.87 ± 0.04
5	4.6 ± 0.33	0.38 ± 0.04	2.1 ± 0.04	74.12 ± 0.04	80.06 ± 0.04
6	3.4 ± 0.05	0.87 ± 0.05	23.1 ± 0.05	68.75 ± 0.05	70.61 ± 0.05
7	8.8 ± 0.87	0.15 ± 0.12	16.3 ± 0.12	72.40 ± 0.12	72.73 ± 0.12
8	1.8 ± 0.18	0.44 ± 0.09	30.6 ± 0.09	78.03 ± 0.09	85.62 ± 0.09
9	2.7 ± 0.24	0.69 ± 0.03	25.0 ± 0.03	86.14 ± 0.03	81.29 ± 0.03
10	2.5 ± 0.03	0.51 ± 0.00	13.5 ± 0.00	99.10 ± 0.00	97.42 ± 0.00
11	4.2 ± 0.12	0.35 ± 0.06	14.9 ± 0.06	91.10 ± 0.06	93.87 ± 0.06
12	3.6 ± 0.68	0.67 ± 0.01	4.0 ± 0.01	91.75 ± 0.01	99.46 ± 0.01
13	4.5 ± 0.09	0.62 ± 0.17	9.7 ± 0.17	80.86 ± 0.17	83.06 ± 0.17
14	2.1 ± 0.27	0.72 ± 0.03	18.1 ± 0.03	77.74 ± 0.03	78.51 ± 0.03
15	10.2 ± 1.05	0.12 ± 0.03	26.5 ± 0.03	77.38 ± 0.03	89.15 ± 0.03
16	1.8 ± 0.43	0.83 ± 0.03	6.0 ± 0.03	97.62 ± 0.03	93.95 ± 0.03
17	8.3 ± 0.51	0.32 ± 0.00	24.3 ± 0.00	72.51 ± 0.00	
18	12.8 ± 0.17	0.20 ± 0.06	19.3 ± 0.06	72.48 ± 0.06	
19	10.1 ± 0.48	0.76 ± 0.01	9.1 ± 0.01	69.32 ± 0.01	
20	9.2 ± 0.99	0.17 ± 0.17	21.7 ± 0.17	19.49 ± 0.17	
21	19.2 ± 1.14	0.22 ± 0.03	30.6 ± 0.03	71.04 ± 0.03	
22	11.7 ± 0.23	0.19 ± 0.03	20.2 ± 0.03	75.70 ± 0.03	
23	1.4 ± 0.32	0.89 ± 0.03	25.7 ± 0.03	91.32 ± 0.03	
Avg	6.4 ± 4.60	0.46 ± 0.25	18.47 ± 9.00	79.36 ± 16.24	86.33 ± 8.60

Solving Classification Problems Using Infix Form Genetic Programming

Mihai Oltean and Crina Groşan

Department of Computer Science,
Faculty of Mathematics and Computer Science,
Babeş-Bolyai University, Kogălniceanu 1
Cluj-Napoca, 3400, Romania.
{moltean,cgrosan}@cs.ubbcluj.ro

Abstract. A new evolutionary technique, called Infix Form Genetic Programming (IFGP) is proposed in this paper. The IFGP individuals are strings encoding complex mathematical expressions. The IFGP technique is used for solving several classification problems. All test problems are taken from PROBEN1 and contain real world data. IFGP is compared to Linear Genetic Programming (LGP) and Artificial Neural Networks (ANNs). Numerical experiments show that IFGP is able to solve the considered test problems with the same (and sometimes even better) classification error than that obtained by LGP and ANNs.

1 Introduction

Classification is a task of assigning inputs to a number of discrete categories or classes [3,9]. Examples include classifying a handwritten letter as one from A-Z, classifying a speech pattern to the corresponding word, etc.

Machine learning techniques have been extensively used for solving classification problems. In particular Artificial Neural Networks (ANNs) [3] have been originally designed for classifying a set of points in two distinct classes.

Recently Genetic Programming (GP) techniques [1,4] have been used for classification purposes. GP has been developed as a technique for evolving computer programs. Originally GP chromosomes were represented as trees and evolved using specific genetic operators [4]. Later, several linear variants of GP have been proposed. Some of them are Linear Genetic Programming (LGP) [1], Grammatical Evolution (GE) [6], Gene Expression Programming (GEP) [2]. In all these techniques chromosomes are represented as strings encoding complex computer programs.

Some of these GP techniques have been used for classification purposes. For instance, LGP [1] has been used for solving several classification problems in PROBEN1. The conclusion was that LGP is able to solve the classification problems with the same error rate as a neural network.

A new evolutionary technique, called Infix Form Genetic Programming (IFGP), is proposed in this paper. IFGP chromosomes are strings encoding complex mathematical expressions using infix form. This departs from GE [6] and

M.R. Berthold et al. (Eds.): IDA 2003, LNCS 2810, pp. 242–253, 2003.

GADS [7] approaches which store expressions using Polish [4] form. An interesting feature of IFGP is its ability of storing multiple solutions of a problem in a chromosome.

In this paper IFGP is used for solving several real-world classification problems taken from PROBEN1.

The paper is organized as follows. Several machine learning techniques (Neural Networks and Linear Genetic Programming) for solving classification problems are briefly described in section 2. Section 3 describes in detail the proposed IFGP technique. The benchmark problems used for comparing these techniques are presented in section 4. Comparison results are presented in section 5.

2 Machine Learning Techniques Used for Classification Task

Two of the machines learning techniques used for solving classification problems are described in this section.

2.1 Artificial Neural Networks

Artificial Neural Networks (ANNs) [3,9] are motivated by biological neural networks. ANNs have been extensively used for classification and clustering purposes. In their early stages, neural networks have been primarily used for classification tasks. The most popular architectures consist of multi layer perceptrons (MLP) and the most efficient training algorithm is backpropagation [3].

ANNs have been successfully applied for solving real-world problems. For instance, the PROBEN1 [8] data set contains several difficult real-world problems accompanied by the results obtained by running different ANNs architectures on the considered test problems. The method applied in PROBEN for training was RPROP (a fast backpropagation algorithm - similar to Quickprop). The parameters used were not determined by a trial-and-error search. Instead they are just educated guesses. More information about the ANNs parameters used in PROBEN1 can be found in [8].

2.2 Linear Genetic Programming

Linear Genetic Programming (LGP) [1] uses a specific linear representation of computer programs.

Programs of an imperative language (like ***C***) are evolved instead of the tree-based GP expressions of a functional programming language (like ***LISP***)

An LGP individual is represented by a variable-length sequence of simple ***C*** language instructions. Instructions operate on one or two indexed variables (registers) r or on constants c from predefined sets. The result is assigned to a destination register.

An example of LGP program is the following:

```
void LGP_Program(double v[8])
{
...
v[0] = v[5] + 73;
v[7] = v[4] - 59;
if (v[1] > 0)
if (v[5] > 21)
v[4] = v[2] *v[1];
v[2] = v[5] + v[4];
v[6] = v[1] * 25;
v[6] = v[4] - 4;
v[1] = sin(v[6]);
if (v[0] > v[1])
v[3] = v[5] * v[5];
v[7] = v[6] * 2;
v[5] = [7] + 115;
if (v[1] ≤ v[6])
v[1] = sin(v[7]);
}
```

A linear genetic program can be turned into a functional representation by successive replacements of variables starting with the last effective instruction. Variation operators are crossover and mutation. By crossover continuous sequences of instructions are selected and exchanged between parents. Two types of mutations are used: micro mutation and macro mutation. By micro mutation an operand or operator of an instruction is changed. Macro mutation inserts or deletes a random instruction.

Crossover and mutations change the number of instructions in chromosome.

The population is divided into demes which are arranged in a ring. Migration is allowed only between a deme and the next one (in the clockwise direction) with a fixed probability of migration. Migrated individuals replace the worst individuals in the new deme. The deme model is used in order to preserve population diversity (see [1] for more details).

3 IFGP Technique

In this section the proposed *Infix Form Genetic Programming* (IFGP) technique is described. IFGP uses linear chromosomes for solution representation (i.e. a chromosome is a string of genes). An IFGP chromosome usually encodes several solutions of a problem.

3.1 Prerequisite

We denote by F the set of function symbols (or operators) that may appear in a mathematical expression. F usually contains the binary operators $\{+, -, *, /\}$. By *Number_of_Operators* we denote the number of elements in F. A correct mathematical expression also contains some terminal symbols. The set of terminal symbols is denoted by T. The number of terminal symbols is denoted by *Number_of_Variables*.

Thus, the symbols that may appear in a mathematical expression are $T \cup F \cup \{`(', `)'\}$. The total number of symbols that may appear in a valid mathematical expression is denoted by *Number_of_Symbols*.

By C_i we denote the value on the ith gene in a IFGP chromosome and by G_i the symbol in the ith position in the mathematical expression encoded into an IFGP chromosome.

3.2 IFGP Algorithm

A steady-state [10] variant of IFGP is employed in this paper. The algorithm starts with a randomly chosen population of individuals. The following steps are repeated until a termination condition is reached. Two parents are chosen at each step using binary tournament selection [5]. The selected individuals are recombined with a fixed crossover probability p_c. By recombining two parents, two offspring are obtained. The offspring are mutated and the best of them replaces the worst individual in the current population (only if the offspring is better than the worst individual in population). The algorithm returns as its answer the best expression evolved for a fixed number of generations.

3.3 IFGP Individual Representation

In this section we describe how IFGP individuals are represented and how they are decoded in order to obtain a valid mathematical expression.

Each IFGP individual is a fixed size string of genes. Each gene is an integer number in the interval [0 .. *Number_Of_Symbols* - 1]. An IFGP individual can be transformed into a functional mathematical expression by replacing each gene with an effective symbol (a variable, an operator or a parenthesis).

Example

If we use the set of functions symbols $F = \{+, *, -, /\}$, and the set of terminals $T = \{a, b\}$, the following chromosome

C = 7, 3, 2, 2, 5

is a valid chromosome in IFGP system.

3.4 IFGP Decoding Process

We will begin to decode this chromosome into a valid mathematical expression. In the first position (in a valid mathematical expression) we may have either a variable, or an open parenthesis. That means that we have *Number_Of_Variables* + 1 possibilities to choose a correct symbol on the first position. We put these possibilities in order: the first possibility is to choose the variable x_1, the second possibility is to choose the variable x_2 ... the last possibility is to choose the closed parenthesis ')'. The actual value is given by the value of the first gene of the chromosome. Because the number stored in a chromosome gene may be larger than the number of possible correct symbols for the first position we take only the value of the first gene *modulo* [6,7] number of possibilities for the first gene.

Generally, when we compute the symbol stored in the ith position in expression we have to compute first how many symbols may be placed in that position. The number of possible symbols that may be placed in the current position depends on the symbol placed in the previous position. Thus:

(i) if the previous position contains a variable (x_i), then for the current position we may have either an operator or a closed parenthesis. The closed parenthesis is considered only if the number of open parentheses so far is larger than the number of closed parentheses so far.
(ii) if the previous position contains an operator, then for the current position we may have either a variable or an open parenthesis.
(iii) if the previous position contains an open parenthesis, then for the current position we may have either a variable or another open parenthesis.
(iv) if the previous position contains a closed parenthesis, then for the current position we may have either an operator or another closed parenthesis. The closed parenthesis is considered only if the number of open parentheses so far is larger than the number of closed parentheses.

Once we have computed the number of possibilities for the current position it is easy to determine the symbol that will be placed in that position: first we take the value of the corresponding gene modulo the number of possibilities for that position. Let p be that value (p = C_i *mod Number_Of_Possibilities*). The pth symbol from the permitted symbols for the current is placed in the current position in the mathematical expression. (Symbols that may appear into a mathematical expression are ordered arbitrarily. For instance we may use the following order: x_1, x_2, ..., +, -, *, /, '(', ')'.)

All chromosome genes are translated but the last one. The last gene is used by the correction mechanism (see below).

The obtained expression usually is syntactically correct. However, in some situations the obtained expression needs to be repaired. There are two cases when the expression needs to be corrected:

The last symbol is an operator (+, -, *, /) or an open parenthesis. In that case a terminal symbol (a variable) is added to the end of the expression. The added symbol is given by the last gene of the chromosome.

The number of open parentheses is greater than the number of closed parentheses. In that case several closed parentheses are automatically added to the end in order to obtain a syntactically correct expression.

Remark: If the correction mechanism is not used, the last gene of the chromosome will not be used.

Example

Consider the chromosome C = 7, 3, 2, 0, 5, 2 and the set of terminal and function symbols previously defined ($T = \{a, b\}$, F = {+, -, *, /}).

For the first position we have 3 possible symbols (a, b and '('). Thus, the symbol in the position C_0 ***mod*** $3 = 1$ in the array of possible symbols is placed in the current position in expression. The chosen symbol is b, because the index of the first symbol is considered to be 0.

For the second position we have 4 possibilities (+, -, *, /). The possibility of placing a closed parenthesis is ignored since the difference between the number of open parentheses and the number of closed parentheses is zero. Thus, the symbol '/' is placed in position 2.

For the third position we have 3 possibilities (a, b and '('). The symbol placed on that position is an open parenthesis '('.

In the fourth position we have 3 possibilities again (a, b and '('). The symbol placed on that position is the variable a.

For the last position we have 5 possibilities (+, -, *, /) and the closed parenthesis ')'. We choose the symbol on the position 5 **mod** $5 = 0$ in the array of possible symbols. Thus the symbol '+' is placed in that position.

The obtained expression is $E = b\ /\ (a+$.

It can be seen that the expression E it is not syntactically correct. For repairing it we add a terminal symbol to the end and then we add a closed parenthesis. Now we have obtained a correct expression:

$$E = b\ /\ (a + a).$$

The expression tree of E is depicted in Fig. 1.

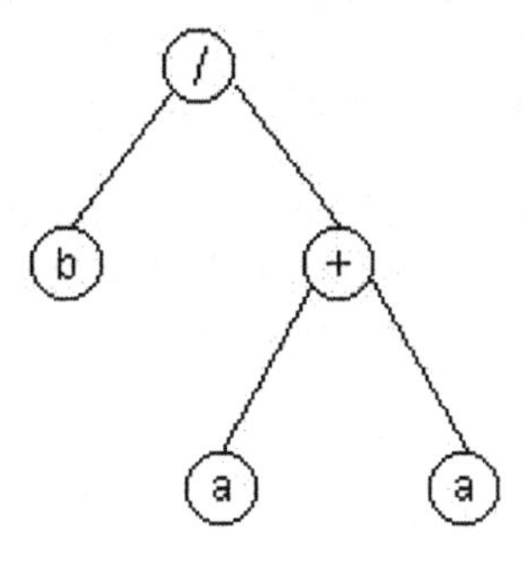

Fig. 1. The expression tree of $E = b\ /\ (a + a)$.

3.5 Using Constants within IFGP Model

An important issue when designing a new GP technique is the way in which the constants are embaded into the proposed model.

Fixed or ephemeral random constants have been extensively tested within GP systems [4]. The interval over which the constants are initially generated is usually problem-dependent. For a good functionality of the program a priori knowledge about the required constants is usually needed.

By contrast, IFGP employs a problem-independent system of constants.

We know that each real number may be written as a sum of powers of 2.

In the IFGP model each constant is a power of 2. The total number of constants is also problem-independent. For instance, if we want to solve problems using double precision constants, we need 127 constants: 2^{-63}, 2^{-62},..., 2^{-1}, 2^0, 2^1,..., 2^{63}. Particular instances of the classification problems may require fewer constants. As can be seen in section 5 a good solution for some classification problems can be obtained without using constants. However, if we do not know what kind of constants are required by the problems being solved it is better the use the double precision system of constants (as described above).

Within the IFGP chromosome each constant is represented by its exponent. For instance the constant 2^{17} is stored as the number 17, the constant 2^{-4} is stored as the number - 4 and the constant $1 = 2^0$ is stored as the number 0. These constants will act as terminal symbols. Thus the extended set of terminal symbols is $T = \{x_1, x_2, \ldots$, -63, -62, .., -1, 0, 1, ..., 62, 63$\}$.

3.6 Fitness Assignment Process

In this section we describe how IFGP may be efficiently used for solving classification problems.

A GP chromosome usually stores a single solution of a problem and the fitness is normally computed using a set of fitness cases.

Instead of encoding a single solution, an IFGP individual is allowed to store multiple solutions of a problem. The fitness of each solution is computed in a conventional manner and the solution having the best fitness is chosen to represent the chromosome.

In the IFGP representation each sub-tree (sub-expression) is considered as a potential solution of a problem. For example, the previously obtained expression (see section 3.4) contains 4 distinct solutions (sub-expressions):

$E_1 = a,$
$E_2 = b,$
$E_3 = a + a,$
$E_4 = b \;/\; (a + a).$

Now we will explain how the fitness of a (sub)expression is computed.

Each class has associated a numerical value: the first class has the value 0, the second class has the value 1 and the mth class has associated the numerical value $m-1$. Any other system of distinct numbers may be used. We denote by o_k the number associated to the kth class.

The value $v_j(E_i)$ of each expression E_i (in an IFGP) chromosome for each row (example) j in the training set is computed. Then, each row in the training set will be classified to the nearest class (the class k for which the difference $|v_j(E_i) - o_k|$ is minimal). The fitness of a (sub)expression is equal to the number of incorrectly classified examples in the training set. The fitness of an IFGP chromosome will be equal to the fitness of the best expression encoded in that chromosome.

Remarks:

(i) Since the set of numbers associated with the problem classes was arbitrarily chosen it is expected that different systems of number to generate different solutions.

(ii) When solving symbolic regression or classification problems IFGP chromosomes need to be traversed twice for computing the fitness. That means that the complexity of the IFGP decoding process it is not higher than the complexity of other methods that store a single solution in a chromosome.

3.7 Search Operators

Search operators used within the IFGP model are recombination and mutation. These operators are similar to the genetic operators used in conjunction with binary encoding [5]. By recombination two parents exchange genetic material in order to obtain two offspring. In this paper only two-point recombination is used [5]. Mutation operator is applied with a fixed mutation probability (p_m). By mutation a randomly generated value over the interval [0, *Number_of_Symbols*–1] is assigned to the target gene.

3.8 Handling Exceptions within IFGP

Exceptions are special situations that interrupt the normal flow of expression evaluation (program execution). An example of exception is *division by zero* which is raised when the divisor is equal to zero.

GP techniques usually use a *protected exception* handling mechanism [4,6]. For instance if a division by zero exception is encountered, a predefined value (for instance 1 or the numerator) is returned. This kind of handling mechanism is specific for Linear GP [1], standard GP [4] and GE [6].

IFGP uses a new and specific mechanism for handling exceptions. When an exception is encountered (which is always generated by a gene containing a function symbol), the entire (sub) tree which has generated the exception is mutated (changed) into a terminal symbol. Exception handling is performed during the fitness assignment process.

4 Data Sets

Numerical experiments performed in this paper are based on several benchmark problems taken from PROBEN1 [8]. These datasets were created based on the datasets from the UCI Machine Learning Repository [11].

Used problems are briefly described in what follows.

Cancer

Diagnosis of breast cancer. Try to classify a tumor as either benignant or malignant based on cell descriptions gathered by microscopic examination.

Diabetes

Diagnosis diabetes of Pima Indians based on personal data and the results of medical examinations try to decide whether a Pima Indian individual is diabetes positive or not.

Heartc

Predicts heart disease. Decides whether at least one of four major vessels is reduced in diameter by more than 50%. The binary decision is made based on personal data such as age sex smoking habits subjective patient pain descriptions and results of various medical examinations such as blood pressure and electro cardiogram results.

This data set was originally created by Robert Detrano from V.A. Medical Center Long Beach and Cleveland Clinic Foundation.

Horse

Predicts the fate of a horse that has colic. The results of a veterinary examination of a horse having colic are used to predict whether the horse will survive will die or should be euthanized.

The number of inputs, of classes and of available examples, for each test problem, are summarized in Table 1.

Table 1. Summarized attributes of several classification problems from PROBEN1

Problem	Number of inputs	Number of classes	Number of examples
cancer	9	2	699
diabetes	8	2	768
heartc	35	2	303
horse	58	3	364

5 Numerical Experiments

The results of several numerical experiments with ANNs, LGP and IFGP are presented in this section.

Each data set is divided in three sub-sets (training set -50%, validation set - 25 %, and test set – 25%) (see [8]).

The test set performance is computed for that chromosome which had minim validation error during the search process. This method, called *early stopping*, is a good way to avoid overfitting [8] of the population individuals to the particular training examples used. In that case the generalization performance will be reduced.

In [1] Linear GP was used to solve several classification problems from PROBEN1. The parameters used by Linear GP are given in Table 2.

Table 2. Linear GP parameters used for solving classification tasks from PROBEN1

Parameter	**Value**
Population size	5000
Number of demes	10
Migration rate	5%
Classification error weight in fitness	1.0
Maximum number of generations	250
Crossover probability	90%
Mutation probability	90%
Maximum mutation step size for constants	±5
Maximum program size	256 instructions
Initial maximum program size	25 instructions
Function set	{+, -, *, /, *sin*, *exp*, if >, if ≤}
Terminal set	{0,..,256} ∪ {input variables}

The parameters of the IFGP algorithm are given in Table 3.

Table 3. IFGP algorithm parameters for solving classification problems from PROBEN1

Parameter	**Value**
Population Size	250
Chromosome length	30
Number of generations	250
Crossover probability	0.9
Crossover type	Two-point Crossover
Mutation	2 mutations per chromosome
Number of Constants	41 $\{2^{-20}, \ldots, 2^{0}, 2^{20}\}$
Function set	{+, -, *, /, *sin*, *exp*}
Terminal set	{input variables} ∪ The set of constants.

The results of the numerical experiments are presented in Table 4.

From Table 4 it can be seen that IFGP is able to obtain similar performances as those obtained by LGP even if the population size and the chromosome length used by IFGP are smaller than those used by LGP. When compared to ANNs we can see that IFGP is better only in 3 cases (out of 12).

We are also interested in analysing the relationship between the classification error and the number of constants used by the IFGP chromosomes. For this purpose we will use a small population made up of only 50 individuals. Note that this is two magnitude orders smaller than those used by LGP. Other

Table 4. Classification error rates of IFGP, LGP and ANN for some date sets from PROBEN1. LGP results are taken from [1]. ANNs results are taken from [8]. Results are averaged over 30 runs

Problem	IFGP–test set			LGP–test set			NN–test set	
	best	mean	stdev	best	mean	stdev	mean	stdev
cancer1	1.14	2.45	0.69	0.57	2.18	0.59	1.38	0.49
cancer2	4.59	6.16	0.45	4.02	5.72	0.66	4.77	0.94
cancer3	3.44	4.92	1.23	3.45	4.93	0.65	3.70	0.52
diabetes1	22.39	25.64	1.61	21.35	23.96	1.42	24.10	1.91
diabetes2	25.52	28.92	1.71	25.00	27.85	1.49	26.42	2.26
diabetes3	21.35	25.31	2.20	19.27	23.09	1.27	22.59	2.23
heart1	16.00	23.06	3.72	18.67	21.12	2.02	20.82	1.47
heart2	1.33	4.40	2.35	1.33	7.31	3.31	5.13	1.63
heart3	12.00	13.64	2.34	10.67	13.98	2.03	15.40	3.20
horse1	23.07	31.11	2.68	23.08	30.55	2.24	29.19	2.62
horse2	30.76	35.05	2.33	31.87	36.12	1.95	35.86	2.46
horse3	30.76	35.01	2.82	31.87	35.44	1.77	34.16	2.32

IFGP parameters are given in Table 3. Experimental results are given in Table 5.

Table 5. Classification error rates of IFGP (on the test set) using different number of constants. Results are averaged over 30 runs

Problem	IFGP–41 constants			IFGP–0 constants			IFGP–81 constants		
	best	mean	stdev	best	mean	stdev	best	mean	stdev
cancer1	1.14	2.45	0.60	1.14	3.18	1.06	1.14	2.41	0.77
cancer2	4.02	6.16	0.91	4.02	6.14	0.81	4.59	6.24	0.99
cancer3	3.44	5.17	1.10	2.87	5.07	1.50	2.87	5.15	1.07
diabetes1	22.39	25.74	2.19	21.87	26.04	1.76	21.87	25.34	2.08
diabetes2	26.04	29.91	1.65	25.00	29.21	2.21	25.52	29.82	1.30
diabetes3	21.87	25.34	1.76	19.79	24.79	1.91	22.91	25.88	3.60
heart1	17.33	24.44	3.76	18.67	23.28	3.33	18.66	25.28	3.64
heart2	1.33	6.97	4.07	1.33	4.97	3.16	1.33	7.06	4.60
heart3	12.00	14.00	2.51	10.67	15.42	3.40	9.33	15.11	4.25
horse1	26.37	32.16	3.12	25.27	30.69	2.49	27.47	31.57	1.91
horse2	30.76	35.64	2.37	30.76	35.49	2.81	31.86	35.71	2.23
horse3	28.57	34.13	3.14	28.57	35.67	3.90	28.57	34.90	3.53

From Table 5 it can be seen that the best results are obtained when the constants are not used in our IFGP system. For 8 (out of 12) cases the best result obtained by IFGP outperform the best result obtained by LGP. That does not mean that the constants are useless in our model and for the considered test problems. An explanation for this behaviour can be found if we have a look at the parameters used by IFGP. The population size (of only 50 individuals) and the chromosome length (of only 30 genes) could not be enough to obtain a perfect convergence knowing that some problems have many parameters (input variables). For instance the '*horse*' problem has 58 attributes and a chromosome of only 30 genes could not be enough to evolve a complex expression that contains sufficient problem's variables and some of the considered constants. It is expected that longer chromosomes will increase the performances of the IFGP technique.

6 Conclusion and Future Work

A new evolutionary technique, Infix Form Genetic Programming (IFGP) has been proposed in this paper. The IFGP technique has been used for solving several classification problems. Numerical experiments show that the error rates obtained by using IFGP are similar and sometimes even better than those obtained by Linear Genetic Programming.

Further numerical experiments will try to analyse the relationship between the parameters of the IFGP algorithm and the classification error for the considered test problems.

References

1. Brameier, M., Banzhaf, W.: A Comparison of Linear Genetic Programming and Neural Networks in Medical Data Mining. IEEE Transactions on Evolutionary Computation (2001) 5:17–26.
2. Ferreira, C.: Gene Expression Programming: a New Adaptive Algorithm for Solving Problems, Complex Systems (2001), 13(2):87–129.
3. Haykin S.: Neural Networks, a Comprehensive Foundation. second edition, Prentice-Hall, Englewood Cliffs (1999).
4. Koza, J. R.: Genetic Programming: On the Programming of Computers by Means of Natural Selection. Cambridge, MA: MIT Press (1992).
5. Goldberg, D.E.: Genetic Algorithms in Search, Optimization, and Machine Learning. Addison-Wesley, Reading, MA (1989).
6. O'Neill, M., Ryan, C.: Grammatical Evolution. IEEE Transaction on Evolutionary Computation, Vol 5, (2001) 349–358.
7. Paterson, N.R., Livesey, M.J.: Distinguishing Genotype and Phenotype in Genetic Programming. In: Genetic Programming: Proceedings of the first annual conference, Koza,John R; Goldberg, David E; Fogel, David B; Riolo, Rick (Eds), The MIT Press (1996) 141–150.
8. Prechelt, L.: PROBEN1 – A Set of Neural Network Problems and Benchmarking Rules, Technical Report 21, University of Karlsruhe (1994).
9. Russell, S., Norvig, P.: Artificial Intelligence: A Modern Approach, Prentice Hall, Englewood Cliffs, NJ (1994).
10. Syswerda, G.: Uniform Crossover in Genetic Algorithms. In: Proc. 3rd Int. Conf. on Genetic Algorithms, Schaffer, J.D. (eds.), Morgan Kaufmann Publishers, San Mateo, CA (1989) 2–9.
11. UCI Machine Learning Repository, Available from www.ics.uci.edu/~mlearn/MLRepository.html

What Is Fuzzy about Fuzzy Clustering? Understanding and Improving the Concept of the Fuzzifier

Frank Klawonn and Frank Höppner

Department of Computer Science
University of Applied Sciences Braunschweig/Wolfenbüttel
Salzdahlumer Str. 46/48
D-38302 Wolfenbüttel, Germany
{f.klawonn,f.hoeppner}@fh-wolfenbuettel.de

Abstract. The step from the well-known c-means clustering algorithm to the fuzzy c-means algorithm and its vast number of sophisticated extensions and generalisations involves an additional clustering parameter, the so called fuzzifier. This fuzzifier controls how much clusters may overlap. It also has some negative effects causing problems for clusters with varying data density, noisy data and large data sets with a higher number of clusters. In this paper we take a closer look at what the underlying general principle of the fuzzifier is. Based on these investigations, we propose an improved more general framework that avoids the undesired effects of the fuzzifier.

1 Introduction

Clustering is an exploratory data analysis method applied to data in order to discover structures or certain groupings in a data set. Fuzzy clustering accepts the fact that the clusters or classes in the data are usually not completely well separated and thus assigns a membership degree between 0 and 1 for each cluster to every datum.

The most common fuzzy clustering techniques aim at minimizing an objective function whose (main) parameters are the membership degrees and the parameters determining the localisation as well as the shape of the clusters. Although the extension from deterministic (hard) to fuzzy clustering seems to be an obvious concept, it turns out that to actually obtain membership degrees between zero and one, it is necessary to introduce a so-called fuzzifier in fuzzy clustering. Usually, the fuzzifier is simply used to control how much clusters are allowed to overlap. In this paper, we provide a deeper understanding of the underlying concept of the fuzzifier and derive a more general approach that leads to improved results in fuzzy clustering.

Section 2 briefly reviews the necessary background in objective function-based fuzzy clustering. The purpose, background and the consequences of the additional parameter in fuzzy clustering – the fuzzifier – is examined in section 3.

M.R. Berthold et al. (Eds.): IDA 2003, LNCS 2810, pp. 254–264, 2003.

Based on these considerations and on a more general understanding, we propose an improved alternative to the fuzzifier in section 4 and outline possible other approaches in the final conclusions.

2 Objective Function-Based Fuzzy Clustering

Clustering aims at dividing a data set into groups or clusters that consist of similar data. There is a large number of clustering techniques available with different underlying assumptions about the data and the clusters to be discovered. A simple and common popular approach is the so-called c-means clustering [4]. For the c-means algorithm it is assumed that the number of clusters is known or at least fixed, i.e., the algorithm will partition a given data set $X = \{x_1, \ldots, x_n\} \subset R^p$ into c clusters. Since the assumption of a known or a priori fixed number of clusters is not realistic for many data analysis problems, there are techniques based on cluster validity considerations that allow to determine the number of clusters for the c-means algorithm as well. However, the underlying algorithm remains more or less the same, only the number of clusters is varied and the resulting clusters or the overall partition is evaluated. Therefore, it is sufficient to assume for the rest of the paper that the number of clusters is always fixed.

From the purely algorithmic point of view, the c-means clustering can be described as follows. Each of the c clusters is represented by a prototype $v_i \in R^p$. These prototypes are chosen randomly in the beginning. Then each data vector is assigned to the nearest prototype (w.r.t. the Euclidean distance). Then each prototype is replaced by the centre of gravity of those data assigned to it. The alternating assignment of data to the nearest prototype and the update of the prototypes as cluster centres is repeated until the algorithm converges, i.e., no more changes happen.

This algorithm can also be seen as a strategy for minimizing the following objective function:

$$f = \sum_{i=1}^{c} \sum_{j=1}^{n} u_{ij} d_{ij} \tag{1}$$

under the constraints

$$\sum_{i=1}^{c} u_{ij} = 1 \qquad \text{for all } j = 1, \ldots, n \tag{2}$$

where $u_{ij} \in \{0, 1\}$ indicates whether data vector x_j is assigned to cluster i ($u_{ij} = 1$) or not ($u_{ij} = 0$). $d_{ij} = \| x_j - v_i \|^2$ is the squared Euclidean distance between data vector x_j and cluster prototype v_i.

Since this is a non-trivial constraint nonlinear optimisation problem with continuous parameters v_i and discrete parameters u_{ij}, there is no obvious analytical solution. Therefore an alternating optimisation scheme, alternatingly optimising one set of parameters while the other set of parameters is considered as fixed, seems to be a reasonable approach for minimizing (1). The above mentioned c-means clustering algorithm follows exactly this strategy.

It should be noted that choosing the (squared) Euclidean distance as a measure for the distance between data vector u_{ij} and cluster i is just one choice out of many. In this paper we are not interested in the great variety of specific cluster shapes (spheres, ellipsoids, lines, quadrics,...) that can be found by choosing suitable cluster parameters and an adequate distance function. (For an overview we refer to [2,5].) Our considerations can be applied to all cluster shapes. In this paper we concentrate on the assignment of data to clusters specified by the u_{ij}-values, especially in fuzzy clustering where the assumption $u_{ij} \in \{0,1\}$ is relaxed to $u_{ij} \in [0,1]$. In this case, u_{ij} is interpreted as the membership degree of data vector x_j to cluster i. Especially, when ambiguous data exist and cluster boundaries are not sharp, membership degrees are more realistic than crisp assignments. However, it turned out that the minimum of the objective function (1) under the constraints (2) is still obtained, when u_{ij} is chosen in the same way as in the c-means algorithm, i.e. $u_{ij} \in \{0,1\}$, even if we allow $u_{ij} \in [0,1]$. Therefore, an additional parameter m, the so-called fuzzifier [1], was introduced and the objective function (1) is replaced by

$$f = \sum_{i=1}^{c}\sum_{j=1}^{n} u_{ij}^{m} d_{ij}. \tag{3}$$

Note that the fuzzifier m does not have any effects, when we use hard clustering. The fuzzifier $m > 1$ is not subject of the optimisation process and has to be chosen in advance. A typical choice is $m = 2$. We will discuss the effects of the fuzzifier in the next section. The fuzzy clustering approach with the objective function (3) under the constraints (2) and the assumption $u_{ij} \in [0,1]$ is called probabilistic clustering, since due to the constraints (2) the membership degree u_{ij} can be interpreted as the probability that x_j belongs to cluster i.

This still leads to a nonlinear optimisation problem, however, in contrast to hard clustering, with all parameters being continuous. The common technique for minimizing this objective function is similar as in hard clustering, alternatingly optimise either the membership degrees or the cluster parameters while considering the other parameter set as fixed.

Taking the constraints (2) into account by Lagrange functions, the minimum of the objective function (3) w.r.t. the membership degrees is obtained at [1]

$$u_{ij} = \frac{1}{\sum_{k=1}^{c}\left(\frac{d_{ij}}{d_{kj}}\right)^{\frac{1}{m-1}}}, \tag{4}$$

when the cluster parameters, i.e. the distance values d_{ij}, are considered to be fixed. (If $d_{ij} = 0$ for one or more clusters, we deviate from (4) and assign x_j with membership degree 1 to the or one of the clusters with $d_{ij} = 0$ and choose $u_{ij} = 0$ for the other clusters i.)

If the clusters are represented by simple prototypes $v_i \in R^p$ and the distances d_{ij} are the squared Euclidean distances of the data to the corresponding cluster prototypes as in the hard c-means algorithm, the minimum of the objective function (3) w.r.t. the cluster prototypes is obtained at [1]

$$v_i = \frac{\sum_{j=1}^{n} u_{ij}^m x_j}{\sum_{j=1}^{n} u_{ij}^m}, \qquad (5)$$

when the membership degrees u_{ij} are considered to be fixed. The prototypes are still the cluster centres. However, using $[0,1]$-valued membership degrees means that we have to compute weighted cluster centres. The fuzzy clustering scheme using alternatingly equations (4) and (5) is called fuzzy c-means algorithm (FCM). As mentioned before, more complicated cluster shapes can be detected by introducing additional cluster parameters and a modified distance function. Our considerations apply to all these schemes, but it would lead too far to discuss them in detail. However, we should mention that there are alternative approaches to fuzzy clustering than only probabilistic clustering. Noise clustering [3] maintains the principle of probabilistic clustering, but an additional noise cluster is introduced. All data have a fixed (large) distance to the noise cluster. In this way, data that are near the border between two clusters, still have a high membership degree to both clusters as in probabilistic clustering. But data that are far away from all clusters will be assigned to the noise cluster and have no longer a high membership degree to other clusters. Our investigations and our alternative approach fit also perfectly to noise clustering. We do not cover possibilistic clustering [8] where the probabilistic constraint is completely dropped and an additional term in the objective function is introduced to avoid the trivial solution $u_{ij} = 0$ for all i, j. However, the aim of possibilistic clustering is actually not to find, but even to avoid the global optimum of the corresponding objective function. This global optimum of the possibilistic objective function is obtained, when all clusters are identical [9], covering only the best data cluster and neglecting all other data. Essentially, the philosophy of possibilistic clustering is to initialise possibilistic clustering with the result of a probabilistic cluster analysis and then find the nearest local minimum of the possibilistic objective function, avoiding the undesired global minimum in this way. Therefore, possibilistic clustering does often lead to non-satisfactory, when it is not initialised by probabilistic clustering. However, the probabilistic initialisation suffers from the before mentioned problems.

3 Understanding the Fuzzifier

The fuzzifier controls how much clusters may overlap. Figure 1 illustrates this effect by placing two cluster centres on the x-axis at 0 and 1. The leftmost plot shows the membership functions for the two clusters, when the fuzzifier is set to $m = 2$, the middle one for $m = 1.5$ (and the rightmost will be discussed later). It is well known and can be seen easily that for a larger m the transition from a high membership degree from one cluster to the other is more smooth than for a smaller m. Looking at the extremes, we obtain crisp $\{0,1\}$-valued membership degrees for $m \to 1$ and equal membership degrees to all clusters for $m \to \infty$ with all cluster centres converging to the centre of the data set. Note that we have chosen a non-symmetric range around the cluster centres, so that images do not look symmetric, although the membership functions are complimentary.

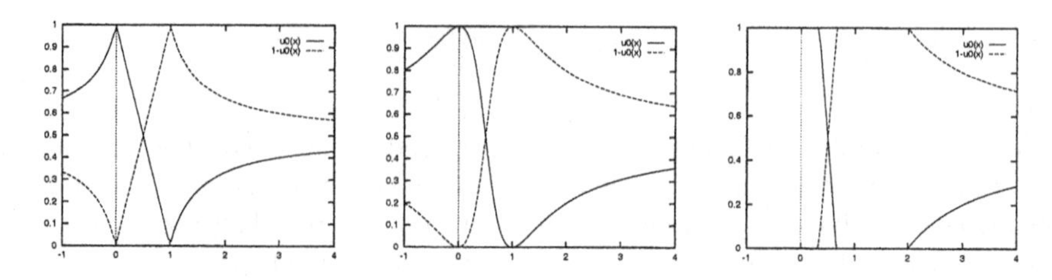

Fig. 1. Effects of the fuzzifier (from left to right): $m = 2$, $m = 1.5$, $\beta = 0.5$

The update equation (4) for the membership degrees derived from the objective function (3) can lead to undesired or counterintuitive results, because zero membership degrees never occur (except in the extremely rare case, when a data vector coincides with a cluster centre). No matter, how far away a data vector is from a cluster and how well it is covered by another cluster, it will still have nonzero membership degrees to all other clusters and therefore influence all clusters. This applies also in the case of noise and of possibilistic clustering, although the effect is weaker in these cases.

Figure 2 shows an undesired side-effect of the probabilistic fuzzy clustering approach. There are obviously three clusters. However, the upper cluster has a much higher data density than the other two. This single dense cluster attracts all other cluster prototypes so that the prototype of the left cluster is slightly drawn away from the original cluster centre and the prototype we would expect in the centre of the lower left cluster migrates completely into the dense cluster. In the figure we have also indicated for which cluster a data vector has the highest membership degree. This undesired effect can be reduced to a certain degree, if we introduce a special volume parameter in the clustering scheme [6].

Another counterintuitive effect of probabilistic fuzzy clustering occurs in the following situation. Assume we have a data set that we have clustered already. Then we add more data to the data set in the form of a new cluster that is far away from all other clusters. If we recluster this enlarged data set with one more cluster as the original data set, we would expect the same result, except that the new data are covered by the additional cluster, i.e., we would assume that the new cluster has no influence on the old ones. However, since we never obtain zero membership degrees, the new data (cluster) will influence the old clusters.

This means also that, if we have many clusters, clusters far away from the centre of the whole data set tend to have their computed cluster centres drawn into the direction of the centre of the data set.

These effects can be amended, when a small fuzzifier is chosen. The price for this is that we end up more or less with hard clustering again and even neighbouring clusters become artificially well separated, although there might be ambiguous data between these clusters.

As can be seen in figure 1, the membership degrees tend to increase again, when we move far away from all clusters. This undesired effect can be amended

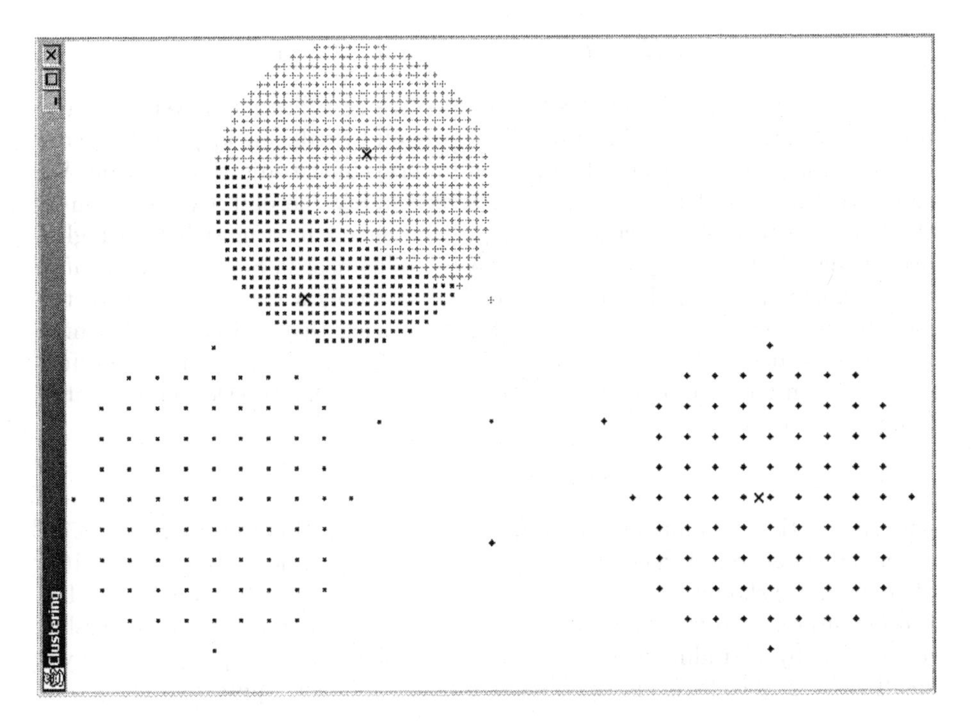

Fig. 2. Clusters with varying density

by applying noise clustering. Nevertheless, even in the case of noise clustering, noisy data, no matter how far away they are from all other clusters, will still have nonzero membership degrees to all clusters.

In order to propose an alternative to the fuzzifier approach, we examine more closely what impact the fuzzifier has on the objective function. When we want to generalise the idea of deterministic or hard clustering to fuzzy clustering, using the original objective function (1) of hard clustering simply allowing the u_{ij} values to be in $[0,1]$ instead of $\{0,1\}$, still leads to crisp partitions, as we have already mentioned before. In order to better understand why, let us consider the following situation. We fix the cluster prototypes, i.e. the distance values d_{ij}, for the moment – we might even assume that we have already found the best prototypes – and want to minimize the objective function (1) by choosing the appropriate membership degrees taking the constraints (2) into account. Due to these constraints we cannot set all u_{ij} to zero. In order to keep the objective function small, a good starting point seems to be to choose $u_{i_0j} = 1$, if $d_{i_0j} \leq d_{ij}$ for all $i = 1, \dots, c$ and $u_{ij} = 0$ otherwise. This means that the only non-zero membership degree for a fixed j is multiplied by the smallest possible distance value d_{ij}. When we try to further reduce the resulting value of the object function by decreasing an u_{i_0j}-value that was set to one, we have to increase another u_{ij}-value to satisfy the constraint (2). When we reduce u_{i_0j} by ε and increase u_{ij} by ε instead, the change in the objective function will be

$$\Delta = \varepsilon \cdot d_{ij} - \varepsilon \cdot d_{i_0 j} = \varepsilon(d_{ij} - d_{i_0 j}).$$

Since $d_{i_0 j} \leq d_{ij}$, this can only lead to an increase and therefore never to an improvement of the objective function. The trade-off by reducing $u_{i_0 j}$ and therefore increasing u_{ij} always means a bad pay-off in terms of the objective function. We can turn the pay-off into a good one, if we modify the objective function in the following way: A reduction of a $u_{i_0 j}$-value near 1 by ε must have a higher decreasing effect than the increment of a u_{ij}-value near 0. Since the factor $d_{i_0 j}$ of $u_{i_0 j}$ is smaller than the factor d_{ij} of u_{ij}, we apply a transformation to the membership degrees in the objective function, such that a decrease of a high membership degree has a stronger decreasing effect than the increasing effect caused by an increase of a small membership value. One transformation, satisfying this criterion, is

$$g : [0,1] \to [0,1], \qquad u \mapsto u^m$$

with $m > 1$ that is commonly used in fuzzy clustering. However, there might be other choices as well. Which properties should such a transformation satisfy? It is obvious that g should be increasing and that we want $g(0) = 0$ and $g(1) = 1$. If g is differentiable and we apply the above mentioned reduction of the membership degree $u_{i_0 j}$ by ε, trading it in for an increase of the membership degree u_{ij} by a small value ε, the change in the objective function is now approximately

$$\Delta \approx \varepsilon \cdot g'(u_{ij}) \cdot d_{ij} - \varepsilon \cdot g'(u_{i_0 j}) \cdot d_{i_0 j} = \varepsilon(d_{ij} \cdot g'(u_{ij}) - d_{i_0 j} \cdot g'(u_{i_0 j})). \quad (6)$$

In order to let this decrease the objective function, we need at least $g'(u_{ij}) < g'(u_{i_0 j})$, since $d_{i_0 j} \leq d_{ij}$. More generally, we require that the derivative of g is increasing on $[0,1]$. $g(u) = u^m$ with $m > 1$ definitely has this property. Especially, for this transformation we have $g'(0) = 0$, so that it always pays off to get away from zero membership degrees. Figure 3 shows the identity (the line which does not transform the membership degrees at all), the transformation u^2 (the lower curve) and another transformation that also satisfies the requirements for g, but has a nonzero derivative at 0, so that it only pays off to have a nonzero membership degree, if the distance values d_{ij} is not too large in comparison to $d_{i_0 j}$.

Let us take a closer look which effect the transformation g has on the objective function. Assume, we want to minimize the objective function

$$f = \sum_{i=1}^{c} \sum_{j=1}^{n} g(u_{ij}) d_{ij} \quad (7)$$

under the constraints (2) w.r.t. the values u_{ij}, i.e., we consider the distances as fixed. The constraints lead to the Lagrange function

$$L = \sum_{i=1}^{c} \sum_{j=1}^{n} g(u_{ij}) d_{ij} + \sum_{j=1}^{n} \lambda_j \left(1 - \sum_{i=1}^{c} u_{ij}\right)$$

and the partial derivatives

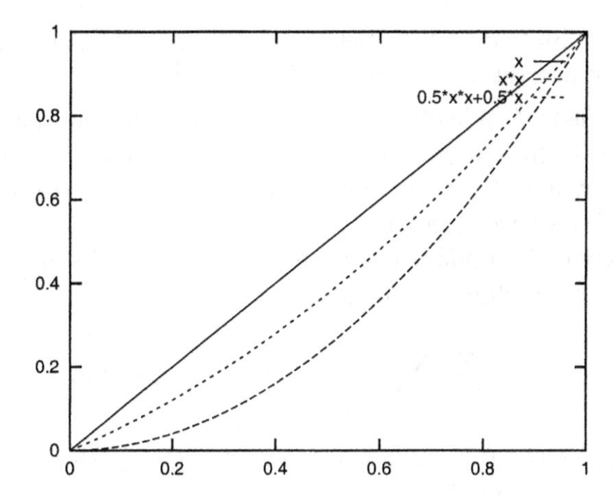

Fig. 3. Transformations for probabilistic clustering

$$\frac{\partial L}{\partial u_{ij}} = g'(u_{ij})d_{ij} - \lambda_j. \tag{8}$$

At a minimum of the objective function the partial derivatives must be zero, i.e. $\lambda_j = g'(u_{ij})d_{ij}$. Since λ_j is independent of i, we must have $g'(u_{ij})d_{ij} = g'(u_{kj})d_{kj}$ for all i, k at a minimum. This actually means that these products must be balanced during the minimization process. In other words, the minimum is not reached unless the Δ-values in (6) are all zero.

4 An Alternative for the Fuzzifier

Taking into account the analysis carried out in the previous section, we propose a new approach to fuzzy clustering that replaces the transformation $g(u) = u^m$ by another transformation. In principle, we can think of any differentiable function satisfying the requirements stated in the previous section. However, when we want to maintain the computationally more efficient alternating optimisation scheme for fuzzy clustering with explicit update equations for the membership degrees, we easily run into problems for general functions g. From (8) we can immediately see that we will need the inverse of g' in order to compute the u_{ij}. Therefore, we restrict our considerations here to quadratic transformations. Since we require $g(0) = 0$ and $g(1) = 1$ and an increasing first derivative, the choices reduce to quadratic functions of the form $g(u) = \alpha u^2 + (1 - \alpha)u$ with $0 \leq \alpha \leq 1$.

Instead of α we use the parameter

$$\beta = \frac{g'(0)}{g'(1)} = \frac{1 - \alpha}{1 + \alpha}.$$

We can easily compute $\alpha = \frac{1-\beta}{1+\beta}$. Let us assume that $d_{i_0 j}$ is the distance of data vector x_j to the nearest cluster and d_{ij} is the distance of x_j to another cluster further away. Then β indicates the lower bound that the quotient $\frac{d_{i_0 j}}{d_{ij}}$ must exceed, in order to have a nonzero membership degree of x_j to the cluster that lies further away. For $\beta = 0$ we obtain standard fuzzy clustering with fuzzifier $m = 2$ and for $\beta \to 1$ we approach crisp clustering.

We now derive the update equations for our new clustering approach. We have to minimize the objective function

$$f = \sum_{i=1}^{c} \sum_{j=1}^{n} \left(\frac{1-\beta}{1+\beta} u_{ij}^2 + \frac{2\beta}{1+\beta} u_{ij} \right) d_{ij}$$

under the constraints (2) as well as $0 \leq u_{ij} \leq 1$. Computing the partial derivatives of the Lagrange function

$$L = \sum_{i=1}^{c} \sum_{j=1}^{n} \left(\frac{1-\beta}{1+\beta} u_{ij}^2 + \frac{2\beta}{1+\beta} u_{ij} \right) d_{ij} + \sum_{j=1}^{n} \lambda_j \left(1 - \sum_{i=1}^{c} u_{ij} \right)$$

and solving for u_{ij} we obtain

$$u_{ij} = \frac{1}{1-\beta} \left(\frac{(1+\beta)\lambda_j}{2d_{ij}} - \beta \right) \tag{9}$$

if $u_{ij} \neq 0$. Using $\sum_{k:u_{kj} \neq 0} u_{kj} = 1$, we can compute

$$\lambda_j = \frac{2(1 + (\hat{c}-1)\beta)}{(1+\beta) \sum_{k:u_{kj} \neq 0} \frac{1}{d_{kj}}}$$

where $\hat{c}$ is the number of clusters to which data vector x_j has nonzero membership degrees. Replacing λ_j in (9), we finally obtain the update equation for

$$u_{ij} = \frac{1}{1-\beta} \left(\frac{1 + (\hat{c}-1)\beta}{\sum_{k:u_{kj} \neq 0} \frac{d_{ij}}{d_{kj}}} - \beta \right). \tag{10}$$

We still have to determine which u_{ij} are zero. Since we want to minimize the objective function (7), we can make the following observation. If $u_{ij} = 0$ and $d_{ij} < d_{tj}$, then at a minimum of the objective function, we must have $u_{tj} = 0$ as well. Otherwise we could reduce the value by setting u_{tj} to zero and letting u_{ij} assume the original value of u_{tj}. This implies the following. For a fixed j we can sort the distances d_{ij} in decreasing order. Without loss of generality let us assume $d_{1j} \geq \ldots \geq d_{cj}$. If there are zero membership degrees at all, we know that for minimizing the objective function the u_{ij}-values with larger distances have to be zero. (10) does not apply to these u_{ij}-values. Therefore, we have to find the smallest index i_0 to which (10) is applicable, i.e. for which it yields a positive value. For $i < i_0$ we have $u_{ij} = 0$ and for $i \geq i_0$ the membership degree u_{ij} is computed according to (10) with $\hat{c} = c + 1 - i_0$.

With this modified algorithm the data set in figure 2 is clustered correctly, i.e. the cluster centres correspond to the centres of the data clusters. The parameter β was set to $\beta = 0.5$, the corresponding transformation g is the curve in the middle in figure 3. $\beta = 0.5$ turned out to be a good choice for most of our data sets and can be used as a rule of thumb similar to the choice of the fuzzifier $m = 2$. Especially, when our modified approach is coupled with noise clustering, most of the undesired effects of fuzzy clustering can be avoided and the advantages of a fuzzy approach can be maintained. The diagram on the right-hand side of figure 1 shows the membership degrees for two cluster with centres at 0 and 1, when we use our approach with $\beta = 0.5$. The membership degree one is not only assumed in the cluster centre, but also in a region around the cluster centre. The undesired effect that membership degrees tend to $1/c$ or $1/2$ in the case of two clusters for data that are far away from all clusters is still present. However, when we introduce a noise cluster, data lying far away from all clusters will not only have small membership degrees to the clusters as in probabilistic clustering, but membership degree one to the noise cluster and zero to all other clusters.

When using our modified algorithm, the update equations for the cluster prototypes remain the same – for instance, in the case of FCM as in (5) – except that we have to replace u_{ij}^m by

$$g(u_{ij}) = \alpha u_{ij}^2 + (1-\alpha)u_{ij} = \frac{1-\beta}{1+\beta}u_{ij}^2 + \frac{2\beta}{1+\beta}u_{ij}.$$

5 Conclusions

We have proposed a new approach to fuzzy clustering that overcomes the problem in fuzzy clustering that all data tend to influence all clusters. From the computational point of view our algorithm is slightly more complex than the standard fuzzy clustering scheme. The update equations are quite similar to standard (probabilistic) fuzzy clustering. However, for each data vector we have to sort the distances to the clusters in each iteration step. Since the number of clusters is usually quite small and the order of the distances tends to converge quickly, this additional computational effort can be kept at minimum. Looking at the computational costs of computing $\alpha u_{ij}^2 + (1-\alpha)u_{ij}$ instead of u_{ij}^m, we are even faster, when the fuzzifier m is not an integer number, since in this case the power function is need explicitly. But even for the case $m = 2$, we only need one extra multiplication, one subtraction and one addition, when we use

$$\alpha u_{ij}^2 + (1-\alpha)u_{ij} = (\alpha \cdot u_{ij}) \cdot u_{ij} + u_{ij} - (\alpha \cdot u_{ij}).$$

As a future work, we will extend our approach not only to quadratic functions as in this paper or to exponential functions [7], but to more general transformations g and apply the balancing scheme induced by (6) directly to compute the membership degrees.

References

1. Bezdek, J.C.: Pattern Recognition with Fuzzy Objective Function Algorithms. Plenum Press, New York (1981)
2. Bezdek, J.C., Keller, J., Krishnapuram, R., Pal, N.R.: Fuzzy Models and Algorithms for Pattern Recognition and Image Processing. Kluwer, Boston (1999)
3. Davé, R.N.: Characterization and Detection of Noise in Clustering. Pattern Recognition Letters **12** (1991) 657–664
4. Duda, R., Hart, P.: Pattern Classification and Scene Analysis. Wiley, New York (1973)
5. Höppner, F., Klawonn, F., Kruse, R., Runkler. T.: Fuzzy Cluster Analysis. Wiley, Chichester (1999)
6. Keller, A., Klawonn, F.: Adaptation of Cluster Sizes in Objective Function Based Fuzzy Clustering. In: Leondes, C.T. (ed.): Intelligent Systems: Technology and Applications vol. IV: Database and Learning Systems. CRC Press, Boca Raton (2003), 181–199
7. Klawonn, F., Höppner, F.: An Alternative Approach to the Fuzzifier in Fuzzy Clustering to Obtain Better Clustering Results. In: Proceedings EUSFLAT03. Zittau (2003) (to appear)
8. Krishnapuram, R., Keller, J.: A Possibilistic Approach to Clustering. IEEE Trans. on Fuzzy Systems **1** (1993) 98–110
9. Timm, H., Borgelt, C., Kruse, R.: A Modification to Improve Possibilistic Cluster Analysis. IEEE Intern. Conf. on Fuzzy Systems, Honululu (2002)

A Mixture Model Approach for Binned Data Clustering

Allou Samé, Christophe Ambroise, and Gérard Govaert

Université de Technologie de Compiègne
Département Génie Informatique
HEUDIASYC, UMR CNRS 6599
BP 20529, 60205 Compiègne Cedex
{same, ambroise, govaert}@utc.fr

Abstract. In some particular data analysis problems, available data takes the form of an histogram. Such data are also called binned data. This paper addresses the problem of clustering binned data using mixture models. A specific EM algorithm has been proposed by Cadez et al.([2]) to deal with these data. This algorithm has the disadvantage of being computationally expensive. In this paper, a classification version of this algorithm is proposed, which is much faster. The two approaches are compared using simulated data. The simulation results show that both algorithms generate comparable solutions in terms of resulting partition if the histogram is accurate enough.

1 Introduction

Generally, data are collected or transformed into frequencies located in disjoined areas of $\mathbb{R}^p$ either because the measurement instruments have finite resolution or to reduce them because of their large size. Data in this form are often called binned data and the considered areas are called bins.

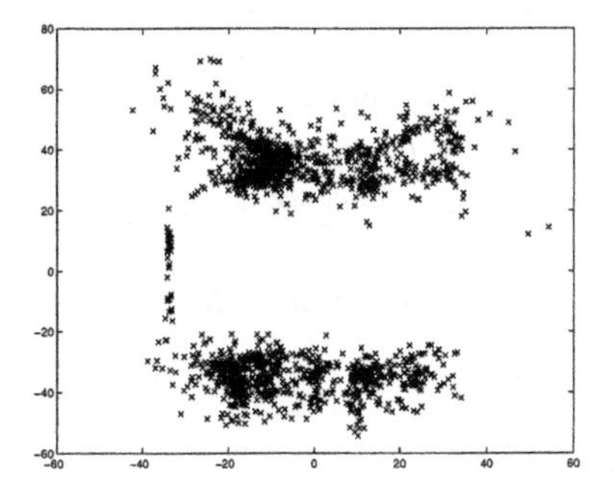

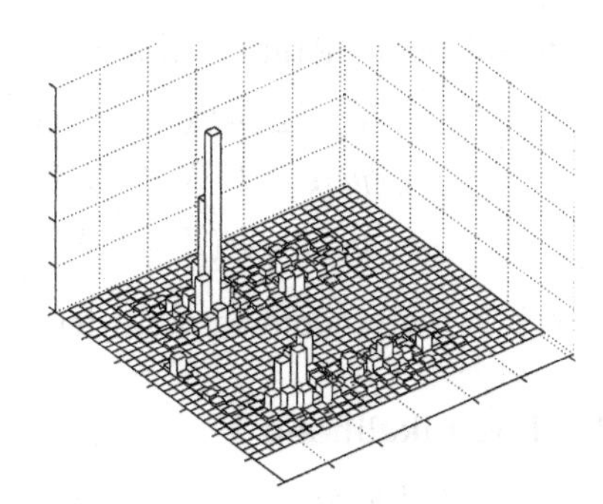

Fig. 1. Real data resulting from acoustic emission for flaw detection in a tank (left) and corresponding binned data (right).

More particularly, in the framework of a flaw detection problem using acoustic emission, we have been brought to classify, under real time constraints, a set of

M.R. Berthold et al. (Eds.): IDA 2003, LNCS 2810, pp. 265–274, 2003.

points located in the plane (fig. 1). This plane represent a tank and each point of the plane represents the localization of an acoustic emission. The clustering CEM algorithm [3], applied using a Gaussian diagonal mixture model assumption, provides a satisfactory solution but cannot react in real time, when the size of the data becomes very large (more than 10000 points). In this paper, we propose to estimate the model parameters and the partition from grouped data, an adapted version of the CEM algorithm.

In the second section, we briefly present the Cadez and al. [2] approach for parameters estimation with the EM algorithm. The third section is devoted to a new clustering approach adapted to binned data and a detailed experimental study is summarized in the fourth section.

2 Clustering of Binned Data via the EM Algorithm

The application of the EM algorithm to binned data has initially been introduced by Dempster Laird and Rubin [5], and then developed in the one-dimensional case by McLachlan and Jones [1]. The generalization to the multidimensional case is due to Cadez and al.[2].

In this work, we suppose that $(\mathbf{x}_1, \ldots, \mathbf{x}_n)$ is an independent sample resulting from a mixture density defined on $\mathbb{R}^p$ by

$$f(\mathbf{x}; \Phi) = \sum_{k=1}^{K} \pi_k f_k(\mathbf{x}; \theta_k),$$

where $\Phi = (\pi_1, \ldots, \pi_K, \theta_1, \ldots, \theta_K)$ and $\pi_1, \ldots, \pi_K$ denote the proportions of the mixture and $\theta_1, \ldots, \theta_K$ the parameters of each component density.

In addition, we consider a partition $(\mathcal{H}_1, \ldots, \mathcal{H}_v)$ of $\mathbb{R}^p$ into v bins and suppose that all available knowledge from the sample is the set of frequencies n_r of the observations $\mathbf{x}_i$ belonging to the bins $\mathcal{H}_r$ indicated by a vector $\mathbf{a} = (n_1, \ldots, n_v)$.

The following notations are also used:

$$\begin{aligned} p_r &= P(X \in \mathcal{H}_r; \Phi) &&= \int_{\mathcal{H}_r} f(\mathbf{x}; \Phi) d\mathbf{x}, \\ p_{r/k} &= P(X \in \mathcal{H}_r | Z = k, \Phi) &&= \int_{\mathcal{H}_r} f_k(\mathbf{x}; \theta_k) d\mathbf{x}, \\ p_{k/r} &= P(Z = k | X \in \mathcal{H}_r, \Phi) &&= \frac{\pi_k \int_{\mathcal{H}_r} f_k(\mathbf{x}; \theta_k) d\mathbf{x}}{p_r}, \end{aligned}$$

where Z is the random variable denoting the class associated to X.

2.1 Log-Likelihood

The associated log-likelihood is written

$$\begin{aligned} L(\Phi; \mathbf{a}) &= \log P(\mathbf{a}; \Phi) \\ &= \sum_{r=1}^{v} n_r \log(p_r) + C, \end{aligned}$$

where $C = \log \frac{n!}{\prod_{r=1}^{v} n_r!}$.

2.2 EM Algorithm for Binned Data

To maximize the log-likelihood, the EM algorithm maximizes iteratively the expectation of the complete data knowing the available data:

$$Q(\Phi, \Phi^{(q)}) = E(L_c(\Phi)|\mathbf{a}, \Phi^{(q)}),$$

where $L_c(\Phi)$ is the log-likelihood of the complete data. To apply this algorithm, it is thus necessary to define the complete data $((\mathbf{x}_1, z_1), \ldots, (\mathbf{x}_n, z_n))$. In this case, we have

$$L_c(\Phi) = \sum_{i=1}^{n}\sum_{k=1}^{K} z_{ik} \log \pi_k f_k(\mathbf{x}_i; \theta_k),$$

where $z_{ik} = \begin{cases} 1 \text{ if } z_i = k \\ 0 \text{ else} \end{cases}$. We can note that the complete data, and thus the complete log-likelihood for binned data are the same as for classical data. The two situations differ only with regard to the observed data: the vector $(\mathbf{x}_1, \ldots, \mathbf{x}_n)$ for the classical situation and the vector $\mathbf{a} = (n_1, \ldots, n_v)$ of the frequencies for the situation studied hereby.

We can now define the two steps of the algorithm:

E-STEP. This step consists in calculating $Q(\Phi, \Phi^{(q)})$. It can be shown that

$$Q(\Phi, \Phi^{(q)}) = \sum_{k=1}^{K}\sum_{r=1}^{v} n_r p_{k/r}^{(q)} \left(\log(\pi_k) + \frac{1}{p_{r/k}^{(q)}} \int_{\mathcal{H}_r} \log(f_k(\mathbf{x}; \theta_k)) f_k(\mathbf{x}; \theta_k^{(q)}) d\mathbf{x} \right).$$

M-STEP. This step consists in maximizing the expression $Q(\Phi, \Phi^{(q)})$. The parameter estimation is given by:

$$\pi_k^{q+1} = \frac{\sum_{r=1}^{v} n_r p_{k/r}^{(q)}}{\sum_\ell \sum_{r=1}^{v} n_r p_{\ell/r}^{(q)}} = \frac{\sum_{r=1}^{v} n_r p_{k/r}^{(q)}}{n},$$

$$\mu_k^{(q+1)} = \frac{\sum_{r=1}^{v} \frac{n_r}{p_r^{(q)}} \int_{\mathcal{H}_r} \mathbf{x} f_k(\mathbf{x}; \theta_k^{(q)}) d\mathbf{x}}{\sum_{r=1}^{v} \frac{n_r}{p_r^{(q)}} \int_{\mathcal{H}_r} f_k(\mathbf{x}; \theta_k^{(q)}) d\mathbf{x}},$$

$$\Sigma_k^{(q+1)} = \frac{Q_k^{(q+1)}}{\sum_{r=1}^{v} n_r \frac{p_{r/k}^{(q)}}{p_r^{(q)}}},$$

where

$$Q_k^{(q+1)} = \sum_{r=1}^{v} \frac{n_r}{p_r^{(q)}} \int_{\mathcal{H}_r} (\mathbf{x} - \mu_k^{(q+1)})(\mathbf{x} - \mu_k^{(q+1)})' f_k(\mathbf{x}; \theta_k^{(q)}) d\mathbf{x}.$$

In the following, we will use the name "binned EM" for this algorithm.

The main difficulty in the implementation of binned EM, contrary to EM algorithm for classical data lies in the evaluation at each iteration, for each component k and each bin r of the integrals $\int_{\mathcal{H}_r} f_k(\mathbf{x};\theta_k^{(q)})d\mathbf{x}$, $\int_{\mathcal{H}_r} \mathbf{x} f_k(\mathbf{x};\theta_k^{(q)})d\mathbf{x}$ and $\int_{\mathcal{H}_r}(\mathbf{x}-\mu_k^{(q+1)})(\mathbf{x}-\mu_k^{(q+1)})' f_k(\mathbf{x};\theta_k^{(q)})d\mathbf{x}$. In the Gaussian diagonal mixture model case that we consider here, these integrals can be computed by simply using the one-dimensional normal cumulative distribution function.

From probabilities $p_{k/r} = P(Z = k|X \in \mathcal{H}_r, \Phi) = \frac{\pi_k \int_{\mathcal{H}_r} f_k(\mathbf{x};\theta_k)d\mathbf{x}}{p_r}$ corresponding to the parameters Φ obtained at convergence, a partition of data can easily be derived by assigning each bin $\mathcal{H}_r$ to the component which maximizes this posterior probability.

3 A Clustering Approach for Binned Data

The clustering method introduced in this part is based on the maximization of a criterion which measures the compatibility between the partition and class parameters with data. We are inspired by criteria which have already been proposed in a situation where the observed data are observations and not frequencies. Among these criteria, let us quote the classification likelihood criteria introduced by Symons [6] and then developed by Celeux and Govaert within the framework of the so-called CEM algorithm (Classification EM) [3,4]. This criterion is expressed as follow:

$$C(z_1,\ldots,z_n,\Phi) = \log p(\mathbf{x}_1,\ldots,\mathbf{x}_n,z_1,\ldots,z_n;\Phi) = \sum_{i=1}^{n}\sum_{k=1}^{K} z_{ik}\log \pi_k f_k(\mathbf{x}_i;\theta_k),$$

where the classes z_i as well as the parameters Φ are supposed to be unknown.

3.1 A Criterion for Binned Data

The method that we propose for the binned data is based on the simultaneous estimation of the mixture parameters and the partition maximizing the criterion

$$L(\Phi;z_1,\ldots,z_n,\mathbf{a}) = \log p(\mathbf{a},z_1,\ldots,z_n;\Phi).$$

Since there is no available information about the exact position of the observations into the bins, we assume that in a same bin, observations have the same label i.e. all the observations of a given bin belong to the same class. Thus, the labels z_i can be replaced by the labels z^r $(1 \le r \le v)$ of the bins if the bin membership of the $\mathbf{x}_i$ is known. Assuming this condition, we get:

$$L(\Phi;z^1,\ldots,z^v,\mathbf{a}) = \sum_{r=1}^{v}\log\left(\pi_{z^r}\int_{\mathcal{H}_r} f_{z^r}(\mathbf{x};\theta_{z^r})d\mathbf{x}\right) + C,$$

where the $z^1,\ldots,z^v$ are the labels of the bins and $C = \log\frac{n!}{\prod_{r=1}^{v} n_r!}$.

3.2 Maximization of $L(\Phi; z^1, \ldots, z^v, \mathbf{a})$

The maximization of $L(\Phi; z^1, \ldots, z^v, \mathbf{a})$ with respect to $(\Phi, z^1, \ldots, z^v)$ is performed similarly to the iterative CEM algorithm i.e. on the basis of an initialization $\Phi^{(0)}$ and then alternating the two following steps until convergence:

Step 1. Computation of $\left(z^{1\,(q+1)}, \ldots, z^{v\,(q+1)}\right) = \underset{(z^1,\ldots,z^v)}{argmax}\ L(\Phi^{(q)};\ z^1, \ldots, z^v, \mathbf{a})$

Step 2. Calculation of $\Phi^{(q+1)} = \underset{\Phi}{argmax}\ L(\Phi; z^{1\,(q+1)}, \ldots, z^{v\,(q+1)}, \mathbf{a})$

Let us detail the 2 steps of this algorithm.

Step 1. This step consists in calculating

$$\left(z^{1\,(q+1)}, \ldots, z^{v\,(q+1)}\right) = \underset{(z^1,\ldots,z^v)}{argmax} \sum_r \log\left(\pi_{z^r}^{(q)} \int_{\mathcal{H}_r} f_{z^r}(\mathbf{x}; \theta_{z^r}^{(q)}\, d\mathbf{x}\right)$$

It can be easily verified for all $r = 1, \ldots, v$ that

$$z^{r\,(q+1)} = \underset{1\leq k\leq K}{argmax}\left(\log(\pi_k^{(q)} \int_{\mathcal{H}_r} f_k(\mathbf{x}; \theta_k^{(q)}) d\mathbf{x})\right).$$

Step 2. In this step, the EM algorithm is used with the initialization $\Phi^{(q)}$ for the maximization of $L(\Phi; z^{1\,(q+1)}, \ldots, z^{v\,(q+1)}, \mathbf{a})$ because it is not straightforward.

The complete data are $((\mathbf{x}_1, z_1^{(q+1)}), \ldots, (\mathbf{x}_n, z_n^{(q+1)}))$ and thus the complete likelihood takes the form

$$L_c(\Phi) = \sum_{i=1}^{n}\sum_{k=1}^{K} z_{ik}^{(q+1)} \log \pi_k f_k(\mathbf{x}_i; \theta_k).$$

The EM algorithm, in this situation, then consists in maximizing iteratively

$$Q(\Phi, \Phi^*) = E(L_c(\Phi)|\mathbf{a}, o_{\mathbf{a}}, \Phi^*),$$

where the notation Φ^* corresponds to the parameter vector which has been obtained at the preceding iteration.

E-STEP. This step consists in calculating $Q(\Phi, \Phi^*)$. It can be easily shown that

$$Q(\Phi, \Phi^*) = \sum_{k=1}^{k}\sum_{r=1}^{v} z_k^{r\,(q+1)} n_r \left(\log \pi_k + \frac{1}{p^*_{r/k}} \int_{\mathcal{H}_r} \log f_k(\mathbf{x}; \theta_k) f_k(\mathbf{x}; \theta_k^*) d\mathbf{x}\right),$$

where $z_k^{r\,(q+1)} = \begin{cases} 1 \text{ si } z^{r\,(q+1)} = k \\ 0 \text{ sinon} \end{cases}$

M-STEP. This step consists in maximizing $Q(\Phi, \Phi^*)$. If we denote $\Phi^{**} = (\pi_1^{**}, \dots, \pi_K^{**}, \mu_1^{**}, \dots, \mu_K^{**}, \Sigma_1^{**}, \dots, \Sigma_K^{**})$ the parameter vector which maximize $Q(\Phi, \Phi^*)$, we get:

$$\pi_k^{**} = \frac{\sum_{r=1}^{v} z_k^{r\,(q+1)} n_r}{n},$$

$$\mu_k^{**} = \frac{\sum_{r=1}^{v} z_k^{r\,(q+1)} \frac{n_r}{p_{r/k}^*} \int_{\mathcal{H}_r} \mathbf{x} f_k(\mathbf{x}; \theta_k^*) d\mathbf{x}}{\sum_{r=1}^{v} z_k^{r\,(q+1)} \frac{n_r}{p_{r/k}^*} \int_{\mathcal{H}_r} f_k(\mathbf{x}; \theta_k^*) d\mathbf{x}},$$

$$\Sigma_k^{**} = \frac{Q_k^{**}}{\sum_{r=1}^{v} z_k^{r\,(q+1)} \frac{n_r}{p_{r/k}^*} \int_{\mathcal{H}_r} f_k(\mathbf{x}; \theta_k^*) d\mathbf{x}},$$

where

$$Q_k^{**} = \sum_{r=1}^{v} z_k^{r\,(q+1)} \frac{n_r}{p_{r/k}^*} \int_{\mathcal{H}_r} (\mathbf{x} - \mu_k^{**})(\mathbf{x} - \mu_k^{**})' f_k(\mathbf{x}; \theta_k^*) d\mathbf{x}.$$

Since this algorithm is related to EM and CEM algorithms, we will name it "binned EM-CEM".

4 Numerical Experiments

This part is devoted to the comparison of the two algorithms proposed for binned data clustering. We consider simulated data, with respect to a mixture of two Gaussian densities with diagonal variance matrix, in dimension 2. The effect of the number of bins on the precision of the results is observed using different experimental settings. Three parameters control the simulation:

- the shape of the classes (spherical or elliptic shapes),
- the overlapping degree of the true partition,
- the sample size.

4.1 Evaluation Criterion

The quality of a partition $P = (P_1, \dots, P_K)$ produced by the clustering algorithms is measured by computing the percentage of misclassification compared to the true partition $P^{true} = (P_1^{true}, \dots, P_K^{true})$ obtained by maximizing posterior probabilities $p_{k/r}$ with the true parameters.

The quality of the partition given by the classical CEM algorithm applied to observations and not to bins is in our case the ideal quality which one hopes to approach, since the information given by the points is more precise than the one given by the bins.

4.2 Strategy

The adopted strategy for simulation consists in initially drawing n points according to a mixture of two bi-dimensional Gaussian, to transform these points into a two-dimensional histogram containing v bins and finally to apply the CEM clustering algorithm to the original data (observations) and the two binned data clustering algorithms to the binned data. For a given simulation, the three algorithms are ran 30 times to take into account the dependence to initialization and the solution which provides the greatest likelihood is selected. For each of the three obtained partitions, the percentage of misclassification is computed.

4.3 Experimental Parameters

- The shape parameters of the classes are specified through the variance matrices. The four shape situations retained are:
  - situation 1: $\Sigma_1 = \Sigma_2 = diag(1,1)$,
  - situation 2: $\Sigma_1 = diag(1,1)$, and $\Sigma_2 = diag(1/5, 1/5)$,
  - situation 3: $\Sigma_1 = diag(1/4, 1)$ and $\Sigma_2 = diag(1/4, 1)$
  - situation 4: $\Sigma_1 = diag(1/4, 1)$ and $\Sigma_2 = diag(1, 1/4)$.

  Figure 2 illustrates the various shape situations that are considered.
- The class overlap parameter (overlap is measured by the theoretical percentage of misclassified observations) corresponds to the distance between the class centers of the mixture. The first center has been fixed at $\mu_1 = (-2, 0)$ and we retain for each of the preceding shape models three centers μ_2 approximately corresponding to 5%, 15% and 25% of theoretical error:

	overlapping		
	≈5%	≈15%	≈25%
situation 1 : μ_2	(1.2;0)	(0;0)	(-0.7;0)
situation 2 : μ_2	(0.1;0)	(-0.9;0)	
situation 3 : μ_2	(-0.4;0)	(-1;0)	(-1.3;0)
situation 4 : μ_2	(1.5;0)	(-1.5;0)	

  Notice that situations 2 and 4 do not have exactly 25% of theoretical error but are close to.
- The various selected sizes of sample are n=100, n=500 and n=1000.

The domain of the histogram has been fixed in practice to $[-10; 10] \times [-10; 10]$. It contains all the simulated observations for each considered situation. The various experiments have been performed for a number of bins on the domain varying from 10×10 bins to 100×100 bins by step of 10.

4.4 Results

Figures 3 and 4 represent the percentage of error obtained with the CEM, binned EM and binned EM-CEM algorithms, as functions of the number of bins per dimension in situations 2 and 4, for an initial overlap of 15% of theoretical error for a sample size of $n = 1000$. A fast decrease of the percentages of error given by binned EM and binned EM-CEM is observed until 30 bins per dimension. Let us note that in this phase, the binned EM algorithm performs better than its concurrent binned EM-CEM algorithm. From 30 bins per dimension to greater

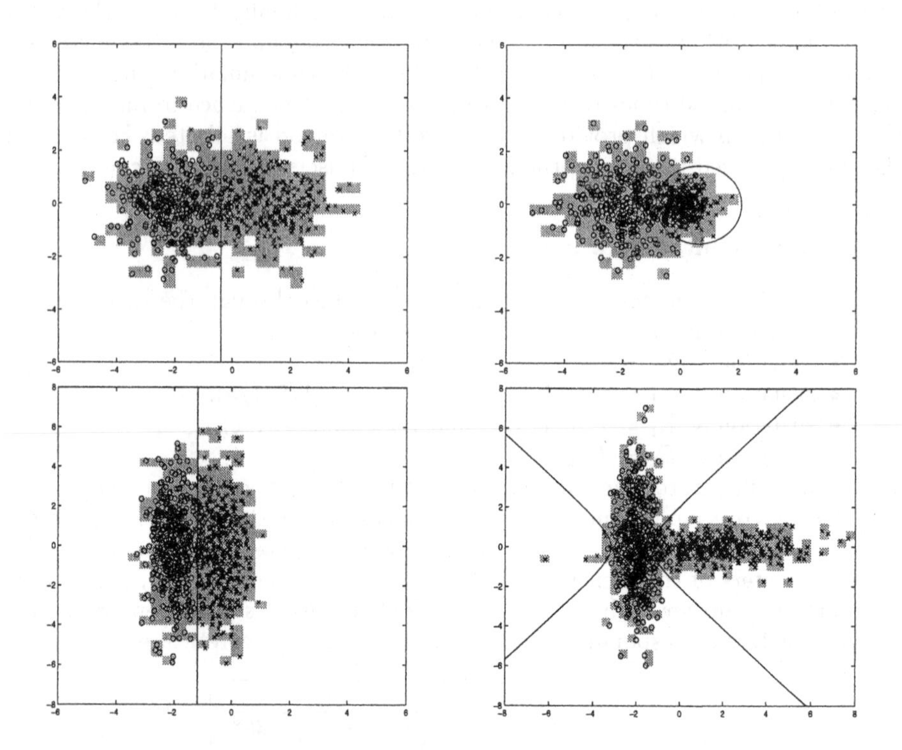

Fig. 2. Various situations of classes considered

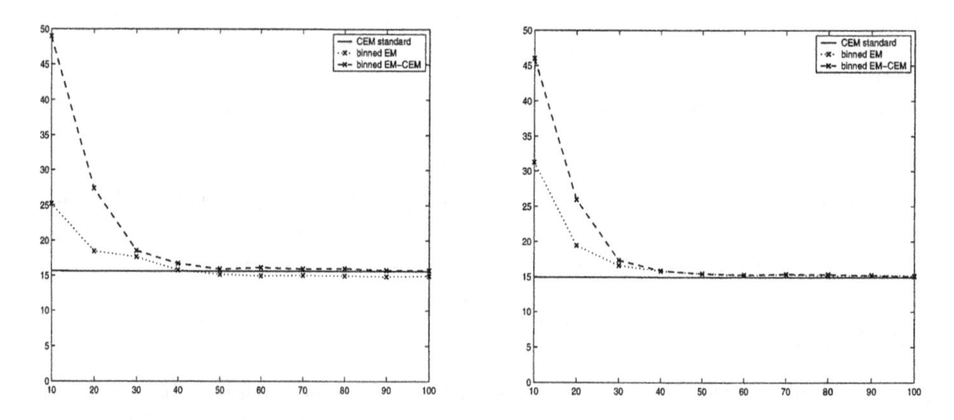

Fig. 3. Variation of the percentage of error obtained with CEM algorithm, binned EM and binned EM-CEM as a function of the number of bins per dimension for situation 2 (on the left) and situation 4 (on the right), with an overlap leading to a 15% of theoretical error for $n = 1000$.

values, the percentages of error given by the two classification algorithms dealing with binned data almost not vary any more and are especially close to the one given by the classical CEM algorithm which constitutes our reference.

This shows that when the number of bins is high enough, the algorithms working the binned data and the algorithm using the original observation do produce results (in term of partition) of comparable quality.

This observed behavior is almost the same for the other situations with different shapes, overlapping degree and sample sizes. In rare cases, when the overlap is particularly high (25% of theoretical error), binned EM performs poorly, even when the number of bins per dimension is higher than 30. In practice, we noted that this effect is due to the fact that the parameters of the mixture obtained in these cases, while being very different from true simulation parameters give a better likelihood.

Consequently, the proposed algorithms are more sensitive to the numbers of bins than they are to the sample size, the shape of the classes or the overlapping degree. Figure 4 shows us an example of partition obtained with the binned EM and binned EM-CEM algorithms in situation 2.

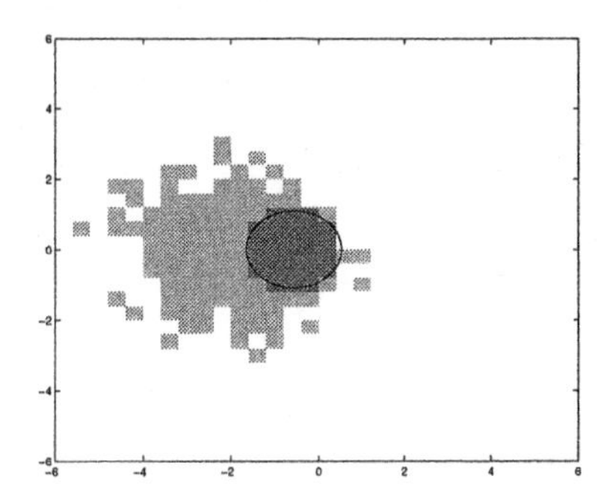

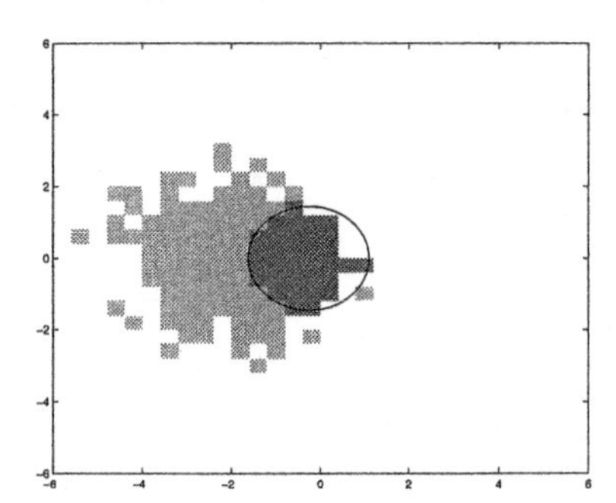

Fig. 4. Example of a classification obtained in situation 2 with 40x40 bins (on the right the one obtained with binned EM and on the left the one obtained with binned EM-CEM) for an overlapping of 15% of theoretical error and for $n = 1000$.

5 Conclusion

Both clustering approaches, using binned EM and binned EM-CEM algorithms, produce partitions using the binned data, which are close to those estimated by CEM algorithm applied to the original observations, when the number of bins is sufficiently large (higher than 30 in the presented simulations). Moreover, the used diagonal Gaussian model (diagonal variance matrices) makes these algorithms faster since they do not require complex computations of integrals.

Each iteration of the proposed algorithm increases the likelihood $L(\Phi^{(q)}; z^{1^{(q)}}, \ldots, z^{v^{(q)}}, \mathbf{a})$ and in practice convergence is observed, but the proof of convergence remains the main prospect of this work.

References

1. McLachlan G. J. and P. N. Jones. *Fitting mixture models to grouped and truncated data via the EM algorithm.* Biometrics, 44(2):571–578, 1988.
2. I. V. Cadez, P. Smyth, G. J. McLachlan and C. E. McLaren. *Maximum likelihood estimation of mixture densities for binned and truncated multivariate data.* Machine Learning 47, 7–34, 2001.
3. Celeux G. and G. Govaert. *A classification EM algorithm for clustering and two stochastic versions.* Computation Statistics and Data analysis, 14, 315–332, 1992.
4. Celeux G. and G. Govaert. *Gaussian parsimonious clustering models.* Pattern Recognition, 28(5), 781–793, 1995.
5. A. P. Dempster, N. M. Laird and D. B. Rubin. *Maximum likelihood from incomplete data via the EM algorithm.* Pattern Recognition, 28(5), J. Royal Stat. Soc. B, 39(1):1–38, 1977.
6. M. J. Symons *Clustering criteria and multivariate normal mixtures.* Biometrics, 37, 35–43, 1981.

Fuzzy Clustering Based Segmentation of Time-Series

Janos Abonyi, Balazs Feil, Sandor Nemeth, and Peter Arva

University of Veszprem, Department of Process Engineering,
P.O. Box 158, H-8201, Hungary,
abonyij@fmt.vein.hu

Abstract. The segmentation of time-series is a constrained clustering problem: the data points should be grouped by their similarity, but with the constraint that all points in a cluster must come from successive time points. The changes of the variables of a time-series are usually vague and do not focused on any particular time point. Therefore it is not practical to define crisp bounds of the segments. Although fuzzy clustering algorithms are widely used to group overlapping and vague objects, they cannot be directly applied to time-series segmentation. This paper proposes a clustering algorithm for the simultaneous identification of fuzzy sets which represent the segments in time and the local PCA models used to measure the homogeneity of the segments. The algorithm is applied to the monitoring of the production of high-density polyethylene.

1 Introduction

A sequence of N observed data, $\mathbf{x}_1, \ldots, \mathbf{x}_N$, ordered in time, is called a time-series. Real-life time-series can be taken from business, physical, social and behavioral science, economics, engineering, etc. Time-series segmentation addresses the following data mining problem: given a time-series T , find a partitioning of T to c segments that are internally homogeneous [1]. Depending on the application, the goal of the segmentation is used to locate stable periods of time, to identify change points, or to simply compress the original time-series into a more compact representation [2]. Although in many real-life applications a lot of variables must be simultaneously tracked and monitored, most of the segmentation algorithms are used for the analysis of only one time-variant variable [3].

The aim of this paper is to develop an algorithm that is able to handle time-varying characteristics of multivariate data: (i) changes in the mean; (ii) changes in the variance; and (iii) changes in the correlation structure among variables. Multivariate statistical tools, like Principal Component Analysis (PCA) can be applied to discover such information [4]. PCA maps the data points into a lower dimensional space, which is useful in the analysis and visualization of the correlated high-dimensional data [5]. Linear PCA models have two particularly desirable features: they can be understood in great detail and they are straightforward to implement. Linear PCA can give good prediction results for simple time-series, but can fail to the analysis of historical data with changes in regime

M.R. Berthold et al. (Eds.): IDA 2003, LNCS 2810, pp. 275–285, 2003.

or when the relations among the variables are nonlinear. The analysis of such data requires the detection of locally correlated clusters [6]. The proposed algorithm does the clustering of the data to discover the local relationships among the variables similarly to mixture of Principal Component Models [4].

The changes of the variables of the time-series are usually vague and do not focused on any particular time point. Therefore, it is not practical to define crisp bounds of the segments. For example, if humans visually analyze historical process data, they use expressions like "this point belongs to this operating point less and belongs to the another more". A good example of this kind of fuzzy segmentation is how fuzzily the start and the end of *early morning* is defined. Fuzzy logic is widely used in various applications where the grouping of overlapping and vague elements is necessary [7], and there are many fruitful examples in the literature for the combination of fuzzy logic with time-series analysis tools [8,2,9,10,16].

Time-series segmentation may be viewed as clustering, but with a time-ordered structure. The contribution of this paper is the introduction of a new fuzzy clustering algorithm which can be effectively used to segment large, multivariate time-series. Since the points in a cluster must come from successive time points, the time-coordinate of the data should be also considered. One way to deal with time is to use time as an additional variable and cluster the observations using a distance between cases which incorporates time. Hence, the clustering is based on a distance measure that contains two terms: the first distance term is based on how the data is in the given segment according to Gaussian fuzzy sets defined in the time domain, while the second term measures how far the data is from hyperplane of the PCA model used to measure the homogeneity of the segments.

The remainder of this paper is organized as follows. Section 2 defines the time-segmentation problem and presents the proposed clustering algorithm. Section 3 contains a computational example using historical process data collected during the production of high-density polyethylene. Finally, conclusions are given in Section 4.

2 Clustering Based Time-Series Segmentation

2.1 Time-Series and Segmentations

A time-series $T = \{\mathbf{x}_k | 1 \leq k \leq N\}$ is a finite set of N samples labelled by time points $t_1, \ldots, t_N$, where $\mathbf{x}_k = [x_{1,k}, x_{2,k}, \ldots, x_{n,k}]^T$. A segment of T is a set of consecutive time points $S(a,b) = \{a \leq k \leq b\}$, $\mathbf{x}_a, \mathbf{x}_{a+1}, \ldots, \mathbf{x}_b$. The c-segmentation of time-series T is a partition of T to c non-overlapping segments $S_T^c = \{S_i(a_i, b_i) | 1 \leq i \leq c\}$, such that $a_1 = 1, b_c = N$, and $a_i = b_{i-1} + 1$. In other words, an c-segmentation splits T to c disjoint time intervals by segment boundaries $s_1 < s_2 < \ldots < s_c$, where $S_i(s_{i-1} + 1, s_i)$.

Usually the goal is to find homogeneous segments from a given time-series. In such a case the segmentation problem can be described as constrained clustering:

data points should be grouped by their similarity, but with the constraint that all points in a cluster must come from successive time points. In order to formalize this goal, a cost function $cost(S(a,b))$ with the internal homogeneity of individual segments should be defined. The cost function can be any arbitrary function. For example in [1,11] the sum of variances of the variables in the segment was defined as $cost(S(a,b))$. Usually, the cost function $cost(S(a,b))$ is defined based on the distances between the actual values of the time-series and the values given by the a simple function (constant or linear function, or a polynomial of a higher but limited degree) fitted to the data of each segment. The optimal c-segmentation simultaneously determines the a_i, b_i borders of the segments and the θ_i parameter vectors of the models of the segments. Hence, the segmentation algorithms minimize the cost of c-segmentation which is usually the sum of the costs of the individual segments:

$$cost(S_T^c) = \sum_{i=1}^{c} cost(S_i)\,. \tag{1}$$

The (1) cost function can be minimized by dynamic programming to find the segments (e.g. [11]). Unfortunately, dynamic programming is computationally intractable for many real data sets. Consequently, heuristic optimization techniques such as greedy top-down or bottom-up techniques are frequently used to find good but suboptimal c-segmentations:

Sliding window: A segment is grown until it exceeds some error bound. The process repeats with the next data point not included in the newly approximated segment. For example a linear model is fitted on the observed period and the modelling error is analyzed [12].

Top-down method: The time-series is recursively partitioned until some stopping criteria is met. [12].

Bottom-up method: Starting from the finest possible approximation, segments are merged until some stopping criteria is met [12].

Search for inflection points: Searching for primitive episodes located between two inflection points [5].

In the following subsection a different approach will be presented based on the clustering of the data.

2.2 Clustering Based Fuzzy Segmentation

When the variance of the segments are minimized during the segmentation, (1) results in the following equation:

$$cost(S_T^c) = \sum_{i=1}^{c} \sum_{k=s_{i-1}+1}^{s_i} \| \mathbf{x}_k - \mathbf{v}_i^x \|^2 = \sum_{i=1}^{c} \sum_{k=1}^{N} \beta_i(t_k) D_{i,k}^2(\mathbf{v}_i^x, \mathbf{x}_k)\,. \tag{2}$$

where $D_{i,k}^2(\mathbf{v}_i^x, \mathbf{x}_k)$ represents the distance how far the $\mathbf{v}_i^x$ = mean of the variables in the i-th segment (center of the i-th cluster) is from the $\mathbf{x}_k$ data point; and

$\beta_i(t_k) = \{0, 1\}$ stands for the crisp membership of the k-th data point is in the i-th segment:

$$\beta_i(t_k) = \begin{cases} 1 \text{ if } s_{i-1} < k \leq s_i \\ 0 \quad \text{otherwise} \end{cases} \tag{3}$$

This equation is well comparable to the typical error measure of standard k-means clustering but in this case the clusters are limited to being contiguous segments of the time-series instead of Voronoi regions in R^n.

The changes of the variables of the time-series are usually vague and do not focused on any particular time point. As it is not practical to define crisp bounds of the segments, Gaussian membership functions are used to represent the $\beta_i(t_k) = [0, 1]$ fuzzy segments of time-series $A_i(t_k)$, which choice leads to the following compact formula for the *normalized degree of fulfillment* of the membership degrees of the k-th observation is in the i-th segment:

$$A_i(t_k) = \exp\left(-\frac{1}{2}\frac{(t_k - v_i^t)^2}{\sigma_i^2}\right), \ \beta_i(t_k) = \frac{A_i(t_k)}{\sum\limits_{j=1}^{c} A_j(t_k)} \tag{4}$$

For the identification of the v_i^t centers and σ_i^2 variances of the membership functions, a fuzzy clustering algorithm is introduced. The algorithm, which is similar to the modified Gath-Geva clustering [13], assumes that the data can be effectively modelled as a mixture multivariate (including time as a variable) Gaussian distribution, so it minimizes the sum of the weighted squared distances between the $\mathbf{z}_k = [t_k, \mathbf{x}_k^T]^T$ data points and the η_i cluster prototypes

$$J = \sum_{i=1}^{c}\sum_{k=1}^{N} (\mu_{i,k})^m D_{i,k}^2(\eta_i, \mathbf{z}_k) = \sum_{i=1}^{c}\sum_{k=1}^{N} (\mu_{i,k})^m D_{i,k}^2(v_i^t, t_k) D_{i,k}^2(\mathbf{v}_i^x, \mathbf{x}_k) \tag{5}$$

where $\mu_{i,k}$ represents the degree of membership of the observation $\mathbf{z}_k = [t_k, \mathbf{x}_k^T]^T$ is in the i-th cluster ($i = 1, \ldots, c$), $m \in [1, \infty)$ is a weighting exponent that determines the fuzziness of the resulting clusters (usually chosen as $m = 2$).

As can be seen, the clustering is based on a $D_{i,k}^2(\eta_i, \mathbf{z}_k)$ distance measure, which consists of two terms. The first one, $D_{i,k}^2(v_i^t, t_k)$, is the distance between the k-the data point and the v_i^t center of the i-th segment in time

$$1/D_{i,k}^2(v_i^t, t_k) = \frac{1}{\sqrt{2\pi\sigma_i^2}} \exp\left(-\frac{1}{2}\frac{(t_k - v_i^t)^2}{\sigma_i^2}\right), \tag{6}$$

where the center and the standard deviation of the Gaussian function are

$$v_i^t = \frac{\sum\limits_{k=1}^{N} (\mu_{i,k})^m t_k}{\sum\limits_{k=1}^{N} (\mu_{i,k})^m}, \ \sigma_i^2 = \frac{\sum\limits_{k=1}^{N} (\mu_{i,k})^m (t_k - v_k^t)^2}{\sum\limits_{k=1}^{N} (\mu_{i,k})^m}. \tag{7}$$

The second term represents the distance between the cluster prototype and the data in the feature space

$$1/D^2_{i,k}(\mathbf{v}^x_i, \mathbf{x}_k) = \frac{\alpha_i\sqrt{\det(\mathbf{A}_i)}}{(2\pi)^{r/2}} \exp\left(-\frac{1}{2}\left(\mathbf{x}_k - \mathbf{v}^x_i\right)^T \left(\mathbf{A}_i\right)\left(\mathbf{x}_k - \mathbf{v}^x_i\right)\right), \quad (8)$$

where α_i represents the *a priori* probability of the cluster and $\mathbf{v}^x_i$ the coordinate of the i-th cluster center in the feature space,

$$\alpha_i = \frac{1}{N}\sum_{k=1}^{N}\mu_{i,k}, \ \mathbf{v}^x_i = \frac{\sum_{k=1}^{N}\left(\mu_{i,k}\right)^m \mathbf{x}_k}{\sum_{k=1}^{N}\left(\mu_{i,k}\right)^m}, \quad (9)$$

and r is the rank of $\mathbf{A}_i$ distance norm corresponding to the i-th cluster. The $\mathbf{A}_i$ distance norm can be defined in many ways. It is wise to scale the variables so that those with greater variability do not dominate the clustering. One can scale by dividing by standard deviations, but a better procedure is to use statistical (Mahalanobis) distance, which also adjusts for the correlations among the variables. In this case $\mathbf{A}_i$ is the inverse of the fuzzy covariance matrix $\mathbf{A}_i = \mathbf{F}_i^{-1}$, where

$$\mathbf{F}_i = \frac{\sum_{k=1}^{N}\left(\mu_{i,k}\right)^m \left(\mathbf{x}_k - \mathbf{v}^x_i\right)\left(\mathbf{x}_k - \mathbf{v}^x_i\right)^T}{\sum_{k=1}^{N}\left(\mu_{i,k}\right)^m}. \quad (10)$$

When the variables are highly correlated, the $\mathbf{F}_i$ covariance matrix can be ill - conditioned and cannot be inverted. Recently two methods have been worked out to handle this problem [14]. The first method is based on fixing the ratio between the maximal and minimal eigenvalue of the covariance matrix. The second method is based on adding a scaled unity matrix to the calculated covariance matrix. Both methods result in invertible matrices, but none of them extract the potential information about the hidden structure of the data.

Principal Component Analysis (PCA) is based on the projection of correlated high-dimensional data onto a hyperplane which is useful for the visualization and the analysis of the data. This mapping uses only the first few p nonzero eigenvalues and the corresponding eigenvectors of the $\mathbf{F}_i = \mathbf{U}_i\Lambda_i\mathbf{U}_i^T$, covariance matrix, decomposed to the Λ_i matrix that includes the eigenvalues of $\mathbf{F}_i$ in its diagonal in decreasing order, and to the $\mathbf{U}_i$ matrix that includes the eigenvectors corresponding to the eigenvalues in its columns, $\mathbf{y}_{i,k} = \Lambda_{i,p}^{-\frac{1}{2}}\mathbf{U}_{i,p}^T\mathbf{x}_k$,

When the hyperplane of the PCA model has adequate number of dimensions, the distance of the data from the hyperplane is resulted by measurement failures, disturbances and negligible information, so the projection of the data into this p-dimensional hyperplane does not cause significant reconstruction error

$$Q_{i,k} = (\mathbf{x}_k - \hat{\mathbf{x}}_k)^T(\mathbf{x}_k - \hat{\mathbf{x}}_k) = \mathbf{x}_k^T(\mathbf{I} - \mathbf{U}_{i,p}\mathbf{U}_{i,p}^T)\mathbf{x}_k. \quad (11)$$

Although the relationship among the variables can be effectively described by a linear model, in some cases it is possible that the data is distributed around some separated centers in this linear subspace. The Hotelling T^2 measure is often used to calculate the distance of the data point from the center in this linear subspace

$$T^2_{i,k} = \mathbf{y}^T_{i,k}\mathbf{y}_{i,k}. \tag{12}$$

The T^2 and Q measures are often used for the monitoring of multivariate systems and for the exploration of the errors and the causes of the errors. The Figure 1 shows these parameters in case of two variables and one principal component.

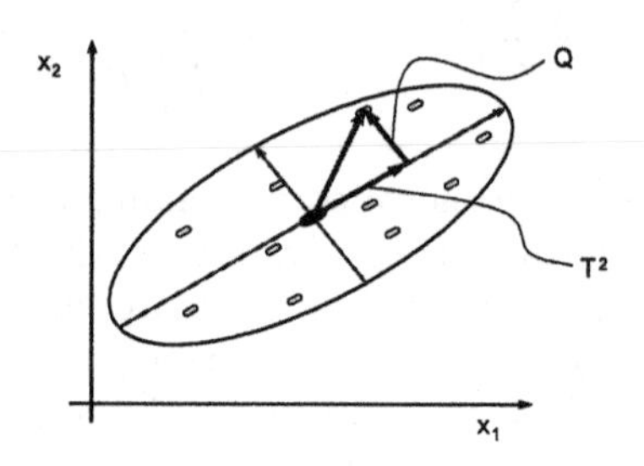

Fig. 1. Distance measures based on the PCA model.

The main idea of this paper is the usage of these measures as measure of the homogeneity of the segments:

Fuzzy-PCA-Q method: The distance measure is based on the Q model error. The $\mathbf{A}_i$ distance norm is the inverse of the matrix constructed from the $n-p$ smallest eigenvalues and eigenvectors of $\mathbf{F}_i$.

$$\mathbf{A}_i^{-1} = \mathbf{U}_{i,n-p}\Lambda_{i,n-p}\mathbf{U}^T_{i,n-p}, \tag{13}$$

Fuzzy-PCA-T^2 method: The distance measure is based on the Hotelling T^2. $\mathbf{A}_i$ is defined with the largest p eigenvalues and eigenvectors of $\mathbf{F}_i$

$$\mathbf{A}_i^{-1} = \mathbf{U}_{i,p}\Lambda_{i,p}\mathbf{U}^T_{i,p}, \tag{14}$$

When p is equal to the rank of $\mathbf{F}_i$, this method is identical to the fuzzy covariance matrix based distance measure, where $\mathbf{F}_i$ is inverted by taking the Moore-Penrose pseudo inverse.

The optimal parameters of the $\eta_i = \{\mathbf{v}^x_i, \mathbf{A}_i, v^t_i, \sigma^2_i, \alpha_i\}$ cluster prototypes are determined by the minimization of the (5) functional subjected to the following constraints to get a fuzzy partitioning space,

$$\mathbf{U} = [\mu_{i,k}]_{c\times N} | \, \mu_{i,k} \in [0,1], \forall i,k; \, \sum_{i=1}^{c} \mu_{i,k} = 1, \forall k; \, 0 < \sum_{k=1}^{N} \mu_{i,k} < N, \forall i \tag{15}$$

The most popular method to solve this problem is the alternating optimization (AO), which consists of the application of Picard iteration through the first-order conditions for the stationary points of (5), which can be found by adjoining the constraints (15) to J by means of Lagrange multipliers [15]

$$\overline{J}(\mathbf{Z},\mathbf{U},\eta,\lambda)=\sum_{i=1}^{c}\sum_{k=1}^{N}(\mu_{i,k})^m\,D^2_{i,k}(\mathbf{z}_k,\eta_i)+\sum_{k=1}^{N}\lambda_i\left(\sum_{i=1}^{c}\mu_{i,k}-1\right) \tag{16}$$

and by setting the gradients of $\overline{J}$ with respect to $\mathbf{Z},\mathbf{U},\eta$ and λ to zero.

Initialization. Given a time-series T specify c, choose a termination tolerance $\epsilon>0$, and initialize the $\mathbf{U}=[\mu_{i,k}]_{c\times N}$ partition matrix randomly.
Repeat for $l=1,2,\ldots$
Step 1. Calculate the η_i parameters of the clusters by (7), (9), and (13) or (14).
Step 2. Compute the $D^2_{i,k}(\eta_i,\mathbf{z}_k)$ distance measures (5) by (6), (8).
Step 3. Update the partition matrix

$$\mu^{(l)}_{i,k}=\frac{1}{\sum_{j=1}^{c}\left(D_{i,k}/D_{j,k}\right)^{2/(m-1)}},1\le i\le c,\,1\le k\le N\,. \tag{17}$$

until $||\mathbf{U}^{(l)}-\mathbf{U}^{(l-1)}||<\epsilon$.

3 Application to Process Monitoring

Manual process supervision relies heavily on visual monitoring of characteristic shapes of changes in process variables, especially their trends. Although humans are very good at visually detecting such patterns, for a control system software it is a difficult problem. Researchers with different background, for example from pattern recognition, digital signal processing and data mining, have contributed to the process trend analysis development and have put emphasis on different aspects of the field, and at least the following issues characteristic to the process control viewpoint are employed: (I.) Methods operate on process measurements or calculated values in time-series which are not necessarily sampled at constant intervals. (II.) Trends have meaning to human experts. (III.) Trend is a pattern or structure in one-dimensional data. (IV.) Application area is in process monitoring, diagnosis and control [3,5,10].

The aim of this example is to show how the proposed algorithm is able to detect meaningful temporal shapes from multivariate historical process data. The monitoring of a medium and high-density polyethylene (MDPE, HDPE) plant is considered. HDPE is versatile plastic used for household goods, packaging, car parts and pipe. The plant is operated by TVK Ltd., which is the largest Hungarian polymer production company (www.tvk.hu) in Hungary. An interesting problem with the process is that it requires to produce about ten product grades according to market demand. Hence, there is a clear need to minimize the time of changeover because off-specification product may be produced during transition. The difficulty of the problem comes from the fact that there are more than ten process variables to consider. Measurements are available in every 15 seconds on process variables $\boldsymbol{x}_k$, which are the ($x_{k,1}$ polymer production intensity (PE), the inlet flowrates of hexene ($C6_{in}$), ethlylene (C_{2in}), hydrogen ($H2_{in}$), the isobutane solvent (IB_{in}) and the catalyzator (Kat), the concentrations of ethylene (C_2), hexene (C_2), and hydrogen (H_2) and the slurry in the reactor ($slurry$), and the temperature of the reactor (T).

The section used for this example is a 125 hours period which includes at least three segments: a product transition around the 25th hour, a "calm" period until the 100th hour, and a "stirring" period of 30 hours. (see Figure 2).

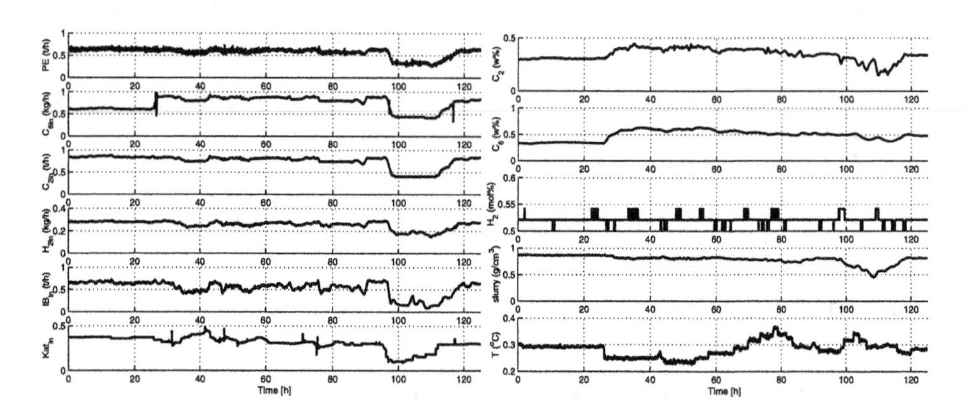

Fig. 2. Example of a time-series of a grade transition.

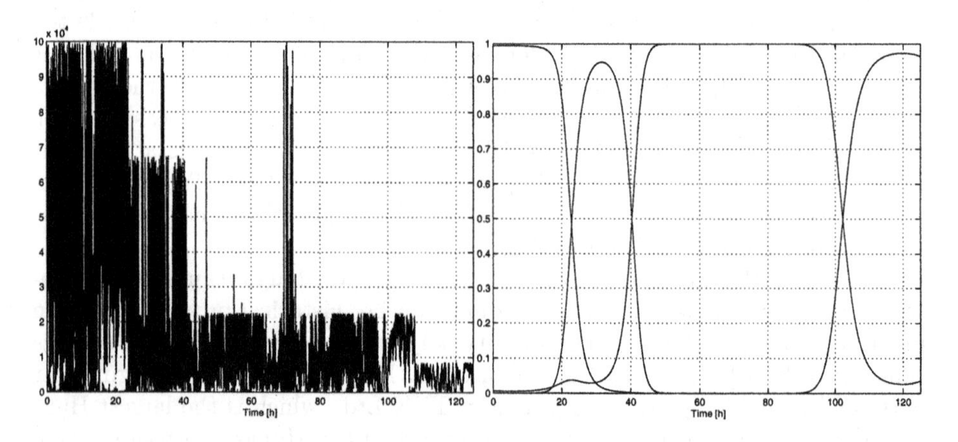

Fig. 3. Fuzzy-PCA-Q segmentation: (a) $1/D_{i,k}(\mathbf{v}_i^x, \mathbf{x}_k)$ distances, (b) fuzzy time-series segments, $\beta_i(t_k)$

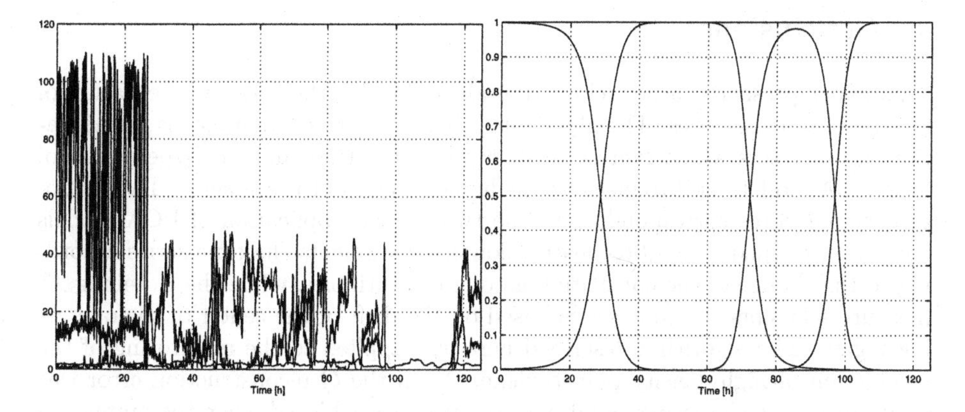

Fig. 4. Fuzzy-PCA-T^2 segmentation: (a) $1/D_{i,k}(\mathbf{v}_i^x, \mathbf{x}_k)$ distances, (b) fuzzy time-series segments, $\beta_i(t_k)$

The distance between the data points and the cluster centers can be computed by two methods corresponding to the two distance measures of PCA: the Q reconstruction error and the Hotelling T^2. These methods give different results which can be used for different purposes.

The *Fuzzy-PCA-Q method* is sensitive to the number of the principal components. With the increase of p the reconstruction error decreases. In case of $p = n = 11$ the reconstruction error becomes zero in the whole range of the data set. If p is too small, the reconstruction error will be large for the entire time-series. In these two extreme cases the segmentation does not based on the internal relationships among the variables, so equidistant segments are detected. When the number of the latent variables is in the the range $p = 3, ..., 8$, reliable segments are detected, the algorithm finds the grade transition about the 25th hours Figure 3).

The *Fuzzy-PCA-T^2 method* is also sensitive to the number of principal components. The increase of the model dimension does not obviously decreases the Hotelling T^2, because the volume of the fuzzy covariance matrix (the product of the eigenvalues) tends to zero (see (8)). Hence, it is practical to use mixture of PCA models with similar performance capacity. This can be achieved by defining p based on the desired accuracy (loss of variance) of the PCA models:

$$\sum_{f=1}^{p-1} diag\left(\Lambda_{i,f}\right) / \sum_{f=1}^{n} diag\left(\Lambda_{i,f}\right) < q \leq \sum_{f=1}^{p} diag\left(\Lambda_{i,f}\right) / \sum_{f=1}^{n} diag\left(\Lambda_{i,f}\right) \qquad (18)$$

As Figure 4 shows, when $q = 0.99$ the algorithm finds also correct segmentation that is highly reflects the "drift" of the operating regions (Figure 3).

4 Conclusions

This paper presented a new clustering algorithm for the fuzzy segmentation of multivariate time-series. The algorithm is based on the simultaneous identification of fuzzy sets which represent the segments in time and the hyperplanes of local PCA models used to measure the homogeneity of the segments. Two homogeneity measures corresponding to the two typical application of PCA models have been defined. The Q reconstruction error segments the time-series according to the change of the correlation among the variables, while the Hotelling T^2 measure segments the time-series based on the drift of the center of the operating region. The algorithm described here was applied to the monitoring of the production of high-density polyethylene. When the Q reconstruction error was used during the clustering, good performance was achieved when the number of the principal components were defined as the parameters of the clusters. Contrary, when the clustering is based on the T^2 distances from the cluster centers, it was advantageous to define the desired reconstruction performance to obtain a good segmentation. The results suggest that the proposed tool can be applied to distinguish typical operational conditions and analyze product grade transitions. We plant to use the results of the segmentations to design intelligent query system for typical operating conditions in multivariate historical process databases.

Acknowledgements. The authors would like to acknowledge the support of the Cooperative Research Center (VIKKK) (project 2001-II-1), the Hungarian Ministry of Education (FKFP-0073/2001), and the Hungarian Science Foundation (T037600). Janos Abonyi is grateful for the financial support of the Janos Bolyai Research Fellowship of the Hungarian Academy of Science. The support of our industrial partners at TVK Ltd., especially Miklos Nemeth, Lorant Balint and dr. Gabor Nagy is gratefully acknowledged.

References

1. K. Vasko, H. Toivonen, Estimating the number of segments in time series data using permutation tests, IEEE International Conference on Data Mining (2002) 466–473.
2. M. Last, Y. Klein, A. Kandel, Knowledge discovery in time series databases, IEEE Transactions on Systems, Man, and Cybernetics 31 (1) (2000) 160–169.
3. S. Kivikunnas, Overview of process trend analysis methods and applications, ERUDIT Workshop on Applications in Pulp and Paper Industry (1998) CD ROM.
4. M. E. Tipping, C. M. Bishop, Mixtures of probabilistic principal components analysis, Neural Computation 11 (1999) 443–482.
5. G. Stephanopoulos, C. Han, Intelligent systems in process engineering: A review, Comput. Chem. Engng. 20 (1996) 743–791.
6. K. Chakrabarti, S. Mehrotra, Local dimensionality reduction: A new approach to indexing high dimensional spaces, Proceedings of the 26th VLDB Conference Cairo Egypt (2000) P089.

7. J. Abonyi, Fuzzy Model Identification for Control, Birkhauser Boston, 2003.
8. A. B. Geva, Hierarchical-fuzzy clustering of temporal-patterns and its application for time-series prediction, Pattern Recognition Letters 20 (1999) 1519–1532.
9. J. F. Baldwin, T. P. Martin, J. M. Rossiter, Time series modelling and prediction using fuzzy trend information, Proceedings of 5th International Conference on Soft Computing and Information Intelligent Systems (1998) 499–502.
10. J. C. Wong, K. McDonald, A. Palazoglu, Classification of process trends based on fuzzified symbolic representation and hidden markov models, Journal of Process Control 8 (1998) 395–408.
11. J. Himberg, K. Korpiaho, H. Mannila, J. Tikanmaki, H. T. Toivonen, Time-series segmentation for context recognition in mobile devices, IEEE International Conference on Data Mining (ICDM'01), San Jose, California (2001) 203–210.
12. E. Keogh, S. Chu, D. Hart, M. Pazzani, An online algorithm for segmenting time series, IEEE International Conference on Data Mining (2001) http://citeseer.nj.nec.com/keogh01online.html.
13. J. Abonyi, R. Babuska, F. Szeifert, Modified Gath-Geva fuzzy clustering for identification of takagi-sugeno fuzzy models, IEEE Transactions on Systems, Man, and Cybernetics 32(5) (2002) 612–321.
14. R. Babuška, P. J. van der Veen, U. Kaymak, Improved covariance estimation for gustafson-kessel clustering, IEEE International Conference on Fuzzy Systems (2002) 1081–1085.
15. F. Hoppner, F. Klawonn, R. Kruse, T. Runkler, Fuzzy Cluster Analysis, Wiley, Chichester, 1999.
16. F. Hoppner, F. Klawonn, Fuzzy Clustering of Sampled Functions. In NAFIPS00, Atlanta, USA, (2000) 251–255.

An Iterated Local Search Approach for Minimum Sum-of-Squares Clustering

Peter Merz

University of Kaiserslautern
Department of Computer Science
Postbox 3049, D-67653 Kaiserslautern, Germany
peter.merz@ieee.org

Abstract. Since minimum sum-of-squares clustering (MSSC) is an NP-hard combinatorial optimization problem, applying techniques from global optimization appears to be promising for reliably clustering numerical data. In this paper, concepts of combinatorial heuristic optimization are considered for approaching the MSSC: An iterated local search (ILS) approach is proposed which is capable of finding (near-)optimum solutions very quickly. On gene expression data resulting from biological microarray experiments, it is shown that ILS outperforms multi–start k-means as well as three other clustering heuristics combined with k-means.

1 Introduction

Clustering can be considered as an optimization problem in which an assignment of data vectors to clusters is desired, such that the sum of squared distances of the vectors to their cluster mean (centroid) is minimal. Let $\mathcal{P}_k$ denote the set of all partitions of X with $X = \{x_1, \dots, x_n\}$ denoting the data set of vectors with $x_i \in I\!\!R^m$ and C_i denoting the i-th cluster with mean $\hat{x}_i$. Thus, the objective is

$$\min_{P_K \in \mathcal{P}_k} \sum_{i=1}^{k} \sum_{x_j \in C_i} d^2(x_j, \hat{x}_i), \quad \text{with} \quad \hat{x}_i = \frac{1}{|C_i|} \sum_{x_j \in C_i} x_j \tag{1}$$

where $d(\cdot, \cdot)$ is the Euclidean distance in $I\!\!R^m$. Alternatively, we can formulate the problem as searching for an assignment p of the vectors to the clusters with $C_i = \{j \in \{1, \dots, n\} \mid p[j] = i\}$. Thus, the objective becomes

$$\min_{p} \sum_{i=1}^{n} d^2(x_i, \hat{x}_{p[i]}). \tag{2}$$

This combinatorial problem is called the minimum sum-of-squares clustering (MSSC) problem and is known to be NP-hard [1]. Although exact optimization algorithms for clustering criteria exist [2,3], their application is limited to small data sets.

M.R. Berthold et al. (Eds.): IDA 2003, LNCS 2810, pp. 286–296, 2003.

Clustering algorithms have received renewed attention in the field of bioinformatics due to the breakthrough of microarrays. This technology allows monitoring simultaneously the expression patterns of thousands of genes with enormous promise to help genetics to understand the genome [4,5]. Since a huge amount of data is produced during microarray experiments, clustering techniques are used to extract the fundamental patterns of gene expression inherent in the data. Several clustering algorithms have been proposed for gene expression profile analysis: Eisen *et al.* [6] applied a hierarchical average link clustering algorithm to identify groups of co-regulated genes in the yeast cell cycle. In [7], the k-means algorithm was used for clustering yeast data and in [8], a comparison of several approaches was made including k-means with average link initialization, which performed well for all tested data sets.

The k-means algorithm is a heuristic which minimizes the sum-of-squares criterion provided an initial assignment/choice of centroids. The number k of clusters is fixed during its run. The k-means heuristic can in fact be regarded as a local search heuristic for this hard combinatorial optimization problem. It is well-known that the effectiveness of k-means highly depends on the initial choice of centroids [9,10].

Since iterated local search (ILS) is known to be highly effective for several combinatorial optimization problems such as the traveling salesman problem [11], the application of ILS to the MSSC appears to be promising.

In this paper, iterated local search for clustering gene expression profiles using the k-means heuristic is proposed. It is shown that instead of repeatedly starting k-means to find better local optima, the ILS framework is much more effective in finding near optimum solutions quickly for the investigated data sets.

The paper is organized as follows. The general ILS framework is described in section 2. In section 3, an ILS approach to minmum-sum-of-squares clustering is introduced. In section 4, distance measures for comparing clustering solutions are provided. The ILS approach is applied for clustering gene expression data in section 5. Section 6 concludes the paper and outlines areas for future research.

2 Iterated Local Search

Per definition, local search heuristics suffer from the fact that they get trapped in local optima with respect to some objective function. In fact, for NP-hard combinatorial problems, there is no known polynomial time algorithm that guarantees to find the optimum solution. Hence, the only practical way is to approach these problems with heuristics. Local search algorithms are improvement heuristics which often produce near-optimum solutions very quickly. To improve these local optimum solutions, additional techniques are required. Iterated Local Search (ILS) [12] is such a technique which is very simple but known to be powerful. A famous example is the iterated Lin-Kernighan heuristic for the traveling salesman problem [11], which is one of the best performing heuristics for the traveling salesman problem to date.

The basic idea behind ILS is as follows: A local search is applied to an initial solution either generated randomly or with some constructive heuristic. Then, repeatedly, the current local optimum is mutated, i. e. slightly modified, and the local search procedure is applied to reach another local optimum. If the new local optimum has a better objective value, the solution is accepted as the new current solution. The outline of the algorithm is provided in Fig. 1.

procedure Iterated-Local-Search(n_{max} : $\mathbb{N}$) : S;
  begin
    Create starting solution s;
    $s :=$ Local-Search(s);
    $n := 0$;
    repeat
      $s' :=$ Mutate(s);
      $s' :=$ Local-Search(s');
      if $f(s') < f(s)$ **then** $s := s'$;
      $n := n + 1$;
    until $n = n_{max}$;
    return s;
  end;

Fig. 1. Iterated Local Search pseudo code.

ILS provides a simple extension to local search. In contrast to Multi-Start Local search (MLS), in which repeatedly local search is applied to newly randomly generated solutions, some information is preserved from previous iterations in ILS.

ILS can be considered as an evolutionary algorithm incorporating local search. In fact, it is a special case of a memetic algorithm [13,14], with a population size of one. In terms of evolutionary algorithms ILS can be regarded as a (1+1)-ES (Evolution Strategy) with additional local search. Compared to common evolutionary algorithms, mutation is aimed at escaping from a local optimum and thus much more disruptive than mutation for neighborhood searching.

3 ILS for Clustering

There are two operators in ILS specific to the problem: (a) the local search, and (b) the mutation operator. Both operators used in our ILS for MSSC are described in the following.

3.1 Initialization and Local Search

In our ILS, local search is performed using the k-means heuristic [15,16]: Given k centroid vectors a_j, for each input vector $x_i \in X$ the nearest centroid s_j is determined and the vector x_i is associated with the corresponding cluster. Afterwards, all centroid vectors are recalculated by calculating the mean of the vectors in each cluster. This procedure is repeated until the partition does no longer change, and thus a local optimum is reached.

The initial solutions in ILS for MSSC are generated by randomly selecting input vectors as initial cluster centers for the k-means.

Variants to the k-means heuristic above may be used within the framework, such as k-means using k-d trees [17,18,19] and X-means [20]. For prove of concept we concentrate on the simple k-means described above.

3.2 The Mutation Operator

The mutation operator used in ILS works as follows: A randomly chosen mean vector a_j $(1 \leq j \leq K)$ is replaced by a randomly chosen input vector x_i $(1 \leq i \leq N)$.

4 Comparing Clustering Solutions

To allow a comparison of clusters produced by the clustering algorithms used in our studies, we define two distance measures between clustering solutions. The first defines a distance between two assignments p and q by the formula

$$D(p,q) = \sum_{i=1}^{n} d^2(\hat{x}_{p[i]}, \hat{x}_{q[i]}). \tag{3}$$

We will refer to this distance measure as the *means distance*. However, the value of this distance metric cannot be interpreted easily. It would be helpful to have a measure indicating how many input vectors have been assigned to different clusters. Furthermore, a measure independent of the values of the centroids allows to compare clustering solutions obtained by other heuristics not using centroids at all.

Therefore, we propose another technique called *matching* to provide a measure for (dis-)similarity of clustering solutions (assignments). Matching works as follows. First, a counter for common vectors is set to zero. Then, for each cluster i of solution A, find the cluster j of solution B containing the most vectors of cluster i. Additionally, find the cluster k from solution A containing most vectors of cluster j. If k equals i, we have a matching and the counter is incremented by the number of vectors contained in cluster i of solution A and also in cluster j of solution B. If we have done the matching for all clusters in solution A, the counter contains the total number of vectors assigned to matching clusters. The distance of solutions A and B is simply the difference of the total number of input vectors and the number of matching vectors. This distance measure is easily interpreted since it provides a measure of how many vectors have been assigned to 'different' clusters in the two solutions. In the following, we refer to this distance as the *matching distance.*

5 Computational Experiments

We compared the iterated local search (ILS) utilizing the mutation operator described above with a multi-start k-means local search (MLS). In the MLS, a predefined number of times a start configuration is randomly generated and the k-means algorithm is applied. The best solution found is reported.

All algorithms were implemented in Java 1.4. Instead of comparing absolute running times, we provide the number of k-means iterations performed by the algorithms to have a measure independent of programming language or code optimizations.

5.1 The Gene Expression Data Sets

The first data set denoted as HL-60 is taken from [21] and contains data from macrophage differentiation experiments. The data consists of 7229 genes and expression levels at 4 time points. We applied a variation filter which discarded all genes with an absolute change in expression level less than or equal to 5. The number of genes which passed the filter was 3230. The vectors were normalized afterwards to have mean 0 and variance 1, as described in [7].

The second data set denoted as HD-4CL is also taken from [21] and contains data from hematopoietic differentiation experiments across 4 cell lines. The data consists of 7229 genes, 17 samples each. We applied the same variation filter as above which discarded all genes with an absolute change in expression level less than or equal to 5. The number of genes which passed the filter was 2046. Afterwards, the vectors were normalized to have mean 0 and variance 1.

The third data set is denoted as Cho-Yeast and is described in [22]. It contains the expression levels of 6565 yeast genes measured at 17 time points over two complete cell cycles. As in [7], we discarded the time points at 90 and 100 min, leading to a 15 dimensional space. A variation filter was used which discarded all genes with an absolute change in expression level less than or equal to 50 and an expression level of max/min < 2.0. The resulting number of genes was 2931. Again, the vectors were normalized afterwards to have mean 0 and variance 1.

To study the capability of ILS to find globally optimum solutions, the fourth and fifth data set were randomly generated with 5000 and 6000 vectors, denoted DS-5000 and DS-6000, respectively. The main vectors were generated with 16 time points: For each cluster, the cluster center $x = (x_1, \ldots, x_{16})$ was generated as follows. x_1 was chosen with a uniform random distribution, and all other x_i by an AR(1) process such that $x_{i+1} = x_i + \epsilon_i$, where ϵ_i is a normally distributed random number. All other cluster members were chosen to be normally distributed 'around' the cluster center. No variation filter was applied and the vectors were normalized as described above.

The number of clusters for the clustering were taken from [7] for Cho-Yeast ($k = 30$), and from [21] for the data sets HL-60 ($k = 12$), HD-4CL ($k = 24$). For the data sets DS-5000 and DS-6000, k was set to the known number of clusters in the experiments: $k = 25$ and $k = 30$, respectively.

5.2 ILS Performance

In the experiments, the ILS algorithm was run with a maximum number of iterations n_{max} set to 2000. Analogously, the MLS was run by calling k-means 2000 times. This way, the number of local searches was set to 2000 in both cases.

The results for MLS and ILS are reported in Table 1. For each data set and

Table 1. Comparison of Iterated Local Search and Multi-Start Local Search

Data Set	Algorithm	No. LS	Iter LS	Best	Avg. Obj.	Excess
HL-60	MLS	2000.0	95896.7	1749.86	1749.90	0.00%
	ILS	164.9	5660.2	1749.85	1749.89	0.00%
HD4CL	MLS	2000.0	61435.9	12476.92	12493.51	0.47%
	ILS	1721.5	26186.2	12434.91	12441.91	0.05%
Cho-Yeast	MLS	2000.0	74632.5	16966.78	16984.32	0.46%
	ILS	2000.0	32652.8	16905.22	16915.77	0.08%
DS-5000	MLS	2000.0	35954.7	3192.35	3497.14	9.55%
	ILS	207.4	2560.4	3192.35	3192.35	0.00%
DS-6000	MLS	2000.0	40929.9	5868.59	6314.74	8.65%
	ILS	272.1	3474.9	5811.82	5811.82	0.00%

algorithm, the average number of local searches (No. LS), the average number of local search iterations (Iter LS), the best objective found (best), the average best objective found (Avg. Obj.), and the average percentage excess over the optimum or best-known solution (Excess) is displayed. For each algorithm, 30 independent runs were performed.

The results show that ILS is clearly superior to MLS. ILS achieves better average and best objective values than MLS for all tested data sets. Moreover, ILS is at least more than two times faster. As can be seen for the data set Cho-Yeast, the number of iterations within k-means is more than two times lower in ILS than in MLS. This is due to the fact that mutation changes a local optimum solution only slightly and thus a new local optimum is reached with much fewer k-means iterations. In case of the largest data sets DS-5000 and DS-6000, the optimum partition is found in all runs with an average number of 207.4 and 372.1 ILS iterations, respectively.

The results of ILS are comparable with those reported in [23]. However, ILS outperforms the memetic algorithms in [23] on the data set Cho-Yeast, and the running times in terms of k-means iterations is lower for data sets DS-5000 and DS-6000.

5.3 Comparison with Other Heuristics

To assess the performance of our ILS approach, we conducted experiments with three other clustering algorithms from the bioinformatics literature. These algorithms use k-means but with a different initialization. The first algorithm denoted 'Average-Link+k-means' is an agglomerative hierarchical clustering algorithm combined with k-means. Starting with each vector in its own cluster, it works by stepwise merging the two clusters with the greatest similarity. The cluster similarity is determined by the average pairwise similarity between the genes in the clusters. The algorithm terminates if a predefined number of clusters is produced and k-means has been applied to the resulting solution. This combination has been shown to be superior to several other clustering approaches [8] for gene expression profiles.

Table 2. Comparison of the solution quality of three heuristics for the MSSC

Data Set	Algorithm	Objective	Excess
HL-60	Average-Link + k-Means	1753.93	0.23 %
	Farthest-Split + k-Means	1818.74	3.94 %
	MST-Cut + k-Means	1756.33	0.37 %
HD-4CL	Average-Link + k-Means	12824.10	3.13 %
	Farthest-Split + k-Means	12881.83	3.59 %
	MST-Cut + k-Means	12666.80	1.86 %
Cho-Yeast	Average-Link + k-Means	17118.25	1.28 %
	Farthest-Split + k-Means	17196.08	1.74 %
	MST-Cut + k-Means	17313.77	2.43 %
DS-5000	Average-Link + k-Means	3865.94	21.1 %
	Farthest-Split + k-Means	4354.12	36.39 %
	MST-Cut + k-Means	4359.50	36.56 %
DS-6000	Average-Link k-Means	6954.48	19.66 %
	Farthest-Split + k-Means	9600.66	65.19 %
	MST-Cut + k-Means	8213.83	41.33 %

The second algorithm denoted 'Farthest-Split + k-Means' is a divisive hierarchical algorithm used in [7] to cluster the yeast data and works as follows. The first cluster center is chosen as the centroid of the entire data set and subsequent centers are chosen by finding the data point farthest from the centers already chosen. If the desired number of clusters is reached, the process is terminated and the k-means heuristic is applied. The third algorithm denoted by 'MST-Cut + k-Means' is a combination of the k-Means heuristic and a heuristic utilizing minimum spanning trees as proposed in [24] for gene expression data clustering. The latter works by calculating the minimum spanning tree for the graph defined by the data set and cutting the tree at its $k-1$ longest edges. The resulting connected components of the tree define the k clusters.

The three algorithms were applied to the same data sets as ILS above. The results are displayed in Table 2. In the table, the objective value (Objective) in terms of the minimum sum-of-squares criterion is displayed as well as the percentage excess over the optimum or best-known objective value (Excess) found by ILS is presented. As the results show, these sophisticated heuristics are not capable of finding the globally optimum solution. The first heuristic 'Average-Link + k-Means' performs reasonably well in all the cases and is only outperformed by 'MST-Cut + k-Means' on the data set HD-4CL. However, all heuristics can not achieve the clustering quality of ILS. In case of the largest instances, all heuristics are even outperformed by the multi-start k-means heuristic. From the combinatorial optimization perspective, these results are not surprising. In many cases, ILS or the more sophisticated memetic algorithms are known to outperform combinations of constructive heuristics and local search [11,25,26].

Since the cluster memberships are more important than the clustering error, the distances to the best-known solutions for the heuristics are shown in Table 3. Both matching distance and means distance are provided. Interestingly, for the

Table 3. Distances to best-known solutions found by ILS

Data Set	Algorithm	Matching Distance	Means Distance
Cho-Yeast	Average-Link + k-Means	852.0	5379.0
	Farthest-Split + k-Means	922.0	5744.0
	MST-Cut + k-Means	1012.0	6202.0
DS-5000	Average-Link + k-Means	3.0	247.0
	Farthest-Split + k-Means	155.0	707.0
	MST-Cut + k-Means	70.0	711.0
DS-6000	Average-Link + k-Means	4.0	656.0
	Farthest-Split + k-Means	428.0	2678.0
	MST-Cut + k-Means	33.0	1314.0

data sets DS-5000 and DS-6000 the heuristics produce solutions which are similar to the optimum solution in terms of cluster memberships. In case of the real data set Cho-Yeast, however, they appear to be far away from the best-known solution in terms of cluster memberships. Hence, to arrive at optimum clustering solutions, it is not sufficient to use these heuristics proposed previously in the bioinformatics literature. ILS appears to be much more robust in finding (near)-optimum solutions to the NP-hard clustering problem. The results by ILS are similar to the results found by memetic algorithms [23], but the algorithmic framework is much simpler. Hence, at least for the data sets tested, the use of sophisticated evolutionary algorithms is not required.

6 Conclusions

An new approach for minimum sum-of-squares clustering has been proposed which is based on concepts of combinatorial heuristic optimization. The approach, called iterated local search (ILS), combines k-means with a simple and powerful mechanism to escape out of local optima. It has been shown on gene expression data from the bioinformatics literature, that ILS is capable of finding optimum clustering solutions for data sets up to 6000 genes. Compared to multi–start k-means, ILS is superior in respect to the best solution found and the average best solution found while requiring less than half of the computation time. Moreover, it is demonstrated that more sophisticated initialization mechanisms for k-means used in the bioinformatics literature such as average linkage clustering, clustering by splitting off the farthest data point, and clustering by partitioning minimum spanning trees does not lead to a clustering quality achieved by ILS. Compared to the recently proposed memetic algorithms for clustering, ILS performs comparable but requires much less implementation efforts.

There are several issues for future research. The estimation of the number of clusters has to be incorporated in the ILS algorithm. Moreover, the application of large data sets with more than 100000 data points should be addressed in further studies. Finally, different local search procedures and k-means variants should be tested within the ILS framework.

References

1. Brucker, P.: On the Complexity of Clustering Problems. Lecture Notes in Economics and Mathematical Systems **157** (1978) 45–54
2. Grötschel, M., Wakabayashi, Y.: A Cutting Plane Algorithm for a Clustering Problem. Mathematical Programming **45** (1989) 59–96
3. Hansen, P., Jaumard, B.: Cluster Analysis and Mathematical Programming. Mathematical Programming **79** (1997) 191–215
4. Zhang, M.: Large-scale Gene Expression Data Analysis: A New Challenge to Computational Biologists. Genome Research **9** (1999) 681–688
5. Brazma, A., Vilo, J.: Gene Expression Data Analysis. FEBS Letters **480** (2000) 17–24
6. Eisen, M., Spellman, P., Botstein, D., Brown, P.: Cluster Analysis and Display of Genome-wide Expression Patterns. In: Proceedings of the National Academy of Sciences, USA. Volume 95. (1998) 14863–14867

7. Tavazoie, S., Hughes, J.D., Campbell, M.J., Cho, R.J., Church, G.M.: Systematic Determination of Genetic Network Architecture. Nature Genetics **22** (1999) 281–285
8. Yeung, K., Haynor, D., Ruzzo, W.: Validating Clustering for Gene Expression Data. Bioinformatics **17** (2001) 309–318
9. Bradley, P.S., Fayyad, U.M.: Refining Initial Points for k-Means Clustering. In: Proc. 15th International Conf. on Machine Learning, Morgan Kaufmann, San Francisco, CA (1998) 91–99
10. Penã, J.M., Lozano, J.A., Larranãga, P.: An Empirical Comparison of Four Initialization Methods for the k-Means Algorithm. Pattern Recognition Letters **20** (1999) 1027–1040
11. Johnson, D.S., McGeoch, L.A.: The Traveling Salesman Problem: A Case Study. In Aarts, E.H.L., Lenstra, J.K., eds.: Local Search in Combinatorial Optimization. Wiley and Sons, New York (1997) 215–310
12. Lourenco, H.R., Martin, O., Stützle, T.: Iterated Local Search. In Glover, F., Kochenberger, G., eds.: Handbook of Metaheuristics. Kluwer Academic Publishers (2003)
13. Moscato, P.: Memetic Algorithms: A Short Introduction. In Corne, D., Dorigo, M., Glover, F., eds.: New Ideas in Optimization. McGraw–Hill, London (1999) 219–234
14. Merz, P., Freisleben, B.: Memetic Algorithms for the Traveling Salesman Problem. Complex Systems **13** (2001) 297–345
15. Forgy, E.W.: Cluster Analysis of Multivariate Data: Efficiency vs. Interpretability of Classifications. Biometrics **21** (1965) 768–769
16. MacQueen, J.: Some Methods of Classification and Analysis of Multivariate Observations. In: Proceedings of the Fifth Berkeley Symposium on Mathemtical Statistics and Probability. (1967) 281–297
17. Alsabti, K., Ranka, S., Singh, V.: An Efficient Space-Partitioning Based Algorithm for the k-Means Clustering. In Zhong, N., Zhou, L., eds.: Proceedings of the 3rd Pacific-Asia Conference on Methodologies for Knowledge Discovery and Data Mining (PAKDD-99). Volume 1574 of Lecture Notes in Artificial Intelligence., Berlin, Springer (1999) 355–359
18. Pelleg, D., Moore, A.: Accelerating Exact k-Means Algorithms with Geometric Reasoning. In Chaudhuri, S., Madigan, D., eds.: Proceedings of the Fifth ACM SIGKDD International Conference on Knowledge Discovery and Data Mining, New York, ACM Press (1999) 277–281
19. Likas, A., Vlassis, N., Verbeek, J.J.: The Global k-Means Clustering Algorithm. Pattern Recognition (**36**)
20. Pelleg, D., Moore, A.: X-means: Extending K-means with Efficient Estimation of the Number of Clusters. In: Proc. 17th International Conf. on Machine Learning, Morgan Kaufmann, San Francisco, CA (2000) 727–734
21. Tamayo, P., Slonim, D., Mesirov, J., Zhu, Q., Kitareewan, S., Dmitrovsky, E., Lander, E.S., Golub, T.R.: Interpreting Patterns of Gene Expression with Self-organizing Maps: Methods and Application to Hematopoietic Differentiation. In: Proceedings of the National Academy of Sciences, USA. Volume 96. (1999) 2907–2912
22. Cho, R.J., Campbell, M.J., Winzeler, E.A., Conway, S., Wodicka, L., Wolfsberg, T.G., Gabrielian, A.E., Landsman, D., Lockhart, D.J., Davis, R.W.: A Genome-wide Transcriptional Analysis of the Mitotic Cell Cycle. Molecular Cell **2** (1998) 65–73

23. Merz, P., Zell, A.: Clustering Gene Expression Profiles with Memetic Algorithms. In et al., J.J.M.G., ed.: Proceedings of the 7th International Conference on Parallel Problem Solving from Nature, PPSN VII. Lecture Notes in Computer Science 2439, Springer, Berlin, Heidelberg (2002) 811–820.
24. Xu, Y., Olman, V., Xu, D.: Clustering Gene Expression Data using a Graph-Theoretic Approach: An Application of Minimum Spanning Trees. Bioinformatics **18** (2002) 536–545
25. Merz, P., Freisleben, B.: Fitness Landscapes, Memetic Algorithms and Greedy Operators for Graph Bi-Partitioning. Evolutionary Computation **8** (2000) 61–91
26. Merz, P., Katayama, K.: Memetic Algorithms for the Unconstrained Binary Quadratic Programming Problem. Bio Systems (2002) To appear.

Data Clustering in Tolerance Space

Chun-Hung Tzeng and Fu-Shing Sun

Computer Science Department
Ball State University
Muncie, IN 47306
{tzeng,fsun}@cs.bsu.edu

Abstract. This paper studies an abstract data clustering model, in which the similarity is explicitly represented by a tolerance relation. Three basic types of clusters are defined from each tolerance relation: maximal complete similarity clusters, representative clusters, and closure clusters. Heuristic methods of computing corresponding clusterings are introduced and an experiment on two real-world datasets are discussed. This paper provides a different view in the study of data clustering, where clusters are derived from a given similarity and different clusters may have non-empty intersection.

1 Introduction

This paper reports a study of tolerance space as a mathematical data-clustering model. The task of data clustering is to group *similar* data together [3]. For numerical (continuous) data, an Euclidean distance function is often used to measure the similarity. For non-numerical (e.g., boolean or categorical) data, the concept of such similarity is vague and sometimes intuitive. Many metrics have been used to measure such similarity (or dissimilarity) such as simple matching coefficient, Jaccard coefficient, and category utility functions [3,5].

Instead of computing clustering directly, this paper studies data similarity first and uses a *reflexive* and *symmetric* binary relation to postulate a similarity. An object should always be similar to itself and, if an object x is similar to another object y, the object y should also be similar to x. A set of elements with such a relation is called a *tolerance space* and the relation is called a *tolerance relation* [9]. Preciseness among similarities is defined by comparing the related binary relations. The concept of data cluster and clustering is inherent in a tolerance space. This paper introduces three basic types of clusters: maximal complete similarity (MCS) clusters, representative clusters, and closure clusters. The MCS clusters have the highest intra-similarity and the closure clusters have the least intra-similarity: each MCS cluster is contained in a representative cluster, which is contained in a closure cluster. Three basic types of clusterings are then defined by the three types of clusters. Except the closure clustering, the clusterings in a tolerance space are in general not partitions. Two clusters may have non-empty intersection, which is natural and common in real-world problems. Although such a non-crisp clustering is called soft clustering in [8], we still use the term "clustering" in this paper.

M.R. Berthold et al. (Eds.): IDA 2003, LNCS 2810, pp. 297–306, 2003.

The closure clustering is uniquely defined for each tolerance relation. However, both MCS and representative clusterings are not unique in general and depend on the input order when computed by an algorithm. To reduce the dependency, this paper introduces a concept of optimal clusterings. Two optimal clusterings of the same type have the same number of clusters, which can always be matched in pairs. The closure clustering of a tolerance space can be computed in a polynomial time. However, the computation of optimal MCS or representative clustering is intractable in general. This paper proposes a heuristic method by use of a density function to compute sub-optimal clusterings and to reduce the dependency on input order. In order to reduce inter-similarity among clusters in a clustering, this paper also proposes a merging process which merges a cluster to another one if the latter one contains a certain percentage of the first one. A similar concept of density and merging has been used in [2].

In the study of data clustering (e.g., [4,7,11]), this paper provides a different view where clusters are derived from a previously given data similarity. For demonstration, this paper introduces an experiment on two real-world datasets. The experiment does not consider the predefined classes when computing clustering. As a result, there is a tolerance relation defined by a Hammming distance which re-discovers the predefined classes for the first dataset. The tolerance relations on the second dataset are also defined by a Hamming distance. For this case, a set of noisy data is clearly identified. In the merged clustering, there are two distinct large clusters, of which each cluster contains mainly records in one of the two original classes, respectively.

The rest of the paper is organized as follows. Section 2 is the introduction of tolerance space. Section 3 introduces the three basic types of data clustering. Section 4 introduces the heuristic method, formulations of noise, and a cluster merging process. Experimental results are discussed in Section 5. Conclusion and future work are in Section 6.

2 Tolerance Relation and Similarity Matrix

Let Ω be a finite set. A **tolerance relation** ξ on Ω is a binary relation with two properties: reflexivity (i.e., $(x,x) \in \xi$ for any $x \in \Omega$) and symmetry (i.e., $(x,y) \in \xi \Rightarrow (y,x) \in \xi$). The pair (Ω,ξ) is called a **tolerance space**. We also use the symbol ξ as the predicate: $\xi(x,y)$ if and only if $(x,y) \in \xi$. If $\xi(x,y)$, we say that x is *ξ-similar* to y. We use *similar* directly if there is no ambiguity. For example, on a metric space (M,d) (including Hamming distance), we have a tolerance relation d_ε for each positive real number $\varepsilon > 0$: $d_\varepsilon(x,y)$ if $d(x,y) < \varepsilon$.

Any undirected graph is a tolerance space (and vice versa), where Ω is the set of all vertices, and two vertices are similar if they are the same vertex or they are adjacent (e.g., Figure 1). Any equivalent relation ξ of a space Ω is a special tolerance relation which has the transitive property: $\xi(x,y)$ and $\xi(y,z) \Rightarrow \xi(x,z)$.

Let ξ and η be two tolerance relations on a space Ω. The tolerance relation ξ is **more precise** than η or η is **less precise** than ξ if $\xi \subset \eta$, denoted by $\xi \leq \eta$. That is, $\xi \leq \eta$ if $\xi(x,y) \Rightarrow \eta(x,y)$ for any $x,y \in \Omega$. It is always $\xi \leq \xi$.

If $\xi \leq \eta$, then any ξ-similar elements are also η-similar. On a metric space (M, d), $d_{\varepsilon_1} \leq d_{\varepsilon_2}$ if $\varepsilon_1 \leq \varepsilon_2$. On the space Ω, the most precise tolerance relation, denoted by Ω^0, is the relation in which each element x is similar only to x itself. We call Ω^0 the **discrete tolerance relation**. On the other hand, the least precise tolerance relation, denoted by Ω^∞, is the relation in which all elements are similar. We call Ω^∞ the **trivial tolerance relation**.

Given a tolerance space (Ω, ξ), we say that an element x can reach an element y in k $(k > 1)$ steps if there are $k-1$ elements $w_1, w_2, ..., w_{k-1}$, so that the following pairs are similar: $(x, w_1), (w_1, w_2), ..., (w_{k-1}, y)$. The **$k-$extended tolerance relation** ξ^k is the relation such that $\xi^k(x, y)$ if x can reach y in no more than k steps. We set $\xi^1 = \xi$. Let ξ^∞ be the union of all ξ^k's; that is, $\xi^\infty(x, y)$ if and only if x can reach y eventually. Actually, ξ^∞ is the transitive closure of ξ and is an equivalent relation. We call ξ^∞ the *closure* of ξ. The extended tolerance relation ξ^k becomes the closure ξ^∞ when k is large enough (e.g., $k \geq |\Omega|$). All ξ^k's form an increasing sequence ($|\Omega| = n$): $\Omega^0 \leq \xi^1 (= \xi) \leq \xi^2 \leq ... \leq \xi^{n-2} \leq \xi^{n-1} = \xi^\infty \leq \Omega^\infty$.

The **similarity matrix** S_ξ of the tolerance space (Ω, ξ) is a function on $\Omega \times \Omega$ defined by

$$S_\xi(x, y) = \begin{cases} 0 & \text{if } x = y, \\ k & \text{if } \xi^k(x, y) \text{ but not } \xi^{k-1}(x, y) \quad (\text{here } \xi^0 = \Omega^0), \\ \infty & \text{for other cases.} \end{cases} \tag{1}$$

If we label the elements of Ω by 1, 2, ..., and n, then S_ξ is a symmetric $n \times n$ matrix, in which all values on the diagonal are zeros. Consider (Ω, ξ) as an undirected graph and the length of each edge is 1. Then $S_\xi(x, y)$ is the shortest distance from x to y, which can be computed by Floyd's algorithm [1]. From $S_\xi(x, y)$, the extended tolerance relation can be derived: $\xi^k(x, y)$ if and only if $S_\xi(x, y) \leq k$; and $\xi^\infty(x, y)$ if and only if $S_\xi(x, y) \neq \infty$. For example, the similarity matrix of the tolerance space on Figure 1 is as follows.

	1	2	3	4	5	6	7
1	0	1	1	2	3	∞	∞
2	1	0	1	2	3	∞	∞
3	1	1	0	1	2	∞	∞
4	2	2	1	0	1	∞	∞
5	3	3	2	1	0	∞	∞
6	∞	∞	∞	∞	∞	0	1
7	∞	∞	∞	∞	∞	1	0

3 Basic Clusters and Data Clusterings

Let (Ω, ξ) be a tolerance space and $k \geq 1$ an integer. We study three basic types of clusters. First consider subspaces in which all elements are ξ^k-similar to each other. Such a subspace is called a *complete similarity* set of order k. A **maximal**

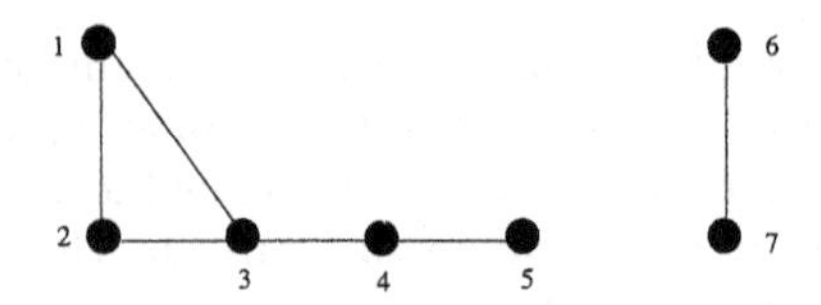

Fig. 1. A tolerance space.

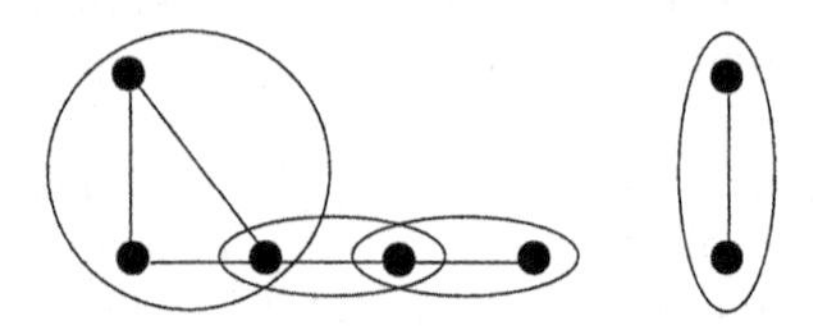

Fig. 2. All MCS clusters of order 1.

complete similarity cluster (an MCS cluster) M of order k is a maximal subset of Ω having the following property: $x, y \in M \Rightarrow \xi^k(x, y)$. The set of all MCS clusters of order k is denoted by $\mathbf{M}_{\xi^k}$. For example, Figure 2 shows all MCS clusters of order 1 in Figure 1. If $k_1 < k_2$, then any member of $\mathbf{M}_{\xi^{k_1}}$ is a complete similar set of order k_2 and is therefore contained in an MCS cluster of order k_2. We represent this property by $\mathbf{M}_{\xi^{k_1}} \leq \mathbf{M}_{\xi^{k_2}}$.

Next, we consider clusters representable by elements. For an element $x \in \Omega$, the **representative cluster** of order k at x is the set $\xi^k(x) = \{y \in \Omega : \xi^k(x, y)\}$, and x is called the *representative*. The set of all representative clusters of order k is denoted by $\mathbf{R}_{\xi^k}$. The cluster $\xi^k(x)$, uniquely determined by x, consists of all elements reachable from x in no more than k steps (e.g., Figure 3). Any cluster in $\mathbf{M}_{\xi^k}$ is contained in a representative cluster of order k (any member can be the representative); therefore, $\mathbf{M}_{\xi^k} \leq \mathbf{R}_{\xi^k}$. Furthermore, since each $\xi^k(x)$ is a complete similarity set of order $2k$, $\xi^k(x)$ is contained in an MCS cluster of order $2k$; that is, $\mathbf{R}_{\xi^k} \leq \mathbf{M}_{\xi^{2k}}$. If $k_1 < k_2$, then $\xi^{k_1}(x) \subseteq \xi^{k_2}(x)$; that is, $\mathbf{R}_{\xi^{k_1}} \leq \mathbf{R}_{\xi^{k_2}}$.

Last, we consider the transitive closure ξ^∞. Since it is an equivalent relation, both the MCS clusters and the representative clusters of ξ^∞ become the partition of the equivalent relation. The member of the partition containing x is $\xi^\infty(x) = \{y \in \Omega : \xi^k(x, y) \text{ for an integer } k \geq 1\}$, which is called the *closure cluster* at x. The corresponding partition is called the **closure partition** of ξ (e.g., Figure 6): $\mathbf{R}_{\xi^\infty} = \{\xi^\infty(x) : x \in \Omega\}$. Note that each MCS or representative cluster (of any order) is always contained in a closure cluster. In summary, we have the following properties about the three types of clusters:

$$\mathbf{M}_{\xi^k} \leq \mathbf{R}_{\xi^k} \leq \mathbf{M}_{\xi^{2k}} \leq \mathbf{R}_{\xi^\infty} \text{ for any integer } k \geq 1. \quad (2)$$

$$\mathbf{M}_{\xi^{k_1}} \leq \mathbf{M}_{\xi^{k_2}} \text{ and } \mathbf{R}_{\xi^{k_1}} \leq \mathbf{R}_{\xi^{k_2}} \text{ for } k_1 < k_2. \quad (3)$$

A **clustering** of a space Ω is a family of subspaces $\mathbf{C} = \{C_i : 1 \leq i \leq m\}$ such that Ω is the union of the subspaces: $\bigcup_{i=1}^m C_i = \Omega$, and any proper subfamily does not have this property; that is, $\mathbf{C}$ is a minimal covering of Ω. Each

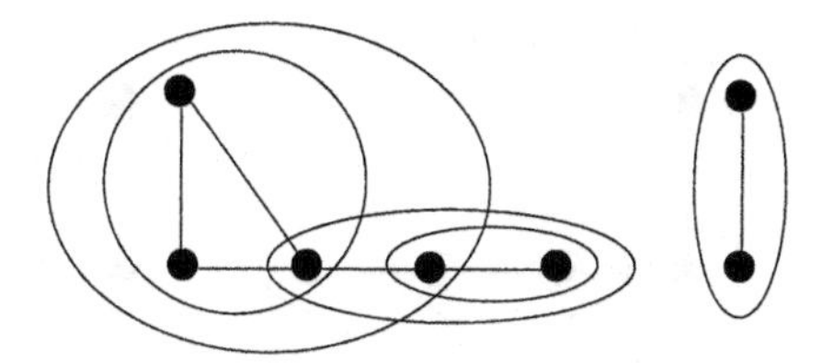

Fig. 3. All representative clusters of order 1.

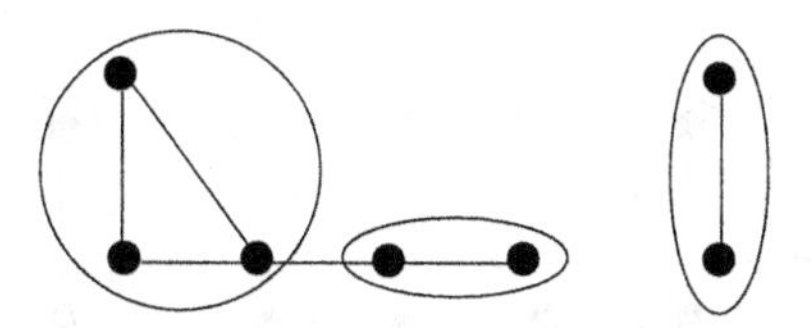

Fig. 4. An MCS clustering.

C_i is called a *cluster*. Each partition of Ω is always a clustering. However, in this paper a clustering is in general not a partition and two clusters may have a non-empty intersection.

In a tolerance space (Ω, ξ), a clustering of the tolerance space is called a **maximal complete similarity clustering** of order k ($\mathbf{M}_{\xi^k}$-clustering) if each cluster is in $\mathbf{M}_{\xi^k}$. A clustering is called a **representative clustering** of order k ($\mathbf{R}_{\xi^k}$-clustering) if each cluster in $\mathbf{R}_{\xi^k}$. The closure partition $\mathbf{R}_{\xi^\infty}$, being a clustering, is called the **closure clustering**. A tolerance space has in general many different MCS clusterings and representative clusterings. However, the closure clustering $\mathbf{R}_{\xi^\infty}$ is always uniquely determined. From the cluster property (2), $\mathbf{R}_{\xi^\infty}$ has the least number of clusters. For example, Figure 4 is an MCS clustering of order 1, Figure 5 is a representative clustering of order 1, and Figure 6 is the closure clustering.

4 Heuristic Search, Noise, and Cluster Merging

Given a tolerance space (Ω, ξ) with n elements, we consider the complexity of clustering. We assume the computation is based on the similarity matrix S_ξ defined in (1), which can be computed by Floyd's Algorithm with a time in $\Theta(n^3)$. Both $\xi^k(x)$ and $\xi^\infty(x)$ can be computed by use of S_ξ with a time in $\Theta(n)$. The closure clustering $\mathbf{R}_{\xi^\infty}$ is computed by repeating the search of closure clusters $\xi^\infty(x)$. Since there are at most n closure clusters, the time of this computation is in $O(n^2)$.

One of typical requirements in data clustering is insensitivity to the input order [3]. To formulate such insensitivity, we introduce a concept of optimal clustering. An $\mathbf{M}_{\xi^k}$-clustering $\mathbf{C} = \{C_1, C_2, ..., C_m\}$ is **optimal** if there is no $\mathbf{M}_{\xi^k}$-clustering with less than m members. That is, m is the minimum size of $\mathbf{M}_{\xi^k}$-clusterings. Optimal $\mathbf{R}_{\xi^k}$-clustering is defined similarly.

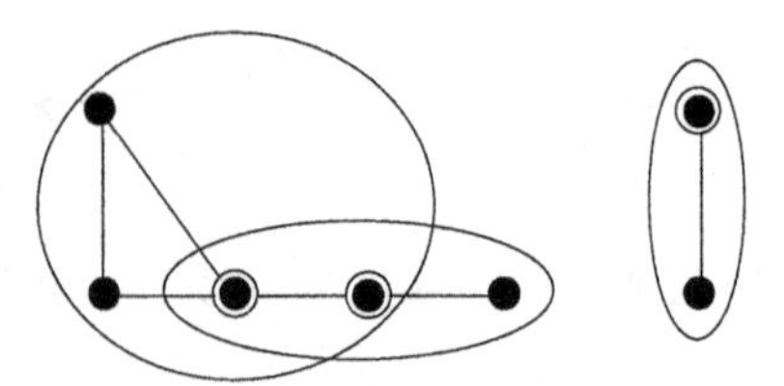

Fig. 5. A representative clustering of order 1.

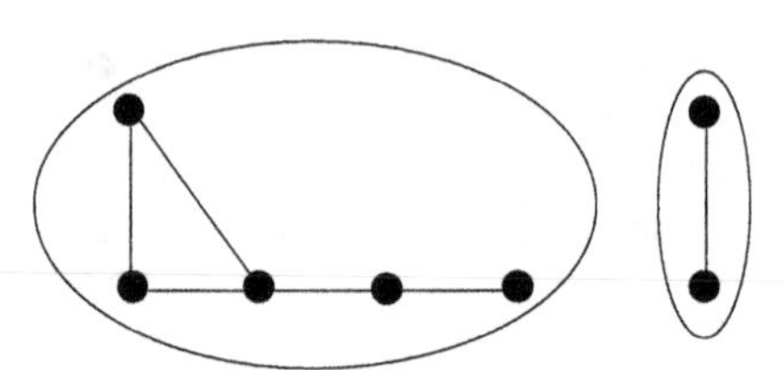

Fig. 6. The closure clustering.

Optimal clusterings are not unique in general. But they have the following property. Suppose $\mathbf{C} = \{C_1, C_2, ..., C_m\}$ and $\mathbf{B} = \{B_1, B_2, ..., B_m\}$ are two optimal clusterings of the same type. By use of the combinatorial lemma in Maak [6], there is a permutation $j_1, j_2, ..., j_m$ of 1, 2, ..., m such that each pair (C_i, B_{j_i}) are related: $C_i \cap B_{j_i} \neq \emptyset$ for $1 \leq i \leq m$. That is, the clusters in two optimal clusterings can be matched in pairs. To achieve certain insensitivity of input order, theoretically, we need to design algorithms to search for optimal clusterings.

To search for optimal clusterings is intractable in general. In the following we propose a heuristic method for sub-optimal clusterings. The goal of the heuristic search is to find a clustering with as few clusters as possible. The idea is to collect large clusters first. For this purpose, we use a *k-density* function: $den_k(x) = |\xi^k(x)|$. According to the density function, we sort the elements of Ω: $x_1, x_2, ..., x_n$, so that $den_k(x_1) \geq den_k(x_2) \geq ... \geq den_k(x_n)$.

To search for a sub-optimal $\mathbf{R}_{\xi^k}$-clustering, choose the cluster $\xi^k(x_1)$ first. Then choose the next cluster $\xi^k(x_i)$ (according to the above order) which is not in the union of previously chosen clusters. Repeat the process until the whole space is covered. Finally, in order to get a minimal covering, scan the chosen clusters backwards and delete the ones which are contained in the union of other collected clusters. The result is called a **sub-optimal** $\mathbf{R}_{\xi^k}$-clustering. Note that the input order has minimal effect in this process (e.g., Figure 5).

It is more difficult to search for a sub-optimal $\mathbf{M}_{\xi^k}$-clustering. We first consider MCS clusters containing a fixed element $x \in \Omega$. Since the MCS clusters are contained in the representative cluster $\xi^k(x)$, we compute the clusters locally. Let $\xi^k(x)$ be represented by an ordered list (any order), x being the first member ($m = den_k(x)$): $\xi^k(x) = (x, y_1, y_2, ..., y_{m-1})$. According to the order we use a backtracking process to search for all MCS clusters of order k containing the first member x and to return an optimal one (i.e., a cluster of the maximum size). This process is intractable in general and hence impractical when $|\xi^k(x)|$ is too large. Therefore, we add a termination strategy to the process, which uses a positive integer m_0 as a threshold. The backtracking process stops and return the currently searched MCS cluster when the process has searched for an MCS cluster of a size equal or larger than m_0. If there is no such cluster, it will return an optimal MCS cluster. Such an MCS cluster is called an m_0-**sub-optimal** MCS cluster at x. In application, users need to adjust the threshold m_0 so that, at one hand, m_0 is not too small and, at the other hand, an m_0-sub-optimal MCS cluster can be returned in a reasonable time.

Now we consider $\mathbf{M}_{\xi^k}$-clustering. Choose a threshold m_0 and compute an m_0-sub-optimal cluster at the first element x_1. Then choose the next x_i which is not in any previously computed clusters and compute an m_0-sub-optimal MCS cluster at x_i. Repeat the process until the whole space is covered. This method is based on a belief that it is more likely to find large MCS clusters at elements of high density. Finally, to minimize the clustering, scan the chosen clusters backwards and delete the ones which are contained in the union of other collected clusters. The resulting clustering is called an m_0-**sub-optimal** $\mathbf{M}_{\xi^k}$-clustering.

To define noisy or exceptional data is possible on a tolerance space. For example, we may use a threshold n_0 (a positive integer) to define a noisy set for each type of clusters: $N_{n_0}^C = \{x| \, |\xi^\infty(x)| < n_0\}$; $N_{n_0}^{R^k} = \{x| \, |\xi^k(x)| < n_0\}$; and $N_{n_0}^M = \{x| \, |M| < n_0, M$ is any optimal MCS cluster containing $x\}$. In computing clusterings, we first delete the noisy elements or put them into a special cluster. If there is a family of clusters with a distinctly small number of elements, we may use a threshold n_0 to separate those clusters. However, not all tolerance relations have such property. For example, each element would be treated as noise (or exceptional data) for the discrete relation and there is no noise at all for the trivial relation.

Except the closure clustering, a data clustering in this paper is not a partition in general. When a sub-optimal clustering has been computed, we may consider a cluster merging as follows. Let p_0 be a positive number $0 < p_0 < 1$. Let C_1 and C_2 be two clusters. If $\frac{|C_1 \cap C_2|}{|C_1|} > p_0$, then we merge the cluster C_1 into C_2. A p_0-**merging** is a process which scan a given clustering starting from the largest cluster and merges the currently scanned cluster into a previously collected cluster if the condition holds. The goal of the merging process is to merge closely related (depending on p_0) clusters together in order to reduce inter-similarity among clusters. However, at the same time, it will loose certain intra-similarity for the merged clusters. When p_0 is small, the merging process will occur more and the merged clustering may become closer to the closure clustering. On the other hand, when p_0 is large, the merging process will occur less.

5 Experiment

In this section, we introduce the experimental results on two datasets from UC Irvine machine learning data repository [10].

Experiment 1. The first dataset is the *Small Soybean Database*, which consists of 47 records with 35 attributes (denoted by A_k , $1 \leq k \leq 35$) and in 4 diagnosis classes: $D1$ (10 records), $D2$ (10 records), $D3$ (10 records), and $D4$ (17 records). The diagnosis class is the 36th attribute, denoted by A_{36}. All the 35 attributes are in categorical values.

Each record is treated as a vector of the first 35 attributes. Let $H(x, y)$ be the Hamming distance between x and y, which is the number of attributes A_k's ($1 \leq k \leq 35$) on which x and y have different values. For an integer $\varepsilon > 0$, consider the tolerance relation: $H_\varepsilon(x, y)$ if $H(x, y) < \varepsilon$. That is, two records are H_ε-similar if they are different at less than ε attributes. The relation H_1 is the discrete tolerance relation, and H_{20} is the trivial tolerance relation. A summary of the result for some ε's is in Table 1.

Table 1. Sub-optimal representative clusterings

ε	no_of_clusters of order 1	no_of_clusters of order 2 (closure)	diagnosis family of closure partition
5	17	**7**	7 singletons
6	11	**4**	$((D1)(D2)(D3)(D4))$
7	7	**3**	$((D1)(D2)(D3\ D4))$
8	6	**3**	$((D1)(D2)(D3\ D4))$

When ε is increased, the number of clusters decreases. For interpretation of the clusterings, we assemble all diagnoses of each cluster into a list, called the *diagnosis set* of the cluster. The *diagnosis family* of a clustering is the list of all diagnosis sets. If all records in a cluster have the same diagnosis, then the diagnosis set is a singleton. All diagnosis sets are singletons for $\varepsilon \leq 6$. Especially, the closure partition of H_6 is the original diagnosis partition. For both $\varepsilon = 7$ and $\varepsilon = 8$, all records of D_3 and D_4 together form a single closure cluster.

Table 2. m_0-sub-optimal MCS clusterings of order 1

ε	m_0	no_of_clusters	no_of_clusters with different diagnoses	no_of_clusters of 0.1-merging
7	2	13	0	4
8	7	11	0	4

In general, MCS clusters have higher intra-similarity than representative clusters of the same order (i.e, the property (2)), which is demonstrated for both H_7 and H_8. Representative clusterings for both cases do not separate records of $D3$ and $D4$ completely (Table 1). However, any m_0-sup-optimal MCS clustering of order 1 for H_7 separates the two classes (Table 2). For the less precise H_8, the m_0-sub-optimal MCS clustering of order 1 can separate the two classes only when m_0 is large enough (e.g., $m_0 = 7$). Therefore, the threshold m_0 is important in this case. Larger threshold returns clusters of higher intra-similarity. For both cases, one 0.1-merging returns the original diagnosis partition of four classes.

Experiment 2. The second dataset is the *1984 United State Congressional Voting Record Database.* The dataset consists of 435 records. Each record corresponds to one congressman's votes on 16 issues. All attributes are boolean

(y or n). The symbol ? represents a missing value. There are 124 records with one missing value, 43 records with 2 missing values, 16 records with 3 missing values, 6 records with 4 missing values, 5 records with 5 missing values, and 4 records with 6 missing values. There is exactly one record with 7, 9, 14, 15, or 16 missing values, respectively. A classification label for each record is provided (*Democrat* or *Republican*). The dataset contains records for 168 Republicans and 267 Democrats.

In the experiment we define a tolerance relation by use of a Hamming distance. At each boolean attribute, if two different records have the same value, the distance at the attribute is defined to be zero. If they have different values, then the distance is defined to be 1. If one or both have ?, then the distance is defined to be $\frac{1}{2}$ (there is no special reason for the value $\frac{1}{2}$). The Hamming distance $H(x,y)$ between two different records x and y is the sum of the distance at the 16 attributes. For each positive number $\varepsilon > 0$, we define a tolerance relation: $H_\varepsilon(x,y)$ if $H(x,y) < \varepsilon$, that is, two records are H_ε-similar if they are different at less than ε votes. Note that $H_{\frac{1}{2}}$ is discrete and H_{16} is trivial. In the following, we summarize the result for H_3 (i.e., $\varepsilon = 3$).

The closure clustering of H_3 consists of 35 clusters, of which a cluster consists of 396 records, 5 clusters consist of only two records, and the other 29 clusters are singletons. We treat the small 34 clusters as noise (i.e., noise threshold $n_0 = 2$), which consists of 39 records in total, 9 percent of the whole dataset. All records containing more than 5 missing attributes (9 records in total) are in this set. The further clustering is computed on the 396 records. The sub-optimal $\mathbf{R}_{H_3}$-clustering computed in the experiment consists of 49 clusters. After one 0.2-merging, the number of clusters is reduced to 5. For each cluster, the numbers of records in each class are in Table 3.

Table 3. A merged $\mathbf{R}_{H_3}$-clustering

Cluster	republican	democrat
1	143	30
2	6	201
3	0	12
4	4	0
5	3	0

There are two large clusters: Cluster 1 consists of mainly republican records (83%), and Cluster 2 consists of mainly democrat records (97%). Note that the total number of records in the merged clustering is 399; that is, the final 5 clusters do not form a partition. Cluster 1 and Cluster 3 share a common record, and Cluster 2 and Cluster 3 share two common records. Except that there are more small clusters, the 7-sub-optimal $\mathbf{M}_{H_3}$-clustering also contains two large clusters with the same property.

6 Conclusion

This paper introduces tolerance space as a mathematical model of similarity in data clustering, in which clustering is inherent in the formulation of similarity.

Our current effort is focusing on both the mathematical theory and implementation issues of tolerance space. Although this paper uses Hamming functions to define tolerance relations in the experiment, we plan to study more tolerance relations, including similarities in existing clustering methods, for different types of problems. In addition to categorical values, other types of data such as numerical and boolean values are to be studied in our model. For scalability, we plan to use both small and huge real-world data as well as synthetic data in future experiments.

References

1. Brassard, G. and Bratley, P. *Fundamentals of Algorithmics*, Prentice Hall, 1996.
2. Ester, M., Kriegel, H.-P., Sander, J., and Xu, X. A Density-Based Algorithm for Discovering Clusters in Large Spatial Databases with Noise, *Proc. 2nd International Conference on Knowledge Discovery and Data Mining*, pages 226–231, Portland, OR, USA, 1996.
3. Han, J. and Kamber, M. *Data Mining: Concepts and Techniques*, Morgan Kaufmann, 2001.
4. Huang, Z. A Fast Clustering Algorithm to Cluster Very Large Categorical Data Sets in Data Mining, *Research Issues on Data Mining and Knowledge Discovery*, 1997.
5. Jain, A. K. and Dube, R. C. *Algorithms for Clustering Data*, Prentice Hall, 1988.
6. Maak, W. *Fastperiodishe Funktionen*, Springer-Verlag, 1967.
7. Ng, R. T. and Han, J. Efficient and Effective Clustering Methods for Spatial Data Mining, *Proc. 20th International Conference on Very Large Data Bases*, pages 144–155, Santiago, Chile, 1994.
8. Pelleg, D. and Moore, A. Mixtures of Rectangles: Interpretable Soft Clustering *Proc. 18th International Conf. on Machine Learning* Morgan Kaufmann, pages 401–408, 2001.
9. Sun, F. and Tzeng, C. A Tolerance Concept in Data Clustering, 4th International Conference on Intelligent Data Engineering and Automated Learning (IDEAL 2003), Hong Kong, March 2003.
10. University of California at Irvine, Machine Learning Repository. http://www.ics.uci.edu/~mlearn/MLRepository.html
11. Zhang, T., Ramakrishnan, R., and Livny, M. BIRCH: An Efficient Data Clustering Method For Very Large Databases, *ACM SIGMOD International Conference on Management of Data*, pages 103–114, Montreal, Canada, 1996.

Refined Shared Nearest Neighbors Graph for Combining Multiple Data Clusterings

Hanan Ayad and Mohamed Kamel

Pattern Analysis and Machine Intelligence Lab,
Systems Design Engineering, University of Waterloo,
Waterloo, Ontario N2L 3G1, Canada
{hanan,mkamel}@pami.uwaterloo.ca

Abstract. We recently introduced the idea of solving cluster ensembles using a Weighted Shared nearest neighbors Graph (*WSnnG*). Preliminary experiments have shown promising results in terms of integrating different clusterings into a combined one, such that the natural cluster structure of the data can be revealed. In this paper, we further study and extend the basic *WSnnG*. First, we introduce the use of fixed number of nearest neighbors in order to reduce the size of the graph. Second, we use refined weights on the edges and vertices of the graph. Experiments show that it is possible to capture the similarity relationships between the data patterns on a compact refined graph. Furthermore, the quality of the combined clustering based on the proposed *WSnnG* surpasses the average quality of the ensemble and that of an alternative clustering combining method based on partitioning of the patterns' co-association matrix.

1 Introduction

Cluster analysis is an unsupervised learning method that constitutes a cornerstone of an intelligent data analysis process. It is used for the exploration of inter-relationships among a collection of patterns, by organizing them into homogeneous clusters. It is called unsupervised learning because unlike classification (known as supervised learning), no *a priori* labelling of some patterns is available to use in categorizing others and inferring the cluster structure of the whole data.

Cluster analysis is a difficult problem because many factors (such as effective similarity measures, criterion functions, algorithms and initial conditions) come into play in devising a well tuned clustering technique for a given clustering problem. Moreover, it is well known that no clustering method can adequately handle all sorts of cluster structures (shape, size and density). In fact, the cluster structure produced by a clustering method is sometimes an artifact of the method itself that is actually imposed on the data rather than discovered about its true structure.

Combining of multiple classifiers has been successful in improving the quality of data classifications. Significant theoretical advances and successful practical

M.R. Berthold et al. (Eds.): IDA 2003, LNCS 2810, pp. 307–318, 2003.

applications have been achieved in this area [1]. An experimental comparison of various combining schemes can be found in [2]. In a recent review of multi-classifier systems [1], the integration of multiple clustering is considered as an example to further broaden and stimulate new progress in the area. However, the theoretical foundation of combining multiple clusterers, is still in its early stages. Some of the important related contributions can be found in [3] and [4].

In fact, combining multiple clusterings is a more challenging problem than combining multiple classifiers. In [5], the reasons that impede the study of clustering combination have been identified as: (1) It is believed that the quality of clustering combination algorithms can not be evaluated as precisely as combining classifiers. A priori knowledge or user's judgement plays a critical role in estimation of clustering performance. This problem represents an obstacle to proposing a mathematical theory to design clustering combination algorithms. (2) As various clustering algorithms produce largely different results due to different clustering criteria, combining the clustering results directly with integration rules, such as sum, product, median and majority vote can not generate a good meaningful result. In [1], combining multiple clusterings has been stated to be a more difficult problem, since cluster labels are symbolic and a correspondence problem must also be solved. Other difficulties include variability of the number and shape of clusters provided by individual solutions, and the desired or the unknown "right" number and shape.

However, as noted in [1,3], cluster ensembles can lead to improved quality and robustness of clustering solutions across a wider variety of data sets, enable *"knowledge reuse"*, and *"distributed clustering"* both in terms of objects or features. Moreover, we believe that one of the potential benefits of cluster ensembles lies in leveraging the ability of cluster analysis to reveal the natural cluster structure through careful analysis and consolidation of multiple different clustering of the input data. Therefore, cluster ensembles have a potential to play a critical role in the field of knowledge extraction from data, distributed learning and decision fusion of multiple unsupervised learners.

Cluster ensembles can be formed in a number of different ways, such as (1) the use of a number of different clustering techniques (either deliberately or arbitrarily selected). (2) The use of a single technique many times with different initial conditions. (3) The use of different partial subsets of features or patterns. In this paper, we use the first approach. As to the selection of the techniques, we use a set of techniques from those proposed in [3] for creating a diverse collection of clustering methods.

2 Shared Nearest-Neighbors Based Combiner

In a recent contribution [6], we introduced the idea of combining multiple different clusterings of a set of data patterns based on a *Weighted Shared nearest neighbors Graph WSnnG*. We compared the performance of the Shared nearest neighbors-based (Snn-based) combiner with the supra-consensus function introduced in [3]. Results were promising and showed the ability of the Snn-based combiner to reveal cluster structures that may be unbalanced.

2.1 Similarity Based on Sharing of Neighbors

Based on the principle of evidence accumulation introduced in [4] and which was inspired by the work in sensor fusion and classifier combination techniques in [2], a voting mechanism can be adopted to determine the similarity between patterns based on their co-occurrence in the given set of clusterings.

An alternative, or rather a complement to direct similarity between patterns is their shared nearest neighbors, an approach to similarity proposed in 1973 by Jarvis and Patrick [7] and recently extended in [8,9]. The philosophy of the approach is that one can confirm the similarity between two patterns by their common (i.e shared) nearest neighbors. For instance, if patterns i and j are similar, and if they are both similar to a set of shared patterns S, then one can have greater confidence in the similarity between i and j, since their similarity is confirmed by the set of their shared neighbors S.

In the proposed combiner, we emphasize the shared neighborhood relationships between patterns. The extracted relationships are used to construct the $WSnnG$ whose vertices correspond to the patterns and the edges represent the links between the patterns based on their neighborhood sharing. Weights are assigned on both edges and vertices, as will be further explained later.

The advantages of $WSnnG$ are: first, we find that the shared nearest neighbors approach to similarity (co-association) can be particularly useful in combining different data clusterings. In combining different clusterings based on co-associations, one faces the issue of transitivity of co-associations in merging patterns. We believe that re-defining co-associations in terms of neighborhood sharing provides a means to carefully analyze the co-associations among patterns, as revealed by the consensus and conflicts among the different clusterings. Therefore, ideas based on this concept, such as those used in this paper, can help in integrating different cluster structures of the data, beyond applying strict consensus or transitivity on co-associations.

Second, we derive from this approach, the definition of the *shared nearest neighbors population* associated with each pattern i as the sum or weighted sum of shared neighbors with pattern i. We use this number in determining the relative weights of patterns. This is analogous to the notion of probability density of a point defined in [8]. The idea is that we assume that if a point has a lot of highly associated shared nearest neighbors, then it is more likely to be a member of a large and/or dense cluster. On the other hand, a point that has fewer weakly associated shared nearest neighbors, then it is more likely to be a member of a small and/or dispersed cluster.

2.2 Problem Formulation

Let $X = \{x_1, x_2, \cdots, x_n\}$ denote a set of n data patterns, and $C = \{C_1, C_2, \cdots, C_r\}$ is a set of r clusterings (cluster ensemble) of the n data patterns. From the given set of clusterings C of the n patterns, we construct a $WSnnG(V, E)$, where $V = \{v_1, v_2, \cdots v_n\}$ is the set of weighted vertices corresponding to the set of patterns. $E = \{e_1, e_2, \cdots, e_m\}$ represents the set of weighted edges connecting the patterns, where m is the number of edges. An edge between any pair of vertices is counted only once, i.e. edge connecting vertex v_i to vertex v_j is not counted separately from edge connecting v_j to v_i.

Since we combine the different clusterings of the ensemble without accessing the data patterns X in their feature space, we will refer, from now on, to a pattern by its index, that is, we use i rather than x_i. Before building the $WSnnG$, the following preprocessing steps are performed.

A. Computation of co-associations (direct-similarities):
Similar to the voting mechanism described in [4], a similarity measure between patterns co-occurring in the same cluster is computed as

$$co - assoc(i, j) = \frac{votes_{ij}}{r}$$

where $votes_{ij}$ is the number of times patterns i and j are assigned to the same cluster among the r clusterings. That is $0 \leq co - assoc(i, j) \leq 1.0$.

B. Sparsification of the nearest neighbors:
Based on the co-associations between patterns, a list of nearest neighbors P^i to each pattern i is defined as the list of patterns whose co-associations with i satisfy a selected $vote_{Thresh}$, where $0 < vote_{Thresh} \leq 1.0$. That is, $\forall(i, j), i \in P^j$ and $j \in P^i \iff co - assoc(i, j) \geq vote_{Thresh}$.

A number of factors can play a role in choosing the value of the $vote_{Thresh}$, such as the size of the ensemble, knowledge about the relative performance of the individual clustering techniques. Other issue such as the number of clusters relative to the size of the data. Increasing the value of $vote_{Thresh}$ corresponds to imposing stricter consensus among the clusterings, while decreasing it, corresponds to relaxes this condition and leads to revealing of more nearest neighbors. A majority vote corresponds to $vote_{Thresh} = 0.5$. In the experiments described, we maintain a majority vote threshold, as we find it suitable for the diverse ensemble that we use. The ensemble consist of most of the techniques originally proposed in [3].

The construction of the $WSnnG$, is described in the next two subsections. After its construction, the graph is then partitioned using the graph partitioning package METIS [10], in which the underlying algorithms described in [10,11, 12] are based on the state-of-the-art multilevel paradigm that has been shown to produce high quality results and scale to very large problems. METIS can partition a graph in the presence a balancing constraint. The idea is that each vertex has a weight associated with it, and the objective of the partitioning algorithm is to minimize the edge-cut subject to the constraint that the weight of the vertices is equally distributed among the clusters.

2.3 Basic WSnnG

The computation of the weights in the basic WSnnG [6] are described as follows.

i. Evaluation of neighborhood sharing: For each pair of patterns i and j, determine the size (count) of their shared nearest neighbors, i.e., the size of the intersection $|P^i \cap P^j|$ of their respective nearest neighbors lists P^i and P^j. Using the Jaccard measure of similarity, we compute the strength σ_{ij} of the neighborhood sharing between the two objects i and j as

$$\sigma_{ij} = 2 \times \frac{|P^i \cap P^j|}{|P^i| + |P^j|} \tag{1}$$

If σ_{ij} satisfies a strength threshold $\sigma_{ijThresh}$, where this threshold is evaluated dynamically for each pair of objects i and j, and is given as, $\sigma_{ijThresh} = \mu \times max(|P^i|, |P^j|)$. The parameter μ determines the required ratio of overlap. In the reported experiments $\mu = 0.5$, indicating that objects must share at least 50% of their nearest neighbors in order to satisfy the strength threshold.

Then

- Create an edge between i and j with weight σ_{ij}
- Increment shared nearest neighbors populations θ_i, θ_j for objects i and j

ii. Assignment of weights to the vertices of the graph: The weight ω_i of each object i is computed as the ratio of a balancing factor l to the shared nearest neighbors population θ_i associated with object i. That is, $\omega_i = \frac{l}{\theta_i}$ The idea of the relative weights ω_i is to in reflect in the graph the varying cluster sizes.

This graph is characterized by the variability of the size of each pattern's nearest neighbors. This type of graph, however, can grow into a significantly large one reaching an $O(n^2)$. For instance, in such cases where n is large and the number of clusters is small, this leads to large lists of nearest neighbors P^i to each pattern i. Furthermore, the weighting scheme used is binary, it is sensitive to the number of shared nearest neighbors as shown in Equation 1, rather than their corresponding co-associations (values of votes ratios).

2.4 Refined WSnnG

In the refined graph, we want to use only the k-nearest neighbors of each pattern, thus ensuring a compact graph of size $O(kn)$ and consequently reducing the computational cost. In addition, we use a refined weighting of the edges and vertices that is sensitive to the corresponding vote ratios instead of the binary scheme used in the basic graph.

We refer to the list of k-nearest neighbors of pattern i as L^i. A characteristic of the lists P^i (defined in step B. of Section 2.2) that is not found in L^i is their of *mutuality*. We have, $\forall(i,j), i \in P^j \iff j \in P^i$, whereas this does not hold for the lists L^i.

Therefore, two algorithmic options present themselves in dealing with the issue of mutuality, while building the $WSnnG$ based on k-nearest neighbors. It is possible that one chooses to connect two patterns i and j only when they are mutual neighbors. Alternatively, one can choose to connect them whenever either $i \in L^j$ or $j \in L^i$. In this paper, we choose the second option because it reveals more relationships at smaller values of k and leads to faster convergence.

The weights assigned to the vertices are meant to reflect an estimate of the probability density of the patterns, thus estimating the probability distribution of the clusters to which the patterns belong. In fact, this is an important aspect of the solution that we propose for accommodating natural cluster structures that may be unbalanced. In the refined *WSnnG*, we use the weighted sum of the shared nearest neighbors as the density estimates of the patterns. Again, at this stage, two options present themselves. One option is to compute the weight of pattern i based only neighborhood sharing with the elements of its pruned k-nearest neighbors list L^i. The second option is to use all the elements of the P^i list. In this paper, we choose the second option in order to maintain a stability to the partitioning solutions at different values of k. This is achieved due to the global computation of the weights assigned to the vertices. Therefore, choosing k becomes less of an issue. The computations of the weights are described below followed by a description of the algorithm *WSnnG-Builder* in Algorithm 1.

i. Evaluation of neighborhood sharing: Let $S^{ij} = \{s_1^{ij}, s_2^{ij} \cdots s_{n^{ij}}^{ij}\} = P^i \cap P^j$, be the set of shared nearest neighbors between pattern i and j, and $W^{ij} = \{w_1^{ij}, w_2^{ij}, ..., w_{n^{ij}}^{ij}\}$ is a corresponding list of weights. n^{ij} is the number of shared patterns between patterns i and j. The weight $w_p^{ij} \in W^{ij}$ is computed as the product of ϑ_p^i and ϑ_p^j where ϑ_p^i is the $co-assoc(i, s_p^{ij})$, and ϑ_p^j is the $co-assoc(j, s_p^{ij})$. That is, the corresponding actual vote ratios. The formulas for computing w_p^{ij} and the total weight w^{ij} are as follows.

$$w_p^{ij} = \vartheta_p^i \times \vartheta_p^j.$$

$$w^{ij} = \sum_{p=1}^{n^{ij}} w_p^{ij}. \tag{2}$$

Let W^i and W^j be the lists of corresponding weights to P^i and P^j, as determined from the vote ratios (in the preprocessing steps of Section 2.2). Using on the extended Jaccard measure [13], we compute the strength of neighborhood sharing σ_{ij} as

$$\sigma_{ij} = \frac{w^{ij}}{||W^i||^2 + ||W^j||^2 - w^{ij}} \tag{3}$$

ii. Assignment of weights to the vertices of the graph: The weight ω_i of each pattern i is computed as the ratio of a balancing factor l to the weighted sum of the shared nearest neighbors population θ_i, where $\theta_i = \sum_{j=1}^{n^i} w^{ij}$, and n_i is size of the list P^i. Finally, ω_i is given by $\omega_i = \frac{l}{\theta_i}$.

Notice that we use only the k-nearest neighbors L^i of each pattern i, to connect edges between the vertices of the graph, but we use the nearest neighbors P^i to evaluate the strength of neighborhood sharing, and to evaluate the patterns' weights (density estimates). Consequently, the size of the graph is reduced from a worst case complexity of $O(n^2)$ (Basic Graph) to $O(kn)$.

Algorithm 1 WSnnG-Builder

```
for all pattern i in the dataset do
  for all pattern j ∈ P^i, such that the symmetric pair (j, i) was not processed do
    compute w^{ij} as given by Equation 2
    θ_i ← θ_i + w^{ij}
    θ_j ← θ_j + w^{ij}
    if j ∈ L^i OR i ∈ L^j then
      connect patterns i and j with an edge of strength σ_{ij} as given by Equation 3
    end if
  end for
  ω_i ← l/θ_i
  assign the weight ω_i to pattern i.
end for
```

Also, notice that we eliminated the μ parameter that was used in the basic graph. This is done to avoid filtering of relationships, since we only consider a limited number of neighbors. We rely instead on the added sensitivity to the weighting scheme in determining the strength of the links, and weights of patterns. However, one may still choose to use a threshold and eliminate weak relationships.

Basically, the idea is to extract from the relationships revealed by the ensemble, representative interconnections, and corresponding weights, in a concise graph. An important characteristic of the graph is to be flexible enough to capture different clusters distributions, which may naturally occur in the data, making it more difficult to clustering techniques to reveal them, and also causing variability in the clustering generated using different techniques.

2.5 Partitioning of the WSnnG

The $WSnnG$ is partitioned into q clusters, so as to minimize the edge-cut subject to the constraint that the weights on the vertices are equally distributed among the q clusters. The issue of how to determine q is not addressed in this work. An estimate of q is rather assumed to be available for the combiner and the individual clustering techniques forming the ensemble. By assigning relative weights ω_i to the vertices based on their corresponding weighted sum of neighborhood sharing, we embed in the graph information about the sizes of neighborhood associated with each pattern, thus accommodating for varying cluster sizes and structure imbalances. The weights σ_{ij} on the edges, are based on the strength of the link connecting patterns i and j. Stronger links are less likely to be cut, again subject to the balance constraint.

3 Experimental Analysis

In [3], an ensemble of diverse techniques was proposed to be used as a portfolio of methods across a wide variety of data from different domains and dimensionality. We use most of the techniques of this portfolio, totalling to 9 different techniques.

They are the graph partitioning and k-means, each using 4 different similarity measures (Euclidean, cosine, extended Jaccard and correlation), in addition to hypergraph partitioning. We used ClusterPack, a cluster analysis package by Alexander Strehl available at http://strehl.com/ to generate the ensemble.

3.1 Quality Evaluation

We use the ensemble of clusterings to build a refined $WSnnG$, at variable values of K in the range [10-100], in order to study the behavior of the method. For each K, we generate the combined clustering based on the graph partitioning. In order to study the merit of our approach, we compare it to solution generated by the partitioning of the original co-association matrix, which is the cluster ensemble method called CSPA (Cluster-based Similarity Partitioning Algorithm) in [3]. The partitioning is generated using METIS on the original co-association matrix (direct similarities based on votes of co-associations). We also compare the quality of the combined clustering to the average quality of the individual clustering techniques forming the ensemble.

We use the F-measure [14], which combines both precision and recall to evaluate the quality of the different clustering solutions. The F-measure is an external criteria based on additional information not given to the clusterers, which is the labelled categorization assigned externally by humans.

3.2 Artificial Data

We artificially generated 3 datasets of 2 dimensional points. The datasets are shown in Figure 1. Dataset 2D-2C consists of two clusters of points of sizes (50, 150) sampled from two Gaussian distributions with different means and co-variance matrices. Dataset 2D-3C consists of three clusters of points of sizes (50, 75, 150) also sampled from three Gaussian distributions using different means, and co-variance matrices. Dataset 2C-NonConvex consists of points forming two non-convex clusters of sizes (120, 300). The clusters were generated by concatenating several smaller Gaussian clusters with means moving along half circles.

From the results shown in Figure 2, we notice the superiority of the combined clustering based on the refined $WSnnG$ over the ensemble's mean and the CSPA clustering. In addition, we find an improvement in terms of the discovered cluster distribution relative to the natural clusters inherent in the data. In contrast to the combined clustering based on the CSPA method which always generate equal size clusters, thus imposing well balanced clusters, our approach thrives to discover the true probability distributions of the clusters.

The clustering of the datasets 2D-2C, 2D-3C and 2C-NonConvex based on the CSPA, discovered clusters of sizes as follows (100, 100), (91, 91, 93) and (210, 210), respectively. The WSnnG solutions for different K discovered clusters of sizes around (44, 156), (27, 78, 170) and (108, 312) respectively. By comparing these values to the original cluster sizes, we find that the WSnnG solutions represent better estimates of the cluster sizes.

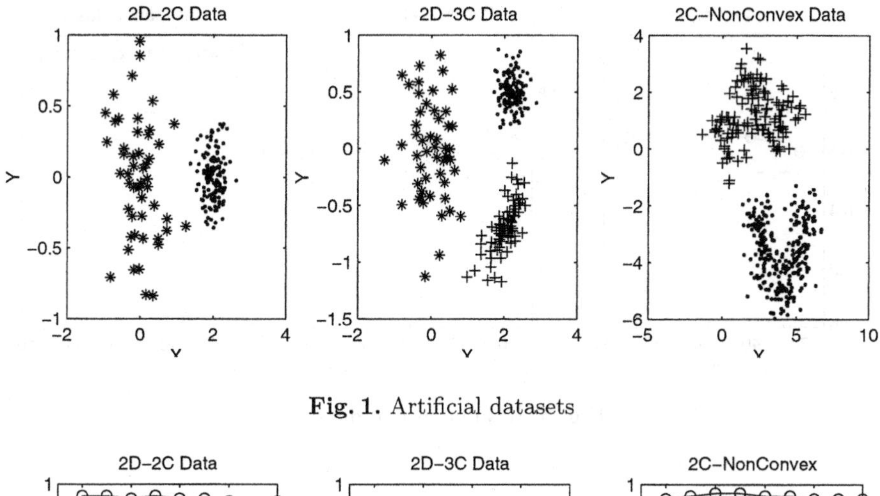

Fig. 1. Artificial datasets

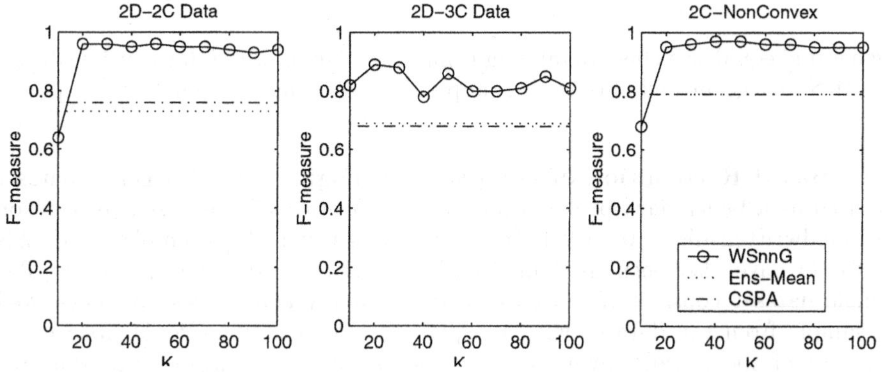

Fig. 2. Artificial data results. The ensemble's mean (Ens-Mean) and the CSPA are not function of K, they have constant values. We plot them on the same plot of WSnnG vs K to facilitate the comparison. With the 2D-3C and the 2C-NonConvex datasets, the Ens-Mean and the CSPA overlap.

3.3 Real Data

We experimented with datasets from the UCI machine learning repository available at http://www.ics.uci.edu/ mlearn/MLRepository.html. The datasets selected are generally used in classification and pattern recognition problems. The results are shown in Figure 3, and further discussed below.

Ecoli Dataset. The Ecoli Dataset is used in classification problems for predicting the cellular localization sites of proteins [15]. The dataset consists of 336 instances represented by 7 numeric predictive attributes. There are 8 classes of unbalanced distribution, with class sizes ranging from 2 to 143 patterns per cluster (ppc). From the results shown in Figure 3, we find that the quality of the combined clustering based on the $WSnnG$ surpasses both the ensemble's average and combined clustering based on CSPA. In addition, the CSPA method

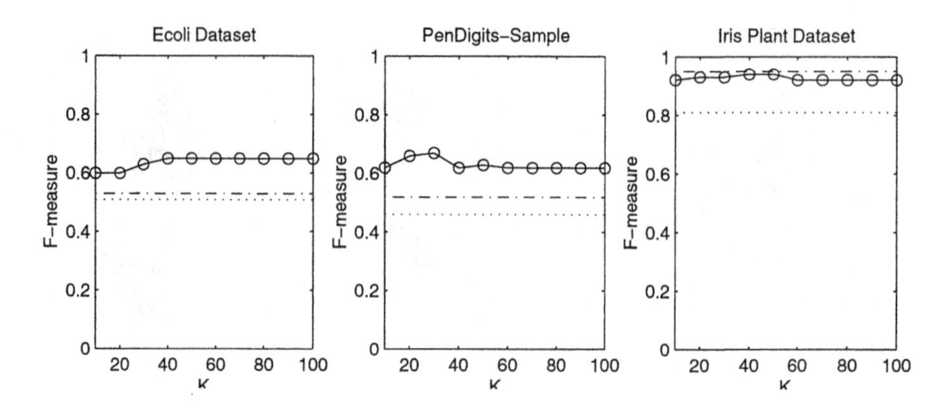

Fig. 3. Results on real datasets

generated equal size clusters ranging from 41 to 43 ppc, while the solutions based on $WSnnG$ generated cluster sizes approximately ranging from 1 to 105 ppc.

Pen-Based Recognition of Handwritten Digits. The Pen Digits dataset is used as a benchmark problem for the classification of pen-based recognition of handwritten digits 0 to 9 from 16 spatial features [16]. Samples from this datasets have also been used in experiments with cluster ensembles [3]. The whole dataset consists of ten classes of of roughly equal sizes. We generated a sample from the dataset with unequal class distribution. PenDigits-Sample consists of 490 instances with varying class sizes ranging from 10 to 150 ppc. From the results shown in Figure 3, the F-measure of the clusterings based on the $WSnnG$ surpasses both the ensemble's mean and the CSPA-based clustering. We also find that while the combined clustering based on CSPA consists of equal size clusters ranging from 48 to 50 ppc, the $WSnnG$ clusters were approximately ranging from 1 to 114 ppc.

Iris Plant Dataset. The Iris plant dataset is one of the best known datasets in the pattern recognition literature. The dataset contains 3 classes of 50 instances each, where each class refers to a type of Iris plant, and the data is represented by 4 attributes. One class is linearly separable from the other 2; the latter are not linearly separable from each other. As opposed to the other datasets used in this paper, this is an example of a well balanced dataset with patterns equally distributed among the clusters. The challenge for the individual clustering methods of the ensemble is rather due to the interleaving of two classes. In this particular example, the CSPA performed well, since the natural clusters are well balanced. The $WSnnG$ solutions also generated balanced clusters of comparable performance to CSPA while achieving an improvement over the ensemble's mean.

4 Conclusions

We introduced in this paper a compact and refined version of the *WSnnG*. The shared nearest neighbors approach to similarity provides a complement to direct similarities and is a more reliable approach to analyze and extract complex similarity relationships among patterns. We believe that this approach can be particularly useful in combining clusterings. As shown in the experiments, the quality of combined clustering based on the refined *WSnnG* surpasses the ensemble's mean, and enhances the combined clustering based the CSPA method, in terms of the F-measures and the cluster distributions revealed.

The WSnnG-based approach thrives to discover the natural cluster structure, using the relationships revealed by the ensemble of individual clustering techniques. We argue that this approach can successfully capture and flexibly adapts to natural cluster distributions that may be unbalanced. We believe that it can provide a new tool to solve cluster ensembles for the purpose of enhancing the clustering quality while preserving fidelity to the true clusters inherent in the data. In this paper, experiments have shown promising results in terms of capturing the significant inter-relationships among patterns on a compact and refined graph.

In Future work, we want to further analyze and extend the *WSnnG*. Moreover, we want to explore its applicability in the domain of very high dimensional data, overlapping clusters, and multi-class patterns. Finally, it is worth noting that in this work, we emphasized the underlying model for combining a given number of clusterings, while the issue of which clustering techniques to use, as an ensemble, is not emphasized. Nonetheless, the study of different ensembles and which combination of clustering techniques would produce the best results with the proposed combiner, will be addressed in future work.

Acknowledgements. This work was partially funded by an NSERC strategic grant.

References

1. J. Ghosh. Multiclassifier systems: Back to the future. In J. Kittler F. Roli, editor, *Multiple Classifier Systems: Third International Workshop, MCS 2002, Proceedings. LNCS 2364*, pages 1–15, Cagliari, Italy, June 2002. Springer.
2. K. Kittler, M. Hatef, R.P.W Duin, and J. Matas. On combining classifiers. *IEEE Transactions on Pattern Analysis and Machine Intelligence*, 20(3):226–239, March 1998.
3. A. Strehl and J. Ghosh. Cluster ensembles – a knowledge reuse framework for combining partitionings. In *Conference on Artificial Intelligence (AAAI 2002)*, pages 93–98, Edmonton, July 2002. AAAI/MIT Press.
4. A. Fred and A.K. Jain. Data clustering using evidence accumulation. In *Proceedings of the 16th International Conference on Pattern Recognition. ICPR 2002*, volume 4, pages 276–280, Quebec City, Quebec, Canada, August 2002.
5. Y. Qian and C. Suen. Clustering combination method. In *International Conference on Pattern Recognition. ICPR 2000*, volume 2, pages 732–735, Barcelona, Spain, September 2000.

6. H. Ayad and M. Kamel. Finding natural clusters using multi-clusterer combiner based on shared nearest neighbors. In *Multiple Classifier Systems: Fourth International Workshop, MCS 2003, Guildford, Surrey, United Kingdom, June 11–13. Proceedings. To Appear*, 2003.
7. R.A. Jarvis and E.A. Patrick. Clustering using a similarity measure based on shared nearest neighbors. *IEEE Transactions on Computers*, C-22(11):1025–1034, November 1973.
8. L. Ertoz, M. Steinbach, and V. Kumar. Finding clusters of different sizes, shapes, and densities in noisy, high dimensional data. Technical report, Department of Computer Science, University of Minnesota, 2002.
9. L. Ertoz, M. Steinbach, and V. Kumar. A new shared nearest neighbor clustering algorithm and its applications. In *Workshop on Clustering High Dimensional Data and its Applications at 2nd SIAM International Conference on Data Mining*, Arlington, VA, 2002.
10. George Karypis and Vipin Kumar. A fast and high quality multilevel scheme for partitioning irregular graphs. Technical Report TR 95–035, Department of Computer Science and Engineering, University of Minnesota, 1995.
11. George Karypis and Vipin Kumar. Multilevel k-way partitioning scheme for irregular graphs. Technical Report TR 95–064, Department of Computer Science and Engineering, University of Minnesota, 1995.
12. George Karypis and Vipin Kumar. Multilevel algorithms for multi-constraint graph partitioning. In *Conference on High Performance Networking and Computing. Proceedings of the 1998 ACM/IEEE conference on Supercomputing*, San Jose, CA, 1998.
13. A. Strehl and J. Ghosh. Relationship-based clustering and visualization for high dimensional data mining. *INFORMS Journal on Computing*, pages 1–23, 2002.
14. M. Steinbach, G. Karypis, and V. Kumar. A comparison of document clustering techniques. In *KDD Workshop on Text Mining*, 2000.
15. Paul Horton and Kenta Nakai. A probablistic classification system for predicting the cellular localization sites of proteins. *Intelligent Systems in Molecular Biology*, pages 109–115, 1996.
16. F. Alimoglu. Combining multiple classifiers for pen-based handwritten digit recognition. Master's thesis, Institute of Graduate Studies in Science and Engineering, Bogazici University, 1996.

Clustering Mobile Trajectories for Resource Allocation in Mobile Environments*

Dimitrios Katsaros[1], Alexandros Nanopoulos[1], Murat Karakaya[2], Gokhan Yavas[2], Özgür Ulusoy[2], and Yannis Manolopoulos[1]

[1] Department of Informatics,
Aristotle University, Thessaloniki, 54124, Greece
{dimitris, alex, manolopo}@skyblue.csd.auth.gr

[2] Department of Computer Engineering
Bilkent University, Bilkent 06800, Ankara, Turkey
{muratk, gyavas, oulusoy}@cs.bilkent.edu.tr

Abstract. The recent developments in computer and communication technologies gave rise to Personal Communication Systems. Due to the nature of the PCS, the bandwidth allocation problem arises, which is based on the notion of *bandwidth-on-demand.* We deal with the problem of how to predict the position of a mobile client. We propose a new algorithm, called *DCP*, to discover user mobility patterns from collections of recorded mobile trajectories and use them for the prediction of movements and dynamic allocation of resources. The performance of the proposed algorithm is examined against two baseline algorithms. The simulation results illustrate that the proposed algorithm achieves recall that is comparable to that of the baseline algorithms and substantial improvement in precision. This improvement guarantees very good predictions for resource allocation with the advantage of very low resource consumption.

1 Introduction

The recent developments in computer and communication technologies gave rise to Personal Communication Systems (PCS), which ensure ubiquitous availability of services. Unlike ordinary static networks (e.g., public telephone network), PCS allows the dynamic relocation of mobile terminals. This *network mobility* gives rise to some new and important problems. Among them are the *location management* and the *bandwidth allocation* problems. Location management consists of two issues, namely the *location* and *paging* procedures. The former allows the system to keep the user's location knowledge (exact or approximate) in order to be able to locate him. The paging procedure consists of sending paging messages in all the locations (cells) where the mobile client could be located. The bandwidth allocation issues arise due to the nature of the PCS, which is

* This research has been funded through the bilateral program of scientific cooperation between Greece and Turkey. (*Γ.Γ.Ε.Τ.* and from *TÜBITAK* grant no 102E021.)

M.R. Berthold et al. (Eds.): IDA 2003, LNCS 2810, pp. 319–329, 2003.

based on the notion of *bandwidth-on-demand*, that is, allocate network resources only after there exists a request for them.

The issue of location management has mainly focused on the procedure of database updating and relatively little (e.g., [6,7,1]) has been done in the area of location prediction. Location prediction is a dynamic strategy in which the PCS tries to proactively estimate the position of the mobile client. Location prediction could provide substantial aid in the decision of resource allocation in cells. Instead of blindly allocating excessive resources in the cell-neighborhood of a mobile terminal, we could selectively allocate resources to the most "probable-to-move" cells. Thus, we could cut down the resource wastage without introducing an increase in the *call drop ratio*.

Related work. Location prediction is addressed in the following works: [6,7, 5,11,1,4,13]. Nevertheless, they exhibit one or some of the following shortcomings: a) They assume the existence of user movement patterns (by only applying pattern matching techniques), without providing a method to discover them in the first place [7]. b) They rely on knowledge of the probability distribution of the mobile's velocity and/or direction of movement [5,1], which assumes the existence of a sophisticated and costly equipment, e.g., GPS. c) They are susceptible in small deviations in client trajectories, which cause significant degradation in prediction accuracy because they do not distinguish between regular and random movements of a mobile [6]. d) They do not exploit the valuable historical information that in any case is gathered (logged) from the PCS and regards the daily movements of clients from cell to cell (intercell movements), spanning large time intervals. Finally, the works described in [4,13,2] address different problems and their proposed solutions are not applicable in our problem. The work in [4,2] deal with cache relocation and data relocation, respectively, and the work in [13] aims at designing a better paging area for each mobile client for each time region and thus, the focus of the work is the estimation of the time-varying location probability. The only relevant work to ours is that in [11], which proposes recording the "cell-to-cell" transition probabilities in order to make predictive resource reservation. The deficiencies of this method are shown by the experimental results of our work.

In this work, we deal with the problem of location prediction for efficient resource allocation. Our contributions are:

- We identify the importance of exploiting historical movement data in the context of predictive resource allocation.
- We develop a new clustering algorithm for discovering regularities in movements of mobile clients to address the large volumes of historical data.
- The design of the proposed algorithm considers the important factor of randomness in the mobile trajectories, which can result in noise within trajectories themselves or to outliers (trajectories that do not follow any regularities). This is addressed by the used distance measure and by the method of representing the clusters.
- The performance of the proposed algorithm is examined against two baseline algorithms, based on simulation. The first of them (EXH) allocates resources

to every neighboring cell and the second one (TM) performs allocation based on cell-to-cell transition probabilities. The simulation results illustrate that the proposed algorithm achieves recall that is comparable to that of the baseline algorithms and substantial improvement in precision (40% over EXH and 25% over TM). This improvement guarantees very good predictions for resource allocation with the advantage of very low resource consumption.

The rest of this article is organized as follows: Section 2 describes the network model assumed in the present study. In Section 3 we give the problem and present its solution in Section 4. Finally, in Section 5 we present the experimental results and in Section 6 we conclude this paper.

2 Preliminaries

In the present work, we consider a wireless PCS (see Figure 1) with architecture similar to those used in *EIA/TIA IS-41* and *GSM* standards [9]. The PCS serves a geographical area, called the *coverage area*, where *mobile users* or *terminals (MU)* or *(MT)* can move. The coverage area served by the PCS is partitioned into a number of non-overlapping regions, called *cells*. At the heart of the PCS lies a fixed backbone (wireline) network. A number of fixed hosts are connected to this network. Each cell is served by one *base station* (BS), which is connected to the fixed network and it is equipped with wireless transmission and receiving capability.

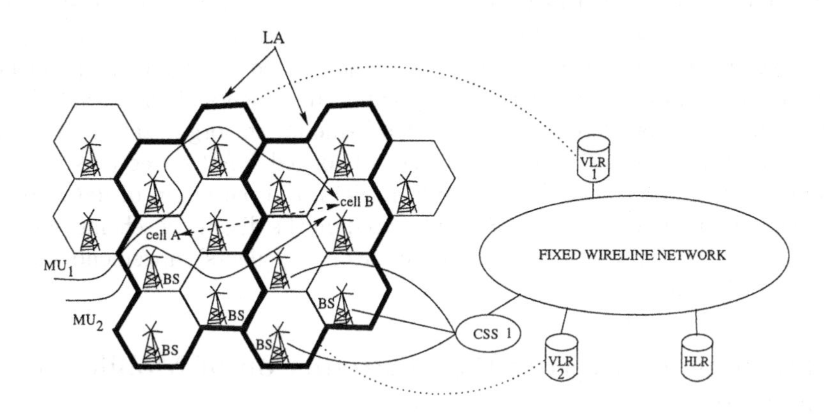

Fig. 1. A typical architecture for a PCS.

Following the settings in [13,4] where each mobile client can report its trajectories to the fixed network, we collect for each user trajectories of the form $T = \langle (id_1, t_1), \ldots, (id_k, t_k) \rangle$. The id_i represents the ID of the cell that the client entered at time t_i ($1 \leq i \leq k$). The characteristic of this sequence is that for every i ($1 \leq i \leq k-1$) the id_i and id_{i+1} correspond to neighboring cells. Following related work on session formation, each such trajectory T is divided into subsequences based on a threshold for the difference $t_{i+1} - t_i$ (i.e., when a client stays in a cell for a sufficiently long time).

The procedure of trajectory recording ends up with a host of the fixed network collecting (logging) the movements T of individual mobile clients within the coverage area. Let these trajectories be called *User Actual Paths (UAP).* UAPs are an invaluable source of information, because the "movement of people consists of random movement and regular movement and the majority of mobile users has some regular daily (hourly, weekly, ...) movement patterns and follow these patterns more or less every day" [7]. In this work, we focus on discovering patterns from multiple users, because it better fits the needs of resource allocation problem.

3 Problem Formulation

Let $\mathcal{D}$ be a collection of logged user actual paths (UAPs), each one of the form $U = \langle c_1, \ldots, c_n \rangle$, where c_i is the ID of the i-th cell in the two dimensional grid of the coverage area U. The discovery of patterns in clients' movement can be attained by organizing the logged UAPs according to their *in-between similarity.* The similarity considers the network topology [7]. Such a procedure identifies groups of similar UAPs. These groups represent a kind of aggregation of large numbers of UAPs into a set of much less patterns, which correspond to the expected regularities which will be called *user mobility patterns* (UMPs).

Having a set of discovered UMPs, let a client that up to the current time point has visited cells described in UAP $C = \langle c_1, \ldots, c_t \rangle$. According to the similarity of UAP C with the discovered UMPs, the movement of the client can correspond to one (or possibly more) discovered regularity. Based on this correspondence, the mobile system can predict the forthcoming movements of the client.

In summary, what is required first is a scheme for the organization of similar UAPs into UMPs. Due to the large sizes of logged UAP collections and the possibility of the existence of noise and outliers, the paradigm of data mining will be followed for this purpose. Second, based on the discovery scheme, we need a method for the dynamic prediction of forthcoming clients' movements and the allocation of resources.

4 Dynamic Clustering-Based Prediction of Mobile-Client Movements

Let $A = \langle a_1, \ldots, a_m \rangle$ and $B = \langle b_1, \ldots, b_n \rangle$ be two cell sequences. To measure their similarity we are based on the well known metric of *edit distance*. However, to reflect the geometry of the cell grid, we use weighted edit distance measures [7], considering that it suffices to limit the effect of temporal deviation from a specific cell only within the cells *around* it. We extend [7] so that, for a cell c, we examine all the cells with a city-block distance from c, denoted as CB[1], less than or equal

[1] Given a cell c_1 with coordinates (x_1, y_1) and c_2 with (x_2, y_2), $CB(c_1, c_2) = \max\{|x_1 - x_2|, |y_1 - y_2|\}$.

to t_d. Henceforth, for two UAPs A and B, the weighted edit distance is denoted as $d_e(A, B)$. The selection of t_d is value is discussed in Section 5.

The UAPs and the aforementioned weighted edit distance form a metric space which, however, may not be Euclidean. Although several clustering algorithms have been proposed for Euclidean spaces (e.g., [3] and the references therein), we have to focus on algorithms that can be used for general metric spaces. Hierarchical agglomerative algorithms belong to the latter category, because they are not based on coordinates of an Euclidean space. Plain hierarchical clustering uses one medoid object for each cluster. Other approaches extended this case and represent a cluster with more objects, e.g., [8] which represents clusters by a subset of distant objects within them (this approach was also adopted in [3, 10]).

For a cluster c, let c.rep denote the set of representative UAPs. Given two clusters c and c', their distance $d(c, c')$ is defined as

$$d(c, c') = \min_{\forall u \in c.rep, v \in c'.rep} \{d_e(u, v)\}$$

A general scheme of hierarchical algorithm to cluster UAPs is given Figure 2. For an efficient implementation of HIERCLUSTER, we use a heap to maintain the distances between the clusters. Following the approach of [3,10], we sample the initial collection of UAPs, to be able to efficiently process, with the clustering algorithm, the selected sample of UAPs in main memory. The merging of representative points can be done in several ways. Based on [3], when merging two clusters c and c' into a new cluster n, we initially set n.rep = c.rep $\cup$ c'.rep. If $|n|$.rep $\geq r$, then the subset of the r most distant members of n.rep can be found.[2] The aforementioned approach is denoted as FU (selects the Farthest UAPs). Although FU captures the shape of a cluster, it may be sensitive to outliers (i.e., UAPs that do not to belong to clusters and tend to form isolated small neighborhoods), because they can become representatives (since projection cannot be applied) and impact the correct cluster detection.

Following the approach of [10] when merging two clusters, if $|n|$.rep $\geq r$, then the medoid representative can be found and then the procedure selects the $r-1$ nearest neighbors of it. Thus, the medoid and its nearest neighbors form the n.rep set. We denote this method as NN. Its objective is to reduce the sensitivity of the clustering algorithm against the outliers, since it reduces the probability of outliers to become representatives.

The result of HIERCLUSTER is the set of UMPs, each one consisting of a group of representatives. Given a mobile client that currently follows a UAP, the purpose of prediction is to first find the UMPs that best match the client's current UAP. Then, based on the matched UMPs, the most probable next movements are predicted, and resource allocation is applied for them. Let A be a client's UAP (its recently crossed cells), where $A = \langle a_1, \ldots, a_k \rangle$. A is examined

[2] The shrinking and projections (towards the center) of the selected subset of representatives can be applied for Euclidean spaces (they use the coordinates), thus not for the problem examined herein.

```
Algorithm HIERCLUSTER
Input  The collection D of UAPs
       The number k of required UMPs
       The max number r of repr.
Output The discovered UMPs
1. Preprocess(D)
2. while c > k {
3.     extract heap root with two clusters c, c'
       for which d(c, c') ≤ min_{x,y≠c,c'} {d(x, y)}
4.     merge c and c' into a new cluster n
5.     merge c.rep and c'.rep to form n.rep
6.     update the heap
7.     c = c − 1
8. }
9. return the set of discovered clusters
end of HIERCLUSTER
```

```
Procedure PREPROCESS
1. foreach u ∈ D {
2.     form a c_u cluster
3.     set c_u.rep = {u}
4.     build the heap
5. }
6. set c = |D|
end of PREPROCESS
```

Fig. 2. The clustering algorithm.

against each UMPs B to find those with distance $d_m(A, B) < t_m$. In such a case we denote that B is matched against A. The calculation of d_m must take into account that a UAP A can be part (i.e., subsequence) of a UMP B. The solution of this problem, which is based on dynamic programming, can be found in [12]. Therefore, assuming that $B = \langle b_1, \ldots, b_j, \ldots b_n \rangle$, the remaining cells in the part $\langle b_j, \ldots, b_n \rangle$ of B can be used for the prediction of the next movements of the client.

The number of predictions that can be made from each matched UMP is set as a parameter. Assume that the remaining part of a matched UMP consists of p cells $\langle b_j, \ldots, b_n \rangle$ (i.e., $n - j = p$). We can use up to a maximum number m ($m \leq p$) cells from this part, by maintaining the order of their appearance, i.e., cells b_j, b_{j+1}, b_{j+m-1} can be predicted as the next movements of the client. The aforementioned cells are denoted as the *predicted trajectory* of the client. Evidently, one may expect that cell b_j has larger probability of correct prediction than b_{j+1}, b_{j+1} than b_{j+2} and so on. For the tuning of m it has to be considered that the smaller m is the more accurate predictions are made. However, the total number of correct predictions is going to be smaller. The experimental results in Section 5 elaborate further on this issue.

Finally, it has to be noticed that the current UAP of a mobile client can be matched by one or more UMPs (i.e., representatives) in each cluster. Thus, one may select the best matched UMP (with the minimum d_m) or all the matched UMPs. We have examined both options and we choose the latter one, because we found that it leads to better performance, since the number of correct predictions increases without significant increase in the total number of predictions made (that corresponds to resource consumption).

Symbol	Definition	Default
m	max length of predicted trajectory	2
r	max number of representatives	20
t_d	cut-off CB distance for d_e	2
L	average length of UAPs	5
c	corruption factor	0.4
o	outlier percentage	30%

Fig. 3. Symbol table.

5 Performance Study

This section presents the empirical evaluation of the proposed method (DCP), which is compared against TM-k and EXH. TM-k makes predictions based on the cell-to-cell transition matrix, where k denotes the number of the most probable-to-visit cells. As described, EXH is the method that allocates resources to every neighboring cell. We first describe the simulation model for the mobile system considered. Next, we give the results for the sensitivity of DCP, by comparing it with TM-k and EXH. Finally, we examine the efficiency of the prediction procedure of DCP with respect to execution time.

To simulate the mobile computing system, we assumed a square-shaped mesh network consisting of 225 cells. To generate UAPs, we first formed a number of UMPs whose length is uniformly distributed with mean equal to L and max equal to $L + 3$. Each UMP is taken as a random walk over the mesh network. We had two types of UAPs, those following a pattern (i.e., a UMP) and those not. The latter are called *outliers*, and their ratio to the number of UAPs that follow a UMP is denoted as o. The formers are formed as follows: First, a UMP is selected at random. Then, with a corruption ratio c, we intervene random paths through the nodes of the selected UMP. The total number of the nodes in these random paths are equal to c times the size of the UMP. The default number of generated UAPs is 100,000. From this set of UAPs we form the train and an evaluation data set. The sampling for the clustering algorithm is done according to the results in [3,10]. The performance measures we used are the following: a) *recall*, which represents the number of correctly predicted cells divided by the total number of requests and b) *precision*, which represents the number of correctly predicted cells divided by the total number of predictions made. Figure 3 summarizes the symbols used henceforth, and, for each one, gives the default value used throughout the performance evaluation.

We saw that the DCP has some parameters, namely the cluster-merging method, the number m of maximum predictions made at each time, and the number r of representatives used for clusters, which require tuning. Similar tuning for the k is required also for the TM-k algorithm. We performed extensive experiments in order to select appropriate values for these parameters. Due to the limited space we do not present here the experimental results for the tuning of the algorithms.

5.1 Sensitivity Experiments

In this series of experiments, we examine the sensitivity of DCP with respect to the pattern length L, the percentage c of corruption, and the ratio o of outliers. DCP is compared with TM-2 and EXH.

Initially we examined the impact of L. In the left part of Figure 4 it can be clearly observed that, for very small L values (i.e., 2), the recall of all methods is reduced significantly. Evidently, the first cell-visit in a UAP may not be predicted by any method. Therefore, when $L = 2$, there is not enough room for correct predictions (even for EXH). With increasing L, recall is increased. EXH clearly attains the best recall in all cases.

However, it has to be considered that EXH is, in fact, an upper-bound for the recall of *any* prediction scheme, because it is not dynamic and bases its prediction on all possible forthcoming cells visits (EXH may miss only the first visit in each UAP). Based on this fact, the dynamic methods, i.e., DCP and TM-2, achieve only around 10% smaller recall than EXH. The recall of DCP and TM-2 are comparable with each other. However, the good performance of DCP is not to the cost of resource consumption.

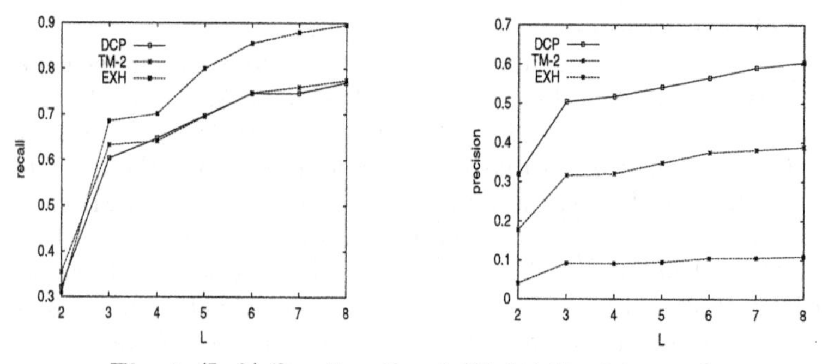

Fig. 4. (Left) Recall vs L and (Right) Precision vs L.

This is shown in the right part of Figure 4, which illustrates precision. Clearly, EXH has the lowest precision (around 0.1), since it allocates every possible cell. This indicates the ineffectiveness of EXH in realistic cases, due to its high resource consumption. TM-2 is second best, but significantly outperformed by DCP in all cases. Since L is the only factor that affects EXH (for the reasons explained earlier), we do not further examine this method, because it achieves constant recall (80%) and precision (1%) with respect to all the following factors. This helps for a more clear comparison between the two dynamic methods, i.e., DCP and TM-2.

Next, we examined the impact of corruption in the data set, measured through parameter c. The results are illustrated in Figure 5 (c is given as a percentage). Focusing on recall (left part of Figure 5), both methods are not affected significantly by increasing c, and their recall is comparable. The precision of DCP (right part of Figure 5) for very low c values is very high, reaching 90% for the case of no corruption. However, with increasing c, the precision of DCP reduces, since the clustering, performed by it, is affected. However, for c larger than 40% the reduction is less significant.

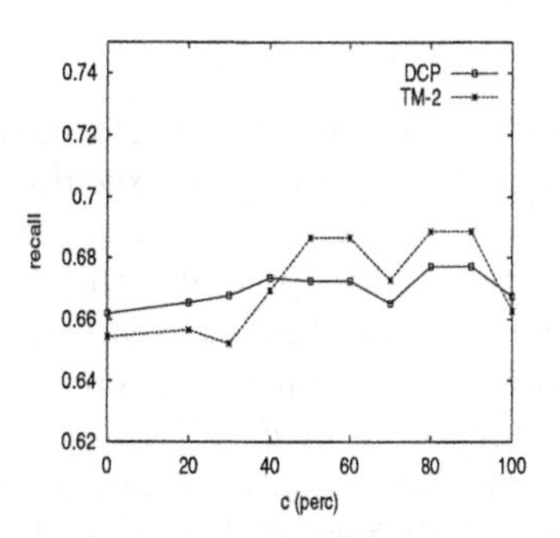

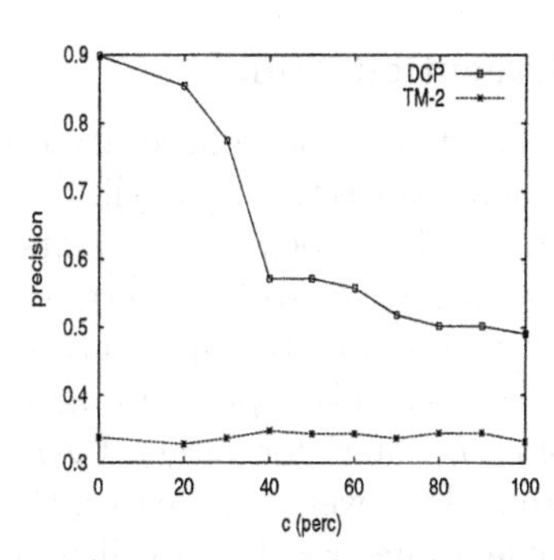

Fig. 5. Comparison of DCP and TM-2 w.r.t. c: (Left) Recall. (Right) Precision.

This is explained from the fact that before this value, DCP performs almost unhindered matchings of UAPs against patterns, reaching the aforementioned very high values of precision. However, these cases does not correspond to realistic ones, because the absence of corruption is not probable[3]. But for larger values of c, DCP manages to maintain a good value of precision (around 50%) without being affected by further increase in c. Evidently, DCP outperforms TM-2, which is less affected by c but does not attain precision larger than 35%.

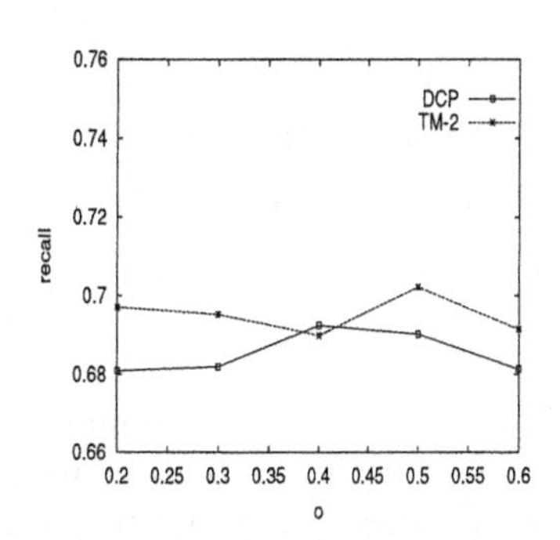

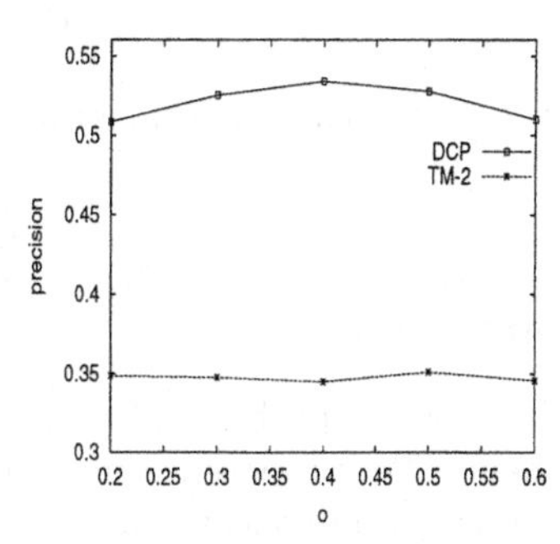

Fig. 6. Comparison of DCP and TM-2 w.r.t. o: (Left) Recall. (Right) Precision.

Finally, we examined the impact of outliers in the data set. The results are illustrated in Figure 6 with respect to the outlier percentage o. The recall (left part of Figure 6) of both methods is not affected significantly by increasing o,

[3] This is the reason why throughout the rest experiments we do not use c smaller than 40%.

where DCP and TM-k attain comparable recall values. The precision of both methods (right part of Figure 6.b) is also not affected significantly. For DCP this indicates that the filtering used by its clustering procedure is effective in reducing the impact of outliers (TM-k does not use clustering, thus is not directly affected by outliers). However, DCP compares favorably with TM-2 in all cases.

5.2 Execution Time

For DCP, the on-line computation of predicted trajectories is performed each time a client moves to a new cell. For this reason, we examined the execution time required for this procedure. Results are depicted in Figure 7 (time is measured in ms) with respect to the average UAP length L. More precisely, the depicted times correspond to the application of prediction procedure at each cell, where the UMPs have been pre-computed. As shown, execution time increases with increasing L, since for larger UAPs the computation of d_m distance requires more steps. However, in all cases, the prediction requires only few ms. Assuming that a client that reaches a cell may stay in it for a period that is much larger than few ms, the prediction time is a negligible overhead.

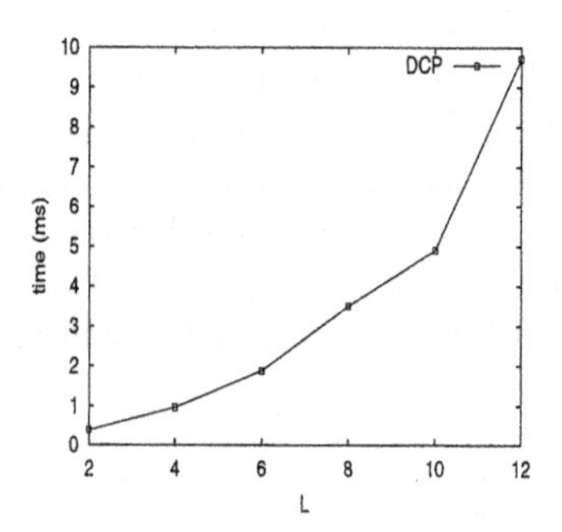

Fig. 7. Execution time (in ms) per cell of the prediction procedure of DCP.

6 Conclusions

We considered the problem of predictive resource allocation in cellular mobile environments. We proposed a new algorithm called *DCP*, to discover user mobility patterns from collections of recorded mobile trajectories and to use them for the prediction of movements and dynamic allocation of resources.

The design of the proposed algorithm considers the important factor of randomness in the mobile trajectories, which can result in noise within trajectories themselves or to outliers (trajectories that do not follow any regularities).

Using simulation, the performance of *DCP* was compared against that of *EXH* and *TM*. Our experiments showed that for a variety of trajectory length, noise and outliers, *DCP* achieves an very good tradeoff between prediction recall and precision.

References

1. A. Aljadhai and T. Znati. Predictive mobility support for QoS provisioning in mobile wireless environments. *IEEE Journal on Selected Areas in Communications*, 19(10):1915–1930, 2001.
2. W.-C. Chen and M.S. Chen. Developing data allocation schemes by incremental mining of user moving patterns in a mobile computing system. *IEEE Transactions on Knowledge and Data Engineering*, 15(1):70–85, 2003.
3. S. Guha, R. Rastogi, and K. Shim. CURE: An efficient clustering algorithm for large databases. In *Proceedings of the ACM Conference on Management of Data (ACM SIGMOD'98)*, pages 73–84, 1998.
4. S. Hadjiefthymiades and L. Merakos. Using proxy cache relocation to accelerate Web browsing in wireless/mobile communications. In *Proceedings of the World Wide Web Conference (WWW'01)*, pages 26–35, 2001.
5. B. Liang and Z. Haas. Predictive distance-based mobility management for PCS networks. In *Proceedings of the IEEE Conference on Computer and Communications (IEEE INFOCOM'99)*, pages 1377–1384, 1999.
6. G.Y. Liu and M.Q. Gerald. A predictive mobility management algorithm for wireless mobile computing and communications. In *Proceedings of the IEEE International Conference on Universal Personal Communications*, pages 268–272, 1995.
7. T. Liu, P. Bahl, and I. Chlamtac. Mobility modeling, location tracking, and trajectory prediction in wireless ATM networks. *IEEE Journal on Selected Areas in Communications*, 16(6):922–936, 1998.
8. R. Michalski, R. Stepp, and E. Diday. A recent advance in data analysis: Clustering objects into classes characterized by conjuctive concepts. In *Progress in Pattern Recognition*, volume 1. North Holland Publishing, 1983.
9. S. Mohan and R. Jain. Two user location strategies for Personal Communication Systems. *IEEE Personal Communications Magazine*, pages 42–50, First Quarter 1994.
10. A. Nanopoulos, Y. Theodoridis, and Y. Manolopoulos. C^2P: Clustering with closest pairs. In *Proceedings of the 27th Conference on Very Large Data Bases (VLDB'01)*, pages 331–340, 2001.
11. S. Rajagopal, R.B. Srinivasan, R.B. Narayan, and X.B.C. Petit. GPS-based predictive resource allocation in cellural networks. In *Proceedings of the IEEE International Conference on Networks (IEEE ICON'02)*, pages 229–234, 2002.
12. P.H. Sellers. An algorithm for the distance between two finite sequences. *Journal of Algorithms*, pages 359–373, 1980.
13. H.-K. Wu, M.-H. Jin, J.-T. Horng, and C.-Y. Ke. Personal paging area design based on mobile's moving behaviors. In *Proceedings of the IEEE Conference on Computer and Communications (IEEE INFOCOM'01)*, pages 21–30, 2001.

Fuzzy Clustering of Short Time-Series and Unevenly Distributed Sampling Points

Carla S. Möller-Levet[1], Frank Klawonn[2], Kwang-Hyun Cho[3], and Olaf Wolkenhauer[4]

[1] Control Systems Centre, Department of Electrical Engineering and Electronics, UMIST, Manchester, U.K.
c.moller-levet@postgrad.umist.ac.uk

[2] Department of Computer Science, University of Applied Sciences, D-38302 Wolfenbüttel, Germany.
f.klawonn@fh-wolfenbuettel.de

[3] School of Electrical Engineering, University of Ulsan, Ulsan, 680-749, Korea.
ckh@mail.ulsan.ac.kr

[4] Department of Computer Science, University of Rostock, Albert Einstein Str. 21, 18051 Rostock, Germany.
wolkenhauer@informatik.uni-rostock.de

Abstract. This paper proposes a new algorithm in the fuzzy-c-means family, which is designed to cluster time-series and is particularly suited for short time-series and those with unevenly spaced sampling points. Short time-series, which do not allow a conventional statistical model, and unevenly sampled time-series appear in many practical situations. The algorithm developed here is motivated by common experiments in molecular biology. Conventional clustering algorithms based on the Euclidean distance or the Pearson correlation coefficient are not able to include the temporal information in the distance metric. The temporal order of the data and the varying length of sampling intervals are important and should be considered in clustering time-series. The proposed short time-series (STS) distance is able to measure similarity of shapes which are formed by the relative change of amplitude and the corresponding temporal information. We develop a fuzzy time-series (FSTS) clustering algorithm by incorporating the STS distance into the standard fuzzy clustering scheme. An example is provided to demonstrate the performance of the proposed algorithm.

1 Introduction

Microarrays revolutionize the traditional way of one gene per experiment in molecular biology [1], [2]. With microarray experiments it is possible to measure simultaneously the activity levels for thousands of genes. The appropriate clustering of gene expression data can lead to the classification of diseases, identification of functionally related genes, and network descriptions of gene regulation, among others [3], [4].

M.R. Berthold et al. (Eds.): IDA 2003, LNCS 2810, pp. 330–340, 2003.

Time course measurements are becoming a common type of experiment in the use of microrarrays, [5], [6], [7], [8], [9]. If a process is subject to variations over time, the conventional measures used for describing similarity (e.g. Euclidean distance) will not provide useful information about the similarity of time-series in terms of the cognitive perception of a human [10]. An appropriate clustering algorithm for short time-series should be able to identify similar shapes, which are formed by the relative change of expression as well as the temporal information, regardless of absolute values. The conventional clustering algorithms based on the Euclidean distance or the Pearson correlation coefficient, such as hard k-means (KM) or hierarchical clustering (HC) are not able to include temporal information in the distance measurement. Fig. 1 shows three time-series with different shapes to illustrate this point. An appropriate distance for the three expression profiles would identify g_2 as more similar to g_3 than to g_1, since the deviation of shape across time of g_3 from the shape of g_2 is less than that of g_1. That is, the deviation of expression level of g_1 from g_2 in the transition of the first to the second time point is one unit per one unit of time, while the deviation of expression level of g_3 from g_2 in the transition of the third to the fourth time point is one unit per seven units of time. The Euclidean distance and the Pearson correlation coefficient do not take into account the temporal order and the length of sampling intervals; for these metrics both g_1 and g_3 are equally similar to g_2. In this paper we introduce a new clustering algorithm which is able to use the temporal information of uneven sampling intervals in time-series data to evaluate the similarity of the shape in the time domain.

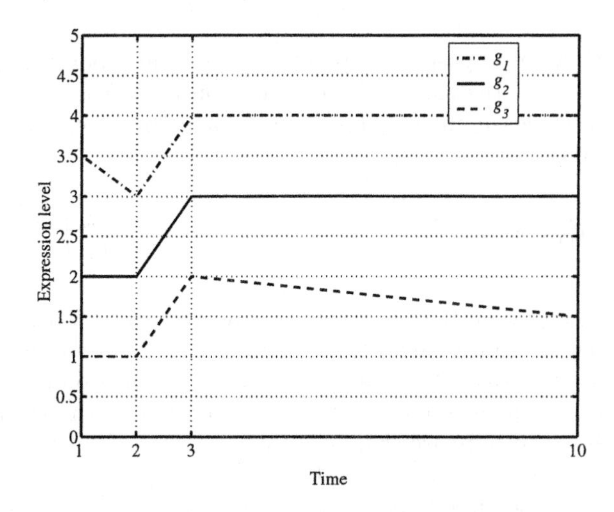

Fig. 1. Three unevenly sampled time-series with different shapes

This paper is organized as follows: Section 2 defines the objective and basic concepts of the short time-series (STS) distance based on the requirements of short time-series clustering. In Section 3, the fuzzy short time-series (FSTS) algorithm is introduced as a modification of the standard fuzzy c-means algorithm (FCM). Section 4 presents an artificial data set to illustrate and compare the performance of the proposed algorithm with FCM, KM and single linkage HC. Finally, conclusions are made in Section 5 summarizing the presented research.

2 Short Time-Series Distance

This section presents a measure of similarity for microarray time-series data. The performance of the distance is illustrated by means of simple tests for which temporal information is a key aspect.

The objective is to define a distance which is able to capture differences in the shapes, defined by the relative change of expression and the corresponding temporal information, regardless of the difference in absolute values. We approach the problem by considering the time-series as piecewise linear functions and measuring the difference of slopes between them. Considering a gene expression profile $x = [x_0, x_1, \ldots, x_{n_t}]$, where n_t is the number of time points, the linear function $x(t)$ between two successive time points t_k and $t_{(k+1)}$ can be defined as $x(t) = m_k t + b_k$, where $t_k \leq t \leq t_{(k+1)}$, and

$$m_k = \frac{x_{(k+1)} - x_k}{t_{(k+1)} - t_k} \tag{1}$$

$$b_k = \frac{t_{(k+1)} x_k - t_k x_{(k+1)}}{t_{(k+1)} - t_k} \ . \tag{2}$$

The STS distance we propose corresponds to the square root of the sum of the squared differences of the slopes obtained by considering time-series as linear functions between measurements. The STS distance between two time-series x and v is defined as:

$$d^2_{\mathrm{STS}}(x, v) = \sum_{k=0}^{n_t - 1} \left(\frac{v_{(k+1)} - v_k}{t_{(k+1)} - t_k} - \frac{x_{(k+1)} - x_k}{t_{(k+1)} - t_k} \right)^2 \ . \tag{3}$$

To evaluate the performance of this distance in comparison with the Euclidean distance and the Pearson correlation coefficient, two tests are performed. The objective of the first test is to evaluate the ability to incorporate temporal information into the comparison of shapes. The objective of the second test is to evaluate the ability to compare shapes regardless of the absolute values.

For the first test, let us consider the time-series shown in Fig. 1. Table 1 illustrates the corresponding STS distance, Euclidean distance, and the Pearson correlation coefficient between g_2 and g_1, and g_2 and g_3, respectively. The results show that the STS distance is the unique distance metric which reflects the temporal information in the comparison of shapes.

For the second test, let us consider a linear transformation of the absolute values of the time-series shown in Fig. 1. These modified series are shown in

Table 1. STS distance, Euclidean distance, and Pearson correlation coefficient between g_2 and g_1, and g_2 and g_3 in Fig. 1

	Euclidean distance	STS distance	Pearson correlation coefficient
(g_2, g_1)	2.29	0.500	0.904
(g_2, g_3)	2.29	0.071	0.904

Fig. 2(a). Since the STS and the Euclidean distance are both sensitive to scaling, a z-score standardization of the series is required for them to neglect absolute values [11]. The z-score of the ith time point of a gene x is defined in (4), where $\overline{x}$ is the mean and s_x the standard deviation of all the time points $x_1, \ldots, x_n$ in vector x

$$z_i = \frac{(x_i - \overline{x})}{s_x} . \tag{4}$$

The time-series after standardization are shown in Fig. 2(b).

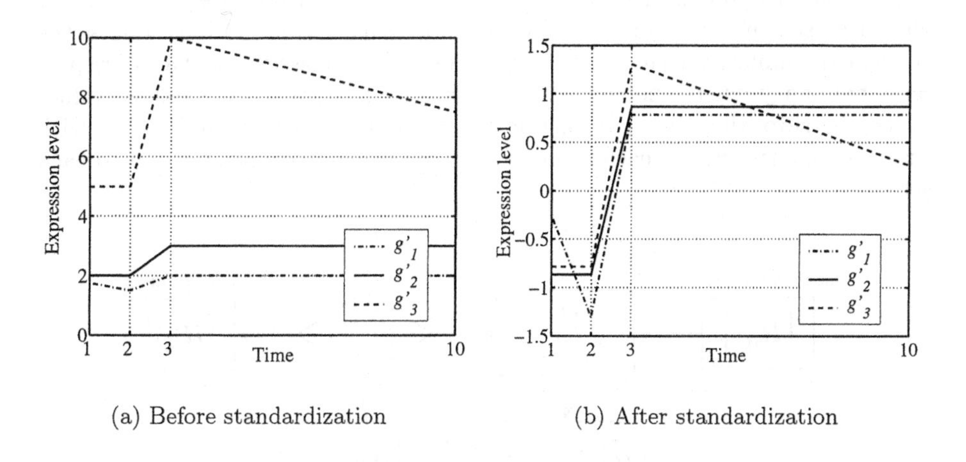

(a) Before standardization (b) After standardization

Fig. 2. Three unevenly sampled time-series data with different shapes, which correspond to linear transformations of the time-series in Fig. 1

Table 2 summarizes the STS distance, the Euclidean distance, and the Pearson correlation coefficient between g'_2 and g'_1, and g'_2 and g'_3. The results show that the STS distance is the unique distance measure which can properly capture temporal information, regardless of the absolute values.

Table 2. STS distance, the Euclidean distance, and the Pearson correlation coefficient between g_2' and g_1', and g_2' and g_3' in Fig. 2(b)

	Euclidean distance	STS distance	Pearson correlation coefficient
(g_2, g_1)	0.756	1.103	0.904
(g_2, g_3)	0.756	0.386	0.904

3 Fuzzy Short Time-Series Clustering Algorithm

This section introduces the FSTS clustering algorithm as a new member of the fuzzy c-means (FCM) family [12], [13], [14]. We present the minimization of the standard objective function and the resulting cluster prototypes.

There are a wide variety of clustering algorithms available from diverse disciplines such as pattern recognition, text mining, speech recognition and social sciences amongst others [11], [15]. The algorithms are distinguished by the way in which they measure distances between objects and the way they group the objects based upon the measured distances. In the previous section we have already established the way in which we desire the "distance" between objects to be measured; hence, in this section, we focus on the way of grouping the objects based upon the measured distance. For this purpose we select a fuzzy clustering scheme, since fuzzy sets have a more realistic approach to address the concept of similarity than classical sets [16], [14]. A classical set has a crisp or hard boundary where the constituting elements have only two possible values of membership, they either belong or not. In contrast, a fuzzy set is a set with fuzzy boundaries where each element is given a degree of membership to each set.

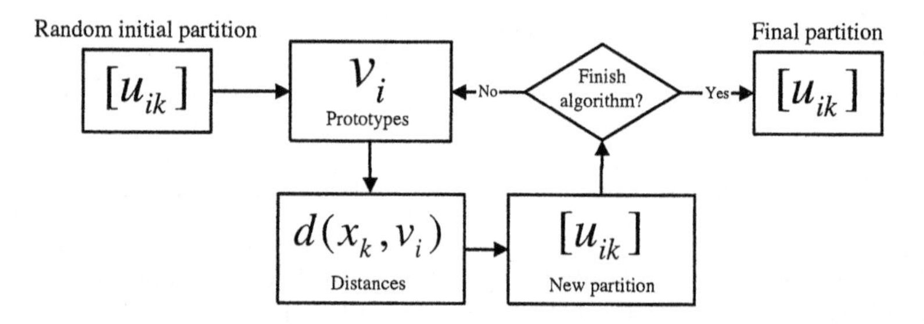

Fig. 3. Diagram of the iteration procedure for the FCM clustering algorithms. Considering the partition of a set $X = [x_1, x_2, \dots, x_{n_g}]$, into $2 \leq n_c < n_g$ clusters, the fuzzy clustering partition is represented by a matrix $U = [u_{ik}]$, whose elements are the values of the membership degree of the object x_k to the cluster i, $u_i(x_k) = u_{ik}$

Fuzzy clustering is a partitioning-optimization technique which allows objects to belong to several clusters simultaneously with different degrees of membership to each cluster [12], [13]. The objective function that measures the desirability of partitions is described in (5), where n_c is the number of clusters, n_g is the number of vectors to cluster, u_{ij} is the value of the membership degree of the vector x_j to the cluster i, and $d^2(x_j, v_i)$ is the squared distance between the vector x_j and the prototype v_i and w is a parameter (usually set between 1.25 and 2), which determines the degree of overlap of fuzzy clusters.

$$J(x,v,u) = \sum_{i=1}^{n_c} \sum_{j=1}^{n_g} u_{ij}^w \, d^2(x_j, v_i) \, . \tag{5}$$

Fig. 3 illustrates the iteration steps of the FCM algorithm, the representative of the fuzzy clustering algorithms. In order to use the STS distance following the conventional fuzzy clustering scheme, we need to obtain the value of the prototype v_k that minimizes (5), when (3) is used as the distance. Substituting (3) into (5) we obtain

$$J(x,v,u) = \sum_{i=1}^{n_c} \sum_{j=1}^{n_g} u_{ij}^w \sum_{k=0}^{n_t-1} \left(\frac{v_{i(k+1)} - v_{ik}}{t_{(k+1)} - t_k} - \frac{x_{j(k+1)} - x_{jk}}{t_{(k+1)} - t_k} \right)^2 . \tag{6}$$

The partial derivative of (6) with respect to v_{ik} is:

$$\frac{\partial J(x,v,u)}{\partial v_{ik}} =$$

$$\sum_{j=1}^{n_g} u_{ij}^w \frac{\partial}{\partial v_{ik}} \left(\left(\frac{v_{(k+1)} - v_k}{t_{(k+1)} - t_k} - \frac{x_{(k+1)} - x_k}{t_{(k+1)} - t_k} \right)^2 + \left(\frac{v_k - v_{(k-1)}}{t_k - t_{(k-1)}} - \frac{x_k - x_{(k-1)}}{t_k - t_{(k-1)}} \right)^2 \right) =$$

$$\sum_{j=1}^{n_g} u_{ij}^w \left[\frac{2\left(v_{ik} - v_{i(k+1)} - x_{jk} + x_{j(k+1)}\right)}{\left(t_k - t_{(k+1)}\right)^2} \right] - \left[\frac{2\left(v_{i(k-1)} - v_{ik} - x_{j(k-1)} + x_{jk}\right)}{\left(t_k - t_{(k-1)}\right)^2} \right] =$$

$$\sum_{j=1}^{g} 2u_{ij}^w \frac{\left(a_k v_{i(k-1)} + b_k v_{ik} + c_k v_{i(k+1)} + d_k x_{j(k-1)} + e_k x_{jk} + f_k x_{j(k+1)}\right)}{\left(t_k - t_{(k+1)}\right)^2 \left(t_k - t_{(k-1)}\right)^2} \tag{7}$$

where

$$a_k = -\left(t_{(k+1)} - t_k\right)^2 \qquad b_k = -(a_k + c_k) \qquad c_k = -\left(t_k - t_{(k-1)}\right)^2$$
$$d_k = \left(t_{(k+1)} - t_k\right)^2 \qquad e_k = -(d_k + f_k) \qquad f_k = \left(t_k - t_{(k-1)}\right)^2 .$$

Setting (7) equal to zero and solving for v_{ik} we have

$$a_k v_{i(k-1)} + b_k v_{ik} + c_k v_{i(k+1)} = -\frac{\sum_{j=1}^{n_g} u_{ij}^w \left(d_k x_{j(k-1)} + e_k x_{jk} + f_k x_{j(k+1)}\right)}{\sum_{j=1}^{n_g} u_{ij}^w}$$

$$a_k v_{i(k-1)} + b_k v_{ik} + c_k v_{i(k+1)} = m_{ik} \tag{8}$$

where

$$m_{ik} = -\frac{\sum_{j=1}^{n_g} u_{ij}^w \left(d_k x_{j(k-1)} + e_k x_{jk} + f_k x_{j(k+1)}\right)}{\sum_{j=1}^{n_g} u_{ij}^w} .$$

Equation (8) yields an undetermined system of equations. We know the relations of the prototype values among the time points, but not the absolute value at each time point. That is, we know the shape but not the absolute level. If we add two fixed time points at the beginning of the series with a value of 0, and solve the system for any n_t, the prototypes can be calculated as

$$v(i,n) = \sum_{r=2}^{n-3} m_{ir} \prod_{q=1}^{r-1} c_q \left[\prod_{q=r+1}^{n-1} a_q + \prod_{q=r+1}^{n-1} c_q + \sum_{p=r+3}^{n} \prod_{j=p-1}^{n-1} c_j \prod_{j=r+1}^{p-2} a_j \right] / \prod_{q=2}^{n-1} c_q + m_{i(n-1)} \prod_{q=1}^{n-2} c_q / \prod_{q=2}^{n-1} c_q + m_{i(n-2)} \prod_{q=1}^{n-3} c_q (a_{(n-1)} + c_{(n-1)}) / \prod_{q=2}^{n-1} c_q \quad (9)$$

where $1 \leq i \leq n_c$, $3 \leq n \leq n_t$ (since $v(i,1) = 0$ and $v(i,2) = 0$), $m_{i1} = 0$ and $c_1 = 1$.

The same scheme of the iterative process as for the FCM, described in Fig. 3 is followed, but the distance and the prototypes are calculated using (3) and (9), respectively. The same three user-defined parameters found in the FCM algorithm; the number of clusters n_c, the threshold of membership to form the clusters α, and the weighting exponent w are also found in the proposed FSTS algorithm. Fig. 4 illustrates the pseudocode of the proposed FSTS clustering algorithm.

4 Illustrative Example

This section presents a simple artificial data set to illustrate and compare the performance of the proposed FSTS clustering algorithm in terms of the cognitive perception of a human. Four groups of five vectors are created where each group has the same parameters of linear transformation between time points, as shown in Table 3. That is, for the group i, $1 \leq i \leq 4$, $x_{j(k+1)} = m_{ik} x_{jk} + b_{ik}$ with $0 \leq k < (n_t - 1)$ and $1 \leq j \leq 5$. The values of m and b were obtained randomly for each group.

Table 3. Artificial profile $x = [x_0, x_1, \ldots, x_{n_t}]$. A group of vectors with a similar shape can be obtained by changing the initial value

Time points	Value
x_0	initial value
x_1	$m_1 x_0 + b_1$
x_2	$m_2 x_1 + b_2$
$\vdots$	$\vdots$
x_{n_t}	$m_{(n_t-1)} x_{(n_t-1)} + b_{(n_t-1)}$

The resulting artificial data set, shown in Fig. 5(a), was clustered using FCM, FSTS, KM and HC algorithms, respectively. All the algorithms were able to

STEP 1: Initialization

n_g	:	number of genes
n_t	:	number of time points
X	:	gene expression matrix (GEM) $[n_g \times n_t]$
n_c	:	number of clusters
w	:	fuzzy weighting factor
a	:	threshold for membership
ϵ	:	termination tolerance

STEP 2: Initialization of the partition matrix

Initialize the partition matrix randomly, $U^{(0)}$ $[n_c \times n_g]$.

STEP 3: Repeat for $l = 1, 2, ...$

3.1 Compute the cluster prototypes:

$v(i,1)^{(l)} = 0$,

$v(i,2)^{(l)} = 0$,

For $v(i,n)^{(l)}$ use Equation (9) $1 \leq i \leq n_c, \quad 3 \leq n \leq n_t$.

3.2 Compute the distances:

$$d^2_{\mathrm{STS}}(x_j, v_i) = \sum_{k=0}^{n_t-1} \left(\frac{v^{(l)}_{i(k+1)} - v^{(l)}_{ik}}{t_{(k+1)} - t_k} - \frac{x_{j(k+1)} - x_{jk}}{t_{(k+1)} - t_k} \right)^2$$

$$1 \leq i \leq n_c, \quad 1 \leq j \leq n_g.$$

3.3 Update the partition matrix:

if $d_{\mathrm{STS}ij} > 0$ for $1 \leq i \leq n_c$, $1 \leq j \leq n_g$,

$$u^{(l)}_{ij} = \frac{1}{\sum_{q=1}^{n_c} (d_{\mathrm{STS}ij}/d_{\mathrm{STS}qj})^{1/(w-1)}},$$

otherwise $u^{(l)}_{ij} = 0$ if $d_{\mathrm{STS}ij} > 0$, and $u^{(l)}_{ij} \in [0,1]$ with $\sum_{i=1}^{n_c} u^{(l)}_{ij} = 1$.

Until $|U^{(l)} - U^{(l-1)}| < \epsilon$.

Fig. 4. Pseudo code of the FSTS clustering algorithm.

identify the four clusters shown in Fig. 5(b) successfully. The clustering parameters for both fuzzy algorithms which yield successful results were $w = 1.6$ and $\alpha = 0.4$.

The second test used a subset of the artificial data set shown in Fig. 5(a). The original data set was "resampled" selecting 10 time points randomly out of the 20 original time points. The resulting data set is shown in Fig. 6(a). In this case, only the FSTS algorithm is able to identify the four clusters successfully, while FCM and HC identify two clusters and the other two mixed, as shown in Fig. 7, and KM does not produce consistent results. The clustering parameters for the FSTS algorithm were $w = 1.2$ and $\alpha = 0.6$. Different parameters were

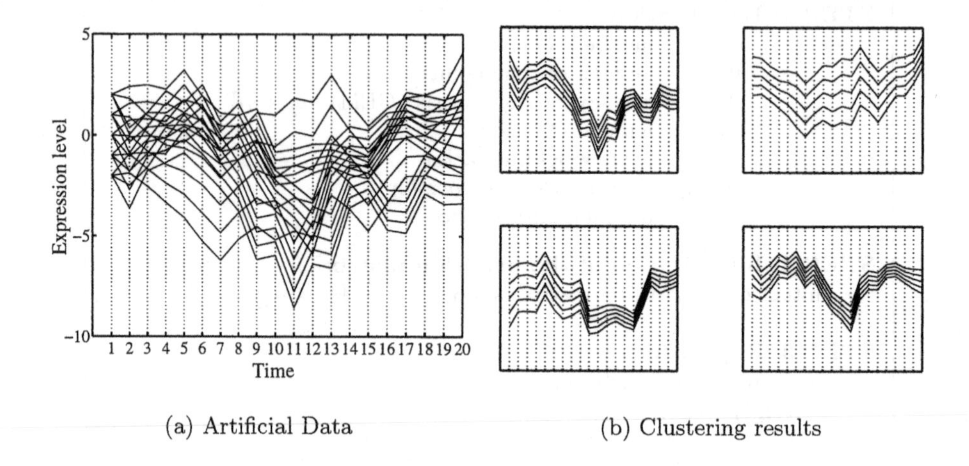

(a) Artificial Data (b) Clustering results

Fig. 5. Artificial data set and clustering results for FCM, FSTS, HK and HC algorithms

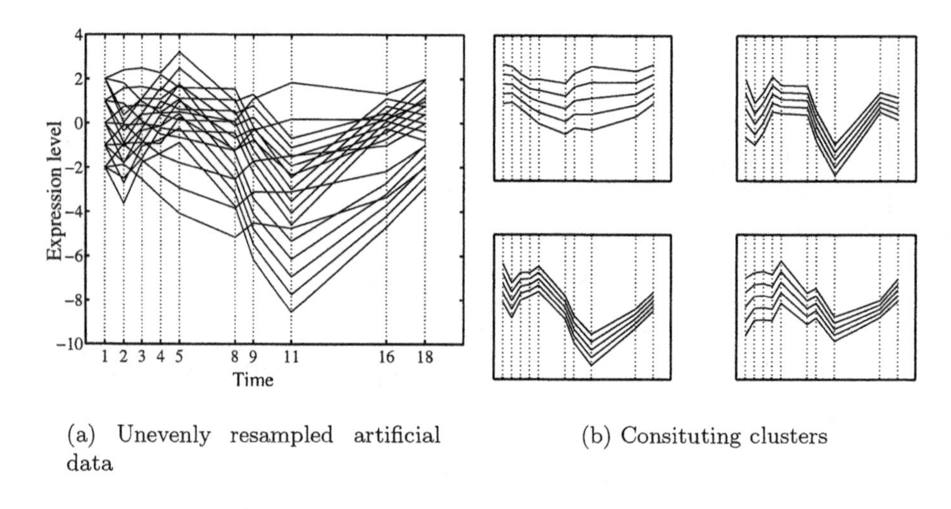

(a) Unevenly resampled artificial data (b) Consituting clusters

Fig. 6. Unevenly resampled artificial data set and the constituting clusters

tested for the FCM, $1.2 < w < 2.5$ and $0.3 < \alpha < 0.6$ (54 combinations) giving unsuccessful results. Finally the algorithms were evaluated using the three time-series data presented in Section 2. The objective is to cluster g_1, g_2 and g_3 in two clusters. The FSTS algorithm is the unique method capable of grouping g_2 with g_3 separated from g_1 consistently. FCM, HK, and HC do not have consistent results since they find g_2 as similar to g_1 as to g_3 as described in Section 2.

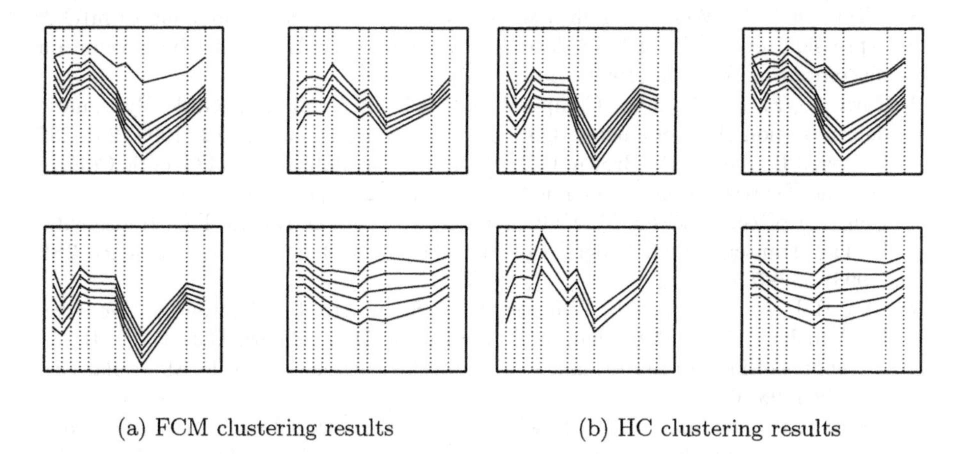

(a) FCM clustering results (b) HC clustering results

Fig. 7. Clustering results for FCM and HC algorithms

5 Conclusions

The FSTS clustering algorithm was presented as a new approach to cluster short time-series. The algorithm is particularly well suited for varying intervals between time points, a situation that occurs in many practical situations, in particular in biology. The FSTS algorithm is able to identify similar shapes formed by the relative change and the temporal information, regardless of the absolute levels. Conventional clustering algorithms, including FCM, KM, or HC, are not able to properly include the temporal information in the distance metric. We tackled the problem by considering the time-series as piecewise linear functions and measuring the difference of slopes between the functions. We illustrated the algorithm with an artificial data set. The FSTS algorithm showed better performance than the conventional algorithms in clustering unevenly sampled short time-series data.

Acknowledgements. This work was supported in part by grants from ABB Ltd. U.K., an overseas Research Studentship (ORS) award, Consejo Nacional de Ciencia y Tecnologia (CONACYT), and by the Post-doctoral Fellowship Program of Korea Science & Engineering Foundation (KOSEF).

References

1. Brown, P.O., Botstein, D.: Exploring the new world of the genome with DNA microarrays. Nature Genetics supplement **21** (1999) 33–37
2. Duggan, D.J., Bittner, M., Chen, Y., Meltzer, P., Trent, J.M.: Expression profiling using cDNA microarrays. Nature **21** (1999) 10–14

3. D'Haeseleer, P., Wen, X., Fuhrman, S., Somogyi, R.: Linear modeling of mRNA expression levels during CNS development and injury. In Pacific Symposium on biocomputing, Hawaii. (1999) 41–52
4. Tavazoie, S., Huges, J.D., Campbell M.J., Cho R.J., Church, G.M.: Systematic determination of genetic network architecture. Nature Genetics **22** (1999) 281–285
5. DeRisi, J.L., Iyer, V.R., Brown, P.O.: Exploring the Metabolic and Genetic Control of Gene Expression on a Genomic Scale. Science **278** (1997) 680–686
6. Chu, S. DeRisi, J., Eisen, M. Mulholland, J. Botstein, D. Brown, P.O. Herskowitz, I.: The Transcriptional Program of Sporulation in Budding Yeast. Science **282** (1998) 699–705
7. Cho, R.J., Campbell, M.J., Winzeler, E.A., Steinmetz, L., Conway, A., Wodicka, L., Wolfsberg, T.G., Gabrielian, A.E., Landsman, D., Lockhart, D.J., Davis, R.W.: A Genome-Wide Transcriptional Analysis of the Mitotic Cell Cycle. Molecular Cell **2** (July 1998) 65–73
8. Eisen, M.B., Spellman, P.T., Brown, P.O., Botstein, D.: Cluster analysis and display of genome-wide expression patterns. Proc. Natl. Acad. Sci. **95** (Nov 1998) 14863–14868
9. Spellman, P.T., Sherlock, G., Zhang, M.Q., Iyer, V.R., Anders, K., Eisen, M.B., Brown, P.O., Botstein, D., Futcher, B.: Comprehensive Identification of Cell Cycle-regulated Genes of Yeast *Saccharamyces cerevisiae* by Microarray Hybridization. Molecular Biology of the Cell **9** (1998) 3273–3297
10. Höppner, F.: Learning Temporal Rules from State Sequences. In IJCAI Workshop on Learning from Temporal and Spatial Data, Seattle, USA (2001) 25–31
11. Everitt, B.: Cluster Analysis. Heinemann Educational Books, London, England (1974)
12. Bezdek, J.: Pattern Recognition with Fuzzy Objective Function Algorithms. Plenum Press, New York (1981)
13. Höppner, F., Klawonn, F., Krause, R., Runkler, T.: Fuzzy Cluster Analysis. John Wiley & Sons. Chichester, England (1999)
14. Wolkenhauer, O.: Data Engineering: Fuzzy Mathematics in System Theory and Data Analysis. John Wiley and Sons, New York (2001)
15. Jain, A.K., Dubes, R.C.: Algorithms for Clustering Data. Prentice Hall, Englewood Cliffs, NJ (1998)
16. Zadeh, L.A.: Fuzzy sets. Information and Control **8** (1965) 338–352

Combining and Comparing Cluster Methods in a Receptor Database

Elena V. Samsonova[1,2], Thomas Bäck[2,3], Margot W. Beukers[1], Ad P. Ijzerman[1], and Joost N. Kok[2]

[1] Leiden University, Leiden / Amsterdam Center for Drug Research, Einsteinweg 55, PO Box 9502, 2300RA Leiden, The Netherlands
{beukers, ijzerman}@chem.leidenuniv.nl

[2] Leiden University, Leiden Institute for Advanced Computer Science, Niels Bohrweg 1, 2333CA Leiden, The Netherlands
{baeck,elena.samsonova,joost}@liacs.nl

[3] NuTech Solutions GmbH, Martin-Schmeisser-Weg 15, 44227 Dortmund, Germany

Abstract. Biological data such as DNA and protein sequences can be analyzed using a variety of methods. This paper combines phylogenetic trees, experience-based classification and self-organizing maps for cluster analysis of G protein-coupled receptors (*GPCRs*), a class of pharmacologically relevant transmembrane proteins with specific characteristics. This combination allows to gain additional insights.

1 Introduction

Advances in deciphering the genomes of various organisms have led to the inclusion into biological databases of large numbers of proteins, whose amino acid sequence is predicted based on their corresponding genes. With the only known characteristic of these proteins being their amino acid sequence, biologists are facing an even greater puzzle than the elucidation of the genome itself. It is well known from biochemistry that the properties of a protein with respect to participation in chemical and biophysical processes are largely defined by its three-dimensional structure. Although the laws governing protein folding are understood to a certain extent, attempts to compute a protein conformation based on its amino acid sequence have so far been unsuccessful. Without a three-dimensional structure, potential functions of predicted proteins often remain unclear. As biological experiments aimed at elucidating these functions are lengthy and complex, a computational approach becomes attractive.

One of the most popular ways to predict unknown properties of proteins is classification. Classification methods group similar proteins together according to some pre-determined criteria. When proteins with unknown properties (also called *orphans*) are found close to well-studied ones, the orphan properties are assumed to be similar to the known properties of their neighbors. Subsequent

M.R. Berthold et al. (Eds.): IDA 2003, LNCS 2810, pp. 341–351, 2003.

biological experiments, now narrowed down to very specific conditions, can test the prediction.

To this date many attempts at automated classification of proteins have been made, with an average result of 70% (e.g. [1], [2]) of proteins correctly classified into families. These methods can be roughly divided into those based on sequence homology determined by alignment (e.g. [3], [4], [5]) and on a combination of protein properties (e.g. [6], [1], [2], [7]). Methods of the first group are more intuitive for non-mathematicians as they work with protein sequences directly and do not use any kind of numeric transformations making it easy to check the results manually. Perhaps for this reason they are favored by biologists and widely used in practice. Methods of the second group, however, allow classification according to a broad range of protein properties, including but not limited to sequence homology.

This paper clarifies one of the methods of the second group (self-organizing maps) with one of the most popular methods of the first group (phylogenetic trees) bringing it in perspective with familiar classifications. Combination of these methods provides more information than each of them alone highlighting areas of potentially incorrect classification and improving overall selectivity (see section 5 for details).

In addition, we would like to bring the G protein-coupled receptor database, GPCRDB, a publicly available data set to the attention of the IDA community because it is an interesting test bed for IDA methods. It has a compact size (around 3,000 data elements) and yet high diversity and it contains both very similar and very different elements thus avoiding potential biases in data. Moreover, biologists and pharmacologists worldwide are highly interested in these proteins as they represent the largest class of drug targets. Although most medicines interact with GPCRs, many targets have not yet been exploited as they belong to the category of orphan receptors. Hence, classification of these proteins would be a major step forward in their exploitation as a drug target.

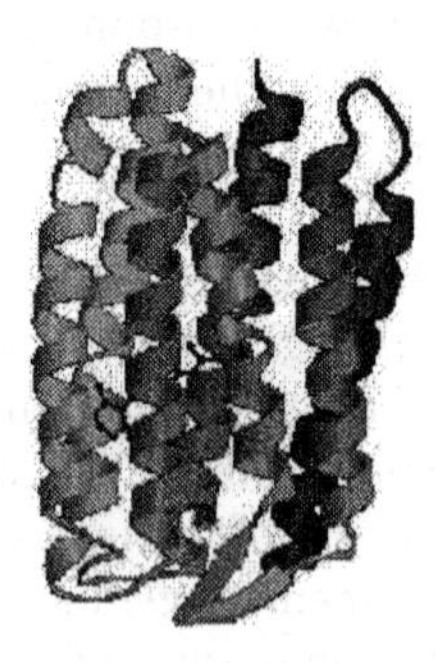

Fig. 1. Schematic representation of a crystal structure of bovine rhodopsin

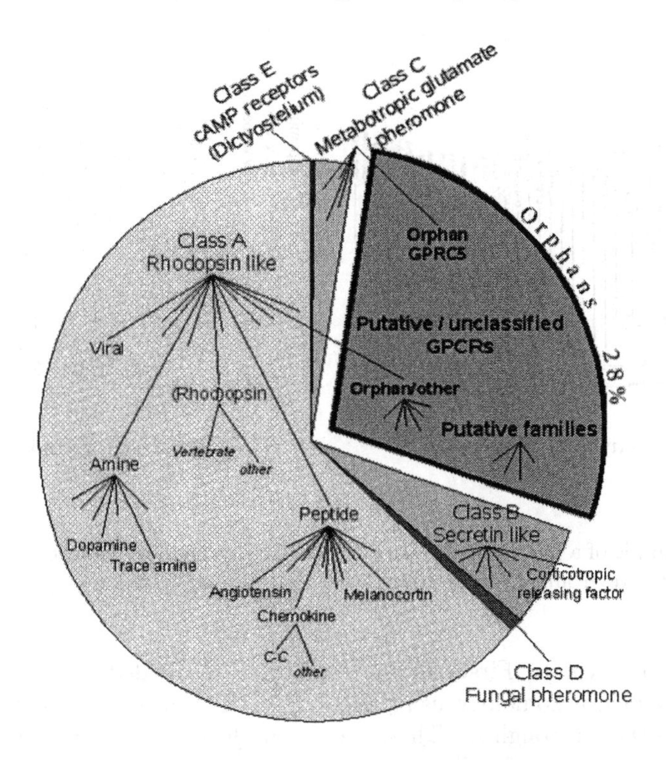

Fig. 2. Receptor classification scheme of GPCRDB

2 Data Set

The data set used in this study is a group of proteins with distinct properties: G protein-coupled receptors (*GPCRs*). They are the "docking site" for many of the body's own hormones and neurotransmitters, such as noradrenaline, dopamine and serotonin, and transmit the "signals" these ligands exert. These cell membrane proteins have a typical structure with seven serpentine-like transmembrane domains, co-determined by the biochemical properties of cell membranes (see figure 1). Moreover, the GPCRs' signal transduction mechanism is based on binding G proteins inside a cell and thus separates them from other transmembrane receptors.

A wealth of information on GPCRs is collected in the previously mentioned publicly available database, GPCRDB (www.gpcr.org), which presents an ideal data set for data mining studies. The GPCRDB [8] is mainly derived from a protein database SWISS-PROT [9] adding various annotations to the SWISS-PROT records as well as providing additional primary and secondary data on GPCRs. However, so far only protein sequences are being used for classification which presents an additional challenge for GPCRs as they often show little homology with each other. GPCRDB uses a mix of computerized and manual

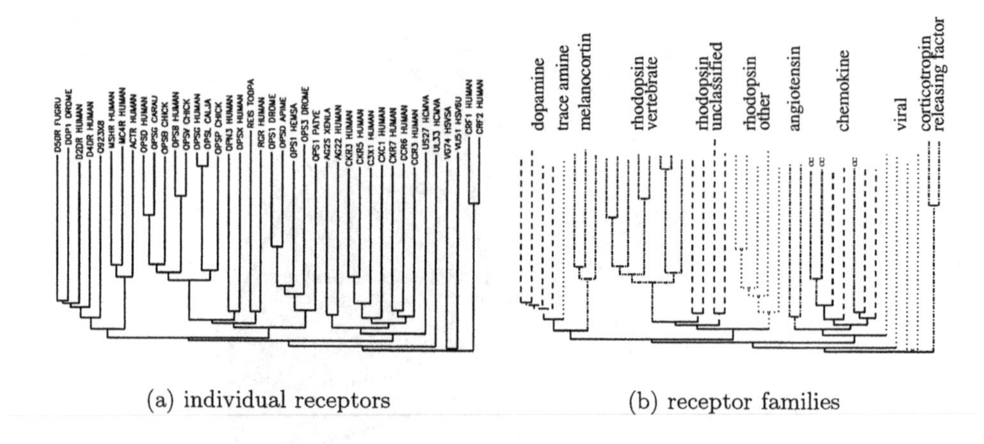

Fig. 3. Example of a phylogenetic tree of 40 GPCRs: (a) individual receptors are listed, (b) receptors are grouped into families according to their function

classification methods. First, all the receptors are divided into major classes based on sequence homology using the WHAT IF computer program [10] (see figure 2, classes A through E). These classes are then augmented with a manually created tree structure of well-studied receptors dividing them into families and sub-families based on ligand types or on other relevant criteria. E.g., in figure 2 *viral, (rhod)opsin, amine* and *peptide* are families of class A, and *dopamine* and *trace amine* are sub-families of *amine*. For each family and sub-family a *consensus profile* is calculated, which may be described as an "average" protein sequence based on the sequences in the family or sub-family in question. The remaining receptors (*orphans*) are compared to the consensus profiles of each of the families and sub-families within its class in an attempt to classify them within the existing structure. This step is also based on sequence homology and is performed with WHAT IF. It allows classification of some orphans but still leaves a large group unclassified indicating that they may form separate families of their own. Figure 2 gives an overview of the resulting classification. Note that orphans can be found on each level of the hierarchy reflecting the degree of confidence of their location. Here we look at the groups with the least confidence degree found on the top level (*putative families* and *putative / unclassified GPCRs*) and on the first level of the hierarchy (*class A - orphan / unclassified* and *class C - orphan GPRC5*). These receptors comprise as much as 28% of all the GPCRs in GPCRDB indicating a great need for innovative classification methods.

3 Phylogenetic Tree

Classification methods that are based on sequence homology often visualize relationships between proteins in a form of an unrooted binary tree in which branch lengths represent distances (see figure 3a). Such a tree often allows to visually locate clusters of receptors that show a higher homology between themselves than between them and the rest of the receptors in the sample. Such clusters appear as sub-trees of the main tree (see figure 3b, *dopamine, melanocortin, trace amine, rhodopsin vertebrate, rhodopsin other, angiotensin* and *corticotropin releasing factor* families). However, receptors that have low sequence homology cannot be clearly grouped using a phylogenetic tree, such as *rhodopsin unclassified* and *viral* families in figure 3b. Also in some cases the method fails to detect sub-families such as is the case for the *chemokine* family in figure 3b where the *CC* sub-family cannot be isolated in the phylogenetic tree. It has also been shown (see [11]) that given a set of phylogenetic trees, one is not guaranteed to find a supertree that maintains correct branch lengths with respect to degree of sequence homology between any two elements of the supertree.

Thus it becomes clear that although phylogenetic trees allow for a clear visualization of the results, they present problems for aggregation of clusters into superclusters (or trees into supertrees). Also, classification based on sequence homology determined by alignment is successful in discovering many protein families, especially when they exhibit high sequence homology. However, it is less effective in cases when proteins with low sequence homology have similar functions.

4 Kohonen Self-Organizing Map

We use Kohonen self-organizing maps (*SOM*, see [12]) to discover clusters and create a graphical representation of the results. SOM tries to preserve distances between the data vectors it is trained on, and creates a graphical map of the data space. Regions containing closely related elements are colored in light shades of grey, and regions separating clusters are colored dark. This technique allows to see the number, size and selectivity of the clusters at a glance, as well as the relationships between the clusters (just as the data elements themselves, closely related clusters are located close together on the map). One of the strengths of self-organizing maps as compared to phylogenetic trees lies in the possibility to use any kind of mathematical modeling technique to convert protein sequences into characteristic numeric fingerprints and analyze them from different points of view (e.g. [13], [14]).

It is well known from literature that the quality of maps created by the SOM algorithm heavily depends on the parameters used for training. Unfortunately, there is no straightforward method to determine optimal parameters in advance, so that several dozens of attempts are recommended for finding acceptable settings (see [12]). Due to the high computational cost of this approach, the present work does not claim to present fully optimized SOMs. Instead, we focus on combining and comparing SOMs with phylogenetic trees and demonstrating the advantages and opportunities opened by this approach.

5 SOM Based on Sequence Homology

We compare the two methods by using the same distance: create a self-organizing map based on sequence homology data and compare it with a phylogenetic tree. Both methods are based on a data matrix (available upon request) that lists pairwise sequence homology scores of all the proteins under consideration. As mentioned in section 1, a sequence homology score for a pair of proteins indicates the degree of similarity of their amino acid sequences and is calculated by use of a sequence alignment algorithm. In the present work we used ClustalW [15] to obtain sequence homology data.

As explained before, figure 3 shows a phylogenetic tree constructed using a random selection of GPCRs belonging to several families taken from the different branches of the manual classification system in GPCRDB. In figure 4a we present an aggregated phylogenetic tree of receptor families derived from the tree in figure 3b by substituting sub-trees by single branches whose length equals to the average of the branch lengths within the respective family in figure 3b. Figure 4b shows an excerpt of the GPCRDB classification in figure 2 presenting selected families in a way similar to a phylogenetic tree for the sake of comparison.

Figure 4c shows the self-organizing map built on sequence homology data with family clusters highlighted. Note that the family *chemokine* is split into separate sub-clusters - *chemokine CC* and *other* - according to the GPCRDB classification whereas this split does not appear in the phylogenetic tree. The *rhodopsin* family is split into *vertebrates* and *other* not showing a separate cluster for *unclassified*, members of this group being spread among the other *rhodopsin* receptors. This placing does not contradict the phylogenetic tree which also fails to place these receptors on a sub-tree of their own. Similarly, the *viral* family is split into separate clusters on the SOM as well as on the phylogenetic tree.

One of the main contributions of this paper is the combination of the two methods into one visualization. The map can be overlayed with a tree in order to highlight differences in classification. The overlay is constructed in three-dimensional perspective view, such that the map is positioned at an angle to the plane of view, so to speak "on the floor". The tree is then constructed vertically and "upside down" placing the branches in the center of each cluster and subsequently connecting them with horizontal bars according to the tree configuration. In a phylogenetic tree branch lengths carry important information, so correct lengths are retained. On the other hand, lengths of horizontal bars are irrelevant, and these are stretched as required in order to maintain the shape of the tree. Note that due to different branch height, horizontal bars may no longer appear horizontal in this case. Construction of a classification tree where only the hierarchy of the branches and not their lengths is relevant, is even simpler. In this case branch height is chosen so as to present a clear picture.

Overlaying the map with the phylogenetic tree (figure 4d) and with the GPCRDB classification tree (figure 4e) reveals differences in cluster placement with respect to either of the trees. Although the SOM tends to place clusters of closely related families near each other, it does not always succeed in doing so, partially due to the fixed geometric space it is constrained by. Thus *dopamine*

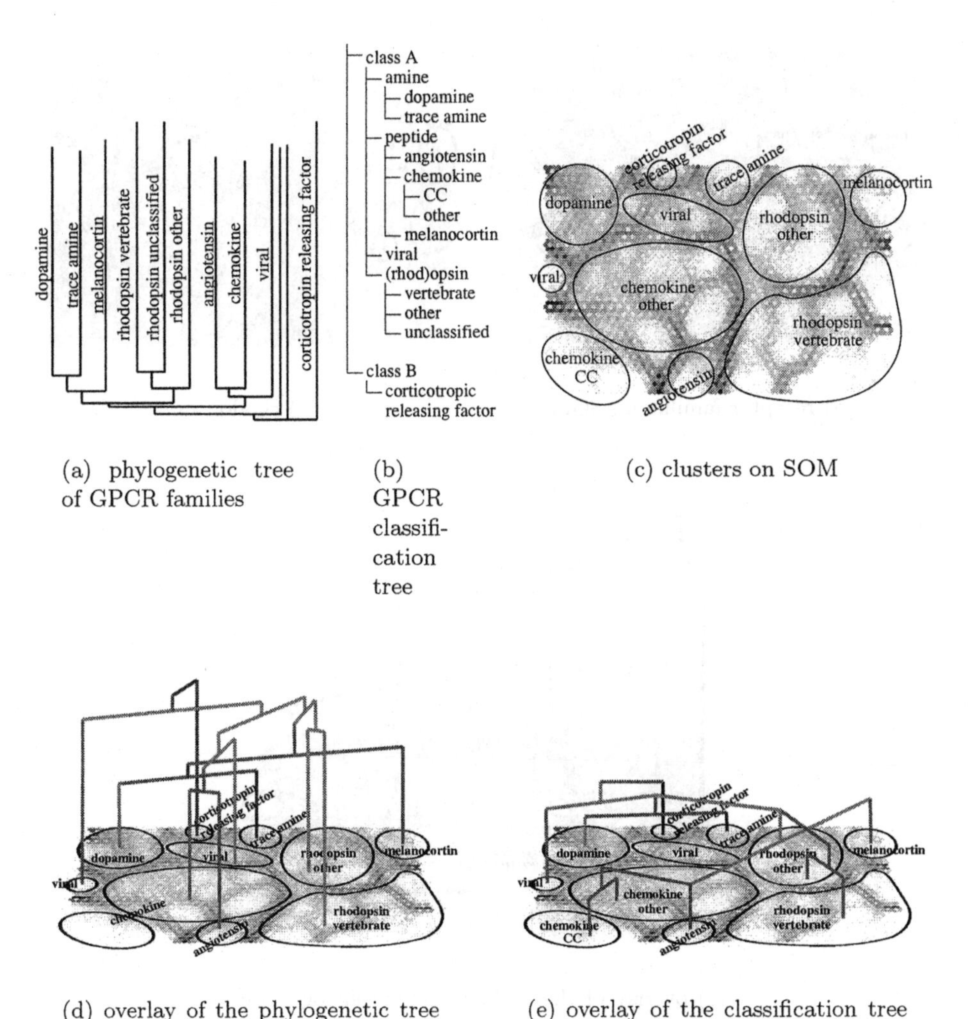

(a) phylogenetic tree of GPCR families

(b) GPCR classification tree

(c) clusters on SOM

(d) overlay of the phylogenetic tree over the SOM

(e) overlay of the classification tree over the SOM

Fig. 4. Overlaying a phylogenetic tree over a self-organizing map. *(a)* Aggregated phylogenetic tree of the receptor families constructed from the tree in figure 3b. *(b)* Selected receptor families according to the classification of GPCRDB. *(c)* Self-organizing map (*SOM*) of these families based on sequence homology with the family clusters highlighted. SOM parameters are: Gaussian neighborhood, 40x30 units, hexagonal representation; pass one: initial learning rate $\alpha_1 = 0.2$, initial radius $r_1 = 30$, training length $I_1 = 100,000$ iterations; pass two: $\alpha_2 = 0.02$, $r_2 = 3$, $I_2 = 300,000$ iterations. *(d)* SOM *(c)* overlayed with the phylogenetic tree *(a)*. *(e)* SOM *(c)* overlayed with the GPCRDB classification tree *(b)*

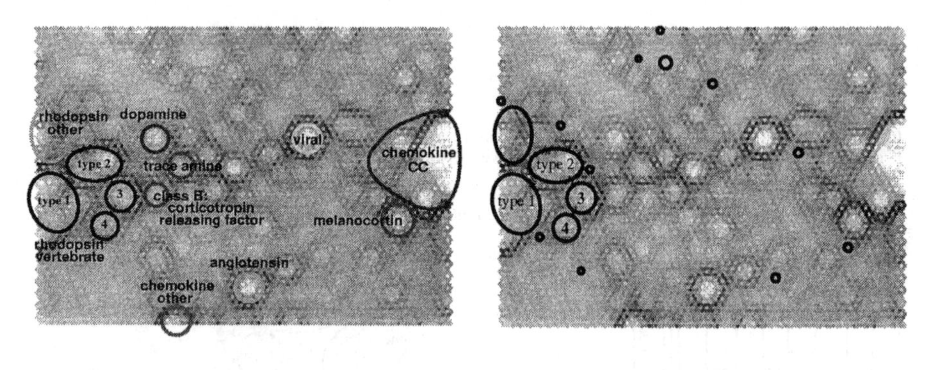

(a) receptor families on SOM (b) rhodopsin

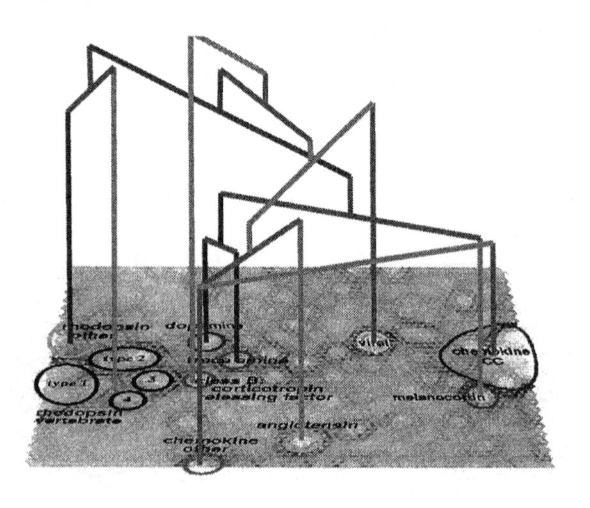

(c) overlay of the SOM with the phylogenetic tree

Fig. 5. Receptor families on the SOM based on the third neighbor pairs of amino acids in protein sequence. SOM parameters are: Gaussian neighborhood, 60x50 units, hexagonal representation; pass one: initial learning rate $\alpha_1 = 0.2$, initial radius $r_1 = 30$, training length $I_1 = 100,000$ iterations; pass two: $\alpha_2 = 0.02$, $r_2 = 3$, $I_2 = 300,000$ iterations. *(a)* Main family clusters for selected families. *(b)* All clusters for the *rhodopsin* family. *(c)* Overlay of the SOM *(a)* with the phylogenetic tree from figure 4a

and *melanocortin* families are no longer neighboring on the SOM as they are on the phylogenetic tree. On the other hand, the SOM proves more selective in identifying sub-families as becomes apparent from the separation of the *chemokine* family into two major sub-families *CC* and *other*.

6 SOM Based on a Characteristic Protein Fingerprint

As mentioned in section 4, a SOM can be trained on any kind of characteristic fingerprint of the protein sequences. In the previous section we used pairwise sequence homology scores to construct protein fingerprints. The disadvantage of such comparative methods lies in the fact that vector length is necessarily equal to the number of elements in the data set meaning that the latter directly and significantly influences training time of the SOM (see [12]). Alternatively, a protein fingerprint can be based on the properties of the protein sequence itself rather than on its relation to other sequences. This approach makes vector length independent of the number of data elements in the training set allowing to construct much larger SOMs.

An example of such a characteristic fingerprint is a distribution of the third neighbor pairs of amino acids in a protein sequence. The idea behind it is as follows: as mentioned in the introduction, GPCRs are known to have seven α-helices traversing a cell membrane. An α-helix is a spiral structure with 3.6 amino acids per turn so that every third and fourth amino acid are found next to each other in space along the helix axis and consequently may exhibit similar properties with respect to the surrounding cell membrane, the ligand binding pocket of the receptor and helical interactions. Therefore prevalence of certain pairs may co-determine three-dimensional orientation of the α-helices with respect to each other which in turn plays a role in the ability of the receptor to bind certain types of ligands. Thus a specific distribution of third or fourth neighbor pairs of amino acids in a protein sequence may exhibit a correlation with receptor function. In this example we use the distribution of the third neighbor pairs.

Figure 5 presents a SOM built on the entire data set of GPCRDB based on the fingerprint described above. In figure 5a we locate clusters on the map exclusively containing receptors of the selected families. Here, as in figure 4b, *rhodopsin* and *chemokine* families are split into distinct sub-families *rhodopsin vertebrate* and *other*, and *chemokine CC* and *other* respectively. In addition, *rhodopsin vertebrate* is further split into types 1 through 4 which do not appear on the phylogenetic tree in figure 3a but do appear in the GPCRDB classification (not shown). This method thus appears to be more selective than the one based on sequence homology scores. As the number of false positives in these clusters is consistently zero, there is a clear correlation between these results and the GPCRDB function-based classification which leads us to conclude that the third neighbor pairs of amino acids in a protein sequence co-determine protein function.

On the other hand, in figure 5b we see that not all the receptors of the *rhodopsin* family are classified within the main clusters. This "spillage" demonstrates that the current fingerprint method is not fully selective and consequently a distribution of the third neighbor pairs of amino acids in a protein sequence is not the only factor determining protein function.

Figure 5c presents an overlay of the SOM from figure 5a with the phylogenetic tree of receptor families from figure 4a. As with the overlays in figures 4d and 4e, there are differences in cluster placement with respect to the phylogenetic tree. Thus, *chemokine CC* and *melanocortin* appear to be direct neighbors on the SOM whereas this is not the case on the phylogenetic tree. On the other hand, the *chemokine CC* and *other* families are placed far from each other on the SOM whereas the phylogenetic tree even fails to make a distinction between them. Once again, overlaying the tree over the SOM proves helpful in finding differences in classification.

Varying and combining fingerprint construction methods allows to study different protein sequence properties with respect to their correlation with the GPCRDB classification. This, in turn, may provide new insights into factors co-determining protein function.

7 Conclusions

Combining self-organizing maps with phylogenetic trees as well as with other classification types (e.g. a hierarchical classification of GPCRDB) proves beneficial on several levels. Users familiar with one type of visual representation become introduced to another type in relation to familiar material. Differences in results produced by various methods are highlighted providing additional insights into the data being analyzed. The flexibility of self-organizing maps with respect to the aspects of the data set expressed as the data fingerprint allows to discover major factors determining protein classification. Finally, a self-organizing map presents a convenient interface for interactive representations of classifications presenting the whole data set in a compact form. Made "clickable", the image can easily accommodate zooming functions and pop-up windows offering detailed views, often found in applications of this type. A phylogenetic tree overlayed over such a map may serve as an additional guide to the users unfamiliar with this representation.

In conclusion, the combination of self-organizing maps with phylogenetic trees will be of use not only to GPCRs but also to other protein families and other classification problems where phylogenetic trees have been the major classification instrument to this date.

References

1. R. C. Graul and W. Sadée, Evolutionary relationships among G protein-coupled receptors using a clustered database approach, AAPS Pharmacological Sciences, vol. 3, no. 2, p. E12, 2001.
2. R. Karchin, K. Karplus, and D. Haussler, Classifying G protein-coupled receptors with support vector machines, Bioinformatics, vol. 18, no. 1, pp. 147–159, 2002.
3. S. F. Altschul, W. Gish, W. Miller, E. W. Meyers, and D. J. Lipman, Basic Local Alignment Search Tool, Journal of Molecular Biology, vol. 215, no. 3, pp. 403–410, 1990.
4. M. Vingron and M. S. Waterman, Sequence alignment and penalty choice. review of concepts, case studies and implications., Journal of Molecular Biology, vol. 235, no. 1, pp. 1–12, 1994.

5. M. S. Waterman, Parametric and ensemble sequence alignment algorithms, Bulletin of Mathematical Biology, vol. 56, no. 4, pp. 743–767, 1994.
6. E. A. Ferrán and P. Ferrara, Clustering proteins into families using artificial neural networks, Computer Applications in Biosciences (CABIOS), vol. 8, no. 1, pp. 39–44, 1992.
7. J. Lazović, Selection of amino acid parameters for fourier transform-based analysis of proteins, Computer Applications in Biosciences (CABIOS), vol. 12, no. 6, pp. 553–562, 1996.
8. F. Horn, E. Bettler, L. Oliveira, F. Campagne, F. E. Cohen, and G. Vriend, GPCRDB information system for G protein-coupled receptors, Nucleic Acids Research, vol. 31, no. 1, pp. 294–297, 2003.
9. A. Bairoch and R. Apweiler, The SWISS-PROT protein sequence database and its supplement TrEMBL in 2000, Nucleic Acids Research, vol. 28, no. 1, pp. 45–48, 2000.
10. G. Vriend, WHAT IF: a molecular modeling and drug design program, Journal of Molecular Graphics, vol. 8, pp. 52–56, 1990.
11. D. Bryant, Optimal agreement supertrees, in *Computational Biology* (O. Gascuel and M.-F. Sagot, eds.), vol. 2066 of *LNCS, Lecture Notes in Computer Science*, pp. 24–31, Springer-Verlag, 2001.
12. T. Kohonen, J. Hynninen, J. Kangas, and J. Laaksonen, *SOM_PAK: the self-organizing map program package*, second ed., 1995.
13. J. Hanke and J. G. Reich, Kohonen map as a visualization tool for the analysis of protein sequences: multiple alignments, domains and segments of secondary structures, Computer Applications in Biosciences (CABIOS), vol. 12, no. 6, pp. 447–454, 1996.
14. P. Somervuo and T. Kohonen, Clustering and visualization of large protein sequence databases by means of an extension of the self-organizing map, in *DS 2000* (S. Arikawa and S. Morishita, eds.), vol. 1967 of *LNAI, Lecture Notes in Artificial Intelligence*, pp. 76–85, Springer-Verlag, 2000.
15. F. Jeanmougin, J. D. Thompson, M. Gouy, D. G. Higgins, and T. J. Gibson, Multiple sequence alignment with Clustal X, Trends in Biochemical Sciences, vol. 23, pp. 403–405, 1998.

Selective Sampling with a Hierarchical Latent Variable Model

Hiroshi Mamitsuka

Institute for Chemical Research, Kyoto University
Gokasho Uji, 611-0011, Japan
mami@kuicr.kyoto-u.ac.jp

Abstract. We present a new method which combines a hierarchical stochastic latent variable model and a selective sampling strategy, for learning from co-occurrence events, i.e. a fundamental issue in intelligent data analysis. The hierarchical stochastic latent variable model we employ enables us to use existing background knowledge of observable co-occurrence events as a latent variable. The selective sampling strategy we use iterates selecting plausible non-noise examples from a given data set and running the learning of a component stochastic model alternately and then improves the predictive performance of a component model. Combining the model and the strategy is expected to be effective for enhancing the performance of learning from real-world co-occurrence events. We have empirically tested the performance of our method using a real data set of protein-protein interactions, a typical data set of co-occurrence events. The experimental results have shown that the presented methodology significantly outperformed an existing approach and other machine learning methods compared, and that the presented method is highly effective for unsupervised learning from co-occurrence events.

1 Introduction

In this paper, we focus on two-mode and co-occurrence events, such as product pairs in purchasing records and co-occurred words in texts. A data set of such co-occurrence events is one of the most frequently seen types of data sets in real world, and learning from co-occurrence events is a fundamental issue in intelligent data analysis. One of the most important property of co-occurrence events is that they are only positive examples, i.e. unsupervised examples in nature. An effective existing approach for learning from co-occurrence events is then based on model-based clustering [1]. In general, model-based clustering has a latent variable corresponding to a latent cluster and obtains one or more clusters of co-occurrence events by clustering co-occurrence events through estimating model parameters. Then, when two events are newly given, we can assign a latent cluster to the events with a likelihood, which can be computed by the estimated model parameters.

M.R. Berthold et al. (Eds.): IDA 2003, LNCS 2810, pp. 352–363, 2003.

We present a new methodology to improve the predictive ability of the existing model-based clustering approach for co-occurrence events. Our method combines a hierarchical latent variable model and a selective sampling strategy for unsupervised learning. The hierarchical latent variable model for co-occurrence events was presented in [2] and has two types of latent variables hierarchically. In general, an event has its own existing background knowledge, and in most cases, the background knowledge can be given as a set of classes, to one or more of which an event belongs. With the background knowledge, we can build a hierarchical latent variable model, in which one of the two types of latent variables and the other latent variable correspond to an existing class of events and a latent cluster of the classes, respectively. Estimating the parameters of the latent variable model results in performing clustering class pairs, instead of clustering event pairs, and we can obtain co-occurrence classes after estimating the parameters. Using the estimated model, arbitrary two events are predicted to be co-occurrence events if the two events belong to one or more pairs of two co-occurrence classes. The space complexity of the model is only roughly linear in the number of given events, and the parameters can be estimated by time-efficient EM (Expectation-Maximization) algorithm. We expect that the hierarchical model improves the predictive performance of existing approaches for co-occurrence events, since the model enables us to use existing classes of co-occurrence events.

We further focus on a selective sampling strategy for unsupervised learning and combine the hierarchical latent variable model and the strategy by employing the learning algorithm of the hierarchical model as the component subroutine of the strategy. The selective sampling was originally proposed for unsupervised noisy data in [3] and has been empirically shown that it is effective to improve the predictive performance of its component subroutine for noisy data. It repeats selecting plausible non-noise candidates by using previously obtained component models and running a component subroutine on the selected set of examples to obtain a new component model alternately. The final hypothesis is given by combining all obtained stochastic models, and the final likelihood for an example is obtained by averaging the likelihoods computed by all stochastic models. We can expect that the selective sampling also improves the predictive performance of existing approaches for real-world co-occurrence events, which is inevitably noisy.

The purpose of this paper is to empirically evaluate the effectiveness of the presented methodology using real data of co-occurrence events, comparing with an existing approach and other machine learning techniques, including a variety of simpler probabilistic models and support vector machines (SVMs). In our experiments, we focus on a real protein-protein interaction data set, a typical co-occurrence data set. We note that predicting unknown protein-protein interactions is an important and challenging problem in current molecular biology. We used the data set to perform five-fold cross-validation and generated synthetic negative examples for each of the five runs in the cross-validation to evaluate the ability of our method for discriminating positive from negative examples as the prediction accuracy of our method.

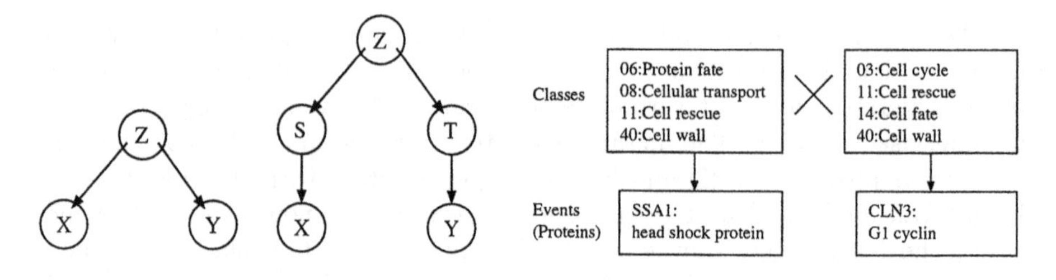

Fig. 1. Graphical models of (a) AM and (b) HAM, and (c) an example of co-occurrence events and their classes

Experimental results have shown that the presented method drastically improved the performance obtained by an existing approach and other methods tested. The component hierarchical model significantly outperformed other methods compared, and the selective sampling strategy further improved the prediction accuracy obtained by the component stochastic model. Over all, we have empirically shown that our method is highly effective for learning from real-world co-occurrence events, and we expect that our method is useful for predicting other numerous co-occurrence events found in a variety of application domains.

2 Methods

In this paper, we use the following notation. We denote a variable by a capitalized letter, e.g. X, and the value of a corresponding variable by that same letter in lower case, e.g. x. Now let X and Y be observable random variables taking on values $x_1, ..., x_L$ and $y_1, ..., y_M$, respectively, each of which corresponds to a discrete value (i.e. event). More concretely, x and y correspond to the first and the second event of an example of co-occurrence events. Let N be the total number of given examples (co-occurrence events). Let S and T be latent random variables taking on values $s_1, ..., s_U$ and $t_1, ..., t_V$, respectively, each of which corresponds to an existing latent class (background knowledge) of observable events. More concretely, for given co-occurrence events, s corresponds to a class taken by the first event and t corresponds to that by the second event. Let Z be a discrete-valued latent variable taking on values $z_1, ..., z_K$, each of which corresponds to a latent cluster. Let $n(x, y)$ be a binary value defined as follows: If x and y co-occur in a given data set then $n(x, y)$ is 1, otherwise it is 0. Let $v_{x,y}(s, t)$ be also a binary value defined as follows: If x and y belong to classes s and t respectively in given data then $v_{x,y}(s, t)$ is 1, otherwise it is 0.

In our experiments, we used protein-protein interactions as co-occurrence events. In this case, a protein corresponds to an event x (and y), and a protein class corresponds to an event class s (and t).

2.1 Hierarchical Latent Variable Model for Co-occurrence Events

Hierarchical Aspect Model (HAM). We first explain an existing model-based clustering approach for modeling co-occurrence events. The approach, which is called aspect model (AM for short), has been applied to some fields, including collaborative filtering recommender systems and semantic analysis in natural language processing [1]. An AM for X and Y with K clusters has the following form:

$$p(x, y; \theta) = \sum_{k}^{K} p(x|z_k; \theta)p(y|z_k; \theta)p(z_k; \theta).$$

The graphical model of AM is shown in Figure 1 (a).

In order to use existing classes of co-occurrence events, we extend the model to a hierarchical latent variable model for X and Y with S, T and K as follows:

$$p(x, y; \theta) = \sum_{l,m,k}^{U,V,K} p(x|s_l; \theta)p(y|t_m; \theta)p(s_l|z_k; \theta)p(t_m|z_k; \theta)p(z_k; \theta).$$

We call the model HAM, standing for *hierarchical aspect model* [2], and a graphical model of HAM is shown in Figure 1 (b).

Figure 1 (c) shows an example of co-occurrence events and their classes, which is derived from the real data of protein-protein interactions used in our experiments. In the figure, each of two events, i.e. proteins named as SSA1 and CLN3, belongs to a different set of four classes. We note that estimating the parameters of HAM results in clustering class pairs to obtain co-occurrence classes, instead of clustering event pairs as done in AM. Thus newly given two events will be predicted as co-occurrence events if the two events belong to co-occurrence classes. In addition, the larger the number of co-occurrence classes to which the two events belong, the more confidently the two events will be predicted as co-occurrence events. In the example shown in Figure 1 (c), we can consider 16 ($= 4 \times 4$) possible combinations of class pairs, and the proteins, SSA1 and CLN3, are predicted as co-occurrence events more strongly if a larger number of class pairs are co-occurrence classes. We emphasize that this model naturally combines co-occurrence events and their existing classes and is applicable to any co-occurrence events found in a variety of real-world applications.

Estimating Model Parameters. When the number of clusters K and training data D are given, a possible criterion for estimating probability parameters of HAM is the maximum likelihood (ML): $\theta^{ML} = arg \max_\theta \log p(D; \theta)$, where $\log p(D; \theta) = \sum_i \sum_j n(x_i, y_j) \log p(x_i, y_j; \theta)$.

We use a time-efficient general scheme, EM (Expectation-Maximization) algorithm [4], to obtain the ML estimators of HAM. The algorithm starts with initial parameter values and iterates both an expectation step (E-step) and a maximization step (M-step) alternately until a certain convergence criterion is satisfied. The following equations derived by us are consistent with those of a hierarchical mixture model shown in [5].

In E-step, we calculate the latent class conditional probability, namely the probability that a given example (x, y) is generated by a cluster z and latent classes s and t given parameters θ:

$$w(z_k, s_l, t_m|x_i, y_j; \theta) = \frac{p(x_i|s_l;\theta)p(y_j|t_m;\theta)p(s_l|z_k;\theta)p(t_m|z_k;\theta)p(z_k;\theta)}{\sum_{k,l,m} p(x_i|s_l;\theta)p(y_j|t_m;\theta)p(s_l|z_k;\theta)p(t_m|z_k;\theta)p(z_k;\theta)}.$$

In M-step, we take a corresponding summation over $w(z_k, s_l, t_m|x_i, y_j; \theta)$ with $n(x_i, y_j)$:

$$\theta_{x_i|s_l} \propto \sum_{j,k,m} n(x_i, y_j)\ v_{x_i,y_j}(s_l, t_m)\ w(z_k, s_l, t_m|x_i, y_j; \theta_{old}).$$

$$\theta_{y_j|t_m} \propto \sum_{i,k,l} n(x_i, y_j)\ v_{x_i,y_j}(s_l, t_m)\ w(z_k, s_l, t_m|x_i, y_j; \theta_{old}).$$

$$\theta_{s_l|z_k} \propto \sum_{i,j,m} n(x_i, y_j)\ v_{x_i,y_j}(s_l, t_m)\ w(z_k, s_l, t_m|x_i, y_j; \theta_{old}).$$

$$\theta_{t_m|z_k} \propto \sum_{i,j,l} n(x_i, y_j)\ v_{x_i,y_j}(s_l, t_m)\ w(z_k, s_l, t_m|x_i, y_j; \theta_{old}).$$

$$\theta_{z_k} \propto \sum_{i,j,l,m} n(x_i, y_j)\ v_{x_i,y_j}(s_l, t_m)\ w(s_l, t_m|x_i, y_j; \theta_{old}).$$

The required memory size of the EM procedure depends only on the parameters of the model, i.e. $p(x|s), p(y|t), p(s|z), p(t|z)$ and $p(z)$. Note that K, i.e. the size of latent clusters, can be given as a certain constant, and the size of $p(s|z)$ and $p(t|z)$ is linear in the number of given events. Note further that the size of $p(x|s)$ and $p(y|t)$ is also practically linear in the number of events, since the number of classes to which an event belongs is limited to a certain small number. Overall, the total space complexity of HAM is roughly linear in the number of given events. In this sense, the HAM is a space-efficient model.

2.2 Selective Sampling with Hierarchical Aspect Models (SsHAM)

In order to further improve the predictive accuracy of HAM, we employ a selective sampling strategy for unsupervised learning [3]. It has been shown that the performance advantage of the selective sampling has more pronounced for noisier data sets.

The selective sampling uses a learning algorithm of an arbitrary stochastic model as a component subroutine, assuming that given an input x, the subroutine returns its likelihood $L(x)$. It further assumes that given a set of examples including x, x is regarded as an outlier (i.e. a noise) if $L(x)$ is the lowest among the likelihoods given to the example set. Under these assumptions, it repeats selecting plausible non-noise examples from a given whole set of examples and running a component learning algorithm on the selected examples alternately. In

Input: A set of given examples: E
Learning algorithm of hierarchical aspect model: A
Number of iterations: I
Number of examples selected at each iteration: N_s
Number of noise candidates to be removed at each iteration: N_n
Initialization: 0. Randomly choose N_s examples as E_0 from E
For $t = 0, ..., I-1$
1. Run A on E_t with random initial parameter values and obtain a trained model.
2. For each $(x, y) \in E$, compute its likelihood $L_t(x, y)$ using the trained model.
3. For each $(x, y) \in E$, compute average $\bar{L}(x, y)$ by $\frac{\sum_t L_t(x,y)}{t+1}$.
4. Select $\langle(x_1^*, y_1^*), ..., (x_{N_n}^*, y_{N_n}^*)\rangle$, whose $\bar{L}(x_1^*, y_1^*), ..., \bar{L}(x_{N_n}^*, y_{N_n}^*)$ are the smallest, and remove them from E as $E'_{t+1} = remove(E, \langle(x_1^*, y_1^*), ..., (x_{N_n}^*, y_{N_n}^*)\rangle)$.
5. Randomly choose N_s examples from E'_{t+1} as E_{t+1}.

Output: Output all obtained component stochastic models.

Fig. 2. Algorithm: Selective sampling with hierarchical aspect models (SsHAM).

the example selection, the strategy first removes examples whose likelihoods are the lowest as noises (i.e. outliers) from all examples and then selects examples from the remaining examples randomly as a new set of training examples. The likelihood for a newly given test example is computed by averaging over the likelihoods computed by all models obtained as the final output. We combine the strategy and HAM by using the learning algorithm of HAM as a subroutine of the selective sampling strategy, and we call this combination 'SsHAM', standing for *selective sampling for hierarchical aspect models.* The pseudocode of SsHAM is shown in Figure 2.

2.3 Methods Compared in Our Experiments

In order to compare our method with other methods in our experiments, we generate a training data set using the data of co-occurrence events and a set of classes to which an event belongs. The data set is a table, in which each record corresponds to an example of co-occurrence events, i.e. a pair (x, y) that satisfies $n(x, y) = 1$, each attribute corresponds to a class pair and attribute values are all binary and represented by $v_{x,y}(s, t)$. The data set used in our experiments is a one-class data set, and we can run any unsupervised learning method on this set. We note that the complexity of the attribute size of the data set is quadratic in the size of event classes. Thus we cannot run a learning method whose computational complexity rises drastically with increasing the size of attributes. For example, learning Baysian belief networks in general falls into this type of unsupervised learning methods.

Simple Probabilistic Models. We first tested a simple method, which computes the probability that arbitrary two classes appear in given data. We consider three variations in using the probabilities: summing over them (SSum), averaging over them (SAve) and selecting the maximum (SMax). The procedure of the method can be summarized as follows:

1. For all s and t, $p(s,t)$ is computed as follows:

$$p(s,t) = \frac{q(s,t)}{\sum_{s,t} q(s,t)},$$

where $q(s,t) = \sum_{i,j} n(x_i, y_j) \cdot v_{x_i,y_j}(s,t)$.

2. For any new events x and y, $\hat{p}(x,y)$, i.e. the probability that they are co-occurrence events, is calculated as follows:

$$\text{SSum:} \quad \hat{p}(x,y) = \sum_{x \in s, y \in t} p(s,t).$$

$$\text{SAve:} \quad \hat{p}(x,y) = \frac{\sum_{x \in s, y \in t} p(s,t)}{\sum_{x \in s, y \in t} 1}.$$

$$\text{SMax:} \quad \hat{p}(x,y) = \max_{x \in s, y \in t} p(s,t).$$

One-Class Support Vector Machine. We then tested 'one-class support vector machine (one-class SVM, hereafter OC.SVM)' proposed by [6]. Learning OC.SVM is unsupervised learning, and thus it is trained by positive examples only. In our experiments, we used LIBSVM ver.2.36, which has an option to run the OC.SVM and can be downloaded from http://www.csie.ntu.edu.tw/~cjlin/libsvm/. In prediction, for each example, LIBSVM gives a binary output, indicating whether the example belongs to the class of training data or not.

Support Vector Classifier. We finally tested a support vector classifier (hereafter, SVC), a supervised learning approach with significant success in numerous real-world applications. When we use the SVC in our experiments, we randomly generated negative training examples, which has the same size as that of positive training examples, not to overlap with any positive training examples and test examples. We then run both SVM^{light} [7] and the LIBSVM, and the two softwares outputted approximately the same performance results for all cases in our experiments.

For both OC.SVM and SVC, we used linear kernels in our experiments. We tested other types of kernels, such as a polynomial kernel, within the limits of main memory, but no significant improvements were obtained.

3 Experimental Results

3.1 Data

In our experiments, we focus on the data set of protein-protein interactions as a typical co-occurrence data set. We focus on yeast proteins, by which high-throughput bio-chemical experiments have been actively done these few years, and used the MIPS [8] comprehensive yeast genome database, which contains the recent results of the experiments. From the MIPS data base, we obtained 3,918 proteins and 8,651 physical protein-protein interactions, each protein of which is one of the 3,918 proteins.

We further used the MIPS database to obtain a set of protein classes. We focus on the functional classification catalogue of the MIPS database. The information of the catalogue takes a tree structure, in which each node corresponds to a protein class and a protein falls into one or more nodes. To obtain a set of classes of proteins from the tree, the problem is how we cut the tree, using given training data. We employ the idea by [9] which cuts a thesaurus based on the MDL (Minimum Description Length) principle [10] to cut the tree for classifying proteins. By adjusting a parameter in the MDL, we obtained some sets of protein classes which range from 30 to 400 in their size. The 30 classes correspond to the nodes which have a root as their parent, and the set of 400 classes is the largest set under the condition that each class has one or more proteins.

3.2 Five-Fold Cross Validation

The evaluation was done by five-fold cross validation. That is, we split the data set into five blocks of roughly equal size, and in each trial four out of these five blocks were used as training data, and the last block was reserved for test data. The results were then averaged over the five runs.

To evaluate the predictive ability of each of the methods, we used a general manner for supervised learning. That is, for each test data set, we generated negative examples, the number of which is the same as that of positive examples, and examined whether each method can discriminate positive from negative examples. In this 50:50 setting, a prediction accuracy obtained by a random guessing (and a predictor, which outputs only one label) is maximally 50%, and we can check the performance of each of the methods by how it is better than 50%. We used two types of negative examples, both of which are randomly generated not to overlap with any positive example. The first type of examples are randomly generated from the 3,918 proteins. The second type of examples are randomly generated from a set of proteins, which are used in positive examples, i.e. the proteins contained in the 8,651 protein-protein interactions. Hereafter we call the two test data sets containing the first and second type of negative examples as Data1 and Data2, respectively.

In our experiments, we tested eight methods, i.e. SsHAM, HAM, SVC, OC.SVM, SSum, SAve, SMax and AM (aspect model), with varying the size of protein classes from 30 to 400, for Data1 and Data2. Throughout the experiments, we fixed $K = 100$ for AM and $K = 100$, $I = 10$, $N = N_n + N_s$ and $\frac{N_n}{N} = 0.1$ for SsHAM (and HAM). All our experiments were run on a Linux workstation with Intel Xeon 2.4GHz processor and 4G bytes of main memory. The actual computation time of HAM was within two minutes for all cases. Thus the computational burden of HAM is extremely small.

Table 1. Average prediction accuracies and t-values for Data1.

#classes	SsHAM	HAM	SVC	OC.SVM	SSum	SAve	SMax	AM
30	**79.5**	79.0	64.2	57.9	57.9	56.9	56.7	
		(1.76)	**(24.7)**	**(29.5)**	**(24.6)**	**(26.8)**	**(28.8)**	
100	**79.1**	78.3	67.0	59.3	59.8	58.4	58.1	
		(6.72)	**(23.1)**	**(26.8)**	**(22.0)**	**(29.8)**	**(24.9)**	
200	**78.5**	77.6	68.1	58.8	58.9	58.9	58.7	47.6
		(5.39)	**(19.0)**	**(21.6)**	**(18.8)**	**(21.5)**	**(18.8)**	**(41.0)**
300	**77.8**	76.6	68.7	61.3	63.2	64.6	64.2	
		(10.8)	**(20.7)**	**(22.9)**	**(16.8)**	**(22.0)**	**(14.0)**	
400	**77.5**	76.2	68.6	61.6	62.7	64.8	64.3	
		(8.20)	**(19.1)**	**(25.7)**	**(18.2)**	**(25.9)**	**(15.3)**	

Table 2. Average prediction accuracies and t-values for Data2.

#classes	SsHAM	HAM	SVC	OC.SVM	SSum	SAve	SMax	AM
30	67.0	**67.2**	59.7	52.6	53.2	53.0	52.7	
		(1.71)	**(23.7)**	**(47.0)**	**(39.2)**	**(68.5)**	**(85.5)**	
100	67.4	**67.5**	61.3	53.2	53.5	53.8	52.5	
		(0.14)	**(15.4)**	**(31.5)**	**(41.4)**	**(38.9)**	**(54.9)**	
200	**68.0**	67.6	62.6	52.8	53.3	55.2	53.1	49.4
		(1.64)	**(14.8)**	**(32.2)**	**(28.8)**	**(29.2)**	**(36.3)**	**(63.0)**
300	**66.6**	66.4	62.9	55.4	56.3	60.0	58.4	
		(0.84)	**(11.2)**	**(53.3)**	**(43.3)**	**(20.0)**	**(34.5)**	
400	**66.1**	65.8	62.6	55.3	55.4	59.1	58.4	
		(0.45)	**(11.6)**	**(34.3)**	**(45.4)**	**(19.3)**	**(24.2)**	

3.3 Average Prediction Accuracies

We first evaluated our method with other methods by average prediction accuracies obtained for Data1 and Data2. All of the methods except for SVC are trained by positive examples only, and predictions of all of methods are done for each of both positive and negative examples as a likelihood or a score. The prediction accuracy is obtained by first sorting examples according to their predicted likelihoods (or scores) and then computing an accuracy at a threshold which maximally discriminates positive from negative examples. The average prediction accuracy is obtained by averaging the accuracy over five runs in cross-validation.

We further used the 't' values of the (pairwise) mean difference significance test for statistically comparing the accuracy of SsHAM with that of another method. The t values are calculated using the following formula: $t = \frac{|ave(D)|}{\sqrt{\frac{var(D)}{n}}}$, where we let D denote the difference between the accuracies of the two methods for each data set in our five trials, $ave(W)$ the average of W, $var(W)$ the variance of W, and n the number of data sets (five in our case). For $n = 5$, if t is greater than 4.604 then it is more than 99% statistically significant that SsHAM achieves a higher accuracy than the other.

Tables 1 and 2 show the average prediction accuracies and t-values (in parentheses) for SsHAM, HAM, SVC, OC.SVM, SSum, SAve, SMax and AM. As shown in the tables[1], SsHAM greatly outperformed the other methods, being statistically significant for all cases in both Data1 and Data2. For Data1, the maximum average prediction accuracy of SsHAM achieved roughly 80%, whereas that of other methods except HAM was less than 70%. We emphasize that the performance of SsHAM should be compared with other unsupervised learning approaches, because the accuracy of SVC may be improved as increasing the size of negative examples or changing the types of negative examples. Surprisingly, for Data1, the maximum average prediction accuracy of other unsupervised learning methods except HAM was lower than 65%, and for each case, the performance difference between SsHAM and one of other unsupervised learning approaches except HAM and AM reached approximately 15 to 20%. Moreover, the performance advantage of SsHAM over AM, an existing model-based clustering method, was roughly 30%. These results indicate that our methodology is highly effective for learning from co-occurrence events, unsupervised data in nature. This performance advantage of SsHAM over other methods was further confirmed for Data2. In particular, for Data2, the prediction accuracies obtained by other unsupervised methods except HAM was less than 60% except for one case, but the highest prediction accuracy of SsHAM reached 68%. This result also shows that our method is especially effective for learning from co-occurrence events.

We next check the predictive performance of HAM and the sampling strategy separately. For the two data sets, HAM significantly outperformed other methods compared, and SsHAM improved the performance of HAM in eight out of all ten cases, being statistically significant in four out of the eight cases. More exactly, SsHAM significantly outperformed HAM in four out of five cases for Data1, but for Data2, SsHAM outperformed HAM in only three cases and the performance difference between them was statistically insignificant in all cases. From this result, we can say that the advantage of the selective sampling strategy varies, depending on a given data set. Another possible explanation for this result is that the training data set used this time (particularly Data2) may not contain so many noises (outliers), because it has been shown that the selective sampling strategy is especially effective for noisy data.

3.4 ROC Curves

We then examined ROC (Receiver Operating Characteristic) curves obtained by each of the methods. An ROC curve is drawn by plotting 'sensitivity' against '1 - specificity'. The sensitivity is defined as the proportion of correctly predicted

[1] Approximate class sizes are shown. Exact sizes are 30, 88, 202, 297 and 398.

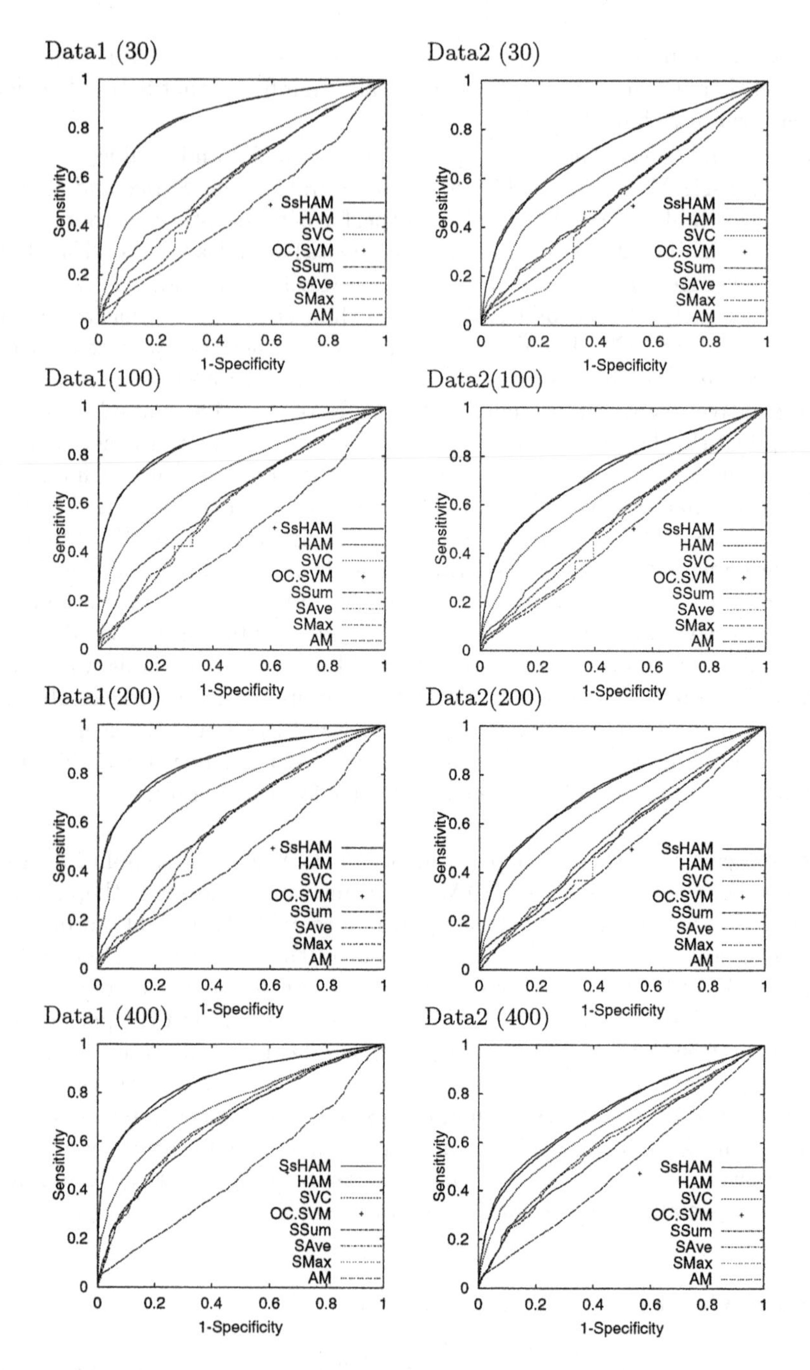

Fig. 3. ROC curves.

examples out of positive examples, and the specificity is defined as the proportion of correctly predicted examples out of negative examples.

Figure 3 shows the ROC curves obtained by all of methods tested for Data1 and Data2, for 30, 100, 200 and 400 protein classes (The sizes are shown in the parentheses.). The experimental findings obtained by the average prediction accuracies are confirmed from these curves.

4 Concluding Remarks

We have presented a new methodology which combines a selective sampling strategy and a hierarchical latent variable model, for the problem of modeling and predicting co-occurrence events. We have empirically shown that the presented method is highly effective for learning from co-occurrence events, using real data sets of protein-protein interactions, i.e. typical and real-world co-occurrence data sets. We believe that the presented method is successfully applicable to other numerous co-occurrence events in a variety of real-world applications as well as the protein-protein interactions used in our experiments.

References

1. Hofmann, T.: Unsupervised learning by probabilistic latent semantic analysis. Machine Learning **42** (2001) 177–196
2. Mamitsuka, H.: Hierarchical latent knowledge analysis for co-occurrence data. In: Proceedings of the Twentieth International Conference on Machine Learning, Morgan Kaufmann (2003)
3. Mamitsuka, H.: Efficient unsupervised mining from noisy data sets. In: Proceedings of the Third SIAM International Conference on Data Mining, SIAM (2003) 239–243
4. Dempster, A.P., Laird, N.M., Rubin, D.B.: Maximum likelihood from incomplete data via the EM algorithm. Journal of the Royal Statistical Society: Series B **39** (1977) 1–38
5. Bishop, C.M., Tipping, M.E.: A hierarchical latent variable model for data visualization. IEEE Transactions on Pattern Analysis and Machine Intelligence **20** (1998) 281–293
6. Schölkopf, B., et al.: Estimating the support of a high-dimensional distribution. Neural Computation **13** (2001) 1443–1471
7. Joachims, T.: Making large-scale SVM learning practical. In Schölkopf, B., Burges, C., Smola, A., eds.: Advances in Kernel Methods – Support Vector Learning. MIT Press (1999)
8. Mewes, H.W., et al.: MIPS: A database for genomes and protein sequences. Nucleic Acids Research **30** (2002) 31–34
9. Li, H., Abe, N.: Generalizing case frames using a thesaurus and the MDL principle. Computational Linguistics **24** (1998) 217–244
10. Rissanen, J.: Modeling by shortest data description. Automatica **14** (1978) 465–471

Obtaining Quality Microarray Data via Image Reconstruction

Paul O'Neill, George D. Magoulas, and Xiaohui Liu

Department of Information Systems and Computing,
Brunel University, Uxbridge, Middlesex, UB8 3PH, UK
Paul.O'Neill@brunel.ac.uk

Abstract. This paper introduces a novel method for processing spotted microarray images, inspired from image reconstruction. Instead of the usual approach that focuses on the signal when removing the noise, the new method focuses on the noise itself, performing a type of interpolation. By recreating the image of the microarray slide, as it would have been with all the genes removed, the gene ratios can be calculated with more precision and less influence from outliers and other artefacts that would normally make the analysis of this data more difficult. The new technique is also beneficial, as it does not rely on the accurate fitting of a region to each gene, with its only requirement being an approximate coordinate. In experiments conducted the new method was tested against one of the mainstream methods of processing spotted microarray images. Our method is shown to produce much less variation in gene measurements. This evidence is supported by clustering results that show a marked improvement in accuracy.

1 Introduction

Microarray technology is progressing at an amazing rate, the number of genes that can be processed is ever increasing, along with a multitude of techniques that can be applied to analyse the various stages of data [15][16]. Problems arising with real world data analysis are well known and can include many things such as noise from a variety of sources, missing data, inconsistency and of course the presence of outliers. These are all problems that microarrays suffer from in abundance. No matter how carefully a microarray experiment is conducted, it is always certain that there will be sources of external errors within the slide. These errors can come in many forms such as technical errors like the random variation in the scanning laser intensity, inaccurate measurement of gene expressions and a wide range of artefacts such as hair, dust or even fingerprints on the slide. Another main source of errors are more biological in nature with possible contamination of the cDNA solution and inconsistent hybridisation. If this did not cause enough problems the technology is still very expensive and so this often means that measurements cannot be repeated or they are at most duplicated. This is not an ideal situation for data analysis, however, it does emphasise, that

M.R. Berthold et al. (Eds.): IDA 2003, LNCS 2810, pp. 364–375, 2003.

as much work as possible should go into the pre-processing of the data to obtain the best possible results for further analysis.

Currently one of the mainstream uses of gene expression data is that of clustering [7]. Generally this utilises methods such as Hierarchical clustering to group or order the genes based on their similarity or dissimilarity, an example of which can be seen in work by Gasch et al. [8]. Based on this, biologist can make suppositions about the functions of previously unknown genes or possibly study the interactions between genes such as the discovery of gene pathways. However all these forms of analysis are very susceptible to noise, highlighting again the need to make sure that the pre-processing of the data is as reliable as possible. Although many people have focused on the post processing of the microarray data [4][7], comparatively there has still been very little work done looking into improved methods of analysing the original image data [17],[13]. In this paper we will propose a novel method for analysing the image data that inherently deals with slide variation and a large degree of slide artefacts and then contrast this to GenePix Pro [3], a software package that is often used for analysis. We contrast them through a series of experiments designed to look at variation within genes, between repeated genes and between clustering results. These tests will be conducted on a large dataset that contains a high number of repeated genes. As a guide to this paper, the next section takes a closer look at the methods used in the pre-processing microarrays and possible sources of error. Following this, we explain both the new method and existing techniques for processing cDNA scanned images. After this we describe the dataset that was used during this paper and outline how we plan to test and compare the different methods. Finally we present the results from each of the experiments and finish with a summary of our findings and a brief look at how this work can be extended.

2 Techniques for Analysing Microarray Data

Although there are many varied techniques for analysing microarray data, generally they have several similarities. They all require knowledge of a central point for the spot they are going to analyse and need to know how the slide was laid out originally. At this point they then all define some sort of boundary around the gene, in GenePix [3] this is a circle of varying diameter, while other techniques use the partitioning of the pixels values by use of a histogram [4],[9] and growing a region from the centre [1]. For a comparison of these techniques and more details about their implementation please refer to Yang et al. [17], who have compared them all in detail.

For this paper we will focus on comparing our method to that of GenePix, a software package commonly used by biologists. As shown in Fig. 1, GenePix first fits a circle to the gene itself using a simple threshold to calculate the diameter and then the median value of these pixels is used as the intensity of spot signal. The technique samples the surrounding noise by placing four rectangular regions in the diagonal space between this and adjacent spots. Again the median value of all the pixels within these regions is taken as the noise. The final stage is to subtract the noise from the signal measurement and once this is done for the red and green channels of the image, the ratio of the spot can be calculated. This process makes the assumption there is little variation both within the gene spot and in the surrounding background noise and that an exact region can be fitted to the gene. Unfortunately, this is not always the case as various types of artefacts are commonly found. In the next section we describe a technique that deals with this problem.

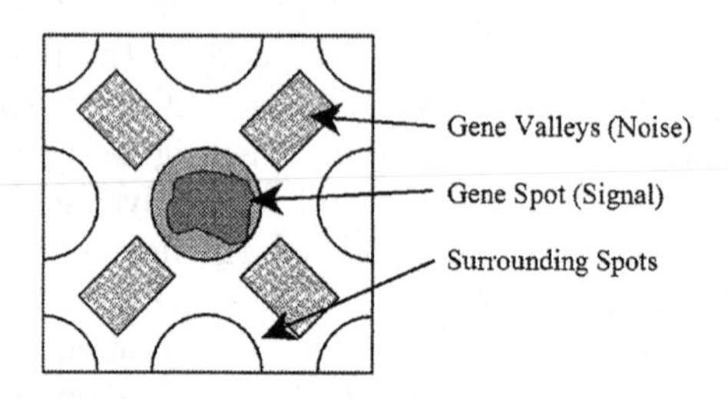

Fig. 1. GenePix sampling method.

3 A New Analysis Method Based on Image Reconstruction

In our attempt to develop a resilient technique that will be able to deal with the types of artefacts that are commonly found both in and around the gene spots, we decided to look at ways of recreating the noise that is found behind the spot itself. Rather than relying on a median sample of the surrounding pixels, we will recreate by a process of interpolation what we believe the noise behind the spot to have been. The Image Reconstruction Technique (IRT) is unlike other methods that focus on the signal of the gene spots. Instead we focus on the noise from start to finish. The result of this process is an image of the noise that can be subtracted from the original to leave us with the remaining signal. The genes can then be calculated in a normal fashion by measuring the intensity of the remaining pixels for each gene. This method is advantageous as it inherently deals with slide variation and a large degree of slide artefacts. Due to its nature it does not necessitate the exact fitting of a boundary around the spot, which is very important if this technique is to be automated.

The algorithm we propose requires that we can successfully reconstruct the noise in the part of the image where the gene resides. The technique used is a simplified version of the method proposed by A. Efros et al. [5]; a well-established technique for image tiling and hole filling. For example if there is a hair on an image that we wish to remove, these techniques can take a mask of the artefact and then recreate the image, as it would have been. Applying this to microarray images, we mask out all the genes so that we can recreate the noise. The original version of the algorithm proposed in [5] proved to be very time consuming. For an individual gene this would not be a problem, but in order to scale this technique up to process thousands of genes on multiple slides the algorithm needs to be as fast as possible.

```
Function ReconstructSpot(Image,GeneInfo,SampleSize,MatchSize)
  CreateEmptyList SpotPixels
  CreateEmptyList SamplePixels
  For each pixel within the square defined by SampleSize
      If InSpot(GeneInfo,pixel.x,pixel.y) Then SpotPixels.Add(pixel)
      If InSample(GeneInfo, pixel.x,pixel.y) Then SamplePixels.Add(pixel)
  End For Each
  While SpotPixels Not empty Do
    PixelToFill=GetPixelToFill(SpotPixels, MatchSize)
    BestMatch=GetBestMatch(PixelToFill,SamplePixels)
    PixelToFill=BestMatch
  End While
End
```

Fig. 2. Pseudo function used to fill in gene spots.

To this end we tested several variations inspired by [5] and came up with the following process (see Fig. 2) that should be applied to every gene in each channel of the slide. The main difference between this and the method proposed in [5] is that we do not use the Markov random field model in selecting pixels. Instead, for any given pixel that we need to fill, we compute and score candidate pixels based on a simple distance score of the surrounding pixels within a given window.

We now use the example shown in Fig. 3 to explain the process outlined in Fig. 2 further. The function first of all creates two lists, one with a set of surrounding background pixels (*Background Noise*) within a square (*Sample Window*) centred on the gene spot with dimensions *SampleSize* and the other a set of pixels in the spot that need to be reconstructed. Then the list of pixels to be filled is sorted in order, based on how many surrounding pixels exist around each pixel within a square of dimensions *MatchSize* (e.g. *Sample Windows* A or B). The pixel with the most neighbours is selected (e.g. pixel p_A in window A). This is then scored against all of the pixels within the background area. Scoring is based on how closely the surrounding pixels present within a square (*Match Window*), defined by *MatchSize*, match one another. For example when matching pixel p_A, the window of pixels around it would be compared to the corresponding background pixels (for example, those in windows C, D, E and so on).

Various measures such as the Euclidean or the Manhattan distance can be used for the comparisons. Following some preliminary testing we have found that the average Manhattan distance between all noise pixel intensities $p_X(i,j)$ gave

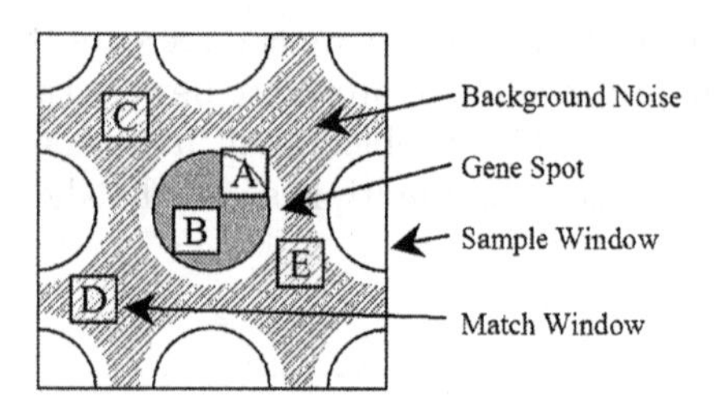

Fig. 3. IRT sampling method.

the best performance (where i,j are the relative coordinates of the pixels within each window centred around pixel p). For example, this was calculated between all noise pixel intensities $p_A(i, j)$ in a window A of width *MatchSize* and the corresponding pixel intensities of a background window of similar width, such as the pixels $p_C(i, j)$ of window C. This calculation was repeated for all background pixels and then the intensity value of the pixel with minimum average distance was used to reconstruct the intensity of $p_A(i, j)$. This procedure is then repeated for all pixels within the gene spot.

4 Experimental Study

With the theory behind both the new and existing technique laid out, it is then necessary to compare them empirically so that we can draw conclusions about their performance. To this end this section presents a series of three tests that are all carried out on a large real world dataset. The first two tests look at the performance of the algorithms based on absolute error and the standard deviation of biological repeats. The third test looks at scoring clustering results in an attempt to assess the impact of these techniques on later processing of the data.

4.1 Dataset

The slides used in this paper came from two experiments that were conducted using the human gen1 clone set slides [10]. The experiments were designed to contrast the effects of two cancer inhibiting drugs (PolyIC and LPS) on two different cell lines, one normal (Control) and one cancerous (HeLa) over a series of time points. Altogether this gives us a total of 48 microarray slides for which we have the uncompressed image data and the initial GenePix results. Each slide in this dataset consists of 24 gene blocks that have 32 columns and 12 rows. The first row of each odd numbered block is the Lucidea scorecard [14] consisting of a set of 32 predefined genes that can be used to test the accuracy of the experimental process. The remaining 11 rows of each block are the human genes. The even numbered blocks are biological repeats of each of the proceeding odd blocks. This is important as it means that on each slide there are a large number of repeats, 24 repeats of each of the 32 scorecard genes and duplicates of each of the remaining 4224 (32×11×12) human genes.

The scanned images of the spotted microarray slides are stored in the standard tiff format. Some scanning software will store the red and the green channels separately while others such as the GenePix software store the two channels and preview images in a single composite file. In either case there are essentially two tiff images that we are interested in. These both have exactly the same properties and represent the two channels of the scanned image, Cy5 (red) and Cy3 (green). Each image is stored as a uncompressed 16bit grey level bitmap with each pixel taking a value between 0 and 65535. It is generally accepted that any pixels at the extremes of this scale (i.e. 0 or 65535) should be ignored as these are beyond the range of the scanner to accurately measure.

Along with the scanned images another useful resource is the GenePix result files that store the initial analysis from the software provided with the scanner. These files store a list of all of the genes along with detailed information about each spot. This information is in the form of spot co-ordinates, spot diameter, its ratio and various statistical measures such as the standard deviation of the pixels within the spot. Although GenePix can work in an automated mode when locating all the spots, it also allows for manual intervention. In the case of the GenePix results we compare our algorithm to, all spots on every slide have been visually inspected and manually corrected if necessary. This gives us a high degree of confidence that the GenePix results are a good representation of the programs best capabilities.

4.2 Description of Experiments

The method that has been proposed relies on the fact that it can reconstruct the area behind a spot. Therefore the first line of testing was to see if this could be achieved. There is no way that you can truly know what the background would have been behind the gene spot and therefore it is impossible to verify the technique on the gene spot locations. However, if we instead look at blank areas of the slide then this is no longer the case and we can easily test how well the reconstruction methods work. This can be achieved by locating a blank portion of the slide, usually between the blocks of genes, and then, by cutting a hole in the slide and recreating the background, we can then verify the recreated sample against what we know to have been there previously. From this we can calculate the absolute error of the reconstruction and then compare this to the other processing techniques such as GenePix.

As we stated before, the dataset we are working on contains a high number of repeats for each slide. First of all there are 24 repeats of 32 control genes on each slide, followed by duplicates of 4224 human genes. For this paper we will only be using the control data as we have the most repeats for these. One simple way to use these genes is to simply calculate the standard deviation of each of the 32 genes for its 24 repeats. Theoretically the same gene in different areas of the slide should give us the same ratio, so therefore the lower the standard deviation between repeated genes the better we can say a technique is.

The final method of testing we propose is the validation of clustering results. Normally this would not be achievable as often there is not enough substantial domain knowledge to be able to calculate a score for clusters. However, in this dataset we are lucky to have a large number of repeats for a subset of the genes. These genes are the 32 control genes that make up the Lucidea scorecard found in the first row of every block. Using the 48 experimental slides we now have enough data to enable us to use standard clustering algorithms to blindly group these genes together. Then once we have the resulting clusters, we would expect to find all twenty-four repeats of the same gene clustered together. As each of the 32 genes in the scorecard are meant to be unique we can also expect to find 32 unique clusters. As we now know how to expect the genes to all clusters, it is possible to score the results. A useful metric for providing this score is something that can often be found in medical statistics, weighted kappa (WK). WK is used to rate the agreement between the classification decisions made by two observers [2]. In this case observer 1 is the cluster output from the algorithm being tested and observer 2 is the known arrangement in which the genes should be clustered. The formulae for calculating this metric are as follows:

$$K_w = \frac{p_{o(w)} - p_{e(w)}}{1 - p_{e(w)}} \tag{1}$$

$$p_{e(w)} = \frac{1}{N_k^2} \sum_{i=1}^{k} \sum_{i=1}^{k} w_{ij} Row(i) Col(j) \tag{2}$$

$$p_{o(w)} = \frac{1}{N_k} \sum_{i=1}^{N} \sum_{i=1}^{N} w_{ij} Count_{ij} \tag{3}$$

$$w_{ij} = 1 - \frac{|i-j|}{N-1} \text{ where } 1 \leq i, j \leq k \tag{4}$$

$$N_k = \sum_{i=1}^{k} \sum_{i=1}^{k} Count_{ij} = \sum_{i=1}^{k} Row(i) \sum_{i=1}^{k} Col(i). \tag{5}$$

where K_w is the WK agreement; $p_{o(w)}$ and $p_{e(w)}$ represent the observed weighted proportional agreement and the expected weighted proportional agreement, respectively; $Row(i)$ and $Col(i)$ are row and column totals respectively; the sum of all of the cells is denoted by N_k and $Count_{ij}$ is the count for a particular combination of classifications; and k is the number of classes (i.e. we have two classes: the algorithms either agree or disagree; thus $k = 2$). This metric will score the output clusters against the expected clusters and give us a value between -1 and 1, the higher the value the better the arrangement with 1 being total agreement. Based on this metric we can now score how well the results produced via the image reconstruction and GenePix compares to the original arrangement of genes.

5 Results

The first experiment aimed to test out the image reconstruction in blank areas of the slide, for example the area that is found between the 24 blocks of genes. For this experiment 3 slides were picked at random and 1056 spots of random size were positioned within the blank regions for both the Cy5 and Cy3 channels. After reconstructing each spot, the difference between the recreated background and the true background was compared to the difference between the GenePix noise estimation and the true pixel values. If the image reconstruction was closer to the true background then this was counted as a positive improvement.

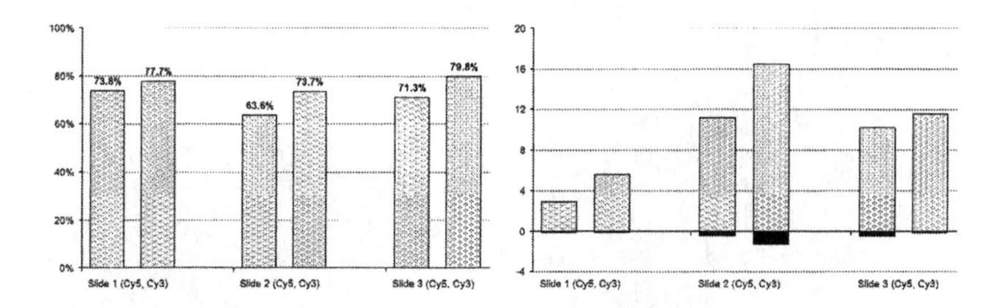

Fig. 4. Left: The percentage of pixels out of 1056 test spots that showed a positive improvement over GenePix. **Right:** The gained improvement in pixel values (positive axis) against the loss in erroneous pixels (negative axis).

The left-hand graph in Fig. 4 shows what percentage of spots showed improved results. On average 73% of spots were improved, although it would be more interesting if we can see how those spots, where GenePix estimated the background better, compared to those which benefited from using the image reconstruction. On the right in Fig. 4 is a graph that shows just this. The positive y-axis of the graph shows how much each pixel within all the spots was improved when using the image reconstruction technique. This is contrasted to the negative portion of the y-axis that shows the improvement that was gained when using GenePix for those spots that showed negative results. It is clear to see from these two graphs that image reconstruction is having a significant effect on how the noise is estimated for any given spot value. Using the image reconstruction techniques not only improves nearly three quarters of the spots processed, but also this improvement is in orders of magnitude higher than the negative effects which are seen in just over one quarter of the spots processed.

After testing the image reconstruction technique it was then necessary to discover what effect applying this method would have on the gene ratios. This experiment analyses the standard deviation between the 24 repeats of the scorecard genes on each of the 48 slides. This is based on the assumption that if a gene sample is repeated on a slide then it is reasonable to expect its ratio to be the same. Therefore if a given technique is producing a lower standard deviation between repeated measures of the same gene it must be analysing that gene more successfully. Fig. 5 shows the standard deviation of each of the 32 genes for the 48 experiments. It is clear from this that the image processing technique we present on the right has significantly reduced the standard deviation. It should be noted that it also allowed us to measure the ratios of an extra 2908 genes. To summarise these results, the mean standard deviation is reduced from 0.91 to 0.53 and the maximum standard deviation is reduced from 6.23 to 2.43. This marks a significant improvement in the reliability of repeated genes.

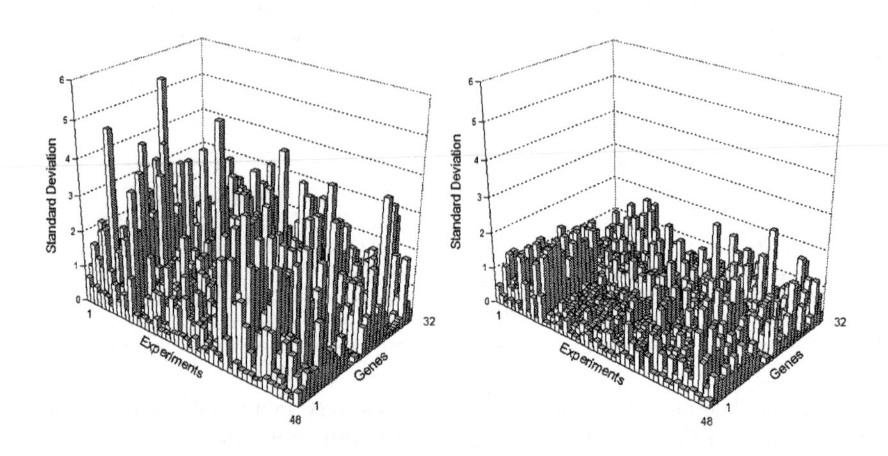

Fig. 5. The standard deviation of 32 genes repeated 24 times across 48 experiments. On the left are the results for GenePix contrasted to the image reconstruction techniques on the right.

The previous two experiments have indicated that the image reconstruction is an improvement, how can we be sure that the biological features of the genes have remained intact? In this final experiment we attempt to assess this by looking at clustering results generated from analysis produced by both techniques. For this we used the average linkage hierarchical clustering based on Pearson correlation and 'cutree' which are both provided in the freeware statistical package 'R'. The results are scored using weighted kappa [2] that as explained before scores agreement between the results and the known clusters with 0 being a random arrangement and 1 the perfect result.

In Fig. 6 you can see the weighted kappa results for datasets with various degrees of missing data, along with the same results tabulated in table 1 and the number of usable genes in table 2. Missing data refers to the number of observations that were missing for any given sample, for example if we had a measurement for a gene in all 48 slides it is complete, but if we could only measure it in 36 of those slides then there was 25% missing data. We filtered genes based on this, so that we tested genes with less than 40% missing observations, then those with less than 20% and finally only those that were complete. In this experiment you can see the results varied as would be expected, whereby

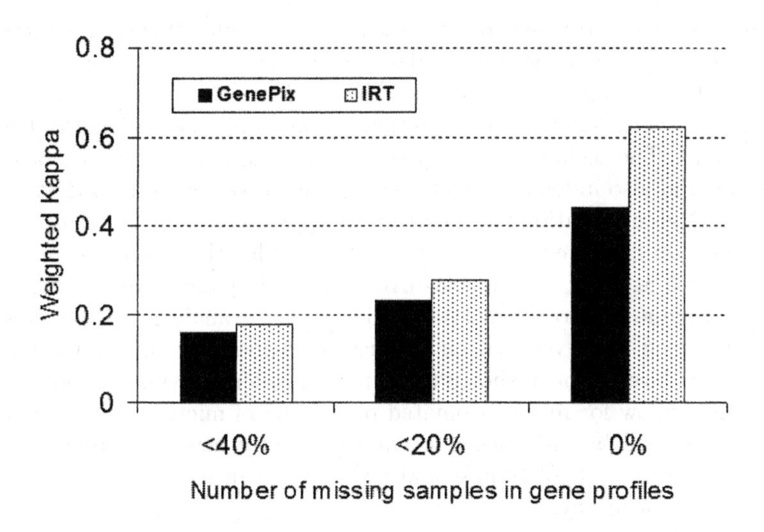

Fig. 6. Weighted Kappa of resultant clusters using hierarchical clustering.

the more missing data we had, the lower the weighted kappa score, indicating increased error. In all cases the gene ratios produced using the image reconstruction technique (IRT) have shown better results than that of GenePix. It is clear to see that when there is no missing data there is a remarkable improvement when using the image reconstruction, however with missing data the results are generally very poor with WK scores of less than 0.3 and only a marginal improvement shown by our technique. This seems to suggest that missing data strongly affects comparisons between gene profiles and we believe the effects of this are something that should be studied in greater detail.

Table 1. WK of repeated genes

Missing Samples	GenePix	IRT
<40%	0.157836	0.177876
<20%	0.231482	0.277247
=0%	0.437749	0.621469

Table 2. Number of Genes Used

Missing Samples	GenePix	IRT
<40%	624	644
<20%	455	457
=0%	299	210

6 Conclusion

This paper proposed a new method of approaching microarray image analysis. Instead of focusing on the signal as with most techniques, we looked towards being able to reconstruct the noise accurately in a way that we could demonstrate to be more robust. To this end we presented the results of several experiments run on real microarray experimental data to demonstrate the performance of this technique at various processing stages. The results obtained have shown a great improvement over a software package that is commonly used, GenePix Pro from Axon systems. The technique greatly reduces the variance between repeats on a given slide, and also in looking at clustering results we have shown that not only is the biological information left intact but it appears to be more accurate. In the future we would like to extend our testing to include other mainstream methods that exist such as Spot [17] and ScanAlyze [6] and QuantArray [9]. We plan to test the technique on another datasets and include the analysis of other clustering techniques along with verification against biological domain knowledge. This technique has already been shown to be more resilient to spot boundary fitting and this may allow for more automated processing of microarrays in the future. To this end it will be interesting to analyse exactly how it compares to other techniques in terms of robustness and whether it can it be used automatically without loss of accuracy.

Acknowledgements. The authors would like to thank Paul Kellam and Antonia Kwan from the Department of Immunology and Molecular Pathology, University College London for providing the datasets used in this study and the EPSRC (GR/P00307/01) and BBSRC (100/BIO14300;100/EGM17735) for funding this work.

References

1. Adams, R., Bischof, L. Seeded region growing. IEEE Transactions on Pattern Analysis and Machine Intelligence, (1994), Vol. 16 641–647
2. Altman, D. G. Practical Statistics for Medical Research. Chapman and Hall, (1997)
3. Axon Instruments Inc.: GenePix Pro Array analysis software. http://www.axon.com
4. Chen, Y., Dougherty, E.R., and Bittner, M.L. Ratio-based decisions and the quantitative analysis of cDNA microarray images. J. Biomed. Optics, (1997), Vol. 2 364–374
5. Efros, A. A., and Leung, T. K. Texture Synthesis by Non-parametric Sampling. IEEE International Conference on Computer Vision, (1999)
6. Eisen, M. B. ScanAlyze, (1999) http://rana.lbl.gov/index.htm?software
7. Eisen, M. B., Spellman, P.T., Brown, P.O., and Botstein, D. Cluster analysis and display of genome-wide expression patterns. Proc. Natl. Acad. Sci., (1998) 14863–14868
8. Gasch, A.P., Spellman, P.T., Kao, C.M., Carmel-Harel, O., Eisen, M.B., Storz, G., Botstein, D. and Brown, P.O. Genomic Expression Programs in the Response of Yeast Cells to Environmental Changes. Molecular Biology of the Cell., (2000), Vol. 11 4241–57, 2000.
9. GSI Lumonics. QuantArray analysis software. http://www.gsilumonics.com
10. "Human Gen1 Clone Set Array", HGMP. Available, http:// www.hgmp.mrc.ac.uk/Research/Microarray/HGMP-RC_Microarrays/desc ription_of_arrays.jsp

11. Kooperberg, C.,. Fazzio, T. G, Delrow, J. J., and Tsukiyama, T. Improved Background Correction for Spotted DNA Microarrays. Journal of Computational Biology, (2002), Vol. 9 55–66
12. Newton, M.A., Kendziorski, C.M., Richmond, C.S., Blattner, F.R., and Tsui, K.W. On differential variability of expression ratios:Improving statistical inference about gene expression changes from microarray data. J. Comp. Biol., (2000), Vol. 8 37–52
13. Quackenbush J. Computational analysis of microarray data. Nat Rev Genet, (2001), Vol. 2, No. 6 418–27
14. Samartzidou, H., Turner, L., Houts, T., Frome, M., Worley, J., and Albertsen, H. Lucidea Microarray ScoreCard: An integrated analysis tool for microarray experiments. Life Science News, (2001)
15. The Chipping Forcast I, Nature Genetics Supplement, (1999) Available, http://www.nature.com/cgi-taf/DynaPage.taf?file=/ng/journal/v21/n1s/index.html
16. The Chipping Forcast II, Nature Genetics Supplement, (2002) Available, http://www.nature.com/cgi-taf/dynapage.taf?file=/ng/journal/v32/n4s/index.html
17. Yang, Y. H., Buckley, M. J., Dudoit, S., Speed, T. P. Comparison of methods for image analysis on cDNA microarray data. Journal of Computational and Graphical Statistics, (2002), Vol 11. 108–136

Large Scale Mining of Molecular Fragments with Wildcards

Heiko Hofer[1], Christian Borgelt[2], and Michael R. Berthold[1]

[1] Tripos Inc., Data Analysis Research Lab,
601 Gateway Blvd., Suite 720, South San Francisco, CA 94080, USA
heiko.hofer@e-technik.tu-chemnitz.de
berthold@tripos.com

[2] School of Computer Science, Otto-von-Guericke-University of Magdeburg,
Universitätsplatz 2, 39106 Magdeburg, Germany
borgelt@iws.cs.uni-magdeburg.de

Abstract. The main task of drug discovery is to find novel bioactive molecules, i.e., chemical compounds that, for example, protect human cells against a virus. One way to support solving this task is to analyze a database of known and tested molecules with the aim to build a classifier that predicts whether a novel molecule will be active or inactive, so that future chemical tests can be focused on the most promising candidates. In [1] an algorithm for constructing such a classifier was proposed that uses molecular fragments to discriminate between active and inactive molecules. In this paper we present two extensions of this approach: A special treatment of rings and a method that finds fragments with wildcards based on chemical expert knowledge.

1 Introduction

The computer-aided analysis of molecular databases plays an increasingly important role in drug discovery. One of its major goals is to build classifiers that predict for a novel molecule whether it will be active or inactive, for example, w.r.t. the protection of human cells against a virus. Such a classifier can then be used to focus the expensive and time-consuming real chemical tests on the most promising candidates.

The classification model we are considering here is a set of *discriminative fragments*. Such fragments are parts of molecules that are frequent in the set of active molecules and rare in the set of inactive ones and thus discriminate between the two classes. The rationale underlying this approach is that discriminative fragments may be the key substructures that determine the activity of compounds and therefore can be used to predict the activity of molecules: If a novel molecule contains at least one discriminative fragment it is classified as active, otherwise it is classified as inactive.

Recently, several algorithms have been suggested for finding discriminative fragments efficiently. In [2] an approach was presented that finds linear fragments using a method that is based on the well-known Apriori algorithm for association rule mining [3]. However, the restriction to linear chains of atoms

M.R. Berthold et al. (Eds.): IDA 2003, LNCS 2810, pp. 376–385, 2003.

is limiting in many real-world applications, since substructures of interest often contain rings or branching points. The approach presented in [1] does not suffer from this restriction. It is based on a transfer of the Eclat algorithm for association rule mining [6] and thus avoids the expensive reembedding of fragments. Another approach that can find arbitrary fragments was proposed in [4]. The main differences between these approaches are the ways in which fragments of interest are generated, an issue that will become important when we study the algorithm of [1] in more detail below.

In this paper we present two extensions of the algorithm suggested in [1]. Due to a symmetry problem that makes it impossible to suppress all redundant search, this algorithm runs into problems if fragments contain a lot of rings. Our first extension solves this problem by finding rings in a preprocessing step and then treating them as units in the mining algorithm, thus making the search much more efficient. The second extension is inspired by the fact that certain atom types can sometimes be considered as equivalent from a chemical point of view. To handle such cases, we introduce wildcard atoms into the search.

As a test case for our extended algorithm we use a publicly available database of the National Cancer Institute, namely the NCI-HIV database (DTP AIDS Antiviral Screen) [5], which has also been used in [1] and [2]. This database is the result of an effort to discover new compounds capable of inhibiting the HI virus (HIV). The screen utilized a soluble formazan assay to measure protection of human CEM cells from HIV-1. Compounds able to provide at least 50% protection to CEM cells were retested. Compounds providing at least 50% protection on retest were listed as moderately active (*CM*). Compounds that reproducibly provided 100% protection were listed as confirmed active (*CA*). Compounds not meeting these criteria were listed as confirmed inactive (*CI*). The dataset consists of 41,316 compounds, of which we have used 37,171. For the remaining 4,145 no structural information was available at the time the experiments reported here were carried out. Each compound has a unique identification number (NSC number), by which we refer to it. Out of the 37,171 compounds we used, 325 belong to class *CA*, 896 to class *CM*, and the remaining 35,950 to class *CI*. We neglected the 896 moderately active compounds in our experiments. The size of this database stresses the need for a truly efficient and scalable algorithm. The method described in [1], for example, could only be applied to this data set by using small initial fragments as "seeds" for the underlying search algorithm.

2 Molecular Fragment Mining

As stated above, the goal of molecular fragment mining is to find discriminative fragments in a database of molecules, which are classified as either active or inactive. To achieve this goal, the algorithm presented in [1] represents molecules as attributed graphs and carries out a depth first search on a tree of fragments. Going down one level in this search tree means extending a fragment by adding a bond and maybe an atom to it. For each fragment a list of embeddings into the available molecules is maintained, from which the lists of embeddings of its extensions can easily be constructed. As a consequence, expensive reembeddings of fragments are not necessary. The support of a fragment, that is, the number of

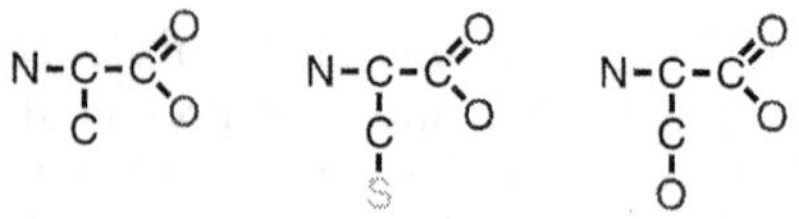

Fig. 1. The amino acids clycin, cystein and serin.

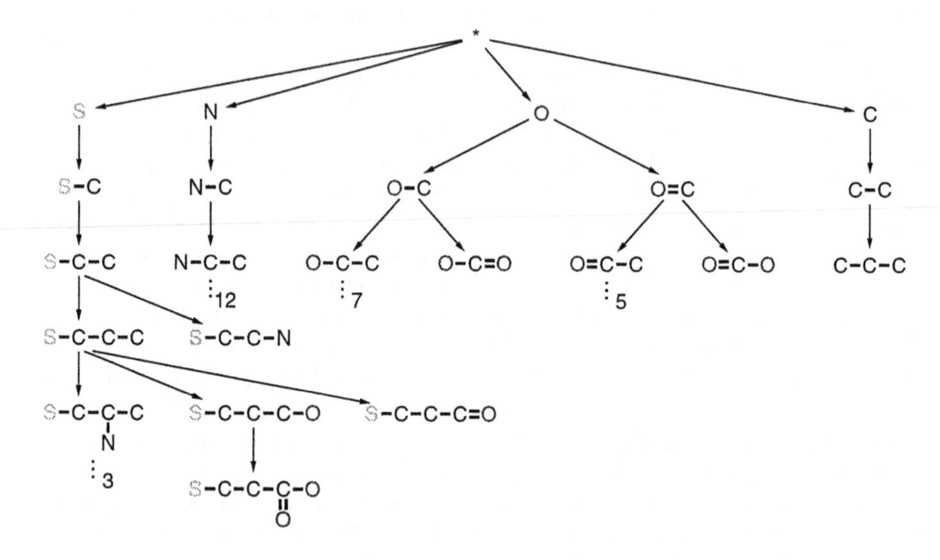

Fig. 2. The tree of fragments for the amino acids example.

molecules it is contained in, is then determined by simply counting the number of different molecules the embeddings refer to. If the support of a fragment is high in the set of active molecules and low in the set of inactive molecules it is reported as a discriminative fragment.

The important ingredients of the algorithm are different search tree pruning methods, which can be grouped into three categories: *size based pruning*, *support based pruning*, and *structural pruning*. Among these the latter is the most important and most complicated one. It is based on a set of rules that defines a local order of the extensions of a fragment, which is used to avoid redundant searches. (See [1] for details.)

Since the search tree we traverse in our modified algorithm differs slightly from the one used in [1], we illustrate it with a simple example. Figure 1 shows the amino acids clycin, cystein and serin with hydrogens and charges neglected. The upper part of the tree (or forest if the empty fragment at the root is removed) that is traversed by our algorithm for these molecules is shown in Figure 2. The first level contains individual atoms, the second connected pairs and so on. The dots indicate subtrees that are not depicted in order to keep the figure understandable. The numbers next to these dots state the number of fragments in these subtrees, giving an indication of the total size of the tree.

The order, in which the atoms on the first level of the tree are processed, is determined by their frequency of occurrence in the molecules. The least frequent atom type is considered first. Therefore the algorithm starts on the left by embedding a sulfur atom into the example molecules. That is, the molecules are searched for sulfur atoms and their locations are recorded. In our example there is only one sulfur atom in cystein, which leads to one embedding of this (one atom) fragment. This fragment is then extended (depth first search) by a single bond and a carbon atom (`-C`), which produces the fragment `S-C` on the next level. All other extensions of fragments that are generated by going down one level in the tree are created in an analogous way.

If a fragment allows for more than one extension (as is the case, for instance, for the fragments `O-C` and `S-C-C-C`), we sort them according to the local ordering rules mentioned above (see [1] for details). The main purpose of this local order is to prevent certain extensions to be generated, in order to avoid redundant searches. For instance, the fragment `S-C-C-C-O` is not extended by adding a single bond to a nitrogen atom at the second carbon atom, because this extension has already been considered in the subtree rooted at the left sibling of this fragment.

Furthermore, in the subtree rooted at the nitrogen atom, extensions by a bond to a sulfur atom are ruled out, since all fragments containing a sulfur atom have already been considered in the tree rooted at the sulfur atom. Similarly, neither sulfur nor nitrogen are considered in the tree rooted at the oxygen atom, and the rightmost tree contains fragments that consist of carbon atoms only.

Although the example molecules considered here do not contain any rings, it is clear that we have to handle them in the general case. While the original algorithm treated extensions to new atoms and extensions closing rings basically alike, only handling them through an ordering rule to avoid redundant search, our new algorithm processes them in two separate phases: Even though extensions closing rings are allowed in any step, once a ring has been closed in a fragment, the possible future extensions are restricted to closing rings as well. While this does not lead to performance improvements, it has certain advantages w.r.t. the order in which fragments are generated.

Up to now we only described how the search tree is organized, i.e., the manner in which the candidates for discriminative fragments are generated and the order in which they are considered. However, in an application this search tree is not traversed completely—that would be much too expensive for a real world database. Since a discriminative fragment must be frequent in the active molecules and extending a fragment can only reduce the support (because only fewer molecules can contain it), subtrees can be pruned as soon as the support falls below a user-defined threshold (support based pruning). Furthermore, the depth of the tree may be restricted, thus limiting the size of the fragments to be generated (size based pruning).

Discriminative fragments should also be rare in the inactive molecules, defined formally by a user-specified upper support threshold. However, this threshold cannot be used to prune the search tree: Even if a fragment does not satisfy this threshold, its extension may (again extending a fragment can only reduce the support), and thus it has to be generated. Therefore this threshold is only used

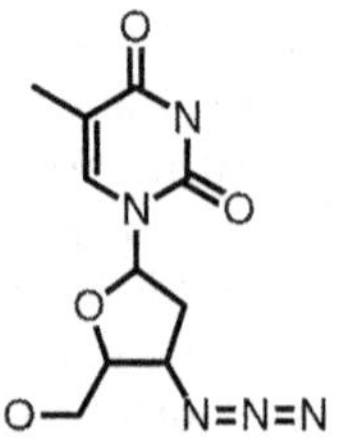

Fragment 1		Fragment 2	
freq. in CA:	15.08%	freq. in CA:	9.85%
freq. in CI:	0.02%	freq. in CI:	0.00%

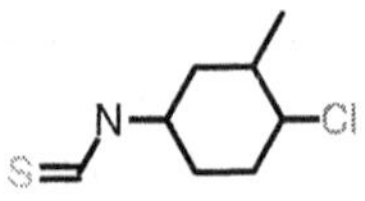

Fig. 3. Discriminative fragments.

to filter the fragments that are frequent in the active molecules. Only those fragments that satisfy this threshold (in addition to the minimum support threshold in the active molecules) are reported as discriminative fragments.

In [1] the original form of this algorithm was applied to the NCI-HIV database with considerable success. Several discriminative fragments could be found, some of which could be related to known classes of HIV inhibitors. Among the results are, for example, the two fragments shown in Figure 3. Fragment 1 is known as AZT, a well-known nucleoside reverse transcriptase inhibitor, which was the first drug approved for anti-HIV treatment. AZT is covered by 49 of the 325 (15.08%) compounds in *CA* and by only 8 of the 35,950 (0.02%) compounds in *CI*. Fragment 2 is remarkable, because it is not covered by any compound of the 35,950 inactive compounds, but is covered by 31 molecules in *CA* (9.85%), making it a very good classification model. (Fragment 2 was not reported in [1]).

3 Rings

An unpleasant problem of the search algorithm as we presented it up to now is that the local ordering rules for suppressing redundant searches are not complete. Two extensions by identical bonds leading to identical atoms cannot be ordered based on locally available information alone. A simple example demonstrating this problem and the reasons underlying it can be found in [1]. As a consequence, a certain amount of redundant search has to be accepted in this case, because otherwise incorrect support values would be computed and it could no longer be guaranteed that all discriminative fragments are found.

Unfortunately, the situation leading to these redundant searches occurs almost always in molecules with rings, whenever the fragment extension process enters a ring. Consequently, molecules with a lot of rings, like the steroid shown in Figure 4—despite all clever search tree pruning—still lead to a prohibitively high search complexity: The original algorithm considers more than 300,000 fragments for the steroid, a large part of which are redundant.

To cope with this problem, we add a preprocessing step to the algorithm, in which the rings of the molecules are found and marked. One way to do this is to

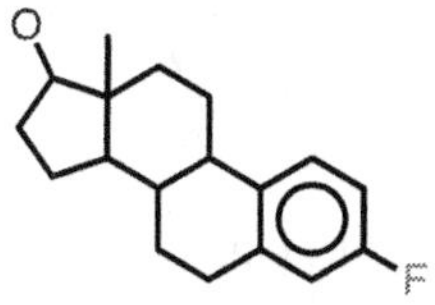

Fig. 4. Example of a steroid, which can lead to problems due to the large number of rings.

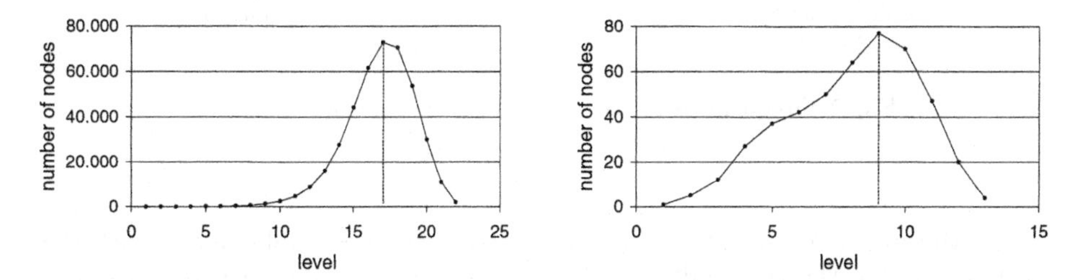

Fig. 5. Number of nodes per level for the algorithms with and without ring extensions.

use the normal algorithm to find a substructure covering a given ring structure as well as a given molecule, where the search neglects the atom and bond types. In our current implementation we confined ourselves to rings with five and six atoms, because these are most interesting to chemists, but it is obvious that the approach can easily be extended to rings of other sizes.

In the search process itself, whenever an extension adds a bond to a fragment that is part of a ring, not only this bond, but the whole ring is added (all bonds and atoms of the ring are added in one step). As a consequence, the depth of the search tree is reduced (the basic algorithm needs five or six levels to construct a ring bond by bond) and many of the redundant searches of the basic algorithm are avoided, which leads to enormous speed-ups. For example, in this way the number of fragments generated for the steroid shown in Figure 4 reduces to 93.

On the NCI-HIV database our implementation of the basic algorithm (without one step ring extensions) generates 407,364 fragments in *CA* that have a support higher than 15%, for which it needs 45 minutes.[1] Using ring extensions reduces the number of fragments to 456 which are generated in 13 seconds.

A more detailed illustration of the gains resulting from one step ring extensions is provided by the diagrams shown in Figure 5. They display the number of nodes per level in the search tree for the two versions of the algorithm. As can be seen in the left diagram, the search tree of the basic algorithm has a width of approximately 80,000 fragments and a depth of 22 levels. In contrast to this, the width of the search tree using one step ring extensions is approximately 80 fragments, almost a factor of 1000 less.

[1] Our experiments were run using a Java implementation on a 1GHz Xeon Dual-Processor machine with 1GB of main memory using jre1.4.0

Fig. 6. Aromatic and Kekulé representations of benzene.

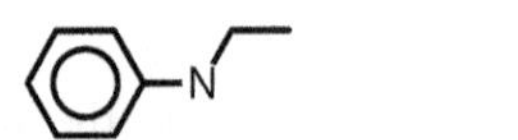

Fragment 3
basic algorithm
freq. in CA: 22.77%

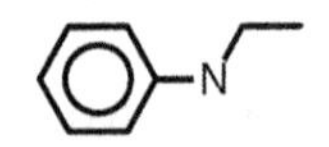

Fragment 4
with ring extensions
freq. in CA: 20.00%

Fig. 7. Frequencies in *CA* with and without ring extensions.

Furthermore, the shapes of the search trees are remarkable. While with the basic algorithm the tree width grows roughly exponentially with the depth of the tree, until a maximum is reached, with the ring extension algorithm the tree width grows only roughly linearly with the depth. As a side note, these diagrams also justify the use of a depth first search in our algorithm: The typical and extremely high width to depth ratio of the tree makes a breadth first search disadvantageous or, for the basic algorithm, infeasible.

Besides a considerable reduction of the running time, treating rings as units when extending fragments has other advantages as well. In the first place, it makes a special treatment of atoms and bonds in rings possible, which is especially important for aromatic rings. For example, benzene can be represented in different ways as shown in Figure 6. On the left, all six bonds are represented as aromatic bonds, while the other two representations—called *Kekulé structures*—use alternating single and double bonds. A chemist, however, considers all three representations as equivalent. With the basic algorithm this is not possible, because it would require treating single, double, and aromatic bonds alike, which obivously leads to undesired side effects. By treating rings as units, however, these special types of rings can be identified in the preprocessing step and then be transformed into a standard aromatic ring.

Secondly, one step ring extensions make the support values associated with a fragment more easily interpretable. To see this, consider the two fragments shown in Figure 7. The left was found with the basic algorithm, the right with the ring extension algorithm. At first glance it may be surprising that the two fragments look alike, but are associated with different support values (74 compounds, or 22.77%, for the basic algorithm and 65 compounds, or 20.00%, for the ring extension algorithm). However, the reason for this is very simple. While with the basic algorithm the `N-C-C` chain extending from the aromatic ring may be a real chain or may be a part of another ring, it must be a real chain with the ring extension algorithm. The reason is that with the latter it is not possible to step into a ring bond by bond, because rings are always added in one step. Therefore the ring extension algorithm does not find this fragment, for instance, in the two compounds shown in Figure 8. Although it may appear as a disadvantage at

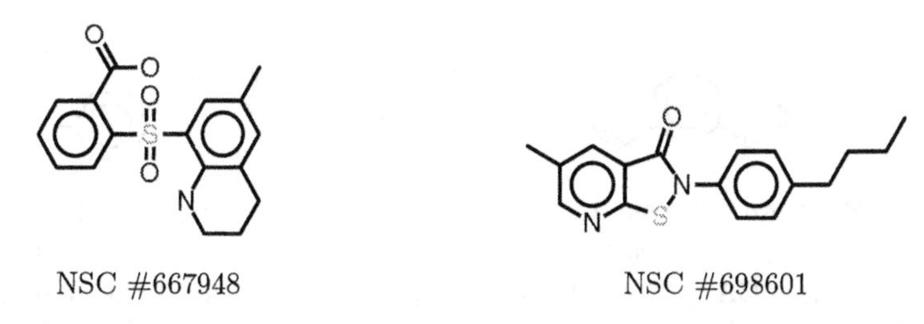

Fig. 8. Compounds that contain Fragment 3 but not Fragment 4.

first sight, this behavior is actually highly desirable by chemists and biologists, because atoms in rings and atoms in chains are considered to be different and thus should not be matched by the same fragment. A fragment should either contain a ring or not, but not parts of it, especially if that part can be confused with a chain.

4 Fragments with Wildcards

Chemists and biologists often regard fragments as equivalent, even though they are not identical. Acceptable differences are, for instance, that an aromatic ring consists only of carbon atoms in one molecule, while in another ring one carbon atom is replaced by a nitrogen atom. An example of such a situation from the NCI-HIV database is shown in Figure 9. A chemist would say that the two molecules shown on the left have basically the same structure and that the difference consists only in the oxygen atom attached to the selenium atom. However, the original algorithm will not find an appropriate fragment because of this mismatch in atom-type [1].

More generally, under certain conditions certain types of atoms can be considered as equivalent from a chemical point of view. To model this equivalence we introduce wildcard atoms into the search, which can match atoms from a user-specified range of atom types. As input we need chemical expert knowledge to define equivalence classes of atoms, possibly conditioned on whether the atom is part of a ring or not (that is, certain atom types may be considered as equivalent in a ring, but as different in a chain). In the example shown in Figure 9 we may introduce a wildcard atom into the search that can match carbon as well

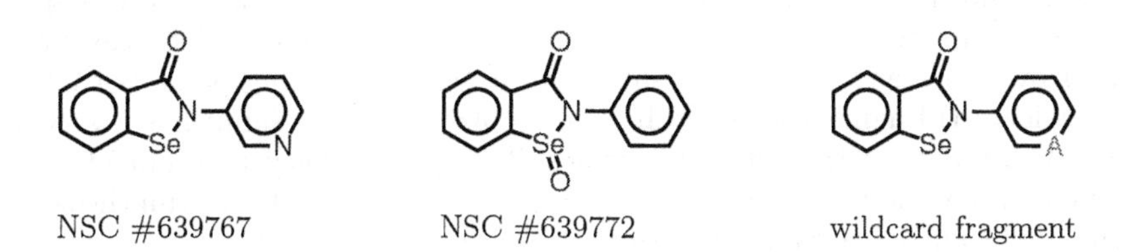

Fig. 9. Matching molecules with a fragment with a wildcard atom.

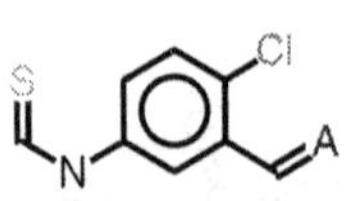

Wildcard fragment 1			Wildcard fragment 2		
A:	O	N	B:	O	S
freq. in CA:	5.5%	3.7%	freq. in CA:	5.5%	0.01%
freq. in CI:	0.0%	0.0%	freq. in CI:	0.0%	0.00%

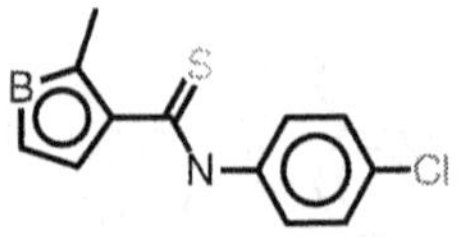

Fig. 10. Fragments with wildcards in the NCI-HIV database.

as nitrogen atoms in rings. This enables us to find the fragment on the right, where the green A denotes the wildcard atom.

Formally, wildcard atoms can be introduced into the search process by adding to each node a branch for each group of equivalent atoms, provided that there are at least two molecules that contain different atoms from the group at the position of the wildcard atom. In the later search process, this branch has to be treated in a special way, because it can be pruned if—due to further extensions—the molecules matching the fragment all contain the same atom for the wildcard atom. In our implementation we restrict the number of wildcard atoms in a fragment to a user-defined maximum number.

We applied the wildcard atom approach to the NCI-HIV database, restricting the number of wildcard atoms to one. With this approach the two wildcard fragments shown in Figure 10 are found (among others). The left fragment has a wildcard atom indicated by a red *A* on the right, which may be replaced by either an oxygen or a nitrogen atom. The (non-wildcard) fragment that results if A is replaced with an oxygen atom has a support of 5.5% in the active compounds, whereas it does not occur in the inactive ones. The fragment with A replaced with nitrogen has a support of 3.7% in the active compounds, and also does not occur in the inactive ones. Consequently, the wildcard fragment has a support of 9.2% in the active compounds. This demonstrates one of the advantages of wildcard fragments, that is, they can be found with much higher minimum support thresholds, which can cut down the output of irrelevant fragments.

However, wildcard fragments can also make it possible to find certain interesting fragments in the first place. This is demonstrated with the wildcard fragment shown in the right in Figure 10. It contains a wildcard atom denoted by a red B in the five-membered aromatic ring on the left, which may be replaced by either an oxygen or a sulfur atom. The interesting point is that the (non-wildcard) fragment that results from replacing B with sulfur has a very low support of 0.01% in the active molecules. This (non-wildcard) fragment would not be found without wildcard atoms, because it is not possible to run the search with a sufficiently low minimum support threshold.

5 Conclusions

This paper adressed two limitations of the molecular fragment mining algorithm introduced in [1]. The first extension of the original algorithm deals with the frequent occurrence of rings and their effect on the width of the generated search tree. By preprocessing the molecular database we can identify rings before the actual search starts and then treat them as one entity during the search. This not only speeds up the computation tremendously, but also finds chemically more meaningful fragments since chemists tend to think of rings as one molecular unit. The second extension addresses the need to find fragments that behave chemically similar but are different from a graph point of view. We have proposed a mechanism that allows the user to specify equivalence classes of atoms. Fragments can then contain a certain maximum number of such *wildcard* atoms. Using this method we not only find chemically meaningful fragments, but we can also detect fragments that would otherwise fall below the support threshold.

The two extensions presented here make the original algorithm described in [1] directly applicable to much larger databases and allow the chemist to introduce expert knowledge about similarities on an atomic level, as demonstrated on the NCI-HIV dataset. As a result, the method presented here has been used very successfully, among others, at Tripos Receptor Research (Bude, UK) for the prediction of synthetic success.

References

1. C. Borgelt and M.R. Berthold. Mining Molecular Fragments: Finding Relevant Substructures of Molecules. Proc. IEEE Int. Conf. on Data Mining (ICDM 2002, Maebashi, Japan), 51–58. IEEE Press, Piscataway, NJ, USA 2002
2. S. Kramer, L. de Raedt, and C. Helma. Molecular Feature Mining in HIV Data. Proc. 7th ACM SIGKDD Int. Conf. on Knowledge Discovery and Data Mining (KDD 2001, San Francisco, CA), 136–143. ACM Press, New York, NY, USA 2001
3. R. Agrawal, T. Imielienski, and A. Swami. Mining Association Rules between Sets of Items in Large Databases. *Proc. Conf. on Management of Data*, 207–216. ACM Press, New York, NY, USA 1993
4. M. Kuramochi and G. Karypis. An Efficient Algorithm for Discovering Frequent Subgraphs. Technical Report TR 02-026, Dept. of Computer Science/Army HPC Research Center, University of Minnesota, Minneapolis, USA 2002
5. O. Weislow, R. Kiser, D. Fine, J. Bader, R. Shoemaker, and M. Boyd. New Soluble Formazan Assay for HIV-1 Cytopathic Effects: Application to High Flux Screening of Synthetic and Natural Products for AIDS Antiviral Activity. *Journal of the National Cancer Institute*, 81:577–586. Oxford University Press, Oxford, United Kingdom 1989
6. M. Zaki, S. Parthasarathy, M. Ogihara, and W. Li. New Algorithms for Fast Discovery of Association Rules. *Proc. 3rd Int. Conf. on Knowledge Discovery and Data Mining (KDD'97)*, 283–296. AAAI Press, Menlo Park, CA, USA 1997

Genome-Wide Prokaryotic Promoter Recognition Based on Sequence Alignment Kernel

Leo Gordon[1], Alexey Ya. Chervonenkis[1,2], Alex J. Gammerman[1], Ilham A. Shahmuradov[1], and Victor V. Solovyev[3]

[1] Department of Computer Science, Royal Holloway, University of London, Egham, Surrey, TW20 0EX, UK,
leo@cs.rhul.ac.uk

[2] Institute of Control Science, Profsoyuznaja 65, Moscow, Russia,
chervnks@ipu.rssi.ru

[3] Softberry Inc., 116 Radio Circle, Suite 400, Mount Kisco, NY, 10549, USA,
victor@softberry.com

Abstract. In this paper an application of Sequence Alignment Kernel for recognition of prokaryotic promoters with transcription start sites (TSS) is presented. An algorithm for computing this kernel in square time is described. Using this algorithm, a "promoter map" of *E.coli* genome has been computed. This is a curve reflecting the likelihood of every base of a given genomic sequence to be a TSS. A viewer showing the likelihood curve with positions of known and putative genes and known TSS has also been developed and made available online.
Although the visual analysis of the promoter regions is very intuitive, we propose an automatic genome-wide promoter prediction scheme that simplifies the routine of checking the whole promoter map visually. Computational efficiency and speed issue are also discussed.

1 Introduction

A *promoter region* is an area on the chromosome which determines conditions and the startpoint of a gene's transcription. A prokaryotic *promoter region* occupies several hundred base pairs upstream of *transcription start site (TSS)* and a smaller area downstream of TSS, and may serve for transcription of a single gene as well as of a group of genes (an operon).

In the last few years complete genomes of many prokaryotic organisms, including *Escherichia coli*, were sequenced [1]. The current information on the gene content of these genomes, obtained by both experimental and computational methods, is relatively complete. However, we are still far from genome-wide promoter/TSS maps of these organisms and the software able to accurately produce this information is of great importance.

Promoter recognition, the computational task of finding the promoter regions on a DNA sequence, is very important for determining the transcription units

M.R. Berthold et al. (Eds.): IDA 2003, LNCS 2810, pp. 386–396, 2003.

responsible for specific pathways (because gene prediction alone cannot provide the solution) and for analysis of gene/operon regulation.

Promoter regions contain *binding motifs* (which are relatively short DNA sequences) interacting with *RNA polymerase* and special regulatory proteins that control the transcription [2]. There are various approaches to DNA signal recognition that are used to recognise prokaryotic *promoter regions*, from very basic to quite complex ones [3,4,5,6,7,8,9,10,11,12,13,14,15], but most of them rely on *binding motif* model. That means, they emphasise the importance of *binding motifs* and essentially neglect the sequence data between them. However, as the number of known promoters increased considerably, the accuracy of those methods generally dropped [16]. Our own experiments on the largest set of *E. coli* promoters available to date (669 positive examples) shows greater variability of the *binding motifs* than was reported previously [17,3,18]. This *might be* due to too rigid constraints that such models impose on the data.

In our alternative approach we abandon the *binding motif* model, where an intermediate step of extracting features of an "ideal promoter" has to be performed. Instead, we have proposed a scheme, in which a specially computed alignment score defines the similarity between two DNA sequences.

In the rest of this section we briefly discuss the motivation for using sequence alignment for recognition, outline the algorithm for its calculation and describe the data set that we have assembled for training and testing our method. In the second part of the paper we describe the *promoter map* and the viewer that we have put online for visual examination of the TSS likelihood curve computed for the whole genome of *E.coli*. Then we discuss the automatic genome-wide promoter prediction scheme that simplifies the routine of checking the whole promoter map visually. We further describe the speed problem that arises when our method is applied genome-wide and propose ways to solve it.

1.1 Using Alignment for Recognition

Sequence alignment is a classical tool in bioinformatics. It is used to trace two or more sequences to the common origin by highlighting the similar parts of the sequences in question.

The global pairwise alignment algorithm of Needleman and Wunsch [19] is a dynamic programming algorithm that takes two sequences as its input and finds the operations (mutations of one letter into another, insertions and deletions of single letters) that transform one sequence into another at the minimum cost (so-called *alignment score*). An additional feature called "affine gaps" [20] allows for a complex operation, insertion or deletion of whole fragments, to be considered as being "simpler" than the insertion or deletion of its constituents and thus having smaller cost.

The *alignment score* itself (though it is only a side result) is a valuable measure that could be used in recognition. However, our experiments have shown that it does not give good recognition when used as the distance in Nearest Neighbours algorithm. On the other hand, it does not necessarily satisfy the necessary conditions for a kernel function to be used in Support Vector Machine or other kernel-based recognition methods. Fortunately, a modification of the alignment algorithm does provide a valid kernel function, which we call Sequence Alignment Kernel (SAK).

1.2 Sequence Alignment Kernel (the Core Algorithm)

The input to the algorithm are two sequences, R and Q, over the alphabet A, C, G, T. The parameters of the algorithm are:

- a matrix $Swap(x, y)$, which defines the score corresponding to a single point mutation of letter x into letter y or vice versa (the matrix is symmetric);
- a vector $Gap(x)$, which defines the score corresponding to a single point deletion or insertion of letter x
- a scalar $GapStart$, which defines the cost of starting the gap ($GapStart = 1$ means that the "affine gap" mechanism is off)

The main difference of SAK from conventional global alignment is that we consider all possible alignment paths, giving each the probability-like score. The sum of those scores is proven to yield a valid kernel [21]. The benefit of using the SAK scheme is its square time complexity $O(|R| * |Q|)$, where $|R|$ and $|Q|$ are the lengths of input sequences. It coincides with time complexity of Needleman-Wunsch algorithm. Computing all possible alignment scores separately and then adding them together would take exponential time.

We apply a model, where all the single mutations/insertions/deletions have a probabilistic score associated with them. To find the score of one full alignment path we have to multiply all the single scores along the path (and operation). When we want to find the score of any two different full paths, we add the scores (or operation). Consequently, if we want to find the score of the alignment as such, irrespective to the path, we have to add the scores of all possible full paths.

Now let us draw all possible full alignment paths of two given sequences, $R = ACGTC$ and $Q = ACCT$ (Fig. 1). All the alignment paths share the same beginning and end points. The trick we use to attain square complexity is by noting that for any two alignment paths that have a common prefix part we do not have to multiply this part twice. It is enough to do it once and multiply by the sum of the scores of the differing suffixes. Ultimately, we get the following recursive formulae:

$$
\begin{aligned}
&p^d_{0,0} = 1 \\
&p^h_{0,0} = p^v_{0,0} = 0 \\
&p^d_{i,0} = p^h_{i,0} = p^v_{i,0} = 0, \ i \neq 0, \\
&p^d_{0,j} = p^h_{0,j} = p^v_{0,j} = 0, \ j \neq 0, \\
&p^h_{i,j} \leftarrow Gap(Q_j) \cdot (p^h_{i,j-1} + p^d_{i,j-1} \cdot StartGap), \\
&p^v_{i,j} \leftarrow Gap(R_i) \cdot (p^v_{i-1,j} + p^d_{i-1,j} \cdot StartGap), \\
&p^d_{i,j} \leftarrow Swap(R_i, Q_j) \cdot (p^h_{i-1,j-1} + p^v_{i-1,j-1} + p^d_{i-1,j-1}).
\end{aligned}
$$

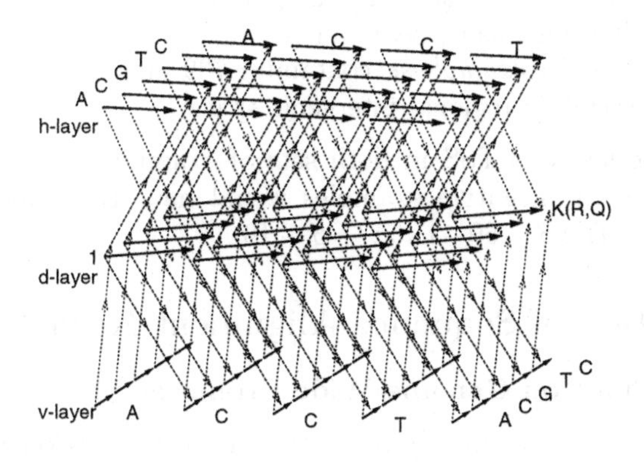

Fig. 1. A Sequence Alignment scheme that accounts for affine gaps

The end result, the kernel value of strings R and Q is the sum of scores of all paths $K(R,Q) = p^d_{|R|,|Q|}$.

Sequence Alignment Kernel is used in conjunction with Dual Support Vector Machine [22], which makes the predictions. This method was compared with several different promoter region predicting methods, showing superior performance in *false negative rate* and *sensitivity*, and close to the best in *specificity*. For more details of the algorithm and results of comparison - see [23].

Sequence Alignment Kernel method is preferable in cases when we have a sufficient number of known promoter regions, but might not know anything about their composition.

1.3 Source and Composition of Data

As the primary source of data we used the sequenced genome of *Escherichia coli* (strain K-12, substrain MG1655) [1]. We have put together a list of 669 experimentally confirmed σ^{70} promoters with known TSS positions from RegulonDB [24] and PromEC [25] databases.

Then *promoter regions* $[TSS - 60 \cdots TSS + 19]$ were taken as the *positive examples.*

As there is no experimentally confirmed negative data (i.e. the positions that are **confirmed not to be** TSS), we had to take the risk and pick the *negative examples* from random places of the same chromosome. We have checked that approximately 81% of known TSS are located in the intergenic non-coding regions and 19% in the coding regions. So two different *negative example* sets were prepared:

(a) *coding negative example set* containing 709 subsequences, 80 letters each, from the coding regions (genes) and
(b) *non-coding negative example set* containing 709 subsequences, 80 letters each, from the non-coding regions (intergenic spacers).

The hypothetical non-TSS in both sets of examples is located in the 61st position, so that the negative examples would have the same format as the positive ones: $[nonTSS - 60 \cdots nonTSS + 19]$.

2 Application of Sequence Alignment Kernel Method

2.1 Escherichia Coli Genome-Wide Promoter Map

We construct every example E - be it positive, negative or unknown test example, as a sequence of the fixed length $|E| = 80$, "off-centered" around the potential TSS with 60 bases upstream and 19 downstream of it. Classifying a particular example means finding out whether the base at 61st position of this example **is** a TSS or not.

To find out whether a given sequence S **contains** a promoter, we simply scan the sequence S with a sliding window, classifying every example E that fits in the window, using our core method. When applying SVM, we take the value of the signed distance between E and the separating hyperplane (which was produced during the training phase) and use it as the likelihood measure that E is a *promoter region* (positive example). The likelihood curve computed for a sequence of DNA upsteam of a gene should show peaks in the area where potential TSS could be.

We have computed the likelihood curve for the full genome of *E.coli*, which has 9,278,442 positions in total on the two strands. For completing the curve for most of the genome all the available 669 *positive* and 709 *coding negative examples* were used for training[1]. A sliding window $E(p) = [p - 60 \cdots p + 19]$ was taken as the example for which the prediction (the *distance from the separating hyperplane*) was made. This distance makes the ordinate of the corresponding point p on the likelihood curve.

To facilitate the visual analysis of the likelihood curve we have the positions of all known and putative genes and all known promoters marked on the same graph, which we call the *promoter map*. A viewer of *E.coli promoter map* is available on-line [26].

Figure 2 shows a sample screenshot of a region of the *promoter map* which contains 5 known genes: 3 on the forward strand (shown as convex arcs above 0-line) and 2 on the reverse (shows as concave arcs under 0-line). The curve shown corresponds to the forward strand, so the "reverse" genes are shown only for reference. The peaks upstream of the genes **b0005** and **talB** clearly show the most probable positions of TSS.

[1] Exceptions were made for windows that intersected with one of either *positive* or *negative* examples, to avoid snooping. When computing the curve for those areas, the "dangerous" example was excluded from the training set and the SVM was retrained.

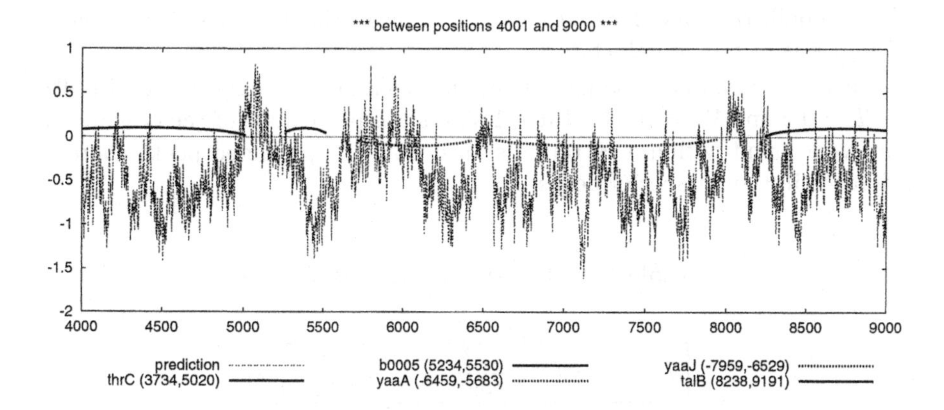

Fig. 2. On the screenshot of *E.coli promoter map*, the peaks in the intervals $5000 \cdots 5200$ and $8000 \cdots 8100$ correspond to the intergenic regions that are likely to contain TSS.

2.2 Genome-Wide Promoter Recognition

The *promoter map* may be convenient for visual inspection, but some processing is definitely needed to get automatic predictions of TSS for the full genome. The 0-line (i.e. the original "separating hyperplane" in terms of SVM), which is optimal for separating the relatively small toy-sets of promoters from non-promoters, would be inappropriate on the full genome scale. If we assume everything above this line to be promoter regions, we would end up with too many false positives (non-promoters taken for promoters). For practical reasons it would be better to keep the false positives as low as possible, even at the expense of loosing some good predictions.

Let us introduce the following constraints:

1. we shall only predict in the non-coding regions,
2. because the curve is quasi-continuous and has "peaks" (local maxima), our prediction of TSS in a given region will be at the peak(s),
3. we shall ignore the peaks that are lower than a certain threshold T,
4. we shall produce either
   a. exactly one prediction per non-coding region - at the highest peak, which is above the threshold T ("highest peak" method), or
   b. as many predictions as there are peaks above the threshold T, but there'll be a certain "insensitivity zone" W around each peak chosen, so that no two peaks can be closer than W ("all peaks" method).

Let's compare the two methods. For assessment we used a simple leave- one-out procedure. For every non-coding region where exactly one TSS is known we trained the SVM on the full *coding negative example set* and the full *positive example set* except for the TSS that belongs to the region we were assessing.

Then we built the curve for that region only, measuring how close our prediction (if any) was to the actual TSS.

There were 353 non-coding regions that contained exactly one known TSS. For $T = 0.2$ and $W = 70$ the Table 1 shows the percentage of the original TSS that were found within a certain distance from the predicted peaks. Note the difference in the number of peaks.

Table 1. Performance of the original curve

$\lvert TSS - peak \rvert$	% of known TSS	
	"highest peak"	"all peaks"
< 10	30.6%	40.5%
< 20	39.7%	54.7%
< 30	47.3%	65.2%
< 40	50.1%	71.4%
< 50	56.1%	77.6%
< 60	61.8%	84.1%
< 70	67.1%	88.7%
peaks	348	812

We may have one more step of improvement if we assume that the gene map on the genome is available. It gives us the histogram of distances between TSS and corresponding Translation Initiation Sites (TIS), which is not uniform at all - see Fig. 3. The exponential distribution density function $0.23 \times e^{-\frac{x}{60}}$ gives quite a close approximation. If we "weigh" the original curve by this function, starting from TIS (zero distance) and asymptotically approaching zero in the ∞, we'll achieve better recognition (less number of peaks and greater percentage of hits) - see Table 2.

Table 2. Performance of the weighted curve

$\lvert TSS - peak \rvert$	% of known TSS	
	"highest peak"	"all peaks"
< 10	38.5%	42.3%
< 20	53.8%	59.0%
< 30	59.8%	65.7%
< 40	66.9%	73.7%
< 50	70.0%	76.3%
< 60	73.1%	79.8%
< 70	75.4%	82.1%
peaks	317	312

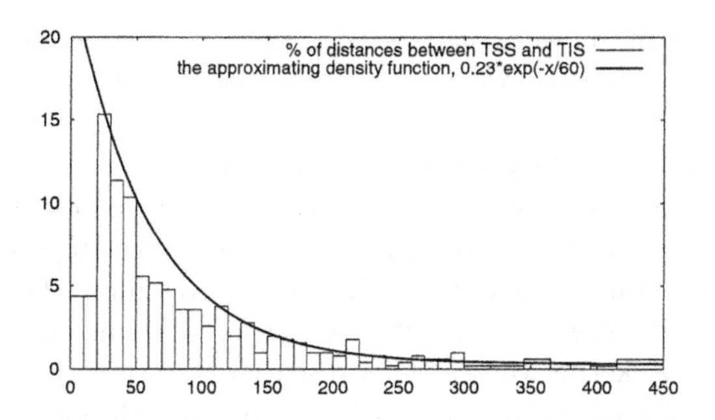

Fig. 3. The histogram of distances from Transcription Start Sites to corresponding Translation Initiation Sites can be approximated by exponential distribution and used to improve the predictions.

In our ViGen [26] promoter map viewer it is possible to set the threshold value T and insensitivity zone W and see the predictions.

3 Discussion

When running the series of experiments to compare different methods, the issue of time was not really crucial, as the computation of the symmetrical 1378×1378 *kernel matrix* takes 5 minutes and application of SVM takes a few seconds on a 1Ghz Linux PC.

However, the calculation of the genome-wide *promoter map* requires the calculation of $4,639,221 \times 2^{\star} \times 1378$ kernel values, which would take 49 days on an average 1Ghz Linux PC, definitely off-line. For comprarison, the nuclear genome of Arabidopsis thaliana is $117,276,964 \times 2$ letters long. If we'd try to do the same kind of *promoter map* for Arabidopsis, we would need 3.4 years on the same computer!

Fortunately, the separate *kernel values* can be calculated independently and concurrently. We used this property to do the off-line calculations for E.coli genome concurrently on 8 workstations [2] by splitting the whole $[1 \cdots 4,639,221 \times 2]$ interval into 8 subintervals, proportional to speeds of the computers. As the result, the calculation took 9 days. Because this problem can be so easily parallelized, Grids [27] can be used to do the task in much shorter time.

An alternative speed-up option would be to use an array processor instead of the traditional one, and compute all the overlapping frames in parallel. Suppose the size of the sliding window equals to w. Then even the most obvious way of parallelizing the algorithm would allow to compute w values simultaneously, yielding w-fold reduction in computation time.

[*] both forward and reverse strand are treated as separate strings

[2] the CPU speeds ranging from 300MHz to dual 2Ghz

Also, the amount of work can be significantly reduced if, instead of the genome-wide *promoter map*, we are only interested in finding the curve for intergenic regions.

Even though our current prediction is not binary, and reflects the intuitively preceivable likelihood measue, the statistical significance of each prediction is not being measured. It would be therefore very interesting to substitute the "traditional" SVM for Transductive Confidence Machine [28], which supplies *confidence* and *credibility* measures with every prediction, and compare the results using the same kernel values. This research is currently in progress.

Another direction of research would be to optimize the criteria for automatic genome-wide promoter recognition.

4 Conclusion

In this paper we have described a new method for recognition of *promoter regions* based on Sequence Alignment Kernel. This method does not require explicit feature extraction, which is a benefit in cases of poorly studied genomes.

This core method gives likelihood estimates for each position of the fixed length window that slides along the sequence, which enables us to draw a nice curve for the whole genome. Mapping the beginnings and the ends of known and putative genes along this curve might give the experimenter an intuitive feeling of where to look for potential promoters. We have developed an on-line viewer that visualizes the genome-wide *promoter map* of E.coli [26].

However, the end users are generally interested in more automation. To make the results more practical, we have analyzed some properties of the curve itself and proposed several heuristics to help recognize the promoters automatically throughout the genome.

Acknowledgements. This work is supported by BBSRC grant No.111/BIO14428, "Pattern recognition techniques for gene identification in plant genomic sequences" and EPSRC grant GR/M14937, "Predictive complexity: recursion-theoretic variants". We also wish to thank Spanish Ministerio de Educacion, Cultura y Deporte and School of Telecommunications, Universidad Politecnica de Madrid (UPM) for their support through the grant SAB2001-0057.

The authors are grateful to Heladia Salgado, member of RegulonDB [24] database support group, and Ruti Hershberg, member of PromEC [25] database support group for the data for the experiments and other useful information. We also would like to thank Jose Carlos Gonzalez of UPM for useful discussions and Aleksei Kirillov and Konstantin Skaburskas from the University of Tartu for their help in parallel calculations.

References

1. Blattner, F.R., Plunkett, G., Bloch, C.A., Perna, N.T., Burland, V., Riley, M., Collado-Vides, J., Glasner, J.D., Rode, C.K., Mayhew, G.F., Gregor, J., Davis, N.W., Kirkpatrick, H.A., Goeden, M.A., Rose, D.J., Mau, B., Shao, Y.: The complete genome sequence of *Escherichia coli* k-12. Science **277** (1997) 1453–1462 Genbank/EMBL/DDBJ record: U00096.
2. De Haseth, P.L., Zupancic, M.L., Record Jr., M.T.: RNA polymerase-promoter interactions: the comings and goings of RNA polymerase. Journal of Bacteriology **180** (1998) 3019–3025
3. Staden, R.: Computer methods to locate signals in nucleic acid sequences. Nucleic Acids Res. **12** (1984) 505–519
4. Lukashin, A.V., Anshelevich, V.V., Amirikyan, B.R., Gragerov, A.I., Frank-Kamenetskii, M.D.: Neural network models for promoter recognition. J Biomol. Struct.Dyn. **6** (1989) 1123–1133
5. O'Neill, M.C.: Training back-propagation neural networks to define and detect DNA-binding sites. Nucleic Acids Res. **19** (1991) 313–318
6. O'Neill, M.C.: *Escherichia coli* promoters: neural networks develop distinct descriptions in learning to search for promoters of different spacing classes. Nucleic Acids Res. **20** (1992) 3471–3477
7. Mahadevan, I., Ghosh, I.: Analysis of *E.coli* promoter structures using neural networks. Nucleic Acids Res. **22** (1994) 2158–2165
8. Alexandrov, N.N., Mironov, A.A.: Application of a new method of pattern recognition in DNA sequence analysis: a study of *E.coli* promoters. Nucleic Acids Res. **18** (1990) 1847–1852
9. Pedersen, A.G., Baldi, P., Brunak, S., Chauvin, Y.: Characterization of prokaryotic and eukaryotic promoters using hidden markov models. In: Proceedings of the 1996 Conference on Intelligent Systems for Molecular Biology. (1996) 182–191
10. Bailey, T., Hart, W.E.: Learning consensus patterns in unaligned DNA sequences using a genetic algorithm (web) http://citeseer.nj.nec.com/172804.html.
11. Rosenblueth, D.A., Thieffry, D., Huerta, A.M., Salgado, H., Collado-Vides, J.: Syntactic recognition of regulatory regions in *Escherichia coli.* Computer Applications in Biology **12** (1996) 415–422
12. Leung, S.W., Mellish, C., Robertson, D.: Basic gene grammars and dna-chartparser for language processing of *escherichia coli* promoter dna sequences. Bioinformatics **17** (2001) 226–236
13. Bailey, T.L., Elkan, C.: Unsupervised learning of multiple motifs in biopolymers using expectation maximization. Machine Learning **21** (1995) 51–80
14. Tompa, M.: An exact method for finding short motifs in sequences, with application to the ribosome binding site problem. In: Seventh International Conference on Intelligent Systems for Molecular Biology. (1999) 262–271
15. Kent, J.: Improbizer motif discovery program with web interface (web) http://www.cse.ucsc.edu/˜kent/improbizer /improbizer.html.
16. Horton, P.B., Kanehisa, M.: An assesment of neural network and statistical approaches for prediction of *E.coli* promoter sites. Nucleic Acids Res. **20** (1992) 4331–4338
17. Hawley, D.K., McClure, W.R.: Compilation and analysis of escherichia coli promoter dna sequences. Nucleic Acids Res. **11** (1983) 2237–2255
18. O'Neill, M.C.: *Escherichia coli* promoters. I. Consensus as it relates to spacing class, specificity, repeat substructure, and three-dimensional organization. Journal of Biological Chemistry **264** (1989) 5522–5530

19. Needleman, S.B., Wunsch, C.D.: A general method applicable to the search for similarities in the amino acid sequence of two proteins. J. Mol. Biol. **48** (1970) 433–453
20. Gotoh, O.: An improved algorithm for matching biological sequences. J. Mol. Biol. **162** (1982) 705–708
21. Watkins, C.: Dynamic alignment kernels. In Smola, A.J., Bartlett, P.L., Schölkopf, B., Schuurmans, D., eds.: Advances in Large Margin Classifiers. MIT Press, Cambridge, MA (2000) 39–50
22. Vapnik, V.N.: Statistical learning theory. Wiley, New York (1998)
23. Gordon, L., Chervonenkis, A.Y., Gammerman, A.J., Shahmuradov, I.A., Solovyev, V.V.: Sequence Alignment Kernel for recognition of promoter regions. To appear in Bioinformatics (2003)
24. Salgado, H., Santos-Zavaleta, A., Gama-Castro, S., Millan-Zarate, D., Blattner, F.R., Collado-Vides, J.: Regulondb (version 3.0): transcriptional regulation and operon organization in *Escherichia coli* K-12. Nucleic Acids Res. **28** (2000) 65–67 http://www.cifn.unam.mx /Computational_Genomics/regulondb/.
25. Hershberg, R., Bejerano, G., Santos-Zavaleta, A., Margalit, H.: Promec: An updated database of *Escherichia coli* mRNA promoters with experimentally identified transcriptional start sites. Nucleic Acids Res. **29** (2001) 277 http://bioinfo.md.huji.ac.il/marg/promec/.
26. Gordon, L.: VIsualiser of GENes – *E.coli* gene and TSS map together with promoter prediction curve – web interface (web) http://nostradamus.cs.rhul.ac.uk/~leo/vigen/.
27. Foster, I., Kesselman, C.: Computational grids. In: The Grid: Blueprint for a New Computing Infrastructure. Morgan-Kaufman (1999)
28. Gammerman, A., Vovk, V.: Prediction algorithms and confidence measures based on algorithmic randomness theory. Theoretical Computer Science **287** (2002) 209–217

Towards Automated Electrocardiac Map Interpretation: An Intelligent Contouring Tool Based on Spatial Aggregation

Liliana Ironi and Stefania Tentoni

IMATI – CNR, via Ferrata 1, 27100 Pavia, Italy

Abstract. The interpretation of cardiac potential maps is essential for the localization of the anomalous excitation sites associated with electrical conduction pathologies, such as arrythmia, and its automation would make clinical practice both realistic and of great impact on health care. Conventional contouring tools allow us to efficiently visualize patterns of electrical potential distribution but they do not facilitate the automated extraction of general rules necessary to infer the correlation between pathopysiological patterns and wavefront structure and propagation. The Spatial Aggregation (SA) approach, which aims at interpreting a numeric input field, is potentially capable of capturing structural information about the underlying physical phenomenon, and of identifying its global patterns and the causal relations between events. These characteristics are due to its hierarchical strategy in aggregating objects at higher and higher levels. However, when dealing with a complex domain geometry and/or with non uniform data meshes, as in our context, the original version of SA may unsoundly perform contouring. Isocurve entanglement and/or segmentation phenomena may occur due to metric-based adjacency relations and a scarse density of isopoints. This paper discusses the problems above, and presents definitions and algorithms for a sound representation of the spatial contiguity between isopoints and for the construction of isocurves.

1 Introduction

Electrocardiology is a stimulating and promising application domain for Intelligent Data Analysis (IDA). Most research effort aimed at the automated interpretation of electrocardiograms (ECG's) [1,2,3,4], for which an interpretative rationale is well established. However, a major drawback of ECG's, which record electrical signals from nine sites only on the body surface, is their poor spatial covering. Some arrythmias, but also conduction anomalies caused by infarcts and ischemia, can be missed by ECG's.

Thanks to the latest advances in technology, a far more informative technique, namely Body Surface Mapping (BSM), is becoming more widely available. In BSM, the electrical potential is measured on the entire chest surface over a complete heart beat, and a great deal of spatial information about the heart electrical activity is then available, though the ability to relate visual features

M.R. Berthold et al. (Eds.): IDA 2003, LNCS 2810, pp. 397–408, 2003.

to the underlying complex phenomena still belongs to very few experts [5]. Besides body surface maps, also epicardial or endocardial maps, which are by far more precise in locating the anomalous conduction sites, are becoming available, either measured by endocavitary probing or obtained non invasively from body surface data through mathematical inverse procedures.

As a rationale for the interpretation of electrocardiographic maps is progressively being defined, the goal of bringing such techniques to the clinical practice gets closer. At this stage, numerical simulations are crucial to highlight the physical mechanisms that link the observable patterns (effects) to the underlying bioelectrical phenomena (causes). In this context, an important role can be played by IDA methodologies for spatial/temporal reasoning, that could 1) support the expert electrocardiologist in identifying salient features in the maps and help him in the definition of an interpretative rationale, and 2) achieve the long-term goal of delivering an automated map interpretation tool to be used in a clinical context. Such a tool would be of great impact on health care.

Currently, the search for features that can be correlated with significant abnormalities mainly relies on the analysis of contour maps of the cardiac electrical potential, and of the activation time. Then, the first step towards the realization of such an intelligent tool consists in the definition and implementation of contouring algorithms capable of preserving the spatial relations between geometrical objects at different abstraction levels. Conventional contouring methods [6] would allow us to visualize the electrical potential distribution but it would not facilitate the automated extraction of features and general rules necessary to infer the causal relationships between pathophysiological patterns and wavefront structure and propagation. Our interest in the Spatial Aggregation (SA) computational paradigm [7,8] has been motivated by such issues. The hierarchical structure of the abstract objects constructed by the SA computational approach to represent and interpret a numeric input field is a promising ground (i) to capture spatial and temporal adjacencies at multiple scales, (ii) to identify causal relations between events, and (iii) to generate high-level explanations for prediction and diagnosis.

The SA approach has been successfully applied to different domains and tasks, [9,10,11,12,13], among those we mention the weather data analysis problem [14] as it presents some similarities with the cardiac map interpretation: global patterns and structures are identified by looking at the isocurves and reasoning about their spatial relations. However, in the case of contouring, which is often the first level for spatial reasoning, the originally proposed SA method has some major drawbacks that limit its use in many applications. A first limitation deals with the way topological contiguity of isopoints is taken into account: the current methods use a kind of neighborhood graph that fails to adequately represent such property. Secondly, the strong adjacency relation, upon which isocurve abstraction from the set of given data points is performed, is based on a metric criterion [14]. This approach proves unsound when the domain geometry is discretized with a non uniform mesh, which is quite a frequent situation.

With respect to our application problem, we deal with the 3D geometries of heart and chest, and data are given on the nodes of non uniform quadrangular meshes: either planar 2D elements when transversal or longitudinal sections of the myocardium are considered, or 3D surface elements when the epicardial surface is explored. Through examples related to the construction of ventricular isochrone maps, herein we discuss such problems, and present definitions and algorithms for a sound aggregation and classification of isopoints within the SA framework.

2 Data Pre-processing and Isopoint Set Generation

Figure 1A illustrates a 3D model ventricular geometry. Its discretization was carried out by 15 horizontal sections, 30 angular sectors on each section, and 6 radial subdivisions of each sector. A numerical simulation was carried out on this mesh to simulate the excitation wavefront propagation in the anisotropic ventricular tissue [15], and for each node $\mathbf{x}_i$ the activation time u_i, i.e. the instant at which the excitation front reaches the node, was computed. In this paper, given the current limitation of SA at dealing with more than two dimensions, we bound our study to the contouring problem in 2D. Nevertheless, this is meaningful from the application point of view as 3D processes are suitably studied also by transversal/longitudinal sections of the 3D domain. For example, to investigate the wavefront propagation, the contour maps of the activation time (isochrones) need to be built and analyzed not only on the ventricular surfaces, but also on sections of the wall (Figure1B). An activation isochrone map delivers a lot of information about the wavefront propagation: each contour line aggregates all and none but the points that share the same excitation state, and subsequent isocurves are nothing but subsequent snapshots of the travelling wavefront.

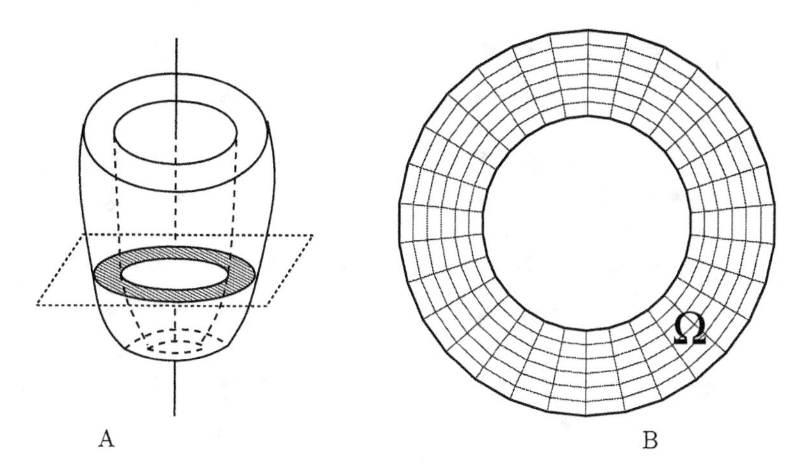

Fig. 1. A - Model ventricle geometry; B - A horizontal section of the discretized ventricular wall.

The knowledge necessary to successfully perform the relation-based abstraction processes must be derived and organized from the available input data in a first pre-processing step. Input data for a 2D contouring problem consist of:

1. a mesh which approximates the geometrical domain Ω by means of a set of elements $\{E_i\}_{i=1..NE}, \cup_i E_i \simeq \Omega$, and of nodes $\{\mathbf{x}_i\}_{i=1..N}$, vertices of the elements;
2. the scalar values $\{u_i\}_{i=1..N}$ of a function $u(\mathbf{x})$ either computed or measured at the mesh nodes;
3. a couple of scalar values $(\bar{u}_0, \Delta u)$ to uniformly scan the range of $u(\mathbf{x})$ and generate isocurves for the levels $\bar{u}_0$, $\bar{u}_0 + \Delta u$, $\bar{u}_0 + 2\Delta u$, .., $\bar{u}_0 + n\Delta u$.

Depending on the irregularities in the shape of the domain Ω as well as on the need to adequately capture the spatial course of u, the given mesh is rarely uniform: elements, usually triangles or quadrangles, can widely vary in size and shape.

The optimal choice of the scan step Δu is usually domain-specific, and is critical in so far as it should be small enough to capture the function main changes over the geometric domain, but large enough to produce a contour map that is readable and not burdened with an unnecessary mass of minor, possibly artifactual, details.
We generate isopoints for the required levels by assuming a piecewise-linear approximation $\tilde{u}(\mathbf{x})$ for the function $u(\mathbf{x})$, with $\tilde{u}(\mathbf{x_i}) = u(\mathbf{x_i})$ for each mesh node $\mathbf{x_i}$. The isopoint set, $\mathcal{I}$, is therefore generated by comparing, for each element, the u values at the vertices with the required levels, and by interpolating from the nodal values when necessary. Figure 2 lists the situations that can occur in the case of a quadrangular mesh, which is more complex to be dealt with than a triangular one.

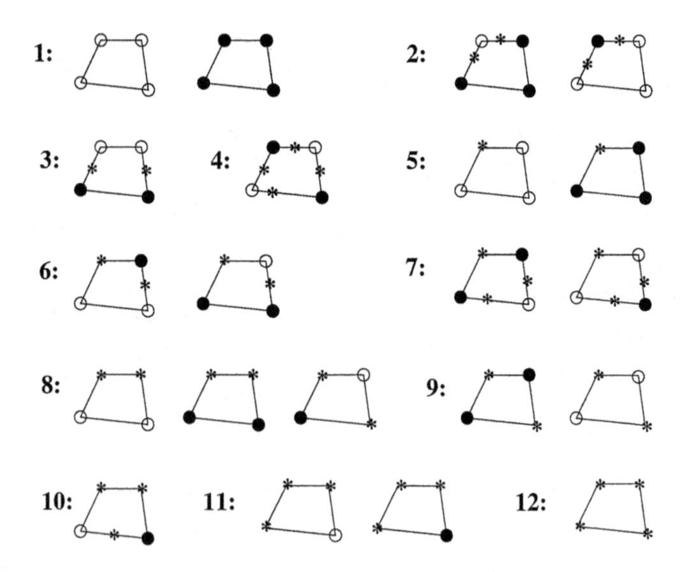

Fig. 2. Quadrangular element: possible locations of isopoints for level $\bar{u}$. A black-filled circle denotes a value lower than $\bar{u}$, an unfilled circle a value greater than $\bar{u}$, a star a value equal to $\bar{u}$.

During the generation of the set $\mathcal{I}$, we also incrementally build two maps, τ and γ, which state the inter and intra-element topological adiacency, respectively. The former, τ, maps each isopoint $\mathbf{x}$ that is added in $\mathcal{I}$ to the set of mesh elements that share $\mathbf{x}$, that is:

$$\forall \mathbf{x} \in \mathcal{I}, \quad \tau : \quad \mathbf{x} \to \tau(\mathbf{x}) = \{E_i : \ E_i \ni \mathbf{x}, \ i = 1..NE\}.$$

This inter-element mapping plays a crucial role in the definition of a strong adjacency relation between isopoints, but is unable to treat complex and even ambiguous situations that may occur with a quadrangular mesh. To this end, we introduce the map γ that is defined for each couple of newly generated isopoints $\mathbf{x}, \mathbf{y}$ belonging to the same element E: it flags their intra-element topological adjacency in accordance with the result of the comparison of their value $\bar{u} = u(\mathbf{x}) = u(\mathbf{y})$ with the nodal values u_i $(i = 1,..,4)$. Thus, it holds $\gamma(\mathbf{x}, \mathbf{y}) = 1$ if the segment $\mathbf{xy} = \{\mathbf{z} \in E \mid \mathbf{z} = \theta\mathbf{x} + (1-\theta)\mathbf{y}, \ \theta \in [0,1]\}$ divides E in two regions where the values of $\tilde{u}$ are in agreement with the nodal values u_i. In other words, $\gamma(\mathbf{x}, \mathbf{y}) = 1$ means that $\mathbf{x}, \mathbf{y}$ belong to the same isocurve that crosses the element. With reference to Figure 2, we have [1]:

Cases 1,5: The element E contains either none or just one isopoint, and then, no intra-element relation applies to these cases.

Cases 2,3,6,8: The element E contains two isopoints $\mathbf{x}$, $\mathbf{y}$ such that the segment $\mathbf{xy}$ leaves the nodal values less than $\bar{u}$ and greater than $\bar{u}$ on opposite sides, respectively. Thus, $\gamma(\mathbf{x}, \mathbf{y}) = 1$.

Case 4: The element E contains four isopoints $\mathbf{y}_{ij}$, one of them on each side $\mathbf{x}_i\mathbf{x}_j$; then, set $\gamma(\mathbf{y}_{12}, \mathbf{y}_{23}) = \gamma(\mathbf{y}_{34}, \mathbf{y}_{41}) = 1$, and $\gamma(\mathbf{y}_{23}, \mathbf{y}_{34}) = \gamma(\mathbf{y}_{41}, \mathbf{y}_{12}) = 0$.

Case 7: The element E contains three isopoints, one of them coincides with one of its vertices $\mathbf{x}_i$, $i = 1..4$. If both the isopoints $\mathbf{x}, \mathbf{y}$ are different from $\mathbf{x}_i$, $\gamma(\mathbf{x}, \mathbf{y}) = 1$, otherwise $\gamma(\mathbf{x}, \mathbf{y}) = 0$.

Case 9: The opposite vertices, $\mathbf{x}_{i_1}$ and $\mathbf{x}_{i_3}$, of the element E are isopoints, and both the other two opposite vertices, $\mathbf{x}_{i_2}$ and $\mathbf{x}_{i_4}$ assume a value either greater or lower than $\bar{u}$, thus $\gamma(\mathbf{x}_{i_1}, \mathbf{x}_{i_3}) = 0$.

Cases 10 - 12 : The element E contains either three or four isopoints for which $\gamma(\mathbf{x}, \mathbf{y}) = 1$ in all cases. Let us observe that case 12 describes an isoregion of value $\bar{u}$.

Cases from 1 to 4 are the most frequent ones; cases 5 to 7 are less frequent but not rare, while cases 8 to 12 are very unlikely. Case 4, $u_{i_k} < \bar{u} < u_{i_l}$, $k = 1,3$ and $l = 2,4$, which corresponds to the presence of a saddle point of u located inside the element, deserves particular attention. In fact this situation is intrinsically ambiguous with respect to which isopoints truly belong to the same curve. Figure 3 illustrates such ambiguity: four new isopoints, one on each side, are generated, and two distinct $\bar{u}$-valued isopolylines are expected to cross the element leaving the saddle point in between. The choice between a) and

[1] in the following, (i_1, i_2, i_3, i_4) is a permutation of $(1, 2, 3, 4)$

b), that marks the two possible ways of connecting the isopoints, is arbitrary at this level of abstraction, and is fixed by γ. We observe that such ambiguity might be naturally removed at a higher level of abstraction if an inconsistent interpretation would occur.

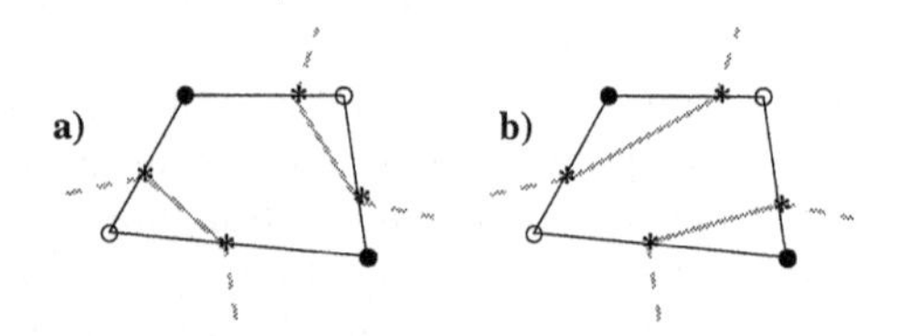

Fig. 3. Saddle point: ambiguity in assessing intra-element isopoint adjacency.

In the case of a triangular mesh, intra-element adjacency of isopoints is never ambiguous since for any two isopoints belonging to the same element their associated segment divides the element into regions where the values of $\tilde{u}$ are in agreement with the nodal values u_i ($\gamma = 1$).

3 Representation of the Spatial Contiguity between Isopoints

The spatial contiguity between isopoints is encoded in a *neighborhood graph* (n-graph), and is made explicit through an aggregation procedure on the isopoint set $\mathcal{I}$. With respect to the contouring task, the n-graph should respect the topological adjacencies of isopoints. Spatial contiguity between isopoints is not well expressed through an absolute metric distance criterion, but it should rather be explicated at topological level. A natural way to spatially aggregate isopoints consists in building a triangulation upon them. However, let us remark that the distribution of the isopoints can be very disuniform in space even if the original mesh nodes were not. A Delaunay triangulation [16] is generally a rather good choice due to its regularity; but, it just expresses context-free spatial contiguity between isopoints as it ignores that isopoints express the value of a function whose course on Ω affects their spatial proximity. As a consequence, the resulting n-graph may fail to represent the actual spatial contiguity between isopoints. It may happen - depending on the spatial course of u - that isopoints which truly belong to the same isocurve are not connected through the n-graph, whereas isopoints of two topologically non-adjacent isocurves are connected. Figure 4 exemplifies such an occurrence: A and B will not be attributed to the same curve since they are not contiguous within the n-graph, unlike C and D.

As mentioned in [14,17] topological contiguity losses may realistically occur when isopoints are not “dense” enough on each isocurve. Since this latter property depends on the size of the mesh elements, this is not a problem when the input data are from a uniform mesh with a sufficiently small element diameter, but it needs care when such conditions are not met, which often occurs with

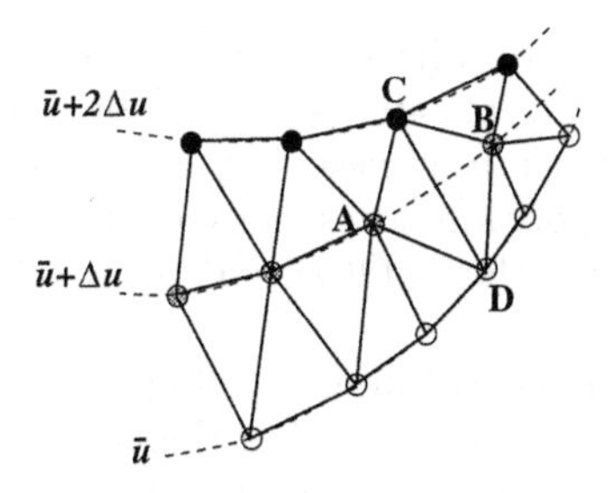

Fig. 4. Isopoint contiguity loss. Unfilled nodes denote a value $\bar{u}$; grey nodes have value $\bar{u} + \Delta u$; black nodes a value $\bar{u} + 2\Delta u$; true isolines are represented by dashed lines; solid lines mark contiguities by the n-graph.

a generic mesh. It has been proved in [14] that if the set $\mathcal{I}$ satisfies the *closeness condition*, then the n-graph $\mathcal{D}$ generated by a Delaunay triangulation of $\mathcal{I}$ guarantees that no contiguity loss occurs. Such a condition is satisfied when $d < \sqrt{2}e$, where d is the maximum distance between two consecutive isopoints on the same isocurve and e the minimum distance between any two isopoints on different isocurves. But, unfortunately, it is objectively difficult to verify such a condition [14] within a contouring context as it explicitely relies on curves whose construction is the final goal.

The n-graph plays a crucial role in the isocurve generation phase, as it establishes spatial contiguities upon which subsets of isopoints are binded into isocurves. As a matter of fact, an inadequate n-graph might induce, by effect of possible isopoint contiguity loss, undesired curve segmentation. While it is objectively difficult to generate a triangulation that *a priori* guarantees from the risk of topological contiguity losses, it is however necessary to identify risky situations and adopt some strategy to avoid the occurrence of such events. The adequacy of the candidate n-graph $\mathcal{D}$ generated by Delaunay triangulation to represent the spatial contiguity between isopoints must be verified before isopoint classification is performed. The algorithm given below identifies possible contiguity loss situations, and locally amends $\mathcal{D}$ by adding new connections where necessary.

N-graph generation algorithm:

1. Generate $\mathcal{D}$ by Delaunay triangulation of $\mathcal{I}$.
2. For each node $\mathbf{x} \in \mathcal{D}$
    consider the set $\mathcal{A}_{\mathcal{D}}(\mathbf{x})$ of its adjacent nodes within $\mathcal{D}$;
    for each node $\mathbf{y} \in \mathcal{A}_{\mathcal{D}}(\mathbf{x})$
        if $|u(\mathbf{x}) - u(\mathbf{y})| > \Delta u$ then
            augment $\mathcal{D}$ with any new connections between couples of nodes $\mathbf{z}, \mathbf{w} \in \mathcal{A}_{\mathcal{D}}(\mathbf{x}) \cap \mathcal{A}_{\mathcal{D}}(\mathbf{y})$ such that $|u(\mathbf{z}) - u(\mathbf{w})| \leq \Delta u$
    endfor
endfor

As new connections are introduced where "false" contiguities have locally taken the place of true ones, the resulting n-graph $\mathcal{D}$ provides for a better local spatial coverage of any region where adjacent points with unexpectedly far u values are detected. The algorithm introduces only a minimum number of extra connections, which guarantees from unnecessary computational overload. It can be proved that the overall algorithm keeps the same time complexity of the Delaunay triangulation algorithm, that is $O(n \log n)$ where n is the number of isopoints in $\mathcal{I}$.

Let us remark that the algorithm may end up with an n-graph that does not correspond to a triangulation any more, but this is irrelevant with respect to the abstraction processes that are subsequently performed.

4 Construction of the Isocurves

Isocurves, which are abstracted as new higher level spatial objects, redescribe the classes resulting from the application of a *strong adjacency relation* on contiguous elements of $\mathcal{I}$ as represented by the n-graph.

The success of the abstraction process relies on the appropriateness of both the n-graph, and the definition of strong adjacency relation. To make this issue clear, let us consider the definition of strong adjacency between isopoints given in [14,17].

Definition D1. Let $\mathcal{A}_{\mathcal{D}}(\mathbf{x})$ be the set of the adjacent nodes to $\mathbf{x} \in \mathcal{I}$ within the n-graph $\mathcal{D}$. $\forall \mathbf{x}, \mathbf{y} \in \mathcal{I}$ and such that $\mathbf{y} \in \mathcal{A}_{\mathcal{D}}(\mathbf{x})$, $\mathbf{x}$ and $\mathbf{y}$ are *strongly adjacent* if and only if $u(\mathbf{x}) = u(\mathbf{y}) \;\wedge\; d(\mathbf{x}, \mathbf{y}) < \delta^*$, where $d(\mathbf{x}, \mathbf{y})$ is the euclidean distance between $\mathbf{x}$ and $\mathbf{y}$, and $\delta^* > 0$ is a suitable nearness radius.

The definition of strong adjacency D1 establishes spatial adjacencies between aggregate objects using a metric distance, and spatial nearness is assessed when such a distance is within a user-defined radius. The limits of this approach when applied to high level objects, such as curves, have been discussed in [17]. However, in a contouring context, analogous limits can hold at the lowest level of abstraction as well, that is when the objects are points which represent the values of a function.

The choice of a proper nearness radius δ^* is quite critical as it depends on the spatial course of u and the scan levels. If definition D1 is used, a few problems may arise in the classification of isopoints, and lead to incorrect isocurve abstraction: roughly speaking, a large δ^* may be responsible for curve entanglement (Figure 5A); whereas a small one may lead to curve segmentation, and intermediate values may even cause both phenomena (Figure 5B). For the isopoint set illustrated in the figure, a smaller value of δ^* would completely eliminate entanglement but it would leave many points in singleton classes, and produce a diffuse artifactual fragmentation of the isocurves. Let us observe that the entanglement problem may appear not only in presence of a saddle point but also where a single curve makes a sharp bend with a high curvature.

The solution to the mentioned problems is provided by a new definition of strong adjacency relation. Such a relation exploits the available knowledge related to the mesh data, and grounds on the inter and intra-element topological adjacency properties of isopoints. It allows us to tackle, satisfactorily and automatically, the most general case of non uniform meshes, both triangular and quadrangular ones.

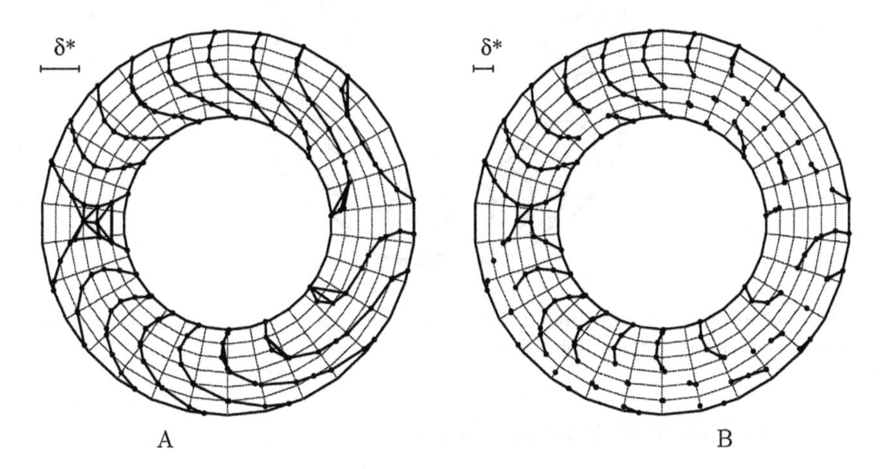

Fig. 5. Activation isochrone map on a horizontal section of a model ventricular wall obtained by the application of metric-based strong adjacency D1. A large and a small nearness radius δ^* produce (A) isocurve entanglement, and (B) isocurve segmentation, respectively.

Given the n-graph $\mathcal{D}$, we define a strong adjacency relation on $\mathcal{I}$ in the following way:

Definition D2. $\forall \mathbf{x}, \mathbf{y} \in \mathcal{I}$ and such that $\mathbf{y} \in \mathcal{A}_{\mathcal{D}}(\mathbf{x})$, we say that $\mathbf{x}$ and $\mathbf{y}$ are *strongly adjacent* if and only if

$$(1)\ u(\mathbf{x}) = u(\mathbf{y})\ \wedge\ (2)\ \tau(\mathbf{x}) \cap \tau(\mathbf{y}) \neq \emptyset\ \wedge\ (3)\ \gamma(\mathbf{x}, \mathbf{y}) = 1.$$

Conditions (2) and (3) in the definition D2 allow us to ground the relation on the topological adjacency properties of the isopoints at element level, which are available from the input mesh data and the pre-processing step. The definition of adjacency grounded on higher level knowledge, namely topological properties rather than a metric distance, makes the classification abstraction process more robust with respect to the given task as it no longer depends on the particular mesh size and shape.

Figure 6 shows the result of isopoint classification according to definition D2 of strong adjacency: the use of a topological criterion allows us to a priori avoid the entanglement and segmentation problems.

The proposed approach has been successfully applied to the construction of isochrones on sets of transversal and longitudinal sections of the myocardium. This allows us to explore the wavefront propagation in the heart wall. The data we considered can be assumed noise-free as they result from numerical simulations. In presence of measurement noise, it could be advisable to pre-process the data by applying suitable filtering procedures.

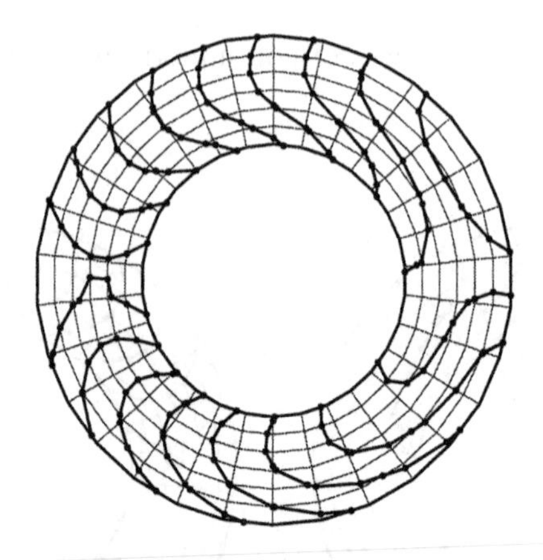

Fig. 6. Activation isochrone map on the same section as in Figure 5. Isocurves are obtained in accordance with definition D2.

5 Discussion and Future Work

This paper deals with a critical revision of the SA approach that revealed to be a necessary step towards the ultimate goal of automating electrocardiac map analysis. More precisely, the work here described focuses on the construction of isocurves from a non uniform numeric input field, and it gives a significant contribution to SA as to the contouring problem by providing new algorithms and definitions for soundly building neighborood and adjacency relations upon 2D complex domain geometry. As a matter of fact, previous SA contouring methods failed in correctly capturing spatial adjacencies between isopoints. Robust contouring is a must in view of the next spatial object abstraction processes aimed at the identification and extraction of salient features and structural information about the underlying physical phenomena.

An alternative way to solve the unsoundness problem of SA in performing contouring could be its integration with conventional contouring tools. More precisely, we could exploit the former ones to efficiently and soundly construct isocurves, and, afterwards, consider them as primitive spatial objects upon which the SA approach is applied to build spatial relations. But, in this way of proceeding, the hierarchical structure of the built spatial objects would be impoverished of the lower-level spatial aggregates with a consequent loss of information which could result to be relevant for a specific task, for example for the explanations of causal relations between events or for the identification of local features within isocurves that can be used to define new kinds of relations among them.

Future work will deal with the identification and definition of spatial relations between isocurves, as well as between possible higher-level aggregated objects. Such relations should reveal adjacencies and interactions between the spatial objects, and, then, underlie the feature extraction process. To this end, a crucial issue deals with the specification of the neighborhood graphs that properly represent the spatial relations between geometrical objects at different levels of abstraction.

Besides Electrocardiology, many other application domains, potentially all those where numeric fields are explored by visualization techniques, would benefit from our SA contouring approach, which throws the basis for further automated imagistic reasoning within the SA framework. As examples, let us mention fluidodynamics, thermal diffusion, and, in particular, weather data analysis where the characterization of pressure troughs and of the propagation of storm fronts presents some analogies with that one of excitation wavefront.

Further work needs to be done as regards the specific application problem. The interpretation of activation isochrone maps currently grounds on features such as contour shapes, the directions along which the front velocity is higher, the minimum and maximum regions. The contour shapes and highest velocity directions reflect the conduction anisotropy, and the fiber architecture of the cardiac tissue. Minimum and maximum correspond to the points where the front emerges and extinguishes, respectively. Such features can be correlated to the expected activation sequence, and used in a diagnostic context: deviations from the expected patterns would highlight ectopic sites associated with ventricular arrythmias. Besides isochrone maps, sequences of isopotential maps can be considered to extract temporal information about electrical phenomena, such as current inflows and outflows.

Future extensions of the current work will deal with the ventricular/epicardial surface maps, and then with the extraction of salient features. This implies to adapt the current methods to 3D analysis.

References

1. Bratko, I., Mozetic, I., Lavrac, N.: Kardio: A Study in Deep and Qualitative Knowledge for Expert Systems. MIT Press, Cambridge, MA (1989)
2. Weng, F., Quiniou, R., Carrault, G., Cordier, M.O.: Learning structural knowledge from the ECG. In: ISMDA-2001. Volume 2199., Berlin, Springer (2001) 288–294
3. Kundu, M., Nasipuri, M., Basu, D.: A knowledge based approach to ECG interpretation using fuzzy logic. IEEE Trans. Systems, Man, and Cybernetics **28** (1998) 237–243
4. Watrous, R.: A patient-adaptive neural network ECG patient monitoring algorithm. Computers in Cardiology (1995) 229–232
5. Taccardi, B., Punske, B., Lux, R., MacLeod, R., Ershler, P., Dustman, T., Vyhmeister, Y.: Useful lessons from body surface mapping. Journal of Cardiovascular Electrophysiology **9** (1998) 773–786
6. Watson, D.F.: Contouring: A Guide to the Analysis and Display of Spatial Data. Pergamon, Oxford (1992)
7. Yip, K., Zhao, F.: Spatial aggregation: Theory and applications. Journal of Artificial Intelligence Research **5** (1996) 1–26
8. Bailey-Kellogg, C., Zhao, F., Yip, K.: Spatial aggregation: Language and applications. In: Proc. AAAI-96, Los Altos, Morgan Kaufmann (1996) 517–522
9. Joskowicz, L., Sacks, E.: Computational kinematics. Artificial Intelligence (1991) 381–416
10. Yip, K.: Structural inferences from massive datasets. In Ironi, L., ed.: QR97, IAN-CNR (1997) 215–220

Study of Canada/US Dollar Exchange Rate Movements Using Recurrent Neural Network Model of FX-Market

Ashwani Kumar[1], D.P. Agrawal[2], and S.D. Joshi[3]

[1] ABV Indian Institute of Information Technology and Management, Gwalior, India-474003, ashwani_iiitm@yahoo.co.in

[2] ABV Indian Institute of Information Technology and Management, Gwalior, India-474003

[3] Indian Institute of Technology, Delhi, India-110016

Abstract. Understanding exchange rate movements has long been an extremely challenging and important task. Unsatisfactory results produced by time series regression models have led to the claim by several authors that in foreign exchange markets, past movements of the price of a given currency have no predictive power in forecasting future movements of the currency price. In this paper, we build a recurrent neural network model for FX-market to explain exchange rate movements. Asset prices are discovered in the marketplace by the interaction of market design and agents' behaviour. The interaction is simulated by integrating 1) the FX-market mechanism; 2) an economic framework; and 3) the embedding of both tasks in neural network architectures. The results indicate that both macroeconomic and microeconomic variables are useful to forecast exchange rate changes. Results from regression model based on neural-fuzzy forecasting system are also included for comparison.

1 Introduction

Time series prediction is an important area of forecasting in which past observations of the prediction variable and other explanatory variables are collected and analyzed to develop a model describing the underlying relationship. The model is then used to extrapolate the time series into the future. One of the most important and widely used time series models is the autoregressive integrated moving average (ARIMA) model. The popularity of the ARIMA model is due to its statistical properties as well as the well known Box-Jenkins methodology [1] in the model building process. The major limitation of ARIMA models is the pre-assumed linear form of the model. That is, a linear correlation structure is assumed among the time series values and therefore, no nonlinear patterns can be captured by the ARIMA model. The approximation of linear models to complex real-world problems is not always satisfactory. Recently, neural networks (NN)

M.R. Berthold et al. (Eds.): IDA 2003, LNCS 2810, pp. 409–417, 2003.

and fuzzy inference systems (FIS) have been extensively studied and used in time series forecasting [2,3]. The major advantage of these approaches is their flexible nonlinear modeling capability. With these soft computing tools, there is no need to specify a particular model form. Rather, the model is adaptively formed based on the features presented from the data. The data-driven time series modeling approaches (ARIMA, NN, FIS) are suitable for many empirical data sets where no theoretical guidance is available to suggest an appropriate data generating process, or when there is no satisfactory explanatory model that relates the prediction variable to other explanatory variables. Our focus in this paper is on the study of exchange rate movements under the shadow of a long history of unsatisfactory results produced by time series models. Several authors have claimed that foreign exchange markets are efficient in the sense that past movements of the price of a given currency have no prediction value in forecasting future [4,5]. These results have helped popularize the view that the exchange rate is best characterized by a random walk. Recent papers that develop models based on long-term relationship between the real exchange rate and macroeconomic variables have found some encouraging results [6,7,8]. Macroeconomic models, however, provide no role for any market microstructure effects to enter directly into the estimated equation. These models assume that information is widely available to all market participants and that all relevant and ascertainable information is already reflected in the exchange rates. For the spot FX trader, what matters is not the data on any of the macroeconomic fundamentals, but information about demand for currencies extracted from purchases and sales orders, i.e., order flow. The work done at the Bank of Canada[9] demonstrates the strong link between macroeconomic and microeconomic variables, and real exchange rates. In this paper we examine whether introducing a market microstructure variable (that is, order flow) into a set of observations of macroeconomic variables (interest rate, crude oil price) can explain Canada/US dollar exchange rate movements. Our work looks at FX-market structure (or the trading process) : the arrival and dissemination of information; the generation and arrival of orders; and the market architecture which determines how orders are transformed into sales. Prices are discovered in the marketplace by the interaction of market design and participant behaviour. It is presumed that certain FX traders manage to get information that are not available to all the other traders and in turn, the market efficiency assumption is violated atleast in the very short term. It may be that, without these market microstructure frictions, markets would be efficient, but trading frictions impede the instantaneous embodiment of all information into prices.

2 FX-Market Mechanism

Basically, a market consists of various agents, which make their buying/selling decisions in accordance with their objective function (e.g., profit maximization). By summing up the individual decisions, the aggregated demand and supply is generated. Both aggregates can be used to describe an equilibrium condition for the market. If the aggregated demand is equal to the aggregated supply, a market equilibrium is reached. Vice versa, if both aggregates are unequal, the market is in disequilibrium state. Since the aggregated demand as well as the aggregated supply are functions of the market price, a situation of a market disequilibrium can be corrected by adjusting the market price level. Therefore, one can anticipate upcoming price shifts by comparing these aggregates. For example if the aggregated demand is higher than the supply, a positive price shift will bring the market back into an equilibrium state [10,11].

The market model discussed above can be represented by a block diagram of Fig.1. The inputs $x^1, \cdots, x^n$ are the macroeconomic/microeconomic indicators. The decision behaviour of an agent a can be modeled by a single neuron.

$$z_t^a = \sigma\left(\sum_{i=1}^{n} w_t^a x_t^{(i)} + D^a \pi_{t+1} - \theta^a\right) \tag{1}$$

$$\pi_{t+1} = \lambda n\left(\frac{p_{t+1}}{p_t}\right); t = 1, 2, \cdots, T \tag{2}$$

The external inputs $x_t^{(i)}$and the feedback π_{t+1} (price shift; p_t: FX-rate) are weighted by adjusting the weights $w_i^a andD^a$, respectively. By summing up the weighted inputs and comparing the sum with a threshold value θ_a, the agents opinion is formed. Afterwards the result is transformed through a nonlinearity (e.g., tanh(.)) in order to generate the neurons output signal z_t^a which is analogous to the action carried out by the agent (e.g., buying / selling). A neural network consisting of A such neurons for agents $a = 1, 2, \cdots, A$, describe the decision making process of agents. For simplicity, we assume that 1) no transaction costs are involved; 2) each agent is allowed to carry out one trading activity at a time, i.e.,z_t^a can either be positive or negative; 3) agents have the same time horizon T, and 4) agents have the same risk aversion characteristic. Typically, the trading activities of a single agent are driven by a profit-maximization task, according to

$$max\left[\frac{1}{T}\sum_t \pi_{t+1} z_t^a\right] \tag{3}$$

The market clearance condition, formulated as$\sum_a z_t^a = 0$describes an equilibrium condition for the FX-market. Agents $a = 1, \cdots, A$interact with each other through the market clearance condition.

If we have reached an equilibrium, i.e., the excess demand/supply is equal to zero and market is cleared, there is no need to adjust the market price in order to balance demand and supply, Formally, in case of market equilibrium,

$$\sum_a z_t^a = 0 \Rightarrow \pi_{t+1} = 0 \tag{4}$$

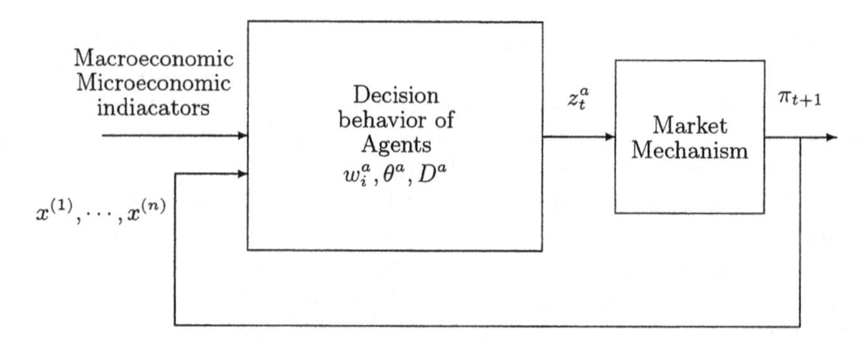

Fig. 1. Block diagram representation of FX-market model

In case the market clearance condition is violated, adjustments in market price will balance demand and supply. The price shift required to rebalance the market can be derived by

$$\sum_t z_t^a \neq 0 \Rightarrow \pi_{t+1} = C \sum_a z_t^a \tag{5}$$

where C is a positive constant. If there is excess demand, i.e., $\sum_a z_t^a > 0$ the FX-rate p_t tends to rise to re-balance the market $\pi_{t+1} > 0$ Since the excess demand tends to decrease when the price level pt is marked up, the partial derivative of the excess demand with respect to price shift is negative. Therefore

$$\frac{\delta \left(\sum_a z_t^a\right)}{\delta \pi_{t+1}} < 0 \tag{6}$$

3 Recurrent Neural Network Model for FX-Market

Now we integrate the elementary trading mechanism of the agents, their decision behaviour and objective function, and the market mechanism discussed in Section II into a neural network architecture [10,11]. The proposed architecture distinguishes between a training and a test phase. The training phase is done by a feedforward neural network (Fig.2) while we use a fix-point recurrent network during the test phase (Fig.3). The weight matrix between the external inputs $x^{(i)}$ and the agents layer is sparse and spreads the information among the agents such that their decision behaviour is heterogeneous. This assumption clearly represents the underlying dynamics of the real world. In our experiments, the sparseness of weight matrix is initialized randomly, i.e., weights which should be turned to zero are randomly selected.

During the training phase of the network (Fig.2), the agents try to maximize their profits using both external input information $x^{(i)_t}$ and the upcoming price shift . Simultaneous to agents profit maximization, the market mechanism adjusts the corresponding price level in order to vanish existing excess

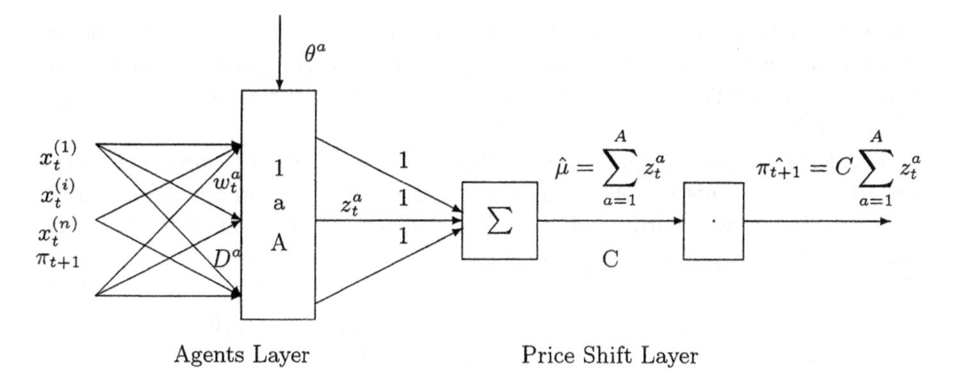

Fig. 2. Feedforward neural network for training phase

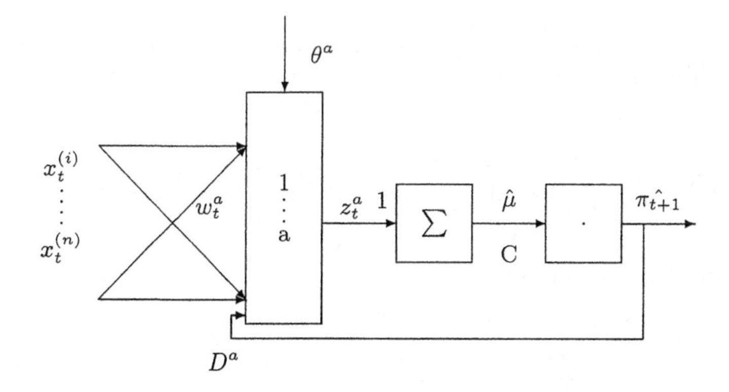

Fig. 3. Fix-point recurrent neural network for testing phase

demands/supplies. This is done indirectly by influencing the agents decision making using the related price shift π_{t+1} . The agents are provided with information concerning upcoming price movements by connecting an input terminal π_{t+1} to the agents layer. Since we have to guarantee that the partial derivative of excess demand with respect to price shift is negative (refer (6)) we only allow negative weights $D^a < 0$. The neural network of Fig.2 is trained with respect to the following error function :

$$E = \frac{1}{T} \sum_{T=1}^{T} \left(\pi_{t+1} - C \sum_{a=1}^{A} z_t^a \right)^2 - \lambda \sum_{a=1}^{A} \pi_{t+1} z_t^a \tag{7}$$

The first element in (7) is related to the market clearance condition. We use squared error function to minimize the divergence between the observed price shift π_{t+1} and the estimated price shift $\hat{\pi}_{t+1} = C \sum_a z_t^a$ which is derived from the agents decision making process. The second element in (7) describes the neural network implementation of agents objective function, i.e., profit maximization. It should be noted that the solutions are not sensitive to the choice of the penalty parameter λ.

The learning parameters are $w_i^a, \theta^a and D^a$. Parameter C varies between 0 and 1(0<C $\leq$1). This is not taken as a learning parameter in our algorithm. It is a separate tunable parameter (C is initialized sufficiently small to guarantee convergence of fix-point recurrence). From Eqn.(7), we obtain

$$\frac{\delta E}{\delta w_i^a} = -\left[C(\pi_{t+1} - \hat{\pi_{t+1}}) + \frac{1}{2}\lambda\pi_{t+1}\right]\left[(1-(z_t^a)^2)x_t^{(i)}\right] \quad (8)$$

$$\frac{\delta E}{\delta \theta^a} = -\left[C(\pi_{t+1} - \hat{\pi_{t+1}}) + \frac{1}{2}\lambda\pi_{t+1}\right][(1-(z_t^a)^2)] \quad (9)$$

Parameter D^a is tuned with respect to the following error function :$E\prime = \mu - \hat{\mu}$; $\mu = 0$ (Equilibrium condition)

$$\hat{\mu} = \sum_{a=1}^{A} z_t^a = \sum_{a=1}^{A} \tanh\left(\sum_{a=1}^{A} w_i^a x_t^{(i)} - \theta^a + D^a\pi_{t+1}\right); D^a < 0 \quad (10)$$

From this equation, we obtain

$$\frac{\delta E\prime}{\delta D^a} = -\frac{1}{2}[\pi_{t+1}][(1-(z_t^a)^2)] \quad (11)$$

Equations (9)-(11) provide the error derivatives for the gradient-descent back-propagation training of the neural network of Fig.2.

4 Simulation Results

For our empirical study, the database (www.economagic.com; www.bankofcanada.com) consists of monthly average values of FX-rate from January 1974 to November 2001 (335 data points) to forecast monthly price shifts of the FX-rate. Number of agents working in the market is assumed to be 40. As further decision of the empirical study, we had to assemble a database which serves as a raw input data for the agents taking part in the FX-market. The selection of raw inputs was guided by Gradojevic and Yang [9]. Besides the Canada/US dollar FX-rate itself, the agents may consider exports, imports, interest rate (Canada), interest rate (US), and crude oil price. We composed a data set of 5 indicators for the period of our empirical study. We trained the neural network until the convergence using incremental training. Following parameters were used in our model

- Excess demand constant C=0.25
- Discount rate λ=0.05
- Sparsity of parameter matrix$w_i^a = 0.35$
- Sparsity of parameter matrix $\theta^a = 0.8$
- Learning rate for training phase I $(parameterw_i^a, \theta^a) = 0.01$
- Learning rate for training phase II $(parametersD^a) = 0.01$
- Number of samples used for training=300

Figure 4 shows the simulation results for prediction of price-shift π (price=FX-rate).

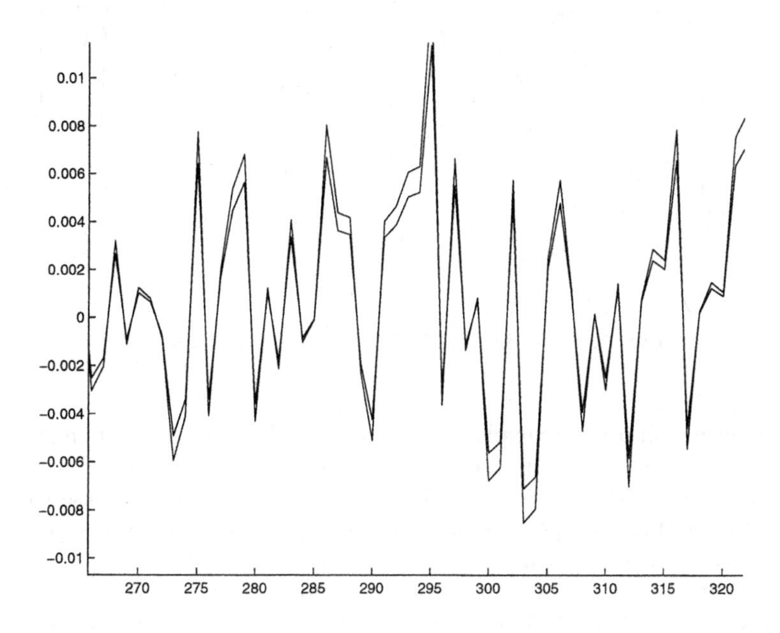

Fig. 4. Price-shift prediction using FX-market model

For evaluation of the performance of recurrent neural network model of the FX-market, we built a regressor model for forecasting using neuro-fuzzy strategy(fuzzy inference system with backpropagation tuning) . As inputs of the neuro-fuzzy network, we utilized the database of our multiagent approach. The results obtained using the neurofuzzy method (NFIS) of [12], are shown in Fig 5. As expected, the regressor model has poor forecasting power for the FX-rate.

5 Conclusions

The goal of this paper was to study the predictability of exchange rates with macroeconomic and microeconomic variables. A variable from the field of microstructure (order flow) is included in a set of macroeconomic variables (interest

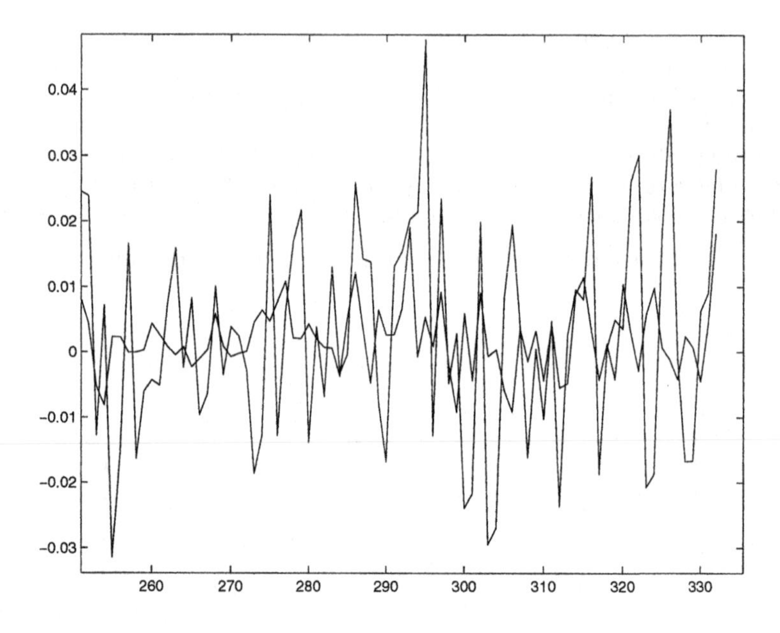

Fig. 5. Price-shift prediction using regression model

rate, and crude oil price) to explain Canada/US dollar exchange rate movements. Fuzzy Inference Systems with backpropagation tuning are employed to estimate a nonlinear relationship between exchange rate movements and these variables. The results confirm the poor performance of the time series regression models for exchange rate forecasting.

The prime focus of the paper is on a recurrent neural network model for FX-market to explain exchange rate movements. Asset prices are discovered in the marketplace by the interaction of market design and agents behaviour. The interaction is simulated by integrating 1) the FX-market mechanism; 2) an economic framework; and 3) the embedding of both tasks in neural network architectures. The results indicate that both macroeconomic and microeconomic variables are useful to forecast exchange rate changes. The inclusion of other microeconomic and macroeconomic variables could improve these findings. The power of this approach should be tested on other currencies.

References

1. Box, G., G.Jenkins, and G.Reinsel, Time Series Analysis : Forecasting and Control, 3rd ed., Englewood Cliffs, NJ: Prentice Hall, 1994.
2. Ashwani Kumar, D. P. Agrawal and S.D. Joshi, Neural Networks and Fuzzy Inference Systems for Business Forecasting, Proceedings of the International Conference on Fuzzy Systems and Knowledge Discovery, FSKD02, Singapore, Vol. 2, p. 661–666

3. Ashwani Kumar, D. P. Agrawal and S.D. Joshi, A GA-based method for generating TSK Fuzzy Rules from Numerical Data, Accepted for presentation and publication in the proceedings of the IEEE International Conference on Fuzzy Systems, FUZZ_IEEE2003, May, USA
4. Medeiros, M.C., A.Veiga, and C.E.Pedreira, Modeling Exchange Rates : Smooth Transitions, Neural Networks and Linear Models, IEEE Trans. Neural Networks, vol.12, no.4, 2001, pp.755–764.
5. White, H., and J.Racine, Statistical Inference, The Bootstrap, and Neural-Network Modeling with Application to Foreign Exchange Rates, IEEE Trans. Neural Networks, vol.12, no.4, 2001, pp.657–673.
6. Charron, M., A Medium-Term Forecasting Equation for the Canada-US Real Exchange Rate, Economic Analysis and Forecasting Division Canada, April 2000.
7. Amano, R.A., and S.van Norden, Terms of Trade and Real Exchange Rates : The Canadian Evidence, Journal of International Money and Finance 14(1) : 83–104
8. Amano, R.A., and S.van Norden, Exchange Rates and Oil Prices, Review of International Economics 6(4) : 683–94.
9. Gradojevic, N., and J.Yang, The Application of Artificial Neural Networks to Exchange Rate Forecasting : The Role of Market Microstructure Variables, Working Paper 2000–23, Bank of Canada, December 2000.
10. Zimmermann, G., and R.Neuneier, Multi-Agent Market Modeling of Foreign Exchange Rates, Advances in Complex Systems, vol.4, no.1 (2001) 29–43.
11. Zimmermann, H.G., R.Neuneier, and R.Grothmann, Multiagent Modeling of Multiple FX-Markets by Neural Networks, IEEE Trans. Neural Networks, vol.12, no.4, 2001, pp.735–743.
12. Ashwani Kumar, D. P. Agrawal and S.D. Joshi, Rapid Prototyping for Neural-Fuzzy Inference System Modeling based on Numerical Data" communicated for review to Soft Computing A Springer Journal.

Gaussian Mixture Density Estimation Applied to Microarray Data

Christine Steinhoff[1]*, Tobias Müller[1,3]*, Ulrike A. Nuber[2], and Martin Vingron[1]

[1] Max-Planck-Institute for Molecular Genetics, Computational Mol. Biology Human Molecular Genetics, Ihnestraße 63–73, D-14195 Berlin, Germany

[2] Department of Bioinformatics, Biocenter, University of Würzburg, Am Hubland, D-97074 Würzburg, Germany

{Christine.Steinhoff, Tobias.Mueller, Nuber, Martin.Vingron}@molgen.mpg.de

Abstract. Several publications have focused on fitting a specific distribution to overall microarray data. Due to a number of biological features the distribution of overall spot intensities can take various shapes. It appears to be impossible to find a specific distribution fitting all experiments even if they are carried out perfectly. Therefore, a probabilistic representation that models a mixture of various effects would be suitable. We use a Gaussian mixture model to represent signal intensity profiles. The advantage of this approach is the derivation of a probabilistic criterion for expressed and non-expressed genes. Furthermore our approach does not involve any prior decision on the number of model parameters. We properly fit microarray data of various shapes by a mixture of Gaussians using the EM algorithm and determine the complexity of the mixture model by the Bayesian Information Criterion (BIC). Finally, we apply our method to simulated data and to biological data.

1 Introduction

Microarray experiments can be used to determine gene expression profiles of specific cell types. In this way, genes of a given cell type might be categorized into active or inactive. Typically, one would like to infer a probability for each gene whether it is expressed in a given experiment or not. This has mainly been addressed by selecting an arbitrary threshold and defining spot intensities above this threshold as "high signal" and below as "low signal". However, this approach does not yield a probabilistic interpretation. As a probabilistic model we propose a mixture of normal distributions. Thus we derive not only an overall description of the data but also a probabilistic framework. There are many methods described in literature to fit various kinds of distributions to ratios of microarray data [1,2]. Only few attempts have been published in which a single distribution is fitted to the signal intensities rather than to the ratios of intensities. Hoyle et al. [3] propose to fit a lognormal distribution with a tail that is

* These authors contributed equally to this work

M.R. Berthold et al. (Eds.): IDA 2003, LNCS 2810, pp. 418–429, 2003.

close to a power law. This is - to our knowledge - the only publication in which single sample-intensities are being approximated. The use of one specific parameterized distribution however poses problems when overall intensities occur in various shapes, especially if they show non-uniformly shaped tails. These effects can occur for different reasons and cannot be captured by normalization procedures in all cases. Hoyle et al. [3] used only the highest expressed genes for the fitting procedure. This simultaneously reduces the probability to observe data having complex shaped, multi-modal distribution, because it focuses only on the highest expressed genes. Therefore, it provides no proper overall model. In [4] a lognormal distribution is fitted to the data in an iterative way. The dataset is iteratively reduced by truncating the data above an arbitrary threshold. Thus, they derive a model which only captures not expressed genes or low expressed genes while the high signal intensities remain unfitted. In this paper we present examples of microarray data which are unlikely to be properly fitted by a single lognormal distribution as described by Hoyle et al. [3]. One example is displayed in figure 1. The data appears to be a mixture of different distributions and can lead to multi-modal shapes of various kinds. Different reasons may account for this: (1) When using microarrays with a small number of probes there might be genes being over-represented at a specific intensity-range. (2) Signal saturation effects could be another reason. It is not advisable for all biological settings to remove this effect by stretching upper intensity range. Or, it might be not possible to repeat the experiment. (3) Also, specific effects which can not be localized and captured by normalization might lead to varying shapes. (4) If one microarray-design is based on different oligo selection procedures the resulting intensity distribution could show more than one mode. Thus, we assume that there is no unimodal single overall distribution which can explain microarray experiments in general. Even though scanner effects or local effects could be excluded by repeating experiments, this might be difficult and expensive in some cases. Although we observe a strong bias in the data it might be possible to answer specific biological questions reliably. This is the case if one is not particularly interested in the quantification of expression levels. An example for this could be the selection of a set of genes that are highly expressed in a specific tissue. Furthermore, it might be desirable to give a probability for a gene to be highly expressed. The main intention of this paper is to infer a robust probabilistic framework that captures inhomogeneous datasets. The paper is organized as follows: In Section 2 we introduce the statistical concepts that we use to fit a mixture model. In Section 3 we describe the analyzed biological as well as the simulated datasets. In Section 4 we show the results of different datasets and compare our method with the established one [3]. Finally, we discuss the merits of our approach and its biological implications.

2 Statistics of Mixture Models

In this section we introduce the statistical concepts that are necessary to formulate the estimation problem of *Gaussian mixture models* (GMMs). First, we briefly review the method of maximum-likelihood followed by a short description of the basic *Expectation-Maximization* (EM) algorithm [5]. Finally we apply this algorithm to the estimation problem of GMMs. For a detailed introduction and discussion of this problem see [6]. [7] is an excellent tutorial of the EM algorithm where also some applications are given.

2.1 Maximum-Likelihood

First, we recall the definition of the maximum-likelihood estimation problem. We are given a set of density functions $p(\cdot|\Theta)$ indexed by a parameter vector Θ. We also have a dataset $\mathcal{D} = \{x_1, \dots, x_N\}$ of size N, supposedly drawn independently and identically from one of these density functions. Therefore, we get the following expression for the likelihood of the parameters given the data $p(\mathcal{D}) = \prod_{i=1}^{N} p(x_i|\Theta) = \mathcal{L}(\Theta|\mathcal{D})$. The likelihood $\mathcal{L}(\cdot)$ is considered as a function of the parameters Θ where the dataset $\mathcal{D}$ is fixed. In the maximum-likelihood problem, the goal is to find the parameter vector Θ that maximizes $\mathcal{L}$. That is, we wish to find $\Theta^\star$ where $\Theta^\star = \operatorname{argmax}_\Theta \mathcal{L}(\Theta|\mathcal{D})$. Equivalently, one looks for the maximum of $\log \mathcal{L}(\Theta|\mathcal{D})$ because this term is often analytically easier to manipulate. Depending on the form of the density $p(x|\Theta)$ this optimizing problem often turns out to be very hard but in some cases it can be solved easily and analytically. For many problems, however, it is not possible to find such analytical expression, and one needs to use more elaborate techniques.

2.2 The Expectation Maximization Algorithm

The EM algorithm is one such elaborate technique. It is a general method for finding the maximum-likelihood estimates of an underlying distribution from a given dataset when the data is incomplete. Let us assume that we are given the data $\mathcal{Z} = (\mathcal{X}, \mathcal{Y})$ where $\mathcal{X}$ denotes the *observed* data and $\mathcal{Y}$ denotes the *unobserved* data. In the sequel we model $\mathcal{Y}$ as a random variable. Therefore, also the *complete-data* log-likelihood $\mathcal{L}(\Theta|\mathcal{X}, \mathcal{Y})$ constitutes a random variable. The EM algorithm starts with an initial guess of the parameter vector $\Theta^{(0)}$ (INIT-STEP). Then the expected complete-data log-likelihood

$$Q(\Theta, \Theta^{(i-1)}) = \int_y \log p(\mathcal{X}, y|\Theta) p(y|\mathcal{X}, \Theta^{(i-1)})\, dy \tag{1}$$

is calculated. For a given parameter vector $\Theta^{(i-1)}$ the function $Q(\cdot, \Theta^{(i-1)})$ is real valued. The evaluation of the expectation $Q(\Theta, \Theta^{(i-1)})$ is called the E-STEP. In the next step the expected complete-data log-likelihood is maximized $\Theta^{(i)} = \operatorname{argmax}_\Theta Q(\Theta, \Theta^{(i-1)})$ over all possible values for Θ (M-STEP). The E-STEP and the M-STEP are repeated until convergence is attained. Applying Jensen's inequality shows that in each iteration the log-likelihood is increased. Furthermore many authors have shown [8,9] that under general regularity conditions, the solution from the EM algorithm will converge to a local maximum of the log-likelihood.

2.3 Gaussian Mixture Models

Let us start with a mixture of M densities

$$p(x|\theta) = \sum_{i=1}^{M} \alpha_i p_i(x|\theta_i) \tag{2}$$

with the parameter vector $\Theta = (\alpha_1, \ldots, \alpha_M, \theta_1, \ldots, \theta_M)$ and $\sum_i \alpha_i = 1$. This means, we assume that we are dealing with M component densities mixed together with M mixing coefficients α_i. Thus, we obtain the following so called *incomplete-data-log-likelihood* for the data $\mathcal{D}$: $\log \mathcal{L}(\Theta|\mathcal{D}) = \log \prod_{i=1}^{N} p(x_i|\Theta) = \sum_{i=1}^{N} \log \sum_{j=1}^{M} \alpha_j p_j(x_i|\theta_j)$ which is difficult to optimize because it contains the log of the sum in the right hand side expression.

The key trick is to consider the data $\mathcal{D}$ as incomplete and posit the existence of unobserved data items $Y = \{y_i\}_{i=1}^{N}$, where each y_i inform us from which component each item x_i was generated. That is $y_i = k$ means that x_i was generated by the k-th mixture density. Surprisingly, the introduction of the missing values simplifies the expression for the log-likelihood: $\log \mathcal{L}(\Theta|\mathcal{D}, \mathcal{Y}) = \log \mathbb{P}(X, Y|\Theta) = \sum_{i=1}^{N} \log \mathbb{P}(x_i|y_i)\mathbb{P}(y) = \sum_{i=1}^{N} \log p_{y_i}(x_i|\theta_{y_i})\alpha_{y_i}$. Since we model the unobserved data $\mathcal{Y}$ as a random vector, we first have to determine the distribution of $\mathcal{Y}$. This can be done by an application of Bayes' rule:

$$p(y_i|x_i, \Theta^\star) = \frac{\mathbb{P}(y_i)p(x_i|y_i, \Theta^\star)}{\sum_k \mathbb{P}(y_k)p(x_i|y_k, \Theta^\star)} = \frac{\alpha^\star_{y_i} p(x_i|\Theta^\star)}{\sum_k \alpha^\star_{y_k} p_{y_k}(x_i|\Theta^\star_k)} ,$$

where we initially have to guess the parameter $\Theta^\star$ of the mixture model. The mixing parameters $\alpha^\star_i$ can be thought of as prior probabilities for density i. Our guesses $\Theta^\star = (\alpha^\star_1, \ldots, \alpha^\star_M, \theta^\star_1, \ldots, \theta^\star_M)$ are based on an initial run of a k-means clustering procedure.

Because the unobserved data-points $\mathcal{Y} = (y_1, \ldots, y_N)$ are modeled to be independent of each other we get $p(y|\mathcal{D}, \Theta^\star) = \prod_{i=1}^{N} p(y_i|x_i, \Theta^\star)$, which is the desired marginal density for performing the EM algorithm, see equation (1).

If we plug in the density of the unobserved data $\mathcal{Y}$ in the term of the expression of the EM-algorithm (1) one can show that the expression takes the form

$$Q(\Theta, \Theta^\star) = \sum_{l=1}^{M} \sum_{i=1}^{N} \log(\alpha_l) p(l|x_i, \Theta^\star) + \sum_{l=1}^{M} \sum_{i=1}^{N} \log(p_l(x_i|\theta_l)) p(l|x_i, \Theta^\star) . \tag{3}$$

The evaluation of this equation completes the E-STEP.

In the M-STEP we have to find the maximum of $Q(\Theta, \Theta^\star)$ in Θ, which can be derived by calculating the roots of the derivative of $Q(\Theta, \Theta^\star)$. Since both summands are not related, see equation (3), this can be done independently from each other. As maximum-likelihood estimates for α_l we get $\alpha_l = 1/N \sum_{i=1}^{N} p(l|x_i, \Theta^\star)$. This means in particular that the term for α_l is independent of the structure of the density $p(l|x_i, \Theta)$. Clearly, this is not the case for the maximum-likelihood estimators for the parameters θ_l. These are the roots of the equation

$$\frac{\delta}{\delta\theta_l} \sum_{l=1}^{M} \sum_{i=1}^{N} \log\left(p_l(x_i|\theta_l)\right) p(l|x_i, \theta^\star) \ . \tag{4}$$

Note that the density $p(l|x_i, \theta^\star)$ depends on the initial guess $\theta^\star$ and not on θ_l. This is also the reason why we can derive in some cases a closed formula for the maximum-likelihood estimators for θ_l.

In our case we want to derive the most likely mixture of Gaussians. Therefore we consider a parameter vector $\Theta = (\mu, \sigma)$, where $\mu = (\mu_1, \ldots, \mu_M)$ and $\sigma = (\sigma_1, \ldots, \sigma_M)$ and densities $p_l(x|\mu_l, \sigma_l) = (2\pi\sigma_l^2)^{-1/2} \exp(-\frac{(x_i-\mu_l)^2}{2\sigma_l^2})$. As maximum-likelihood estimates we get

$$\mu_l = \frac{\sum_{i=1}^{N} x_i p(l|x_i, \theta^\star)}{\sum_{i=1}^{N} p(l|x_i, \theta^\star)} \quad \text{and} \quad \sigma_l^2 = \frac{\sum_{i=1}^{N} (x_i - \mu_l)^2 p(l|x_i, \theta^\star)}{\sum_{i=1}^{N} p(l|x_i, \theta^\star)}$$

for the means and for the variances. In summary, the estimates of the new parameters in terms of the old parameters are as follows:

$$\alpha_l^{new} = \frac{1}{N} \sum_{i=1}^{N} p(l|x_i, \Theta^\star), \mu_l^{new} = \frac{\sum_{i=1}^{N} x_i p(l|x_i,\theta^\star)}{\sum_{i=1}^{N} p(l|x_i,\theta^\star)}, \sigma_l^{2\,new} = \frac{\sum_{i=1}^{N} (x_i - \mu_l^{new})^2 p(l|x_i,\theta^\star)}{\sum_{i=1}^{N} p(l|x_i,\theta^\star)} .$$

In our application we ran the EM algorithm with a number of different initial guesses derived from an initial randomized k-means procedure. Thus the EM algorithm yields a series of estimates from which we choose that with the highest likelihood.

2.4 Application to Microarray Data

Our goal is to model the entity of microarray data which do not follow a specific distribution but rather show shapes which might be based on several distributions. In fact what we observe by looking at the whole set of data is a mixture of many of distributions. In our case we assume a mixture of normal distributions. The maximum-likelihood parameters of this probabilistic model are estimated by an application of the EM algorithm.

The EM algorithm is based on an initial guess of the parameters that are to be estimated. We derive these initial rough estimates by an application of a K-means clustering procedure where the initial centers are chosen randomly. Then the initial guesses of the EM algorithm are given by the sample mean and sample variance of each K-means cluster, where the mixing parameter is just the relative cluster size.

The EM algorithm only guarantees the convergence to a local maximum of the likelihood, but we are interested in finding a global maximum. Thus we repeat the initial K-means procedure as well as the following EM steps many times. From the resulting estimates we choose the most likely ones in order to get a better approximation of the global maximum of the likelihood.

Since we do not know the number of clusters this parameter has to be estimated too. Clearly, if we increase the number of parameters we will simultaneously achieve higher likelihoods, but we also run into the problem of overfitting. To overcome this problem it is common practice to look at the *penalized* likelihood where additional parameters lower the likelihood of the data [10]. The Bayesian Information Criterion (BIC) is a widely used criterion to select a model out of a sequence of models involving a different number of parameters. Here, the model with the highest penalized likelihood $BIC = 2 * \log \mathcal{L}(\Theta|\mathcal{D}) - (\log N) * d$ is chosen, where N denotes the size of the data and d the number of parameters in the fitted model. This criterion was first described by Schwarz ([11]). In fact, BIC is proportional to the Akaike information criterion (AIC). However, using BIC fewer high complex models will be accepted and thus strict penalization of a high number of distributions in our mixture model will be achieved. We used the EM algorithm for a number of distribution mixtures k and computed for each model the BIC. The model with maximal BIC was chosen and in fact there was always a clear maximum.

3 Datasets

We analyze two different types of datasets to judge the performance of our fitting procedure. These comprise one dye-swap experiment and one simulated dataset.

We first analyze data derived from a human X chromosome-specific cDNA array with 4799 spotted cDNA sequences (dataset 1). IMAGE clones were obtained from the resource center of the German human genome project (`http://www.rzpd.de/`). RNA is isolated from B-cell rich lymphoblastoid cells of a healthy male person and a mentally retarded male patient and labeled in reverse transcriptase reaction with Cy3- or Cy5-dUTP. Labeled cDNAs are co-hybridized on the array and a repeat experiment is performed with fluorescent dyes swapped. Fluorescence intensities of Cy3 and Cy5 are measured separately at 532 and 635 nm with a laser scanner (428 Array Scanner, Affymetrix). Image processing is carried out using Microarray Suite 2.0 (IPLab Spectrum Software, Scanalytics, Fairfax, VA, USA). Raw spot intensities are locally background subtracted. The intensity values of each channel of the co-hybridization experiment are then subjected to variance stabilizing normalization as described in [12]. As a second dataset we draw 10^4 samples from a mixture of normal densities. A necessary condition for the proposed estimation procedure is that the underlying model parameters can at least be reestimated. We show that our implementation of the EM algorithm in fact can properly infer the underlying model. As an example we show the performance sampling for the following mixture model: $0.6N(0.5, \sqrt{0.6}) + 0.3N(1.5, \sqrt{0.4}) + 0.1N(3.2, \sqrt{0.5})$ (dataset 2a). Since there are discussions that log-microarray data could follow a normal distribution in low intensities and could show a gamma distribution like tail in high intensities, we also show, that a mixture of a normal and a gamma distribution can be approximated quite well by a mixture of normal distributions. Furthermore, the normal distribution of the underlying model can be properly reestimated.

As an example we sample once more 10^4 samples from the following model: $\alpha_1 N(\mu, \sigma) + \alpha_2 F(r, a)$, where $\alpha = [0.6, 0.4]$, $\mu = 0.5$, $\sigma = 1.2$, $r = 2$, $a = 3$ and F is a gamma distribution with parameters r and a (dataset 2b).

4 Results

In dataset 1 we first normalized the two co-hybridized samples. Normalized values of the same targets were averaged and used in the analysis. While normalizing all data points, empty spots and plant control genes were excluded from the mixture fitting. We fitted the data of each channel using mixtures of 1 up to 5 normal distributions. As an example the result from one target sample is shown in table 1. It appears that there is an optimum according to BIC using

Table 1. Mixtures of normal distributions: The table gives an overview of fitting 1-5 mixtures of normal distributions via the EM-algorithm. The first column indicates the number of normal distributions used for the fit. The second, 7th and 12th columns give the resulting BIC value divided by 10^4 (BIC'). The third, 8th and 13th columns display the log-likelihood divided by the length of data. The columns labeled with μ, σ, α depict the estimated mean, standard deviation and mixing probability resp.

K	Set 1 BIC'	LL/N	μ	σ	α	Set 2a BIC'	LL/N	μ	σ	α	Set 2b BIC'	LL/N	μ	σ	α
1	1.5785	−1.6643	0.94	1.28	1.00	2.8926	−2.8926	1.08	1.03	1.00	5.5672	−2.7827	2.69	3.91	0.676
2	1.5118	−1.5912	0.43	0.82	0.69	2.6642	−1.3298	0.60	0.58	0.65	4.9658	−2.4806	0.80	1.32	0.67
			3.09	1.31	0.31			1.87	1.04	0.35			6.47	4.58	0.33
3	1.4971	−1.5731	0.89	0.53	0.38	2.6604	−1.3265	0.50	0.54	0.60	4.9224	−2.4575	0.66	1.25	0.64
			2.26	0.78	0.53			1.58	0.60	0.33			4.80	2.86	0.27
			4.48	0.95	0.09			3.26	0.67	0.07			10.52	5.29	0.09
4	1.4934	−1.5665	0.86	0.51	0.37	2.6620	−1.3259	0.45	0.50	0.45	4.9160	−2.4530	0.59	1.22	0.63
			2.12	0.69	0.45			1.28	0.77	0.39			3.73	2.05	0.20
			3.53	0.95	0.16			1.90	1.15	0.12			8.21	3.31	0.14
			5.75	0.25	0.02			3.42	0.39	0.03			14.52	5.84	0.03
5	1.4960	−1.5665	0.78	0.48	0.29	2.6644	−1.3258	0.44	0.48	0.37	4.9185	−2.4528	0.09	1.11	0.25
			1.73	0.69	0.29			0.99	0.76	0.38			0.94	1.20	0.39
			2.32	0.72	0.25			1.65	0.77	0.17			3.82	1.98	0.18
			3.54	0.97	0.15			2.92	0.86	0.06			7.94	3.39	0.15
			5.75	0.25	0.02			3.43	0.28	0.02			14.23	5.92	0.03

a mixture of 4 normal distributions. The resulting optimal fit is displayed in figure 1. In the high intensity range it appears as if there was a signal saturation effect which in fact can be confirmed from the scanner analysis. To determine expressed genes in the patient's tissue it is not necessary to exclude this experiment. Furthermore, especially in clinical settings it might be impossible to repeat experiments due to lack of additional patient's tissue material. In figure 1 we show that empty spots are clearly separated, even from clones with lowest intensities which suggests, that non-expressed genes display an overall higher signal than empty spots after normalization. That is what one might expect from the biochemical background. In figure 1 it is apparent that the saturation

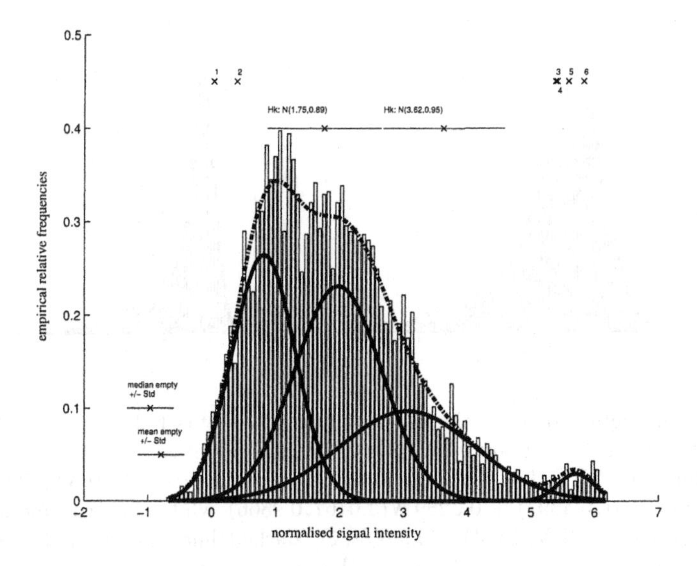

Fig. 1. The figure displays the optimal fitting of dataset 1. In the high intensity-range a probable saturation effect is being captured by one normal distribution given by $N(5.75, 0.25)$. The whole mixture model has been estimated to be: $Y = 0.3682\,N(0.8573, 0.5104) + 0.449\,N(2.1163, 0.686) + 0.1638\,N(3.5254, 0.9481) + 0.019\,N(5.7473, 0.2539)$ (dashed line). Each single distribution is marked by solid lines. In the lower left part of the first figure mean +/- standard deviation and median +/- standard deviation of empty spots which have been excluded from the mixture fitting are being displayed. In the upper part of the figure the means and standard deviations as estimated by the EM algorithm for housekeeping genes is shown. In the very upper part location of housekeeping genes is marked with x followed by a number: (1) G6PD, (2) TAX1BP1, (3) ATP50, (4) EEF1A1, (5) RPL37A and (6) RPS4X.

effect is clearly described by one additional normal distribution. One would expect that for the non-saturated data three distributions would optimally fit. In order to get an impression of the expression level we also fitted housekeeping gene intensities separately. Housekeeping genes were obviously not differentially expressed and were best fitted using 2 normal distributions.

For the highest and lowest expressed housekeeping genes present on the array (G6PD, TAX1BP1 (both low) and EEF1A1, ATP50, RPL37A and RPS4X (high)) we used GeneCards^{TM} (`http://bioinfo.weizmann.ac.il/cards/`) to judge their expression in normal human tissues based on quantifying ESTs from various tissues in Unigene clusters. Thus, from the biological background it is apparent that G6PD and TAX1BP1 are expressed at a very low level while we expect a high expression for EEF1A1, ATP50, RPL37A and RPS4X in cells of lymphoid organs. These expectations are in accordance with our data.

To determine the expression of X chromosomal genes in patient's lymphoblastoid cells we would suggest to model non-expressed genes with a normal distribu-

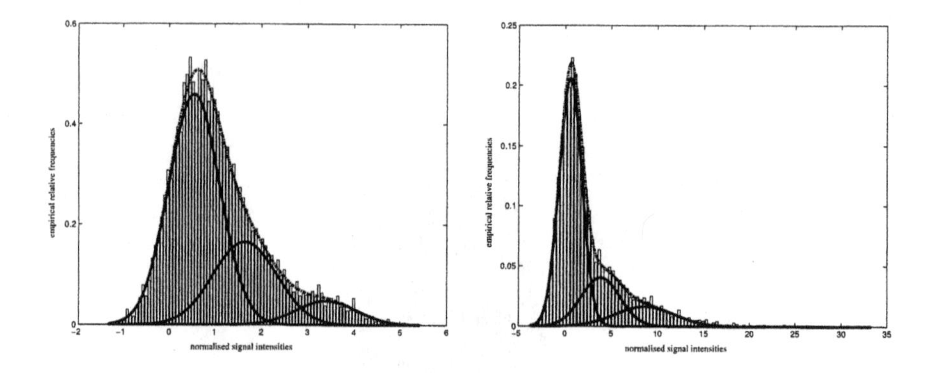

Fig. 2. The figure displays the optimal fitting of dataset 2a and 2b. The figure on the left hand side displays the optimal fitting of dataset 2 part a using the EM algorithm and BIC. The optimal fit is given by $Y = 0.2620\,N(1.4991, 0.2935) + 0.4120\,N(1.9231, 0.4391) + 0.3259\,N(3.0167, 0.9866)$ while the true model was $Y = 0.3\,N(1.5, 0.3) + 0.4\,N(2, 0.4) + 0.3\,N(3, 1)$ (dashed line). Each single distribution is marked by solid lines. The figure on the right hand side displays the optimal fitting of dataset 2 part b using the EM algorithm. The optimal fit is given by $Y = 0.6271\,N(0.5867, 1.2198) + 0.2022\,N(3.7347, 2.0450) + 0.1447\,N(8.2126, 3.3050) + 0.026\,N(14.5245, 5.8399)$ while the true model was $Y = 0.6\,N(0.5, 1.2) + 0.4\,F(2, 3)$, where $F(2, 3)$ is a gamma distributions with parameter $r = 2$; $a = 3$ (dashed line). Each single distribution is marked by solid lines.

tion given by $N_1 = N(0.86, 0.51)$. Now, we can easily retrieve a list of probabilities for each gene to belong to each of the mixture components. If we interpret the normal distribution given by $N_1 = N(0.86, 0.51)$ as non - expressed genes we can infer a probability measure for each gene belonging to N_1 and thus being not expressed or belonging to the mixture of expressed genes.

We simulated a dataset of normal distributions (dataset 2a) to demonstrate that the EM algorithm can in fact properly reestimate the underlying model. This is displayed on the left hand side in figure 2. The number of normal distributions in the mixture as well as the parameters α, μ, σ are almost exactly reestimated. It has been proposed that high intensities in log scale might follow a gamma distribution while low intensities follow a normal distribution. We therefore simulated a mixture of a normal and a gamma distribution (dataset 2b). In fact, this mixture can be fitted very well and the underlying normal distribution is being reestimated properly as shown in the right hand side of figure 2 and in table 1. For each dataset we fit a power law for (A) the highest expressed 500 genes as described in [3], (B) all genes. This is displayed in figure 3. While fitting the highest expressed 500 genes the use of a power law seems to be a good approximation for simulated data of a mixture of normal distributions as well as for simulated data of a mixture of a normal and a gamma distribution, the fits perform very badly in the case of biological data. The fit is even worse in the case of dataset (1) in which we have a signal saturation effect. In part (B)

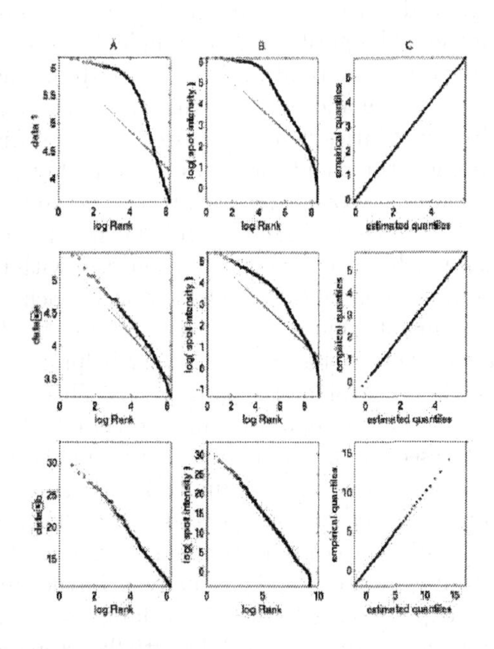

Fig. 3. Overview of fitting results. In column A the logarithm of normalized intensities (y-axis) of each dataset is plotted against log-rank of top 500 spots (x-axis). The solid line marks the fit to Zipf's law. In column B the logarithm of normalized intensities (y-axis) of each dataset is plotted against log-rank of all spots (x-axis). In column C estimated quantiles resulting from the optimal fit to a mixture of normal distributions (x-axis) is plotted against the empirical quantiles of each dataset 1 (first row), 2a (second row) and 2b (third row).

of figure 3 fitting of the whole dataset to a power law is shown. As one expects the fitting gets worse due to the fact that in low intensities an approximation using a lognormal distribution would be appropriate [3]. On the other hand the quantile-quantile-plots comparing empirical quantiles show a very good overall fit to the data. This is displayed in part (C) of figure 3.

5 Discussion and Conclusion

In our study we propose a new robust probabilistic description of microarray datasets. This description comprises various effects, these are avoidable experimental effects as well as immanent effects. The motivation of our approach is to incorporate in our probabilistic model effects of microarray data like saturation, local effects and others. This mixture of effects in the biological data is modeled by a mixture of normal densities. The parameters of the model are estimated by an application of the EM-algorithm. The complexity of the mixture model is determined by the Bayesian Information Criterion (BIC). There are several biological questions that can be addressed by this approach. The proposed model provides genes with probabilities to be expressed at a specific level. Often one is interested in defining a threshold for genes to be highly expressed, as is the case in tissue classification. Typically, an arbitrary and convenient threshold is chosen like the highest x-percent. Using the probabilistic framework we can however choose the genes which are assigned to the densities with the highest means. The underlying idea of using a mixture model is that to each fitted density there is a biological or experimental effect of the associated data. These effects are a priori unknown. In this context, however, it is very helpful if a priori knowledge is available, like using housekeeping or spike-in genes. This is done in our biological datasets. In this case we can rephrase a more precise biological question like: Which genes are expressed at the same level as the highly expressed controls? That means in particular that we are not guided by one single spike-in gene, but we make our decision based on the associated density. Other questions focus on quality checks in series of microarray experiments. This might be based on the comparison of fitted mixture models. And finally, a further important question addresses differential expression. For two given datasets two models are fitted. The question, whether a certain gene is differentially expressed or not, can be formulated in a straightforward way in terms of the mixture densities. There have been many attempts addressing the problem of fitting a mixture model to microarray data. However, typically a mixture model is being fitted to the logarithm of the expression ratios. Ghosh and Chinnaiyan [1] propose an algorithm in which first the number of measured spot intensities is extremely reduced by preprocessing steps. For this procedure one either needs additional biological knowledge or one has to choose an arbitrary threshold which yields obviously truncated data. Initially, they use hierarchical agglomerative clustering and reduce the dimension of the data by principal component analysis. In another approach McLachlan et al. [6] focus on a classification of samples (for example tissues) based on a mixture model describing the ratios of preprocessed intensities. Again they use only a subset of genes for their fitting procedure rather than fitting the whole dataset. Medvedovic and Sigaganesan [13] use log ratios of a selected subset of genes for which significant results already were shown. Their clustering procedure is based on a Bayesian infinite mixture model and similar clusters are derived from the posterior distribution given by a Gibbs sampler. However, there is one approach described in the literature which is based directly on expression intensities [3]. In this paper the authors focus on the very right tail of the expression intensities that represent very highly expressed genes to fit a Zipf-law. Although they suggest that the intensities follow a mixture of a log normal and a power-law, they do not take advantage of this special structure in their fitting procedure. Therefore, they run into the problem to define precisely the threshold above which the Zipf-law has to be fitted. First, based on simulated data we demonstrate that in case of "perfect data" a power law approximation performs well, but this is also the case for the proposed mixture model of normal densities. Secondly, we focus on biological data. This kind of data often shows some effects. As is carried out in [3] the fit of the Zipf-law to this kind of data is very bad. Clearly, such a simple model can not comprise effects of various kinds. There is a need for a more complex model. We do not think that microarray data can be modeled to follow a specific class of distribution because already the selection bias of clones has an effect on the shape of the distribution. A mixture model based approach can capture this kind of data and therefore is the appropriate model for describing microarray data. Summarizing, in this paper it is demonstrated that it is natural to model directly the intensities and not the ratio of intensities. This kind of data is optimally described by a mixture of log normal distributions, which can be fitted by the EM-algorithm efficiently. We prove that the dimension of the model can be derived from the Bayesian information criterion very well. Using this probabilistic framework the underlying biological question can be answered in a straightforward manner including associated probabilities.

References

1. Ghosh, D., Chinnaiyan, A.M.: Mixture modelling of gene expression data from microarray experiments. Bioinformatics **18** (2002) 275–86
2. Li, W., Yang, Y.: Zipf's law in importance of genes for cancer classification using microarray data. J Theor Biol **219** (2002) 539–551
3. Hoyle, D.C., Rattray, M., Jupp, R., Brass, A.: Making sense of microarray data distributions. Bioinformatics **18** (2002) 576–84
4. Beissbarth, T., Fellenberg, K., Brors, B., Arribas-Prat, R., Boer, J., Hauser, N.C., Scheideler, M., Hoheisel, J.D., Schutz, G., Poustka, A., Vingron, M.: Processing and quality control of dna array hybridization data. Bioinformatics **16** (2000) 1014–22
5. Dempster, A.P., Laird, N.M., Rubin, D.B.: Maximum likelihood from incomplete data via the EM algorithm. JRSSB **39** (1977) 1–38
6. McLachlan, G., Bean, R., Peel, D.: A mixture model-based approach to the clustering of microarray expression data. Bioinformatics **18** (2002) 413–22
7. Bilmes, J.A.: A gentle tutorial on the EM algorithm and its application to parameter estimation for gaussian mixture and hidden markov models (1998)
8. Wu, C.F.J.: On the convergence of the EM algorithm. Ann. Stat. **11** (1983) 95–103
9. Boyles, R.A.: On the convergence of the EM algorithm. JRSS B **45** (1983) 47–50
10. McLachlan, G., Peel, D.: Finite Mixture Models. Wiley Series in Probability and Statistics. Wiley, New York (2000)
11. Schwarz, G.: Estimating the dimension of a model. Ann. Stat. **6** (1978) 461–64
12. Huber, W., Heydebreck, A.V., Sultmann, H., Poustka, A., Vingron, M.: Variance stabilization applied to microarray data calibration and to the quantification of differential expression. Bioinformatics **18 Suppl 1** (2002) S96–S104
13. Medvedovic, M., Sivaganesan, S.: Bayesian infinite mixture model based clustering of gene expression profiles. Bioinformatics **18** (2002) 1194–206

Classification of Protein Localisation Patterns via Supervised Neural Network Learning

Aristoklis D. Anastasiadis, George D. Magoulas, and Xiaohui Liu

Department of Information Systems and Computing, Brunel University, UB8 3PH, Uxbridge, United Kingdom
{Aristoklis.Anastasiadis, George.Magoulas, Xiaohui.Liu}@brunel.ac.uk

Abstract. There are so many existing classification methods from diverse fields including statistics, machine learning and pattern recognition. New methods have been invented constantly that claim superior performance over classical methods. It has become increasingly difficult for practitioners to choose the right kind of the methods for their applications. So this paper is not about the suggestion of another classification algorithm, but rather about conveying the message that some existing algorithms, if properly used, can lead to better solutions to some of the challenging real-world problems. This paper will look at some important problems in bioinformatics for which the best solutions were known and shows that improvement over those solutions can be achieved with a form of feed-forward neural networks by applying more advanced schemes for network supervised learning. The results are evaluated against those from other commonly used classifiers, such as the K nearest neighbours using cross validation, and their statistical significance is assessed using the nonparametric Wilcoxon test.

1 Introduction

An area of protein characterization that it considered particularly useful in the post-genomics era is the study of protein localization. In order to function properly, proteins must be transported to various localization sites within a particular cell. Description of protein localization provides information about each protein that is complementary to the protein sequence and structure data. Automated analysis of protein localization may be more complex than the automated analysis of DNA sequences; nevertheless the benefits to be derived are of same importance [1]. The ability to identify known proteins with similar sequence and similar localization is becoming increasingly important, as we need structural, functional and localization information to accompany the raw sequences. Among the various applications developed so far, the classification of protein localization patterns into known categories has attracted significant interest. The first approach for predicting the localization sites of proteins from their amino acid sequences was an expert system developed by Nakai and Kanehisa [8,9]. Later, expert identified features were combined with a probabilistic model, which could learn its parameters from a set of training data [4].

M.R. Berthold et al. (Eds.): IDA 2003, LNCS 2810, pp. 430–439, 2003.

Better prediction accuracy has been achieved by using standard classification algorithms such as K nearest neighbours (KNN), the binary decision tree and a naïve Bayesian classifier. The KNN achieved the best classification accuracy comparing to these methods in two drastically imbalanced dataset namely the E.coli and Yeast [5]. E.coli proteins were classified into 8 classes with an average accuracy of 86%, while Yeast proteins were classified into 10 classes with an average accuracy of 60%. Recently, genetic algorithms, growing cell structures, expanding range rules and feed-forward neural networks were comparatively evaluated for this problem, but no improvements over the KNN algorithm were reported [2].

This paper is not about proposing another classification algorithm, but rather about conveying the message that some existing algorithms, if properly used, can lead to better solutions to this challenging real-world problem. The paper advocates the neural network-based approach for classifying localization sites of proteins. We investigate the use of several supervised learning schemes [11] to improve the classification success of neural networks. The paper is organized as follows. First, we briefly describe the methods that we use to predict the localisation sites. Next, we introduce the datasets and the evaluation methodology that are used in the paper, and presents experimental results by the neural network approach when compared with the best existing method.

2 Classification Methods

2.1 The K Nearest Neighbours Algorithm

Let $X = \{x^i = (x_1^i, \ldots, x_p^i), i = 1, \ldots, N\}$ be a collection of p-dimensional training samples and $C = \{C_1, \ldots, C_M\}$ be a set of M classes. Each sample x^i will first be assumed to posses a class label $L_i \in \{1, \ldots, M\}$ indicating with certainty its membership to a class in C. Assume also x^S be an incoming sample to be classified. Classifying x^S corresponds to assigning it to one of the classes in C, i.e. deciding among a set of M hypotheses: $x^S \in C_q, q = 1, \ldots, M$. Let Φ^S be the set of the K nearest neighbours of x^S in X. For any $x^i \in \Phi^S$, the knowledge that $L_i = q$ can be regarded as evidence that x^S increases our belief that also belongs to C_q. However, this piece of evidence does not by itself provide 100% certainty. The K nearest neighbours classifier (KNN), as suggested by Duda and Hart [3], stores the training data, the pair (X, L). The examples are classified by choosing the majority class among the K closest examples in the training data, according to the Euclidean distance measure. In our experiments we set $K = 7$ for the E.coli proteins and $K = 21$ for the yeast dataset; these values have been found empirically to give the best performance [5].

2.2 Multilayer Feed-Forward Neural Networks and Supervised Learning

In a multilayer Feed-forward Neural Network (FNN) nodes are organised in layers and connections are from input nodes to hidden nodes and from hidden nodes to output nodes. In our experiments we have used FNNs with sigmoid hidden and output nodes. The notation $I-H-O$ is used to denote a network architecture with I inputs, H hidden layer nodes and O outputs nodes. The most popular training algorithm of this category is the batch Back-Propagation; a first order method that minimizes the error function using the steepest descent.

Adaptive gradient-based algorithms with individual step-sizes try to overcome the inherent difficulty of choosing the appropriate learning rates. This is done by controlling the weight update for every single connection during the learning process in order to minimize oscillations and maximize the length of the step-size. One of the best of these techniques, in terms of convergence speed, accuracy and robustness with respect to its parameters, is the Rprop algorithm [11]. The basic principle of Rprop is to eliminate the harmful influence of the size of the partial derivative on the weight step, considering only the sign of the derivative to indicate the direction of the weight update. The Rprop algorithm requires setting the following parameters: (i) the learning rate increase factor $\eta^+ = 1.2$; (ii) the learning rate decrease factor $\eta^- = 0.5$; (iii) the initial update-value is set to $\Delta_0 = 0.1$; (iv) the maximum weight step, which is used in order to prevent the weights from becoming too large, is $\Delta_{max} = 50$[11].

2.3 Ensemble-Based Methods

Methods for creating ensembles focus on creating classifiers that disagree on their decisions. In general terms, these methods alter the training process in an attempt to produce classifiers that will generate different classifications. In the neural network context, these methods include techniques for training with different network topologies, different initial weights, different learning parameters, and learning different portions of the training set (see [10] for reviews and comparisons).

We have investigated the use of two different methods for creating ensembles. The first one consists of creating a simple neural network ensemble with five networks where each network uses the full training set and differs only in its random initial weight settings. Similar approaches often produce results as good as Bagging [10]. The second approach is based on the notion of diversity, where networks belonging to the ensemble can be said to be diverse with respect to a test set if they make different generalisation errors on that test set [12]. Different patterns of generalisations can be produced when networks are trained on different training sets, or from different initial conditions, or with different numbers or hidden nodes, or using different algorithms [12]. Our implementation consisted of using four networks and belongs to the so called level 3 diversity [12], which allows to weight the outputs of the networks in such a way that either the correct answer is obtained or at least that the correct output is obtained often enough that generalisation is improved (see [12] for details).

3 Experimental Study

3.1 Description of Datasets

The datasets used have been submitted to the UCI Machine Learning Data Repository by Murphy and Aha [7] and are described in [4,8,9]. They include an E.coli dataset with 336 proteins sequences labelled according to 8 localization sites and a Yeast data set with 1484 sequences labelled according to 10 sites.

In particular, protein patterns in the E.coli data set are organized as follows: 143 patterns of cytoplasm (cp), 77 of inner membrane without signal sequence (im), 52 of periplasm (pp), 35 of inner membrane with uncleavable signal sequence (imU), 20 of outer membrane without lipoprotein (om), 5 of outer membrane with lipoprotein (omL), 2 of inner membrane with lipoprotein (imL) and 2 patterns of inner membrane with cleavable signal sequence (imS).

Yeast proteins are organized as follows: there are 463 patterns of cytoplasm (CYT), 429 of nucleus (NUC), 244 of mitochondria (MIT), 163 of membrane protein without N-terminal signal (ME3), 51 of membrane protein with uncleavable signal (ME2), 44 of membrane protein with cleavable signal (ME1), 35 of extracellular (EXC), 30 of vacuole (VAC), 20 of peroxisome (POX) and 5 patterns of endoplasmic reticulum (ERL).

3.2 Evaluation Methods

Cross Validation. In k-fold cross-validation a dataset D is randomly split into k mutually exclusive subsets $D_1, \ldots, D_k$ of approximately equal size. The classifier is trained and tested k times; each time $t \in \{1, 2, \ldots, k\}$ is trained on all $D_i, i = 1, \ldots, k$, with $i \neq t$ and tested on D_t. The cross-validation estimate of accuracy is the overall number of correct classifications divided by the number of instances in the dataset.

We used cross-validation to estimate the accuracy of the classification methods. Both data sets were randomly partitioned into equally sized subsets. The proportion of the classes is equal in each partition as this procedure provides more accurate results than a plain cross-validation does [6]. In our experiments we have partitioned the data into 4 equally sized subsets for the E.coli dataset and into 10 equally sized subsets for the Yeast dataset, as proposed in previous works [4,5].

The Wilcoxon Test of Statistical Significance. The Wilcoxon signed rank test is a nonparametric method. It is an alternative to the paired t-test. This test assumes that there is information in the magnitudes of the differences between paired observations, as well as the signs. Firstly, we take the paired observations, we calculate the differences and then we rank them from smallest to largest by their absolute value. After adding all the ranks associated with positive and negative differences giving the T_+ and T_- statistics respectively. Finally, the probability value associated with this statistic is found from the appropriate table. In our experiments, we analyzed the statistical significance by implementing the Wilcoxon rank sum test as proposed in [13]. All statements refer to a significance level of 10%, which corresponds to $T_+ < 8$ for the E.coli dataset and $T_+ < 14$ for the Yeast dataset.

4 Results

4.1 Classifying E.coli Patterns Using a Feed-Forward Neural Network

We conducted a set of preliminary experiments to find the most suitable FNN architecture in terms of training speed. We trained several networks with one and two hidden layers using the Rprop algoritrhm, trying various combinations of hidden nodes, i.e. 8, 12, 14, 16, 24, 32, 64, 120 hidden nodes. Each FNN architecture was trained 10 times with different initial weights. The best available architecture found was a 7-16-8 FNN that was further used in our experiments. 100 independent trials were performed with two different termination criteria, $E_{MSE} < 0.05$ and $E_{MSE} < 0.015$, in order to explore the effect of the network error on the accuracy of the classifications. We have conducted two experiments as recommended in previous works [2,4,5,8,9]. In the first experiment the neural networks are tested on the entire dataset. Table 1 shows the classification success achieved on each class.

Table 1. The accuracy of classification of E.coli proteins for each class.

No of Patterns	Class	KNN (%)	FNN $E_{MSE} < 0.05$ (%)	FNN $E_{MSE} < 0.015$ (%)
77	im	75.3	82.5	89
143	cp	98.6	98.1	99
2	imL	0.0	85.0	91.8
5	omL	80.0	100.0	100
35	imU	65.7	68.0	88.3
2	imS	0.0	6.5	49
20	om	90.0	86.6	93.2
52	pp	90.3	88.7	93.6
Mean		**62.5**	**76.9**	**88.0**

The results of the FNN show the mean value over 100 trials. As we can observe the FNNs outperforms KNN in every class. It is very important to highlight the neural network classification success in inner membrane with cleavable signal sequence (imS) and in inner membrane with lipoprotein (imL) classes. In the first case, FNNs trained with $E_{MSE} < 0.05$ exhibit a 6% success while the other methods have 0%. In the second case, the FNNs have 85% success while the other two methods only 50%. Results for FNNs become even better when trained with an $E_{MSE} < 0.015$. This obviously causes an increase to the average number of epochs required to converge (approximately 2000 epochs are needed) but leads to significant improvement: the average of overall successful predictions out of 100 runs was 93% with a standard deviation of 0.33. All FNNs results satisfy the conditions of the Wilcoxon test showing that the improvements achieved by the FNNs are statistically significant (e.g. comparing FNN performance when trained with an $E_{MSE} < 0.05$ against the the KNN gives $T_+ = 7$).

In order to identify common misclassifications, we calculated the confusion matrix for the FNN that exhibited the best training time for an $E_{MSE} < 0.015$. These results are shown in Table 2. The neural network achieved high percentage of classification compared to other methods (cf. with Table 5 in [5]). These results also show that fast convergences achieved by the FNN by no means affect its classification success.

Table 2. Confusion matrix for E.coli proteins with FNN.

No of Patterns	Class	cp	imL	imS	imU	im	omL	om	pp
143	cp	142	0	0	0	0	0	0	1
2	imL	0	2	0	0	0	0	0	0
2	imS	0	0	1	1	0	0	0	0
35	imU	0	0	0	31	4	0	0	0
77	im	2	0	0	6	69	0	0	0
5	omL	0	0	0	0	0	5	0	0
20	om	0	0	0	0	0	0	19	0
52	pp	3	0	0	0	0	0	0	49

In the second experiment, we performed a leave-one-out cross validation test by randomly partitioning the dataset into 4 equally sized subsets, as suggested in [4,5]. Each time three subsets are used for training while the remaining one is used for testing. Table 3 gives the best results for each method using the leave-one-out cross-validation showing that the FNN when trained with an $E_{MSE} < 0.015$ exhibits better performance. Lastly, it is important to mention that the performance of KNN algorithm is significantly improved when leave-one-out cross-validation is used but it still lacks in performance compared to the best FNN.

4.2 Classifying Yeast Patterns Using a Feed-Forward Neural Network

We conducted a set of preliminary experiments as with the E.coli dataset in order to find the most suitable architecture. An 8-16-10 FNN architecture exhibited the best performance. 100 FNNs were trained with the Rprop algorithm using different initial weights to achieve an $E_{MSE} < 0.045$. As previously we conducted two experiments following the guidelines of [4,5]. The first experiment

Table 3. Best performance for each method with leave-one-out cross-validation for E.coli proteins.

Cross Validation	Partition	KNN (%)	FNN $E_{MSE} < 0.015$ (%)
	0	89.3	91.7
	1	95.2	88.1
	2	88.1	84.5
	3	69.1	88.1
Mean		**82.4**	**88.1**

concerns testing using the whole dataset. The FNN outperforms significantly the other methods. The average neural network performance out of 100 runs is 67%; the worst-case performance was 64% (which is still an improvement over other methods) and the best one 69%. 10000 epochs were required on average. The result of the KNN is 59.5%.

In an attempt to explore the influence of the error goal on the success of the classifications we slightly increased the error goal to a value of $E_{MSE} < 0.05$. This results to a significant reduction on the average number of epochs; 3500 epochs are needed in order to reach convergence. Table 4 shows the classification success achieved on each class. The results of the FNN represent the average of 100 trials. The neural network outperforms the other methods in almost every class. It is very important to highlight the neural network classification success in the POX and ERL classes. With regards to the results of the Wilcoxon test of statistical significance, the improvements achieved by the neural networks are statistically significant giving $T_{+} = 7$ when compared against the KNN. The overall average classification of the FNNs is 64%, which shows that a small variation in the value of the error goal might affect the classification success. This might provide an explanation for the unsatisfactory performance that FNNs exhibited in the experiments reported in [2].

To identify the misclassifications in the Yeast dataset we have created the confusion matrix for the FNN that exhibited the fastest convergence to an $E_{MSE} < 0.045$. The results are shown in Table 5. It is important to highlight the significant improvement to classify the localisation sites in each class (cf. with Table 6 in [5]).

The second experiment involves the use of leave-one-out cross validation method with 10 equally sized partitions. The results are shown in Table 6. The KNN algorithm improves its generalization success. The performance of the FNNs is also improved, achieving average classification success of 70% (an $E_{MSE} < 0.045$ was used and an average of 3500 epochs). The results of the Wilcoxon test gives $T_{+} = 0$ showing that the improved performance achieved by the FNNs is statistically significant in all partitions when compared against the results of the KNN algorithm.

Table 4. The accuracy of classification of Yeast proteins for each class.

No of Patterns	Class	KNN (%)	FNN E_{MSE} <0.05 (%)
463	cyt	70.7	66.7
5	erl	0.0	99.6
35	exc	62.9	62.7
44	me1	75.0	82.9
51	me2	21.6	47.8
163	me3	74.9	85.6
244	mit	57.8	61.3
429	nuc	50.7	57.7
20	pox	55.0	54.6
30	vac	0.0	4.1
Mean		**46.8**	**62.3**

Table 5. The confusion matrix of Yeast proteins for each class using a neural network.

No of Patterns	Class	cyt	erl	exc	me1	me2	me3	mit	nuc	pox	vac
463	cyt	335	0	0	1	0	7	24	96	0	0
5	erl	0	5	0	0	0	0	0	0	0	0
35	exc	4	0	23	2	2	0	2	2	0	0
44	me1	2	0	1	0	0	1	1	0	0	0
51	me2	5	0	2	3	28	4	7	2	0	0
163	me3	8	0	0	0	0	145	1	8	0	1
244	mit	46	0	0	2	4	10	155	27	0	0
429	nuc	121	0	0	0	48	13	24	223	0	0
20	pox	6	0	0	0	0	0	2	1	11	0
30	vac	11	0	2	0	1	6	2	7	0	1

4.3 Classifying Protein Patterns Using Ensemble-Based Techniques

In this section we report results using the Simple Network Ensemble (SNE) method and the Diverse Neural Networks (DNN) method. Table 7 shows the results of the two ensemble-based methods on the E.coli dataset. The overall classification success of the SNE method was 93.5%, while the overall classification success of the DNN method was 96.8%.

We decided to concentrate on the DNN method and created an ensemble for the Yeast dataset. The results are shown in Table 8. The DNN ensemble significantly outperforms all other methods tested so far [2,4,5].

5 Conclusions

In this paper we have explored the use of a neural network approach for the prediction of localisation sites of proteins in the E.coli and Yeast datasets. Numerical results using supervised learning schemes, with and without the use of

Table 6. Best performance for each method using leave-one-out cross-validation for Yeast proteins.

Cross Validation	Partition	KNN (%)	FNN $E_{MSE} < 0.045$ (%)
	0	55.8	68.5
	1	59.2	69.1
	2	61.0	69.2
	3	65.8	69.0
	4	48.6	69.6
	5	62.3	69.5
	6	68.5	69.8
	7	58.9	69.8
	8	56.9	69.7
	9	58.2	70.5
Mean		**59.5**	**69.5**
Std. Dev		**5.49**	**0.56**

Table 7. Accuracy of classification for E.coli proteins using ensemble-based techniques.

No of Patterns	Class	SNE E_{MSE} <0.015 (%)	DNN $E_{MSE} < 0.015$ (%)
77	im	87.0	92.22
143	cp	98.0	100
2	imL	100	100
5	omL	100	100
35	imU	77.2	94.3
2	imS	50.0	50
20	om	80.0	100
52	pp	91.4	96.15
Mean		**85.45**	**91.7**

Table 8. Accuracy of classification for Yeast proteins using diverse neural networks

No of Patterns	Class	DNN E_{MSE} <0.045 (%)
463	cyt	80.5
5	erl	100
35	exc	74.3
44	me1	100
51	me2	70.6
163	me3	92.65
244	mit	71.4
429	nuc	70.1
20	pox	55.0
30	vac	20.0
Mean		**73.5**

cross validation, are better than the best previous attempts. This is mainly because of the use of more advanced training schemes. We investigated the use of network ensembles exhibiting level 3 diversity. Our future work focuses on exploring the use of network ensembles with higher degrees of diversity, taking into account the drastically imbalanced nature of these datasets.

References

1. Boland M.V. and Murphy R.F.: After sequencing: quantitative analysis of protein localization. IEEE Engineering in Medicine and Biology. (1999) 115–119
2. Cairns, P. Huyck, C. Mitchell, I. Wu, W.: A Comparison of Categorisation Algorithms for Predicting the Cellular Localization Sites of Proteins. IEEE Engineering in Medicine and Biology.(2001) 296–300
3. Duda, R.O. and Hart.: Pattern Classification and Scene Analysis. John Wiley and Sons,(1973)
4. Horton, P., and Nakai, K.: A probabilistic classification system for predicting the cellular localization sites of proteins. Proceedings of the Fourth International Conference on Intelligent Systems for Molecular Biology.(1996) 109–115
5. Horton, P., and Nakai, K.: Better Prediction of Protein Cellular Localization Sites with the k Nearest Neighbors Classifier. Proceedings of Intelligent Systems in Molecular Biology.(1997) 368–383
6. Kohavi, R.: A study of cross-validation and bootstrap for accuracy estimation and model selection. IInternational Joint Conference on Artificial Intelligence.(1995) 223–228
7. Murphy, P. M., and Aha, D. W.: UCI repository of machine learning databases. http://www.ics.uci.edu/mlearn,(1996)
8. Nakai, K. and Kanehisa, M.: Expert system for predicting protein localization sites in gram-negative bacteria. PROTEINS. **11**(1991) 95–110
9. Nakai, K. and Kanehisa, M.: A knowledge base for predicting protein localization sites in eukaryotic cells. Genomics. **14** (1992) 897–911
10. Opitz D. and Maclin R.: Popular Ensemble Methods: An Empirical Study", Journal of Articial Intelligence Research.**11** (1999) 169–198
11. Riedmiller, M. and Braun, H.: A direct adaptive method for faster backpropagation learning: The RPROP algorithm. Proceedings International Conference on Neural Networks, (1993) 586–591
12. Sharkey A. J.C. and Sharkey N.E.:Combining diverse neural nets.The Knowledge Engineering Review.**12** (1997) 231–247
13. Snedecor, G., and Cochran, W.: Statistical Methods. 8th edn.Iowa State University Press, (1989)

Applying Intelligent Data Analysis to Coupling Relationships in Object-Oriented Software

Steve Counsell[1], Xiaohui Liu[2], Rajaa Najjar[1], Stephen Swift[2], and Allan Tucker[2]

[1] School of Computer Science and Information Systems,
Birkbeck, University of London, Malet Street, London WC1E 7HX, UK.
steve@dcs.bbk.ac.uk

[2] Department of Information Systems and Computing,
Brunel University, Uxbridge, Middlesex UB8 3PH, UK.

Abstract. Class coupling lies at the heart of object-oriented (OO) programming systems, since most objects need to communicate with other objects in any OO system as part of the system's functions. Minimisation of coupling in a system is a goal of every software developer, since overly-coupled systems are complex and difficult to maintain. There are various ways of coupling classes in OO systems. However, very little is known about the different forms of coupling and their relationships. In this paper, three data analysis techniques, namely, Bayesian Networks, Association Rules and Clustering were used to identify coupling relationships in three C++ systems. Encouraging results were shown for the Bayesian Network approach, re-inforcing existing knowledge and highlighting new features about the three systems. Results for the other two techniques were disappointing. With association rules, it was clear that only a very general relationship could be discovered from the data. The clustering approach produced inconsistent results, casting doubt on whether such a technique can provide any insight into module discovery when applied to these type of systems.

1 Introduction

A key feature of any object-oriented (OO) system is the amount of coupling it contains. Coupling is defined generally and informally as the inter-relatedness of the components in a software system. In an OO sense, the components are the system's classes and the coupling between those classes represents their inter-dependencies, i.e., the extent to which a class depends on other classes in a system to carry out its services. For example, in a C++ system, a parameter passed to a method of class X whose type is class Y represents parameter coupling between those two classes. If class X also inherits properties from class Y as part of an inheritance hierarchy, then the two classes are coupled via inheritance [15]. There are various other ways that two OO classes may be coupled, yet in the software engineering community we know very little about the way

M.R. Berthold et al. (Eds.): IDA 2003, LNCS 2810, pp. 440–450, 2003.

OO systems are constructed in terms of coupling inter-dependencies (on an empirical basis), or how one form of coupling may depend on, or trade-off, against another. Information of this type would be invaluable to developers when anticipating changes to existing systems or, indeed, when developing systems from scratch.

In this paper, an investigation of coupling patterns and trends in three large C++ systems is described. The systems comprised a compiler, a class library and a framework, all developed using the C++ language. Data representing five different forms of coupling was collected from each system, organised for convenience into coupling matrices, and then examined using intelligent data analysis (IDA) to identify associations therein.

The aims of the research in this paper are three-fold. Firstly, to attempt to shed light on empirical software engineering problems using IDA; the analysis tools available (and described in this paper) significantly widen the scope of research in the empirical software engineering field; search-based software engineering is an exciting and emerging discipline. Secondly, to learn about OO practice so that similar studies on other systems may be carried out, thereby informing general OO practice. Finally, to show how the complexity of industrial-sized OO systems can be tackled using IDA techniques. The results described herein would have taken significantly longer to produce had less intelligent forms of analysis been used.

In the next section, we describe the motivation behind the investigation. In Section 3, we describe the methods used to analyse the datasets and Section 4 documents the results. Finally, we draw some conclusions and discuss future work (Section 5).

2 Motivation

In the software engineering community and more specifically in the empirical software engineering community, a significant amount of work has been carried out into OO coupling per se, its different forms and, in particular, on metrics for measuring coupling [3]. An acknowledged software engineering principle is that coupling should be kept to a minimum, since both class reuse and maintenance becomes unwieldy and difficult when there are too many class dependencies. In theory, too much system coupling detracts from system cohesion (itself the subject of a number of studies) and is generally harmful [4]. There is also evidence (but mostly anecdotal) to suggest that a high proportion of faults in software are invested at the interface level and occur when objects interact with one another. Firmer evidence shows that use of friend coupling in C++ is a major source of faults in C++ software [4]. Finally, there is empirical evidence which strongly associates use of friends with that of inheritance [8], but the exact strength of this association is unknown and difficult to quantify with empirical evidence alone.

The work in this paper thus attempts to build on results found in previous work, but with the important and valuable addition of IDA. For example, it has been shown in earlier work [7] that a compiler-based C++ system tends to engage in significant amounts of coupling through the return types of its methods. Methods tend to take primitive values (integers, strings etc.) as parameters and return objects. Yet we have no way of knowing what other forms of coupling are associated with return type coupling and the strength of that coupling.

It has also been shown in earlier work that framework systems conform strongly to OO principles (in terms of encapsulation, inheritance, etc) [6]. They should therefore contain relatively low amounts of coupling and be more balanced in their use of coupling (i.e., not emphasise one form of coupling too much). Pure empirical analysis can only show limited evidence of this. Use of intelligent data analysis, by its very nature, significantly broadens the scope and potential for analysis of association and trade-off of coupling forms. A further objective of the work in this paper was to investigate the application of intelligent data analysis to problems in the OO arena.

2.1 Data

The C++ language was chosen for this study because it enjoys widespread use in the IT industry for developing small, medium and large-scale systems across a variety of application domains. Other OO languages (and languages generally) would have been valid as a basis for the study. However, an important part of the research in this paper was to augment our knowledge of the three C++ systems accrued through previous studies. The three C++ systems analysed as part of the study were:

1. Rocket, a compiler, consisting of 32.4 thousand non-comment source lines (KNCSL) and containing 322 classes.
2. The GNU C++ Class Library, consisting of 53.5 KNCSL and containing 96 classes.
3. ET++, a user interface framework, consisting of approximately 56.3 KNCSL and containing 508 classes.

The three systems were chosen from a range of publicly available C++ systems. A compiler, library and framework were chosen to cover as wide a range of these systems and sizes as possible. This improves the chances of the results being generalisable to OO systems. For each of the three systems analysed, a coupling matrix was manually collected from the program source code. A relationship was defined in terms of zero or more couplings between each pair of classes according to whether that relationship was through Aggregation (A), Friend (F), Inheritance (N), Parameter (P) or Return type (R). We note that in C++ systems, all these forms of coupling are used frequently. Table 1 shows summary data for the three systems including thousands of non-comment source lines (KNCSL), number of classes, the number of relationships scored and a value for sparseness calculated by dividing the number of relationships by the total number of possible relationships, i.e., $s = 1-(r/(n(n-1)))$ where s is sparseness, r is number of relationships and n is the number of classes.

Table 1. Summary data for the three systems.

System	KNCSL	Classes	No. Relationships	Sparseness (%)
Rocket	32.4	216	709	98.5
GNU	53.5	105	127	98.8
ET++	56.3	207	1035	97.6

3 Methods

In the following sections, we describe the methods used to analyse the datasets representing the three C++ systems.

3.1 Bayesian Networks

A Bayesian Network (BN) models a domain probabilistically using a Directed Acyclic Graph (DAG) and conditional probability tables which allow us to store a joint probability distribution by making conditional independence assumptions about variables in the domain. A DAG is a directed graph that contains no more than one directed path from one node to another. A thorough discussion of BNs can be found in [14]. A BN consists of the following:

1. A DAG, S, where the set of nodes represents variables in the domain. This is made up of a set of nodes, X, representing variables within the domain, where $|X| = N$, and directed links between them representing conditional dependencies between a node, x_i, and its parents, π_i.
2. A set of conditional probability distributions where each node, x_i, is assigned a Conditional Probability Table (CPT) which determines $p(x_i \mid \pi_i)$, the probability of its different possible instantiations given the instantiations of its parents.

There has been a lot of research into learning BNs from data [5,10,12]. Most methods make use of a scoring function and a search procedure. In our experiments we treat the different possible relationships between classes as variables in a BN. Therefore, each BN will only contain 5 binary variables: A, F, N, P, and R, which are present or not present in a class. This means an exhaustive search can be applied to the BN learning process. We use the log likelihood scoring metric as implemented in Bayesware (http://bayesware.com).

3.2 Association Rules

An Association Rule [1] is a form of relationship. However, an Association Rule differs from a Bayesian Network in that it relates values of a field to values of another field, as opposed to relating the whole field to another. The most common use of Association Rules regards shopping basket data. For example, 95% of people, who buy bread, also buy milk. This is useful for marketing and especially relevant now due to loyalty cards, which can lead to huge databases being formed.

As with structure discovery within Bayesian Networks, there are many types of algorithms for discovering Association Rules. The algorithms for generating Association Rules are quite large and complex, and hence will be outlined rather than listed.

Formal Definition

- Let I be a set of items
- Let T be a subset of I, called Transactions
- Let D be a set of Transactions (T)
- We say T contains X if $X \subseteq I$
- An association rule is an implication of the form $X \Rightarrow Y$, where X and Y are transactions, and $X \cup Y = \emptyset$
- The rule $X \Rightarrow Y$ holds with confidence $c\%$ if $c\%$ of the transactions in D that contain X also contain Y
- The rule $X \Rightarrow Y$ has support $s\%$ if $s\%$ of the transactions in D contain X $\cup Y$
- The algorithms rely on the minimum support and minimum confidence being defined as part of the implementation

The algorithm apriori works by generating all large 1-itemsets (sets of items that are of size 1, and have minimum support). In each further pass, the previous (k-1) large itemset is used as a seed to create the k-itemsets. This process is repeated until there are no large k-itemsets. Association Rules were implemented using the software downloaded from http://fuzzy.cs.uni-madgeburg.de/~borgelt.

3.3 Bunch

The software tool Bunch [13] is a Hill-Climbing based clustering tool designed towards grouping software components together into modules based upon their coupling. The system tries to ensure that components grouped together are highly coupled with low coupling existing between modules.

The Hill-Climbing method is described in the following algorithm and attempts to maximise the coupling based metric (called MQ) defined in the following equation:

$$MQ(G) = \sum_{i=1}^{|G|} CF(g_i)$$

$$CF(g_i) = \frac{2\mu_i}{2\mu_i + 0.5 \sum_{j=1, j \neq i}^{|G|} \epsilon_{ij} + \epsilon_{ji}}$$

where G is a partition or clustering arrangement of $\{1, ..., n\}$ and g_i is the ith group or cluster of G. Here, μ_i is the number of coupling links between elements in g_i and ϵ_{ij} is the number of coupling links between elements in g_i and elements in group j.

Algorithm: The Bunch Hill-Climbing Method

Input to the algorithm: The Coupling Information for the System Being Grouped.

```
 1. Let G be a Random Module Grouping
 2. Let Complete = False
 3. While Not Complete
 4.        Let Better = False
 5.        Let Moves be the Set of all Possible Small Changes
 6.        While Moves is Not Empty or Not Better
 7.               Try the Next Element in Moves
 8.               If the Move was Better Then
 9.                      Let Better = True
10.                      Update G with current Element in Moves
11.               End If
12.        Wend
13.        If Not Better Then Complete = True
14. Wend
```

Output from the algorithm: The Grouping G.

The algorithm works by considering all small steps (in this case moving a component from one group to another) that are possible given the current modularisation under consideration. If a better solution (according to the MQ metric) is found then the process is repeated from the new configuration, otherwise the procedure is terminated when there are no more small moves left to do. The creation or deletion of new clusters and a few other implementation details have been omitted but is clearly and fully documented in [13].

4 Results

To analyse the results from this study, we will look at each technique in turn, drawing conclusions from the data analysis where appropriate. We also support our findings with previous empirical studies which have highlighted features of the three systems.

4.1 Bayesian Network Analysis

It was interesting to note from the Rocket system that two types of class were identified. Firstly, there were classes where neither Inheritance (N) or Return type (R) were observed, but whose methods had a number of Parameters (P).

This type of class would be typically found in a system such as a compiler to carry out mundane functions. The second type of class on the other hand did observe N, P and R. This latter type of class comprises the majority of classes in Rocket and is an observation supported by a previous study [7].

A further observation about the Rocket system was due to the C++ friend facility [15]. The Friend facility (F) is considered harmful to code in terms of its robustness and design because it causes a violation of OO encapsulation principles. One of the chief strengths of the OO approach is its data encapsulation (i.e., protection of data from illegal outside interference).

The Bayesian network also identified two other types of class in Rocket concerning F. The observed features of the first type of class were that in the presence of F and no P relationships in the methods of the class, there is a tendency for no R either. The observed features of the second type of class were that in the presence of F *and* P, you are more likely to have an R.

A simple, yet interesting explanation may account for this. The first type of class contains friends used for operator overloading. The second type of class contains friends in the more destructive case (i.e., where the friend is actually another class and is a true violation of encapsulation). Operator overloading tends to occur in classes where there are very few object-based parameters.

For the ET++ system, very few relationships were observed. However, a similar relationship was identified between F, N and R. In the case where no F was observed, an inverse relationship seemed to exist between R and N. From an OO perspective, this would suggest that these two types of coupling are almost mutually exclusive. For the GNU system, an almost identical Bayesian network was discovered to that of the Rocket network, the difference being a link between F and aggregation relationships (A).

An interesting feature of all three networks is that the ET++ network is a subset of the GNU network, in turn a subset of the Rocket network; this implies that there may be general relationships applicable to all OO systems, in addition to relationships specific to a particular system. This feature is shown diagramatically in Figure 1.

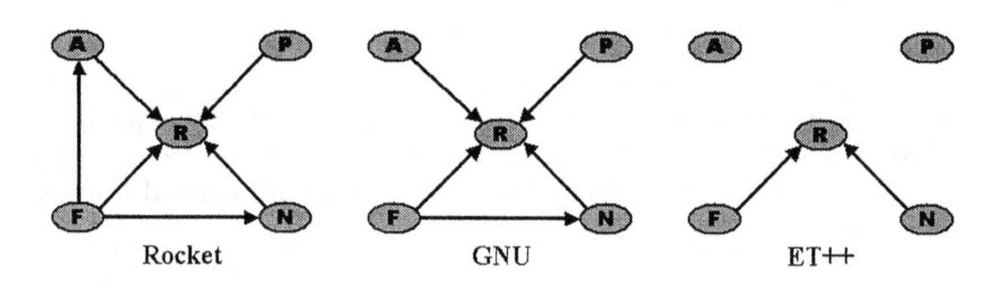

Fig. 1. The discovered Bayesian networks for each dataset.

4.2 Association Rules

Due to the number of single individual relationships present within the data, the association rules were run with a support of 10% and a confidence of 50%. Table 2 shows the results of running association rules on each of the datasets.

Table 2. Association rules for the three systems.

System	Rules	Support (%)	Confidence (%)
Rocket	P $\Leftarrow$	50.4	50.4
	R $\Leftarrow$ A	13.5	58.2
	P $\Leftarrow$ A	11.7	50.3
GNU	P $\Leftarrow$ R	27.6	85.4
	R $\Leftarrow$ P	27.6	57.4
ET++	P $\Leftarrow$	59.1	59.1
	P $\Leftarrow$ A	13.6	56.4
	P $\Leftarrow$ R	25.4	61.6

The first noticeable result from all of the systems is the small number of rules that have been generated. Ignoring singleton rules, it can be seen that each system has produced just two rules, even though the levels for the support and confidence were very low. Confidence cannot be set lower than 50% since this would produce rules that are false more often than true. Similarly, setting the support lower than 10% would result in rare rules occurring that do not reflect the overall relationships within the data. From an OO perspective, very little can be gleaned from the results for any of the systems. For ET++, the association rules imply that when given a coupling relationship of aggregation or return type, we get a parameter (although the supporting confidence for both these association rules is quite low). For the GNU system, the 85% confidence level suggests a strong probability of getting a parameter in the presence of a return type. The 57% confidence level suggests that the reverse is true, but not to the same degree. From an OO perspective, one would generally expect a strong relationship between the parameters passed to methods (functions) and the values returned. For Rocket, support and confidence for any relationships is too weak to be of any real value.

4.3 Bunch

The results for the clustering experiment will be compared against each other using the weighted kappa metric [2]. This metric can be used to compare how similar two clustering arrangements are [11]. The metric returns a value of +1 indicating an identical set of clusters and -1 for a totally dissimilar set of clusters. A value of zero would be expected for two random arrangements.

Bunch was run ten times on each dataset. An additional dataset (labelled B) was also used which consisted of all of the other five coupling matrices OR'ed together for each system, thus creating six datasets. This new dataset was created to see if the results of a combined matrix would be similar to the ones it was created from. Additionally, this type of matrix is very similar to those used in many modularisation methods, i.e., a link is any coupling relationship. Sixty experiments were therefore run, and the weighted-kappa for each possible comparison was taken.

Table 3. Clustering results: ET++ system.

ET++	A	B	F	N	P	R
A	0.583	0.178	0.020	0.017	-0.002	0.048
B	0.178	0.683	0.087	0.196	-0.002	0.092
F	0.020	0.087	0.109	0.008	0.000	0.012
N	0.017	0.196	0.008	0.829	-0.001	0.060
P	-0.002	-0.002	0.000	-0.001	0.004	-0.001
R	0.048	0.092	0.012	0.060	-0.001	0.622

Table 4. Clustering results: GNU system.

GNU	A	B	F	N	P	R
A	0.058	0.044	0.011	-0.001	0.025	0.030
B	0.044	0.225	0.021	0.143	0.137	0.040
F	0.011	0.021	0.022	0.001	0.013	0.013
N	-0.001	0.143	0.001	0.205	0.115	0.001
P	0.025	0.137	0.013	0.115	0.219	0.041
R	0.030	0.040	0.013	0.001	0.041	0.068

Tables 3 to 5 contain the average weighted kappa results for each of the comparisons across the six binary matrices for the three datasets. The leading diagonal within each of the tables is an indication of how consistent the clustering results are between repeated runs, using the same binary matrix.

For the ET++ results, it can be seen that the consistency varies greatly; for the N (inheritance) matrix, the consistency is high. For B (combined), A (aggregation) and R (return type) the agreement is good. However, for the parameter matrix, the agreement is no better than random. Results between all the different matrices are very poor and are all below 0.2. For the GNU dataset. all the results including the intra-method comparison are very poor and all lie below 0.23. For the Rocket dataset, the inheritance results are highly consistent; the P, B and R results are reasonably consistent. The results for A (aggregation) are poorly consistent and for F (friends) no better than random. Results between all the different matrices are poor/very poor and all lie below 0.3.

Table 5. Clustering results: Rocket system.

Rocket	A	B	F	N	P	R
A	0.356	0.142	0.008	0.004	0.100	0.093
B	0.142	0.693	0.055	0.057	0.282	0.209
F	0.008	0.055	0.069	0.003	0.023	0.019
N	0.004	0.057	0.003	0.873	0.004	0.005
P	0.100	0.282	0.023	0.004	0.668	0.132
R	0.093	0.209	0.019	0.005	0.132	0.735

5 Conclusions

In this paper, three well-known IDA techniques were applied to three real-world software engineering datasets. Encouraging results were shown for the Bayesian Network approach, reinforcing existing knowledge and highlighting new features about these systems. Results for the other two techniques were disappointing; the clustering approach firstly, produced inconsistent results between repeated runs on the same data matrix and secondly, showed huge variation between the clustering results produced using different measures of coupling. With the association rule results, it is clear that only a very general relationship can be discovered from this type of data (i.e., only two rules per dataset). In order to get a large number of varying rules from the datasets, it would be necessary to reduce the support and confidence levels to an unacceptably low value.

The BN approach has proved somewhat more successful than the other two IDA techniques when applied to the datasets within this paper. One explanation for this is that the relationships within the coupling datasets are more complex than simple pairwise associations. The BN is capable of modelling non-linear complex relations involving more than two variables. Indeed most of the relationships discovered (see Fig. 1) by the BN involved at least three variables. Association rules can also handle more than two variables per relationship. However, it appears that the sparseness of the data as well as the ability of BNs to handle more complex inferences, imply that BNs are more useful in extracting interesting rules from the datasets explored.

The research in this paper was invaluable in supporting and extending previous OO studies. The research also higlighted a number of outstanding research questions in the OO field. Future work will focus on application of the three IDA techniques to more C++ systems and possibly Java systems as well. We also intend to investigate the use of Bayesian Networks classifiers [9] to identify different types of component such as those found for class types in this paper.

Acknowledgments. The authors acknowledge the help of Jim Bieman at the Department of Computer Science, Colorado State University, U.S., for access to two of the systems studied (Rocket and ET++).

References

1. R. Agrawal. Fast algorithms for mining association rules. In *Proceedings of the 20th VLDB Conference, Santiago, Chile*, pages 487–499, 1994.
2. D. Altman. *Practical Statistics for Medical Research.* Chapman and Hall, 1997.
3. L. Briand, J. Daly, and J. Wust. A unified framework for coupling measurement in object-oriented systems. *IEEE Transactions on Software Engineering*, 25(1):91–121, 1999.
4. L. Briand, P. Devanbu, and W. Melo. An investigation into coupling measures for C++. In *Proceedings of the 19th International Conference on Software Engineering (ICSE'97), Boston, USA*, pages 412–421, 1997.
5. G. Cooper and E. Herskovitz. A bayesian method for the induction of probabilistic networks from data. *Machine Learning*, 9:309–347, 1992.
6. S. Counsell, G. Loizou, R. Najjar, and K. Mannock. On the relationship between encapsulation, inheritance and friends in C++ software. In *Proceedings of International Conference on Software and Systems Engineering and their Applications, ICSSEA'02, Paris, France (no page numbers)*, 2002.
7. S. Counsell, E. Mendes, and S. Swift. Comprehension of object-oriented software cohesion: the empirical quagmire. In *Proceedings of the 10th International Workshop on Program Comprehension (IWPC'2002), Paris, France*, pages 33–42, 2002.
8. S. Counsell and P. Newson. Use of friends in C++ software: an empirical investigation. *Journal of Systems and Software*, 53:15–21, 2000.
9. N. Friedman and M. Goldszmidt. Building classifiers using bayesian networks. In *AAAI/IAAI, Vol. 2*, pages 1277–1284, 1996.
10. D. Heckerman, D. Geiger, and D. M. Chickering. Learning bayesian networks: The combination of knowledge and statistical data. In *KDD Workshop*, pages 85–96, 1994.
11. P. Kellam, X. Liu, N. Martin, N. Orengo, S. Swift, and A. Tucker. Comparing, contrasting and combining clusters in viral gene expression data. In *Proceedings of the IDAMAP 2001 Workshop, London, UK*, pages 56–62, 2001.
12. W. Lam and F. Bacchus. Learning bayesian belief networks: An approach based on the MDL principle. *Computational Intelligence*, 10:269–293, 1994.
13. B. Mitchell. *A Heuristic Search Approach to Solving the Software Clustering Problem, Ph.D. Thesis, Drexel University, Philadelphia, USA*. 2002.
14. J. Pearl. *Probabilistic Reasoning in Intelligent Systems: Networks of Plausible Inference.* Morgan Kaufman, 1988.
15. B. Stroustrup. Adding classes to the C language: An exercise in language evolution. *Software – Practice and Experience*, 13:139–161, 1983.

The Smaller the Better: Comparison of Two Approaches for Sales Rate Prediction

Martin Lauer[1], Martin Riedmiller[1], Thomas Ragg[2], Walter Baum[3], and Michael Wigbers[3]

[1] Universität Dortmund, Lehrstuhl Informatik I, 44221 Dortmund, Germany
{lauer, riedmiller}@ls1.cs.uni-dortmund.de

[2] quantiom bioinformatics, Ringstrasse 61, 76356 Weingarten, Germany

[3] Axel-Springer-Verlag, 20355 Hamburg, Germany

Abstract. We describe a system to predict the daily sales rates of newspapers. We deduce a mathematical modeling and its implementation, a data cleaning approach, and a way to augment the training sets using similar time series. The results are compared with a neural prediction system currently in use.

1 Introduction

Every newspaper publisher is faced with the problem of daily determining the number of copies and distributing them to the retail traders. Two aspects have to be balanced out to optimize the economical success: the number of unsold copies should be minimal to reduce the cost of production, and the sellout rate should be minimal to maximize the number of sold copies. Thus, a good sales rate prediction is necessary to optimize both antagonistic aspects.

The objective of our work was to optimize the sales rate prediction of the *BILD-Zeitung* for each of its 110 000 retailers. The *BILD-Zeitung* is the most highly circulated newspaper in Germany. 4.1 million copies are daily sold, and the return quota of unsold copies currently amounts to 15% of all printed copies.

In 1992, the *Axel-Springer-Verlag* as publisher started to develop a newspaper regulation system which uses neural networks to predict the future sales rates. Compared to the formerly used prediction tool based on moving averages, the neural modeling was able to significantly reduce the return quota without increasing the number of sellouts [3].

Despite ongoing improvements, we started to replace the neural prediction system to further reduce the error rate. In this paper we want to show some facets of the new modeling, and we want to compare both approaches on the sales rates of three wholesaler areas with more than 900 retail traders.

Predicting newspaper sales rates is a challenging task since the set of retailers is very heterogeneous. The sales level ranges from, on average, one copy per day up to 400 copies and the training history from 20 weeks up to five years. Trends and seasonal ups and downs are more or less observable, and annual

M.R. Berthold et al. (Eds.): IDA 2003, LNCS 2810, pp. 451–461, 2003.

closing differs from retailer to retailer. However, the modeling must cover all these individualities without becoming unmanageable.

The domain of sales rate prediction has been investigated before in a number of publications. Thiesing and Vornberger [7] compared the prediction error of multi layer perceptrons with the error of the moving average approach while Heskes [1] examined the behavior of multi-task learning and a hierarchical Bayesian modeling for a set of retail traders.

2 Modeling of Sales Rates

The prediction task we are faced with can be described as follows: Given the past sales rates of a retailer, we should predict the sales figures of next week. If we denote the past sales with $X_1, \ldots, X_n$ and next week's sales rates with $X_{n+1}, \ldots, X_{n+6}$ we can describe the task as to find a mapping from $X_1, \ldots, X_n$ to $X_{n+1}, \ldots, X_{n+6}$. Due to the randomness of sales rates, we cannot expect to find an exact prediction but we need to get an approximation as good as possible.

For reasons of convenience we firstly want to focus on a single weekday, Monday say, and thus predict X_{n+1} given $X_1, \ldots, X_n$. The modeling of the other weekdays applies analogously.

Let us assume that the sales figure at time t is distributed normally with mean μ_t and variance σ^2: $X_t \sim N(\mu_t, \sigma^2)$. The temporal ups and downs are modeled by the change of the means μ_t over time. Thus, we simplify the modeling assuming stochastic independence between different X_t of an individual trader.

In the case of time series without trends, i.e. $\mu_t = \mu_{t'}$ for all t, t', the best estimator of the future sales is the empirical mean from the past data. But, since almost all retail traders show long-term or middle-term trends, such a modeling is not sufficient. We could instead try to estimate the past trend and extrapolate to the future, but, as well, since extrapolation is not well done by most regression algorithms, this is not a preferable approach.

An alternative is to model the time series of differences $Z_t := X_t - X_{t-1}$. The prediction probability becomes:

$$\begin{aligned} P(Z_{n+1}|Z_n, Z_{n-1}, \ldots, Z_2) &= \int P(Z_{n+1}|X_n) \cdot P(X_n|Z_n, \ldots, Z_2) dX_n \\ &\propto \int P(Z_{n+1}|X_n) \cdot P(Z_n, \ldots, Z_2|X_n) \cdot P(X_n) dX_n \\ &\sim N\left(\delta_{n+1} + d - z, \frac{n+1}{n}\sigma^2\right) \end{aligned} \tag{1}$$

with $\delta_t := \mu_t - \mu_{t-1}$, $d := \sum_{t=2}^{n} \left(\frac{t}{n}\delta_t\right)$ and $z := \sum_{t=2}^{n} \left(\frac{t}{n} Z_t\right)$. Notice that the expectation value in (1) is independent of the sales level μ_t but only depends on the temporal differences in past $Z_2, \ldots, Z_n$, the trend in past $\delta_2, \ldots, \delta_n$, and the future trend δ_{n+1}. Thus the estimator becomes independent of level shifts and avoids extrapolation. In the stationary case the expectation value of Z_{n+1} simplifies to $-z$ which can directly be computed from the past data.

Although the values of d and z are independent of μ_t and X_t we can rewrite their definitions as: $z = \frac{1}{n}(\sum_{t=1}^{n} X_t) - X_n$ and $d = \frac{1}{n}(\sum_{t=1}^{n} \mu_t) - \mu_n$. z can be described as the difference between the average sales and the current sales figure indicating the deviation of the current sales figure. The value of d reveals the systematic shift of the sales level in past. Thus, $z - d$ is the unbiased estimator of the deviation of the current sales figure from its expectation value.

The prediction variance $\frac{n+1}{n}\sigma^2$ reflects that the accuracy increases the more past data are considered. It does not depend on the stationarity of the time series since it considers the past trend given by d. Notice that the composition of the expectation value gives us hints on the construction of appropriate features.

The stochastic modeling from this section is the basis of the new prediction system. The difference between the current and the future sales figure is predicted given information on the development in the past. To represent the past, we used four features like indicators of trend and curvature. Two of them are directly derived from the value of z. Further two features are used to get an approximation for the value of d. Furthermore, we used two features to predict the future trend. The latter ones are explained in the next section.

3 Determining the Future Trend

As long as the time series are stationary or the time series are influenced by a constant growth rate, the term δ_{n+1} in equation (1) is constant and can easily be modeled without extra effort. But, unfortunately, this is seldom the case and thus, we have to introduce predictors of future trend.

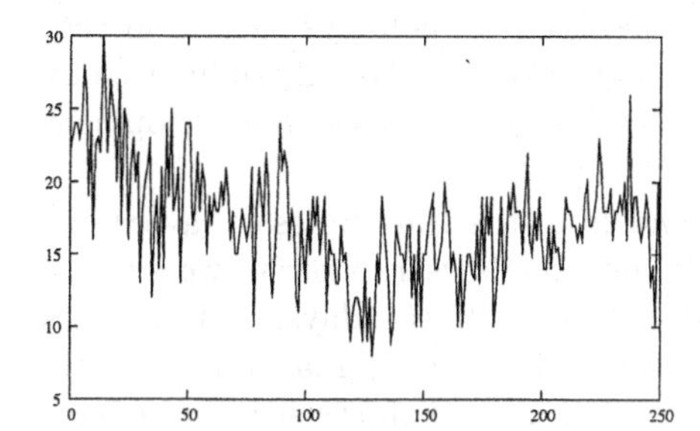

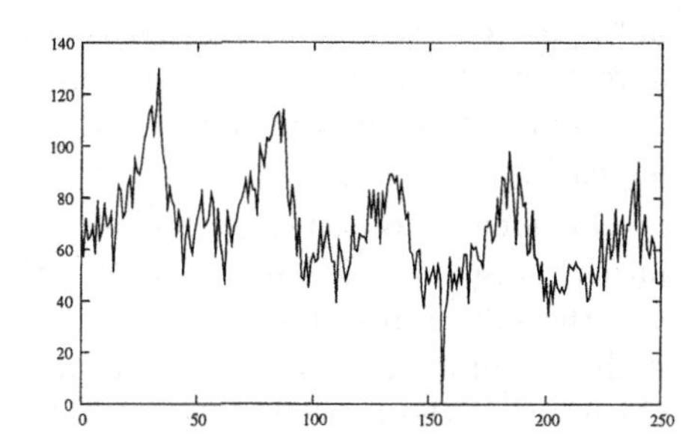

Fig. 1. The sales rates on Monday of two retail traders over 250 weeks (five years). The left retailer reveals a changing long-term trend while the right retailer is strongly influenced by seasonal changes.

Figure 1 shows two time series influenced by trends. In the left figure the noise level is relatively large compared to the trend. Thus, it is very difficult to extract the trend from the time series. But, however, the change from week to week is small and therefore the error caused by the trend remains small.

The right side of figure 1 shows a retailer with a large seasonal influence. Obviously, the changes from week to week are larger than the noise so that ignoring the trend causes a large error. Figure 2 shows the seasonal development for the same retailer in several years. The development of sales rates reveals a typical structure although every year shows individual variation.

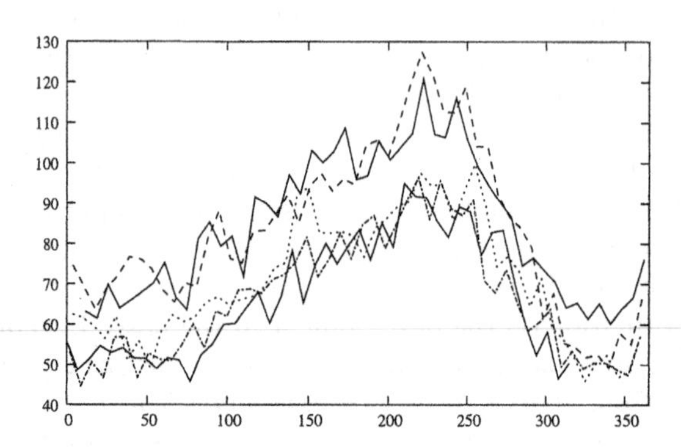

Fig. 2. The sales rates of a retailer with large seasonal changes. Every line represents the development in one year. The abscissa indicates the days of a year.

Heskes [1] used features on the development one year ago to model seasonal ups and downs. This modeling is not adequate for our application due to the fact that we have to predict sales outlets with down to 20 weeks of history. Thus, the data of last year's development are not available or even do not exist. Secondly we identify some shortcomings in the simple look back for one year: a) the seasonal development is similar but not identical; red letter days like Easter and Pentecost as well as weather conditions and school holidays vary from year to year, and b) the sales rates of an individual retailer are strongly influenced by noise so that an increase may occur even when the general seasonal development decreases.

Our idea is to extract the seasonal trend from the time series by averaging out the noise and the special characteristics of a single year. Thereto we average over the sales rates of a certain set of sales outlets, all weekdays, and all years of the training set. Thus, we get time series with a high signal-to-noise ratio which reveal the typical seasonal development. Figure 3 shows seasonal curves for an urban surrounding and a holiday area. The typical ups and downs can be recognized very clearly. The noise level is very low, especially in the holiday area. Notice that even the urban area reveals typical seasonal effects like the break down at Christmas and the decrease in spring. We can use the local derivative of the seasonal curves as feature to indicate the degree of seasonal ascend.

The set of sales outlets that we use to compute the seasonal curves is individual for each retailer. It consists of the sales outlets with the most similar behavior measured by the correlation of the sales rates in past. The number of similar retailers, k, is a fixed number determined from experiments. We used two sets with different k to get one feature that is more specific (small k) and one with a better signal-to-noise ratio (large k).

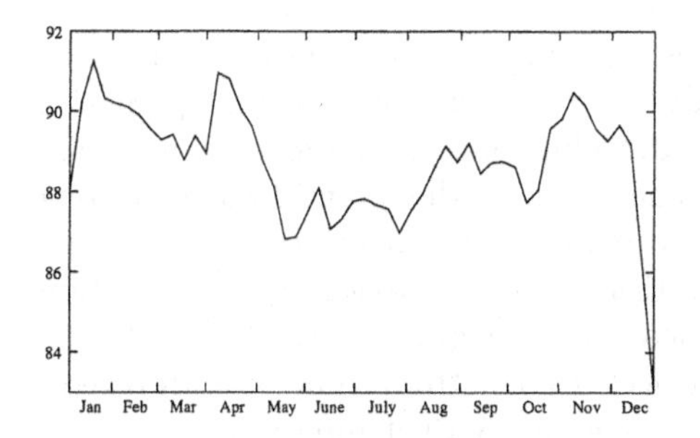

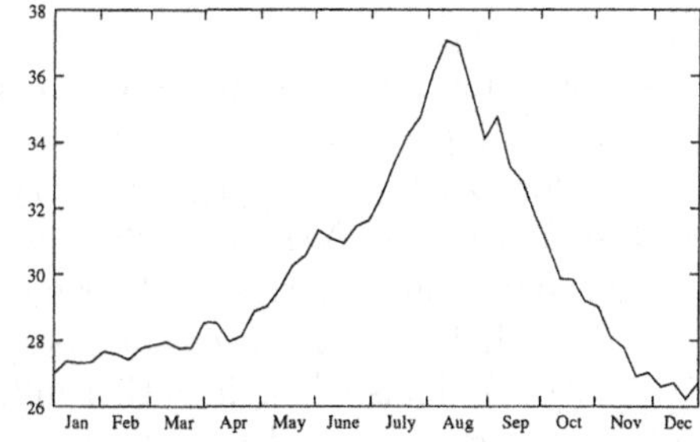

Fig. 3. Seasonal development curves over one year for two sets of retailers: for 133 sales outlets in an urban surrounding (left) and for 880 sales outlets in a holiday area (right).

4 Neural Modeling

We investigated several models to learn the dependency between input patterns and the future sales. The key question was whether the future sales depend non-linearly from the past or not. Thereto we used feed forward neural networks of different topology to compare the prediction accuracy on a validation set.

The number of input neurons was always six derived from the modeling in section 2. The target was the expected difference between the current and the future sales figure. We used neural networks with a linear output neuron and up to four hidden neurons with sigmoidal activation function in a single hidden layer. For reasons of better generalization we applied Bayesian learning [2] to optimize the weight decay parameter. The prediction error was evaluated using the square error function which is the appropriate error measure according to the assumption of normally distributed sales figures.

The empirical results showed that the use of a non-linear model increased the prediction error. Even a single hidden neuron reduced the accuracy significantly. These results indicate that the best model for the given patterns is a linear regression. The features can be seen as a preprocessing to linearize the original task. Notice that already the stochastic modeling in (1) points out that the influencing variables have to be combined linearly.

Surprisingly, even for linear regression the use of a weight decay parameter (ridge regression) combined with Bayesian learning improved the forecast compared to standard linear regression. Thus, although we use a very small model with only six input features, even the linear model without regularization showed a slight tendency to overfit.

5 Data Cleaning

Many time series contain a small number of strange data points due to erroneous sales figures or external events like holidays of neighboring traders or red-letter days. E. g. the right sales outlet in figure 1 reveals a pathological sales figure in week 156. In this section we want to investigate in which way outliers distort the regression and how to improve the prediction removing some of these points.

We can incorporate an outlier in the modeling of section 2 altering a certain sales figure X_t to a very large (or very small) value. This point does not belong to the distribution $N(\mu_t, \sigma^2)$ which would model an inconspicuous sales figure.

Assuming stationarity of the time series, i.e. $\mu_t = \mu_{t'}$ for all t, t', equation (1) gives that the only necessary best feature of length n is the value of $z := \sum_{i=2}^{n} \left(\frac{i}{n} Z_i\right)$. Thus, a single outlier influences $n+1$ patterns: one pattern reveals a large target value and n patterns are influenced by their inputs.

Figure 4 shows the pattern set originating from a stationary time series with a single small outlier. The target value is the difference to the future sales figure while the input value is the optimal feature z with history $n = 2$. Three patterns are influenced by the outlier marked with A (the jump from the normal level to the outlying small value), B (the return from the outlier to the normal level), and C (the pattern generated one time step after the return to normal level). Notice that the influence of these three patterns annihilates, i.e. the regression line induced from these three points is approximately the optimal regression. One can show that for every history length n of the feature z, the patterns influenced by a single outlier induce the optimal regression. Thus, an outlier does not distort the result in general but it may stabilize the regression.

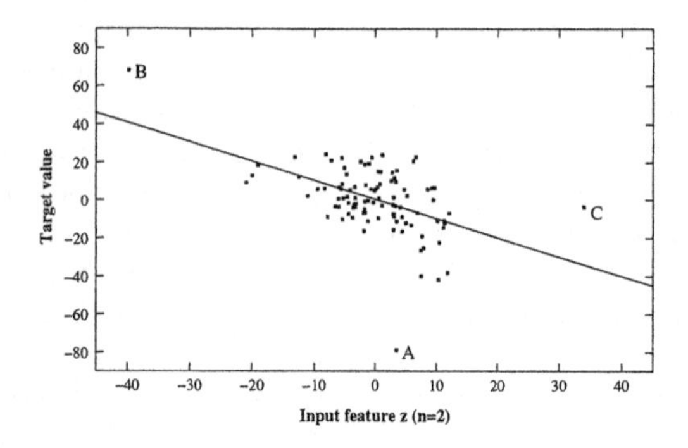

Fig. 4. The pattern set for a stationary time series with a single small outlier. The patterns are indicated by dots. The patterns influenced by the outlier are marked with A, B, and C. The solid line represents the optimal regression.

Since the absolute value of a pattern is no sufficient criterion to classify its contribution to the regression, we need to use other criteria. Rousseeuw [5] gives a survey of outlier tests and outlier robust estimation. Unfortunately, experiments showed that approaches like the Huber error function, the least-trimmed-sum-of-squares estimator, as well as outlier rejection methods based on the hat-matrix do not improve the prediction in our framework.

Instead, we investigated a two-steps approach based on an initial estimate given the complete training set which allows us to determine the utility of each pattern. We used the absolute size of the residuals on the training set as criterion to optimize the training set. The training patterns with the largest residuals are removed, and the regression model is retrained from the cleaned data.

Figure 5 compares the resulting errors on a validation set for a subset of retailers. The solid line shows the effect of random deletion: scaling down the training set increases the error. Removing the patterns with the largest target values (dashed line) slightly reduces the error for a small deletion rate but increases the error considerably for larger deletion rates. The deletion of large target values can be seen as a form of classical outlier rejection. In contrast, the two-steps algorithm based on the magnitude of the residuals (dash-dot line) results in a reduction of the prediction error for a wide range of deletion rates. In further experiments we used a fixed deletion rate of 5%.

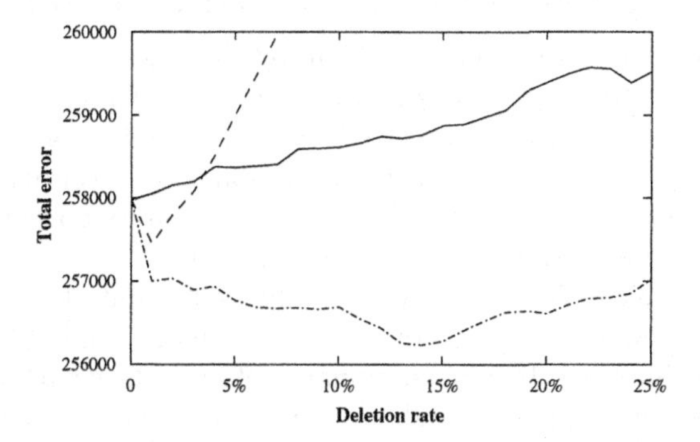

Fig. 5. Comparison of random deletion from the training set (solid line), deletion of patterns with large target values (dashed line) and removing patterns with large residuals (dash-dot line). The abscissa indicates the deletion rate (0% is the complete training set), the ordinate shows the total error on a validation set.

6 Augmenting the Training Set

As mentioned in section 4, even the linear regression model showed a slight tendency to overfit. Beneath the use of regularization approaches like Bayesian learning, a possible way to increase the prediction accuracy is to enlarge the training set. The solid line in figure 5 exemplifies the degree of possible improvement.

For every retailer we had historical data from 20 to 250 weeks which is an amount of data sufficient to get a rough estimate. However, it remains a high potential for improving the forecast by an enlargement of the training set.

Our idea was to use the data of similar sales outlets to augment the training set. Two main questions arose: (i) how can we measure the similarity of several retailers, and (ii) how can we assimilate the data of similar but not identical traders to avoid distorting the regression.

Since we use a linear regression model, the data of other retailers are most helpful if their linear relationship is high. This can be measured in terms of correlation which is invariant to level shift and variance. Since the length of the history of sales differs from retailer to retailer we have to eliminate the influence of different length. Thereto we used the mapping:

$$\rho \mapsto 2 \cdot F_{t_{n-2}} \left(\rho \frac{\sqrt{n-2}}{\sqrt{1-\rho^2}} \right) - 1 \tag{2}$$

where $F_{t_{n-2}}$ denotes the cumulative distribution function of the t-distribution with $n-2$ degrees of freedom, ρ denotes the correlation and n the length of history. Transformation (2) up weights the correlations for large n to compensate the stronger random influence. It is derived from the distribution of the term $\rho\frac{\sqrt{n-2}}{\sqrt{1-\rho^2}}$ given by the t-distribution with $n-2$ degrees of freedom [6].

After choosing a certain set of similar retailers we have to assimilate the additional patterns to avoid distorting the regression. Notice that the sales level and the noise level differ from retailer to retailer. Therefore we have to scale the patterns of each retailer individually to a common level. Otherwise, the influence of each sales outlet on the regression would depend on its noise level since noisy patterns typically lie on the edge of the pattern set.

Secondly, we have to consider that, by adding extra data from similar traders, the data of the primary trader become more and more negligible although they are the most specific data. To avoid dominance of secondary patterns from similar traders we introduced explicit weights on the data. The primary patterns were up weighted by assigning big weights while the secondary patterns were down weighted using small weights.

The explicit weights change the objective of regression. Instead of minimizing the standard mean squared error:

$$MSE := \frac{1}{2} \sum_{all\ patterns\ i} (y(x_i|\theta) - t_i)^2 \tag{3}$$

we minimize the weighted mean squared error:

$$WMSE := \frac{1}{2} \sum_{all\ patterns\ i} w_i(y(x_i|\theta) - t_i)^2 \tag{4}$$

where x_i denotes the pattern input, t_i the pattern target, w_i the weight assigned to pattern i, and $y(\cdot|\theta)$ the regression function parameterized by the parameter vector θ. The minimizing parameter vector can either be found using gradient descend or modifying the approach of pseudoinverse: computing the partial derivatives of the weighted mean squared error leads to:

$$X^T W X \theta = X^T W t \tag{5}$$

where t denotes the column vector of targets, W a diagonal matrix with the i-th entry representing the weight w_i, and X the matrix of pattern inputs with the

i-th row representing x_i. We can derive θ by multiplying (5) with $(X^TWX)^{-1}$ from the left.

The determination of the weights and the number of similar retail traders needs careful empirical investigation. Finally we used up to 150 times as much patterns as primary patterns were available. The forecast improved significantly while the training time still remained below five minutes for a single retailer on a 1.8 GHz PC.

7 Comparison of Two Approaches

We tested the approach against the system currently in use on three wholesaler areas: a) 133 sales outlets from a big city, b) 285 traders from a rural area, and c) 552 retailers from a holiday area. The test interval was one year, the training data were taken from up to five years. The seasonal development in the area a) and c) is shown in figure 3.

The system in use is based on a multilayer perceptron with one hidden layer. The network is trained using the RPROP algorithm [4] and regularized by early stopping and a weight decay term. The training patterns are preprocessed to equalize weekday specific differences and filtered using a classical outlier rejection method. The main properties of both approaches are compared in table 1.

Table 1. The properties of the neural prediction system currently in use compared to the new approach.

	system in use	new approach
number of features	51	6
topology of the neural network	51 – 4 – 2 sigmoidal hidden units	6 – 1 linear
learning algorithm	RPROP (gradient descend)	pseudoinverse
regularization techniques	weight decay, early stopping small initial weights	Bayesian learning
training set	all patterns of the primary retailer	weekday specific patterns + patterns of similar weekdays + patterns of similar retailers
data cleaning	remove patterns with very large and very small target values	residual dependent rejection

The test errors for both approaches and a simple moving average predictor are given in table 2. Notice that the new approach outperforms the system in use in all three areas although the system in use works quite well in the big city. Although an improvement of 0.09 points in the urban area seems to be negligible, it leads to a significant cost reduction since the number of sold copies is very high.

The differing error level in the three areas is primarily caused by differing sizes of the retail traders: the sales outlets are on average larger in the urban

Table 2. Development of the total relative prediction errors. The prediction error is defined as the sum of the absolute values of the individual prediction errors divided by the total sum of sales.

	urban area	rural area	holiday area
best moving average	8.58%	12.71%	14.37%
system in use	8.15%	12.20%	13.56%
new prediction approach	8.06%	11.74%	12.63%

area than in the rural and holiday area and therefore the relative prediction error remains smaller. We can assume as a rule of thumb that the variance of random noise equals the average sales level of a retailer (Poisson model). Under this assumption, the resulting prediction errors of the new prediction approach are very close to the minimal possible errors induced by purely random noise.

Comparing the results month by month helps to understand the effects. Figure 6 plots the relative test error for the holiday area for three approaches: the system in use, the new approach, and a simplified version of the new approach. The system in use has a bad performance above all in the time of strong seasonal volatility, e.g. in September and October while its accuracy is comparable in winter and spring. In contrast, the new modeling works well even in the time of seasonal changes.

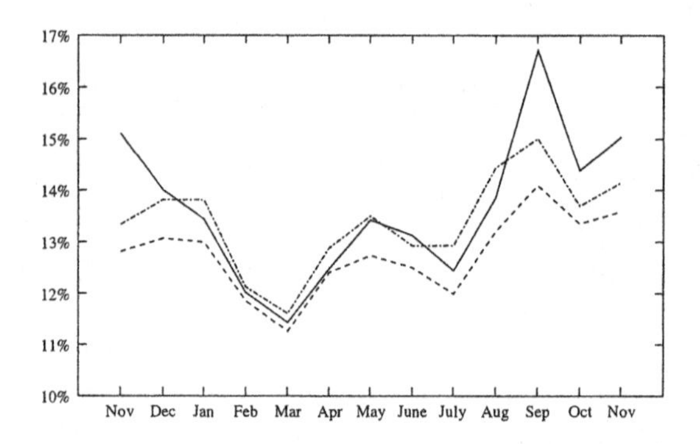

Fig. 6. Average relative prediction error in the holiday area. Notice that the test interval starts and ends in November. The solid line indicates the test error of the system in use, the dashed line shows the test error for the new approach and the dash-dot line shows the error for the new approach without training set augmentation, data cleaning and seasonal features.

Comparing the modeling of both approaches shows notable differences. The system in use is based on a large feature vector and a non-linear modeling. Therefore it needs strict regularization to avoid overfitting. The features allow to distinguish different situations very accurately but the training sets are too small to learn the subtle distinctions.

Vice versa, the new approach is based on a very small modeling with a set of six features and linear regression. The features are built to linearize the problem, and the training sets are augmented to exploit the data as much as possible. Despite using a small modeling, elaborate, data dependent techniques like Bayesian learning and data cleaning are used to optimize the results. The reduction of model complexity was compensated by an improvement of generalization and accurracy so that the total prediction error decreased significantly.

The new approach will soon be used for sales rate prediction in practice. Our ongoing work is aimed at further optimizing the modeling and extending the approach to other types of publications like weekly newspapers and magazines.

References

1. Tom Heskes. Solving a huge number of similar tasks: a combination of multi-task-learning and a hierarchical Bayesian approach. In *Proceedings of the 15th International Conference on Machine Learning*, pages 233–241, 1998.
2. David J. C. MacKay. A practical Bayesian framework for backprop networks. *Neural Computation*, 4(3):448–472, 1992.
3. Thomas Ragg, Wolfram Menzel, Walter Baum, and Michael Wigbers. Bayesian learning for sales rate prediction for thousands of retailers. *Neurocomputing*, 43(1-4):127–144, 2002.
4. Martin Riedmiller and Heinrich Braun. A direct adaptive method for faster back-propagation learning: the RPROP algorithm. In *Proceedings of the IEEE International Conference on Neural Networks*, pages 586–591, 1993.
5. Peter J. Rousseeuw and Annick M. Leroy. *Robust Regression and Outlier Detection.* John Wiley & Sons, 1987.
6. Werner A. Stahel. *Statistische Datenanalyse.* Vieweg, 1995.
7. Frank M. Thiesing and Oliver Vornberger. Sales forecasting using neural networks. In *Proceedings of the International Conference on Neural Networks*, pages 2125–2128, 1997.

A Multiagent-Based Constructive Approach for Feedforward Neural Networks

Clodoaldo A.M. Lima, André L.V. Coelho, and Fernando J. Von Zuben

Department of Computer Engineering and Industrial Automation (DCA)
Faculty of Electrical and Computer Engineering (FEEC)
State University of Campinas (Unicamp)
{moraes, coelho, vonzuben}@dca.fee.unicamp.br

Abstract. In this paper, a new constructive approach for the automatic definition of feedforward neural networks (FNNs) is introduced. Such approach (named MASCoNN) is multiagent-oriented and, thus, can be regarded as a kind of hybrid (synergetic) system. MASCoNN centers upon the employment of a two-level hierarchy of agent-based elements for the progressive allocation of neuronal building blocks. By this means, an FNN can be considered as an architectural organization of reactive neural agents, orchestrated by deliberative coordination entities via synaptic interactions. MASCoNN was successfully applied to implement nonlinear dynamic systems identification devices and some comparative results, involving alternative proposals, are analyzed here.

1 Introduction

Hybrid intelligent systems [5,10] comprehend all classes of methodologies wherein two or more intelligent techniques are glued together to create specialized tools and improved solutions to complex computational problems. Hybrid solutions are always motivated by the absence of a single procedure capable of dealing with all the computational aspects involved in the problem. Complementary attributes of distinct conceptual devices or practical tools may then be coherently explored.

In this paper, we focus on the combination of multiagent-based systems and connectionist models into the same conceptual framework. Although the aggregation of symbolic and subsymbolic systems has already been reported elsewhere [4,18], our approach follows a different research direction, not so common in the literature [13,17]. Essentially, the purpose is to model a feedforward neural network (FNN) as an architectural organization of neural agents, orchestrated by deliberative coordination entities via synaptic interactions. As a constructive methodology [8,9], MASCoNN (standing for "Multiagent Systems-based Constructive Neural Network") employs a two-level hierarchy of agent-based elements for the progressive assembling of customized FNNs. In the following, we briefly review similar approaches, present MASCoNN architectural components and training procedure, and analyze its performance when applied to the task of identification of multivariate nonlinear dynamic systems [11,12].

M.R. Berthold et al. (Eds.): IDA 2003, LNCS 2810, pp. 462–473, 2003.

2 Related Approaches

Amongst the several learning techniques already available, two classes can be clearly distinguished [4,5,10]: (i) *symbolic algorithms*, which basically try to process information (extracted knowledge) by generating rules of decision on how to classify relevant attributes into meaningful groups; and (ii) *subsymbolic algorithms*, wherein implicit models are automatically created to provide functional mappings (or clusters) that emulate data interdependencies.

In this work, the focus is on the progressive construction of customized FNNs by means of a multiagent-based hierarchical framework. In this context, Scherer and Schlageter [15] have already conceived a multiagent-oriented distributed problem solving architecture (PREDICTOR), joining together neural networks and expert systems modules for macro-economic forecasting. Their approach gears toward the application of the multiagent metaphor as an abstraction to encapsulate distinct intelligent systems (such as knowledge-based systems and neural networks) into the same cooperative framework in order to deal with complex predictive tasks. The focus is on the sequential and loosely-coupled integration of symbolic and subsymbolic subsystems into the same architectural model (i.e. the subsystems are clearly identifiable and might be properly separated after combined). In MASCoNN, however, our perspective is more compliant with the philosophy followed by the works of Pham [13] and Śmieja [17]. Based on very adaptable neural agents, the intention is to generate a parsimonious and dedicated solution, capable of incorporating from the very specific to the most generic aspects of the problem in hand.

Pham [13] has proposed the neurOagent approach, which is based on macro-connectionist concepts and stipulates two types of information integration. The first encompasses the association and symbol levels, where a neuro-symbolic fully-integrated processing unit provides learning capabilities and distributed inference mechanisms. The second comprehends the knowledge level, which concerns with the knowledge-based modeling. In neurOagent, the basic unit of modeling is an assembly of neurons and not a single neuron, whereas the framework constitutes three organization scopes: from the cellular to the connectionist to the macro-connectionist.

Śmieja [17] has extended the prior work of Selfridge [16] on the conceptualization of the Pandemonium system. By means of the "divide-and-conquer" principle, each of the former Selfridge's daemons has been replaced by an adaptive NN module dedicated to a particular subtask. In this new version, the knowledge distribution among the NN modules (agents) is granted since there is the requirement that they be reflective. Typically, a reflective agent senses the world and produces an output based on its current internal mapping function, which may be self-corrected through time (via a training procedure) in order to increase the agent self-confidence. Śmieja's version of Pandemonium has the following design features: (a) Multiple NN modules are recruited to solve a particular (huge) task; (b) These modules are not interconnected, may be independently trained and simulated, and their number may dynamically change; (c) Each module may take part in different module groups in such a way as to compose different Pandemonium systems for the realization of distinct tasks; (d) Each Pandemonium is defined in terms of its input/output pairs and is in charge of allocating available information to the individual modules (decomposition) and then to recombine produced outputs to compose a unique task solution (typically by choosing more self-confident NN modules).

MASCoNN borrows features from both neurOagent, in the modeling of the neuronal structures, and from Pandemonium, in the organizational viewpoint.

3 MASCoNN: Constructive Neural Networks from a Multiagent Systems Perspective

The main aim underlying MASCoNN is to properly combine the advantages offered by both neurOagent and Pandemonium: skillful representation, manipulation, and generation of knowledge (agents); high degrees of plasticity, generalization and approximation (NNs); and full potential to adapt and learn (both).

In what follows, we present MASCoNN by focusing on the issues that comprise its macro (architectural) and micro (agent) organizational levels, as well as its training procedure.

3.1 Hierarchical Organization

MASCoNN is influenced by Śmieja's Pandemonium in the following aspects: (i) it also makes use of the "divide-and-conquer" principle to tackle complex problems; (ii) the number of agents may dynamically change; and (iii) the activities are processed in two levels, encompassing the *decision* (or *coordination*) top level and the *operational* (or *competition*) bottom level. In MASCoNN, the agent units in the coordination level are called *decision agents* (DAs) whose fixed number (set in advance) is proportional to the amount of tasks to be accomplished by the whole multiagent system (MAS). Each DA output is the linear combination of the outcomes of all agents working in its associated competition level, which are known as *working agents* (WAs). The number of WAs can vary "on the fly" (initially, there is only one) and is set in accordance with the difficulty of the task at hand. So, at each moment, a DA may coordinate one or more WAs (see Fig. 1). The working agents are trained as they are recruited and can also be retrained whenever novel peers appear to tackle the same task (in order to stimulate competitive behavior among them). Moreover, WAs may be reused in other task achievements (associated with other DAs) provided that they be properly retrained for such endeavor.

In MASCoNN, each WA embodies properties of a typical neuron in the hidden layer of a constructive FNN. By this means, different WAs may produce distinct nonlinear behaviors when performing the same functional task, something that may be executed by choosing disparate types of activation functions. Such flexibility on how *heterogeneous* or *homogeneous* are the WA elements is a great improvement brought by MASCoNN, that can be used in advance by the designer or left to be dynamically resolved by each DA. The idea is to allocate novel WAs in a given competition level for possibly identifying important aspects not yet captured by the former components (i.e. residual learning).

There is no communication between WAs in the same or different groups. However, every DA communicates with its WA group via synaptic interactions in a way such as to stimulate their competitive (first step) and cooperative (second step) behaviors. From these competitive/cooperative operations should emerge self-organizing phenomena for bringing about automatically-defined FNNs.

Firstly, WAs compete in order to find the most relevant aspects underlying the environment they are embedded in. Such environment is typically the representation of the problem under consideration in the form of input/output data. Contrary to other constructive approaches (such as CasCor [3]), the WAs receive no outcome furnished by the other peers ("blind" competition). The competitive phase finishes when no WA can bring more improvements on its task fulfillment (i.e. each automatically converges to a particular subtask). Afterwards, the WAs enter the cooperative stage to be coordinated by the DA. At this phase, the WA outputs are combined to provide the DA's "world model", viewed as the whole group knowledge about the task. If this model is still inadequate (according to some criterion, such as an index of performance), then a new WA is inserted into the group, and the cycle starts again. After generating the "world model", each DA verifies the WAs that have positively contributed to the task accomplishment, and cooperated better with their fellows (typically by evaluating how complementary are their solutions). Such performance assessment may incur the pruning [14] of some WAs. In this case, the remaining WAs are retrained just for the reinforcement of their abilities.

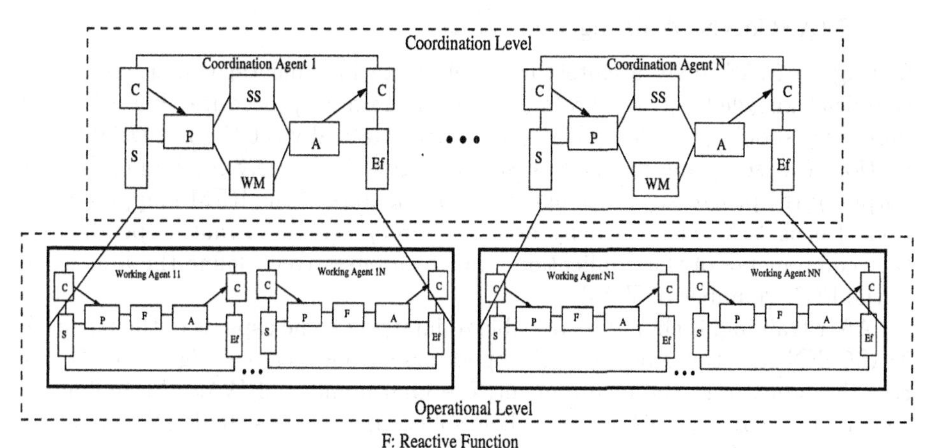

Fig. 1. MASCoNN hierarchical arrangement of coordination and working agents.

3.2 MASCoNN Agent Structure

From Fig. 1, one can notice that each DA is composed of four modules:

Perception Module (PM): at each execution cycle, it receives an input/output pair from the *Sensors* (S) which serves as observed data for comparing the value produced by the *World Model* unit (combination of WA outputs) with the expectation generated by the *Score System* (performance criteria). It should detect environmental regularities (or anomalies) and relationships between WA outputs.

World Model Module (WMM): offers the best estimation of the environmental state from the perspective of the whole group it coordinates.

Action Module (AM): according to the world model, this element plans the new actions to be followed by the WA group.

Score System Module (SSM): judges what is good or bad; rewards/punishes WAs for their results; and continually performs an estimate on how long is the distance from the current group state until the desired environmental state. It is also liable for dynamically spawning/recruiting new WAs.

Conversely, each WA, modeled as a reactive agent (i.e., its effectors' outputs are given as a direct function of its current sensors' inputs), comprises three modules, namely the *Perception Module* (PM) and *Action Module* (AM), as mentioned above, and the *Reflexive Function Module* (RFM), which projects the input data along the direction plan that generates the closest output to the desired one. The reinforcement interplay (reward/punishment) from a DA to its WAs is done via a virtual device, called *Channel*.

3.3 MASCoNN Training Procedure

Setting an MASCoNN architecture implies the incremental construction of a (implicit) knowledge base through a proper training procedure (Fig. 2). Such procedure should provide a means to (a) automatically set the DA weight connections (i.e. searching for the best set of weights for linearly combining the WA outputs); (b) progressively configure the parameters of the RFM mapping function of each WA, which usually assumes a very generic format, depending on the requirements of the application; and (c) progressively define the dimension of the FNN (number of WAs).

After training, it should be possible to interpret some aspects of the resulting MASCoNN structure, including the operators automatically designed to preprocess the input from the environment, the final number of WAs, the individual contributions of each WA for the whole solution, as well as the FNN generalization capability (see Section 5).

4 An Agent-Based Approach for Multidimensional Nonlinear Mapping

As an agent-based perspective on the automatic construction of FNNs, MASCoNN allows the synthesis of more compact and efficient solutions to a wide

Step 1: DA perceives the environment;
Step 2: While not achieving a proper environment model
  2.1: DA performs competition among WAs;
  2.2: DA performs cooperation among WAs;
Step 3: DA performs pruning of deficient WAs;
Step 4: DA performs reinforcement training.

Fig. 2. MASCoNN training procedure.

variety of information processing tasks. Also, being an alternative to synthesize NN models, the input-output behavior can always be interpreted in terms of a multidimensional nonlinear mapping.

The universal approximation capability expressed by a wide range of NN architectures provides theoretical motivations to consider these tools as potential candidates to deal with problems such as pattern classification, time-series prediction, nonlinear regression, and identification and control of nonlinear dynamic systems. However, the guarantee that a proper NN exists, no matter the peculiarities of each application, is not enough. Procedures to search for an adequate architecture, with a proper definition of its free parameters, are necessary and sometimes guide to deceptive results. For example, given two architectures A and B, characterized by the fact that A is totally contained in B, the theoretical conclusion that B will always produce a performance equal or superior to A is false, in practice, because the training procedure is prone to local minima.

One of the reasons why the training procedure can be defined as a multimodal optimization strategy is the usual attempt to deal with all aspects of the problem at the same time, employing the whole flexibility of the NN at once to solve the problem. On the other hand, agent-based approaches to synthesize NNs adopt a completely different perspective: Agents will self-organize themselves, so that each one becomes responsible for a subtask, possibly much simpler than the original task. Some kind of composition of the subtasks will further represent the task as a whole. Specifically in the case of multidimensional nonlinear mapping, the following results may be considered useful to justify a multiagent-based implementation: (a) every multidimensional nonlinear mapping can always be described by an additive composition of linearly independent mappings [2]; (b) among the infinite possibilities for the set of linearly independent mappings to be composed, including the possibility of a set with infinite cardinality, there are the sets composed of ridge functions [1], i.e. functions that express their nonlinear shape in just one direction, being constant along the other orthogonal directions; (c) the number of terms in the additive composition, necessary to achieve a given performance, may vary depending on the adequacy of the mappings to be composed [6]. If these mappings can be defined according to the peculiarities of the application, then the number of mappings will be significantly reduced, because each mapping will be designed specifically to complement the information represented by the others.

MASCoNN will be implemented here to take full advantage of these three results. In the experiments to be reported, each WA tries to accomplish the environmental modeling by projecting the data inputs along different vector space directions, as asserted by the *Project Pursuit Learning* (PPL) constructive algorithm [8,9,19]. So, the activation function of each hidden neuron will play the role of a ridge function and the output layer will perform the additive composition. The number of neurons (agents) and the subtask to be attributed to each one will be automatically defined during the execution of the already described MASCoNN training phase.

In the cooperation phase, each WA should contribute to improve the whole neural network performance in tackling the task (e.g., by minimizing an approximation error). For the n-th WA, two sets of parameters are adjusted. First, assuming the projection direction (v_n) is fixed, the idea is to achieve the best approximation function (f_n) for the projected data, according to:

$$\min_{f_n} \frac{1}{N} \sum_{l=1}^{N} (d_l - f_n(v_n^T x_l))^2 + \lambda_n \phi(f_n),$$

where d_l is the approximation residual (i.e. approximation error that persists after the introduction of all neurons with index different from n), λ_n refers to a regularization parameter, and the function ϕ measures the smoothness of f_n. Then, assuming the activation (mapping) function is fixed, the objective is to iteratively discover (via an optimization method with unidimensional search, like Modified Newton's Method [7]) the best projection direction (v_n), given by

$$\min_{v_n} \frac{1}{N} \sum_{l=1}^{N} (d_l - f_n(v_n^T x_l))^2.$$

The iterative update of both f_n and v_n in the cooperative phase continues until the whole process converges. The convergence is proved to exist [8]. In the competitive stage, each neuron tries to review its position in light of the changes that occurred previously, such as the arrival of new WAs. As before, f_n and v_n may also be updated. Finally, in the pruning stage, each DA defines a threshold delimiting the minimal level of contribution to the whole task each of its WAs should attain for not being eliminated from the FNN. Moreover, at this moment, the DA verifies whether there is any WA working (i.e. representing) on noisy data; in such case, those WAs are also removed.

5 MASCoNN for Nonlinear Dynamic Systems Identification

It is possible to employ either static or dynamic NNs for the system identification task [11,12]. Such effort is particularly useful to the design of adaptive controllers.

Typically, there are some considerations about the plant to be identified: (a) the plant is characterized by a discrete time dynamics; (b) the plant may express linear and/or nonlinear behavior; (c) only input-output data are made available; and (d) the plant may be either time-varying (several regimes) or invariant.

Because usually there is no *a priori* knowledge about the system, purely analytical or parametric methods may not be available, as there is no way to predict the flexibility to be imposed to the identification model. The input $u(k)$ and output $y(k)$ values should guide the neural network (MASCoNN, in this study) to properly represent the system transfer function. Such setting is illustrated in Fig. 3 wherein the aim is to minimize the difference between the target output $(y(k+1))$ and the generated output $(\hat{y}(k+1))$, that is, $||\hat{y}(k+1) - y(k+1)||$.

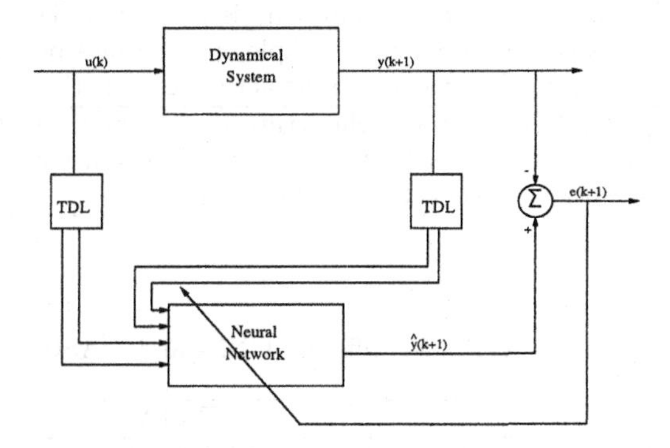

Fig. 3. Nonlinear system identification with NNs (TDL means "tapped delay line").

Contrasted with the identification of linear plants, which is typically done by designing (via the Superposition Principle) a transfer function modeling the plant behavior, identification of multivariate nonlinear dynamic systems are more laborious, since the input-output relation may depend on past state values and the plant may be sensitive to the current points of operation. The extraction of a global NN model from data is also very difficult as the network generalization in some noisy regimes may be hampered (i.e. over- or under-fitting) by the dynamic switching between regimes.

In the following, we compare the behavior of MASCoNN with other conventional neural network (CNN) approaches. As there is only one identification task to be dealt with, we employ only one WA group coordinated by one DA. Table 1 brings the notation used for describing the several FNN architectures (with their supervised training algorithms) applied in the comparative study.

The case study to be considered here was taken from the work of Narendra and Parthasarathy [11] and the plant to be identified is described by a third-order nonlinear difference equation given by:

$$y(k+1) = \frac{y(k) \cdot y(k-1) \cdot y(k-2) \cdot u(k-1)[y(k-2) - 1] + u(k)}{1 + y^2(k-2) + y^2(k-1)}$$

Table 1. FNN models employed in the experiments and their description

NN Model	Description
CNNmlpGRn	CNN with one hidden layer composed of n neurons with identical activation functions $f_j(\cdot) = f(\cdot) = \tanh(\cdot)$ $(j = 1, \cdots n)$ and parameters adjusted via simple gradient algorithm
CNNmlpGCn	CNN with one hidden layer composed of n neurons with identical activation functions $f_j(\cdot) = f(\cdot) = \tanh(\cdot)$ $(j = 1, \cdots n)$ and parameters adjusted via hybrid conjugate gradient algorithm
CNNmlpFKn	CNN with one hidden layer composed of n neurons with identical activation functions $f_j(\cdot) = f(\cdot) = \tanh(\cdot)$ $(j = 1, \ldots, n)$ and parameters adjusted via Extended Kalman Filter
MASCoNN_Her	MASCoNN whose architecture is generated via the training procedure shown in Section 3 employing Hermite polynomials (with six as maximum order) as basic activation functions
MASCoNN_Spl	MASCoNN whose architecture is generated via the training procedure shown in Section 3 employing non-parametric smoothing splines as basic activation functions

Table 2. Comparative results for different CNNs and MASCoNNs for IMNDS

FNN Configuration	Plant			
	Neurons	Stopping Criterion	MSE	
			Training	Test
CNNmlpGC6	6	2000 epochs	0.000678	0.012031
CNNmlpGC10	10	522 epochs	0.000489	0.005785
CNNmlpGC15	15	386 epochs	0.000495	0.008380
CNNmlpFK6	6	....	....	0.012962
CNNmlpFK10	10	....	....	0.009888
CNNmlpFK15	15	....	....	0.007957
MASCoNN_Her	6	MSE = 0.0005	0.000496	0.000126
MASCoNN_Spl	6	MSE = 0.0005	0.000460	0.000818

Table 3. Projection directions achieved by *MASCoNN_Her* WAs

Projections	Coordinates		
	u(k)	y(k)	y(k-1)
1st direction	-0.041526	-0.154477	-0.353771
2nd direction	0.590152	0.407951	0.000369
3rd direction	-0.511678	0.492766	0.001185
4th direction	0.205783	-0.479769	-0.004685
5th direction	0.293872	-0.643541	-0.002054
6th direction	-0.831326	0.406370	-0.018132

The plant output is nonlinearly dependent both on its prior values and on the control input (which are made coupled). The control input is regarded as a uniformly distributed random signal $u(k)$ with 1000 points in the range $[-1, 1]$ for training. For the *CNNmlpFK* model [7], the training set is composed of 10000 samples and the parameters are adjusted online (so, there is no training data). The identification model has five inputs, given by $u(k), u(k-1), y(k), y(k-1)$, and $y(k-2)$, plus one output, $y(k+1)$, and is described as

$$y(k+1) = FNN[u(k), u(k-1), y(k), y(k-1), y(k-2)]$$

where *FNN* refers to a CNN or MASCoNN structure. After concluding the training phase, the model is assessed by applying a sinusoidal signal $u(k)$, whose first 500 samples are generated according to $u(k) = \sin(\frac{2\pi k}{250})$, and the remaining according to $u(k) = 0.8\sin(\frac{2\pi k}{250}) + 0.2\sin(\frac{2\pi k}{25})$. Table 2 brings the comparative results and shows how superior was MASCoNN in face of CNN approaches. Albeit the CNNs have shown concurring error results for training, this is not achieved for testing, even for architectures with high dimensions (such as *CNNmlpGC15*). Just for illustrative purposes, Fig. 4(a) brings the evolution of the mean squared error along the construction of *MASCoNN_Her* and Fig. 4(b) highlights both the plant and *MASCoNN_Her* outcomes for this case study.

As mentioned before, unlike CNNs [18], it is possible to directly extract knowledge from MASCoNN. For sake of explanation, we concentrate here only on the *MASCoNN_Her* setting. Table 3 shows the final projection directions [8] achieved by each of the six progressively created *MASCoNN_Her* WAs whereas Fig. 5 presents their respective activation function formats. From the final projection directions in Table 3, one may note that all WAs "observe" the training data from different viewpoints. This demonstrates that the "divide-and-conquer" principle was satisfied (decomposition of the problem) by employing the cooperative/competitive methodology devised for MASCoNN. Moreover, it is worth to mention the very reduced number of WAs employed in the final MASCoNN architectures (this was observed in all experiments we have conducted on regression, forecasting, and control problems that were not reported here). So, MASCoNN brings support to parsimony in the allocation of resources for generating the solutions, which is mainly due to the orthogonal roles presented by the neurons as a result of the competitive process.

6 Final Remarks

In this paper, we have introduced a novel methodology for the progressive, automatic construction of feedforward neural networks, named MASCoNN. Such approach is compliant with the intelligent hybrid systems standpoint as it stipulates the fusion of multiagent-based and connectionist models into the same framework. A two-level hierarchy composed of coordination and operational levels provides the means for assembling FNNs by employing neural agents to accomplish the high-level tasks. Some advantages of MASCoNN, when compared with conventional approaches, are: (a) it provides a proper constructive approach

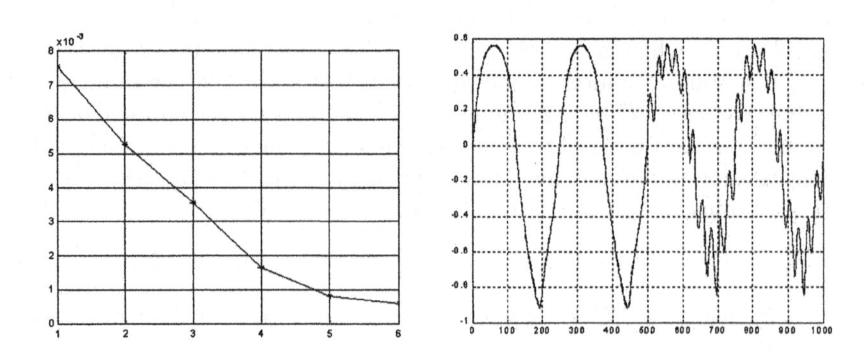

Fig. 4. (a) MSE evolution; (b) MASCoNN_Her output vs. real plant output.

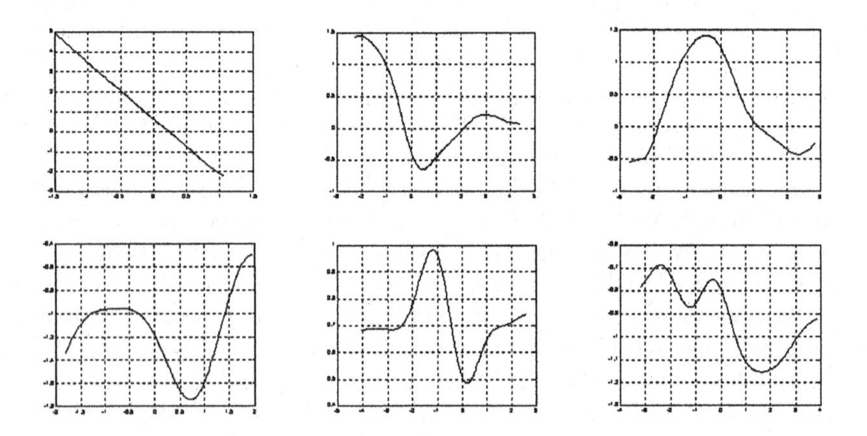

Fig. 5. Final activation functions achieved by *MASCoNN_Her* WAs.

for the handling of very complicated problems by allowing the successive application of differently-shaped, generic activation functions (that is, heterogeneous WAs); (b) it guides to the interpretation of an FNN as a multiagent system, which is more intuitive and facilitates the identification and extraction of knowledge embedded in the architectural elements; (c) it yields a better allocation of resources, bringing support to parsimony; and (d) it provides an extensible architecture that allows that other types of interactions between agents be modeled and introduced.

As future work, we plan to investigate the behavior of MASCoNN with heterogeneous configurations, that is, with WAs with distinct types of activation functions working together. Evolutionary-based techniques shall be also considered as alternative tools for training and combining WAs. Finally, it is possible to extend MASCoNN to encompass other constructive algorithms [9].

Acknowledgement. CNPq/Brazil has sponsored the authors via PhD scholarships #141394/00-5 and #140719/99-7, and via grant #300910/96-7.

References

1. W. Dahmen and C. A. Micchelli. Some remarks on ridge functions. *Approximation Theory and its Applications*, 3(2–3):139–143, 1987.
2. P. J. Davis. *Interpolation & Approximation*. Dover Publications, New York, 1975.
3. S. E. Fahlman and C. Lebiere. The cascade-correlation learning architecture. In D. S. Touretzky, editor, *Advances in Neural Information Processing Systems*, volume 2, pages 524–532. Morgan Kaufmann, 1990.
4. J. Ghosh. Neural-symbolic hybrid systems. In M. Padget et al., editors, *The Handbook of Applied Computational Intelligence*. CRC Press, 2001.
5. S. Goonatilake and S. Khebbal, editors. *Intelligent Hybrid Systems*. Wiley, 1995.
6. W. Härdle. *Applied Nonparametric Regression*. Cambridge University Press, 1990.
7. S. Haykin. *Neural Networks–A Comprehensive Foundation*. Prentice Hall, 1999.
8. J. Hwang, S. Lay, M. Maechler, D. Martin, and J. Schimert. Regression modeling in backpropagation and project pursuit learning. *IEEE Trans. on Neural Networks*, 5(3):342–353, May 1994.
9. T.-Y. Kwok and D.-Y. Yeung. Constructive algorithms for structure learning in feedforward neural networks for regression problems. *IEEE Trans. on Neural Networks*, 8(3):630–645, 1997.
10. L. A. Medskar. *Hybrid Intelligent Systems*. Kluwer Academic Publisher, 1995.
11. K. Narendra and K. Parthasarathy. Identification and control of dynamical systems using neural networks. *IEEE Trans. on Neural Networks*, 1(1):4–27, March 1990.
12. M. Nrgaard, O. Ravn, N. K. Poulsen, P. M. Norgaard, and L. K. Hansen. *Neural Networks for Modelling and Control of Dynamic Systems: A Practitioner's Handbook*. Advanced Textbooks in Control and Signal Processing. Springer, 2000.
13. K. M. Pham. The neurOagent: A neural multi-agent approach for modelling, distributed processing and learning. In Goonatilake and Khebbal [5], chapter 12, pages 221–244.
14. R. Reed. Pruning algorithms–A survey. *IEEE Trans. on Neural Networks*, 4(5):740–747, May 1993.
15. A. Scherer and G. Schlageter. A multi-agent approach for the integration of neural networks and expert systems. In Goonatilake and Khebbal [5], chapter 9, pages 153–173.
16. O. G. Selfridge. Pandemonium: A paradigm for learning. In *Proc. Symp. Held Physical Lab.: Mechanisation Thought Processing*, pages 511–517, London, 1958.
17. F. J. Śmieja. The pandemonium system of reflective agents. *IEEE Trans. on Neural Networks*, 7(1):97–106, January 1996.
18. I. Taha and J. Ghosh. Symbolic interpretation of artificial neural networks. *IEEE Trans. on Knowledge and Data Eng.*, 11(3):448–463, May/June 1999.
19. F. J. V. Zuben and M. Netto. Projection pursuit and the solvability condition applied to constructive learning. In *Proc. of the International Joint Conference on Neural Networks*, volume 2, pages 1062–1067, 1997.

Evolutionary System Identification via Descriptive Takagi Sugeno Fuzzy Systems

Ingo Renners and Adolf Grauel

University of Applied Sciences
Lübecker Ring 2, D-59494 Soest / Germany
Soest Campus, Dept. Mathematics and
Centre Computational Intelligence & Cognitive Systems
phone: +049 2921 378451, fax:+049 2921 378451, renners@fh-swf.de

Abstract. System identification is used to identify relevant input-output space relations. In this article the relations are used to model a descriptive Takagi-Sugeno fuzzy system. Basic terms of system identification, fuzzy systems and evolutionary computation are briefly reviewed. These concepts are used to present the implementation of an evolutionary algorithm which identifies (sub)optimal descriptive Takagi-Sugeno fuzzy systems according to given data. The proposed evolutionary algorithm is tested on the well known gas furnace data set and results are presented.

1 Introduction

By using a broad definition, system identification (and curve fitting) refers to the process of selecting analytical expressions representing the relationships between input and output variables of a data set. Each instance of this dataset consist of independent (predictor) variables $x_1, x_2, \ldots, x_i$ and one dependent (response) variable y. The process of system identification includes three main tasks [9]:

- selecting a pool of independent variables which are potentially related to y
- selecting an analytical expression
- optimizing the parameters of the selected analytical expression

Not stringently these analytical expressions has to be represented by closed form solutions. We used Descriptive Takaki-Sugeno Fuzzy Systems (DTS-FSs) as framework to represent the analytical expression and methods from Evolutionary Computation (EC) to identify a (sub)optimal DTS-FS structure. This structure identification subsumes the task of selecting the inputs and selecting the characteristic analytical expression. Simultaneous the implemented EA optimizes the parameters of the selected DTS-FS.

The sections 2 and 3 briefly introduces the used concepts and nomenclature. Section 4 shortly reviews characteristics of the gas furnace dataset. This is followed by section 5 which presents the implemented EA for system identification and the paper is concluded by some results and a short discussion.

M.R. Berthold et al. (Eds.): IDA 2003, LNCS 2810, pp. 474–485, 2003.

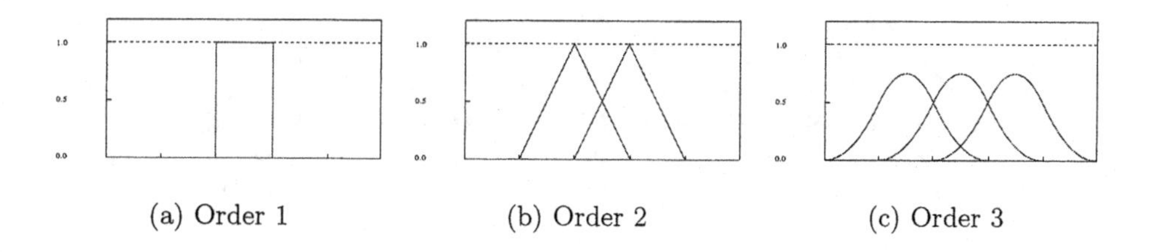

(a) Order 1 (b) Order 2 (c) Order 3

Fig. 1. Univariate b-splines of order 1, 2 and 3 defined over equispaced knots.

2 Fuzzy Rule Based System Basics

Fuzzy logic is an extension to the classical two-valued logic by concepts of fuzzy set theory, introduced by Zadeh in 1965 [17]. A FS make use of fuzzy logic concepts to represent a knowledge base and/or to model interactions and relations of system variables. Knowledge in FSs is represented by linguistic variables with an associated set of linguistic values. Linguistic values are defined by fuzzy sets, where a fuzzy set A in X is a set $A = \{x, \mu_A(x)|x \in X\}$ of ordered pairs which are defined by a Membership Function (MF) $\mu_A(x) \in [0, 1]$ and a linguistic term for labeling.

2.1 Membership Functions

Common used MFs are triangular, trapezoidal, bellshaped and Gaussian functions [7,8]. A desirable property of MFs for many fuzzy systems is the so called partition of unity, meaning that all activation values $\mu_A(x)$ sum up to one. The advantages of MFs forming a partition of unity are:

- advantages in function approximation [16]
- more robust extrapolation behavior [11]
- fuzzy system output lies in a known interval
- simplifies the interpretation of a FS

Especially b-splines, which have been employed and were successfully deployed in surface-fitting algorithms for computer aided design tasks [3], are first class candidates to form MFs in descriptive FSs, because:

- b-splines can form extremely different shapes simply by changing the order or the knot-positions
- b-splines show some for FSs essential characteristics such as positivity and local support and furthermore they form a partition of unity

If we have no partition of unity, the interested reader is referred to [6].

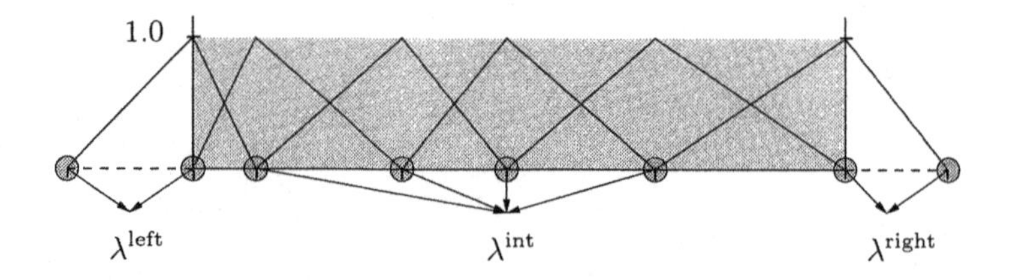

Fig. 2. A knotvector λ with eight knots is used to define six b-splines of order 2.

B-splines are recursively defined over $k+1$ successive arranged knots of a knotvector λ, by

$$
\begin{aligned}
B_j^{k+1}(x) &= \frac{x-\lambda_j}{\lambda_{j+k-1}-\lambda_j} B_j^k(x) + \frac{\lambda_{j+k}-x}{\lambda_{j+k}-\lambda_{j+1}} B_{j+1}^k(x) \\
B_j^1(x) &= 1, \text{if } x \in [\lambda_j, \lambda_{j+1}) \\
&= 0, \text{otherwise,}
\end{aligned}
$$

with x as input value and $B_j^k(x)$ as activation value of the jth b-spline defined over the knots λ_j to λ_{j+k}. The concept of a knotvector is compatible with the construction demands of DTS-FSs, because each knotvector defines a whole set of MFs for one linguistic variable. For the remainder λ^{left}, λ^{int} and λ^{right} refer to sections of the complete knotvector λ. Figure 2 illustrates an exemplary distribution of b-spline MFs, with the grey shaded area fulfilling the partition of unity and thus, representing a valid model input for the considered input dimension. Therefore, the last element of λ^{left} is identical to the minimum valid input value and the first element of λ^{right} is identical to the maximum valid input value. The number of elements of $\lambda^{\text{int}}+k$, with $\lambda^{\text{int}}>0$, is identical to the number of MFs defined over λ.

2.2 Takagi-Sugeno Fuzzy Systems (TS-FS)

Takagi and Sugeno proposed in 1985 [15] a FS with following rule structure: IF x_1 is A_1 AND $\cdots$ AND x_n is A_n THEN $f(x)$, thus each rule output is a function of the input vector and the overall output of a TS-FS is a weighted sum of all rule outputs. For the remainder only zero-th order consequences will be considered, yielding in systems referred as *singleton* FSs.

A FS is called *descriptive* if all rules share the same linguistic terms provided by a global database. Thus, the global semantic of rules is assured and the interpretability of the linguistic values of a linguistic variable can be provided by observing and constraining each set of MFs stored in the global database.

2.3 Learning of Descriptive Takagi-Sugeno (DTS) FSs

By interpreting univariate basis functions (figure 1) as MFs, and n-variate basis functions, built up by multiplying n univariate basis functions, as an antecedent

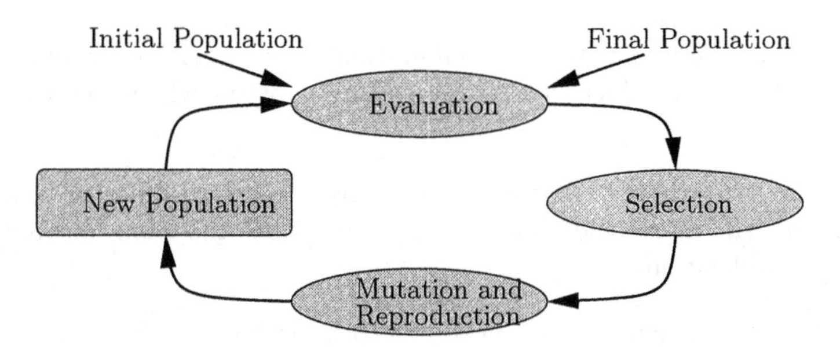

Fig. 3. Basic concept of evolutionary computation.

of a FS rule, Radial Basis Function Networks (RBFNs) has shown to be functional equivalent to zero-th order TS-FSs [10]. As consequence, supervised learning of TS-FS can be performed very efficiently by solving an overdetermined linear system of equations [12]. Furthermore, if the univariate basis functions of a RBFN are defined globally for each possible input n and if the n-variate basis function of all RBFN-neurons are built up by permutating all possible univariate basis function combinations, then a RBFN becomes equivalent to a DTS-FS [2] and the set of rules is called *complete*.

3 Evolutionary Computation

By concerning about a population as a set of solutions, the term Evolutionary Computation (EC) subsumes all population based approaches which perform random combination, alteration of potentially useful structures to generate new solutions and a selection mechanism to increase the proportion of better solutions. The schematic functionality of EC is depicted in figure 3. Thus, EC comprises techniques such as genetic algorithms, evolutionary strategies, evolutionary programming, genetic programming and artificial immune systems or hybrid methodologies resulting from the combination of such techniques. If EC concepts are used in algorithms, those algorithms are referred as Evolutionary Algorithms (EAs).

3.1 Nomenclature of Evolutionary Computation

For simplicity EC make use of terms with their origin in biology:

- individual - a particular biological organism
- fitness (of an individual) - measurement that express the success of an individual to handle its living conditions
- population - group of interbreeding individuals within a given area
- phenotype - refers to the composition of a particular individual
- gene - a gene is a functional entity that encodes a specific feature of an individual

– genotype - refers to a specific combination of genes carried by an individual[1]
– DNA - blueprint which is transcribed into proteins which build the phenotype

Keeping the limitations of nomenclature transfer between different scientific areas in mind, the above mentioned terms has in EC the same meaning beside following modifications:

– individual - particular solution to a certain problem
– fitness (of an individual) - measurement which express the success of an individual to solve the problem
– DNA/genotype - blueprint which is transcribed via function(s) into a phenotype

3.2 Variable Length Coding of Genotypes

Variable length coding in EC becomes inevitable if not only parameter optimization, but also structure optimization of candidate solutions should be performed. This is due to the fact that structure optimization comprises the task to deal with desired ranges and constraints. The more restrictions are postulated and more ranges vary, the more complicated and more memory consuming (using fix length coding) the process of finding valid genotypes becomes. Thus, genetic programming utilizes variable length coding since their debut and it is noteworthy to mention that they mostly use tree-based genotypes. A promising method to create genotypes is by utilizing a genotype-template. Any type of grammar can be used as template and n-ary trees [13], for example, can be used to form the genotype.

4 The Gas Furnace Data

The gas furnace data set consist of 296 input values u(t) representing the flow rate of methane gas in a gas furnace and the according 296 output values y(t) (figure 4) representing the concentration of $C0_2$ in the gas mixture flowing out of the furnace under a steady air supply. The measurements were sampled at a fixed interval of 9 seconds and therefore the 296 samples represents a time series covering 44 minutes and 15 seconds. The data has used extensively as a benchmark example for process identification by many researchers (see [4] for an instructive overview). The new method will be tested on this benchmark example.

The ability to model non-linearity is one of the reasons to use soft computing methods like DTS-FS. At first glance it would be more appropriate to choose a

[1] In EC literature the terms *genotype*, *chromosome* and *DNA* (seldom used) are mostly used interchangeably. In the following the term *genotype* is used to describe a set of genetic parameters that encode a candidate solutions, because this is closer to the biological meaning.

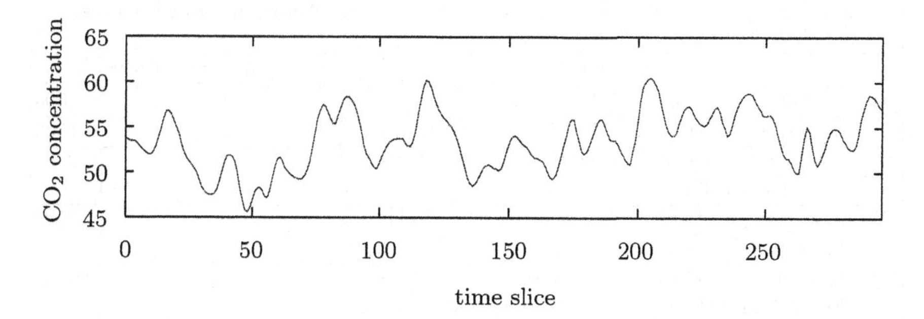

Fig. 4. The gas furnace data.

highly non-linear dataset instead of the approximately linear gas furnace dataset [1]. Nevertheless we used the gas furnace dataset, firstly because of its availability and secondly because the task of feature selection is the main emphasis of this paper. Thus, the expectation is to identify the same used features for optimal linear models (given in table 4) and for the evolutionary optimized DTS-FS. The expected decrease of the MSE for the evolutionary optimized DTS-FS is only small because of the near linear dataset.

Most models work on the gas furnace dataset with a maximal retrospection of $t - 4$. To increase the possible number of inputs for the EA, we decided to use $t - 6$, therefore the resulting dataset consist of 290 samples and the EA has to choose from 13 different inputs, namely u(t-6), u(t-5), u(t-4), u(t-3), u(t-2), u(t-1), u(t) and y(t-6), y(t-5), y(t-4), y(t-3), y(t-2), y(t-1). These 13 different inputs were used to find the concerning optimal linear models by a complete permutation search for different numbers of features. Table 1 lists the optimal linear models and the according MSE values achieved by utilizing all 290 data pattern for training.

5 The Implemented EA for System Identification

For the remaining we utilize a grammar as genotype template and a n-ary tree as genotype representation. By regarding a DTS-FS as a Functional Entity (FE) and each component (i.e. linguistic variables, MFs, etc) of this main FE as subordinate FEs, we can assign FEs to parts of the genotype tree. For example the whole model (the FE DTS-FS) is represented by the complete genotype n-ary tree, thus, the root node of the genotype n-ary tree is labeled "DTS-FS". A linguistic variable is only a part of the whole model and consequently it is represented by a subtree of the genotype n-ary tree and the concerning root node of the subtree is labeled as "linguistic variable" (see figure 6).

By following this approach a possible (not complete) genotype template, represented by a n-ary tree, can be seen in figure 5. The dashed boxed area in figure 5 represents a set of b-spline MFs associated to one linguistic variable.

Table 1. Best possible linear models for different number of used features.

linear model	number of used features	MSE
y(t) = 1.36637 + y(t-1) 0.97471	1	0.5604
y(t) = 2.71727 + y(t-2) -0.857463 + y(t-1) 1.80671	2	0.1502
y(t) = 8.85997 + u(t-3) -0.513869 + y(t-2) -0.51795 + y(t-1) 1.35191	3	0.0688
y(t) = 3.67287 + u(t-6) 0.300255 + u(t-3) -0.509377 + y(t-2) -0.514342 + y(t-1) 1.4456	4	0.0593
y(t) = 3.79162 + u(t-6) 0.232458 + u(t-3) -0.465351 + y(t-4) 0.0675975 + y(t-2) -0.698299 + y(t-1) 1.55972	5	0.0566
y(t) = 3.87595 + u(t-6) 0.234921 + u(t-5) -0.0947678 + u(t-4) 0.141222 + u(t-3) -0.429811 + u(t-2) -0.214547 + u(t-1) 0.193185 + u(t) -0.0673241 + y(t-6) -0.0400303 + y(t-5) 0.05408 + y(t-4) 0.134192 + y(t-3) -0.176416 + y(t-2) -0.586207 + y(t-1) 1.54181	13	0.0555

Each leaf node (a node without children) contains a knot position for order k=1 to order k=3 (i.e. according to a predetermined maximum order $k = 3$). This is justified by the observation that b-splines of different orders have different optimal knotvectors. To increase the possibility that a change in the order (i.e. the variable "order k of MFs") causes a better fitness, a knot-position for each order is stored in each leaf node.

5.1 Constraining the Genotype

As mentioned above we support our grammar in that way, that each node can act as container for decision variables (see figure 8). Regarding the decision variables we implemented tools to constrain variable values. To constrain a decision variable, a variable name, the scope of the constraint and the constraint for itself has to be given. For example all knots of λ^{int} has to be in increasing order and thus, by considering the decision variable $\lambda(k = 2)$, the scope is "sibling" (meaning that all children of the same parent node are considered) and the constraint is ">". Other constrains, detached from decision variables, affect for example the possible number of node children to force any (sub)tree to have nodes with a number of children in a given range. By using this kind of constraint it is possible to restrict for example the number of used features. The fulfillment of each constraint is assured before the genotype is mapped to the phenotype. Details regarding the implementation are beyond the scope of this article, but a short description is given in [13]. Thus, only phenotypes which fits our specifications will be created. It would be also imaginable to penalize candidate solutions which does not exactly fit our demands, but this would be contradictory to the usage of grammar based genotype templates.

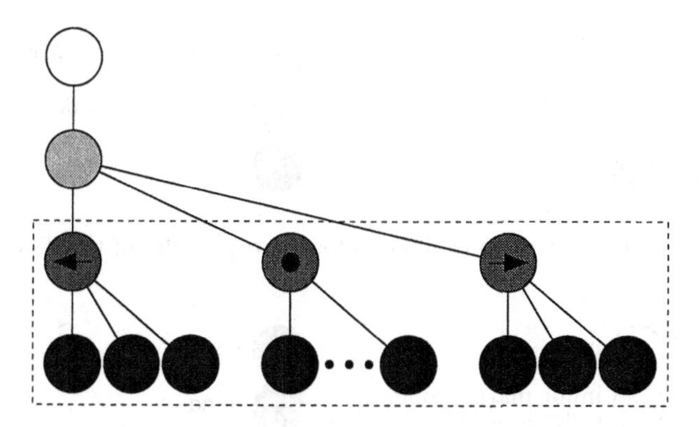

Fig. 5. N-ary tree representing an exemplary genotype template for a DTS-FS genotype.

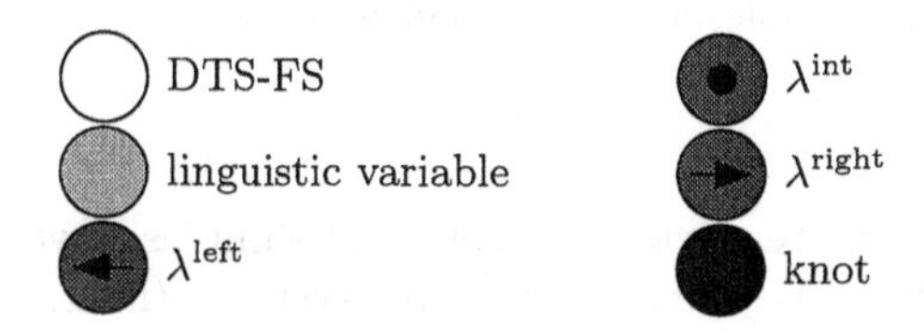

Fig. 6. The functional entity label of the genotype (sub)trees with node as root node.

5.2 Mutation and Crossover

Regarding mutation in n-ary based genotypes, we used two different approaches. Real valued decision variables were changed by evolutionary strategies and the mutation probability of integer based decision variables was set to 0.01. Crossover was done each generation for all individuals by performing a random swap of two interchangeable subtrees of the concerned individuals. If the random chosen crossover-nodes were not interchangeable, crossover was not performed.

6 Results and Discussion

6.1 Results

We computed two evolutionary optimized DTS-FS. For both evolutionary runs we used a population size = 150 with stop generation = 50 and tournament selection with 5 competitors. This yield in 7500 model evaluations which is in most cases marginal regarding a permutation search. We restricted the number of MFs per linguistic variable $\in [1, 8]$ and the order of the b-splines $\in [1, 3]$. During the evolutionary system identification process both DTS-FSs uses the leave-one-out [14] cross-validated MSE as fitness measure. After the system identification process was finished, the MSE was calculated by determing the weights of each

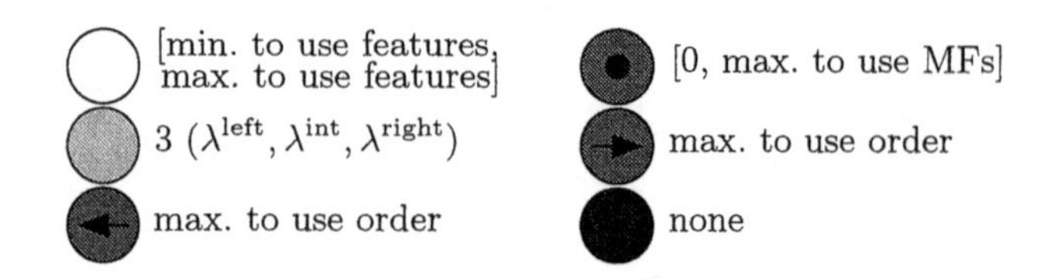

Fig. 7. Valid number of node children for the root nodes of the different FEs.

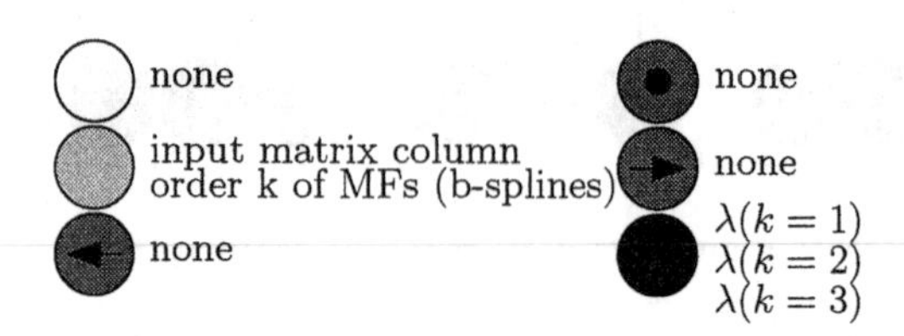

Fig. 8. Variables assigned to different nodes.

n-variate basis function with respect to all 290 data samples. Over-fitting was avoided by restricting the EA to find only small systems (max. 50 n-variate basis functions) and the leave-one-out technique during the system identification process.

The first DTS-FS was restricted to use one or two of the possible 13 inputs. The computed MSE value of this model was 0.143 (the optimal linear model with two inputs yields in a MSE = 0.1502). Figure 10 shows the structure of this optimized DTS-FS and figure 9 shows the prediction error.

The second DTS-FS was restricted to use one, two or three of the possible 13 inputs. The computed MSE value of this model was 0.046 (the optimal linear model with three inputs yields in a MSE = 0.0688). Figure 12 shows the structure of this optimized DTS-FS and figure 11 shows the prediction error.

6.2 Discussion

Because the gas furnace data is only slightly non-linear, the achieved results using DTS-FSs are, as expected, only slightly improved compared to the optimal linear models. Nevertheless the MSE of the DTS-FS with three inputs is also less than the best MSE=0.057 found in literature [5] for non-linear models by using the same inputs. Furthermore the MSE of the identified DTS-FS with three inputs is lower than the linear model utilizing all 13 available features. Moreover the evolutionary process identified the same inputs as optimal as the best linear model did.

The more possible inputs a task shows and the fewer expert knowledge is available, the more important fully data driven system identification becomes. These fully data driven system identification methods has to be capable to model arbitrary non-linear relationships and therefore an approach based on

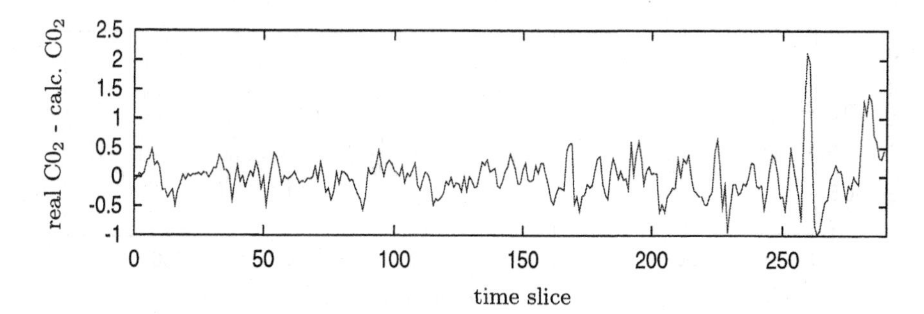

Fig. 9. Error in CO_2 concentration prediction with an optimized DTS-FS using 2 inputs.

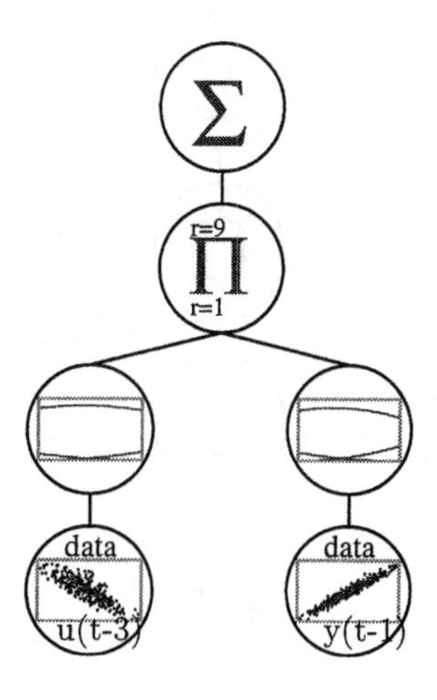

Fig. 10. Architecture of the optimized DTS-FS using 2 linguistic variables with 3 MFs each.

DTS-FSs, which is known as universal approximator, seems very convenient. Furthermore, the more a solution to a problem is afflicted with constraints, the more the need for intelligent genotype encoding methods arises. The presented approach satisfies both requirements. Further investigations will be done on highly non-linear and more constraint problems and especially on grammar based genotype templates.

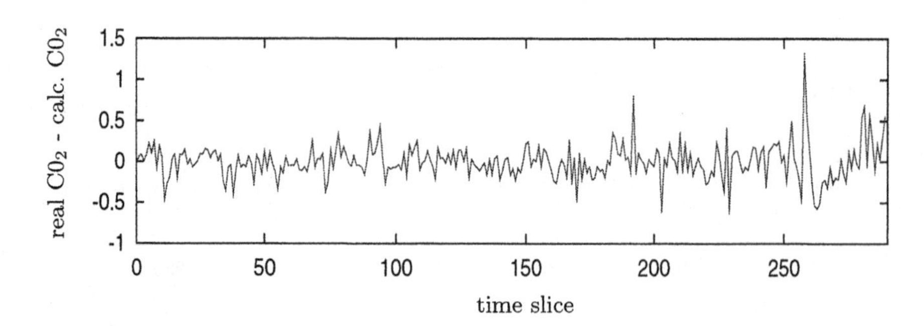

Fig. 11. Error in CO_2 concentration prediction with an optimized DTS-FS using 3 inputs.

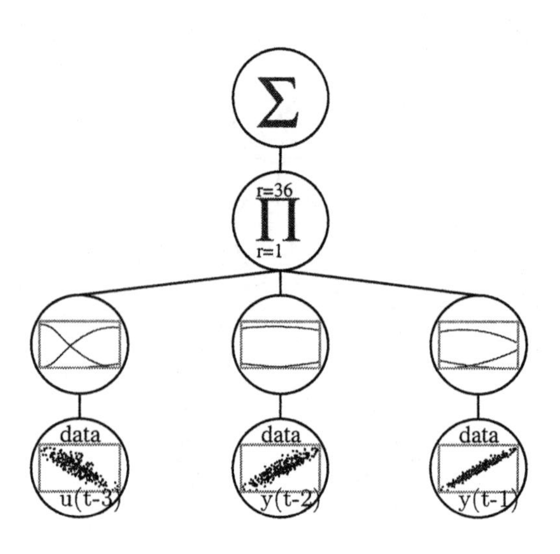

Fig. 12. Architecture of the optimized DTS-FS using 3 linguistic variables with 4,3,3 MFs each.

Acknowledgment. This work has been supported by the Ministry of Science and Research, North-Rhine Westphalia.

References

1. Box, G.E.P. and Jenkins, G.M Time Series Analysis: Forecasting and Control. 2nd ed. San Francisco: Holen-Day, (1970).

2. Cordon, O., Herrera, F., Hoffmann, F. and Magdalena, L. Genetic Fuzzy Systems: Evolutionary Tuning and Learning of Fuzzy Knowledge Bases. *Advances in Fuzzy Systems – Application and Theory Vol. 19*, World Scientific Publishing Co. Pte. Ltd., ISBN 981-02-4016-3, p. 337, (2001).
3. Cox, M.G. Algorithms for Spline Curves and Surfaces. *NPL Report DITC 166/90*, 36 pages, (1990).
4. Farag, W. and Tawfik, A. On Fuzzy Identification and the Gas Furnace Data. *Proc. of the IASTED Int. Conf. about Intelligent Systems and Control*, Honolulu Hawaii, USA, (2000).
5. Gaweda, A.E. Optimal data-driven rule extraction using adaptive fuzzy-neural models. *Dissertation submitted to Faculty of the graduate school of the university of Louisville*, pp. 44–49, (1997).
6. Grauel, A. and Mackenberg, H. Mathematical Analysis of the Sugeno Controller Leading to General Design Rules. *Fuzzy Sets and Systems, 85, pp.165–175*, (1997).
7. Grauel, A. and Ludwig, L.A Construction of Differentiable Membership Functions. *Fuzzy Sets and Systems, 101, pp.215–225*, (1999).
8. Grauel, A. Handbook of Mathematics *Bronstein, Semendjajew & Eds. Musiol and Mühlig, chapter 5.9, Fuzzy Logic, Springer-Verlag, Heidelberg, pp.372–391*, (2003).
9. M. Gulson and A. E. Smith A Hierarchical Genetic Algorithm for System Identification and Curve Fitting with a Supercomputer Implementation. In *IMA Volumes in Mathematics and its Applications – Issue on Evolutionary Computation*, Springer-Verlag (1998).
10. Jang, J.R. and Sun, C.T. Functional equivalence between radial basis function networks and fuzzy inference systems. *IEEE Trans. Neural Networks*, vol. 4, no.1, pp. 156–159, (1993).
11. Nelles, O. Nonlinear System Identification – From Classical Approaches to Neural Networks and Fuzzy Models. *ISBN 3-540-67369-5, Springer*, pp. 284–285, (2001).
12. Renners, I. and Grauel, A. Optimizing lattice-based associative memory networks by evolutionary algorithms. *Information Sciences*, vol. 136, pp. 69–84, (2001).
13. Renners, I. and Grauel, A. Variable length coding of genotypes using n-ary trees. *Proc. of the 9th Int. Conference on Soft Computing – Mendel*, (2003).
14. Sarle, W.S. Frequently asked questions (FAQ) of the neural network newsgroup. *http://www.faqs.org/faqs/ai-faq/neural-nets/part3/*
15. Takagi, T. and Sugeno. M. Fuzzy identification of systems and its application to modeling and control. *IEEE Transactions on Systems, Man, and Cybernetics*, vol. 15(1), pp. 116–132, (1985).
16. Werntges, H.W. Partitions of unity improve neural function approximators. *IEEE Int. Conference on Neural Networks (ICNN)*, vol. 2 pp. 914–918, San Francisco, USA, (1993).
17. Zadeh, L.A. The Concept of a Linguistic Variable and its Application to Approximate Reasoning *Information Sciences – Application to Approximate Reasoning I–III*, (1975).

Minimum Message Length Criterion for Second-Order Polynomial Model Selection Applied to Tropical Cyclone Intensity Forecasting

Grace W. Rumantir[1] and Chris S. Wallace[2]

[1] School of Multimedia Systems
Monash University – Berwick Vic 3806 Australia
[2] School of Computer Science and Software Engineering
Monash University – Clayton Vic 3168 Australia
{grace.rumantir, chris.wallace}@infotech.monash.edu.au

Abstract. This paper outlines a body of work that tries to merge polynomial model selection research and tropical cyclone forecasting research. The contributions of the work are four-fold. First, a new criterion based on the Minimum Message Length principle specifically formulated for the task of polynomial model selection up to the second order is presented. Second, a programmed optimisation search algorithm for second-order polynomial models that can be used in conjunction with any model selection criterion is developed. Third, critical examinations of the differences in performance of the various criteria when applied to artificial *vis-a-vis* to real tropical cyclone data are conducted. Fourth, a novel strategy which uses a synergy between the new criterion built based on the Minimum Message Length principle and other model selection criteria namely, Minimum Description Length, Corrected Akaike's Information Criterion, Structured Risk Minimization and Stochastic Complexity is proposed. The forecasting model developed using this new automated strategy has better performance than the benchmark models SHIFOR (Statistical HurrIcane FORcasting) [4] and SHIFOR94 [8] which are being used in operation in the Atlantic basin.

1 Introduction

Model selection is about finding structure from data. The goal of model selection is to find a model or a set of models from a limited amount of sample data that best explain the structure of the data population. To achieve this goal, two main tasks need to be carried out. The first is in selecting the variables that are useful to be included in the model and the second is in finding the structure of the model. Finding the structure of a model involves two tasks. The first is in determining the form in which the variables collaborate, e.g. Bayesian networks, neural networks, decision tree, polynomials, etc. The second is in determining the degree of influence each variable has in the model.

M.R. Berthold et al. (Eds.): IDA 2003, LNCS 2810, pp. 486–496, 2003.

When dealing with a real-world application domain where observational data are used to build a model like tropical cyclone forecasting, the true model that represents the dependent variable is not known. Hence, no model built using the sample data sets available can be said to be correct. What can be shown is that some models are either consistent or inconsistent with the data based on certain performance criteria.

In the face of the many forms a model can take, it is a common practice in model selection research to make an assumption of the form the model should take. For example, in this work, the form that the model is assumed to have is a second-order polynomial. This is because there already exists such models created based on the real data used to test the new model selection strategy outlined in this work. These models (see [4], [8] and [9]), are currently used to forecast tropical cyclone intensity in the Atlantic basin. Hence, the model to be built using the new model selection strategy can be benchmarked against these models.

In an attempt to find the best selection strategy for building a polynomial forecasting model, this work has tested a wide range of criteria commonly used in the literature, namely Minimum Description Length (MDL) [13], Stochastic Complexity (SC) [14], Structured Risk Minimisation (SRM) [20], Corrected AIC (CAICF) [3], Akaike's Information Criterion (AIC) [1], Bayesian Information Criterion (BIC) [19], Mallows' C_p [10], F-test [11], Adjusted Coefficient of Determination (adjR^2) [7], together with a proposed novel criterion based on the Minimum Message (MML) principle [21] on three different data sets: data generated from known artificial models (reported in [15]), real climatological and atmospheric data from the Atlantic tropical cyclone basin (reported in [16]) and artificial data generated based on the covariance matrix of the real data from the Atlantic basin (reported in [16]). This paper summarises the published body of work by showing the links between the three studies.

2 Background

The form of the model is fixed to be a second-order polynomial which may use the squares of variables and products of two variables. For a given form, once the variables are selected, the coefficients of the variables are determined by computing a least squares solution to the data. Hence, the way in which models differ is in the variables they use. The following is the standardized form of the second-order polynomial model under consideration:

$$y_n = \sum_{p=1}^{P} \gamma_p u_{np} + \sum_{p=1}^{P} \sum_{q \geq p}^{P} \gamma_{pq} u_{np} u_{nq} + \epsilon_n \Leftrightarrow y_n = \sum_{k=1}^{K} \beta_k x_{nk} + \epsilon_n \qquad (1)$$

where for each data item n:

y_n : target variable $\qquad$ x_{nk} : regressor k;

$x_{nk} = u_{np}$ or $x_{nk} = u_{np}u_{nq}$; $q \geq p$

u_{np} : regressor p		β_k : coeff. for regressor k	
γ_p : coeff. for single regressor		K $=2P + P!/2!(P-2)!$	
γ_{pq} : coeff. for compound regressor		ϵ_n : noise/residual/error term	

The values of the error term ϵ is assumed to be uncorrelated, normally and independently distributed $\epsilon_n \sim NID(0, \sigma^2)$.

3 Methods of Model Selection

Model selection is about comparing one model with another. In building a model, a good model selection criterion should serve as a stopping rule of the selection process when the best model has been found amongst all of the models considered.

There are at least two categories of model selection criteria:

1. Criteria which require a separate test data set to decide on a model. These methods require the data set to be divided into 2 parts. The first set of data, the training data set, is used to calculate the parameters of the variables of a model. The second set of data, the test data set, is used to compare and make a choice between two models.
2. Complexity-penalised criteria. These methods require only one data set in deciding on a model. They penalise more complex models, hence seek to find the balance between the complexity of the model and the fit of the model to the data. When comparing two models with different complexities (number of variables, order of polynomial, etc), these methods will choose the more complex model only when the improvement in the fit to the data outweighs the penalty attributed to the increase in model complexity.

Table 1 [15] gives the summary of the model selection criteria considered in this work. The first seven methods belong to the second category and the last five belong to the first category. The model chosen by any of the methods is claimed to be a parsimonious description of the data at hand, therefore has predictive power for future data. This work tests the robustness of each method in support of this claim.

3.1 Minimum Message Length Polynomial Model Selection Criterion

This work proposes a new model criterion based on the Minimum Message Length (MML) principle specifically formulated for the task of polynomial model selection up to the second order as presented in [15]. MML is of the category of complexity-penalised model selection criteria, i.e. the second category described in Section 3 above.

Table 1. Summary of model selection criteria

Method		Ref.	Objective Function
Minimum Message Length	MML	[21]	$-\log f(x\|\theta) + \frac{1}{2}\log\|I(\theta)\| - \log h(\theta)$ $-\frac{k+1}{2}\log 2\pi + \frac{1}{2}\log(k+1)\pi$ $-1 - \log h(\nu,\xi,j,l,J,L)$
Minimum Description Length	MDL	[13]	$-\log f(x\|\theta) + \frac{k}{2}\log n + (\frac{k}{2}+1) + \log k(k+2)$
Stochastic Complexity	SC	[14]	$\frac{n}{2}\log\sum_{i=1}^{n} e_i^2 + \frac{1}{2}\log\|X'X\|$
Structured Risk Minimisation	SRM	[20]	$\frac{1}{n}\sum_{i=1}^{n} e_i^2 / \left(1 - \sqrt{\frac{(k+1)(\log\frac{n}{k+1}+1)-\log\eta}{n}}\right)$
Corrected AIC	CAICF	[3]	$-\log f(x\|\theta) + \frac{1}{2}\log\|I(\theta)\| + k + \frac{1}{k}\log n$
Akaike's Information Criterion	AIC	[1]	$-2(\log f(x\|\theta) - k)$
Bayesian Information Criterion	BIC	[19]	$-2(\log f(x\|\theta) - \frac{1}{2}k\log n)$
Mallows' C_p	C_p	[10]	$\frac{1}{\hat{\sigma}^2}\frac{1}{n}\sum_{i=1}^{n} e_i^2 - (n-2k)$
F-to-enter	F-test	[11]	$\left(\sum_{i=1}^{n} e_i^2\right)_{\text{old}} \geq \left(\sum_{i=1}^{n} e_i^2\right)_{\text{new}}\left(1 + \frac{F_{\text{enter}}}{n-(k+1)}\right)$
F-to-exit			$\left(\sum_{i=1}^{n} e_i^2\right)_{\text{new}} \geq \left(\sum_{i=1}^{n} e_i^2\right)_{\text{old}}\left(1 + \frac{F_{\text{exit}}}{n-(k+1)}\right)$
Adjusted Coeff. of Determination	adjR^2	[7]	$1 - \frac{\sum_{i=1}^{n} e_i^2/(n-k)}{\sum_{i=1}^{n}(y_i-\bar{y})^2/(n-1)}$
Generalized Cross Validation	GCV	[6]	$\frac{1}{(1-k)^2}\sum_{i=1}^{n} e_i^2$
Predictive sum of squares	PRESS	[2]	$\sum_{i=1}^{n} e_i^2$

MML seeks to minimize the length of the joint encoding, the "message", of the model and the data given the model. When comparing between two models, the best model is the one which costs the shortest message length.

For second-order polynomial models that take the form of Equation 1, the cost of encoding the model is composed of two parts: that of the model structure (i.e. which combination of variables) L_s and that of the model parameters L_p. Hence the message length cost function is:

$$L = L_s + L_p + L(x|\theta) \tag{2}$$

This work proposes that the cost of encoding the model structure L_s is composed of three parts: that of the set of single regressors, product of regressors and the combination of regressors:

$$L_s = -\log h(\nu, j) - \log h(\xi, l) - \log \frac{1}{\binom{J+L}{j+l}} \tag{3}$$

$h(\nu, j)$ and $h(\xi, l)$ follow geometric series:

$h(\nu, j) = \nu^j(1-\nu)/(1-\nu^{J+1})$ and $h(\xi, l) = \xi^l(1-\xi)/(1-\xi^{L+1})$

where: ν, ξ : probability of choosing single, product of regressors
j, l: single, product of regressors chosen
J, L: single, product of regressors available

It is assumed that there exists some prior knowledge/expectation about the possible models $\theta = (\beta_k, \sigma)$ which gives a message length $L = -\log h(\theta) = -\log prior(\sigma) prior(\{\beta_k|\sigma\})$. However, the adoption of a discrete message/code string of length L implies that θ itself is regarded as having a prior probability of e^{-L}, which is discrete. MML principle assigns to θ a prior probability $h(\theta) * v(\theta)$ where $v(\theta)$ is the the volume of a region of the search space which includes θ. As shown in [21] the whole message length is minimized when $v(\theta)$is chosen to be proportional to $1/\sqrt{|I(\theta)|}$, where $I(\theta)$ is the Fisher information matrix.

As in Equation 1, the parameters β_k are assumed to be normal. If uniform density prior (i.e. no preference) is chosen for σ, then following [21], the cost of encoding the model parameters L_p takes the form

$$L_p = -\log prior(\sigma) prior(\{\beta_k|\sigma\}) + \frac{1}{2}\log|I(\theta)| - \frac{1}{2}(k+1)\log 2\pi + \frac{1}{2}log(k+1)\pi - 1$$

$$= \frac{1}{\sigma}\prod_{k=1}^{K}\frac{1}{\alpha\sigma\sqrt{2\pi}}e^{-\beta_k^2/2\alpha^2\sigma^2} + \frac{1}{2}\log|I(\theta)| - \frac{1}{2}(k+1)\log 2\pi + \frac{1}{2}log(k+1)\pi - 1 \quad (4)$$

The term $|I(\theta)| = 2N\sigma^{-2(K+1)}|X'X|$ is the expected Fisher information matrix associated with the real-valued parameters of θ. That is, the determinant of the expected second partial differentials of $L(x|\theta)$ (Equation 5) with respect to the model. The last three terms reflect the effect of quantization of $v(\theta)$ (i.e. the discretization of θ) in forming the optimum code [5, pp.59–61] which results in the increase in the message length.

Finally, the cost of encoding the data given the model $L(x|\theta)$ is simply the likelihood function

$$L(x|\theta) = -\log f(y|\sigma, \{\beta_k\}) = -\log\prod_{n=1}^{N}\frac{1}{\sigma\sqrt{2\pi}}e^{-(y_n - \sum_{k=1}^{K}\beta_k x_{nk})^2/2\sigma^2} \quad (5)$$

4 Search Algorithm

A non-backtracking search algorithm has been developed to be used as a common search engine for the different stopping criteria. This algorithm starts with an empty model. Variables are either added to or deleted from the model one at a time. A new variable will be added to the model if it results in the best among the other alternative new models based on the selection criterion being used. After every addition to the model, a variable will be searched to be deleted therefrom. A variable will be deleted from the model if it results in the best new model among the initial model and any other models should any other variable has been chosen. Hence in effect, at each stage of the search process, the model chosen would be the best amongst all of the potential models with the same complexity that are possibly chosen by adding or deleting one variable to or from the existing model. The search terminates when there is no more variable to be added which will result in a better model.

5 Experimental Design and Results

For the task of automated model selection, the first test of the robustness of a criterion is in its ability in recovering the true model that has generated sets of artificial data which have additional noise and irrelevant variables included in them. This first test is done in [15] and outlined in Section 5.1 using a set of true models for the model selection criteria considered.

If a model selection criterion manages to recover true models of artificially generated data, it has the potential of finding a model that is consistent with the real data presented to it within the assumptions and constraints of the model selection strategy in place. This second test is done in [16] and outlined in Section 5.2 in building tropical cyclone intensity forecasting model for the Atlantic basin using the atmospheric and climatological data sets seen to be influential to the tropical cyclone intensity data.

After the above metioned test, we have a set of multivariate real data and have empirically found a polynomial regression model that is so far seen as the right model represented by the data. In [17] and outlined in Section 5.3, we would like to be able to replicate the multivariate data artificially to enable us to run multiple experiments to achieve two objectives. First, to see if the model selection criteria can recover the model that is seen to be the right model. Second, to find out the minimum sample size required to recover the right model.

5.1 Experiments on Artificial Data

Three true models were designed for the experiments from which artificial data were generated. Each true model consists of a target variable and a set of single and compound independent variables. Not all of the variables are necessarily directly or at all connected to the target variable. Each value of an independent variable is chosen randomly from a normal distribution $N(0,1)$. For each model, 6 training and test data sets comprising 500, 1000, 2000, 4000, 6000 and 10000 instances respectively are generated.

The product of two independent variables is calculated from the standardized values of each variable. Each value of the target variable is calculated from the values of all of the independent variables directly linked to it multiplied by the respective link weights plus a noise value which is independently and identically distributed (i.i.d) as Normal $(0,1)$.

The search engine is presented with the data of the target variable and all of the available independent variables and the possible products of the single variables. The performance criteria for the true model discovery task are whether or not a model selection method manage to select a model with the same set of variables and corresponding coefficients as those of the true model, as reflected in the following measures:

1. The number of variables selected in the discovered model
2. How close the coefficients of the discovered model are with those of the true model (model error): $1/K * \sum_{k=1}^{K}(\beta_k - \hat{\beta_k})^2$
3. Model predictive performance (on test data), quantified by

a) Root of the mean of the sum of squared deviations: RMSE $= \sqrt{\frac{1}{n}\sum_{i=1}^{n} e_i^2}$
b) Coefficient of determination: $R^2 = 1 - (\sum_{i=1}^{n} e_i^2) / \sum_{i=1}^{n}(y_i - \bar{y})^2$

Jacobian Orthogonal Transformation (as opposed to straight Gaussian Elimination) is performed on the covariance matrix of the independent variables for the calculation of model coefficients. This ensures that should multicollinearity among variables exists, it is not reflected in the model coefficients. The normalized variances of the single and product of variables are kept to unity by standardizing the product of the standardized single variables.

Results. The results of the experiments suggest that if a model selection procedure is to be fully automated, MML, MDL, CAICF, SRM and SC can reliably converge to the true model (if one exists), or to a reasonably parsimonious model estimate. The models selected by the AIC, BIC, adjR^2 and F-test have *overfitted the training data.* This implies that in these model selection methods, the penalty for chosing a more complex model is too small compared to the reward of better data fit. The methods may need further judgements in deciding on the final model which can take two forms. First, choosing a model half way through the search process just before it chooses a more complex model with worse performance on the test data. Second, pruning out some of the variables with small coefficients. The need for these manual adjustments explains the real reason behind the traditional common practice of specifying beforehand the maximum number of variables for a model (e.g.[18], [11]).

5.2 Experiments on Tropical Cyclone Data

Table 2 [16] gives the summary of the independent variables used to build the TC intensity change forecasting models which can be categorized into three groups:

1. *Persistence:* the TC intensity in knots (5,6)
2. *Climatology:* Julian data (1,2), global position (3,4) and motion in knots (7,8,9)
3. *Synoptic* environmental features: sea sub/surface temperature (12,13–15), shear(16–18), distance from land (25–26), etc.

The target variable is the intensity change 72 hours into the future. The TC track forecasting model CLIPER [12] is used to provide future forecast positions for which the climatological independent variables are calculated.

The work proposes a way to build forecasting models using a set of methods proven in the experimental survey in [15] to be the most robust amongst most of the commonly cited complexity-penalised model selection criteria. The procedure are as follows:

1. Using random subsampling, create m pairs of training and test data sets.As seen in Table 2, temporal relationships between values of a variable over a period of time have been taken into account in the form of a set of related variables.

Table 2. Basic regressors used to build the Atlantic TC intensity change forecasting models. The target variable is the change of intensity (wind speed) 72 hour into the future. To get the *average, at end* and *change* values of a variable, the TC track/location (longitude and latitude) forecast out to 72 hours is required: *average* means the average of the values at the location forecast at 0, 24, 48 and 72 hours, *at end* means the value at 72 hours and *change* means the difference between the current value and the value at 72 hours.

No	Basic Regressor: acronym and explanation
1,2	Julian, JulOff – date: Julian, |Julian - 253|
3,4	LatData, LonData – position: latitude (deg N), longitude (deg W)
5,6	Vmax, Del12V – intensity (in knots): initial, previous 12 hour change
7,8,9	UCurr, VCurr, Speed – motion (in knots): eastward, northward, resultant
10,11	POT, POTEnd – potential intensity: initial, at end
12	DelSST – change of sea surface temperature
13,14,15	SSST, SSSTend, DSSST – sea sub-surface temperature: average, at end, change
16,17,18	UppSpd, Uppend, DUppSpd – windspeed at 200mb: average, at end, change
19,20,21	Stabil, Stabend, DelStab – moist stability 1000mb to 200mb: average, at end, change
22,23,24	200mbT, 200Tend, Del200T – temperature at 200mb: average, at end, change
25,26	DisLand, Closest – distance from land: initial, closest approach
27,28,29	200mbU, 200Uend, Del200U – eastward motion at 200mb: average, at end, change
30	U50 – 50mb Quasi-Biennial Oscillation (QBO) zonal winds
31	RainS – African Western Sahel rainfall index (5W-15W, 10N-20N)
32	RainG – African Gulf of Guinea rainfall index (0W-10W, 5N-10N)
33	SLPA – April-May Caribbean basin Sea Surface Pressure Anomaly
34	ZWA – April-May Caribbean basin Zonal Wind Anomaly at 200mb (12 km)
35	ElNino – Sea surface temperature anomaly in the eastern equatorial pacific
36	SOI – Surface pressure gradient between Darwin and Tahiti

The test data sets are needed not for model development but to see whether or not the performance of a method on unseen data is much different from that on training data.

2. For each of the method, search for forecasting models on each pair of data sets.
3. For each independent variable, count its frequency of being chosen in any of the forecasting models discovered in the previous point. Create a set of new models, each as a collection of variables with at least a certain frequency. Calculate the set of coefficients for each model using all of the available data.
4. Compare the models with increasing complexity based on all of the performance criteria

There are in total 4347 data items available. The experiments are conducted on data sets built using two types of sampling method. Ten training–test data sets are built using 2 : 1 random sampling method. One data set is built so that the training data is taken from the years 1950–1987 and the test data from the years 1988–1994. Convenient separation of data based on consecutiveness is common practice in hurricane intensity change forecasting. SHIFOR (modified by Pike from [4]) and SHIFOR94 [8] have both been built using training data from 1950 to 1987 and test data from 1988 to 1994. The purpose of using these two categories of data sets is to see whether or not the possible changes in atmospheric dynamics from year to year should be taken into account in experimental design.

The performance criteria for model comparison are:

1. Parsimony, reflected in the model cost calculated using the cost functions MML, MDL, CAICF and SRM (on training data)
2. Model predictive performance (on test data) as outlined in Section 5.1

Results. The experiments show that although Stochastic Complexity (SC) has been proven to be a good candidate for automated model discovery using artificial data in [15], it failed to converge into an optimum model before the maximum number of variables set for a model has been reached. Therefore, it is decided not to use SC in the procedure for building forecasting models.

For each of the first 10 data set, there is not too much difference between the predictive performance of MML, MDL, CAICF and SRM on the training data and on the test data. This confirms the findings in [15] and strengthen the believe that all of the available data can be used as training data since overfitting is not a problem for these methods.

The results show that partitioning training and test data based on consecutiveness as commonly done in tropical cyclone research (see [4] and [8]) has resulted in two non-homogeneous data sets; a lot of the regularities learned by a model from the training data set are not present in the test data set resulting in much reduced performance.

One finding of significance to atmospheric research is that there is a strong presence of the seasonal predictors in the new model discovered using the proposed procedure. These predictors have been proven in the literature to have strong influence to the Atlantic cyclone activity.

5.3 Experiments on Data Generated from Covariance Matrix of Tropical Cyclone Data

This work proposes a methodology to replicate real multivariate data using its covariance matrix and a polynomial regression model seen as the right model represented by the data. The covariance matrix of the tropical cyclone data used in the experiments outlined in Section 5.2 is used. The sample data sets generated are then used for model discovery experiments.

The following is the procedure to generate artificial data for the pool of potential regressors and their associated target variable for model discovery:
Step 1: Calculate the covariance matrix of the set of real multivariate data.
Step 2: Generate artificial data for the pool of potential regressors using the eigenvector and eigenvalue of the covariance matrix in the following formula:

$$x_i = \sum_j \sqrt{\lambda_i} q_{ji} n_i \qquad (6)$$

where:

n_i : a unit normal data point generated from $N(0,1)$
q_{ij} : j^{th} element of eigenvector i
λ_i : eigenvalue i

Step 3: Calculate the values for the target variable using the true model. This is done by multiplying the value of the regressor with the link weight connected to the target variable plus a unit normal noise value.
Step 4: For each model selection criterion used in Section 5.2 as the cost function, run a search mechanism to rediscover the true model from the data generate in the previous steps.
Step 5: Compare the models found by each criterion using the performance criteria given in Section 5.1.

Five categories of data sets with increasing level of noise are generated, each category with test data sets comprising 100, 500, 1000, 2000, 4000, 6000 and 10000 data points.

Results. The results of all of the model selection criteria are quite uniform in that they all managed to recover the true model with similar degree of accuracy. It is also observed that all of the criteria managed to recover the true model of 9 regressors found in the experiments in Section 5.2 with a data set as small as 100 data points. Increasing the size of the data set does not significantly increase the accuracy of the results. This finding indicates that most of the potential regressors do not have direct influence to the target variable.

This finding is valuable in that it confirms that the integrated model discovery procedure outline in Section 5.2 indeed has the ability to select a parsimonious subset of variables from a pool of potential regressors in the absence of prior knowledge of the problem domain.

6 Conclusion

A body of work which tries to merge model selection research and tropical cyclone research is outlined. The contributions of the work to model selection research are in the proposal of a new model selection criterion based on the Minimum Message Length (MML) principle and a new strategy in finding forecasting model using a combination of MML, MDL, CAICF and SRM. The contributions of the work to tropical cyclone research are two-fold. First, a new better tropical cyclone forecasting model built using the new MML criterion and the new strategy proposed. Second, findings that the behavior of cyclones in the Atlantic basin has changed significantly between the periods observed suggesting that the practice of partitioning of the data as training and test data sets based on a cronological order that has always been done in building tropical cyclone models is flawed since the training and test data sets do not capture the same features.

References

1. H. Akaike. Information theory and an extension of the Maximum Likelihood principle. In B.N. Petrov and F. Csaki, editors, *Proc. of 2nd Int. Symp. Information Thy.*, pages 267–281, 1973.
2. D.M. Allen. The relationship between variable selection and data augmentation and a method for prediction. *Ann. Inst. Statist. Math.*, 21:243–247, 1974.
3. H. Bozdogan. Model selection and akaike's information criterion (AIC): the general theory and its analytical extensions. *Psychometrika*, 52(3):345–370, 1987.
4. B.R. Jarvinen and C.J. Neumann. Statistical forecasts of tropical cyclone intensity. Technical Report Tech. Memo. NWS NHC-10, National Oceanic and Atmospheric Administration (NOAA), Miami, Florida, 1979.
5. J.H. Conway and N.J.A. Sloane. *Sphere Packings, Lattices and Groups.* Springer–Verlag, New York, 1988.
6. P. Craven and G. Wahba. Smoothing noisy data with spline functions. *Numerische Mathematik*, 31:377–403, 1979.
7. M. Ezekiel. *Methods of Correlation Analysis.* Wiley, New York, 1930.
8. C.W. Landsea. SHIFOR94 – Atlantic tropical cyclone intensity forecasting. In *Proceedings of the 21st Conference on Hurricanes and Tropical Meteorology*, pages 365–367, Miami, Florida, 1995. American Meteorological Society.
9. M. DeMaria and J. Kaplan. A Statistical Hurricane Intensity Prediction Scheme (SHIPS) for the Atlantic basin. *Weather and Forecasting*, 9(2):209–220, June 1994.
10. C. Mallows. Some comments on C_p. *Technometrics*, 15:661–675, 1973.
11. A.J. Miller. *Subset Selection in Regression.* Chapman and Hall, London, 1990.
12. C.J. Neumann. An alternate to the HURRAN tropical cyclone forecasting system. Technical Memo NWS-62, NOAA, 1972.
13. J. Rissanen. Modeling by shortest data description. *Automatica*, 14:465–471, 1978.
14. J. Rissanen. Stochastic complexity. *Journal of the Royal Statistical Society B*, 49(1):223–239, 1987.
15. G.W. Rumantir. Minimum Message Length criterion for second-order polynomial model discovery. In T. Terano, H. Liu, A.L.P. Chen, editor, *Knowledge Discovery and Data Mining: Current Issues and New Applications, PAKDD 2000, LNAI 1805*, pages 40–48. Springer–Verlag, Berlin Heidelberg, 2000.
16. G.W. Rumantir. Tropical cyclone intensity forecasting model: Balancing complexity and goodness of fit. In R. Mizoguchi and J. Slaney , editor, *PRICAI 2000 Topics in Artificial Intelligence, LNAI 1886*, pages 230–240. Springer–Verlag, Berlin Heidelberg, 2000.
17. G.W. Rumantir and C.S. Wallace. Sampling of highly correlated data for polynomial regression and model discovery. In F. Hoffmann, et al., editor, *Advances in Intelligent Data Analysis, LNCS 2189*, pages 370–377. Springer–Verlag, Berlin Heidelberg, 2001.
18. T. Ryan. *Modern Regression Methods.* John Wiley & Sons, New York, 1997.
19. G. Schwarz. Estimating the dimension of a model. *Annals of Statistics*, 6:461–464, 1978.
20. V. Vapnik. *The Nature of Statistical Learning Theory.* Springer, New York, 1995.
21. C.S. Wallace and P.R. Freeman. Estimation and inference by compact coding. *Journal of the Royal Statistical Society B*, 49(1):240–252, 1987.

On the Use of the GTM Algorithm for Mode Detection

Susana Vegas-Azcárate and Jorge Muruzábal

Statistics and Decision Sciences Group
University Rey Juan Carlos, 28936 Móstoles, Spain
{s.vegas,j.muruzabal}@escet.urjc.es

Abstract. The problem of detecting the modes of the multivariate continuous distribution generating the data is of central interest in various areas of modern statistical analysis. The popular self-organizing map (SOM) structure provides a rough estimate of that underlying density and can therefore be brought to bear with this problem. In this paper we consider the recently proposed, mixture-based generative topographic mapping (GTM) algorithm for SOM training. Our long-term goal is to develop, from a map appropriately trained via GTM, a fast, integrated and reliable strategy involving just a few key statistics. Preliminary simulations with Gaussian data highlight various interesting aspects of our working strategy.

1 Introduction

Searching the structure of an unknown data-generating distribution is one of the main problems in statistics. Since modes are among the most informative features of the density surface, identifying the number and location of the underlying modal groups is of crucial importance in many areas (such as, e.g., the Bayesian MCMC approach).

A previously proposed idea is to transform mode detection into a mixture problem following either a parametric [9] or a non-parametric [15] approach. Studies based on kernel estimation [7,16,18] have also provided useful results. While some methods for testing the veracity of modes in $1D$ [8,15] and 2D [16] data are available, tools for making exploration possible in higher dimensions are much needed.

Here we inquire about the potential of *self-organizing maps* (SOM) [11] as an approach to multivariate mode detection. The SOM structure derived from the standard fitting algorithm is often found useful for clustering, visualization and other purposes. However, it lacks a statistical model for the data. Recent approaches to SOM training usually incorporate some statistical notions yielding richer models and more principled fitting algorithms.

Specifically, the *generative topographic mapping* (GTM) [1] is a non-linear latent variable model (based on a constrained mixture of Gaussians) in which its parameters can be determined by maximum likelihood via the *expectation-maximization* (EM) algorithm [6]. GTM attempts to combine the topology-preserving trait of SOM structures with a well-defined probabilistic foundation.

M.R. Berthold et al. (Eds.): IDA 2003, LNCS 2810, pp. 497–508, 2003.

The GTM approach provides a number of appealing theoretical properties and is deemed indeed a major candidate to support our mode detection task.

The remainder of the paper is organized as follows. In section 2 we explore basic properties of the GTM model and discuss the reasons for preferring GTM over the original SOM fitting algorithm to deal with multivariate mode detection. Section 3 first presents the basic tools or key statistics extracted from the trained GTM, then sketches our working strategy for mode detection. A basic test of this tentative strategy is provided in section 4, where 2D, 3D and 10D synthetic data following various Gaussian distributions are used to illustrate the prospect of the approach. Finally, section 5 summarizes some conclusions and suggests future research stemming from the work discussed in the paper.

2 The GTM Model

The GTM [1] defines a non-linear, parametric mapping $\mathbf{y}(\mathbf{x}, \mathbf{W})$ from an L-dimensional latent space ($\mathbf{x} \in \Re^L$) to a D-dimensional data space ($\mathbf{y} \in \Re^D$) where L < D. The transformation $\mathbf{y}(\mathbf{x}, \mathbf{W})$ maps the latent-variable space into an L-dimensional manifold S embedded within the data space. By suitably constraining the model to a grid in latent space, a posterior distribution over the latent grid is readily obtained for each data point using Bayes' theorem.

As often acknowledged [11], the standard SOM training algorithm suffers from some shortcomings: the absence of a cost function, the lack of a theoretical basis to ensure topographic ordering, the absence of any general proofs of convergence, and the fact that the model does not define a probability density. GTM proceeds by optimizing an objective function via the EM algorithm [6]. Since the cost function is of log-likelihood type, a measure is provided on which a GTM model can be compared to other generative models.

But our main interest in working with GTM algorithm instead of the original SOM fitting algorithm to deal with multivariate mode detection is related to the 'self-organization' and 'smoothness' concepts. While the conditions under which the self-organization of the SOM occurs have not been quantified and empirical confirmation is needed in each case, the neighbourhood-preserving nature of the GTM mapping *is an automatic consequence of the choice of a continuous function* $\mathbf{y}(\mathbf{x}, \mathbf{W})$ [2]. In the same way, the smoothness properties of the original SOM are difficult to control since they are determined indirectly by the neighbourhood function, while basis functions parameters of the GTM algorithm explicitly govern the smoothness of the manifold, see below.

Hence, the GTM algorithm seeks to combine the topology preserving properties of the SOM structure with a well defined probabilistic framework. Moreover, since the evaluation of the Euclidean distances from every data point to every Gaussian centre is the dominant computational cost of GTM and the same calculations must be done for Kohonen's SOM, each iteration of either algorithm takes about the same time.

More specifically, GTM training is based on an optimization procedure aimed at the standard Gaussian mixture log-likelihood [2]

$$\ell(\mathbf{W},\beta) = \sum_{n=1}^{N} \ln\{\frac{1}{K}\sum_{i=1}^{K}(\frac{\beta}{2\pi})^{\frac{D}{2}} \exp\{-\frac{\beta}{2} \parallel \mathbf{W}\phi(\mathbf{x}_i) - \mathbf{t}_n \parallel^2\}\},$$

where a generalized linear regression model is chosen for the embedding map, namely

$$\mathbf{y}(\mathbf{x},\mathbf{W}) = \mathbf{W}\phi(\mathbf{x}), \text{with } \phi_m(\mathbf{x}) = \exp\{-\frac{\parallel \mathbf{x} - \mu_m \parallel^2}{2\sigma^2}\},\ m = 1, ..., M.$$

Here $\mathbf{t}_n$ is one of N training points in the D-dimensional data space, $\mathbf{x}_i$ is one of K nodes in the regular grid of a L-dimensional latent space, β is the inverse noise variance of the mixture Gaussian components, $\mathbf{W}$ is the DxM matrix of parameters (weights) that actually govern the mapping and $\phi(\mathbf{x})$ is a set of M fixed (spherical) Gaussian basis functions with common width σ. Some guidelines to deal with GTM training parameters are laid out in [1,2,3,17].

3 Basic Tools

Once we have trained a GTM model, suitable summaries of its structure will be extracted and analyzed to ascertain the modes' location. In particular, we consider *median interneuron distances*, *dataloads*, *magnification factors* and *Sammon's projections.* The first three quantities are introduced in detail in the next sections.

To visualize high-dimensional SOM structures, use of Sammon's projection [14] is customary. Sammon's map provides a useful global image while estimating all pairwise Euclidean distances among SOM pointers and projecting them directly onto 2D space. Thus, since pointer concentrations in data space will tend to be maintained in the projected image, we can proceed to identify high-density regions directly on the projected SOM. Furthermore, by displaying the set of projections together with the connections between immediate neighbours, the degree of self-organization in the underlying SOM structure can be expressed intuitively in terms of the amount of *overcrossing* connections. This aspect of the analysis is rather important as the main problem with the SOM structure, namely *poor organization*, needs to be controlled somehow. As usual, it is crucial to avoid poorly-organized structures (whereby immediate neighbours tend to be relatively distant from each other) but this goal is not so easy when working with high-dimensional data [10,11]. On this matter, Kiviluoto and Oja [10] suggested the use of PCA-initialization instead of random-initialization to obtain suitably organized GTM structures; they also proposed the *S-MAP* method combining GTM and SOM. In addition, since it is not clear how much organization is possible for a given data set, the amount of connection overcrossing lacks an absolute scale for assessment. On the other hand, if overcrossing in Sammon's projection plot is (closely) null, we can proceed with some confidence.

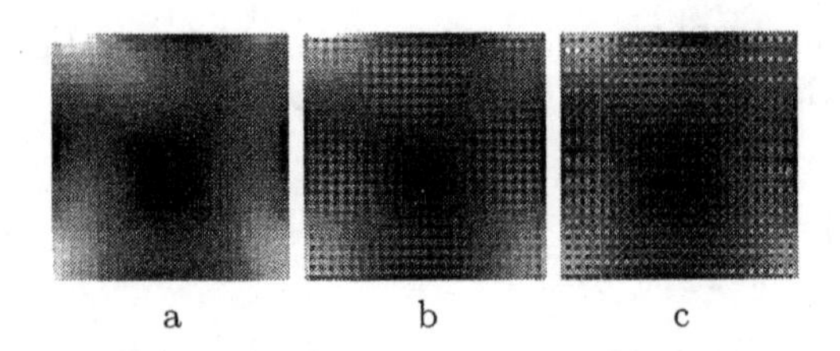

Fig. 1. Data comes from a 2D Gaussian density; (a) MID-matrix showing a higher pointer concentration around the true mode; (b) U-matrix with rectangular topology; (c) U-matrix with hexagonal topology.

3.1 Median Interneuron Distances

Since we are interested in regions with higher pointer (or gaussian centre) density, the relative distance from each pointer to its immediate neighbours on the network will provide a useful bit of information. The inspection of pointer interdistances was pioneered by Ultsch [19], who defined the *unified-matrix* (U-matrix) to visualize Euclidean distances between reference vectors in Kohonen's SOM. Emphasis in the U-matrix is on cluster analysis.

Although modes may be associated with clusters [7], a problem exists with high-dimensional data: "while one cluster might be well-represented by a single multivariate Gaussian, another cluster may required dozens of Gaussian components to capture skewness and yet still be unimodal" [16]. So the relationship between the number of modes and the number of mixture components is not straightforward. In particular, when dealing with multivariate data the mixture may have more modes than mixture components [16].

In this paper, we will work with the alternative *median interneuron matrix* (MID-matrix) proposed in [12]. This is a $\sqrt{K}$x$\sqrt{K}$ matrix whose (i,j) entry, is the *median* of the Euclidean distances between the gaussian centre and all pointers belonging to a star-shaped, fixed-radius neighbourhood containing typically eight units. To facilitate the visualization of higher pointer concentrations, a linear transformation onto a 256-tone gray scale is standard (here the lower the value, the darker the cell). Figure 1 compares the MID-matrix to two variants of the U-matrix. It appears that the latter images provide a more blurry picture regarding pointer concentration.

3.2 Dataloads

The number of data vectors projecting onto each unit, namely the pointer *dataload* $\widehat{\pi}(i,j)$, is the natural estimate of the weight $\pi(i,j)$ obtained from the true underlying distribution f [4],

$$\pi(i,j) = \int_{V(i,j)} f(\mathbf{x})d(\mathbf{x}),$$

where $V(i,j)$ collects all input vectors which are closest to unit (i,j). Again, to easily visualize the dataload distribution over the map, a similar gray image is computed, namely, the DL-matrix. Note that, in this case, darker means higher.

It is important to realize that the "density" of pointers in the trained GTM should serve as an estimate of the density underlying the data. In this ideal case, each neuron would cover about the same proportion of data, that is, a uniform DL-matrix should be obtained. Another interesting way to deal with mode detection and density estimation is to obtain uniformly distributed pointers (over the observed range). Now neurons will present markedly different dataloads, higher densities relating intuitively to larger dataloads. Throughout this paper we focus on the first approach, in which mode detection is based on different pointer concentrations. However, the second idea also seems feasible and is under study.

3.3 Magnification Factors

Since the usual Kohonen algorithm tends to underestimate high probability regions and overestimate low probability areas [3], a concept to express the *magnification* between data and pointer density was needed. A basic theorem related to 1D data was presented by Ritter and Schulten [13]. These authors demonstrated that the limiting pointer density is proportional to the data density raised to a *magnification factor* $\kappa = \frac{2}{3}$. This theorem was later complemented by Bodt, Verleysen and Cottrell [4], who recalled a result from *vector quantization* (VQ) theory, but we shall not be concerned with this problem in the present paper.

Thanks to the topology preserving properties of the GTM, nearby points in latent space will map to nearby points in data space. The concept of *magnification factors* came to represent how the topological map was being locally stretched and compressed when embedded in data space [3]. In the context of the original version of the SOM the topological map is represented in terms of a discrete set of reference vectors, so that a non-continuous expression of magnification factors can also be obtained.

We have focused our study on the GTM model, where local magnification factors can be evaluated as continuous functions of the latent space coordinates in terms of the mapping $\mathbf{y}(\mathbf{x}, \mathbf{W})$ [3]. This constitutes one of our main motivations for working with GTM. We obtain the magnification factors as

$$\frac{dV_y}{dV_x} = \sqrt{\det(\mathbf{J}\mathbf{J}^T)} =_{(L=2)}= \sqrt{\| \frac{\partial \mathbf{y}}{\partial x^1} \|^2 \| \frac{\partial \mathbf{y}}{\partial x^2} \|^2 - (\frac{\partial \mathbf{y}}{\partial x^1}\frac{\partial \mathbf{y}}{\partial x^2})^2}$$

where V_x is the volume of an infinitesimal L-dimensional (L=2) hypercuboid in the latent space and V_y is the volume of its image in the data space. Here $\mathbf{J}$ is the Jacobian of the mapping $\mathbf{y}(\mathbf{x}, \mathbf{W})$ and $\frac{\partial \mathbf{y}}{\partial x^l}$ are the partial derivatives of the mapping $\mathbf{y}(\mathbf{x}, \mathbf{W})$ with respect to the latent variable $x^l = (x_1^l, x_2^l)$.

Using techniques of differential geometry it can be shown that the *directions* and *magnitudes* of stretch in latent space are determined by the eigenvectors and eigenvalues of $\mathbf{J}\mathbf{J}^T$ [3]. Only magnitudes of stretch are considered in this paper. Again, a linear transformation on a gray-scale matrix is presented to visualize

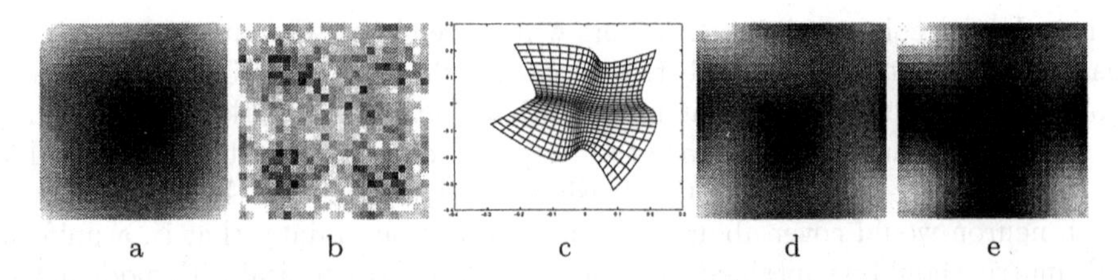

Fig. 2. Data comes from a $N_2(\mathbf{0}, 0.1\mathbf{I}_2)$ density; training parameters are specified in the text; (a) DPO-matrix, showing a map well centered around the mode; (b) DL-matrix, note a tendency towards darker corners; (c) trained map, denser strips are clearly visible; (d) MID-matrix, slightly biased with respect to a); (e) MF-matrix, much coincident with d).

the magnitude of that compression, namely the MF-matrix (as in MID-matrix case, darker means lower). Mode detection analysis based on stretch orientations is postponed for future work.

3.4 A Strategy for Mode Detection

The above summaries of the GTM's self-organizing structure constitute the basis of the following scheme for exploring mode estimation.

1. Train a GTM model [17] until the Sammom's projected pointers show a good level of organization and a (nearly) uniform DL-matrix is obtained. Check also the stability of log-likelihood values.
2. Compute the MID and MF matrices. If the existence of more than one mode is suggested, build subsets of data and return to STEP1 to fit individual GTMs at each unimodal subset.
3. Combining MID-matrix's darkest region, pointer concentration on Sammon's projection and stretch location displayed in the MF-matrix, an approximation to the single mode's location can be performed.

This completes our theoretical presentation. We are now ready to examine some empirical evidence.

4 Simulation Study

As an illustration of the strategy for mode detection presented above, synthetic data samples from several Gaussian distributions with spherical covariance matrices (and often centered at the origin) are considered. Each training set is of size 2500 unless otherwise noted. Since 2-dimensional data allow to examine the GTM directly, a straightforward single Gaussian is tried out first to verify that pointer concentration occurs and can be detected by our summary matrices and

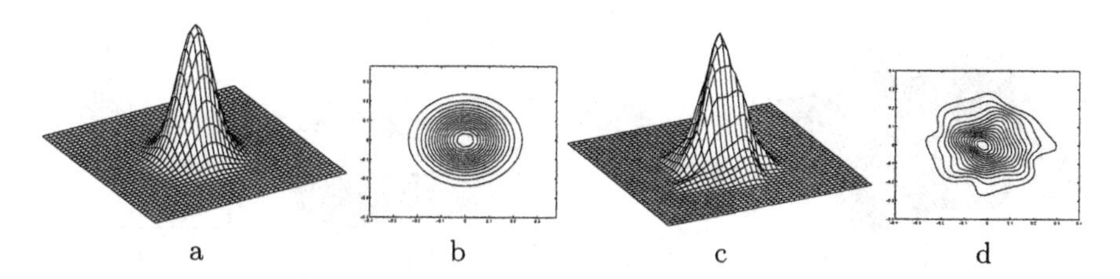

Fig. 3. Data comes from a $N_2(\mathbf{0}, 0.1\mathbf{I}_2)$ density; training parameters are specified in the text; (a) theoretical distribution in 3D; (b) 2D isolines generated from a); (c) mixture distribution obtained after training the GTM; (d) 2D isolines generated from c).

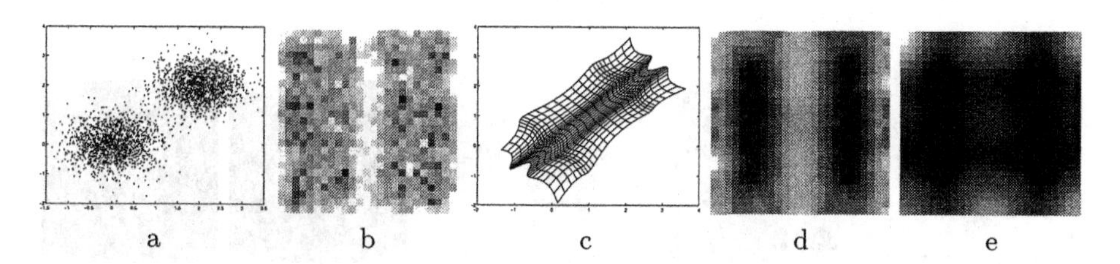

Fig. 4. (a) $N_2(\mathbf{0}, 0.5\mathbf{I}_2) + N_2(\mathbf{2}, 0.5\mathbf{I}_2)$ data set, training parameters: K=625, M=16, $\sigma = 0.4$ and $\tau = 2$; (b) DL-matrix; (c) trained map; (d) MID-matrix; (e) MF-matrix. Note: a) and c) show the data space orientation while b) d) and e) have the latent space one.

statistics, see Figure 2. For the sake of references, we have also compared, in our 2D cases, the final distribution obtained via the GTM,

$$\frac{1}{K} \sum_{i=1}^{K} (\frac{\beta^*}{2\pi})^{\frac{D}{2}} \exp\{-\frac{\beta^*}{2} \parallel \mathbf{y}^*(\mathbf{x}_i, \mathbf{W}) - \mathbf{t} \parallel^2\},$$

where β^* and $\mathbf{y}^*(\mathbf{x}_i, \mathbf{W})$ are the inverse of the variance and the location of the Gaussian centres after training, with the true one, see Figure 3. Whenever synthetic data exhibit a single mode at the origin, a gray-scaled matrix containing the norm of each pointer is also used to test the goodness of our strategy, namely the DPO-matrix, i.e. the Distance from each Pointer to the Origin, see Figure 2.

To obtain a good level of detail in the study, the number of latent points is fixed to 25x25 unless otherwise noted. Note that having a large number of sample points causes no difficulty beyond increased computational cost [1], so smaller maps, say 15x15 or even 10x10, often provide equally useful information. For these many units, the best values for the number of basis functions M and their common width σ seem to be 16 and 0.5 respectively (they tend to provide a well-organized map and a rather uniform DL-matrix). When selecting the number of training cycles, the evaluation of the log-likelihood can be used to monitor convergence. Hence, we have developed `Matlab` code to train the GTM model

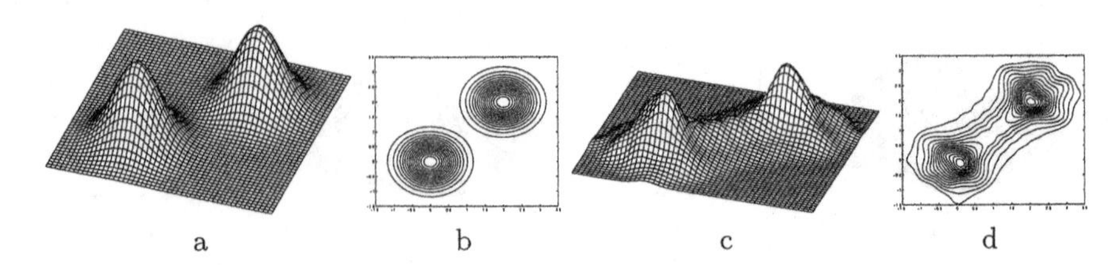

Fig. 5. $N_2(\mathbf{0}, 0.5\mathbf{I}_2)+N_2(\mathbf{2}, 0.5\mathbf{I}_2)$ density, training parameters: K=625, M=16, $\sigma = 0.4$ and $\tau = 2$; (a) theoretical distribution in 3D; (b) 2D isolines generated from a); (c) mixture distribution obtained after training the GTM; (d) 2D isolines generated from c).

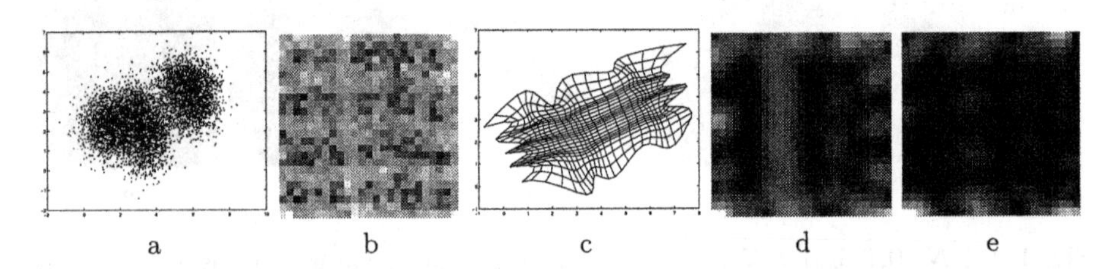

Fig. 6. (a) Carreira data set, N=8000, training parameters: K=625, M=81, $\sigma = 1.0$ and $\tau = 2$; (b) DL-matrix; (c) trained map; (d) MID-matrix; (e) MF-matrix.

until the difference between the log-likelihood values at two consecutive steps is less than a threshold, namely τ. A τ-value smaller than 1.5 is not required for this particular data set, where the training algorithm only needs 20-25 cycles to achieve a good level of convergence.

Special care must be taken with a well-known problem related to the SOM trained structure, namely the *border effect*. By this is meant that units on edges of the network do not stretch out as much as they should [11], which leads to confusing gray-scaled matrices on these map regions, see Figure 2d. Fortunately, these spurious concentrations rarely spread towards the interior of the network, although their traditional presence is somewhat annoying.

To illustrate the strategy in a simple multimodal case, Figure 4 shows an equally weighted mixture of two 2D Gaussians. While all the statistics successfully reveal the existence of two modes, the map does not reflect the support of the true generating distribution. Note also that, somewhat surprisingly, both horizontal and vertical strips are visible in Figure 4e. As a result note the bulky ridge connecting the two modes in Figure 5c.

We now present a more complex multimodal case, first considered by Carreira [5], involving an equally weighted mixture of eight 2D Gaussians with three modes in total, see Figure 7b. The trained map in Figure 6 shows quite informative lines of concentration and stretch in (d) and (e), revealing plausible hori-

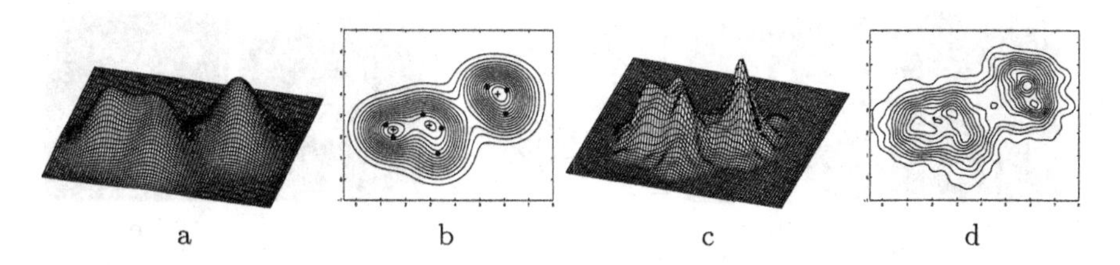

Fig. 7. Carreira data set, training parameters: K=625, M=81, σ = 1.0 and τ = 2; (a) theoretical distribution in 3D; (b) 2D isolines generated from a), component modes are marked with "•", mixture modes with "+"; (c) mixture distribution obtained after training the GTM; (d) 2D isolines generated from c).

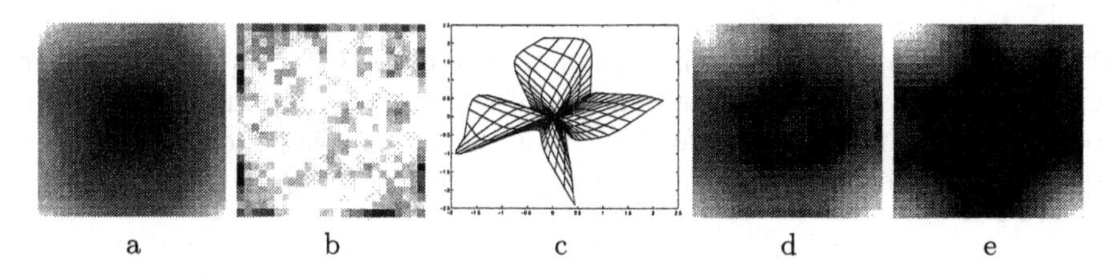

Fig. 8. Data come from a $N_3(\mathbf{0}, \mathbf{I}_3)$ distribution, training parameters: K=625, M=25, $\sigma = 0.5$ and $\tau = 2$; (a) DPO-matrix; (b) DL-matrix; (c) Sammon's projected map; (d) MID-matrix; (e) MF-matrix.

zontal and vertical *split-regions* to separate out the three modes (unfortunately, a strong border effect is also visible in (d)). The fitted density in Figure 7c-d does capture the three modes approximate location (although it tends again to inflate the main gap between modes).

The GTM model in Figure 8 deals with 3D Gaussian data. It reflects how the relationship between adequate organization and uniformly distributed dataloads is already more difficult to obtain. Note that the MID-matrix and MF-matrix show a definite map compression around the center of the map, yet the current pattern is quite different in nature to that found in Figure 2. In these and other higher-dimensional cases, the fitted mixture densities and isolines plots can not be visualized. The main motivation for using SOM structures as an exploratory tool stems from the complexity of the analysis based on a multidimensional fitted density.

When jumping to high dimensional Gaussian data we first note that the probability assigned to the unit radius sphere by the spherical Gaussian distribution goes to zero as the dimension increases. To the extent that lighter concentration should then be expected near the mode, modes in this case may turn out particularly hard to locate [15]. Further, it has been mentioned already that self-organization can be awkward in higher dimensions. Finally, as evidenced in Figure 9, the map may no longer be as centered around the mode as before.

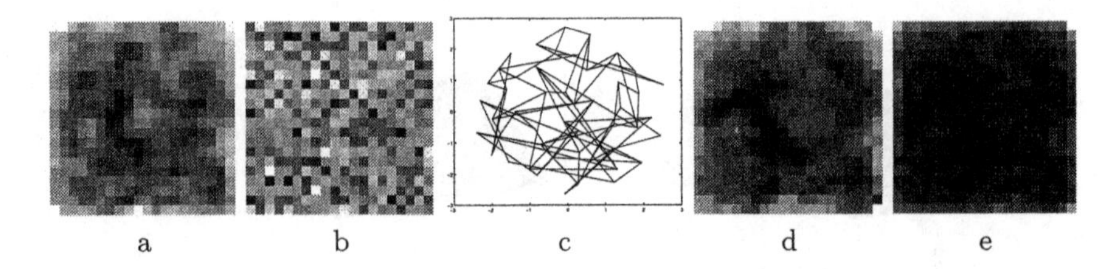

Fig. 9. Data comes from a $N_{10}(\mathbf{0}, \mathbf{I}_{10})$ density, training parameters: K=49, M=144, σ = 1.1, τ = 2 and 200 Sammon steps; (a) DPO-matrix, showing how the mode is shifted a bit with respect to the 2D Gaussian case; (b) DL-matrix, (c) Sammon's projected map; (d) MID-matrix following the pattern in a); (e) MF-matrix.

Overall, this map is only partially organized, but the patterns in plots (d) and (e) here can be seen as diffuse versions of the inner rings found in the corresponding plots in Figure 8. We stress that the shift in mode location has been consistently observed in many runs with this data set.

5 Summary and Discussion

This paper explores the role of the SOM *structure* to deal with the problem of multivariate mode detection. Mode detection differs from cluster detection in that a precise estimation task is faced when tackling unimodal data. The present approach is founded on the tenets that (i) SOM pointers near the modes should lie closer together due to a higher concentration of data in those regions; and (ii), these areas can be highlighted by exploring the structure of pointer interdistances and other summaries whenever sufficient self-organization is achieved. As the GTM training algorithm explicitly controls the smoothness of the map and also leads to a rich, continuous treatment of local magnification factors, it can be seen to provide a potential advantage over the original SOM fitting algorithm. Hence, we have formulated and investigated a strategy for detecting the modes of a multivariate continuous data-generating distribution based on suitable statistics taken from a trained GTM, namely, median interneuron distances, dataloads, magnification factors together with Sammon's projection.

The strategy works as expected with 2D and 3D Gaussian data. While the maps produced by GTM are quite similar to those found by the standard fitting algorithm in some cases (see Figure 2), they differ markedly in others (Figure 4), which suggests that any detection strategy of the present sort must distinguish the origin of the SOM structure. In a sense, this is the price to be paid for exploiting GTM's richer framework.

The paper also highlights the current risks found when fitting SOMs to high-dimensional data, where larger maps are needed to provide enough detail for the strategy to succeed but turn out to be harder to organize. Some unexpected phenomena have been isolated; for example, in the 10D case modes are typically

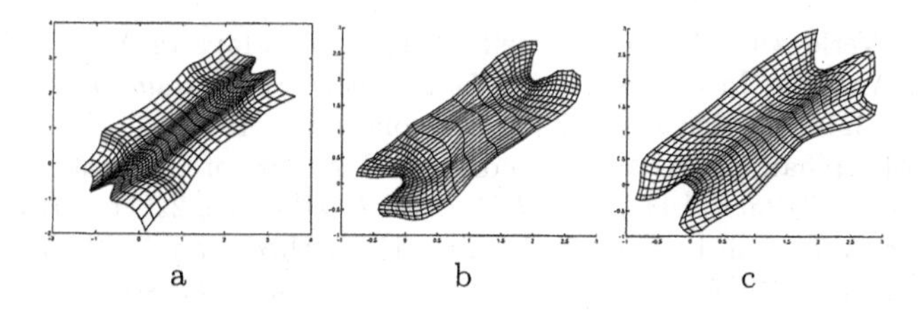

Fig. 10. (a) Map trained via GTM (discussed earlier); (b) the map trained with the convex adjustment reflects more faithfully the sampling density; (c) the map trained with the concave adjustment provides a more uniform distribution of pointers.

anchored at noncentral neurons. Additional research is needed to clarify the setting of GTM parameters with an eye put on better self-organization levels.

Future investigations can proceed in various fronts. New diagnostic values can be based on a suitable combination of MID and MF values intended to stress pointer concentration. Directions of stretch deserve also a close look. On the other hand, an additional smoothing construct can be used in GTM, namely, a Gaussian process prior distribution penalizing larger $\mathbf{W}$ entries. Other prior distributions may also be considered, see e.g. [20].

Finally, note that the present strategy for mode detection is based on the assumption that the density of the reference vectors will be similar in some sense to the density of the training data. As mentioned above, an alternative strategy can proceed on the basis of uniformly spaced pointers with widely different dataloads. To this end, some adjustments to the standard algorithm have been proposed to yield a trained map in line with the desired level of pointer concentration [21]. It remains to be seen whether these adjustments have their counterpart in GTM or else can improve detection accuracy. As a first example of the power of the suggestions made in [21], we finally present the alternative maps obtained in a previously discussed example.

Acknowledgement. We are grateful to Drs. Juan M. Marín and David Ríos for their constructive comments on earlier drafts of this paper. The authors are partially supported by grants from the local CAM Government and the European Council.

References

1. Bishop, C. M., Svensén, M. and Williams, C. K. I. GTM: The Generative Topographic Mapping. *Neural Computation*, 10: 215–235. 1997a.
2. Bishop, C. M., Svensén, M. and Williams, C. K. I. GTM: A principle alternative to the Self-Organizing Map. *Advances in Neural Information Processing Systems*. vol, 9. MIT Press, 1997.
3. Bishop, C. M., Svensén, M. and Williams, C. K. I. Magnification Factors for the SOM and GTM Algorithms. *Proccedings 1997 Workshop on Self-Organizing Maps*, Helsinki, Finland.

4. Bodt, E., Verleysen, M., and Cottrell, M. Kohonen Maps vs. Vector Quantization for Data Analysis. In *Proceedings of the European Symposium on Artificial Neural Networks*, 221–220. Brusseels: D-Facto Publications. 1997.
5. Carreira-Perpiñán, M. A. Mode-Finding for Mixtures of Gaussians Distributions. *IEEE Trans. Pattern Analysis and Machine Intelligence*, 22(11):1318–1323. 2000.
6. Dempster, A. P., Laird, N. M., and Rubin, D. B. Maximum likelihood from incomplete data via the EM algorithm. *Journal of the Royal Statistics Society, B 39*(1), 1–38. 1977.
7. Good, I. J. and Gaskins, R. A. Density Estimation and Bump-Hunting by Penalized Likelihood Method Exemplified by Scattering and Meteorite Data. *Journal of the American Statistical Association*. 75:42–73, 1980.
8. Hartigan, J. A., and Hartigan, P. M. The DIP test of unimodality. *Ann. Statist. 13*, 70–84. 1985.
9. Izenman, A. J., and Sommer, C. J. Philatelic mixtures and multimodal densities. *J. Am. Statist. Assoc. 83*, 941–953. 1988.
10. Kiviluoto, K. and Oja, E. S-map: A Network with a Simple Self-Organization Algorithm for Generative Topographic Mappings. In M. I. Jordan, M. J. Kearns, and S. A. Solla (Eds.), *Advances in Neural Information Processing Systems*. MIT Press 10:549–555. 1998.
11. Kohonen, T. *Self-Organizing Maps*. Springer-Verlag, Berlin, 2001.
12. Muruzábal, J. and Muñoz, A. On the Visualization of Outliers via Self-Organizing Maps. *Journal of Computational and Graphical Statistics*, 6(4): 355–382. 1997.
13. Ritter, H. and Schulten, K. On the Stationary State of Kohonen's Self-Organizing Sensory Mapping. *Biological Cybernetics*, 54:99–106, 1986.
14. Sammon, J. W. A Nonlinear Mapping for Data Structure Analysis. *IEEE Transactions on Computers*, 18(5): 401–409. 1969.
15. Scott, D. W. *Multivariate Density Estimation*. New York: John Wiley and Sons, Inc. 1992.
16. Scott, D. W. and Szewczyk, W. F. The Stochastic Mode Tree and Clustering. To appear in Journal of Computational and Graphical Statistics. 2000
17. Svensén, M. The GTM: Toolbox-User's Guide. 1999. `http://www.ncrg.aston.ac.uk/GTM`.
18. Terrell, G. R. and Scott, D. W. Variable kernel density estimation. *Ann. Statist. 20*, 1236–1265. 1992.
19. Ultsch, A. Self-Organizing Neural Networks for Visualization and Classification. In Opitz, O., Lausen, B., and Klar, R. editors, *Information and Classification*, pages 307–313. Springer-Verlag, Berlin, 1993.
20. Utsugi, A. Bayesian Sampling and Ensemble Learning in Generative Topographic Mapping. *Neural Processing Letters*, 12(3): 277–290. 2000.
21. Zheng, Y. and Greenleaf, J.F. The Effect of Concave and Convex Weight Adjustments on Self-Organizing Maps. *IEEE Transactions on Neural Networks*, 7(1): 87–96, 1996.

Regularization Methods for Additive Models

Marta Avalos, Yves Grandvalet, and Christophe Ambroise

HEUDIASYC Laboratory UMR CNRS 6599
Compiègne University of Technology
BP 20529 / 60205 Compiègne, France
{avalos, grandvalet, ambroise}@hds.utc.fr

Abstract. This paper tackles the problem of model complexity in the context of additive models. Several methods have been proposed to estimate smoothing parameters, as well as to perform variable selection. However, these procedures are inefficient or computationally expensive in high dimension. To answer this problem, the *lasso* technique has been adapted to additive models, but its experimental performance has not been analyzed.
We propose a modified *lasso* for additive models, performing variable selection. A benchmark is developed to examine its practical behavior, comparing it with forward selection. Our simulation studies suggest ability to carry out model selection of the proposed method. The *lasso* technique shows up better than forward selection in the most complex situations. The computing time of modified *lasso* is considerably smaller since it does not depend on the number of relevant variables.

1 Introduction

Additive nonparametric regression model has become a useful statistical tool in analysis of high–dimensional data sets. An additive model [10] is defined by

$$Y = f_0 + \sum_{j=1}^{p} f_j(X_j) + \varepsilon, \qquad (1)$$

where the errors ε are independent of the predictor variables X_j, $\mathbb{E}(\varepsilon) = 0$ and $var(\varepsilon) = \sigma^2$. The f_j, are univariate smooth functions and f_0 is a constant. Y is the response variable.

This model's popularity is due to its flexibility, as a nonparametric method, and also to its interpretability. Furthermore, additive regression circumvents the curse of dimensionality.

Some issues related to model complexity have been studied in the context of additive models. Several methods have been proposed to estimate smoothing parameters [10,8,11,16,4]. These methods are based on generalizing univariate techniques. Nevertheless, the application of these procedures in high dimension is often inefficient or highly time consuming.

Variable selection methods have also been formulated for additive models [10, 5,13,3,16]. These proposals exploit the fact that additive regression generalizes

M.R. Berthold et al. (Eds.): IDA 2003, LNCS 2810, pp. 509–520, 2003.

linear regression. Since nonparametric methods are used to fit the terms, model selection develops some new flavors. We not only need to select which terms to include in the model, but also how smooth they should be. Thus even for few variables, these methods are computationally expensive.

Finally, the *lasso* technique [14] has been adapted to additive models fitted by splines. The lasso (least absolute shrinkage and selection operator) is a regularization procedure intended to tackle the problem of selection of accurate and interpretable linear models. The lasso estimates a vector of regression coefficients by minimizing the residual sum of squares subject to a constraint on the l_1–norm of coefficient vector.

For additive models, this technique transforms a high–dimensional into a low–dimensional hyper–parameter selection problem, which implies many advantages. Some algorithms have been proposed [6,7,1]. Their experimental performance has not been, however, analyzed, and above all, these discriminate between linear and nonlinear variables, but do not perform variable selection.

There are many other approaches to model selection for supervised regression tasks (see for example [9]). Computational costs are a primary issue in their application to additive models.

We propose a modified lasso for additive spline regression, in order to discriminate between linear and nonlinear variables, but also, between relevant and irrelevant variables. We develop a benchmark based on Breiman's work [2], to analyze the practical behavior of the modified lasso, comparing it to forward variable selection. We focus on the situations where the control of complexity is a major problem. The results allow us to deduce conditions of application of each regularization method. The computing time of modified lasso is considerably smaller than the computing time of forward variable selection since it does not depend on the number of relevant variables (when the total number of input variables is fixed).

Section 2 presents lasso applied to additive models. Modifications are introduced in section 3, as well as algorithmic issues. The schema of the testing efficiency procedure and benchmark for additive models are presented in section 4. Section 5 gives simulation results and conclusions.

2 Lasso Adapted to Additive Models

Grandvalet et al. [6,7] showed the equivalence between adaptive ridge regression and lasso. Thanks to this link, the authors derived an EM algorithm to compute the lasso estimates. Subsequently, results obtained for the linear case were generalized to additive models fitted by smoothing splines.

Suppose that one has data $\mathcal{L} = \{(\mathbf{x}, \mathbf{y})\}$, $\mathbf{x} = (\mathbf{x}_1, \dots, \mathbf{x}_p)$, $\mathbf{x}_j = (x_{1j}, \dots, x_{nj})^t$, $\mathbf{y} = (y_1, \dots, y_n)^t$. To simplify, assume that the responses are centered. We remind that the adaptive ridge estimate is the minimizer of the problem

$$\min_{\alpha_1,\dots,\alpha_p} \left\| \mathbf{y} - \sum_{j=1}^{p} \mathbf{x}_j \alpha_j \right\|_2^2 + \sum_{j=1}^{p} \lambda_j \alpha_j^2, \tag{2}$$

subject to

$$\sum_{j=1}^{p} \frac{1}{\lambda_j} = \frac{p}{\lambda} \qquad \lambda_j > 0, \tag{3}$$

and the lasso estimate is given by the solution of the following constrained optimization problem

$$\min_{\alpha_1,\dots,\alpha_p} \left\| \mathbf{y} - \sum_{j=1}^{p} \mathbf{x}_j \alpha_j \right\|_2^2 \quad \text{subject to} \quad \sum_{j=1}^{p} |\alpha_j| \le \tau, \tag{4}$$

where τ and λ are predefined values.

We also remind that the cubic smoothing spline is defined as the minimizer of the penalized least squares criterion over all twice–continuously–differentiable functions. This idea is extended to additive models in a straightforward manner:

$$\min_{f_1,\dots,f_p \in \mathcal{C}^2} \left\| \mathbf{y} - \sum_{j=1}^{p} f_j(\mathbf{x}_j) \right\|_2^2 + \sum_{j=1}^{p} \lambda_j \int_{a_j}^{b_j} f_j''(x)^2 dx, \tag{5}$$

where $[a_j, b_j]$ is the interval for which an estimate of f_j is sought. This interval is arbitrary, as long as it contains the data; $\widehat{f_j}$ is linear beyond the extreme data points no matter what the values of a_j and b_j are. Each function in (5) is penalized by a separate fixed smoothing parameter λ_j.

Let $\mathbf{N}_j$ denote the $n \times (n+2)$ matrix of the unconstrained natural B–spline basis, evaluated at x_{ij}. Let Ω_j be the $(n+2) \times (n+2)$ matrix corresponding to the penalization of the second derivative of $\widehat{f_j}$. The coefficients of $\widehat{f_j}$ in the unconstrained B–spline basis are noted β_j. Then, the extension of lasso to additive models fitted by cubic splines, using the equivalence between (4) and (2)–(3) is given by

$$\min_{\beta_1,\dots,\beta_p} \left\| \mathbf{y} - \sum_{j=1}^{p} \mathbf{N}_j \beta_j \right\|_2^2 + \sum_{j=1}^{p} \lambda_j \beta_j^t \Omega_j \beta_j, \tag{6}$$

subject to

$$\sum_{j=1}^{p} \frac{1}{\lambda_j} = \frac{p}{\lambda} \qquad \lambda_j > 0, \tag{7}$$

where λ is a predefined value. The expression (6) shows that this problem is equivalent to a standard additive spline model, where the penalization terms

λ_j applied to each additive component are optimized subject to constraints (7). This problem has the same solution as

$$\min_{\beta_1,\ldots,\beta_p} \left\| \mathbf{y} - \sum_{j=1}^{p} \mathbf{N}_j \beta_j \right\|_2^2 + \frac{\lambda}{p} \left(\sum_{j=1}^{p} \sqrt{\beta_j^t \Omega_j \beta_j} \right)^2 . \tag{8}$$

The penalizer in (8) generalizes the lasso penalizer $\sum_{j=1}^{p} |\alpha_j| = \sum_{j=1}^{p} \sqrt{\alpha_j^2}$. Note that the writing (6)–(7) can also be motivated from a hierarchical Bayesian viewpoint.

Grandvalet et al. proposed a generalized EM algorithm, where the M-step is performed by backfitting (see Sect. 3.1) to estimate coefficients β_j. This method does not perform variable selection. When after convergence $\widehat{\beta}_j^t \Omega_j \widehat{\beta}_j = 0$, the jth predictor is not eliminated but linearized.

Another algorithm based on sequential quadratic programming was suggested by Bakin [1]. This methodology seems to be more complex than the precedent one and does not perform variable selection either.

3 Modified Lasso

3.1 Decomposition of the Smoother Matrix

Splines are linear smoothers, that is, the univariate fits can be written as $\widehat{\mathbf{f}}_j = \mathbf{S}_j \mathbf{y}$, where $\mathbf{S}_j$ is an $n \times n$ matrix called *smoother matrix*. The latter depends on the smoothing parameter and the observed points $\mathbf{x}_j$, but not on $\mathbf{y}$.

The smoother matrix of a cubic smoothing spline has two unitary eigenvalues corresponding to the constant and linear functions (its projection part), and $n-2$ non–negative eigenvalues strictly smaller than 1 corresponding to different compounds of the non-linear part (its *shrinking* part). Also, $\mathbf{S}_j$ is symmetric, then $\mathbf{S}_j = \mathbf{G}_j + \widetilde{\mathbf{S}}_j$, where $\mathbf{G}_j$ is the matrix that projects onto the space of eigenvalue 1 for the jth smoother, and $\widetilde{\mathbf{S}}_j$ is the *shrinking* matrix [10].

For cubic smoothing splines, $\mathbf{G}_j$ is the *hat* matrix corresponding to the least–squares regression on $(\mathbf{1}, \mathbf{x}_j)$, the smoother matrix is calculated as $\mathbf{S}_j = \mathbf{N}_j(\mathbf{N}_j^t \mathbf{N}_j + \lambda_j \Omega_j)^{-1} \mathbf{N}_j^t$, and $\widetilde{\mathbf{S}}_j$ is found by $\mathbf{S}_j - \mathbf{G}_j$:

$$\widetilde{\mathbf{S}}_j = \mathbf{S}_j - \left(\frac{1}{n} \mathbf{1}\mathbf{1}^t + \frac{1}{||\mathbf{x}_j||_2^2} \mathbf{x}_j \mathbf{x}_j^t \right) . \tag{9}$$

Additive models can be estimated by the backfitting algorithm which consists in fitting iteratively $\widehat{\mathbf{f}}_j = \mathbf{S}_j(\mathbf{y} - \sum_{k \neq j} \widehat{\mathbf{f}}_k)$, $j = 1, \ldots, p$.

Taking into account the decomposition of $\mathbf{S}_j$, the backfitting algorithm can be divided into two steps: 1. estimation of the projection part, $\mathbf{g} = \mathbf{G}(\mathbf{y} - \sum \widetilde{\mathbf{f}}_j)$, where $\mathbf{G}$ is the *hat* matrix corresponding to the least–squares regression on $(\mathbf{1}, \mathbf{x}_1, \ldots, \mathbf{x}_p)$, and 2. estimation of the shrinking parts, $\widetilde{\mathbf{f}}_j = \widetilde{\mathbf{S}}_j(\mathbf{y} - \mathbf{g} - \sum_{k \neq j} \widetilde{\mathbf{f}}_k)$. The final estimate for the overall fit is $\widehat{\mathbf{f}} = \mathbf{g} + \sum \widetilde{\mathbf{f}}_j$.

In addition, linear and nonlinear parts work on orthogonal spaces: $\mathbf{G}\widetilde{\mathbf{f}}_j = \mathbf{0}$ and $\widetilde{\mathbf{S}}_j\mathbf{g} = \mathbf{0}$, so estimation of additive models fitted by cubic splines, using the backfitting algorithm can be separated in its linear and its nonlinear part.

3.2 Isolating Linear from Nonlinear Penalization Terms

The penalization term in (7) only acts on the nonlinear components of $\widehat{\mathbf{f}}$. Consequently, severely penalized covariates are not eliminated but linearized. Another term acting on the linear component should be applied to perform subset selection.

The previous decomposition of cubic splines allows us to decompose the linear and nonlinear parts:

$$\widehat{\mathbf{f}} = \mathbf{g} + \widetilde{\mathbf{f}} = \sum_{j=1}^{p} \mathbf{x}_j \widehat{\alpha}_j + \sum_{j=1}^{p} \widetilde{\mathbf{f}}_j(\mathbf{x}_j) = \mathbf{x}\widehat{\alpha} + \sum_{j=1}^{p} \widetilde{\mathbf{N}}_j \widetilde{\beta}_j, \tag{10}$$

where $\widetilde{\mathbf{N}}_j$ denotes the matrix of the nonlinear part of the unconstrained spline basis, evaluated at x_{ij}, $\widetilde{\beta}_j$ denotes the coefficients of $\widetilde{\mathbf{f}}_j$ in the nonlinear part of the unconstrained spline basis, and $\alpha = (\alpha_1, \ldots, \alpha_p)^t$ denotes linear least squares coefficients.

Regarding penalization terms, a simple extension of (7) is to minimize (with respect to α and $\widetilde{\beta}_j$, $j = 1, \ldots, p$)

$$\left\| \mathbf{y} - \mathbf{x}\alpha - \sum_{j=1}^{p} \widetilde{\mathbf{f}}_j \right\|_2^2 + \sum_{j=1}^{p} \mu_j \alpha_j^2 + \sum_{j=1}^{p} \lambda_j \widetilde{\beta}_j^t \Omega_j \widetilde{\beta}_j, \tag{11}$$

subject to

$$\sum_{j=1}^{p} \frac{1}{\mu_j} = \frac{p}{\mu}, \quad \mu_j > 0 \quad \text{and} \quad \sum_{j=1}^{p} \frac{1}{\lambda_j} = \frac{p}{\lambda}, \quad \lambda_j > 0, \tag{12}$$

where μ and λ are predefined values[1].

When after convergence, $\mu_j \approx \infty$ and $\lambda_j \approx \infty$, the jth covariate is eliminated. If $\mu_j < \infty$ and $\lambda_j \approx \infty$, the jth covariate is linearized. When $\mu_j \approx \infty$ and $\lambda_j < \infty$, the jth covariate is estimated to be strictly nonlinear.

3.3 Algorithm

1. Initialize: $\mu_j = \mu$, $\Lambda = \mu \mathbf{I}_p$, $\lambda_j = \lambda$.
2. Linear components:
   a) Compute linear coefficients: $\alpha = \left(\mathbf{x}^t\mathbf{x} + \Lambda\right)^{-1} \mathbf{x}^t\mathbf{y}$.

[1] Note that $\widetilde{\Omega}_j \equiv \Omega_j$, where $\widetilde{\Omega}_j$ and Ω_j are the matrix corresponding to the penalization of the second derivative of $\widetilde{\mathbf{f}}_j$ and $\mathbf{f}_j$, respectively.

b) Compute linear penalizers: $\mu_j = \mu \frac{\|\alpha\|_1}{p|\alpha_j|}$, $\quad \Lambda = \mathrm{diag}(\mu_j)$.

3. Repeat step 2 until convergence.
4. Nonlinear components:
   a) Initialize: $\widetilde{\mathbf{f}}_j$, $j = 1, \ldots, p$, $\widetilde{\mathbf{f}} = \sum_j^p \widetilde{\mathbf{f}}_j$.
   b) Calculate: $\mathbf{g} = \mathbf{G}\left(\mathbf{y} - \widetilde{\mathbf{f}}\right)$.
   c) One cycle of backfitting: $\widetilde{\mathbf{f}}_j = \widetilde{\mathbf{S}}_j\left(\mathbf{y} - \mathbf{g} - \sum_{k \neq j} \widetilde{\mathbf{f}}_k\right)$, $j = 1, \ldots, p$.
   d) Repeat step (b) and (c) until convergence.
   e) Compute $\widetilde{\beta}_j$ from the final estimates.
   f) Compute nonlinear penalizers: $\lambda_j = \lambda \frac{\sum_{j=1}^p \sqrt{\widetilde{\beta}_j^t \Omega_j \widetilde{\beta}_j}}{p\sqrt{\widetilde{\beta}_j^t \Omega_j \widetilde{\beta}_j^t}}$.
5. Repeat step 4 until convergence.

In spite of orthogonality, projection step 4.(b) is iterated with backfitting step 4.(c) to improve the numerical stability of the algorithm. In order to compute nonlinear penalizers in 4.(f), calculating β_j instead of $\widetilde{\beta}_j$ in 4.(e) is sufficient, since Ω_j is insensitive to the linear components.

An efficient lasso algorithm was proposed by Osborne et al. [12]. It could be used for the linear part of the algorithm, and may be adapted to the nonlinear part.

3.4 Complexity Parameter Selection

The initial multidimensional parameter selection problem is transformed into a 2–dimensional problem. A popular criterion for choosing complexity parameters is cross–validation, which is an estimate of the prediction error (see Sect. 4.1). Calculating the CV function, is computationally intensive. A fast approximation of CV is generalized cross–validation [10,8,14]:

$$\mathrm{GCV}(\mu, \lambda) = \frac{1}{n} \frac{(\mathbf{y} - \widehat{\mathbf{f}})^t (\mathbf{y} - \widehat{\mathbf{f}})}{(1 - df(\mu, \lambda)/n)^2}. \quad (13)$$

The GCV function is evaluated over a 2–dimensional grid of values. The point $(\widehat{\mu}, \widehat{\lambda})$ yielding the lowest GCV is selected. The effective number of parameters (or the effective degrees of freedom) df is estimated by the sum of the univariate effective number of parameters, df_j [10]:

$$df(\mu, \lambda) \approx \sum_{j=1}^p df_j(\mu, \lambda) = \mathrm{tr}\left[\mathbf{x}\left(\mathbf{x}^t\mathbf{x} + \Lambda\right)^{-1}\mathbf{x}^t\right] + \sum_{j=1}^p \mathrm{tr}\left[\widetilde{\mathbf{S}}_j(\lambda_j)\right]. \quad (14)$$

Variable selection methods do not take into account the effect of model selection [17,15]. It is assumed that the selected model is given a priori, which may introduce a bias in the estimator. The present estimate suffers from the same inaccuracy: the degrees of freedom spent in estimating the individual penalization terms estimation are not measured.

4 Efficiency Testing Procedure

Our goal is to analyze the modified lasso behavior and compare it to another model selection method: forward variable selection. Comparison criteria and procedures for simulations are detailed next.

4.1 Comparison Criteria

The objective of estimation is to construct a function $\widehat{f}$ providing accurate prediction of future examples. That is, we want the prediction error

$$\mathrm{PE}(\widehat{f}) = \mathbb{E}_{Y\mathbf{X}}[(Y - \widehat{f}(\mathbf{X}))^2] \tag{15}$$

to be small.

Let $\widehat{\lambda}$ denote complexity parameter selected by GCV and λ^* be the value minimizing PE (Breiman's *crystal ball* [2]):

$$\widehat{\lambda} = \underset{\lambda}{\operatorname{argmin}}\, \mathrm{GCV}\left(\widehat{f}(.,\lambda)\right), \qquad \lambda^* = \underset{\lambda}{\operatorname{argmin}}\, \mathrm{PE}\left(\widehat{f}(.,\lambda)\right). \tag{16}$$

4.2 A Benchmark for Additive Models

Our objective is to define the bases of a benchmark for additive models inspired by Breiman's work [2]: "Which situations make estimation difficult?" and "What parameters are these situations controlled through?"

The number of observations–effective number of parameters ratio, n/df. When n/df is high, the problem is easy to solve. Every "reasonable" method will find an accurate solution. Conversely, a low ratio makes the problem insolvable. The effective number of parameters depends on the number of covariates, p, and on other parameters described next.

Concurvity. Concurvity describes a situation in which predictors are linearly or nonlinearly dependent. This phenomenon causes non–unicity of estimations. Here we control collinearity through the correlation matrix of predictors:

$$\mathbf{X} \sim \mathcal{N}_p(\mathbf{0}, \Gamma), \qquad \Gamma_{ij} = \rho^{|i-j|}. \tag{17}$$

The number of relevant variables. The nonzero coefficients are in two clusters of adjacent variables with centers at fixed l's. The coefficient values are given by

$$\delta_j = \mathbb{I}_{\{\omega - |j-l| > 0\}}, \quad j = 1, \ldots, p, \tag{18}$$

where ω is a fixed integer governing the cluster width.

Underlying functions. The more the structures of the underlying functions are complex, the harder they are to estimate. One factor that makes the structure complex is rough changes in curvature. Sine functions allow us to handle diverse situations, from almost linear to highly zig-zagged curves. We are interested in controlling intra– and inter–covariates curvatures changes, thus define:

$$f_j(\mathbf{x}_j) = \delta_j \sin 2\pi k_j \mathbf{x}_j, \tag{19}$$

and $f_0 = 0$. Consider $\kappa = k_{j+1} - k_j$, $j = 1, \ldots, p-1$, the sum $\sum_{j=1}^{p} k_j$ fixed to keep the same overall degree of complexity.

Noise level. Noise is introduced through error variance

$$Y = \textstyle\sum_{j=1}^{p} f_j + \varepsilon, \qquad \varepsilon \sim \mathcal{N}(0, \sigma^2). \tag{20}$$

In order to avoid sensitivity to scale, the noise effect is controlled through the determination coefficient, which is a function of σ.

5 Results

5.1 Example

Fig. 1 reproduces the simulated example in dimension five used in [7]. The fitted univariate functions are plotted for four values of (μ, λ). The response variable depends linearly on the first covariate. The last covariate is not relevant. The other covariates affect the response, but the smoothness of the dependancies decreases with the coordinate number of the covariates.

For $\mu = \lambda = 1$, the individual penalization terms of the last covariate, μ_5, λ_5, were estimated to be extremely large, as well as the individual nonlinear penalization term corresponding to the first covariate, λ_1. Therefore, the first covariate was estimated to be linear and the last one was estimated to be irrelevant. For high values of μ and λ, the dependences on the least smooth covariates are difficult to capture.

5.2 Comparison

Implementation. The forward selection version of the backward elimination algorithm given in [3] was implemented. Both smoothing parameters and models are selected by GCV. Given that the current model has $q \geq 0$ variables, GCV is computed for each of the $p - q$ models embedding the current model with one more variable. If the $(q+1)$–variable nested model minimizing GCV has smaller GCV value than the q–variable model, then the q–model is replaced with this "best" $(q+1)$–variable nested model and the procedure is continued. Otherwise, the q–variable model is retained and the procedure is stopped. The PE value is estimated from a test set of size 20000 sampled from the same distribution as the learning set.

The GCV criterion is used to select complexity parameters of modified lasso. The GCV function is evaluated over a grid of 5×5 values. A reasonably large

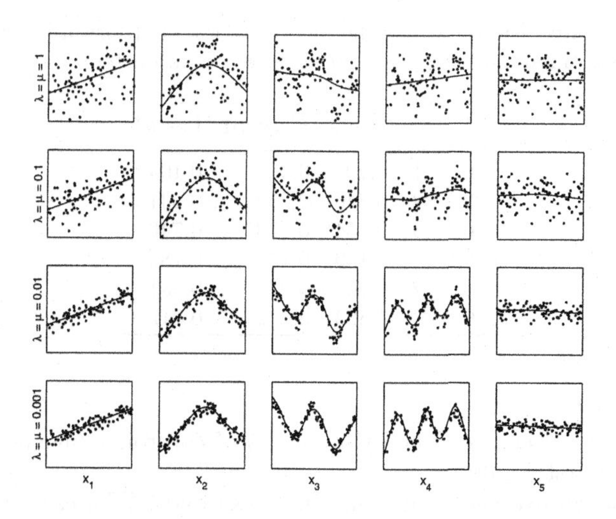

Fig. 1. Modified lasso on simulated data. The underlying model is $y = x_1 + \cos(\pi x_2) + \cos(2\pi x_3) + \cos(3\pi x_4) + \varepsilon$. The covariates are independently drawn from a uniform distribution on $[-1, 1]$ and ε is a Gaussian noise of standard deviation $\sigma = 0.3$. The solid curves are the estimated univariate functions for different values of (μ, λ), and dots are partial residuals.

range was adopted to produce the complexity parameters space, specifically, $\lambda, \mu \in [0.02, 4]$. The log–values are equally spaced on the grid. The PE value of lasso is calculated similarly to the PE of forward selection.

Simulations. Parameters of control described in section (4.2) were fixed as follows. The number of observations and the number of covariates were, respectively, $n = 450$ and $p = 12$. We considered two cases for ρ, low ($\rho = 0.1$) and severe ($\rho = 0.9$) concurvity. With respect to relevant variables, the situations taken into account in simulations were: $\omega = \{1, 2, 3\}$, taken clusters centered at $l = 3$ and $l = 9$. This gave 2, 6 and 10 relevant variables, over a total of 12. No different curvature changes between covariates: $\kappa = 0$ and moderate curvature changes within covariates: $\sum k_j = 9$, were taken into account. We studied a lowly noisy situation: $R^2 = 0.9$.

Table 1 shows estimated PE values for the two methods in comparison, modified lasso using the GCV criteria and forward variable selection, and PE values for the optimal modified lasso, Lasso*, (that is, complexity parameters are the minimizers of PE, they are estimated from an "infinite" test set).

We observe that only in the most difficult situations: 6 or 10 relevant variables ($\omega = 2$ or 3) and severe correlation ($\rho = 0.9$), the lasso applied to additive models had a lower PE than forward selection. Estimated PE values corresponding to modified lasso using the GCV criteria are quite higher than PE values corresponding to optimal modified lasso. This may be because this complexity parameter selection method is not sufficiently accurate. A more extended exper-

Table 1. Estimated PE values for the modified lasso and forward selection and PE values for the optimal modified lasso, for each set of parameters.

ρ	ω	Lasso	Forward	Lasso*
0.1	1	0.109	0.081	0.104
0.1	2	0.346	0.262	0.341
0.1	3	0.754	0.713	0.746
0.9	1	0.199	0.172	0.182
0.9	2	0.907	0.935	0.881
0.9	3	1.823	2.212	0.812

iment, including other described parameters of control, would be necessary to validate these first results.

Table 2 shows average computing time in seconds. Whereas computing time of lasso does not seem to depend on the number of relevant variables, computing time of forward variable selection increases considerably with the number of variables that actually generate the model.

Table 2. Average computing time in seconds needed by modified lasso and forward selection, when 2, 6 and 10 variables generate the model ($\omega = \{1, 2, 3\}$, respectively).

ω	Lasso	Forward
1	9854	4416
2	8707	26637
3	7771	108440

Forward selection never missed relevant variables. However, this is not the best solution when $\rho = 0.9$, considering that variables are highly correlated. Moreover forward variable selection picked two irrelevant variables in the situations $\omega = 1$, $\rho = 0.9$ and $\omega = 2$, $\rho = 0.9$. Linear components of modified lasso were estimated to be highly penalized in all cases, for all variables (we remind that underlying functions are centered sines). Fig. 2 shows inverse of estimated univariate complexity parameters corresponding to the nonlinear components. Modified lasso penalized severely irrelevant variables. Penalization of relevant variables increased with concurvity and with the number of relevant variables.

5.3 Conclusion

We propose a modification of lasso for additive models in order to perform variable selection. For each covariate, we differentiate its linear and its nonlinear part, and penalize them independently. Penalization is regulated automatically from two global parameters which are estimated by generalized cross–validation.

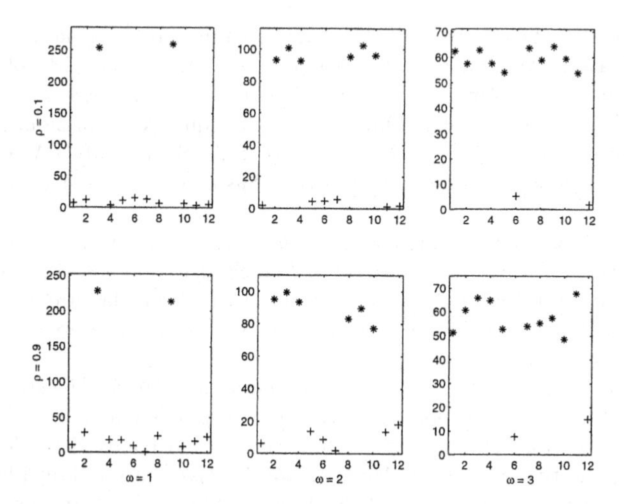

Fig. 2. Inverse of the estimated univariate complexity parameters corresponding to the nonlinear components: $\frac{1}{\lambda_j}$, $j = 1, ..., 12$. All variables are represented in the horizontal axis. Stars $(*)$ are relevant variables and plus $(+)$ are irrelevant ones. First line of graphics corresponds to low concurvity, second one corresponds to severe concurvity. Graphics are ordered along ω from left to right.

We have tested this method on a set of problems, in which complexity was very different.

Results of simulations allow us to conclude that lasso perform better than forward selection in the most complex cases. Whereas computing time of lasso does not depend on the number of relevant variables, computing time of forward variable selection increases considerably with the number of variables that actually generate the model. Performance of modified lasso can be got better by improving the complexity parameter selection method.

References

1. S. Bakin. *Adaptive Regression and Model Selection in Data Mining Problems.* PhD thesis, School of Mathematical Sciences, The Australian National University, Canberra, 1999.
2. L. Breiman. Heuristics of instability and stabilization in model selection. *The Annals of Statistics*, 24(6):2350–2383, 1996.
3. B. A. Brumback, D. Ruppert, and M. P. Wand. Comment on "Variable selection and function estimation in additive nonparametric regression using a data–based prior" by T. S. Shively, R. Khon and S. Wood. *Journal of the American Statistical Association*, 94(447):794–797, 1999.
4. E. Cantoni and T. J. Hastie. Degrees of freedom tests for smoothing splines. *Biometrika*, 89:251–263, 2002.
5. Z. Chen. Fitting multivariate regression functions by interaction spline models. *J. R. Statist. Soc. B*, 55(2):473–491, 1993.

6. Y. Grandvalet. Least absolute shrinkage is equivalent to quadratic penalization. In L. Niklasson, M. Bodén, and T. Ziemske, editors, *ICANN'98*, volume 1 of *Perspectives in Neural Computing*, pages 201–206. Springer, 1998.
7. Y. Grandvalet and S. Canu. Outcomes of the equivalence of adaptive ridge with least absolute shrinkage. In M.S. Kearns, S.A. Solla, and D.A. Cohn, editors, *Advances in Neural Information Processing Systems 11*, pages 445–451. MIT Press, 1998.
8. C. Gu and G. Wahba. Minimizing gcv/gml scores with multiple smoothing parameters via the newton method. *SIAM J. Sci. Statist. Comput.*, 12:383–398, 1991.
9. I. Guyon and A. Elisseeff. An introduction to variable and feature selection. *Journal of Machine Learning Research, Special Issue on Variable/Feature Selection*, 3:1157–1182, 2003.
10. T. J. Hastie and R. J. Tibshirani. *Generalized Additive Models*, volume 43 of *Monographs on Statistics and Applied Probability.* Chapman & Hall, 1990.
11. J. D. Opsomer and D. Ruppert. A fully automated bandwidth selection method for fitting additive models. *J. Multivariate Analysis*, 73:166–179, 1998.
12. M.R. Osborne, B. Presnell, and B.A. Turlach. On the lasso and its dual. *Journal of Computational and Graphical Statistics*, 9(2):319–337, 2000.
13. T. S. Shively, R. Khon, and S. Wood. Variable selection and function estimation in additive nonparametric regression using a data–based prior. *Journal of the American Statistical Association*, 94(447):777–806, 1999.
14. R. J. Tibshirani. Regression shrinkage and selection via the lasso. *Journal of the Royal Statistical Society, B*, 58(1):267–288, 1995.
15. R. J. Tibshirani, and K. Knight. The covariance inflation criterion for adaptive model selection. *Journal of the Royal Statistical Society, B*, 61(3):529–546, 1999.
16. S. N. Wood. Modelling and smoothing parameter estimation with multiple quadratic penalties. *J. R. Statist. Soc. B*, 62(2):413–428, 2000.
17. J. Ye. On measuring and correcting the effects of data mining and model selection. *Journal of the American Statistical Association*, 93:120–131, 1998.

Automated Detection of Influenza Epidemics with Hidden Markov Models

Toni M. Rath[1], Maximo Carreras[2], and Paola Sebastiani[2]

[1] Department of Computer Science
University of Massachusetts at Amherst, MA 01003 USA
trath@cs.umass.edu
[2] Department of Mathematics and Statistics
University of Massachusetts at Amherst, MA 01003 USA
{carreras, sebas}@math.umass.edu

Abstract. We present a method for automated detection of influenza epidemics. The method uses Hidden Markov Models with an Exponential-Gaussian mixture to characterize the non-epidemic and epidemic dynamics in a time series of influenza-like illness incidence rates. Our evaluation on real data shows a reduction in the number of false detections compared to previous approaches and increased robustness to variations in the data.

1 Introduction

Influenza is a contagious disease that affects between 10% and 20% of U.S. residents every year. Although most people regard the flu as a seasonal annoyance, influenza can lead to serious complications such as bacterial pneumonia and, for the elderly or immunocompromised patients, influenza can be fatal — it is estimated that an average of about 20,000 people die from influenza in the U.S. every year, and 114,000 per year have to be admitted to the hospital as a result of influenza. Because early detection of influenza epidemics could have a serious impact on the number of lives saved, great emphasis has been placed on influenza surveillance by monitoring indicators of the spread of epidemics.

The current approach to influenza surveillance is based on Serfling's method [1]. The method uses cyclic regression to model the weekly proportion of deaths from pneumonia and influenza and to define an epidemic threshold that is adjusted for seasonal effects. Because of delays in detection caused by the nature of the data, several authors have proposed to use other indicators of influenza epidemics, such as the proportion of patient visits for influenza like illness (ILI) [2,3] that should provide earlier indications of an influenza epidemic. Although applied to different data, the detection is still based on cyclic regression and suffers from several shortcomings, such as the need for non-epidemic data to model the baseline distribution, and the fact that observations are treated as independent and identically distributed. Although the second issue can be overcome by proper modeling of the time series data and examples are discussed in [4,5], the

M.R. Berthold et al. (Eds.): IDA 2003, LNCS 2810, pp. 521–532, 2003.

first issue is a fundamental obstacle toward the development of an automated surveillance system for influenza.

Recent work in [6] suggested the use of Hidden Markov Models [7] to segment time series of influenza indicators into epidemic and non-epidemic phases. There are two advantages in this approach. The first advantage is that the method can be applied to historical data without the need for distinguishing between epidemic and non-epidemic periods in the data, thus opening the way toward an automated surveillance system for influenza. The second advantage is that the observations are supposed to be independent given knowledge about the epidemic, whereas Serfling's method assumes marginal independence of the data.

For detection of influenza epidemics, the authors in [6] suggest using a mixture of Gaussian distributions to model the ILI rates and demonstrate the detection accuracy using a data set of historical data collected between January 1985 and December 1996 from a sentinel network of 500 general practitioners in France. We analyze the same data set and show that better detection accuracy can be achieved by modeling the data using a mixture of Exponential and Gaussian distributions. The change of the underlying distributions removes the need for explicit modeling of trend and seasonal effects that can introduce systematic bias in the detection accuracy.

The remainder of this paper is structured as follows. The next section describes the traditional approach to detection of influenza epidemics based on Serfling's method. Section 3 reviews the basic concepts of Hidden Markov models. In section 4 we describe the approach proposed by [6] and introduce our model for epidemic detection. The evaluation, which compares these two methods, is described in Section 5. Conclusions and suggestions for further work are in Section 6.

2 Serfling's Method for Epidemic Detection

The current approach to influenza surveillance implemented by the Centers for Disease Control in the U.S. is based on Serfling's method [1] and the objective is to determine an epidemic threshold. The method models the weekly number of deaths due to pneumonia and influenza by the cyclic regression equation

$$y_t = \mu_0 + \theta t + \alpha \sin(2\pi t/52) + \beta \cos(2\pi t/52) + \epsilon_t \tag{1}$$

where y_t is the number of susceptible to deaths from pneumonia and influenza in week t, when there is no epidemic. The parameter μ is the baseline weekly number of deaths, without seasonal and secular trends, and ϵ_t is noise which is assumed to have mean 0 and variance σ^2. The component θt describes secular trend, and the sine-wave component $\alpha \sin(2\pi t/52) + \beta \cos(2\pi t/52)$ models the annual recurrence of influenza epidemics. Assuming that the errors are uncorrelated and normally distributed, the standard least squares method is used to estimate the model parameters $\mu, \theta, \alpha, \beta, \sigma^2$ from non-epidemic data, and to compute confidence bounds about the predicted values. The predicted values are given by

$$\hat{y}_t = \hat{\mu}_0 + \hat{\theta}t + \hat{\alpha}\sin(2\pi t/52) + \hat{\beta}\cos(2\pi t/52)$$

The confidence bounds are then computed as

$$\hat{y}_t \pm t_{\alpha/2} SE(\hat{y}_t),$$

where $SE(\hat{y}_t)$ is the estimated standard error of the prediction, and $t_{\alpha/2}$ is the $(1-\alpha/2)$ percentile of a Student's t distribution. The confidence bounds are used to define a time varying epidemic threshold that is adjusted for trend and seasonal effects.

As an example, Figure 1 describes the epidemic threshold computed by using Serfling's method on the proportion of deaths for pneumonia and influenza recorded in the USA from 122 sentinel cities between January 1998 and May 2002. The Figure is reproduced from the web site of the Centers for Disease Control and Prevention (CDC).

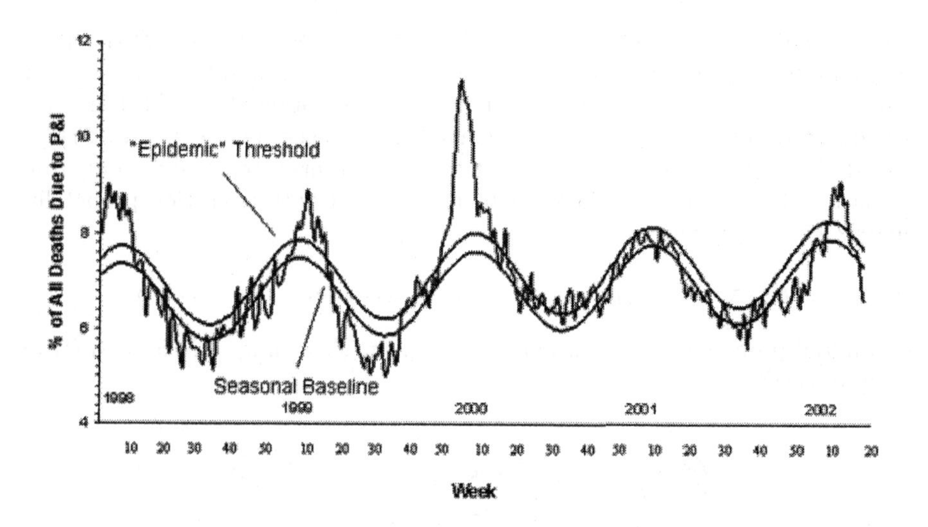

Fig. 1. Weekly proportion of deaths from pneumonia and influenza in 122 sentinel cities in the USA between January 1998 and May 2002 after normalization with the number of deaths for all causes. Source: CDC, Morbidity and Mortality Weekly Report (http://wonder.cdc.gov/mmwr/mmwrmort.asp).

Because of delays in detection due to the nature of the data, Serfling's method has been adapted to monitor either the proportion of influenza like illness or hospital visit data for influenza [2,3]. Although applied to data that should provide earlier indications of epidemics, the detection is still based on cyclic regression and suffers of two shortcomings: the need for non-epidemic data to model the baseline distribution, and the fact that observations are treated as independent and identically distributed. The need for non-epidemic data is a fundamental obstacle toward the development of an automated surveillance system for influenza and it is overcome by the use of Hidden Markov Models to segment time series of influenza indicators into epidemic and non-epidemic phases. This approach was introduced in [6] and is reviewed in the next section.

3 Hidden Markov Models

Hidden Markov Models (HMMs) are a convenient statistical tool for explaining serial dependency in data. They can be described by a sequence of pairs of random variables (Y_t, S_t), $t = 1, ..., n$, satisfying the following conditional independence assumptions:

$$P(y_t | y_1, ..., y_{t-1}, s_1, ..., s_t) = P(y_t | s_t) \tag{2}$$

$$P(s_t | y_1, ..., y_{t-1}, s_1, ..., s_{t-1}) = P(s_t | s_{t-1}), \tag{3}$$

where y_t and s_t are, respectively, the realized values of Y_t and S_t and $P(\cdot)$ denotes a generic probability function. The variable S_t is discrete and takes on one of m values $1, \ldots, m$. Figure 2 shows a graphical representation of an HMM in which the directed arrows define the conditional dependencies. The sequence $\{S_t\}_{t=1}^n$, called the state sequence, is assumed to be unobserved or *hidden* and, from Equation (3), it follows a Markov Chain of order 1 with transition probability matrix $P = (p_{ij})$, where

$$p_{ij} = P(S_t = j | S_{t-1} = i) \quad i, j = 1, \ldots, m; \quad t = 2, \ldots, n$$

and initial probability distribution $\pi = (\pi_1, ..., \pi_m)'$, with $\pi_i = P(S_1 = i)$ for an m-state HMM.

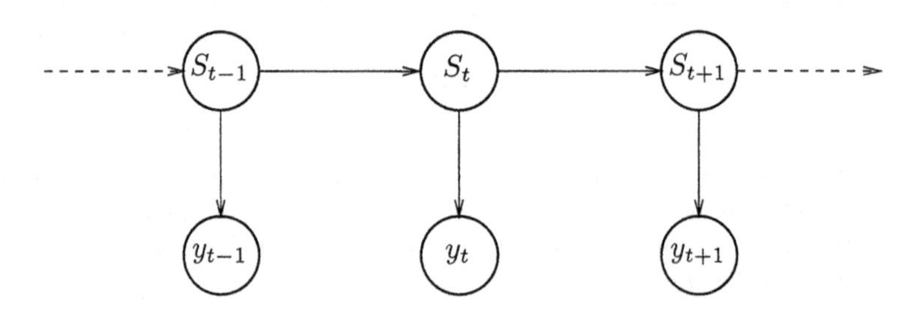

Fig. 2. Illustration of the conditional dependencies in an HMM model.

The conditional distribution of Y_t given $S_t = i$ has a parametric form $f_{it}(y_t; \theta_i)$, where θ_i is a vector of unknown parameters. The notation f_{it} suggests that the conditional density could change with the state as well as with the time. An example of the former is the model in this work: we describe the non-epidemic rates with an Exponential distribution and the epidemic rates with a Gaussian distribution. An example of the latter is the trend and seasonality model for the distribution means in [6].

When applying the model to the data, there are two main objectives:

1. Learning the model form the data; that is, estimating the model parameters from the data. This is achieved by maximum likelihood techniques using the EM-algorithm, known in the HMM literature as the Baum-Welch algorithm [8].
2. Decoding the most likely sequence of hidden states that generated the data. This is done by means of the Viterbi algorithm [9].

The EM-algorithm is the standard technique for computing maximum likelihood estimates of the model parameters from an incomplete data set, see [10]. The method works by alternating an E-step to an M-step until convergence is reached. In the E-step, missing data are replaced by their expected values given the current parameter estimates; in the M-step, maximum likelihood estimates of the parameters are computed using the data completed in the E-step.

4 Epidemic Detection Using Hidden Markov Models

Le Strat & Carrat [6] proposed to detect epidemic and non-epidemic phases of influenza by modeling ILI rates with HMMs using a mixture of Gaussian distributions. They focused on a 2-state model and parameterized the means using Serfling's approach, so that the rates of both the epidemic and non-epidemic phases are modeled as in Equation (1). There are two main shortcomings to this approach:

- The use of a Gaussian distribution for modeling the non-epidemic rates is inappropriate, because it assigns non-zero probability to the occurrence of negative incidence rates.
- The model for the means, although rich in parameters, is quite rigid; for instance, it does not allow for variation in the amplitude and frequency of the sinusoidal pattern. This lack of flexibility raises the question of how much data should be used in the model estimation. For example, modeling data from influenza mortality surveillance has shown that more accurate forecasts are based on a few years historical data, and the CDC influenza surveillance practice is to use five years historical data [11,12,13]. It is clear that using too little data will lead to an unreliable fit that would not extend well to future data. On the other hand, a large amount of data will yield bias and under/overestimation problems.

To address these problems, we propose to use a 2-state HMM, where non-epidemic rates are modeled with an Exponential distribution, and epidemic rates with a Gaussian distribution. Unlike Le Strat & Carrat's model, the means of the densities do not have any seasonal dependency. Non-epidemic rates are always positive and approximately exponentially distributed. As an example, Figure 3 shows a histogram of non-epidemic rates, using a visually determined threshold. The rates are skewed to the right, and follow, approximately, an exponential distribution.

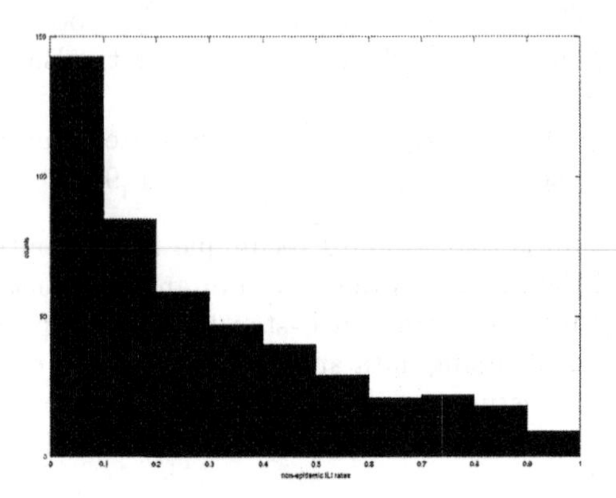

Fig. 3. Histogram of non-epidemic ILI incidence rates using an empirical threshold.

Our model is simple and describes the data well, without the need for modeling seasonality and trend effects. It allows us to capture out-of-season, non-epidemic peaks, which tend to cause false detections in Le Strat & Carrat's periodic model.

5 Evaluation

Le Strat & Carrat [6] tested their model (referred to as *periodic model* in the following) for epidemic detection on a time series of ILI incidence rates in France from 1985 to 1996, collected by a network of sentinel physicians. For reasons of comparability, we performed the evaluation of our model (referred to as *Exponential*) on the same time series. Both models were implemented in *Matlab*[1] and the parameters estimated from the time series using the Baum-Welch/EM estimation algorithm [8]. The updated parameters are then passed to the Viterbi algorithm [9], which extracts the most likely state sequence that generated the observed time series. Table 1 summarizes the classification induced by the two models, and Figure 4 shows the ILI time series as classified by the two models, with the epidemic periods shaded dark.

An ideal classification separates the data into two parts: the non-epidemic, seasonal background pattern with low incidence rates and the epidemic periods with the pronounced peaks. Visual inspection shows that the Exponential model

[1] Source code available from http://ciir.cs.umass.edu/~trath/prj/epdet.html

Table 1. Summary of the classification of the two models

Model	Number of epidemic weeks	Number of non-epidemic weeks
Periodic	159	465
Exponential	124	500

proposed here provides a more cautious classification than the periodic model proposed in [6] and it classifies a smaller number of weeks as epidemic weeks. Furthermore, the periodic model (Figure 4(a)) often tends to detect epidemic periods too early and non-epidemic periods too late, and such mistakes are very minimal with the Exponential model (Figure 4(b)): on average, our approach classified 10.40 weeks per year as epidemic, which is more realistic than 13.31 (over 3 months) obtained with the periodic model. For example, the CDC reports that, in the last 6 years, only during the influenza season 1996-1997 the epidemic threshold was exceeded for more than 13 weeks, and the average number of epidemic weeks per year was about 10, see for example [14]. This number is consistent with our results.

Another advantage of the Exponential model is its classification robustness: in the periodic model, small peaks occurring outside the influenza season trigger classifications (e.g. early classification of the second peak and one extra classification between the last two epidemic seasons in the periodic model). The Exponential model performs much better in these situations.

We attribute these differences in classification performance to the rigid modeling of seasonality in the periodic model, which does not adapt to variations in the data. The effects can be seen in Figure 5, which shows magnifications of the classification results with the periodic model. In Figure 5(a), the beginning of the time series is magnified, where the non-epidemic background pattern varies in a larger range than in the rest of the series. Since the amplitude of the periodic, non-epidemic pattern is a constant that is estimated from the whole time series, the amplitude is underestimated for this part of the data. The consequence is that some non-epidemic incidence rates appear high compared to the estimated seasonal mean, and the corresponding weeks are therefore classified as epidemic. Furthermore, the seasonal model is unable to adapt to bias in the data: Figure 5(b) shows a part of the time series, where a small peak that occurs outside of the influenza season causes a false detection (marked with an arrow). This figure also shows a situation where the non-epidemic seasonal fit falls below zero, which is unrealistic.

Using an Exponential distribution to model non-epidemic incidence rates can lessen these problems. The Exponential distribution is a more adequate model, since it does not allow for negative incidence rates (as in the periodic model) and it is able to explain small out-of-season peaks in the non-epidemic portion of the data. This model prevents the false positive detections that occur in regions where the periodic model does not adequately adjust to the seasonal incidence rate pattern.

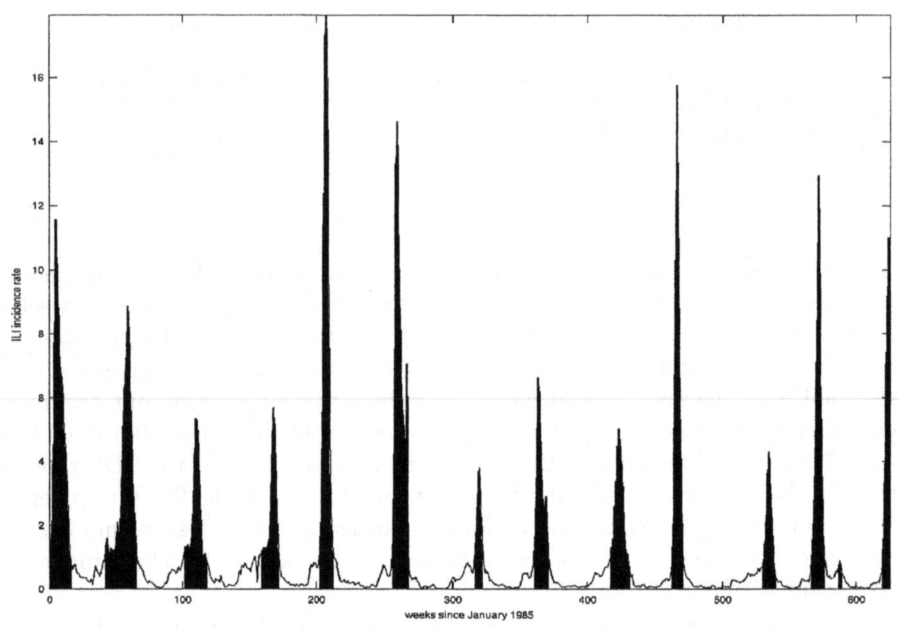

(a) *normal/normal model with periodic means by Le Strat & Carrat,*

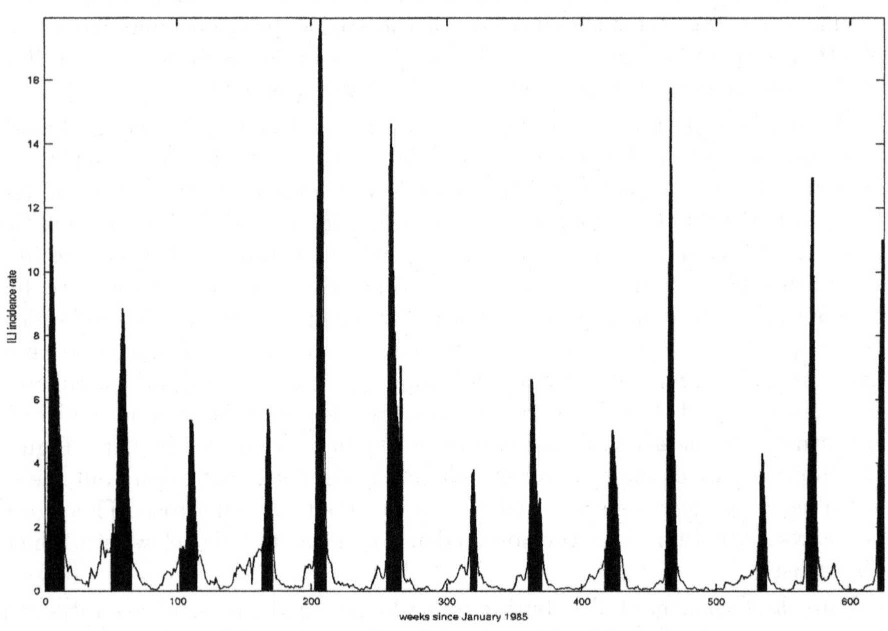

(b) *exp/normal model proposed in this work*

Fig. 4. Time series of weekly ILI rates in France from 1985 to 1996. Dark shaded areas indicate periods that were identified as epidemic by the respective models.

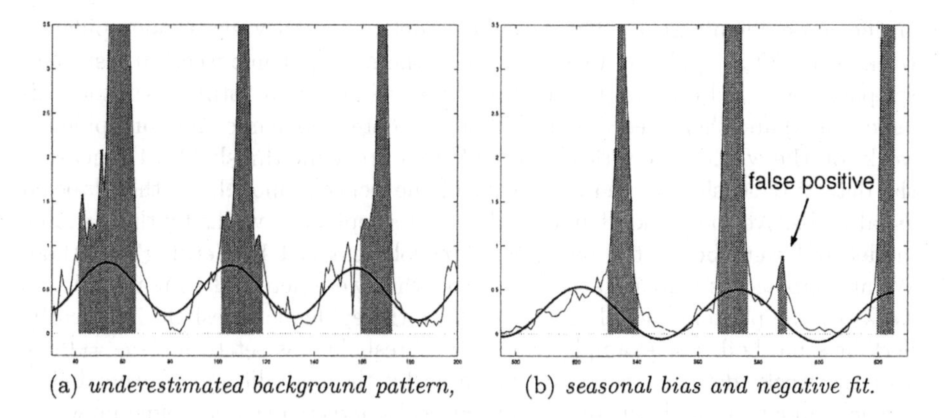

(a) *underestimated background pattern,* (b) *seasonal bias and negative fit.*

Fig. 5. Magnifications of the classification results with the periodic model. The shaded regions indicate the detected epidemic seasons, the sinusoidal pattern represents the estimated mean of the normal distribution which models the non-epidemic seasons.

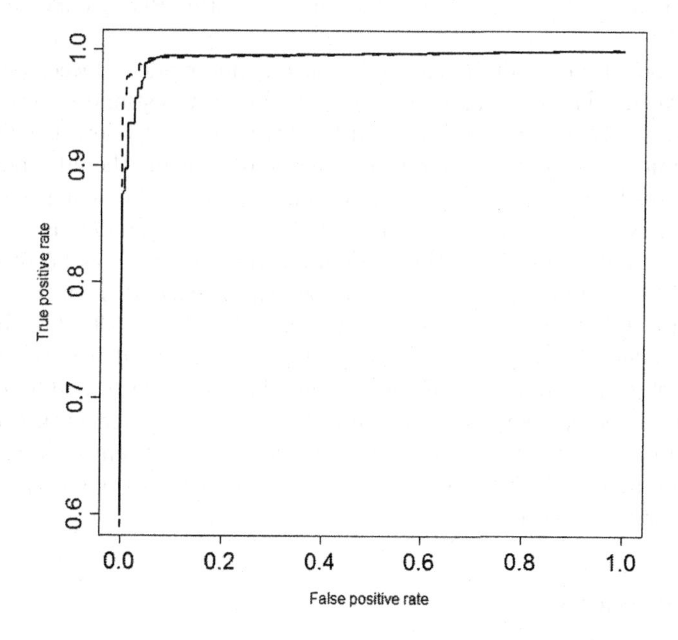

Fig. 6. Receiver Operating Characteristic (ROC) curve plotting the true positive rate against the false positive rate for the different epidemic thresholds. Solid line: ROC curve induced by the periodic model. Dashed line: ROC curve induced by the exponential model.

Because the original data were not labelled into epidemic and non-epidemic weeks, we further compared the classification determined by the two models with the classification that would be induced by an adjusted Serfling's method trained on the non-epidemic weeks. The periodic model labels 465 weeks as non-epidemic weeks and 159 as epidemic weeks. We used the non-epidemic weeks to estimate the parameters in Equation 1 (constrained to avoid negative estimates of the incidence rates) and then used the fitted model to detect epidemic and non epidemic weeks on the whole time series, using different epidemic thresholds. To measure the detection sensitivity and specificity of the periodic model, we then labeled as false positive the weeks that were detected as epidemic weeks by the periodic model and were below the epidemic threshold induced by Serfling's method. We also labeled as false negative the weeks that were detected as non-epidemic weeks by the periodic model and were above the epidemic threshold induced by Serfling's method. For example, when the threshold was set to have $\alpha = 0.05$, 4 of the weeks detected as epidemic weeks by the periodic model were below the epidemic threshold, whereas 24 of the weeks detected as non-epidemic weeks were above the threshold, resulting in a proportion of false positive 4/159=0.02 and a proportion of false negative 24/465=0.0516. Varying the epidemic threshold with $\alpha > 0.5$ results in different false positive and false negative rates that are plotted in the Receiver Operating Characteristic (ROC) curve in Figure 6 (solid line).

The Exponential model labels 500 weeks as non-epidemic weeks and 124 as epidemic weeks. The non-epidemic weeks were used to estimate the parameters of Equation 1. As above, we then used the fitted model to detect epidemic and non-epidemic weeks on the whole time series and to define the false positive and negative rates for different epidemic thresholds. For example, when the threshold was set to have $\alpha = 0.05$, none of the weeks detected as epidemic weeks by the exponential model was below the epidemic threshold, whereas 25 of the weeks detected as non-epidemic weeks were above the threshold, resulting in a false positive rate 0/124=0 and a false negative rate 25/500=0.0500. The dashed line in Figure 6 depicts the tradeoff between false positive and negative rates and shows the slightly superiority of the exponential model with a ROC curve that is closer to the y-axis compared to the ROC curve associated with the periodic model. Figure 7 plots the true positive rates for the Exponential (y-axis) versus the periodic model (x-axis) for $\alpha > 0.5$ and shows the larger true positive rate of the Exponential model.

6 Conclusions

We have proposed a new model for the detection of influenza epidemics in time series of ILI incidence rates using Hidden Markov Models. Unlike previous approaches, we do not model explicitly the periodicity in the data to detect epidemics and non-epidemic weeks. We use a hidden variable to describe the unobservable epidemic status and, conditional on the absence of epidemics, we use an Exponential distribution to model the incidence rates of ILI. ILI incidence rates observed during influenza epidemics are modeled as Gaussian distributions. The exponential distribution provides a better adjustment to the variations in the data and removes the need for constrained parameter estimation that would

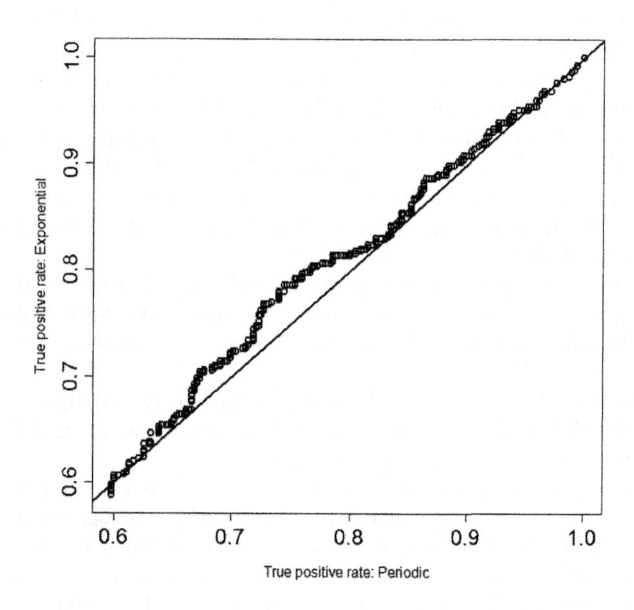

Fig. 7. True positive rates of the Exponential (y-axis) versus the periodic model (x-axis).

be needed with Gaussian models. An evaluation of the proposed model demonstrated the improved performance over previously reported techniques.

Acknowledgments. This work was supported by the Alfred P. Sloan Foundation (Grant 2002-12-1) and by the National Science Foundation (Grant IIS-0113496).

References

1. R. E. Serfling. Methods for current statistical analysis of excess pneumonia-influenza deaths. *Public Health Reports*, 78:494–506, 1963.
2. L. Simenson, K. Fukuda, L. B. Schonberg, and N. J. Cox. The impact of influenza epidemics on hospitalizations. *The Journal of Infectious Diseases*, 181:831–837, 2000.
3. F. C. Tsui, M. M. Wagner, V. Dato, and C. C. H. Chang. Value ICD-9-Coded chief complaints for detection of epidemics. In *Proceedings of the Annual AMIA Fall Symposium*, 2001.
4. L. Wang, P. Sebastiani, J. Tsimikas, and K. Mandl. Automated surveillance for influenza pandemics. Technical report, Department of Mathematics and Statistics. University of Massachusetts at Amherst., 2003. Submitted.

5. P. Sebastiani and K. Mandl. Biosurveillance and outbreak detection using syndromic data. In *Data Mining: Next Generation Challenges and Future Directions*. AAAI/MIT Press, 2003. In press.
6. Y. LeStrat and F. Carrat. Monitoring epidemiologic surveillance data using hidden Markov models. *Statistics in Medicine*, 18 (24):3463–78, 1999.
7. L. R. Rabiner. A tutorial on Hidden Markov Models and selected applications in speech recognition. *Proceedings of the IEEE*, 77:257–285, 1989.
8. L. E. Baum, T. Petrie, G. Soules, and N. Weiss. A maximization technique occurring in the statistical analysis of probabilistic functions of Markov chains. *Annals of Mathematical Statistics*, 41:164–171, 1970.
9. G. D. Forney. The Viterbi algorithm. *Proceedings of the IEEE*, 61:268–278, 1973.
10. A. P. Dempster, D. Laird, and D. Rubin. Maximum likelihood from incomplete data via the EM algorithm. *Journal of the Royal Statistical Society, B*, 39:1–38, 1977.
11. C. P. Farrington, N. J. Andrews, A. D. Beale, and M. A. Catchpole. A statistical algorithm for the early detection of outbreaks of infectious disease. *Journal of the Royal Statistical Society A*, 159:547–63, 1996.
12. L. C. Hutwagner, E. K. Maloney, N. H. Bean, L. Slutsker, and S. M. Martin. Using laboratory-based surveillance data for prevention: An algorithm for detecting salmonella outbreaks. *Emerging Infectious Diseases*, 3:395–400, 1997.
13. L. Stern and D. Lightfoot. Automated outbreak detection: a quantitative retrospective analysis. *Epidemiology and Infection*, 122:103–110, 1999.
14. Centers for Disease Control. Morbidity and mortality tables, 2002. http://wonder.cdc.gov/mmwr/mmwrmort.asp.

Guided Incremental Construction of Belief Networks

Charles A. Sutton, Brendan Burns, Clayton Morrison, and Paul R. Cohen

Department of Computer Science,
University of Massachusetts, Amherst, MA 01002
{casutton, bburns, morrison, cohen}@cs.umass.edu

Abstract. Because uncertain reasoning is often intractable, it is hard to reason with a large amount of knowledge. One solution to this problem is to specify a set of possible models, some simple and some complex, and choose which to use based on the problem. We present an architecture for interpreting temporal data, called AIID, that incrementally constructs belief networks based on data that arrives asynchronously. It synthesizes the opportunistic control of the blackboard architecture with recent work on constructing belief networks from fragments. We have implemented this architecture in the domain of military analysis.

1 Introduction

Reasoning problems in many domains, such as intelligence analysis and common-sense reasoning, seem to require both large amounts of knowledge and reasoning under uncertainty. For example, a military commander may try to intuit an enemy's intended attack based on knowledge of the likely terrain, weather, enemy strength, and other factors, as well as actual sightings of enemy forces. Or, finding a forgotten book can make use of (highly uncertain!) memories of where it was last seen, places where it might have been used, and people who might have wanted to borrow it, as well as data acquired by looking around the office. In both cases, the reasoner is uncertain about features of the world, and can marshal both observations and large amounts of general and domain-specific knowledge to resolve the uncertainty.

Belief networks are a popular method for reasoning with uncertainty, but traditionally they require specifying a complete model of the domain. Such expert systems assume that in advance a designer can winnow the set of possible influences, which can be large, down to the few that are relevant.

This approach has two serious limitations. First, in intelligence domains, part of the uncertainty is knowing how many influences there *are*. We may not know exactly how many units are on a battlefield, for example. It is difficult to design a complete probabilistic model for a domain without knowing how large the domain is.

Second, in common-sense reasoning, any observation (such as a car not starting) has many possible explanations, ranging from plausible (out of gas) to implausible (wrong car) to downright ridiculous (starter motor stolen by aliens to

M.R. Berthold et al. (Eds.): IDA 2003, LNCS 2810, pp. 533–543, 2003.

repair spaceship). To reason with a lot of general knowledge—imagine a probabilistic knowledge base as large as Cyc! [11]—it helps to be able to work with a plausible subset. But if this subset is selected in advance, we cannot handle situations where implausible rules suddenly become plausible, for example, if the car contains unfamiliar belongings or small grey men with large eyes. Knowledge-based systems could be both more robust and generally applicable if they contained implausible knowledge, but were able to ignore it until new observations called for its use.

The *blackboard architecture* [5,13] provides a natural framework for incrementally interpreting large amounts of data. The architecture was designed for problems that require many heterogeneous types of knowledge. Blackboard systems have a set of reasoning modules, called *knowledge sources* (KSs), that develop hypotheses using a global data structure called a *blackboard.*

In this paper, we present AIID, an architecture for incrementally interpreting temporal data. AIID is a blackboard architecture in which hypotheses on the blackboard are nodes in a belief network. It extends traditional blackboard techniques by using a principled method for representing uncertainty, and it extends traditional belief network techniques by incrementally building a model as incoming data warrant.

The principal type of knowledge source in AIID is a *network fragment*, a reusable mini-belief network from which larger networks are constructed. Using fragments, AIID can construct a network incrementally in response to data that arrive asynchronously. Network fragments can be reused across domains. Previous work [6,9,10] has discussed methods for constructing belief networks using network fragments. AIID extends this work by incorporating procedural knowledge sources that programmatically alter the network, and by focusing on how the growth of the network will be controlled.

In sections 2 and 3, we summarize the blackboard architecture and the existing literature on incrementally constructing belief networks. In section 4, we describe AIID, including its blackboard, knowledge sources, and control. In section 5, we illustrate AIID's operation on an example. We conclude by discussing our prototype implementation, in the domain of military analysis.

2 Blackboard Systems

Blackboard systems are knowledge-based problem solvers that work through the collaboration of independent reasoning modules. They were developed in the 1970s and originally applied to signal-processing tasks. The first, HEARSAY-II [5], was used for speech recognition, employing acoustic, lexical, syntactic, and semantic knowledge. Other systems were applied to problems as diverse as interpretation of sonar data, protein folding, and robot control.

Blackboard systems have three main components: the blackboard itself, knowledge sources (KSs), and control. The blackboard is a global data structure that contains hypotheses or partial solutions to the problem. The blackboard is typically organized into sections by levels of abstraction. For example, HEARSAY-II had different levels for phrases, words, syllables, and so forth.

Knowledge sources are small programs that write new hypotheses to the blackboard. Ideally, knowledge sources interact only by posting to the blackboard. Different KSs use different types of knowledge: for example, one might use a syntactic grammar to generate words that are likely to occur next, while another might detect phonemes directly from the acoustic signal. While no single knowledge source could solve the problem, working together they can, and the blackboard model provides an organized means of collaboration.

One focus of the architecture is control [4]. The operation of a blackboard system can be seen as search for hypotheses that explain the data at each level of abstraction, using the KSs as operators. Rather than search bottom-up (i.e., from the data level to the most abstract level) or top-down, blackboard systems can search opportunistically, dynamically rating KSs based on the current data and on the partial solutions that exist so far.

Because hypotheses posted to the blackboard are often uncertain, systems have needed to represent that uncertainty. Heuristic methods have generally been used [4]: for example, HEARSAY-II used a numerical confidence score that ranged from 1 to 100. We are not aware of a blackboard system that represents the links between hypotheses using conditional probabilities.

3 Belief Networks

A belief network is a directed graphical model that represents a probability distribution by a graph that encodes conditional independence relationships, and a set of conditional distributions that give a local distribution for variables based on their parents in the graph. Belief networks are described formally in Russell and Norvig [16].

Belief networks that describe several similar objects often have repetitive structure. For example, in the military domain, every unit has attributes like UNIT-TYPE (e.g., tanks, infantry, artillery) and DIRECT-FIRE-RADIUS. These attributes have relationships that do not depend on the particular unit: for example, tanks can shoot farther than infantry. If we simply have nodes called UNIT-TYPE-FOR-UNIT1, DIRECT-FIRE-RADIUS-FOR-UNIT1, etc., then the humans constructing the network need to specify separate, identical CPTs for each unit, which is impractical because there could be many units, and we do not know in advance how many.

Several authors [6,9,10] have addressed this problem by breaking up large belief networks into smaller subnetworks. Subnetworks have designated input nodes—which have no conditional distribution, requiring that their distribution be specified in a different subnetwork—and resident nodes, which do have CPTs. A standard belief network can be created from subnetworks by unifying the input nodes of one subnetwork with the resident nodes of another. Repetitive structure can be specified once in a subnetwork and instantiated multiple times to exploit redundancies in the domain.

Object-oriented Bayesian networks (OOBNs) [9,15] employ strong encapsulation between subnetworks. Each subnetwork defines a set of output variables, and combinations between subnetworks are made only by connecting the output variables of one subnetwork to the input variables of another. Each subnetwork can be seen as a single cluster node in a higher-level belief network, so that an OOBN defines a single probability distribution over its variables. An attractive feature of OOBNs is that their hierarchical structure can be exploited to guide clustering for the junction-tree algorithm, making inference more efficient. Since the subnetworks are connected by a knowledge engineer, rather than automatically, OOBNs are not a technique for incrementally building models based on incoming evidence.

Network fragments [10] are another approach to constructing modular subnetworks. Unlike OOBNs, nodes can be resident in more than one fragment, so they designate influence combination methods for combining distributions from multiple fragments. So network fragments can combine in more unexpected ways than in OOBNs, which precludes specialized inference algorithms, but can be more flexible for specifying complicated belief networks.

4 Architecture

4.1 The Blackboard

The blackboard of AIID represents the system's current beliefs about the domain. The blackboard contains a possibly disconnected belief network that includes previous observations, background knowledge, and hypotheses about the data. In the military domain, the blackboard contains nodes that include sightings and hypothesized locations of enemy units, locations of key terrain, and hypotheses about the enemy's tactics and strategy. A sample blackboard is shown in figure 1.

As in the subnetwork literature, we use a first-order extension to belief networks to represent multiple similar entities more conveniently, analogous to the extension of propositional logic to predicate logic. Instead of naming the random variables by a single atom, e.g. UNIT-MASS, each node has a *node-type*, for example, UNIT-MASS, and a set of *arguments*, for example, "Tank Regiment 1." Logic variables can be used as arguments in KSs to describe a relationship that does not depend on the particular argument values. The combination of a node-type and arguments uniquely specifies a node on the blackboard.

Information on the blackboard can occur on different temporal scales. For example, we can represent a short meeting between two people as a punctual event, while an activity like "Planning-Attack" takes an extended amount of time. We handle these scales using two temporal representations: a tick-based representation, and an interval representation. At lower levels of the blackboard, where we are considering things like meetings and current locations, each network node is indexed by the time it occurs, and the entire network is a Dynamic Bayesian Network (DBN) [12]. At higher levels of the blackboard, which correspond to

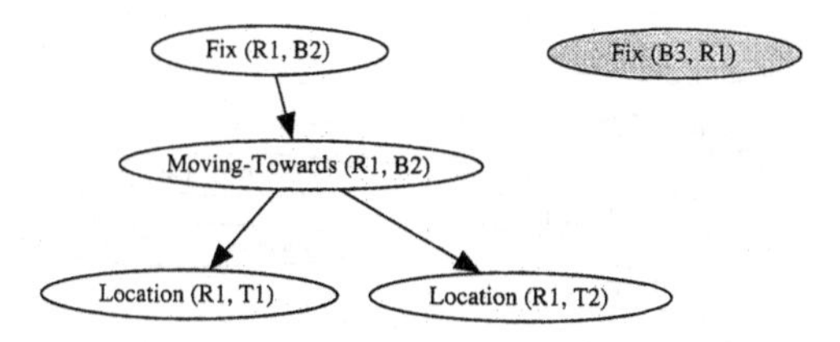

Fig. 1. A sample blackboard in the military analysis domain

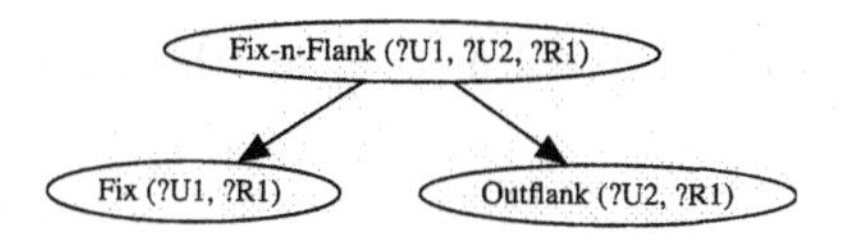

Fig. 2. A sample knowledge fragment for the military analysis domain

long-term actions and intentions, we represent events by the interval in which they occur. Each event has a start-time and an end-time that are explicit nodes in the network. These two representations are integrated as described elsewhere [3].

4.2 Knowledge Sources

A knowledge source is a procedure that modifies the blackboard. Knowledge sources can post new nodes to the blackboard, add edges, alter CPTs, and remove nodes. Every KS has three components, which can be arbitrary procedures: a confidence, a precondition, and an action. The confidence returns a number that indicates how intrinsically useful the KS is. The precondition is run when the blackboard changes and returns true if the KS is applicable. The action is the procedure that actually modifies the blackboard.

As in blackboard systems, KS actions can be full-fledged programs. For example, a KS might use arbitrary heuristics to post simplifying assumptions to make reasoning more tractable. In the military domain, for example, our implementation uses a grouping KS that treats several enemy units as a group if they seem sufficiently close.

Another type of KS cleans up nodes that accumulate from old time steps. Old nodes can slow down inference without greatly affecting current beliefs. Cleanup KSs can remove nodes that are either older than some cutoff or that do not cause a large drop in information about certain nodes of interest. We define the information value of a node in section 4.3.

The most common type of KS is a *network fragment*, which is a belief network that represents a small fragment of knowledge. Example fragments are shown in figures 2 and 6. A node in the fragment *matches* a node on the blackboard when the two nodes have the same type, and their argument lists unify. (Recall that nodes in a KS can have logic variables in their argument lists, and the arguments of a node are distinct from its set of possible outcomes.) By default,

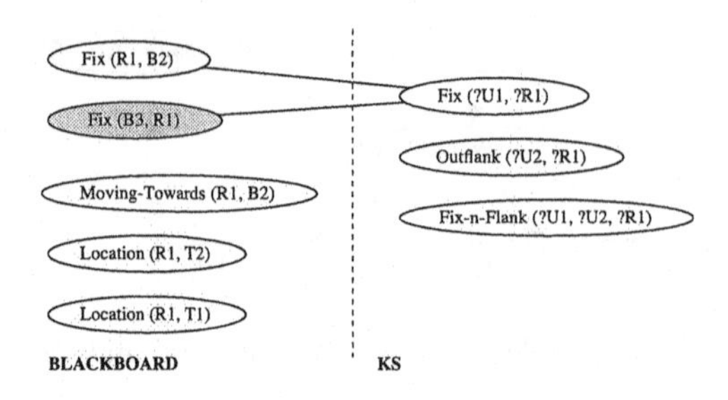

Fig. 3. The bipartite matching problem from matching KS 2 to the blackboard in figure 1

the precondition for a fragment KS is that at least one of the fragment nodes has a match on the blackboard; however, the KS designer can designate certain nodes that must be matched, or write an arbitrary precondition.

Posting Network Fragments. Fragments are posted to the blackboard by a process that resembles unification. A fragment can be posted to the blackboard if three conditions hold. First, each of the fragment nodes must match a node on the blackboard; a new node can be created on the blackboard if necessary. Second, a single unifying assignment must unify the argument lists of all the fragment nodes with their corresponding blackboard nodes—this merely ensures that a logic variable like ?U refers to the same thing throughout the fragment. Third, no two fragment nodes can match the same blackboard node.

We can think of fragment matching as a bipartite matching problem, as shown in figure 3. On the left side of the bipartite graph are all the blackboard nodes; on the right are all the fragment nodes. A blackboard node and a fragment node are linked if they have the same node type. Now, any bipartite matching in this graph describes a way the fragment could be posted to a blackboard. A fragment node unifies with its neighbor in the matching. If it has no neighbor, a new node is posted to the blackboard.

Once a fragment has been matched to the blackboard, it can be posted. An example of a fragment posting is given in figure 4. A fragment is posted to the blackboard in three steps. First, new nodes are posted if they are required by the match. Second, for every pair of fragment nodes that are linked, a corresponding edge is added to the blackboard. Now the nodes on the blackboard have both their original parents $\mathbf{V}_{BB}$ and the new parents that were specified by the fragment, $\mathbf{V}_F$. Third, since every node V in the fragment has both a conditional distribution in the fragment, $\mathrm{P}(V \mid \mathbf{V}_F)$, and a one on the blackboard, $\mathrm{P}(V \mid \mathbf{V}_{BB})$, these two distributions are combined to get $\mathrm{P}(V \mid \mathbf{V}_F, \mathbf{V}_{BB})$. Ways this can be done are given in the next section.

Influence Combination. Since network fragments are themselves belief networks, they specify a complete probabilistic model over their variables. But nodes

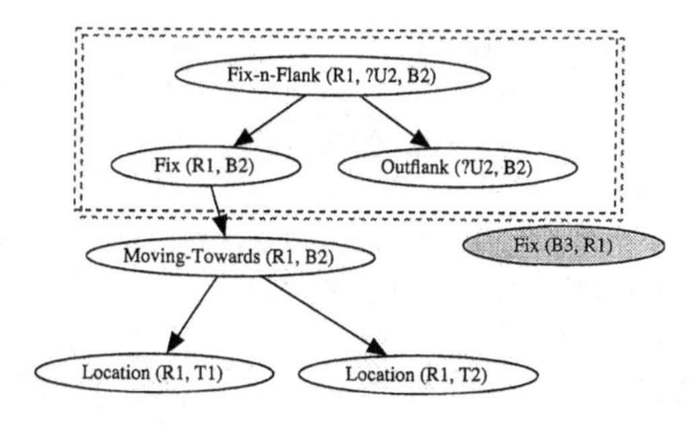

Fig. 4. The blackboard in figure 1 after KS 2 posts

on the blackboard already have a probability distribution. Suppose that some node V has parents $\mathbf{V}_F$ in the fragment and parents $\mathbf{V}_{BB}$ on the blackboard, so that we have probability distributions $\mathrm{P}(V \mid \mathbf{V}_F)$ and $\mathrm{P}(V \mid \mathbf{V}_{BB})$. When the KS posts, the parents of V will be $\{\mathbf{V}_F \cup \mathbf{V}_{BB}\}$, so we must merge these two models to get a CPT for $\mathrm{P}(V \mid \mathbf{V}_F, \mathbf{V}_{BB})$. For example, between figures 7 and 7, the distribution P(ALARM | BURGLAR) must be combined with the distribution P(ALARM | BASEBALL-BREAKS-WINDOW).

There are several ways to do this. Laskey and Mahoney [10] define several *influence combination methods* to combine conditional probability distributions, one of the principal types being parametric causal models like *noisy-or*. The noisy-or model [14,17] allows one to compactly specify a conditional distribution when the parents are independent, stochastic causes of the child. The knowledge engineer can specify which combination method should be used for a given node type.

4.3 Control

Many knowledge sources are applicable at any given time, but only a few can be selected to run. This is both because our implementation of AIID runs on a single processor, so only one KS can be run at a time, and because if too many KSs fire, the blackboard could become too large for probabilistic inference to be tractable. We describe three types of control regimes: a simple one based on KSs' confidence methods, an information-theoretic one based on nodes of interest to the user, and a Bayesian one based on the probability of the blackboard structure given the observed data.

First, KSs can be ordered by confidence. This provides a gross ordering among KSs, but it can be difficult to know in advance, or write a procedure that computes, how useful a KS will be.

Second, certain nodes of the blackboard have more interest to the user. For example, an intelligence analyst may want to know whether two people have communicated, or a military commander may want to know whether a certain attack is a feint. Let $\mathbf{V}$ be a set of these nodes of interest. We can choose to post the knowledge sources that provide the most *information* about $\mathbf{V}$, in the

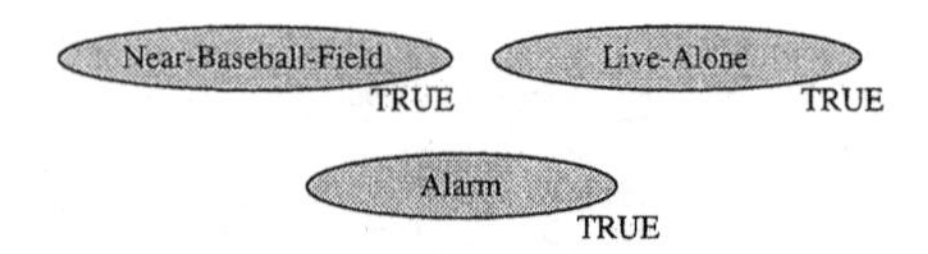

Fig. 5. Initial knowledge in the burglar-alarm example. Darkened nodes are observed, with the observed value printed underneath the node

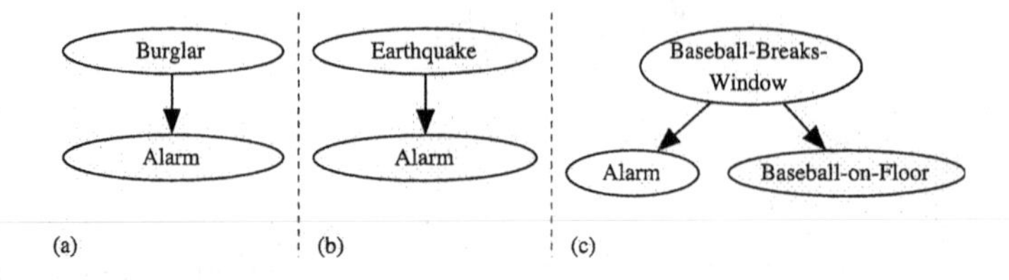

Fig. 6. Three knowledge sources for the burglar-alarm example

sense of reducing its Shannon entropy. The information gained from firing a KS K is given by $I(K) = H(\mathbf{V}) - H_K(\mathbf{V})$, where $H(\mathbf{V})$ is the entropy of $\mathbf{V}$ before K is posted, and $H_K(\mathbf{V})$ is the entropy afterward. We can compute this directly by temporarily posting K, computing the marginal distribution $\mathrm{P}(\mathbf{V})$, and calculating its entropy.

Since this is probably too expensive to use if many KSs are applicable, we can try to approximate this effect. We can get a cheap approximation by simply looking at the distance between where the KS will post and $\mathbf{V}$, that is, the length of the shortest undirected path between a node used by the KS and a member of $\mathbf{V}$. Then we prefer the KSs with the shortest distance, on the assumption that nodes that are closer to $\mathbf{V}$ have more influence on its distribution.

Third, previous work on structure learning in belief networks is applicable to control decisions. Bayesian structure learning is concerned with finding the structure for a belief network B_s which maximizes the posterior probability $\mathrm{P}(B_s \mid D, \xi)$ for some set of observed data D and an environment ξ. For making control decisions, we can select the KS whose application would maximize the posterior probability of the blackboard structure given the observations.

Although there is a closed-form solution for this probability, its calculation is computationally intractable [7]. As a result of this numerous approximation techniques for estimating this probability have been proposed. Heckerman [7] gives a good overview of these approximation techniques, most of which are applicable to the task of control, although some require multiple independent observations, which may not always be available.

5 Example

In this section we illustrate the action of the architecture by a diagnostic example inspired by one from Judea Pearl [16]. Suppose you are upstairs in your house and hear your burglar alarm go off. The alarm might have gone off for many

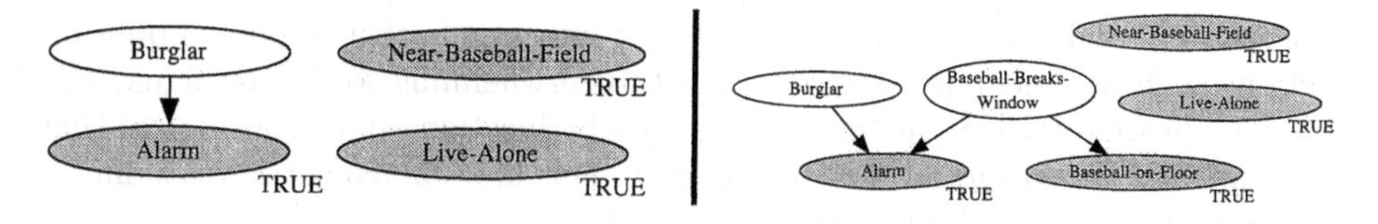

Fig. 7. At left, he blackboard after 6(a) has fired. At right, the blackboard after KS 6(c) has fired

reasons: an earthquake, a forgetful member of the household, a baseball breaking a window, an actual burglar, and many others. Of course you'll investigate, but as you get up to do this, you naturally wonder what you'll find downstairs.

Your initial knowledge might be represented in AIID as figure 5: you have just heard an alarm and you live alone near a baseball field. Since ALARM has just been posted, KSs that use that node are applicable. Some available KSs are listed in figure 6.

Now we have to choose which of those KSs should activate. Suppose we use the Bayesian control regime. In this case, ALARM is observed, so we choose the KS that maximizes its posterior probability. If we assume pessimistically that burglaries are more likely than baseball games, it may be that 6(a) results in the more probable structure, so assume that it is chosen. The result of posting KS 6(a) is shown in figure 7.

Now, suppose that you come downstairs and see a broken window and a baseball on the floor. With this information, KS 6(b) becomes very relevant, so suppose it fires. Now influence combination needs to be performed on the ALARM node. Since BURGLARY and BASEBALL are independent causes of the alarm, it is reasonable to use the noisy-or combination method. At this point P(BASEBALL) is likely to be high and P(BURGLAR) to be low (because of explaining away), so the blackboard now (figure 7) contains a good explanation of the observations.

6 Implementation

We have built a prototype of AIID in the domain of military analysis. We simulate military engagements at the battalion level (roughly a thousand troops), using the Capture the Flag simulator [1,2]. The simulator includes such effects as terrain, fog of war, artillery, combat aviation, and morale.

The data consist of reports about friendly and enemy units, for example, "Unit Red-1 has sound contact with a brigade-sized unit in the east." The prototype does not address the problems of identifying the number and composition of units from individual sightings, which are hard. Rather, our problem is to infer the enemy commander's strategy and, more specifically, the objective of each enemy unit.

The blackboard contains reports about enemy units and hypotheses such as individual unit's objectives, coordinated actions between unit, and theater-wide objectives. We have implemented 30 fragment KSs: for example, one computes the relative combat strength of two opposing units, and others model of military actions such as defeat (i.e., to attack with overwhelming force), outflank, and fix. One procedural KS clusters enemy units by location, and hypothesizes that the units are groups acting in close concert. This kind of geometric reasoning is difficult to implement in a network fragment.

7 Conclusion

We have presented an architecture for solving knowledge-intensive problems under uncertainty by incrementally constructing probabilistic models. The architecture synthesizes ideas from the older literature on blackboard systems and the newer literature on construction of belief networks from fragments. Blackboard systems are a method of incrementally building symbolic models, while the network fragment systems incrementally build probabilistic models. The contribution of the current work is, in making the connection between the two literatures, to point out that both have good ideas the other hasn't used: probabilistic models have a principled method of reasoning under uncertainty, while blackboard systems have focused on controlling the search through the space of possible models.

It would be natural to extend the architecture to handle influence diagrams, the largest requirement being influence combination methods for decision and utility nodes. The most important open questions in this architecture are better methods of evidence combination and evaluating different control methods. We would also like to apply this architecture to other domains, both within intelligence analysis and common-sense reasoning.

References

1. Marc Atkin, Gary W. King, David Westbrook, Brent Heeringa, Andrew Hannon, and Paul Cohen. SPT: Hierarchical Agent Control: A framework for defining agent behavior. In *Proceedings of Fifth International Conference on Autonomous Agents*, pages 425–432, 2001.
2. Marc Atkin, David L. Westbrook, and Paul Cohen. Domain-general simulation and planning with physical schemas. In *Proceedings of the Winter Simulation Conference*, pages 464–470, 2000.
3. Brendan Burns and Clayton Morrison. Temporal abstraction in Bayesian networks. In *AAAI Spring Symposium*, Palo Alto, CA, 2003.
4. Norman Carver and Victor Lesser. The evolution of blackboard control architectures. In *Expert Systems with Applications 7*, pages 1–30, 1994.
5. L.D. Erman, F. Hayes-Roth, V.R. Lesser, and D.R. Reddy. The HEARSAY-II speech understanding system: Integrating knowledge to resolve uncertainty. *ACM Computing Survey*, 12:213–253, 1980.
6. Robert P. Goldman and Eugene Charniak. A language for construction of belief networks. *IEEE Transactions on Pattern Analysis and Machine Intelligence*, 15(3):196–208, 1993.
7. David Heckerman. A tutorial on learning with Bayesian networks. In Michael Jordan, editor, *Learning in Graphical Models*. MIT Press, 1995.

8. Eric Horvitz. *Computation and Action Under Bounded Resources.* PhD thesis, Stanford University, December 1990.
9. Daphne Koller and Avi Pfeffer. Object-oriented Bayesian networks. *Proceedings of the Thirteenth Annual Conference on Uncertainty in Artificial Intelligence (UAI-97)*, pages 302–313, 1997.
10. Kathryn Blackmond Laskey and Suzanne M. Mahoney. Network fragments: Representing knowledge for constructing probabilistic models. In *Proceedings of the Thirteenth Conference on Uncertainty in Artificial Intelligence (UAI-97)*, San Mateo, CA, 1997. Morgan Kaufmann.
11. DB Lenat and RV Guha. *Building large knowledge-based systems: Representation and inference in the Cyc project.* addison Wesley, 1990.
12. Kevin P. Murphy. *Dynamic Bayesian Networks: Representation, Inference and Learning.* PhD thesis, U.C. Berkeley, July 2002.
13. H. Penny Nii. Blackboard systems. In Avron Barr, Paul R. Cohen, and Edward A. Feigenbaum, editors, *The Handbook of Artificial Intelligence: Volume IV*, pages 1–82. Addison-Wesley, 1989.
14. Judea Pearl. *Probabilistic Reasoning in Intelligent Systems: Networks of Plausible Inference.* Morgan Kaufmann, San Mateo, CA, 1988.
15. Avi Pfeffer, Daphne Koller, Brian Milch, and Ken T. Takusagawa. SPOOK: A system for probabilistic object-oriented knowledge representation. In *Proceedings of the 14th Annual Conference on Uncertainty in AI (UAI-99)*, pages 541–550. Morgan Kaufmann, 1999.
16. Stuart Russell and Peter Norvig. *Artificial Intelligence: A Modern Approach.* Addison-Wesley, 2002.
17. Sampath Srinivas. A generalization of the noisy-or model. In *Proceedings of the Ninth Conference on Uncertainty in Artificial Intelligence (UAI-93)*, San Francisco, 1993. Morgan Kaufmann.

Distributed Regression for Heterogeneous Data Sets*

Yan Xing, Michael G. Madden, Jim Duggan, and Gerard J. Lyons

Department of Information Technology,
National University of Ireland, Galway, Ireland
{yan.xing, michael.madden, jim.duggan, gerard.lyons}@nuigalway.ie

Abstract. Existing meta-learning based distributed data mining approaches do not explicitly address context heterogeneity across individual sites. This limitation constrains their applications where distributed data are not identically and independently distributed. Modeling heterogeneously distributed data with hierarchical models, this paper extends the traditional meta-learning techniques so that they can be successfully used in distributed scenarios with context heterogeneity.

1 Introduction

Distributed data mining (DDM) is an active research sub-area of data mining. It has been successfully applied in the cases where data are inherently distributed among different loosely coupled sites that are connected by a network [1,2,3]. Through transmitting high level information, DDM techniques can discover new knowledge from dispersed data. Such high level information not only has reduced storage and bandwidth requirements, but also maintains the privacy of individual records. Most DDM algorithms for regression or classification are within a meta-learning framework that is based on ensemble learning. They accept the implicit assumption that the probability distributions of dispersed data are homogeneous. It means that the data are just geographically or physically distributed across various sites, and there are no differences among the sites. Although a distinction is made between homogeneous and heterogeneous data, it is limited on database schemata [4]. In reality, heterogeneity across the sites is often the rule rather than the exception [5,6]. An example is a health virtual organization having several hospitals, which differ in expertise, skills, equipment and treatment. Another example is a loosely coupled commercial network consisting of several retailers, which adopt different business policies, sell different products and have different price standards. Wirth describes this kind of scenario as "distribution is part of the semantics" [7], which has been seldom discussed and addressed.

* The support of the Informatics Research Initiative of Enterprise Ireland is gratefully acknowledged.

M.R. Berthold et al. (Eds.): IDA 2003, LNCS 2810, pp. 544–553, 2003.

2 Model of Distributed Data across Heterogeneous Sites

Distributed data across homogeneous sites may be regarded as random samples from the same underlying population, even though the probability distribution of the population is unknown. The differences among the data sets are just random sampling errors. From a statistical perspective, the distributed data are identically and independently distributed (IID) [8,9].

A data set stored at the kth site consists of data $\{(y_{ik}, \mathbf{x}_{ik}), i = 1, \ldots, N_k\}$ where the ys are numerical responses for regression problems, N_k is the sample size of the data, $\{k = 1, \ldots, K\}$ and K is the total number of individual sites. If all the sites are homogeneous, IID data of normal distribution with mean θ and variance σ^2 can be expressed as:

$$y_{ik} \overset{IID}{\sim} N\left(\theta, \sigma^2\right) \tag{1}$$

If e_{ik} is the random sampling error at the kth site, and $f(\mathbf{x}_{ik})$ is the real global regression model, then (1) can be rewritten as:

$$\begin{cases} y_{ik} = \theta + e_{ik} = f(\mathbf{x}_{ik}) + e_{ik} \\ e_{ik} \overset{IID}{\sim} N\left(0, \sigma^2\right) \end{cases} \tag{2}$$

When there is heterogeneity across the different sites, distributed data are not IID. The differences across the various sites are not only the sampling errors, but also some context heterogeneities caused by different features of the sites. In practice, it is difficult to determine the exact sources of the context heterogeneities. Considering the earlier example of a hospital domain, the mixture effects of the differences of hospital expertise, equipment, skills and treatment cause context heterogeneity. It is hard or even impossible for us to measure how much of the heterogeneity is caused by one of the above issues. Furthermore, there may be some differences that exist across the hospitals which are unobservable. So it is with other domains such as the network of retailers. In this case, we need to model the context heterogeneity in a reasonable way.

In statistical meta-analysis, a popular way to model unobservable or immeasurable context heterogeneity is to assume that the heterogeneity across different sites is random. In other words, context heterogeneity derives from essentially random differences among sites whose sources cannot be identified or are unobservable [9]. Distributed data across various sites having randomly distributed context heterogeneity is often regarded as conditional IID [5]. This leads to a two-level hierarchical model which describes context heterogeneity with mixture models and employs latent variables in a hierarchical structure [5,10,11].

Assuming that the context of different sites is distributed normally with mean θ and variance τ^2, data y_{ik} at the kth site has normal distribution with mean θ_k and variance σ^2, then the two-level hierarchical model of distributed data across heterogeneous sites is:

$$\begin{cases} \text{between sites level}: & \theta_k \overset{IID}{\sim} N\left(\theta, \tau^2\right) \\ \text{within a certain site}: & (y_{ik}|\theta_k) \overset{IID}{\sim} N\left(\theta_k, \sigma^2\right) \end{cases} \tag{3}$$

If t_k is the random sampling error of context across different sites, and the different level residuals t_k and e_{ik} are independent, then (3) can be rewritten as:

$$\begin{cases} y_{ik} = \theta + e_{ik} + t_k = f(\mathbf{x}_{ik}) + e_{ik} + t_k \\ e_{ik} \overset{IID}{\sim} N(0, \sigma^2), t_k \overset{IID}{\sim} (0, \tau^2) \end{cases} \tag{4}$$

When $\tau^2 = 0$ or $t_k = 0$, (3) will be the same as (1), and (4) will be same as (2). So a distributed scenario with homogeneous sites is only a special case of the generic situation.

In theory of hierarchical modeling, intra-class correlation (ICC) is a concept to measure how much quantity of the total variance of a model is caused by context heterogeneity. ICC is calculated by:

$$ICC = \tau^2 / (\tau^2 + \theta^2) \tag{5}$$

When the various sites are homogeneous, $ICC = 0$. The larger value ICC has, the more quantity of model variance is caused by context heterogeneity.

3 Towards Context-Based Meta-learning

Once distributed data across heterogeneous sites are modeled as (4), the main task of distributed regression is to obtain the global regression model $f(\mathbf{x}_{ik})$.

In hierarchical modeling, statistical linear and nonlinear multilevel model fitting is done by iterative maximum likelihood based algorithms [5,10,11]. Unfortunately, all of these algorithms are designed for centralized data. Even though we can modify the algorithms to a distributed environment, the iteration feature will cause significant burdens on communications among the various sites [12].

Since most existing DDM approaches are within a meta-learning framework, it is necessary for us to extend the state of the art techniques so that they can be successfully used in distributed scenarios with heterogeneous sites.

3.1 Traditional Distributed Meta-learning

When different sites are homogeneous, (1) and (2) apply. The implicit assumption for an ensemble of learners to be more accurate than the average performance of its individual members is satisfied [13,14].

Assuming the base models (learners) generated at different sites are $f_k(\mathbf{x}_{ik})$ and $k = 1, \ldots, K$, then the final ensemble model (meta-learner) is $f_A(\mathbf{x}_{ik}) = E_k[f_k(\mathbf{x}_{ik})]$, where E_k denotes the expectation over k, and the subscript A in f_A denotes aggregation. Distributed meta-learning takes $f_A(\mathbf{x}_{ik})$ as the estimate of the real global model $f(\mathbf{x}_{ik})$.

Meta-learning based distributed regression follows three main steps. Firstly, generate base regression models at each site using a learning algorithm. Secondly, collect the base models at a central site. Produce meta-level data from a separate validation set and predictions generated by the base classifier on it. Lastly, generate the final regression model from meta-level data via combiner (un-weighted or weighted averaging).

3.2 Context-Based Meta-learning

When different sites are heterogeneous, (3) and (4) apply. The variance of data within a certain site is σ^2, but the variance of data from different sites is $(\sigma^2 + \tau^2)$. So distributed data are not IID. The criterion for success of meta-learning is not satisfied. We need an approach to deal with the context variance τ^2 in the meta-learning framework. We call it context-based meta-learning approach.

Global Model Estimation. According to (4), given the context θ_k of the kth site, data within that site can be expressed as:

$$\begin{cases} (y_{ik}|\theta_k) = \theta_k + e_{ik} \\ e_{ik} \overset{IID}{\sim} N(0, \sigma^2) \end{cases} \tag{6}$$

and $\{\theta_k, k = 1, 2, \ldots, K\}$ has the following distribution:

$$\begin{cases} \theta_k = \theta + t_k = f(\mathbf{x}_{ik}) + t_k \\ t_k \overset{IID}{\sim} N(0, \tau^2) \end{cases} \tag{7}$$

So base models $f_k(\mathbf{x}_{ik})$ generated at the local sites are the estimates of $\theta_k, k = 1, \ldots, K$. Given θ_k and $\mathbf{x}_{ik}$, suppose $\hat{\theta}_k$ is the estimate of the real θ_k, then:

$$f_k(\mathbf{x}_{ik}) = \hat{\theta}_k \approx \theta_k = \theta + t_k = f(\mathbf{x}_{ik}) + t_k \tag{8}$$

Then the final ensemble model $f_A(\mathbf{x}_{ik})$ is:

$$f_A(\mathbf{x}_{ik}) = E_k[f_k(\mathbf{x}_{ik})] \approx f(\mathbf{x}_{ik}) + E_k(t_k) \tag{9}$$

Because $E_k(t_k) = 0$, we can use $f_A(\mathbf{x}_{ik})$ to estimate the real global model $f(\mathbf{x}_{ik})$.

Context Residual Estimation. Since there is context heterogeneity, when we use the ensemble model for prediction at a certain kth site, we need to add the value of context residual at that site, which is t_k. From (8) and (9), we have:

$$t_k = \frac{1}{N_k} \sum_{i=1}^{N_k} [f_k(\mathbf{x}_{ik}) - f_A(\mathbf{x}_{ik})] \tag{10}$$

With (10), t_k will never be exactly equal to zero even though the real context residual is zero. So when we get the value of t_k by (10), we use two-tail t-test to check the null hypothesis $H_0 : t_k = 0$, and calculate the context residual of kth site by:

$$t_k = \begin{cases} 0, & H_0 \text{ accepted given } \alpha \\ t_k, & H_0 \text{ rejected given } \alpha \end{cases} \tag{11}$$

where α is the level of significance. When we get all the values of $\{t_k, k = 1, \ldots, K\}$, we can calculate context level variance τ^2.

Algorithm for Context-based Meta-learning. Our context-based algorithm for distributed regression follows seven steps:

1. At each site, use cross-validation to generate a base regression model.
2. Each site transfer its base model to all the other sites.
3. At each site, produce meta-level data from the predictions generated by all the base models on the local data set.
4. At each site, generate the ensemble model from meta-level data via combiner (un-weighted or weighted averaging).
5. At each site, calculate its context residual with (10) and (11).
6. At each site, generate the final regression model of this site by (8).
7. Collect context residuals at a certain site, calculate context level variance.

4 Simulation Experiment

In practice, different distributed scenarios have different levels of intra-class correlation. It ranges from $ICC = 0$, which is the homogeneous case, to $ICC \to 1$, which means variance is mainly caused by context heterogeneity.

In order to evaluate our approach on different values of ICC, we use simulation data sets because it is difficult for us to get real world distributed data sets that can satisfy all our requirements.

4.1 Simulation Data Sets

The three simulation data sets we used are Friedman's data sets [15]. They are originally generated by Friedman and used by Breiman in his oft-cited paper about bagging ([13]). We have made some modifications to the three data sets so that they are compatible with our distributed scenarios.

Friedman #1: there are ten independent predictor variables $x_1, \ldots, x_{10}$ each of which is uniformly distributed over $[0, 1]$. We set the total number of sites is $K = 10$. At kth site, the sample size is 200 for training set and 1000 for testing set. The response is given by:

$$\begin{cases} y_{ik} = 10\sin\left(\pi x_{1ik} x_{2ik}\right) + 20\left(x_{3ik} - 0.5\right)^2 + 10x_{4ik} + 5x_{5ik} + e_{ik} + t_k \\ e_{ik} \overset{IID}{\sim} N\left(0, 1\right),\ t_k \overset{IID}{\sim} N\left(0, \tau^2\right) \end{cases} \tag{12}$$

We adjust the value of τ so that we can get $ICC = 0.0, 0.1, \ldots, 0.9$ respectively.

Friedman #2, #3: these two examples are four variable data with

$$\#2: \begin{cases} y_{ik} = \left(x_{1ik}^2 + \left(x_{2ik} x_{3ik} - \left(1/x_{2ik} x_{4ik}\right)\right)^2\right)^{1/2} + e_{ik} + t_k \\ e_{ik} \overset{IID}{\sim} N\left(0, \sigma_2^2\right),\ t_k \overset{IID}{\sim} N\left(0, \tau_2^2\right) \end{cases} \tag{13}$$

$$\#3 : \begin{cases} y_{ik} = \tan^{-1}\left(\frac{x_{2ik}x_{3ik}-(1/x_{2ik}x_{4ik})}{x_{1ik}}\right) + e_{ik} + t_k \\ e_{ik} \overset{IID}{\sim} N\left(0,\sigma_3^2\right),\ t_k \overset{IID}{\sim} N\left(0,\tau_3^2\right) \end{cases} \tag{14}$$

where x_1, x_2, x_3, x_4 are uniformly distributed as $0 \leq x_1 \leq 100$, $20 \leq (x_2/2\pi) \leq 280$, $0 \leq x_3 \leq 1$ and $1 \leq x_4 \leq 11$. The total site number and sample sizes of training and testing set of each site are the same as #1. The parameters σ_2, σ_3 are selected to give 3 : 1 signal/noise ratios, and τ_2, τ_3 are adjusted for the same purpose as #1.

4.2 Simulation Result

To compare the prediction accuracy of traditional meta-learning approach and our context-based meta-learning approach under different values of ICC, we implement both algorithms. The base learner we use is a regression tree algorithm implemented in Weka [16]; the meta-learner we use is un-weighted averaging and the significance level $\alpha = 0.05$. We evaluate our approach from two angles: the whole virtual organization and its individual sites.

The Whole Organization. From the point of view of the whole organization, the target of DDM is to discover the global trend. Fig. 1 shows the global prediction accuracy under different ICC values.

From Fig. 1, we can see that, as ICC increases, the prediction accuracy of the traditional meta-learning approach decreases, while the prediction accuracy of our approach essentially remains the same. When $ICC = 0$, the prediction accuracy of both approaches are almost the same.

Individual Sites. From the point of view of each individual site, the goal of DDM is to get more accurate model than the local model (base model) only created from its local data. In practice, when $ICC > 0.5$, individual site usually only use local model because the context heterogeneity is too large.

Fig. 2 compares the prediction accuracy of local model, meta-learning model and the model created with our context-based meta-learning approach when $ICC = 0.3$. For those sites ($5^{th}, 8^{th}, 10^{th}$ for #1; 1st, 4th, 5th, 8th for #2; 3rd, 4th, 7th for #3) with relatively larger context residuals, meta-learning behaves worse than our approach, and sometimes even worse than the local models (1st, 4th, 5th, 8th for #2; 4th, 7th for #3). For those sites with very small context residuals, the performance of our approach is comparable with the traditional meta-learning approach. So the overall performance of our approach is the best.

5 Discussion of Our Approach

There are two important issues relating with our approach: number of sites and sample size at individual sites.

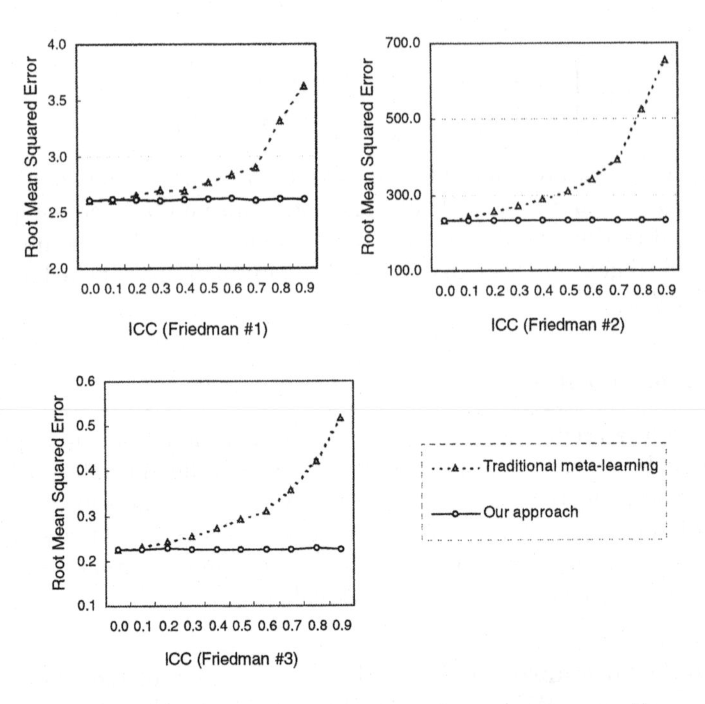

Fig. 1. Global prediction accuracy (average of 20 runs)

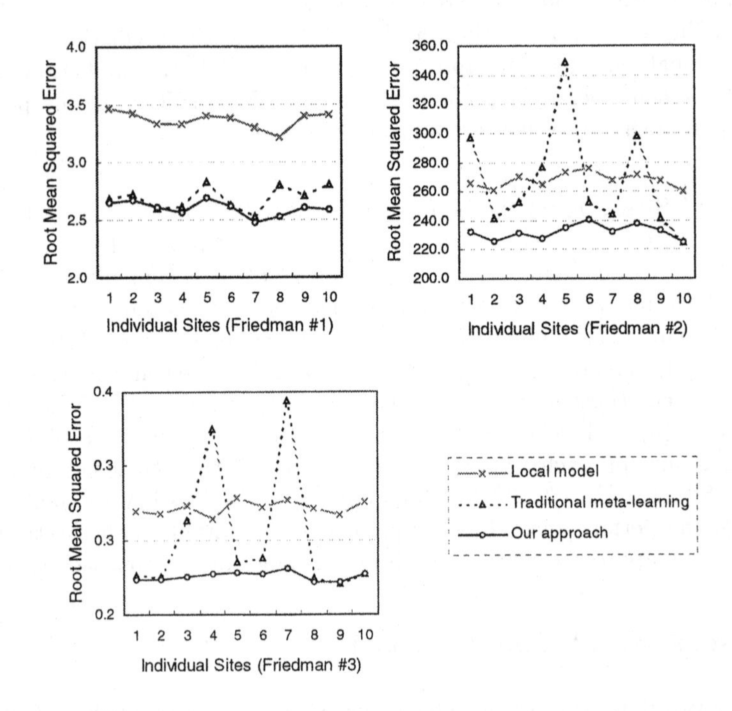

Fig. 2. Prediction accuracy of individual sites when $ICC = 0.30$ (average of 20 runs)

5.1 Number of Sites

In practical distributed scenarios, the number of sites is usually much smaller than the number of data at each individual site. The extreme case is that there are only two sites.

From perspective of statistics, the larger the number of sites is, the more accurate estimations of the quantity of context heterogeneity we can obtain. When the number of sites is extremely low, we usually underestimate the quantity of context heterogeneity [10]. This kind of underestimation will worsen the advantage of our approach. We tried our simulation experiment when the number of sites is 5 and 2 respectively. The results we got demonstrate this.

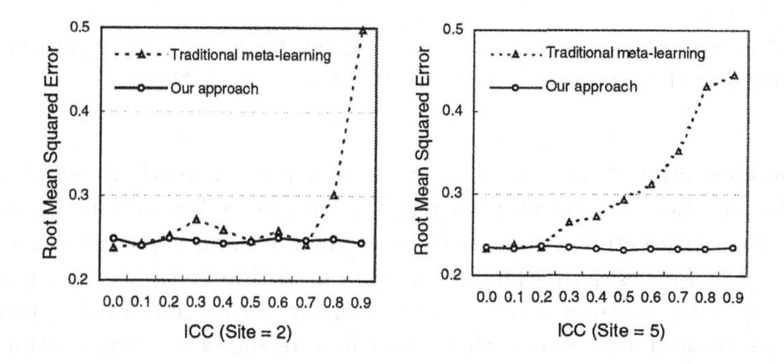

Fig. 3. Comparison of global prediction accuracy when the number of sites is extremely small (Frienman #3, average of 20 runs)

Fig. 3 shows the global prediction accuracy for Friedman #3 data when the number of sites is extremely small. When the number of sites is 5, the advantage of our approach is still obvious. But when the number of sites is 2, the advantage of our approach is less obvious.

5.2 Sample Size of Data at Individual Sites

At each individual site, as the sample size of training data set increases, the accuracy of using $\hat{\theta}_k$ to estimate θ_k in (8) increases. So we can obtain more accurate estimation of the context residual of the site if we have more local data. Thus finally we can get higher prediction accuracy.

Comparing the two graphs in Fig. 4, it can be seen that the advantage of our approach is greater pronounced when there are more training data at each individual site.

5.3 Related Work

Meta-learning is one popular approach of DDM techniques. The most successful meta-learning based DDM application is done by Chan in the domain of credit

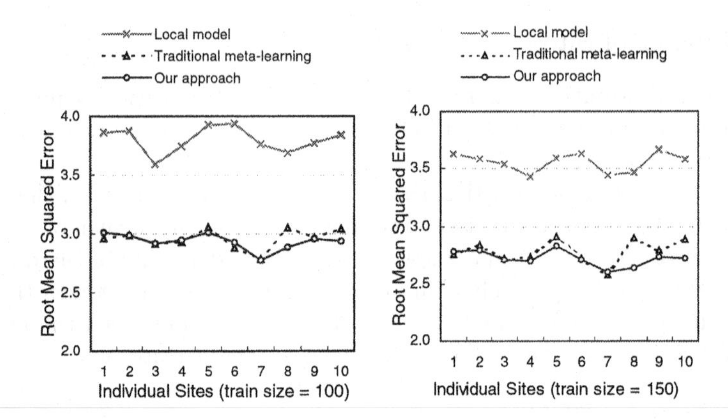

Fig. 4. Comparison of global prediction accuracy with different training size (Frienman #1, testing size=1000 for each site, $ICC = 0.30$, average of 20 runs)

card fraud detection [17] where both database schemata and data distribution are homogeneous. Hershberger and Kargupta introduce collective DDM approaches for scenarios with heterogeneous database schemata and homogeneous data distribution [18]. Most existing DDM applications within meta-learning framework do not explicitly address context heterogeneity across individual sites. Wirth defines distributed scenarios with context heterogeneity as "distribution is part of the semantics" in his work [7], but does not give an approach to solve it. The work we found which explicitly addresses context heterogeneity is done by Páircéir [19], where statistical hierarchical models are used to discover multi-level association rules from dispersed hierarchical data. In our previous work, we use hierarchical modeling with the traditional iterative-based algorithms to address context heterogeneity in the domain of virtual organizations [12] and get some encouraging results. However we also realize that traditional iteration-based algorithms will cause heavy communication traffic among individual sites.

6 Summary and Future Work

Through the analysis of the limitations of distributed meta-learning approach, we model distributed data across heterogeneously distributed sites with two-level statistical hierarchical models, and extend traditional meta-learning approach to suit non-IID distributed data. We successfully use our context-based meta-learning approach on several simulation data sets for distributed regression. We also discuss the important issues related with our approach. The simulation results demonstrate that our context-based meta-learning approach can both deal with heterogeneous data distribution and minimize network traffic among individual sites.

For our future work, we will use some real world distributed data sets to test our approach. Then we plan to extend our approach to distributed classification problems.

References

1. Provost, F.: Distributed data mining: Scaling up and beyond. In Kargupta, H., Chan, P.K., eds.: Advances in Distributed and Parallel Knowledge Discovery. AAAI/MIT Press (2000) 3–27
2. Park, B.H., Kargupta, H.: Distributed data mining: Algorithms, systems, and applications. In Ye, N., ed.: Data Mining Handbook. (2002)
3. Prodromidis, A.L., Chan, P.K., Stolfo, S.J.: Meta-learning in distributed data mining systems: Issues and approaches. In Kargupta, H., Chan, P.K., eds.: Advances in Distributed and Parellel Knowledge Discovery. Chapter 3. AAAI / MIT Press (2000)
4. Kargupta, H., Chan, P.K.: Distributed and parallel data mining: A brief introduction. In Kargupta, H., Chan, P.K., eds.: Advances in Distributed and Parallel Knowledge Discovery. AAAI/MIT Press (2000) xv–xxvi
5. Draper, D.: Bayesian hierarchical modeling. Tutorial on ISBA2000 (2000) http://www.bath.ac.uk/
6. Draper, D.: Inference and hierarchical modeling in the social sciences. Journal of Educational and Behavioral Statistics **20** (1995) 115–147
7. Wirth, R., Borth, M., Hipp, J.: When distribution is part of the semantics: A new problem class for distributed knowledge discovery. In Marinaro, M., Tagliaferri, R., eds.: 5th European Conference on Principles and Practice of Knowledge Discovery in Databases (PKDD'01),Workshop on Ubiquitous Data Mining for Mobile and Distributed Environments. (2001) 56–64
8. Brandt, S.: Data Analysis: Statistical and Computational Methods for Scientists and Engineers. 3nd edn. Springer (1998)
9. Lipsey, M.W., Wilson, D.B.: Practical Meta-Analysis. SAGE Publications (2000)
10. Goldstein, H.: Multilevel Statistical Models. Second edition edn. ARNOLD (1995)
11. Kreft, I., Leeuw, J.D.: Introducing Multilevel Modeling. Sage Publications (1998)
12. Xing, Y., Duggan, J., Madden, M.G., Lyons, G.J.: A multi-agent system for customer behavior prediction in virtual organization. Technical Report NUIG-IT-170503, Department of Information Technology, NUIG (2002)
13. Breiman, L.: Bagging predictors. Machine Learning **24** (1996) 123–140
14. Dietterich, T.G.: Ensemble methods in macine learning. Lecture Notes in Computer Science **1857** (2000) 1–15
15. Friedman, J.H.: Multivariate adaptive regression splines. Annals of Statistics **19** (1991) 1–141
16. Witten, I.H., Frank, E.: Data Mining: Practical Machine Learning Tools and Techniques with Java Implementations. Academic Press (2000)
17. Chan, P.K., Fan, W., Prodromidis, A.L., Stolfo, S.J.: Distributed data mining in credit card fraud detection. IEEE Intelligent System **14** (1999) 67–74
18. Hershberger, D.E., Kargupta, H.: Distributed multivariate regression using wavelet-based collective data mining. Journal of Parallel and Distributed Computing **61** (1999) 372–400
19. Páircéir, R., McClean, S., Scitney, B.: Discovery of multi-level rules and exceptions from a distributed database. In: Sixth ACM SIGKDD International Conference on Knowledge Discovery and Data Mining KDD2000. (2000) 523–532

A Logical Formalisation of the Fellegi-Holt Method of Data Cleaning

Agnes Boskovitz[1], Rajeev Goré[1], and Markus Hegland[2]

[1] Research School of Information Sciences and Engineering
Australian National University, Canberra ACT 0200 Australia

[2] Mathematical Sciences Institute
Australian National University, Canberra ACT 0200 Australia
{agnes.boskovitz, rajeev.gore, markus.hegland}@anu.edu.au

Abstract. The Fellegi-Holt method automatically "corrects" data that fail some predefined requirements. Computer implementations of the method were used in many national statistics agencies but are less used now because they are slow. We recast the method in propositional logic, and show that many of its results are well-known results in propositional logic. In particular we show that the Fellegi-Holt method of "edit generation" is essentially the same as a technique for automating logical deduction called resolution. Since modern implementations of resolution are capable of handling large problems efficiently, they might lead to more efficient implementations of the Fellegi-Holt method.

1 Introduction

Data errors are everywhere. While minor errors might be inconsequential, data errors can often seriously affect data analysis, and ultimately lead to wrong conclusions. Detecting and correcting errors to clean data is a large part of the time-consuming preprocessing stage of any data-intensive project.

Obvious strategies for dealing with potentially erroneous data, like revisiting the original data sources or doing nothing, have clear disadvantages. If the locations of the errors are known, we can discard dirty data, again with clear disadvantages. More practically, statistical techniques can be used to estimate a likely correction. Fortunately, errors can often be located using the data's internal dependencies, or redundancies, as seen in the following (contrived) example.

Example, part 1. (This will be a running example throughout this paper.) A census of school students includes a data record showing a person who is aged 6, has a driver's licence and is in Grade 8 of school. Domain knowledge gives:

Requirement 1: A person with a driver's licence must be in at least Grade 11.
Requirement 2: A person in Grade 7 or higher must be at least 10 years old.

The given data record fails both of the requirements. Since school grade level appears in both of them, it might seem that the record can be corrected by

M.R. Berthold et al. (Eds.): IDA 2003, LNCS 2810, pp. 554–565, 2003.

changing the grade level. However the two requirements cannot be satisfied by changing only the grade level. The difficulty arises because there is an additional requirement that can be deduced from these, namely:

Requirement 3: A person with a driver's licence must be at least 10 years old.

Requirement 3 does not involve school grade level, but is still failed by the data record. In fact two fields must be changed to satisfy all of the requirements. Of course in reality something much stronger than Requirement 3 holds, but Requirement 3 is the logical deduction from the first two Requirements.

In this paper we address the problem of automating the task of correcting record data when each record must satisfy deterministic requirements. After stating our assumptions in Section 2, we describe in Section 3 the elegant contribution of Fellegi and Holt [1], who recognised the need to find "all" deducible requirements to decide where to correct a record. They developed an algorithm for finding deducible requirements, and proved that it finds enough requirements to determine which fields must be changed to correct an incorrect record. The Fellegi-Holt method was used for many years in government statistical agencies, but is less used now because it is too slow for many practical applications [2].

There have been many published refinements to the Fellegi-Holt method [3,4, 5,6,7]. However, they can be confusing, because the method is expressed in terms of records which *fail* requirements, whereas the deducible requirements are about records which *satisfy* requirements. We think a formalisation that helps people to reason about the method itself, especially about the logical deduction of new requirements, should decrease the level of confusion.

Symbolic logic is a well-developed and rigorous framework for formalising reasoning and performing logical deduction. Although the problem of finding "all" deductions is NP-complete, computer implementations for automatically finding logical deductions are being continually improved. In Sections 4 and 5 we formalise the Fellegi-Holt method in logic, giving us two benefits. Firstly we gain a rigorous perspective with which to reason about the problem of data correction and about Fellegi and Holt's solution. Secondly we gain the potential to use automated deduction techniques to speed up the method. In our formalisation, the Fellegi-Holt results become well-known results in classical propositional logic. In particular, Fellegi-Holt deduction is essentially the same as a standard technique called resolution deduction. The vast field of results for resolution deduction translate to Fellegi-Holt deduction, giving a new and potentially faster technique for implementing the Fellegi-Holt method. Section 6 gives the results of some experiments which prove our concept. Although other published formalisations in logic solve the data-correction problem, no other work formalises the Fellegi-Holt method itself in logic.

2 Assumptions and Problem Statement

We assume that we are dealing with data arranged in records. A *record* R is an N-tuple $(R_1, \ldots, R_N)$ where $R_j \in A_j$. Here A_j is the *domain* of the jth *field*, and we assume that it is a finite set. We assume that N is fixed for all records under consideration. The set of all such N-tuples is called the *domain space* $D = A_1 \times \cdots \times A_N = \prod_1^N A_j$.

Example, part 2. The Example has three fields:

$A_1 = \{5, 6, \ldots, 20\}$ (relevant ages), which we relabel $A_{\mathbf{age}}$.
$A_2 = \{\mathrm{Y}, \mathrm{N}\}$ (whether someone has a driver's licence), which we relabel $A_{\mathbf{driver}}$.
$A_3 = \{1, 2, \ldots, 12\}$ (school grade levels), which we relabel $A_{\mathbf{grade}}$.
The domain space is $D = A_{\mathbf{age}} \times A_{\mathbf{driver}} \times A_{\mathbf{grade}}$.
The record in this example is $R = (6, \mathrm{Y}, 8)$.

We assume that potential errors in the data are specified by deterministic *requirements* obtained from domain knowledge, and that the requirements apply to one record at a time. The requirements given in the Example are typical. In the statistical literature, such requirements are usually formalised as *edits*, where an edit consists of the elements of D which are *incorrect* according to the corresponding requirement. Elements of the edit e are *incorrect* according to e or *fail* e. Elements of $D \setminus e$ are *correct* according to e or *satisfy* e.

Example, part 3. The edits corresponding to Requirements 1-3 are:

$$\begin{aligned}
e_1 &= \quad A_{\mathbf{age}} &\times& \quad \{\mathrm{Y}\} &\times& \{1, 2, \ldots, 10\}, \\
e_2 &= \{5, 6, \ldots, 9\} &\times& A_{\mathbf{driver}} &\times& \{7, 8, \ldots, 12\}, \\
e_3 &= \{5, 6, \ldots, 9\} &\times& \quad \{\mathrm{Y}\} &\times& \quad A_{\mathbf{grade}}.
\end{aligned}$$

Since $R \in e_1$, $R \in e_2$ and $R \in e_3$, R is said to fail all three edits.

In our logical formalisation in Section 4, we represent the requirements as logical formulae called *checks*, which represent the *correctness* of a record. To each edit there corresponds a check.

We can now give a formal statement of the data correction problem: given a record R and a set of requirements, find an N-tuple R' which differs from R in as few (weighted) fields as possible and satisfies all of the requirements. Solving the data correction problem involves two steps:

1. *Editing:* the process of testing a data record against edits; and
2. *Imputation:* the correction of incorrect data, which itself takes two steps:
   a) *error localisation:* finding a smallest (weighted) set of fields which can be altered to correct the record, and then
   b) changing those fields so as to preserve the original frequency distributions of the data as far as possible (sometimes also called "imputation").

In this paper we concentrate on error localisation, which is the area in which the Fellegi-Holt method is slow.

3 The Fellegi-Holt (FH) Method

Fellegi and Holt's method hinges on two aspects:

1. *FH edit generation:* a systematic process for generating new edits from a given set of edits (described below in Section 3.1); and
2. *FH error localisation* and *guarantee*: a method of finding a smallest set of fields that is guaranteed to yield a correction (described in Section 3.2).

3.1 FH Generation of New Edits

Fellegi and Holt generate edits in terms of *normal edits*, where a normal edit is an edit e of the form $\prod_{j=1}^{N} A_j^e$, with $A_j^e \subseteq A_j$. The edits in the Example, part 3, are normal edits. Any edit can be written as the union of a set of normal edits.

Given a set E of normal edits where each edit e is written $\prod_{j=1}^{N} A_j^e$, and given a field i, the *FH-generated edit* on E with generating field i is $\mathrm{FHG}(i, E)$:

$$\mathrm{FHG}(i, E) = \prod_{j=1}^{i-1} \bigcap_{e \in E} A_j^e \times \bigcup_{e \in E} A_i^e \times \prod_{j=i+1}^{N} \bigcap_{e \in E} A_j^e. \quad (1)$$

Example, part 4. $e_3 = \mathrm{FHG}(\text{grade}, \{e_1, e_2\})$.

The following proposition tells us that any record that satisfies all edits in E must also satisfy $\mathrm{FHG}(i, E)$, so the edit generation process produces an acceptable edit. The result applies because $\mathrm{FHG}(i, E) \subseteq \bigcup_{e \in E} e$.

Proposition 1 (Soundness). *[1, Lemma 1] Let E be a set of normal edits, i be a field, and R be a record such that for all $e \in E, R \notin e$. Then $R \notin \mathrm{FHG}(i, E)$.*

Given a starting set E of edits, the FH edit generation process can be applied to all subsets of E, on all fields. The newly generated edits can then be added to E and the process repeated until no new edits can be generated. The process will eventually terminate because the domain is finite.

The slowness of this process slows down the Fellegi-Holt method. The process can be sped up by excluding any edit which is a subset of (*dominated* by) another generated edit because the FH error localisation guarantee still applies [3]. We call the unique set of edits left at the end of this process the *set of maximal generated edits*, written $\mathrm{MGE}(E)$, or just MGE when the context is clear.

3.2 FH Error Localisation Method and Guarantee

Given an arbitrary set of edits, the FH error localisation method depends on:

1. If a record R fails a normal edit e, then any correction to R must change at least one field j that is *involved* in e, ie $A_j^e \neq A_j$. Otherwise the change in field j cannot affect R's correctness according to e.

2. Hence if R fails some normal edits then any correction of R must change a set of fields C which includes an involved field from each failed edit. Such a set of fields C is called a *covering set.*
3. There is no guarantee that every covering set of the failed set of edits will yield a correction to R. However if the edits are the failed edits in the *MGE*, then *any* covering set will yield a correction to R (Theorem 1 below).
4. The FH error localisation method finds a smallest covering set, giving a solution to the error localisation problem.

Theorem 1 (Error localisation guarantee). *Let R be a record and let E be a set of normal edits such that the edit $D = \prod_{j=1}^{N} A_j \notin \mathrm{MGE}(E)$. Let $E' = \{e \in \mathrm{MGE}(E) \mid R \in e\}$ be those edits in MGE(E) that are failed by R. Then there exists a covering set of E', and for every covering set C there exists an N-tuple R' which (1) is identical to R on fields outside C, and (2) satisfies all the edits in E. ([1, Theorem 1, Corollaries 1, 2] and [3, Lemma 4, Corollaries 1,2])*

Example, part 5. Both edits in $E = \{e_1, e_2\}$ are failed by the given record, but the smallest covering set of E, namely {grade}, does not yield a correction. The record fails every edit in $\mathrm{MGE}(E) = \{e_1, e_2, e_3\}$, so $E' = \{e_1, e_2, e_3\}$. Any pair of fields is a smallest covering set of the MGE and will yield a correction.

3.3 Overcoming the Slowness of Edit Generation

The Fellegi-Holt method is slow because of the slowness of edit generation. Improvements to the edit generation process have been published over the years, eg [3,4,5], although there is disagreement about whether the "Field Code Forest" algorithm finds the whole MGE [6]. Fellegi and Holt proposed using "essentially new" edit generation, a method that produces fewer edits than the MGE, but which requires much analysis prior to the generation of each new edit.

There has also been work on methods that avoid the large scale edit generation, and instead solve the editing problem record by record. Garfinkel, Liepins and Kunnathur [3,4,5] and Winkler [7] propose a cutting plane algorithm which avoids finding all implied edits, and instead finds only those that are needed for each record. Barcaroli [8] formalises the problem in first order logic, and then uses a systematic search through possible corrections to find a solution. Bruni and Sassano [9] formalise the problem in propositional logic, use a propositional satisfiability solver for edit validation and error detection, and then use integer programming for error localisation and imputation. Franconi et al [10] use disjunctive logic programming with constraints, solving both the editing and imputation problem. In each case, there is no preliminary edit generation but the workload for each record is increased.

In contrast, the Fellegi-Holt idea is to do a large amount of work even before the data is received, with correspondingly less work for each individual record. The Fellegi-Holt method would therefore be best applied to datasets such as population censuses where there is a long time between the finalisation of the questionnaire and data entry. It is during this time that the large scale edit generation might be completed. Once the data is entered the editing itself should take less time than the record-by-record approach. Thus our logical formalisation, which builds on the formalisation of Bruni and Sassano, keeps the full generation of the MGE.

4 Translation of the Fellegi-Holt Method to Logic

In this section we represent the FH method in terms of classical propositional logic. We first represent each of the FH constructs as a corresponding logical construct, and then represent the key error localisation result.

As in [9], we represent individual field values by *atoms*, for example p^6_{age} stands for "age = 6". The set of atoms is $\{p^v_j \mid j = 1, \ldots, N, v \in A_j\}$. In the Example we use "age", "driver" and "grade" to represent 1, 2 and 3 respectively.

Formulae are built from the atoms using the propositional connectives $\neg, \vee, \wedge$ and $\rightarrow$. We represent edits as formulae called *checks*.

Example, part 6. The edits e_1, e_2 and e_3 are represented by the checks

$$\begin{aligned}
\gamma_1 &= \neg\,(& & p^{\text{Y}}_{\text{driver}} \wedge (p^1_{\text{grade}} \vee \cdots \vee p^{10}_{\text{grade}})\,) \\
\gamma_2 &= \neg\,(\,(p^5_{\text{age}} \vee \cdots \vee p^9_{\text{age}}) \wedge & & (p^7_{\text{grade}} \vee \cdots \vee p^{12}_{\text{grade}})\,) \\
\gamma_3 &= \neg\,(\,(p^5_{\text{age}} \vee \cdots \vee p^9_{\text{age}}) \wedge p^{\text{Y}}_{\text{driver}} & &)\,.
\end{aligned}$$

Each N-tuple R is represented by a *truth function* $f_R : \text{Atoms} \rightarrow \{\text{true}, \text{false}\}$ which can be extended in the usual way to truth functions on formulae. In any N-tuple, each component takes exactly one value, so we require every truth function to map to "true" the following sets of checks, called *axioms*:

Axiom 1: $\neg p^v_j \vee \neg p^w_j$, for all $j = 1, \ldots, N$ and for $v \neq w$. (Each field of an N-tuple can take at most one value.)

Axiom 2: $\bigvee_{v \in A_j} p^v_j$, for all $j = 1, \ldots, N$. (Each field of an N-tuple must take at least one value.)

Example, part 7. $R = (6, \text{Y}, 8)$, so $f_R(p^6_{\text{age}}) = f_R(p^{\text{Y}}_{\text{driver}}) = f_R(p^8_{\text{grade}}) =$ true. For all other atoms p, $f_R(p) =$ false. For the checks, $f_R(\gamma_1) = f_R(\gamma_2) = f_R(\gamma_3) =$ false.

If $f_R(\gamma) =$ false then the record R is said to *fail* the check γ or to be *incorrect* according to γ. In the Example, R fails γ_1, γ_2 and γ_3. If $f_R(\gamma) =$ true then the record R is said to *satisfy* the check γ or to be *correct* according to γ. This mimics the situation with the corresponding edits, but note that the set of N-tuples on which a check is "true" is the complement of the corresponding edit.

The following terms from logic will be useful: a *literal* is an atom or a negated ($\neg$) atom; a *clause* is a disjunction ($\vee$) of literals; a clause α *subsumes* a clause β if the literals in α form a subset of the literals in β.

Normal edits are represented by negations of conjunctions ($\wedge$) of clauses, as seen in the Example, part 6. In general, the normal edit $\prod_{j=1}^{N} A^e_j$ is represented by the check $\neg \bigwedge_{j=1}^{N} \bigvee_{v \in A^e_j} p^v_j$. These formulae are complex and difficult to reason about, but there are simpler "semantically equivalent" formulae: two formulae φ and ψ are *semantically equivalent* if for all N-tuples R, $f_R(\varphi) = f_R(\psi)$.

Example, part 8. The checks γ_1, γ_2 and γ_3 from Example, part 6, respectively representing the edits e_1, e_2 and e_3, are semantically equivalent to:

$$\begin{aligned}
\epsilon_1 &= && p^{\mathrm{N}}_{\text{driver}} \vee p^{11}_{\text{grade}} \vee p^{12}_{\text{grade}} \\
\epsilon_2 &= p^{10}_{\text{age}} \vee \cdots \vee p^{20}_{\text{age}} \vee && p^{1}_{\text{grade}} \vee \cdots \vee p^{6}_{\text{grade}} \\
\epsilon_3 &= p^{10}_{\text{age}} \vee \cdots \vee p^{20}_{\text{age}} \vee && p^{\mathrm{N}}_{\text{driver}} .
\end{aligned}$$

In general, the normal edit $e = \prod_{j=1}^{N} A_j^e$ can be represented by the check $\bigvee_{j=1}^{N} \bigvee_{v \in \overline{A_j^e}} p_j^v$, where $\overline{A_j^e}$ is the complement of A_j^e, and so we define a *Fellegi-Holt normal check (FHNC)* ϵ to be a check written in the form $\bigvee_{j=1}^{N} \bigvee_{v \in S_j^\epsilon} p_j^v$, where $S_j^\epsilon \subseteq A_j$. The sets S_j^ϵ may be empty. If they are all empty then the FHNC is the *empty clause*, written $\Box$, which is the check according to which no N-tuple is correct, ie every truth function makes $\Box$ false.

Given a set Σ of FHNCs, the *FH-deduced check* with generating index i is FHD(i, Σ), an exact mimic of the complement of the FHG(i, E) in Equation (1):

$$\mathrm{FHD}(i, \Sigma) = \bigvee \Big\{ p_j^v \mid j \in \{1, \ldots, N\} \setminus \{i\}, v \in \bigcup_{\epsilon \in \Sigma} S_j^\epsilon \Big\} \ \vee \ \bigvee \Big\{ p_i^v \mid v \in \bigcap_{\epsilon \in \Sigma} S_i^\epsilon \Big\} .$$

Note that FHD(i, Σ) is an FHNC.

Example, part 9. FHD(grade, $\{\epsilon_1, \epsilon_2\}) = \epsilon_3$.

As with FH edit generation, any record that satisfies all FHNCs in Σ must also satisfy FHD(i, Σ), so the FH deduction produces acceptable results:

Proposition 2 (Soundness). *Let Σ be a set of FHNCs, i a field, and f a truth function such that for all $\sigma \in \Sigma$, $f(\sigma) = true$. Then $f(\mathrm{FHD}(i, \Sigma)) = true$.*

As with FH edit generation, the FH deduction can be repeatedly applied until no new checks can be generated. We write $\Sigma \vdash_{\mathrm{FH}} \epsilon$ to mean that ϵ is obtained from Σ using one or more steps of FH deduction. As with FH edit generation, checks which represent "dominated" edits can be excluded. This translates to:

1. excluding any subsumed check.
2. excluding any check which is mapped to "true" by every truth function. This will happen if Axiom 2 is part of the check.

We will call the resulting set of FHNCs derived from a starting set Σ of FHNCs the *set of minimal FH-deduced FHNCs*, written MFH(Σ) or just MFH if the context is clear. The field j is an *involved field* of the FHNC $\bigvee_{j=1}^{N} \bigvee_{v \in S_j^\epsilon} p_j^v$ if $S_j^\epsilon \neq \emptyset$. A *covering set* for a record is a set C of fields such that each failed FHNC in the MFH has at least one involved field in C. With these definitions, the error localisation guarantee (Theorem 1) translates to logic as:

Theorem 2 (Translation to logic of Theorem 1, Corollaries 1, 2 of [1], modified in [3]). *Let f be a truth function and let Σ be a set of FHNCs such that $\Box \notin \mathrm{MFH}(\Sigma)$. Let $\Sigma' = \{\epsilon \in \mathrm{MFH}(\Sigma) \mid f(\epsilon) = \text{false}\}$ be those FHNCs in*

MFH(Σ) which f makes false. Then there is a covering set of Σ', and for every covering set C there is a truth function f' such that (1) for all fields j with $j \notin C$ and all $v \in A_j$, $f'(p_j^v) = f(p_j^v)$, and (2) for all $\epsilon \in \Sigma$, $f'(\epsilon) = \text{true}$.

The proof is a faithful translation to logic of Fellegi and Holt's proof.

5 Using Resolution Deduction instead of FH Deduction

Instead of using Fellegi-Holt deduction, we propose using resolution deduction, explained below. In this section we show that the MFH can be obtained using resolution deduction instead of FH deduction, opening the possibility to use the well-developed methods of automated resolution deduction to obtain the MFH.

Resolution deduction is a well-known method of generating formulae from other formulae [11]. Whereas FH deduction applies to FHNCs, resolution deduction applies to clauses. A single step of resolution takes as input two clauses $q \vee l_1 \vee \cdots \vee l_m$ and $\neg q \vee l'_1 \vee \cdots \vee l'_n$, where q is an atom and $l_1, \ldots, l_m, l'_1, \ldots, l'_n$ are literals, and produces the *resolvent* clause $l_1 \vee \cdots \vee l_m \vee l'_1 \vee \cdots \vee l'_n$, by cancelling out q and $\neg q$. The result of resolving the unit clauses q and $\neg q$ is $\Box$.

We define *RA (resolution-axioms)* deduction using resolution and the two Axioms of Section 4. The clause σ is *RA-deducible* from Σ, written $\Sigma \vdash_{\text{RA}} \sigma$, if σ results from $\Sigma \cup \text{Axioms}$ by possibly many resolution steps or $\sigma \in \Sigma \cup \text{Axioms}$.

As with FH-deduction, given a starting set Σ of clauses, we can apply RA-deduction until no new clauses can be generated. The process will eventually terminate because the number of possible clauses is finite. We remove any subsumed clauses or any clause which contains Axiom 2. At the end, we also remove all non-FHNCs. We call the unique resulting set of clauses *the set of minimal RA-deduced FHNCs* and write it as MRA(Σ) or just MRA if the context is clear.

Example, part 10. The FHNC ϵ_3 is RA-deducible from $\{\epsilon_1, \epsilon_2\}$ as follows:

$$\epsilon_1 = p_{\text{driver}}^N \vee p_{\text{grade}}^{11} \vee p_{\text{grade}}^{12} \qquad \epsilon_2 = p_{\text{age}}^{10} \vee \cdots \vee p_{\text{age}}^{20} \vee p_{\text{grade}}^{1} \vee \cdots \vee p_{\text{grade}}^{6}$$

For $v = 1, \ldots, 6$, resolve the axiom $\neg p_{\text{grade}}^{11} \vee \neg p_{\text{grade}}^{v}$ with ϵ_1 on p_{grade}^{11} to obtain 6 resolvents:

$$p_{\text{driver}}^N \vee p_{\text{grade}}^{12} \vee \neg p_{\text{grade}}^{v} . \qquad (*v)$$

For $v = 1, \ldots, 6$, resolve $(*v)$ with the axiom $\neg p_{\text{grade}}^{12} \vee \neg p_{\text{grade}}^{v}$ to obtain 6 resolvents:

$$p_{\text{driver}}^N \vee \neg p_{\text{grade}}^{v} . \qquad (+v)$$

Now resolve $(+1)$ with ϵ_2 on p_{grade}^{1} to obtain:

$$p_{\text{age}}^{10} \vee \cdots \vee p_{\text{age}}^{20} \vee p_{\text{driver}}^N \vee p_{\text{grade}}^{2} \vee \cdots \vee p_{\text{grade}}^{6} . \qquad (\dagger 1)$$

Now resolve $(+2)$ with $(\dagger 1)$ on p_{grade}^{2} to eliminate p_{grade}^{2} from $(\dagger 1)$ to obtain:

$$p_{\text{age}}^{10} \vee \cdots \vee p_{\text{age}}^{20} \vee p_{\text{driver}}^N \vee p_{\text{grade}}^{3} \vee \cdots \vee p_{\text{grade}}^{6} . \qquad (\dagger 2)$$

Continue successively resolving on $p^3_{\text{grade}}, \cdots, p^6_{\text{grade}}$ to eventually obtain

$$\epsilon_3 = p^{10}_{\text{age}} \vee \cdots \vee p^{20}_{\text{age}} \ \vee \ p^N_{\text{driver}} \,. \tag{†6}$$

Example part 10 can be generalised on the number and sizes of fields. Some induction steps yield MFH(Σ) $\subseteq$ MRA(Σ). The reverse direction, MRA(Σ) $\subseteq$ MFH(Σ), requires induction on the number of resolution steps, yielding:

Theorem 3. *If Σ is a set of FHNCs then MFH(Σ) = MRA(Σ).*

Theorem 3 has two consequences. Firstly, automated resolution deduction will find the MFH. Secondly, the set of MRAs guarantees error localisation since Theorem 2 applies, and as with FH-deduction, a smallest covering set is guaranteed to be a solution to the error localisation problem.

The Fellegi-Holt theorems in logical perspective. In effect Fellegi and Holt proved some well-known and important results about resolution [11]. Their Lemma 1 is equivalent to the standard logic result known as soundness, and their Theorem 1 has as a simple consequence the standard logic result known as refutational completeness. In fact, their Theorem 1 (part of our Theorem 2) is a generalisation of a standard logic lemma [11, Lemma 8.14, page 55]. Although we have omitted the details, our logical equivalents of the Fellegi-Holt theorems show that, contrary to the claim of Fellegi and Holt, their Theorem 1 can be proved without using their Theorem 2. These observations support our claim that a recasting of the Fellegi-Holt method into formal logic helps to reason about the method itself.

6 Experiments

We conducted experiments to prove our claim that FH-generation can be done using resolution. We chose OTTER [12], a resolution-based theorem prover, for its reliability and easy availability. Since we were only testing our claim, it did not matter that OTTER is not tuned for propositional deduction. We used OTTER in its default configuration, on a Sun Ultra Sparc 250, and used hyper-resolution (a special type of resolution) as the deduction method, and the "propositional" setting. The clause lists were as follows: "usable" list - the given checks; "set of support" list - Axiom 1; "passive" list - Axiom 2. We used OTTER to generate the MRA for various initial sets of checks.

Our initial tests verified that OTTER does in fact find all the required deduced checks. In particular we used the example analysed in [5] and [6], and obtained the same deduced checks. The CPU time for this small example was 0.7 seconds.

It is hard to compare experimental results since any other implemetations are likely to be out of date, because the FH method has not been used in statistical agencies recently. The only published results that we know of are in [5].

We conducted tests to assess the effect on CPU time of increasing the initial number of checks. We used various combinations of three parameters: number of checks, number of fields and number of values per field. For each such combination we produced 1000–3000 randomly generated sets of edits, in order to observe the variability of the results. Our sets of checks are arranged in two groups, Group A and Group B, whose parameters are listed in Table 1. The sets of checks in Group A contained checks with 13 fields and up to 8 values per field. In Group B, the sets of checks contained checks with 10 fields and up to 10 values per field. Each Group contained sets of checks with 14, 19, 28 and 35 checks. Table 2 gives the CPU times of our results, listing the median times, and the upper 75% and 99% percentiles.

Table 1. Experiment parameters

	Group A	Group B
No. of fields	13	10
No. of values per field	Fields 1 & 2: 2	Fields 1 & 2: 2
	Fields 3 & 4: 3	Field 3: 3
	Fields 5 & 6: 4	Field 4: 4
	Fields 7 & 8: 5	Field 5: 5
	Fields 9 & 10: 6	Field 6: 6
	Fields 11 & 12: 7	Field 7: 7
	Field 13: 8	Field 8: 8
		Field 9: 9
		Field 10: 10

Table 2. CPU time (secs) to generate MRAs

	Group A			Group B		
Initial no. of FHNCs	Median	75% percentile	99% percentile	Median	75% percentile	99% percentile
14	1	1	39	2	5	147
19	2	6	280	3	14	898
28	2	12	2297	1	5	904
35	1	4	1049	1	2	165

As the initial number of checks increases, the computational time first increases and then decreases. The decrease is caused by the interconnectedness of the checks, that is, the checks are more likely to have fields in common which are then eliminated in the deduction process. For example, as the number of randomly generated checks increases, an inconsistency becomes more likely, reducing the MRA to the single clause □. In addition, clauses restricted to one field are also more likely, so that newly generated clauses are more likely to be subsumed. Such situations are unrealistic: contradictory checks are almost always unacceptible as are checks which unduly restrict one field. Thus these experiments demonstrate some characteristics of randomly generated problems as opposed to real problems. The computational times decline because of the increasing proportion of unrealistic sets of checks.

Efficient methods for performing deduction have been studied extensively in the field of automated deduction, and OTTER is by no means state-of-the-art. Systems like Chaff [13] can now perform automated deduction in propositional logic orders of magnitude faster than can OTTER. Simon and del Val [14] state that modern consequence finders can generate and represent huge numbers (10^{70}) of deduced clauses efficiently. With such systems, one would expect that the CPU times will be much lower.

7 Conclusions and Future Work

We have shown the equivalence between Fellegi-Holt deduction and resolution deduction. Fellegi and Holt's work is truly impressive: in apparent isolation from logicians, they reinvented automated propositional deduction and proved the underlying theorems.

The difficulty with the Fellegi-Holt method is the slowness of generating a suitable set of edits for error localisation. As noted by Bruni and Sassano [9], methods like Fellegi-Holt, which generate all implied edits, are not applicable to many problems because the number of implied edits increases exponentially with the number of original edits. The translation to propositional logic cannot avoid this problem, of course, since it is inherent in any NP-complete problem. But modern consequence finders [14] and propositional satisfiability solvers [13] can efficiently handle huge problems which also have the same exponential behaviour. So they are worth investigating for our fomalisation of the Fellegi-Holt method.

We have shown that logic has potential for improving the method by giving a rigorous framework for reasoning about the method, and by demonstrating that the core of the Fellegi-Holt method can be implemented in automated deduction tools based on logic. Our next step will be to use some well-tuned deduction system, or consequence finder, to generate checks efficiently.

Acknowledgements. Thank you to the three anonymous referees for their helpful suggestions, to Matt Gray for his efficient programming assistance, and to Andrew Slater for informative discussions about propositional satisfiability solvers.

References

1. Fellegi, I.P., Holt, D.: A systematic approach to automatic edit and imputation. Journal of the American Statistical Association **71** (1976) 17–35
2. U.N. Statistical Commission, Economic Commission for Europe and Conference of European Statisticians: Work Session on Statistical Data Editing. Helsinki (2002)
3. Liepins, G.E.: A rigorous, systematic approach to automatic editing and its statistical basis. Report ORNL/TM-7126, Oak Ridge National Lab., Tennessee (1980)
4. Liepins, G.E.: Refinements to the Boolean approach to automatic data editing. Technical Report ORNL/TM-7156, Oak Ridge National Lab., Tennessee (1980)
5. Garfinkel, R.S., Kunnathur, A.S., Liepins, G.E.: Optimal imputation of erroneous data: Categorical data, general edits. Operations Research **34** (1986) 744–751
6. Winkler, W.E.: Editing discrete data. Statistical Research Report Series, RR97/04, U.S. Bureau of the Census (1997)
7. Winkler, W.E., Chen, B.C.: Extending the Fellegi-Holt model of statistical data editing. Research Report Series, Statistics #2002-02, U.S. Bureau of the Census (2002)
8. Barcaroli, G.: Un approccio logico formale al probleme del controllo e della correzione dei dati statistici. Quaderni di Ricerca 9/1993, Istituto Nazionale di Statistica , Italia (1993)
9. Bruni, R., Sassano, A.: Errors detection and correction in large scale data collecting. In: Advances in Intelligent Data Analysis, Volume 2189 of Lecture Notes in Computer Science, Springer-Verlag (2001) 84–94
10. Franconi, E., Palma, A.L., Leone, N., Perri, S., Scarcello, F.: Census data repair: A challenging application of disjunctive logic programming. In: 8th International Conference on Logic for Programming, Artificial Intelligence and Reasoning (2001)
11. Nerode, A., Shore, R.A.: Logic for Applications. Springer-Verlag, New York (1997)
12. Argonne National Laboratory, Math. & Comp. Sci. Div.: Otter: An automated deduction system. (Web page) http://www-unix.mcs.anl.gov/AR/otter/
13. SAT Research Group, Electrical Engineering Department, Princeton University: zChaff. (Web page) http://ee.princeton.edu/~chaff/zchaff.php
14. Simon, L., del Val, A.: Efficient consequence finding. In: IJCAI '01: 17th International Joint Conference on Artificial Intelligence, Morgan Kaufmann (2001)

Compression Technique Preserving Correlations of a Multivariate Temporal Sequence

Ahlame Chouakria-Douzal

Sic-Timc-Imag, Université Joseph Fourier Grenoble 1
F-38706, La Tronche, France
Ahlame.Douzal@imag.fr

Abstract. The application of Data Analysis methods to temporal sequences is limited because of the inherent high temporal dimension of the data. One promising solution consists in performing dimensionality reduction before analysing the sequences. We propose a new technique that reduces the temporal dimension of a multivariate sequence while preserving correlations between the sequence's descriptive features, a highly important property for many Data Analysis process. We illustrate the effectiveness of the proposed technique for a multivariate temporal sequence of an intensive care unit application.

1 Introduction

The growing number of industrial, medical and scientific applications using temporal data has increased the importance of temporal sequences analysis techniques. At the same time, the size of the temporal databases often prohibits the use of conventional data analysis methods. A frequently used solution is to reduce the size of temporal sequences before analysing the data. Three of the main dimensionality reduction techniques are based on Discrete Fourier Transform (DFT) [1], Singular-Value Decomposition (SVD) [10] and, more recently, Discrete Wavelet Transform (DWT) [2]. These techniques describe the initial temporal sequence in the spectral (DFT, DWT) or singular-value (SVD) domains, then reduce the dimension of the new descriptive space using spectral or singular-value criteria.

In the field of Data Analysis, the above dimensionality reduction techniques have a number of disadvantages. Temporal information is lost when a sequence is described in a spectral or singular-value domain; in addition, the DFT, DWT reduction processes are limited to a spectral criterion (i.e. the power spectrum). We propose a simple and effictive compression technique based on the extraction of the main local trends. The reduced sequence is described in the temporal domain with preserving the correlations between the sequence's descriptive features, a high important property for many Data Analysis tasks such as: decision or association rules extraction, clustering and classification, discriminant analysis, factorial analysis, feature selection, and so on.

The rest of the paper is organized as follows. In section 2, we present the notion of local variance-covariance initially introduced by Von Neumann, J. [8],

M.R. Berthold et al. (Eds.): IDA 2003, LNCS 2810, pp. 566–577, 2003.

[9]. In section 3, we extend local variance-covariance results to the temporal case. In section 4, we introduce a temporal sequence decomposition operator and the main associated properties. In section 5, we present the new compression technique. Finally, in section 6 we illustrate the effectiveness of the proposed technique by applying it to a multivariate temporal sequence.

2 Local Variance-Covariance

The notion of temporal variance was introduced by Von Neumann, J. [8], [9]. He demonstrated that the conventional measure used to estimate the variance-covariance of data describing some trends will tend to overestimate the realistic deviation. This work was then generalized by Geary, R.C. [5] and Lebart, L. [6], [7] for local variance-covariance estimation in the case of a general proximity structure.

Consider a general proximity structure defined for N observations $w_1, ..., w_N$ where each observation provides the description of a set of features $X_1, ..., X_p$. The proximity structure associates each observation w_i with the set $E_i \subset \{w_1, ..., w_N\}$ of its neighbouring observations. Let $V_L(v_{jl}^L)$ be the local variance-covariance matrix of dimension p corresponding to the N observations and taking into account the proximity structure defined above:

$$v_{jl}^L = \frac{1}{2m} \sum_{i=1}^{N} \sum_{i' \in E_i} (x_{ij} - x_{i'j})(x_{il} - x_{i'l}) , \tag{1}$$

where $m = \sum_{i=1}^{N} Card(E_i)$, $Card(E_i)$ defines the E_i neighbourhood cardinality and v_{jl}^L is the local covariance between the features X_j and X_l. Based on the above, Geary, R.C. [5] and Lebart, L. [6], [7] proposed to compare the local to the classical variance-covariance in order to measure the extent to which the data are dependent on the proximity structure. The closer the local variance-covariance is to the classical variance-covariance, the more the data are independent with respect to the proximity structure (i.e. no trend). On the other hand, the smaller the local variance-covariance compared to the classical variance-covariance, the more the data are dependent on the proximity structure (i.e. presence of trend).

3 Temporal Variance-Covariance

The rest of the paper will focus on a temporal proximity structure. We define a multivariate temporal sequence as a set of successive observations describing a set of features in the time domain. The length of the sequence is defined as the number of observations. Let S be a multivariate temporal sequence of length N and $w_1, ..., w_N$ the observations corresponding to times $t_1, ..., t_N$ and describing p features $X_1, ..., X_p$:

$$S = \begin{pmatrix} w_1 \\ \vdots \\ w_N \end{pmatrix} = \overset{\begin{matrix} X_1 & \dots & X_p \end{matrix}}{\begin{pmatrix} x_{11} & \dots & x_{1p} \\ \vdots & \ddots & \vdots \\ x_{N1} & \dots & x_{Np} \end{pmatrix}}$$

We define the temporal neighborhood E_i of order $k \geq 1$ associated with the observation w_i as the set of observations belonging to the temporal window $[t_{i-k}, t_{i+k}]$ $(0 \leq i \pm k \leq N)$. In the following we consider a temporal neighborhood of order 1.

In the case of a temporal proximity structure, the local variance-covariance matrix $V_T(S)$ (called temporal) corresponding to the sequence S and of the general term $v_{jl}^T(S)$ can be expressed as:

$$v_{jl}^T(S) = \frac{1}{4\,(N-1)} \sum_{i=1}^{N-1} (x_{(i+1)j} - x_{ij})(x_{(i+1)l} - x_{il})\,. \tag{2}$$

4 Definitions and Properties

The proposed technique is based on a "Decomposition" operator and on the main properties of the temporal variance-covariance. We first introduce the "Decomposition" and "Inclusion" operators defined for a set of sequences. We then develop the main properties of "Additivity" and "Growth & Monotonicity".

4.1 Definitions

Decomposition. Let $S = (w_1, ..., w_N)$ be a multivariate temporal sequence of N observations $w_1, ..., w_N$. We decompose S into k sequences $S_1 = (w_1, ..., w_{n_1})$, $S_2 = (w_{n_1}, ..., w_{n_2})$, ..., $S_k = (w_{n_{k-1}}, ..., w_N)$ with, $L_1 = n_1, L_2 = n_2 - n_1 + 1, ..., L_k = N - n_{k-1} + 1$ the respective lengths of $S_1, S_2, ..., S_k$. We note this decomposition as $S = S_1 \oplus S_2 \oplus ..., \oplus S_k$, where $\oplus$ defines the decomposition operator. Note that the observation w_i belongs to both sequences S_i and S_{i+1}.

Inclusion. Let S be a temporal multivariate sequence and $P(S)$ the set of all the sub-sequences extracted from S. We define the inclusion operator in $P(S)$ as:

$$\forall\, S_i, S_j \in P(S) \quad S_i \subset S_j \Rightarrow \exists\, S_k \in P(S) \quad / \quad S_j = S_i \oplus S_k\,.$$

Order Relationship. We define the order relationship $\leq$ in $P(S)$ as:

$$\forall\, S_i, S_j \in P(S) \quad S_i \leq S_j \Leftrightarrow S_i \subset S_j\,. \tag{3}$$

4.2 Properties

Let $S = S_1 \oplus, ..., \oplus S_k$ be a temporal multivariate sequence of length N decomposed into k sequences of respective lengths $L_1, ..., L_k$. The temporal variance-covariance matrix $V_T(S)$ satisfies the following property (see proof in Appendix):

$$V_T(S) = V_T(S_1 \oplus, ..., \oplus S_k) = \sum_{i=1}^{k} \frac{L_i - 1}{N - 1} V_T(S_i) . \qquad (4)$$

In view of the above property, we introduce a new function C_S defined over $P(S)$ which associates with each sub-sequence S_i its contribution $C_S(S_i)$ to the whole temporal variance-covariance of S:

$$C_S : P(S) \longrightarrow R^+$$
$$S_i \longrightarrow C_S(S_i) = \frac{L_i - 1}{N - 1} Tr(V_T(S_i)) .$$

We can easily check that $C_S(S) = Tr(V_T(S))$.

Additivity. Let $S = S_1 \oplus, ..., \oplus S_k$ be a temporal multivariate sequence of length N decomposed into k sequences of respective length $L_1, ..., L_k$. We can easily prove from the result established in formula 4 that the function C_S is additive relative to the operator $\oplus$:

$$C_S(S) = C_S(S_1 \oplus, ..., \oplus S_k) = \sum_{i=1}^{k} C_S(S_i) . \qquad (5)$$

Growth and Monotonicity. Let S be a temporal multivariate sequence of length N. The function C_S satisfies the following property of growth and monotonicity according to the defined order relatioship $\leq$:

$$\forall\, S_i,\, S_j \in P(S) \;\; S_i \leq S_j \Rightarrow C_S(S_i) \leq C_S(S_j) .$$

Indeed, the order relationship definition allows us to write:

$$S_i \leq S_j \Longrightarrow S_i \subset S_j \Longrightarrow \exists\, S_k \subset P(S) \;/\; S_j = S_i \oplus S_k ,$$

according to the additivity property of C_S, we obtain:

$$C_S(S_j) = C_S(S_i \oplus S_k) = C_S(S_i) + C_S(S_k) ,$$

as $C_S(S_k) \geq 0$, then $C_S(S_j) \geq C_S(S_i)$.

5 Temporal Sequence Compression

Any dimensionality reduction, compression process leads to some loss of information. We are interested here in the information provided by temporal correlations. Thus our aim is to reduce the sequence length while preserving the main temporal correlations induced by the descriptives features.

The proposed technique depends on one main parameter: the maximum permissible correlation loss after the reduction process. Let $\alpha_p \in [0,1]$ be such an input parameter. In addition, we introduce a threshold $\alpha_{C_S} \in [0,1]$ that defines a minimum contribution to the whole temporal variance-covariance of S.

To reduce the length of a temporal multivariate sequence, the sequence S will basically be segmented into a set of chronologically ordered sub-sequences. A multiple linear regression is then performed to approximate each extracted sub-sequence by a segment. Finally, depending on the correlation loss of the obtained reduced sequence, the algorithm either adjusts the threshold α_{C_S} and repeats the segmentation and regression steps or outputs the obtained reduced sequence. We will now look in detail at the main steps of the dimensionality reduction process.

5.1 Initialization

The initialization step consists in:

- assigning a value to the correlation loss limit $\alpha_p \in [0,1]$. A typical value of $\alpha_p = 0.1$ means that 90% of the initial correlations are conserved after the reduction process.
- assigning an arbitrary initial value to the minimum contribution threshold $\alpha_{C_S} \in [0,1]$. In the segmentation step, each extracted sub-sequence S_i must then have a contribution $C_S(S_i) \simeq \alpha_{C_S}.C_S(S)$. The initial value of α_{C_S} will be adjusted (increased or decreased) at each algorithm iteration until convergence is achieved (see details below). Although the initial value of α_{C_S} does not affect algorithm results, it may influence computation times. We don't discuss this point in the present work.
- calculating the temporal variance-covariance matrix $V_T(S)$ corresponding to the sequence S.

5.2 Segmentation

The aim of this step is to partition the sequence S into chronologically ordered sub-sequences. No assumptions are made on the number or the length of the sub-sequences. Based on the "Decomposition" operator and the associated properties (i.e. "Additivity" and "Growth & Monotonicity") of the temporal variance-covariance, the segmentation makes it possible to extract only sub-sequences S_i including certain trend (i.e. quantified by the minimum contribution $\alpha_{C_S}(S_i)$ to the whole temporal variance-covariance of S). Semantically, such segmentation ensures that each extracted sub-sequence contains a minimum trend.

Let $S = (w_1, ..., w_N)$ be the sequence to segment, S_1 the first sub-sequence to be extract and $S_1^r = (w_1, ..., w_r)$, $r \in \{1, ..., N\}$ the sub-sequence composed of the r first observations. According to the properties of growth and monotonicity of the temporal contribution function C_S, we can write:

$$\forall r \in [1, N-1] \;\; S_1^r \subset S_1^{r+1} \Rightarrow C_S(S_1^r) \leq C_S(S_1^{r+1}).$$

Based on the above order relationship, the extraction of the sub-sequence S_1 consists in determining the first observation w_r satisfying the following condition:

$$C_S(S_1^r) \leq \alpha_{C_S} . C_S(S) < C_S(S_1^{r+1}) .$$

The first extracted sub-sequence is then $S_1 = (w_1, ..., w_{r-1})$. We perform the same to extract the sub-sequences S_2, S_3,... from the remaining observations $\{w_r, ..., w_N\}$, thus partitioning the entire sequence S.

5.3 Regression

The aim of this step is to perform a multiple linear regression on each extracted sub-sequence in order to determine the best (least squares) approximation surface. After the regression process, each sub-sequence is represented by only two observations with the sub-sequence (segment) time span preserved.

5.4 Correlation Loss Estimation

After the segmentation and regression steps, we estimate the correlation loss corresponding to the reduced sequence. S_1^* denotes the reduced sequence of length $2k$ (k is the sub-sequence number) obtained at the first iteration. Consider the correlation matrices $CorS_1^*(c_{jl}^*)$, $CorS(c_{jl})$ corresponding respectively to S_1^*, S. To estimate the correlation loss, we measure on average of how much initial correlations differ from those correponding to the reduced sequence. Let $e(S, S_1^*)$ be such an estimate:

$$e(S, S_1^*) = \frac{2}{p(p-1)} \sum_{j=1}^{p} \sum_{l=j+1}^{p} \mid c_{jl} - c_{jl}^* \mid ,$$

where $e(S, S_1^*) \in [0, 1]$. We introduce a new compression ratio $\tau_C = 1 - \frac{\text{final length}}{\text{initial length}}$ to measure of how much a sequence was reduced. The correlation loss estimate $e(S, S_1^*)$ is then compared to the α_p limit. Two cases may be distinguished:

- If the correlation loss estimate $e(S, S_1^*)$ is lower than the permissible limit α_p, we increase α_{C_S} by a ratio of $\Delta = \frac{\alpha_p}{e(S, S_1^*)}$ (updated α_{C_S} = old $\alpha_{C_S}.\Delta$) in order to look for the maximum compression ratio τ_C keeping the correlation loss below α_p. We repeat the three above steps (i.e. segmentation, regression, correlation loss estimation) on the sequence S with the new α_{C_S} value, and this until the following condition is satisfied:

$$e(S, S_l^*) \leq \alpha_p < e(S, S_{l+1}^*) .$$

At iteration l, the reduced sequence S_l^* maximizes the compression ratio τ_C for a correlation loss limit of α_p.

- On the other hand, if the correlation loss estimate $e(S, S_1^*)$ is greater than the permissible limit α_p, we decrease α_{C_S} by a ratio of $\Delta = \frac{\alpha_p}{e(S,S_1^*)}$ (updated $\alpha_{C_S} =$ old $\alpha_{C_S}.\Delta$). We repeat the three above steps on the sequence S with the new α_{C_S} value, and this until the following condition is satisfied:

$$e(S, S_l^*) \leq \alpha_p < e(S, S_{l-1}^*) .$$

S_l^* is then the reduced sequence preserving correlations for a limit of $1 - \alpha_p$.

6 Experimental Results

6.1 The Multivariate Temporal Sequence

The experiment is based on monitoring data describing 11 quantitative physiological features, regularly observed over time, of a patient in an intensive care unit. Patient data constitute a multivariate temporal sequence of 5269 observations of 11 numerical features. Figure 1 shows this multivariate temporal sequence. Each curve represents the changes of one descriptive feature with time. This experiment lies within the scope of developing an intelligent monitoring

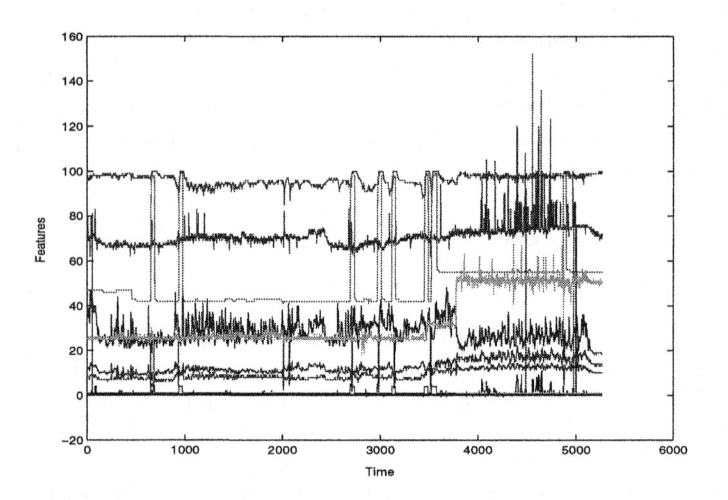

Fig. 1. Initial multivariate sequence with a length of 5269

alarm system in intensive care units. The above physiological data are collected in conjunction with notes of alarm appropriateness (i.e. false alarm, clinically-relevant true alarm, clinically-irrelevant true alarm,...). The aim of such intelligent alarm system is to provide a more accurate alarm system by generating a specific alarm for each recognized abnormal physiological state. Tasks related to temporal feature selection and discriminant analysis (to extract distinctive sub-sequences each expressing an abnormal physiological state) are under study.

Monitoring data are generally very huge. For performance reasons, we propose first to compress these data while preserving the interdependencies among physiological features, a basic information for the following processes of feature selection and discriminant analysis.

6.2 Temporal Sequence Compression

To assess the accuracy and the efficiency of the new compression technique we apply it to the above multivariate sequence. First, we will illustrate the results obtained at each iteration of the compression algorithm. Secondly, we will show and compare the results of the proposed technique when the correlation loss limit is increased from 10% ($\alpha_p = 0.1$) to 70% ($\alpha_p = 0.7$). Table 1 illustrates the iterations of the compression algorithm for a correlation loss limit of 20% (α_p=0.2).

Table 1. Results of the compression algorithm's iterations

Iteration number	Initial α_{CS}	Final length	Correlation loss estimate	Updated α_{CS}	Runtime (s)
1	0.2	10	0.5937	0.0674	38.4690
2	0.0674	10	0.4196	0.0321	34.8280
3	0.0321	26	0.2492	0.0258	37.4370
4	0.0258	52	0.3164	0.0163	34.7190
5	0.0163	72	0.1831	–	38.7500

The compression algorithm converges in five iterations for $\alpha_p = 0.2$. The first iteration (segmentation, regression and correlation loss estimation steps) at $\alpha_{CS} = 0.2$ gives a reduced sequence with a length of 10 (5 sub-sequences) and a correlation loss estimate of 0.5937. As the correlation loss is greater than the prior limit $\alpha_p = 0.2$, we decrease α_{CS} by a ratio of $\Delta = \frac{0.2}{0.5937}$ (updated $\alpha_{CS} = 0.2\frac{0.2}{0.5937} = 0.0674$). The first iteration was carried out in 38.4690 seconds. The second iteration was performed with the new $\alpha_{CS} = 0.0674$ value. At the iteration five, the obtained reduced sequence has a length of 72 (36 sub-sequences) and the stopping condition is reached with a correlation loss of 0.1831 lower than α_p.

The above results show that a compression with $\alpha_p = 0.2$ reduces the length of the sequence from 5269 to 72 observations while preserving 81.69% of the temporal correlations induced by the 11 descriptive features.

Let's now illustrate and compare the results of the proposed algorithm when the correlation loss threshold is increased from 10% ($\alpha_p = 0.1$) to 70% ($\alpha_p = 0.7$). Each row of Table 2 summarizes the compression results for a given correlation loss limit α_p.

Table 2. Compression results for a correlation loss limit increasing from 10% to 70%

α_p	Final length	τ_C (%)	Correlation loss estimate	Runtime (s)
0.1	370	92.97	0.09580	576.8430
0.2	72	98.63	0.1831	184.2030
0.3	26	99.50	0.2840	180.1400
0.4	20	99.62	0.3433	248.1870
0.5	8	99.84	0.4257	105.4070
0.6	4	99.92	0.5973	68.4840
0.7	4	99.92	0.6591	68.7650

For instance, at a correlation loss limit of 10% (α_p=0.1), compression provides a reduced sequence of length 370 with a compression ratio of 92.97%. The effective correlation loss estimate is 9.58% (i.e. a correlation preservation of 90.42%). The algorithm converges in 576.8430 seconds.

We can easily see that the compression ratio τ_C and the correlation loss estimate increase with the correlation loss limit. The reduced sequence obtained at each α_p limit is shown in Figures 2 to 5.

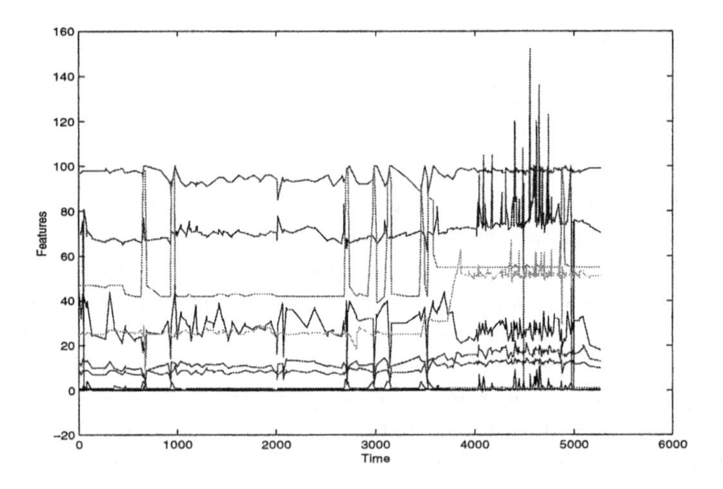

Fig. 2. Reduced sequence with a length of 370 (10% correlation loss)

The comparison of the reduced sequence in Figure 2 with the initial sequence in Figure 1 shows that with only 370 observations we can capture the main variations (crossing points between curves, peak values) simultaneously for the 11 features. Figures 3, 4 and 5 clearly show the loss of correlation due to the increase in α_p.

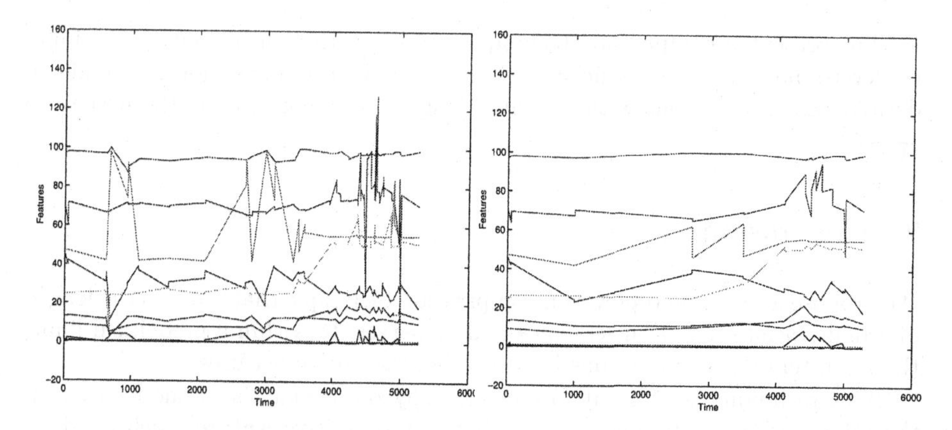

Fig. 3. Left) Reduced sequence of length 72 (20% correlation loss) Right) Reduced sequence with a length of 26 (30% correlation loss)

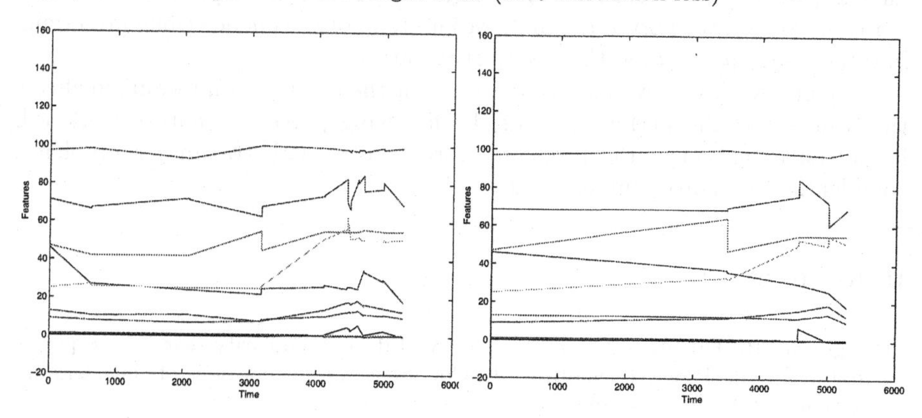

Fig. 4. Left) Reduced sequence with a length of 20 (40% correlation loss) Right) Reduced sequence with a length of 8 (50% correlation loss)

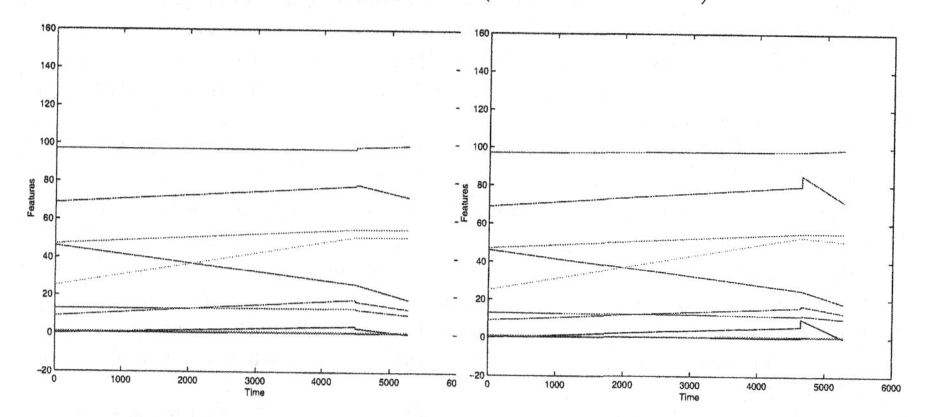

Fig. 5. Left) Reduced sequence with a length of 4 (60% correlation loss) Right) Reduced sequence with a length of 4 (70% correlation loss)

The presented compression technique can be carried out at several α_p limits to determine either the reduced sequence providing a correlation loss under a desired value, or the one with a desired length, or the one giving the best ratio $\frac{\tau_c}{e(s,s^*)}$.

7 Conclusion

We propose a simple and powerful compression technique that reduces the length of a multivariate temporal sequence and preserves, to within a predefined limit, the main temporal correlations between the descriptive features.

This technique is of great interest for any data analysis method based on the information of interdependencies between descritive features such as: decision or association rules extraction, clustering and classification, discriminant analysis, feature selection, factorial analysis... where generally, for performance reasons, we need to shorten the multivariate temporal sequence while preserving interrelationships between the descriptive features.

Ongoing work will evaluate in what extent the compression technique affects the results and the performance of the following processes. Future work will intend an extention of the proposed work to deal with two major problems: confidential and noisy temporal data.

References

1. Agrawal, R., Faloutsos, C. and Swami, A.: Efficient similarity search in sequence databases. In: Proceedings of the 4th Conference on Foundations of Data Organization and Algorithms (1993).
2. Chan, K. and Fu, W.: Efficient time series matching by wavelets. In: Proceedings of the 15th IEEE International Conference on Data Engineering (1999).
3. Chatfield, C.: The analysis of time series. Chapman and Hall (1996).
4. Faloutsos, C., Ranganathan, M. and Manolopoulos, Y.: Fast subsequence matching in time series databases. In: ACM SIGMOD Conference, Minneapolis (1994).
5. Geary, R.C.: The contiguity ratio and statistical mapping. The Incorporated Statistician, 5(3), 115–145 (1954).
6. Lebart, L.: Analyse statistique de la contiguité. Publication de l'ISUP, 18, 81–112 (1969).
7. Lebart, L., Morineau, A. and Piron, M.: Statistiques Exploratoires Multidimensionnelles. Dunod (1995).
8. Von Neumann, J.: Distribution of the ratio of the mean square successive difference to the variance. The Annals of Mathematical Statistics, 12, 4 (1941).
9. Von Neumann, J., Kent, R.H., Bellinson, H.R. and Hart, B.I.: The mean square successive difference to the variance. The Annals of Mathematical Statistics, 153–162 (1942).
10. Wu, D., Agrawal, R., Elabbadi, A., Singh, A. and Smith, T.R.: Efficient retrieval for browsing large image databases. In: Proceedings of the 5th International Conference on Knowledge Information. 11–18 (1996).

Appendix

Temporal Variance-Covariance Contribution

For clarity and without restricting generality, we consider k=2. Let $S = S_1 \oplus S_2$ be a sequence of length N decomposed in two sub-sequences S_1 and S_2 of respective lengths L_1 and L_2. To demonstrate the above property we demonstrate it for the general term $v_{jl}^T(S)$ of the matrix $V_T(S)$. The temporal covariance expression v_{jl}^T associated with S is:

$$v_{jl}^T(S) = \frac{1}{4\,(N-1)} \sum_{i=1}^{N-1} (x_{(i+1)j} - x_{ij})(x_{(i+1)l} - x_{il})\,.$$

We rewrite the above expression as the sum of two parts, the first part is composed of the $L_1 - 1$ first terms and the second is composed of the $L_2 - 1$ remaining terms:

$$\begin{aligned} v_{jl}^T(S) &= \frac{1}{4\,(N-1)} \left[\sum_{i=1}^{L_1-1} (x_{(i+1)j} - x_{ij})(x_{(i+1)l} - x_{il}) \right. \\ &\quad \left. + \sum_{i=L_1}^{N-1} (x_{(i+1)j} - x_{ij})(x_{(i+1)l} - x_{il}) \right] \\ &= \frac{1}{4\,(N-1)} [\frac{4(L_1-1)}{4(L_1-1)} \sum_{i=1}^{L_1-1} (x_{(i+1)j} - x_{ij})(x_{(i+1)l} - x_{il}) + \\ &\quad \frac{4(L_2-1)}{4(L_2-1)} \sum_{i=L_1}^{N-1} (x_{(i+1)j} - x_{ij})(x_{(i+1)l} - x_{il})] \end{aligned}$$

Note that the expression $\frac{1}{4(L_1-1)} \sum_{i=1}^{L_1-1} (x_{(i+1)j} - x_{ij})(x_{(i+1)l} - x_{il})$ defines the temporal covariance v_{jl}^T associated with the sequence S_1, whereas the expression $\frac{1}{4(L_2-1)} \sum_{i=L_1}^{N-1} (x_{(i+1)j} - x_{ij})(x_{(i+1)l} - x_{il})$ defines the temporal covariance associated with the sequence S_2. Therefore:

$$\begin{aligned} v_{jl}^T(S) &= \frac{1}{4\,(N-1)} [4(L_1-1)\, v_{jl}^T(S_1) + 4(L_2-1)\, v_{jl}^T(S_2)] \\ &= \frac{(L_1-1)}{(N-1)}\, v_{jl}^T(S_1) + \frac{(L_2-1)}{(N-1)}\, v_{jl}^T(S_2) \end{aligned}$$

and

$$V_T(S) = \frac{(L_1-1)}{(N-1)}\, V_T(S_1) + \frac{(L_2-1)}{(N-1)} V_T(S_2)\,.$$

Condensed Representations in Presence of Missing Values

François Rioult and Bruno Crémilleux

GREYC, CNRS - UMR 6072, Université de Caen
F-14032 Caen Cédex France
{Francois.Rioult, Bruno.Cremilleux}@info.unicaen.fr

Abstract. Missing values are an old problem that is very common in real data bases. We describe the damages caused by missing values on condensed representations of patterns extracted from large data bases. This is important because condensed representations are very useful to increase the efficiency of the extraction and enable new uses of frequent patterns (e.g., rules with minimal body, clustering, classification). We show that, unfortunately, such condensed representations are unreliable in presence of missing values. We present a method of treatment of missing values for condensed representations based on δ-free or closed patterns, which are the most common condensed representations. This method provides an adequate condensed representation of these patterns. We show the soundness of our approach, both on a formal point of view and experimentally. Experiments are performed with our prototype MVMINER (for Missing Values miner), which computes the collection of appropriate δ-free patterns.

1 Introduction

Context. Missing values in data bases are an old problem that always arises in presence of real data, for instance, in the medical domain. With regard to opinion polls, it is rare that interviewees take the pain to fill entirely the questionnaire. It is a strong problem because many analysis methods (e.g., classification, regression, clustering) are not able to cope with missing values. The deletion of examples containing missing values can lead to biased data analysis. Elementary techniques (e.g., use of the mean, the most common value, default value) are not more satisfactory, because they exaggerate correlations [7]. At last some treatments are devoted to specific databases [8], but it is difficult to apply them in the general case and it is clear that there is not a single, independent and standard method to deal with missing values. We will see that this problem occurs also on condensed representations.

Condensed representations of frequent patterns provide useful syntheses of large data sets [4,12], highlighting the correlations embedded in data. There is a twofold advantages of such an approach. First, it allows to improve efficiency of algorithms for usual tasks such as association rules [1]. Even if this technique is, today, well mastered, the use of condensed representations enables to

M.R. Berthold et al. (Eds.): IDA 2003, LNCS 2810, pp. 578–588, 2003.

achieve extraction of rules in contexts where usual APRIORI-like algorithms fail [3,12]. Second, condensed representations enable multiple uses of frequent patterns [9,6,16] (e.g., strong rules, informative rules or rules with minimal body, non-redundant rules, clustering, classification) which is a key point in many practical applications. These uses are today required by experts on data who know that the whole value embedded in their data can only be acquired by the new developments of such analysis methods.

Motivations. In real data bases, users have to cope with missing values and we are going to see that, unfortunately, condensed representations are no more valid in presence of missing values. This is the starting point of our work. Let us give an example: the left part of Table 1 (called r or the reference table) provides an example of a transactional database composed of 7 transactions (each one identified by its Tid) and 5 items denoted $A \dots E$. For instance, in medical area (e.g., Hodgkin's disease), the item A denotes an item which means `''Bsymptoms''`[1] `= present`, the item B means `mediastinum = enlarged`, and so on. This table is used as the running example throughout the paper. The right part of Table 1 provides the condensed representation based on 0-free sets with an absolute support threshold of 2. For each 0-free, we indicate its closure. These notions are explained in Section 2.

Table 1. Running example of a database without missing values (r)

r

Tid	Items
1	A D
2	C E
3	A B C D E
4	A D
5	A B D E
6	A B D E
7	A B C D E

0-free set	closure	0-free set	closure
A	$\{D\}$	AC	$\{B,D,E\}$
B	$\{A,D,E\}$	AE	$\{B,D\}$
C	$\{E\}$	BC	$\{A,D,E\}$
D	$\{A\}$	CD	$\{A,B,E\}$
E		DE	$\{A,B\}$

For instance, from this condensed representation, we can extract rules such $AE \Rightarrow BD$ with a support of 4 and a confidence of 100%. Now, let us suppose that some parameters used in the definition of `''Bsymptoms''` are not been caught, so we do not know for instance whether `''Bsymptoms'' = present` is true or not. Then, missing values appear and we use the character `'-'` before an item to note such a situation. The left part of Table 2 (called $\tilde{r}$) includes 5 missing values.

How to deal with missing values? Elementary methods remove data with missing values but it may lead to a biased data set from which extracted information is unreliable. Let us see that in our example. The right part of Table 2

[1] Bsymptoms are features of lymphoma, and include fever, drenching night sweats, weight loss more than 10% of body mass in previous 6 months

Table 2. Running example of a database with missing values ($\tilde{r}$)

$\tilde{r}$

Tid	Items
1	A D
2	C E
3	A B C D E
4	$-A$ D
5	A B $-D$ E
6	$-A$ $-B$ D E
7	A B C D $-E$

0-free set	closure	0-free set	closure	0-free set	closure
A		AD		ABC	$\{E\}$
B		AE	$\{D\}$	ABD	$\{E\}$
C	$\{E\}$	BC	$\{E\}$	ABE	$\{D\}$
D		BD	$\{E\}$	ACD	$\{E\}$
E		BE	$\{D\}$	BCD	$\{E\}$
AB		CD	$\{E\}$	$ABCD$	$\{E\}$
AC	$\{E\}$	DE	$\{A, B\}$		

depicts the condensed representation on $\tilde{r}$ achieved with this strategy. This condensed representation contains new patterns which are not present in the condensed representation of reference (extracted on r). We will qualify (see Section 2.3) such patterns of *pollution*. We also note that a lot of items have disappeared from the almost-closures. Consequently, rules such as $AE \Rightarrow BD$ are no longer found. Clearly, missing values are responsible for the removal of relationships and the invention of new groundless associations. Experiments in Section 4 show that this result is general and it is clear that usual condensed representations cannot safely be used in presence of missing values. The aim of this paper is to propose a solution to solve this open problem.

Contributions. The contribution of this paper is twofold. First, we describe the damages caused by missing values in condensed representations. Second, we propose a method of treatment of missing values for condensed representations based on δ-free or closed patterns (which are the most common condensed representations). This method avoids these damages. We show the soundness of the obtained condensed representations, both on a formal point of view and experimentally. We think that this task is important in data mining owing to the multiple uses of condensed representations.

Organization of the paper. This paper is organized as follows: Section 2 briefly reviews the condensed representations of δ-free patterns and we set out in Section 2.3 the effects of missing values on these representations. In Section 3 we describe the corrections (and we give a formal result) that we propose so as to treat missing values. Section 4 presents experiments both on benchmarks and real world data (medical database on Hodgkin's disease).

2 Condensed Representations

We give the necessary material on condensed representations which is required for the rest of the paper and we present the negative effects of missing values.

2.1 Patterns Discovery

Let us consider a transactional database: a database r is a set of transactions t composed of items. In Table 1 $r = \{t_1, \ldots, t_7\}$ where $t_1 = AD$ (note that we use a string notation for a set of items, e.g., AD for $\{A, D\}$), $t_2 = CE$, etc. Let Z be a *pattern* (i.e. a set of items), an association rule based on Z is an expression $X \Rightarrow Y$ with $X \subset Z$ and $Y = Z \backslash X$.

The support of X with respect to a set of transactions r is the number of transactions of r that contain X, i.e. $supp(X, r) = |\{t \in r \mid X \subseteq t\}|$. We note r_X the subset of transactions of r containing X, we have $supp(X, r) = |r_X|$. If r is clear from the context, we will use $supp(X)$ for $supp(X, r)$. The confidence of $X \Rightarrow Y$ is the proportion of transactions containing X that also contain Y [1], i.e. $conf(X \Rightarrow Y) = supp(X \cup Y)/supp(X)$.

2.2 δ-Free Patterns

Due to the space limitation, we give here only the key intuition of δ-free patterns and the background required to understand the effects of missing values. We start by defining δ-strong rules.

Definition 1 (δ-strong rule) *A δ-strong rule is an association rule of the form $X \Rightarrow Y$ that admits a maximum of δ exceptions [13,3].*

The confidence of such a rule is at least equal to $1 - (\delta/supp(X))$.

Definition 2 (δ-free pattern) *A pattern Z is called δ-free if there is no δ-strong rule $X \Rightarrow Y$ (with $X \subset Z$ and $Y = Z \backslash X$) that holds.*

The case $\delta = 0$ (corresponding to 0-free patterns) is important: no rule with confidence equal to 1 holds between proper subsets of Z. For instance, DE is a 0-free pattern because all rules constructed from proper subsets of DE have at least one exception. If $\delta = 1$, DE is not a 1-free set owing to the rule $E \Rightarrow D$ which has only one exception. From a technical perspective, δ-strong rules can be built from δ-free patterns that constitute their left-hand sides [4]. δ-free patterns are related to the concept of almost-closure:

Definition 3 (almost-closure) *Let δ be an integer. $AC(X, r)$, the almost-closure of X in r, gathers patterns Y so that :*

$$supp(X, r) - supp(X \cup Y, r) \leq \delta \qquad (1)$$

Note that if $X \Rightarrow Y$ is a δ-strong rule in r, items of Y belong to $AC(X, r)$. In other words, when an item belongs to $AC(X, r)$, it means that it is present in all the transactions that contain X with a number of exceptions bounded by δ. Following our example given in Table 1, $D \in AC(E, r)$ with $\delta = 1$ (there is only one exception, transaction t_4). δ-freeness satisfies a relevant property (anti-monotonous constraint) and we get tractable extractions for practical mining tasks that are not feasible with APRIORI-like algorithms [4].

A collection of frequent δ-free patterns is a condensed representation of the collection of frequent patterns. If $\delta = 0$, one can compute the support of every frequent pattern. In this case, the almost-closure corresponds to the special case of the *closure.* Such closed patterns have relevant properties to compute informative [2] or non-redundant [16] rules, or achieve clustering [6]. If $\delta > 0$, one can approximate the support of every frequent pattern X with a bounded error: in [3], it is shown that the error is very low in practice. δ-free patterns verify suitable properties to build δ-strong rules with a high value of confidence [3] and rules characterizing classes [5].

2.3 Effects of the Missing Values on the Condensed Representations Based on δ-Free Patterns

Missing values produce effects on δ-free patterns and items of almost-closures. Let us suppose that an item A belongs to the almost-closure of a δ-free X. It means that A is always present with X except a number of exceptions lower than δ. If missing values on A occur, this number of exceptions can only increase and can become higher than δ: then, A comes out of the almost-closure and the $X \cup \{A\}$ pattern becomes free (see Figure 1). In our example, with $\delta = 0$, B and D belong to the closure of AE on r whereas B comes out of this closure on $\tilde{r}$ owing to the missing value on transaction t_6. Moreover, AB and BE become free and such created free patterns are qualified of *pollution.*

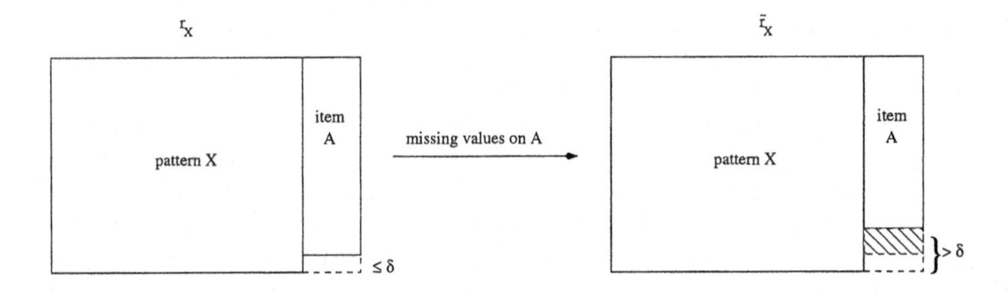

Fig. 1. Missing values on items of almost-closures

3 Corrections of the Missing Values

We present here our corrections on support and almost-closure to deal with missing values. We start by giving the notion of disabled data which is a key step in our corrections.

3.1 Disabled Data

In presence of missing values, support values decrease [14]. For instance, in our example, $supp(DE, r) = 4$ but $supp(DE, \tilde{r}) = 2$. In fact, to compute properly the support of an itemset in $\tilde{r}$, it is necessary to distinguish the transactions of $\tilde{r}$ having at least one missing value among the items of X. These transactions will be temporarily disabled to compute $supp(X, \tilde{r})$ because they do not enable to take a decision to this support. It is shown that this approach allows to retrieve relevant values of support [14].

Definition 4 (disabled data) *A transaction t of $\tilde{r}$ is disabled for X if t contains at least one missing value among the items of X. We note $Dis(X, \tilde{r})$ the transactions of $\tilde{r}$ disabled for X.*

With this approach, the whole data base is not used to evaluate *a* pattern X (owing to the missing values on X) but, as data are only temporarily disabled, finally the whole data base is used to evaluate *all* patterns.

3.2 Correction of the Effect of Missing Values

In case of missing values, we have to take into account, with respect to X, the disabled data of a candidate pattern Y to test if $Y \subseteq AC(X, \tilde{r})$. We propose to redefine the almost-closure of X as follows:

Definition 5 *$AC(X, \tilde{r})$, the almost-closure of X in $\tilde{r}$, gathers patterns Y so that :*

$$supp(X, \tilde{r}) - supp(X \cup Y, \tilde{r}) \leq \delta + Dis(Y, \tilde{r}_X) \tag{2}$$

Of course, this definition makes only sense if there remains at least one transaction in $\tilde{r}$ containing both X and Y.

Let us note that if there is no missing value, $Dis(Y, \tilde{r}_X) = 0$ and $r = \tilde{r}$ and we recognize the usual definition of the almost-closure (see Definition 3). This new definition is fully compatible with the usual one when there are no missing values. Inequality 2 can be seen as a generalization of Inequality 1 (Section 2.2). Inequality 2 can be interpreted as a local relaxation of the constraint on δ (i.e. δ is adjusted for each Y) according to the number of missing values on Y in $\tilde{r}_X$. This definition leads to the important following property:

Property 1 *Definition 5 of the almost-closure is sound in presence of missing values, i.e. $Y \subset AC(X, r) \Rightarrow Y \subset AC(X, \tilde{r})$. Then, by using this definition, the effect of missing values (loss of items from almost-closures and pollution of δ-free patterns (cf. Section 2.3)) is corrected.*

The two following lemmas help to prove this property.

Lemma 1 $supp(X, \tilde{r}) = supp(X, r \backslash Dis(X, \tilde{r})) = supp(X, r) - |Dis(X, \tilde{r})|$

Proof The claim follows directly from Definition 4.

Lemma 2 $supp(X \cup Y, \tilde{r}) = supp(X \cup Y, r) - |Dis(X, \tilde{r})| - |Dis(Y, \tilde{r}_X)|$

Proof The claim follows directly from Lemma 1 and the fact that $|Dis(X \cup Y, \tilde{r})| = |Dis(X, \tilde{r})| + |Dis(Y, \tilde{r}_X)|$.

It is now easy to prove Property 1.
Proof: The effect arises when Y is in $AC(X, r)$ but no more in $AC(X, \tilde{r})$. Consider $Y \subset AC(X, r)$, then $supp(X, r) - supp(X \cup Y, r) \leq \delta$. Adding the terms $|Dis(X, \tilde{r})|$ and $|Dis(Y, \tilde{r}_X)|$ to this inequality, we get:

$$supp(X, r) - |Dis(X, \tilde{r})| - (supp(X \cup Y, r) - |Dis(X, \tilde{r})| - |Dis(Y, \tilde{r}_X)|) \\ \leq \delta + |Dis(Y, \tilde{r}_X)|$$

Lemmas 1 and 2 enable to write $supp(X, \tilde{r}) - supp(X \cup Y, \tilde{r}) \leq \delta + Dis(Y, \tilde{r}_X)$ and we recognize Inequality 2. Thus $Y \subset AC(X, r)$ implies $Y \subset AC(X, \tilde{r})$: Definition 5 is sound in presence of missing values and Property 1 holds.

This property assures to perfectly recover items in the almost-closures in the presence of missing values otherwise they may come out. In our example, this method fully recovers on $\tilde{r}$ the condensed representation of reference (given in the right part of Table 1).

4 Experiments and Results

The purpose of this section is to compare condensed representations achieved by our corrections versus condensed representations obtained without corrections. Experiments are performed with our prototype MVMINER which extracts δ-free patterns with their almost-closures according to our corrections (MVMINER can be seen as an instance of the level-wise search algorithms presented in [10]).

4.1 The Framework of the Experiments

The data base without missing values is the reference data base and it is denoted r. Let us call *reference condensed representation* the condensed representation of reference performed on r. Then, some missing values are randomly introduced in r and the data base with missing values is noted $\tilde{r}$. We are going to discuss of the condensed representation of $\tilde{r}$ obtained by the elementary method (i.e. ignore the missing values) and the one achieved with our corrections (i.e. running MVMINER on $\tilde{r}$) compared to the reference condensed representation. For the following, our method with corrections is simply called MVMINER and the elementary is called "the usual method". Used measures are: pollution of δ-free patterns and recovery of items of almost-closures, differences on the values of support between the usual method and MVMINER, pollution on 0-free patterns according to the rate on missing values.

Experiments have been done both on benchmarks (usual data bases used in data mining and coming from the University of Irvine [11]) and on a real database provided by the European Organization of Research and Treatment of Cancer (EORTC). All the main results achieved are similar and, due to the lack of space, we choose to present here only results on EORTC data base (the results on most data bases are available in [15]). EORTC has the advantage to be a real data base on a problem for which physicians have a great interest. This data base gathers 576 people suffering from Hodgkin's disease, a gland cancer. There are 26 multi-variate attributes, which bring 75 highly correlated binary items.

The protocol of experimentation suggests the use of three parameters: the rate of artificially introduced missing values, the minimum support used in the extraction and the value of δ. Actually, experiments show that only the variation of δ produces changes of tendency [15]. So, we present, below, the results on experiments with 10% of missing values per attribute (randomly introduced), a minimum support of 20%. δ varies from 0 to 20 transactions (i.e. 0 to 3.5%).

4.2 Correction on Almost-Closures

The phenomena described in this section are linked to the effect of missing values, previously described Section 2.3. Figure 2 (on the left) depicts the proportion of δ-free patterns obtained in $\tilde{r}$ by the usual method and *not* belonging to the reference condensed representation (i.e. pollution of δ-free patterns). The usual method undergoes a pollution of 50% as soon as the first values of δ (1.5% or 8 transactions). It means that there are a lot of meaningless patterns which are, obviously, impossible to distinguish from the true patterns. Such a result shows a great interest of MVMINER: we note there is no pollution with MVMINER for any δ value. It is clear that this result was expected owing to the property given in Section 3.2.

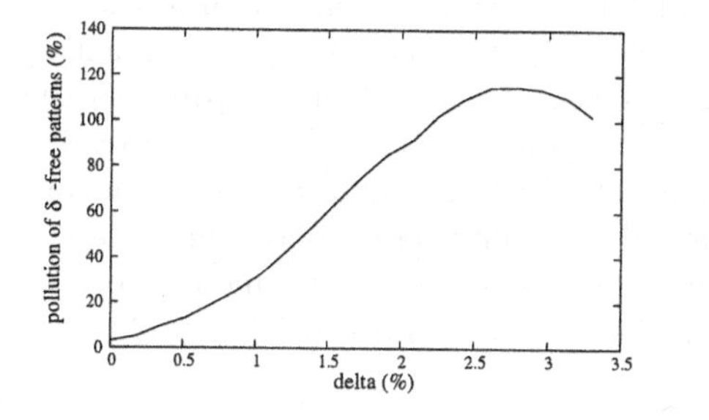

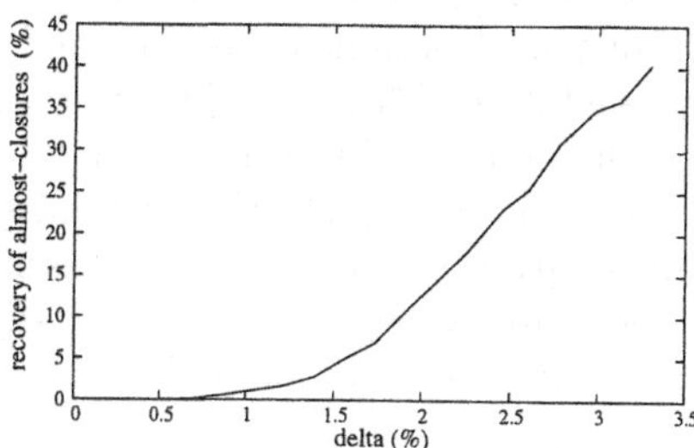

Fig. 2. Drawbacks of the usual method

Figure 2 (on the left) indicates that the level of pollution with the usual method stays low when $\delta = 0$ or is very small and then we can imagine using this method. Figure 2 (on the right) deletes this hope: items of almost-closures are not recovered. Even with high values of δ, only half of the items of almost-closures are found. With corrections, MVMINER fully recovers all items of almost-closures. This shows that it is impossible to trust condensed representations given by the usual method even with a low value of δ and thus highlights again the advantages of MVMINER.

4.3 Support of δ-Free Patterns and Recovery of 0-Free Patterns According to the Rate of Missing Values

Figure 3 (on the left) compares errors on values of the supports of δ-free patterns between the usual method and MVMINER. Error is about twice as low as with MVMINER. Such an improvement may be precious to compute measures of interestingness on rules.

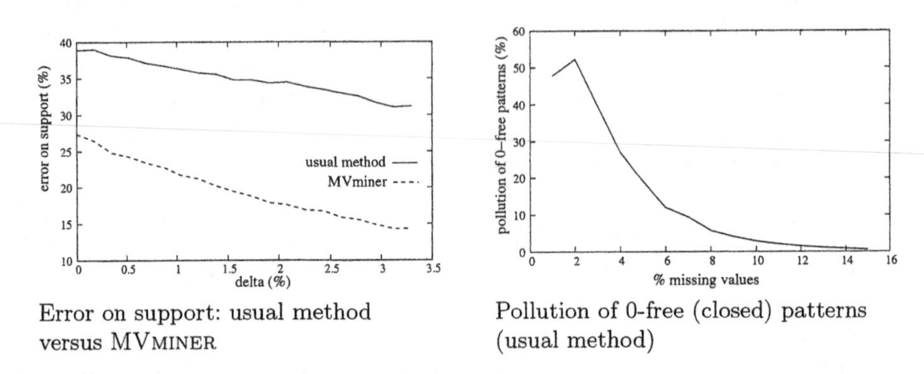

Error on support: usual method versus MVMINER

Pollution of 0-free (closed) patterns (usual method)

Fig. 3.

Let us consider now the pollution brought by the usual method on the condensed representation made up of 0-free patterns (a very often used condensed representation) according to the rate of missing values. This pollution is the proportion of obtained patterns and not belonging to the reference condensed representation. Missing values were introduced on each item according to the percentage indicated on Figure 3. This figure (on the right) shows a high value of pollution as soon as the first missing values are introduced (50% of meaningless patterns). With a lot of missing values, pollution decreases because few patterns are recovered but the condensed representation is groundless. Let us recall that there is no pollution on patterns with MVMINER (cf. Section 4.2).

5 Conclusion and Future Work

We have presented the damages due to missing values on condensed representations based on δ-free patterns and closed patterns which are the most common condensed representations. Without processing, such condensed representations are unreliable in presence of missing values which prevent the multiple uses of extracted patterns. Our analysis clarifies the effects of missing values on δ-free patterns and their almost-closures. We have proposed corrections and a sound new definition of the almost-closure which is a generalization of the usual one and fully compatible with this one. These corrections deal with missing values and recover initial patterns with their almost-closures. The corrections never introduce pollution on patterns. As they check the property of anti-monotonicity, these corrections are effective even in the case of huge, dense and/or highly correlated data. Experiments on real world data and benchmarks show that it is impossible to use trustingly condensed representations without corrections (e.g. high level of pollution on δ-free patterns even with low rates of missing values) and confirm the relevance of the corrections.

A straightforward further work is to use these condensed representations suitable for missing values to determine reliable rules to predict missing values. Such condensed representations have good properties for this task (see for instance [2]). Our collaborations in medicine provide excellent applications.

Acknowledgements. The authors wish to thank Dr M. Henry-Amar (Centre François Baclesse, France) for providing data on Hodgkin's disease and many valuable comments.

References

1. R. Agrawal, T. Imielinski, and A. Swami. Mining association rules between sets of items in large databases. *Proc. of the 1993 ACM SIGMOD International Conference on Management of Data*, pages 207–216, 1993.
2. Y. Bastide, R. Taouil, N. Pasquier, G. Stumme, and L. Lakhal. Mining minimal non-redundant association rules using frequent closed itemsets. *Proceedings of the 6th International Conference on Deductive and Object Databases (DOOD'00)*, pages 972–986, 2000.
3. J.-F. Boulicaut, A. Bykowski, and C. Rigotti. Approximation of frequency queries by means of free-sets. In *Principles of Data Mining and Knowledge Discovery (PKDD'00)*, pages 75–85, 2000.
4. J.-F. Boulicaut, A. Bykowski, and C. Rigotti. Free-sets : a condensed representation of boolean data for the approximation of frequency queries. *Data Mining and Knowledge Discovery journal*, 7(1):5–22, 2003. Kluwer Academics Publishers.
5. B. Crémilleux and J.F. Boulicaut. Simplest rules characterizing classes generated by delta-free sets. *proceedings of the 22nd Annual International Conference of the Specialist Group on Artificial Intelligence (ES'02), Springer, Cambridge, United-Kingdom*, December 2002.
6. N. Durand and B. Crémilleux. Ecclat : a new approach of clusters discovery in categorical data. *22nd Int. Conf. on Knowledge Based Systems and Applied Artificial Intelligence (ES'02)*, pages 177–190, December 2002.
7. J. Grzymala-Busse and M. Hu. A comparison of several approaches to missing attribute values in data mining. *International Conference on Rough Sets and Current Trends in Computing, Banff, Canada*, 2000.
8. S. Jami, X. Liu, and G. Loizou. Learning from an incomplete and uncertain data set: The identification of variant haemoglobins. *The Workshop on Intelligent Data Analysis in Medicine and Pharmacology, European Conference on Artificial Intelligence, Brighton, 23th-28th August*, 1998.
9. H. Mannila and H. Toivonen. Multiple uses of frequent sets and condensed representations (extended abstract). In *Knowledge Discovery and Data Mining*, pages 189–194, 1996.
10. H. Mannila and H. Toivonen. Levelwise search and borders of theories in knowledge discovery. *Data Mining and Knowledge Discovery*, 1(3):241–258, 1997.

11. Set of databases for the data mining. Provided by the university of californy, irvine. *http://www.ics.uci.edu/˜mlearn/MLRepository.html.*
12. N. Pasquier, Y. Bastide, R. Taouil, and L. Lakhal. Efficient mining of association rules using closed itemset lattices. *Information Systems*, 24(1):25–46, 1999.
13. G. Piatetsky-Shapiro. Discovery, analysis and presentation of strong rules. *Knowledge Discovery in Databases*, pages 229–248, 1991.
14. A. Ragel and B. Crémilleux. Treatment of missing values for association rules. *Proceedings of the Second Pacific-Asia Conference on Knowledge Discovery and Data Mining (PAKDD'98), Melbourne, Australia*, pages 258–270, April 1998.
15. F. Rioult. Représentation condensée pour les bases de données adéquate aux valeurs manquantes. *Technical Report, 60 pages, GREYC, Université de Caen*, 2002.
16. Mohammed J. Zaki. Generating non-redundant association rules. *6th ACM SIGKDD International Conference on Knowledge Discovery and Data Mining, Boston*, pages 34–43, 2000.

Measures of Rule Quality for Feature Selection in Text Categorization*

Elena Montañés, Javier Fernández, Irene Díaz, Elías F. Combarro, and José Ranilla

Artificial Intelligence Center
University of Oviedo, Spain
ir@aic.uniovi.es

Abstract. Text Categorization is the process of assigning documents to a set of previously fixed categories. A lot of research is going on with the goal of automating this time-consuming task. Several different algorithms have been applied, and Support Vector Machines have shown very good results. In this paper we propose a new family of measures taken from the Machine Learning environment to apply them to feature reduction task. The experiments are performed on two different corpus (Reuters and Ohsumed). The results show that the new family of measures performs better than the traditional Information Theory measures.

1 Introduction

In the last years, the management and organization of text has become an important task, since the information available is continuously growing.

One of the main tasks in text processing is that of assigning the documents of a corpus into a set of previously fixed categories, what is known as Text Categorization (TC). Formally, TC consists of determining whether a document d_i (from a set of documents D) belongs or not to a category c_j (from a set of categories C), consistently with the knowledge of the correct categories for a set of train documents [17].

Usually, TC is tackled by inducing an automatic classifier from a set of correctly labelled examples using statistical methods [20] or Machine Learning (ML) algorithms [17].

Since the documents (examples) in TC are represented by a great amount of features and most of them could be irrelevant or noisy [16], a previous feature reduction usually improves the performance of the classifiers.

This paper proposes a new family of measures taken from the ML environment. We also compare them with some Information Theory (IT) measures previously applied to feature reduction in [21] and in [8]. The classification is carried out with *Support Vector Machines* (SVM), since they have been shown to perform well in TC [9].

* The research reported in this paper has been supported in part under MCyT and Feder grant TIC2001-3579 and FICYT grant BP01-114.

M.R. Berthold et al. (Eds.): IDA 2003, LNCS 2810, pp. 589–598, 2003.

The rest of the paper is organized as follows. In Section 2 the process of TC is described. In Section 3 the description of the corpus and the experiments are exposed. Finally, in Section 4 some conclusions and ideas for further research are presented.

2 Process of Text Categorization

The process of TC involves document representation, feature reduction and classification. In this section we describe how we perform all these stages. Also, the way of evaluating the model is described.

2.1 Document Representation

All the algorithms applied to TC need the documents to be represented in a way suitable for the induction of the classifier. The most widely used representation is called *vector of words* or *bag of words* (see [16]). It consists of identifying each document with a numerical vector whose components weight the importance of the different words in the document.

In this paper, we adopt the absolute frequency of the word in a document (tf), since it has been widely used in previous works, like [9] and [21]. Another measure, which takes into account the distribution of words in the documents, is $tf \times idf$, but in the collections presented in this work, the words that appear many times in the collection, do not tend to occur in few documents.

2.2 Feature Selection

Feature Selection (FS) is one of the feature reduction approaches most used in TC. The main idea is to select a subset of features from the original one. FS is performed by keeping the words with highest score according to a predetermined measure of the importance of the word.

The motivation of FS is that it is widely accepted, among the Information Retrieval (IR) researchers [16], that only a small percentage of the words are really relevant for the classification. Thereby, FS is performed in this work, not only to reduce the number of the features but also to avoid overfitting and to increase the sparsity of the document vectors. This sparsity helps SVM to separate the positive documents from the rest.

In the rest of the section, the IT measures previously applied for FS in TC are described, as well as the family of measures proposed by us taken from the ML field.

Information Theory Measures. Recently, measures taken from IT have been widely adopted because it is interesting to consider the distribution of a word over the different categories. Among these measures, we find *information gain*

(IG), which takes into account either the presence of the word in a category either its absence, and can be defined as (see, for instance [21])

$$IG(w,c) = P(w)P(c|w)\log(\frac{P(c|w)}{P(c)}) + P(\overline{w})P(c|\overline{w})\log(\frac{P(c|\overline{w})}{P(c)})$$

where $P(w)$ is the probability that the word w appears in a document, $P(c|w)$ is the probability that a document belongs to the category c knowing that the word w appears in it, $P(\overline{w})$ is the probability that the word w does not appear in a document and $P(c|\overline{w})$ is the probability that a document belongs to the category c if we know that the word w does not occur in it. Usually, these probabilities are estimated by means of the corresponding relative frequencies. In the same direction, we can find *expected cross entropy for text* (CET) [11], which only takes into account the presence of the word in a category

$$CET(w,c) = P(w) \cdot P(c|w) \cdot \log\frac{P(c|w)}{P(c)}$$

Yang & Pedersen [21] introduced the statistic χ^2 for feature reduction, which measures the lack of independence between a word and a category. A little modification of this measure is proposed by Galavotti et al. [8], that is called $S - \chi^2$ and defined by

$$S - \chi^2(w,c) = P(w,c) \cdot P(\overline{w},\overline{c}) - P(w,\overline{c}) \cdot P(\overline{w},c)$$

and they have shown to performs better than χ^2.

ML Measures. The measure proposed in this paper is taken from the ML environment. Specifically, it is a measure that has been applied to quantify the quality of the rules induced by a ML algorithm. In order to be able to adopt this measure to feature reduction in TC, we propose to associate to each pair (w,c), the following rule: *If the word w appears in a document, then that document belongs to category c*. From now on, we will denote this rule by $w \to c$. In this way the quantification of the importance of a word w in a category c is reduced to the quantification of the quality of the rule $w \to c$.

Some notation must be introduced in order to define the family of measures. For each pair (w,c), $a_{w,c}$ denotes the number of documents of the category c in which w appears, $b_{w,c}$ denotes the number of documents that contain the word w but they do not belong to the category c, $c_{w,c}$ denotes the number of documents of the category c in which w does not appear and finally $d_{w,c}$ denotes the number of documents which neither belong to category c nor contain the word w.

In general, the rule quality measures are based on the percentage of successes and failures of its application, like the Laplace measure [3].

Besides, there exist other rule quality measures that also take into account the times that the rule is applied (the number of documents of the category in which the word occurs) and the distribution of the documents over the categories. An example of this measure is the *impurity level* (IL) proposed by Ranilla et al. ([14] and [15]). In terms of IR, this measure establishes a difference between a word that only appears in a document of a category and a word that appears in more documents of that category. Intuitively, the second rule has more quality because by itself it is able to cover most of the documents, meanwhile the first one could be noisy. The same occurs in case of failure. Hence, a rule that is applied a considerable number of times (a word that occurs in many documents) with a certain percentage of failures is better than other which has the same percentage of failures, but it is applied to less documents. In order to take into account the distribution of examples over the categories it is necessary to introduce the concept of *inconditional rule* (denoted by $\rightarrow c$, which is the rule that has no antecedent and has the category as consequent). This rule is used as a reference of the rest of the rules of the same category. The measure IL is based on the criterium that Aha [1] uses in IB3, that is, if a rule is applied n times and has m successes, the confident interval of its percentage of success is determined by the formula taken from Spiegel [18]

$$CI_{d,i} = \frac{p + \frac{z^2}{2n} \pm z\sqrt{\frac{p(1-p)}{n} + \frac{z^2}{4n^2}}}{1 + \frac{z^2}{n}}$$

where z is the value of a normal distribution at the significant level α, which in this paper was chosen to be $\alpha = 95\%$ and p is the percentage of success ($\frac{m}{n}$ or $\frac{a_{w,c}}{a_{w,c}+b_{w,c}}$). Then, IL is defined by the degree of overlapping between the confident interval of the rule and the confident interval of its correspondent *inconditional rule*

$$IL(w \rightarrow c) = 100 \cdot \frac{CI_d(\rightarrow c) - CI_i(w \rightarrow c)}{CI_d(w \rightarrow c) - CI_i(w \rightarrow c)}$$

Now, some variants of this measure are proposed, motivated by the fact that χ^2 and $S - \chi^2$ take into account the absence of the term in the rest of the categories. That is, in addition to considering only the pair (w, c), these variants consider the pair $(\overline{w}, \overline{c})$. Hence, counting the successes and failures of the rules we define the percentage of success p_a for IL_a (it is only what it changes)

$$p_a(w \rightarrow c, \overline{w} \rightarrow \overline{c}) = \frac{a_{w,c} + d_{w,c}}{a_{w,c} + b_{w,c} + c_{w,c} + d_{w,c}}$$

On the other hand, since the negative successes, that is, $d_{w,c}$, are typically not considered neither in IR nor in TC, we propose another variant, called IL_{ir}, which consists of removing the parameter $d_{w,c}$ from the previous expression, that is, not to take into account the number of documents that neither contain w nor belong to category c. The expression of the percentage of success for this last measure is

$$p_{ir}(w \rightarrow c, \overline{w} \rightarrow \overline{c}) = \frac{a_{w,c}}{a_{w,c} + b_{w,c} + c_{w,c}}$$

Additional Feature Reduction Techniques. Following [17], we only consider (after eliminating *stop words* and *stemming*, according to Porter [13]) local vocabularies (set of features) rather than global vocabulary. The local vocabulary consists of the words occurring in documents of the category for which the classifier is being constructed and the global vocabulary consists of the words occurring in any document of the corpus. Notice that there will be as many different local vocabularies as there are different categories, while there exists only a global vocabulary (the union of the local ones).

2.3 Classification

The aim of TC is to find out whether a document is relevant to a category or not. Typically, the task of assigning a category to a document from a finite set of m categories is commonly converted into m binary problems, each one consisting of determining whether a document belongs to a fixed category or not (*one-against-the-rest*) [2]. This transformation makes possible the use of binary classifiers to solve the multi-category classification problem [9].

In this paper the classification is performed using SVM [6], since they have been shown to performs fast [7] and well [9] in TC. The key of this performance is that it is a good handler of many features and it deals well with sparse examples. SVM are universal binary classifiers able to find out linear or non-linear threshold functions to separate the examples of a certain category from the rest. They are based on the *Structural Minimization Risk* principle from computational learning theory [19]. We adopt this approach with a linear threshold function since most TC problems are linearly separable [9].

2.4 Evaluation of the Performance

A measure of effectiveness commonly used in IR and called F_1 has been widely adopted in TC [21], which combines with the same relevance the *precision* (P) and the *recall* (R). The former quantifies the percentage of documents that are correctly classified as positives and the latter quantifies the percentage of positive documents that are correctly classified. The following expression defines F_1

$$F_1 = \frac{1}{0.5\frac{1}{P} + 0.5\frac{1}{R}}$$

To compute the global performance over all the categories there exist two different approaches, those are, *macroaveraging* (averaging giving the same importance to each category) and *microaveraging* (averaging giving an importance to each category proportional to the number of documents of each category) [17].

3 Experiments

This section presents a study of the performance of TC with SVM when local selection of the features according to the measures described in section 2.2 is previously applied. First we describe the corpus used (Reuters and Ohsumed) and finally the results are presented.

3.1 The Corpus

Reuters Collection. The Reuters-21578 corpus is a well-know, among IR researchers, set of documents which contains short news related to economy published by Reuters during the year 1987 (it is publicly available at [5]). These stories are distributed on 135 different prefixed categories. Each document is allowed to belong to one or more of these categories. The distribution is quite unbalanced, because there exist empty categories while others contain thousands of documents. The words in the corpus are little scattered, since almost the half (49.91%) of the words only appear in a category. We have chosen the split into train and test documents proposed by Apté (cf. [2,10]). After eliminating some documents which contain no body or no topics we have finally obtained 7063 train documents and 2742 test documents assigned to 90 different categories.

Ohsumed Collection. Ohsumed is a clinically-oriented MEDLINE subset covering all references from 270 medical journals over a 5 year period from 1987 to 1991 (it can be found at [4]). In this work we have restricted our experiments to the MEDLINE documents of 1991 (Ohsumed.91). The stories were classified into at least one of the 15 fixed categories of MeSH structure tree [12]: A, B, C ... Each category is in turn split into subcategories: A1, A2, ..., C1, C2 ... The words in this collection are quite more scattered than in Reuters, since the 19.55% of the words (in average) only appear in a category (49.91% in Reuters). The experimentation was made over what we call OhsumedC, which consists of the subset of Ohsumed.91 formed by the first 20,000 documents of Ohsumed.91 that have abstract. The first 10,000 are labelled as train documents and the rest as test documents. They are split into the 23 subcategories of diseases (category C) of MeSH: C1, C2,...,C23. The distribution of documents over the categories is much more balanced than in Reuters.

3.2 The Results

In this section we summarize the results of the experiments on the collections described in previous section. In all the experiments, a set of filtering levels ranging from 20% to 98% was considered (these levels indicate the percentage of words of the vocabulary that are removed from the representation of the documents). The evaluation is made using *macroaverage* and *microaverage* of F_1. Besides, it is performed a one-tail paired t-*Student* test at a significance level of 95% between $F_1's$ for each pair of measures IL_{ir}-IL, IL_{ir}-IL_a, IL-IL_a, IG-CET, IG-$S-\chi^2$ and CET-$S-\chi^2$.

Table 1 shows the *macroaverage* and the *microaverage* for F_1 of the local reduction for both collections. Also, in table 2 the hypothesis contrasts are presented. In this last table the symbol "+" means that the first measure is significantly better than the second one, the symbol "-" means that the second measure is better than the first one, and finally the symbol "=" means that there are not appreciable differences between both measures.

At sight of tables 1 and 2 we could conclude that for Reuters:

Table 1. *Macroaverage* and *Microaverage* of F_1 for Different Variants

	Macroaverage											
	Reuters						OhsumedC					
$pr(\%)$	IL	IL_{ir}	IL_a	CET	$S-\chi^2$	IG	IL	IL_{ir}	IL_a	CET	$S-\chi^2$	IG
20	58.39	59.18	57.37	57.20	57.29	58.58	46.76	43.45	49.54	45.92	45.86	42.71
40	59.11	59.44	55.83	56.41	59.25	58.56	50.00	44.23	50.57	46.73	46.71	43.61
60	59.57	59.62	53.72	57.17	60.20	59.97	50.95	44.50	50.01	47.39	47.28	45.21
80	60.50	62.11	55.51	61.72	63.03	61.75	51.35	45.06	48.77	48.14	47.40	46.94
85	61.44	64.15	56.11	63.00	61.51	63.40	51.69	45.29	48.93	48.36	47.64	47.11
90	62.94	63.82	56.04	62.98	62.65	64.68	52.43	45.76	50.26	48.79	47.85	47.54
95	62.91	65.98	54.91	64.20	62.58	65.34	52.62	44.90	51.58	48.91	46.94	49.40
98	61.47	66.28	60.15	65.11	65.40	64.30	51.14	42.15	52.27	47.47	46.17	49.66
	Microaverage											
	Reuters						OhsumedC					
$pr(\%)$	IL	IL_{ir}	IL_a	CET	$S-\chi^2$	IG	IL	IL_{ir}	IL_a	CET	$S-\chi^2$	IG
20	82.52	84.92	80.99	83.39	83.35	84.78	54.06	51.75	56.19	53.46	53.47	51.93
40	82.37	84.99	79.42	83.03	83.19	84.96	55.92	51.93	56.20	53.91	53.94	52.31
60	82.58	84.67	78.91	83.03	83.28	85.02	56.62	52.21	55.52	54.49	54.35	53.13
80	82.46	85.07	78.68	83.33	83.40	85.42	57.28	52.11	55.11	54.80	54.49	53.98
85	82.12	84.89	78.63	83.40	83.17	85.53	57.53	51.90	53.45	55.10	54.75	54.19
90	82.69	84.53	77.77	82.99	83.30	85.48	57.95	52.04	54.36	55.42	54.93	54.59
95	82.84	84.35	75.63	83.22	83.57	85.40	57.83	50.71	55.72	55.69	54.48	55.47
98	81.81	83.69	78.40	83.02	83.01	84.76	56.66	47.04	56.24	54.04	51.99	54.81

Table 2. *t-Student* Contrasts of F_1 among Different Variants

	Reuters					
$pr(\%)$	IL_{ir}-IL	IL_{ir}-IL_a	IL-IL_a	IG-CET	IG-$S-\chi^2$	CET-$S-\chi^2$
20	+	=	=	=	=	=
40	=	+	=	+	=	=
60	=	+	+	=	=	=
80	+	+	+	=	=	=
85	+	+	+	=	=	=
90	=	+	+	+	=	=
95	=	+	+	+	+	=
98	+	+	=	=	+	=
	OhsumedC					
$pr(\%)$	IL_{ir}-IL	IL_{ir}-IL_a	IL-IL_a	IG-CET	IG-$S-\chi^2$	CET-$S-\chi^2$
20	=	=	=	=	=	=
40	=	-	-	-	-	=
60	=	-	=	-	-	=
80	=	-	+	-	-	+
85	=	-	+	-	=	+
90	=	-	+	-	=	+
95	=	-	=	=	+	+
98	=	-	=	+	+	+

i) Among IL and its variants, IL_{ir} produces the best *macroaverage* and *microaverage* of F_1 for all filtering levels. In fact, the hypothesis contrasts show that IL_{ir} is significantly better in F_1 for most of the filtering levels. For the rest of the levels there are not noticeable differences.

ii) Among the IT measures, IG offers best *microaverage* of F_1. This means that for the more abundant categories (in Reuters there are few abundant categories, meanwhile there are a lot of small categories) IG performs better. If it is assigned equal weight to each category (see *macroaverage*), IG is only better at some filtering levels. The hypothesis contrasts says that at least IG is not significantly worst than CET and $S-\chi^2$. In fact, it is significantly better for some filtering levels.

For Reuters, the reason why these variants are better may be because this collection have the words little scattered. We found that (in average) the 49.91% of the words only appears in a category. This means that there is a high percentage of words that are specific of most of the categories. Hence, it is advisable to apply a measure which select words which appear in many documents of the category (parameter $a_{w,c}$ high), provided that they do not appear in the rest of the categories (parameter $b_{w,c}$ low) or in a reverse way, that is, words which appears in many documents of the rest of categories (parameter $b_{w,c}$ high) and in few documents of the category in study (parameter $a_{w,c}$ low). This last condition ($b_{w,c}$ high and $a_{w,c}$ low) means to consider the absence of the word. IL_{ir} is, among its variants, the measure which penalizes the most those words which appear in the rest of categories. On the other hand, IG is better than its variants because it takes into account the absence of the word in the category.

At sight of tables 1 and 2 we could conclude that for OhsumedC:

i) IL is the best variant for most of the filtering levels. The hypothesis contrasts allow to argue that there are not noticeable differences between IL and IL_{ir}, but IL is significantly better than IL_a at high filtering levels.
ii) Among the IT measures, CET produces the best results, except for too aggressive filtering levels. The hypothesis contrasts suggest that CET is significantly better than IG for intermediate and high filtering levels and than $S-\chi^2$ for high and aggressive filtering levels.

On the contrary of Reuters, in OhsumedC, the averaged percentage of appearance of words per category is 19.55% (against 49.91% of Reuters). This percentage allow to conclude that the words in OhsumedC are quite more scattered than in Reuters, which means that there are not enough specific words of each category. Hence, for this collection, it is advisable to use measures which primarily select words that appear in many documents of the category in study (parameter $a_{w,c}$ high), although they are present in the rest of the categories (parameter $b_{w,c}$ high). Although it is desirable that $b_{w,c}$ is low, it is unwise to penalize words with high $b_{w,c}$ if this means to give more priority to a word with lower $a_{w,c}$. IL_{ir} penalizes too much words with high $b_{w,c}$ by means of its percentage of success and favours words with high $b_{w,c}$ by means of the number of times the rule is applied, so it is not a good choice in this case. Respect to IL_a, it does not take into account the number of times the rule is applied (it is constant for all the words of a category and equal to the total number of documents), and although it assigns equal relevance to $a_{w,c}$ and $b_{w,c}$, the former property makes this measure unadequate. On the other hand, although IL tends to select words which do not appear in documents of the rest of categories, the number of times the rule is applied is really the issue which makes this measure reinforce words with high $a_{w,c}$. Respect to the IT measures, it is not adequate to apply measures which takes into account the absence of a word (like IG), since OhsumedC do not have enough words of this kind, due to the scattering of the words in this collection. Both CET and $S-\chi^2$ assign more score to words with high $a_{w,c}$ but with $b_{w,c}$ also high, but they differ in the degree of importance assigned to both $a_{w,c}$ and $b_{w,c}$. In this way CET is more desirable for OhsumedC.

Table 3 shows the *t-Student* test among the two selected measures for each corpus.

From tables 1 and 3, we could conclude that for Reuters there is not appreciable differences between IL_{ir} and IG, which means that it is adequate either

Table 3. *t-Student* Contrasts among the Best Variants

	$pr(\%)$	20	40	60	80	85	90	95	98
Reuters	$IL_{ir} - IG$	=	=	=	=	=	=	=	=
OhsumedC	$IL - CET$	+	+	+	+	+	+	+	+

to consider the absence of the word in the category in study or to roughly penalize the words that occur in the documents of the rest of categories. But, IL_{ir} is slightly better than IG for most filtering levels, although IG is better for large categories (see *microaverage*). However, for OhsumedC, IL is better than CET for all filtering levels independent of the size of the categories. In fact, the contrasts show that IL is significantly better than CET for all filtering levels. The reason may be because IL reinforces $a_{w,c}$ in more degree than $b_{w,c}$ more than CET, and this is better for this corpus due to the scattering of the words in the categories.

4 Conclusions and Future Work

In this paper, we propose a family of measures taken from the ML environment (IL, IL_a and IL_{ir}) to perform feature selection in TC, when the classification is carried out with SVM.

The results show that the measures proposed as well as the IT measures for feature selection are dependent of the corpus. For Reuters, which has an unbalanced distribution of the documents along the categories and has the words little scattered, it is better for feature selection to use measures that penalize more those words which occur in documents of the rest of categories (like IL_{ir}) and that take into account the absence of the word in the category in study (like IG). On the other hand, for OhsumedC, whose distribution of documents is more or less uniform and which has the words highly scattered, it is better to reinforce the words that occur in documents of the category in study (like IL and CET) and not to penalize so much those words which appear in the rest of the categories.

We show that, for Reuters, both taking into account the absence of the word in the category in study (IG) and penalizing those words which occur in documents of the rest of categories (IL_{ir}) are good choices for FS. However, for OhsumedC, it is much more adequate to reinforce in more degree those words that appear in documents of the category. That is, for OhsumedC, it is better our proposed measure IL than the IT measure CET, and this improvement is statistically significant.

As future work, we plan to explore the behavior of these new proposed measures when other classification algorithms are applied to TC. We are also interested in finding the optimal levels of filtering, which may depend on properties of the category, like its size.

References

1. D. W. Aha. *A Study of Instance-based Algorithms for Supervised Learning Tasks: Mathematical, Empirical, and Psychological Evaluations.* PhD thesis, University of California at Irvine, 1990.
2. C. Apte, F. Damerau, and S. Weiss. Automated learning of decision rules for text categorization. *Information Systems*, 12(3):233–251, 1994.
3. P. Clark and T. Niblett. The cn2 induction algorithm. *Machine Learning*, 3(4):261–283, 1989.
4. Ohsumed 91 Collection. `http://trec.nist.gov/data/t9-filtering`.
5. Reuters Collection. `http://www.research.attp.com/lewis/reuters21578.html`.
6. C. Cortes and V. Vapnik. Support-vector networks. *Machine Learning*, 20(3):273–297, 1995.
7. S. Dumais, J. Platt, D. Heckerman, and M. Sahami. Inductive learning algorithms and representations for text categorization. In *Proceedings of the International Conference on Information and Knowledge Management*, 1998.
8. L. Galavotti, F. Sebastiani, and M. Simi. Experiments on the use of feature selection and negative evidence in automated text categorization. In *Proceedings of ECDL-00, 4th European Conference on Research and Advanced Technology for Digital Libraries*, pages 59–68, Lisbon, Portugal, 2000. Morgan Kaufmann.
9. T. Joachims. Text categorization with support vector machines: learning with many relevant features. In Claire Nédellec and Céline Rouveirol, editors, *Proceedings of ECML-98, 10th European Conference on Machine Learning*, number 1398, pages 137–142, Chemnitz, DE, 1998. Springer Verlag, Heidelberg, DE.
10. D. D. Lewis and M. Ringuette. A comparison of two learning algorithms for text categorization. In *Proceedings of SDAIR-94, 3rd Annual Symposium on Document Analysis and Information Retrieval*, pages 81–93, Las Vegas, US, 1994.
11. D. Mladenic and M. Grobelnik. Feature selection for unbalanced class distribution and naive bayes. In *Proceedings of 16th International Conference on Machine Learning ICML99*, pages 258–267, Bled, SL, 1999.
12. National Library of Medicine. Medical subject headings (mesh). `http://www.nlm.nih.gov/mesh/2002/index.html`.
13. M. F. Porter. An algorithm for suffix stripping. *Program (Automated Library and Information Systems)*, 14(3):130–137, 1980.
14. J. Ranilla and A. Bahamonde. Fan: Finding accurate inductions. *International Journal of Human Computer Studies*, 56(4):445–474, 2002.
15. J. Ranilla, O. Luaces, and A. Bahamonde. A heuristic for learning decision trees and pruning them into classification rules. *AICom (Artificial Intelligence Communication)*, 16(2):in press, 2003.
16. G. Salton and M. J. McGill. *An introduction to modern information retrieval.* McGraw-Hill, 1983.
17. F. Sebastiani. Machine learning in automated text categorisation. *ACM Computing Survey*, 34(1), 2002.
18. M. R. Spiegel. *Estadística.* McGraw-Hill, (in spanish), 1970.
19. V. Vapnik. *The Nature of Statistical Learning Theory.* Springer-Verlag, 1995.
20. T. Yang. Expert network: effective and efficient learning from human decisions in text categorisation and retrieval. In *Proceedings of SIGIR-94, ACM Int. Conference on Research and Development in Information Retrieval*, pages 13–22, 1994.
21. T. Yang and J. P. Pedersen. A comparative study on feature selection in text categorisation. In *Proceedings of ICML'97, 14th International Conference on Machine Learning*, pages 412–420, 1997.

Genetic Approach to Constructive Induction Based on Non-algebraic Feature Representation*

Leila S. Shafti and Eduardo Pérez

Universidad Autónoma de Madrid, E–28049 Madrid, Spain
{leila.shafti,eduardo.perez}@ii.uam.es

Abstract. The aim of constructive induction (CI) is to transform the original data representation of hard concepts with complex interaction into one that outlines the relation among attributes. CI methods based on greedy search suffer from the local optima problem because of high variation in the search space of hard learning problems. To reduce the local optima problem, we propose a CI method based on genetic (evolutionary) algorithms. The method comprises two integrated genetic algorithms to construct functions over subsets of attributes in order to highlight regularities for the learner. Using non-algebraic representation for constructed functions assigns an equal degree of complexity to functions. This reduces the difficulty of constructing complex features. Experiments show that our method is comparable with and in some cases superior to existing CI methods.

1 Introduction

Most machine learning algorithms based on similarity attain high accuracy on many artificial and real-world domains such as those provided in Irvine databases [1,2]. The reason for high accuracy is that the data representation of these domains is good enough to distribute instances in space in a way that cases belonging to the same class are located close to each other. But for hard concepts due to low-level representation, complex interactions exist among attributes. The *Interaction* means the relationship between one attribute and the concept depends on other attributes [3]. Due to attribute interaction of hard concepts, each class is scattered through the space and therefore a similarity based learning algorithm fails to learn these concepts. Such problems have been seen in real-world domains such as protein secondary structure [4]. Attribute interaction causes problems for almost all classical learning algorithms [5].

Constructive induction (CI) methods have been introduced to ease the attribute interaction problem. Their goal is to automatically transform the original representation space of hard concepts into a new one where the regularity is more apparent [6,7]. This goal is achieved by constructing features from the given set

* This work has been partially supported by the Spanish Interdepartmental Commission for Science and Technology (CICYT), under Grants numbers TIC98-0247-C02-02 and TIC2002-1948.

M.R. Berthold et al. (Eds.): IDA 2003, LNCS 2810, pp. 599–610, 2003.

of attributes to abstract the relation among several attributes to a single new attribute.

Most existing CI methods, such as Fringe, Grove, Greedy3 [8], LFC [9], and MRP [10] are based on greedy search. These methods have addressed some problems. However, they still have an important limitation; they suffer from the local optima problem. When the concept is complex because of the interaction among attributes, the search space for constructing new features has more variation. Because of high variation, the CI method needs a more global search strategy such as Genetic Algorithms [11] to be able to skip local optima and find the global optima solutions. Genetic Algorithms (GA) are a kind of multi-directional parallel search, and more likely to be successful in searching through the intractable and complicated search space [12].

There are only a few CI methods that use genetic search strategy for constructing new features. Among these methods are GCI [13], GPCI [14], and Gabret [15]. Their partial success in constructing useful features indicates the effectiveness of genetic-based search for CI. However, these methods still have some limitations and deficiencies as will be seen later. The most important ones are their consideration of the search space and the language they apply for representing constructed features.

This paper presents a new CI method based on GA. We begin by analyzing the search space and search method, which are crucial for converging to an optimal solution. Considering these we design GABCI, Genetic Algorithm Based CI and present and discuss the results of our initial experiments.

2 Decomposition of Search Space

A CI method aims to construct features that highlight the interactions. For achieving this aim, it has to look for subsets of interacting attributes and functions over these subsets that capture the interaction. Existing genetic-based CI methods, GCI, and GPCI, search the space (of size 2^{2^N} for N input attributes in Boolean domain) of all functions that can be defined over all subsets of attributes. This enormous space has lots of local optima and is more difficult to be explored.

To ease the searching task, we propose to decompose the search space into two spaces: S the space of all subsets of attributes, and F_{S_i} the space of all functions defined over a given subset of attributes S_i (Fig. 1). The decomposition of the search space into two spaces allows a specific method to be adjusted for each search space. This strategy divides the main goal of CI into two easier sub-goals: finding the subset of interacting attributes (*S-Search*) and looking for a function that represents the interaction among attributes in a given subset (F_{S_i}*-Search*).

The genetic-based CI method Gabret also divides the search space into two spaces. It applies an attribute subset selection module before and independently from the feature construction module to filter out irrelevant attributes. However, when high interaction exists among attributes, attribute subset selection module cannot see the interaction among primitive attributes. Therefore, this module

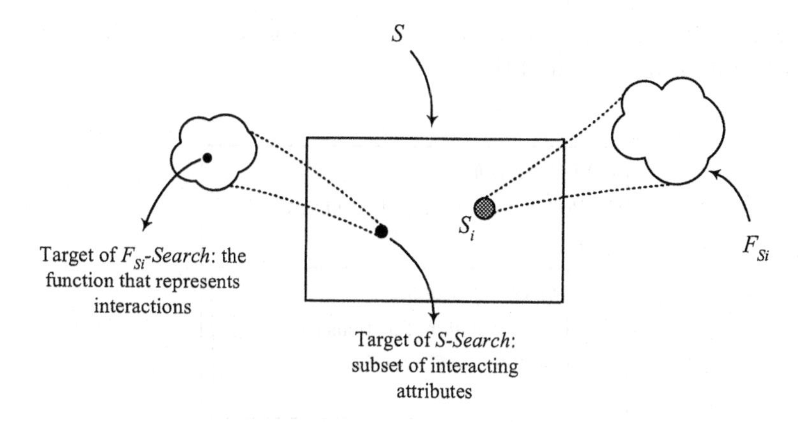

Fig. 1. The decomposition of space of all functions defined over all subsets of attributes. Let S to be the space of all subsets of attributes, $S_i \in S$, and F_{S_i}, the space of all functions defined over subset S_i.

often excludes some relevant attributes from the search space that should be used later for constructing new features.

The two spaces of attribute subsets and functions over subsets are related and each one has a direct effect on the other. In order to find a good subset of attributes it is necessary to see the relation among attributes. An existing interaction among attributes in subset S_i is not apparent unless a function f_{S_i}, that outlines this interaction, is defined over the proper subset of attributes. Similarly, a function can highlight the interaction among attributes only when it is defined over the related attributes. If the chosen subset of attributes is not good enough, the best function found may not be helpful. Therefore, the two tasks, *S-Search* and F_{S_i}*-Search*, should be linked to transfer the effects of each search space into the other.

3 GABCI

Our CI method GABCI consists of two search tasks *S-Search* and F_{S_i}*-Search* as described in Sect. 2. Since both spaces S and F_{S_i} grow exponentially with the number of attributes and have many local optima, GABCI applies GA to each search task. Thus, GABCI is a composition of two GA with two populations: a population of attribute subsets (for *S-Search*) and a population of functions (for F_{S_i}*-Search*). The goal is to converge each population to an optimal solution by means of genetic operators.

Recalling the need to keep the connection between two tasks, we propose to have F_{S_i}*-Search* inside *S-Search*. That is, for each subset S_i in *S-Search*, in order to measure its goodness, GABCI finds the best function defined over this subset using F_{S_i}*-Search*. Then the fitness value of the best individual in F_{S_i}*-Search* determines the fitness of the subset of attributes S_i in *S-Search* (Fig. 2). By this strategy we can maintain the relation between two tasks and their effects to each other, while improving each of them.

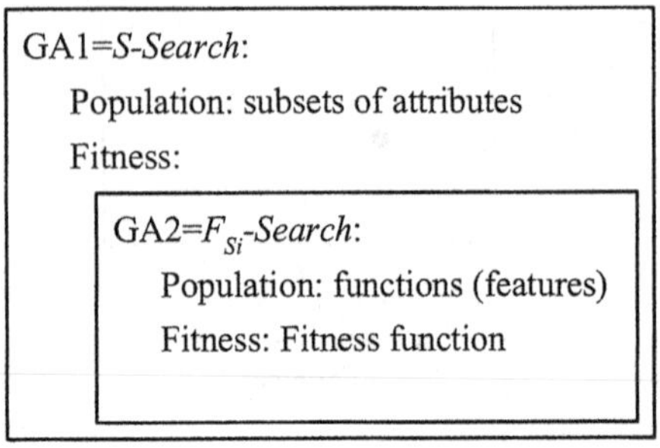

GA1=*S-Search*:
Population: subsets of attributes
Fitness:
GA2=F_{Si}-*Search*:
Population: functions (features)
Fitness: Fitness function

Fig. 2. The GABCI's framework: the fitness of each individual of *S-Search* is determined by performing F_{S_i}-*Search* and calculating the fitness of the best individual of GA2.

Among the several factors that need to be defined carefully when designing a GA, the two main ones are the representation language of individuals and the fitness function. The language has an impact on the way the search is conducted. An inadequate language can make the search fruitless by hiding solutions amongst many local optima. Fitness function has the role of guiding the method toward solution. If this function assigns an incorrect fitness value to individuals the method will be misguided to local optima. In the following, we explain the representation language and the fitness function used for each space in GABCI.

3.1 Individuals Representation

For *S-Search*, we represent individuals (i.e., attribute subsets) by bit-strings; each bit representing presence or absence of an attribute in subset. F_{S_i}-*Search* is more complicated than *S-Search*. Recalling that the aim of F_{S_i}-*Search* is to construct a function over S_i, the given subset of attributes, to represent interactions; the representation language should provide the capability of representing all complex functions or features of interests. This representation has an important role in convergence to optimal solution.

There are two alternative groups of languages for representing features: algebraic form and non-algebraic form. By algebraic form, we mean that features are shown by means of some algebraic operators such as arithmetic or Boolean operators. Most genetic-based CI methods like GCI, GPCI, and Gabret apply this form of representation using parse trees [16]. GPCI uses a fix set of simple operators, AND and NOT, applicable to all Boolean domains. The use of simple operators makes the method applicable to a wide range of problems. However, a complex feature is required to capture and encapsulate the interaction using simple operators. Conversely, GCI and Gabret apply domain-specific operators to reduce complexity; though, specifying these operators properly cannot be performed without any prior information about the target concept.

In addition to the problem of defining operators, an algebraic form of representation produces an unlimited search space since any feature can be appeared in infinite forms. As an example see the following conceptually equivalent functions that are represented differently:

$$X_1 \wedge X_2$$

$$((X_1 \wedge X_2) \vee \overline{X}_1) \wedge X_1$$

$$((((X_1 \wedge X_2) \vee \overline{X}_1) \wedge X_1) \vee \overline{X}_2) \wedge X_2 \ .$$

Therefore, a CI method needs a restriction to limit the growth of constructed functions.

Features can also be represented in a non-algebraic form, which means no operator is used for representing the feature. For example, for a Boolean attribute set such as $\{X_1, X_2\}$ an algebraic feature like $(X_1 \wedge \overline{X}_2) \vee (\overline{X}_1 \wedge X_2)$ can be represented by a non-algebraic feature such as $\langle 0110 \rangle$, where the jth element in $\langle 0110 \rangle$ represents the outcome of the function for j^{th} combination of attributes X_1 and X_2 according to the following table:

X_1	X_2	f
0	0	0
0	1	1
1	0	1
1	1	0

A non-algebraic form is simpler to be applied in a CI method since there is no need to specify any operator. Besides, all features have equal degree of complexity when they are represented in a non-algebraic form. For example two algebraic features like $f(X_1, X_2) = (X_1 \wedge X_2) \vee (\overline{X}_1 \wedge \overline{X}_2)$ and $f'(X_1, X_2) = X_1$ are represented in non-algebraic form as $R(X_1, X_2) = \langle 1001 \rangle$ and $R'(X_1, X_2) = \langle 0011 \rangle$, which are equivalent in terms of complexity. Since for problems with high interaction, CI needs to construct features more like the first feature in example, which are complex in algebraic form, a non-algebraic representation is preferable for these methods. Such kind of representation is applied in MRP [10], a learning method that represents features by sets of tuples using multidimensional relational projection. However, MRP applies a greedy method for constructing features. As the search space for complex problems has local optima, MRP fails in some cases.

Non-algebraic representation also eases the traveling through the search space by GA. Changing one bit in an individual as non-algebraic function by genetic operators can produce a large change in the function. For example if the function is $X_1 \vee X_2$, represented as $\langle 0111 \rangle$, changing the last bit gives $\langle 0110 \rangle$, that is $(X_1 \wedge \overline{X}_2) \vee (\overline{X}_1 \wedge X_2)$, which is very different from its parent in terms of algebraic form of representation. So this form of representation results in more variation after performing genetic operators and can provide more useful features in hard problems.

Thus we apply a non-algebraic form of representation for individuals in F_{S_i}*-Search*. Without losing generality, suppose that attributes have Boolean domain. For each subset S_i of size k of the population in *S-Search*, a population of bit-strings of length 2^k is generated in F_{S_i}*-Search*. The jth bit of the string represents the outcome of the function for j^{th} combination of attributes in the subset S_i.

3.2 Fitness Function

The fitness of each individual S_i in *S-Search* depends on performance of F_{S_i}*-Search* and the fitness value of the best constructed feature. If the fitness value of the best feature F in F_{S_i}*-Search* is $Fitness_{S_i}(F)$, then the fitness of S_i is calculated as:

$$Fitness(S_i) = Fitness_{S_i}(F) + \epsilon \frac{W(S_i)}{|S_i|} \quad (1)$$

where $W(S_i)$ is the number of ones in bit string S_i, and $|S_i|$ is the length of the string. The first term of the formula has the major influence on fitness value of the individual S_i. The last term roughly estimates the complexity of the new feature, to follow Occam's razor bias, by measuring the fraction of attributes participating in the feature. It is multiplied by $\epsilon = 0.01$ to have effect only when two subsets S_j and S_k perform equally well in F_{S_i}*-Search*; that is, $Fitness_{S_j}(F') = Fitness_{S_k}(F'')$. Therefore, between two equally good subsets, this formula assigns a better fitness value to the smaller.

The fitness of individuals in F_{S_i}*-Search* can be determined by a hypothesis-driven or a data-driven evaluation. A hypothesis-driven evaluation can be done by adding the new function to the list of attributes and updating data, to then apply a learner like C4.5 [17] and measure its accuracy. GCI and Gabret apply this kind of evaluation. A hypothesis-driven evaluation relies on the performance of the learner. Therefore, if the learner assigns inaccurately a low or high fitness value to individuals, it will guide the algorithm to a local solution. Another possibility is to apply a data-driven evaluation formula like conditional entropy [18] to measure the amount of information needed to identify the class of a sample in data knowing the value of the new feature. A data-driven approach only depends on data, and therefore, is more reliable than a hypothesis-driven approach for guiding GA. Moreover, in GA computation time of fitness function is very important since this function is called for every individual during each GA generation. The current version of GABCI applies the latter evaluation function as explained below. These fitness functions will be compared empirically in Sect. 4.2.

To prevent the system from constructing features that fits well with training data but does not fit with unseen data (overfitting), we first shuffle and divide the training data into three sets T_1, T_2, and T_3. Then fitness of the new feature,

F, is measured by the following formula:

$$Fitness(F) = \frac{1}{3}\sum_{i=1}^{3} Entropy(T_i|F) \tag{2}$$

where $Entropy(T|F)$ is the conditional Entropy [18]. Figure 3 summarizes GABCI's algorithm.

S-Search:

1. Generate a population of bit-strings of length N representing subsets of attributes S_i.
2. For each individual S_i run F_{Si}-*Search*, assign the best individual of F_{Si}-*Search* to S_i and calculate the fitness according to formula (1).
3. Unless desired conditions are achieved, apply genetic operators, generate a new population, and go to step 2.
4. Add the corresponding function of the best individual to the original set of attributes and update data for further process of learning by applying a classical learning system over the new set of attributes.

F_{Si}-Search:

1. Given the subset S_i, generate a population of bit-strings of length 2^w representing non-algebraic functions, where w is the weight of (i.e. number of ones in) S_i.
2. Calculate the fitness of each non-algebraic function according to formula (2).
3. Unless desired conditions are achieved, apply genetic operators, generate a new population, and go to step 2.
4. Return the best function found over S_i and its fitness value.

Fig. 3. The GABCI's algorithm

4 Experimental Results and Analysis

We compare GABCI with other CI methods based on greedy search by conducting experiments on artificial problems known as difficult concepts with high interactions. These problems are used as prototypes to exemplify real-world hard problems. Therefore, similar results are expected for real-world problems. We also analyze the behavior of GABCI with data-driven and hypothesis-driven fitness functions.

4.1 Comparison with Other Methods

The concepts used for these experiments are defined over 12 Boolean attributes $a_1, \ldots, a_{12}$. These concepts are parity of various degrees, multiplexor-6, n-of-m (exactly n of m attributes are one), Maj-n-m (among attributes in $\{a_n, \ldots, a_m\}$, at least half of them are one), and nm-4-5 (among attributes in $\{a_4, \ldots, a_{10}\}$, 4 or 5 of them are one). A high interaction exists among relevant attributes of these concepts.

Each experiment was run 10 times independently. For each experiment, we used 10% of shuffled data for training and kept the rest (90% of data) unseen for evaluating the final result. When GABCI finished, its performance is evaluated by the accuracy of C4.5 [17] on modified data after adding the constructed feature; using 10% data as training and 90% unseen data as test data.

For implementing GA, we used PGAPACK library [19] with default parameters except those indicated in Table 1.

Table 1. GA's modified parameters

GA Parameter	***S-Search***	F_{S_i}***-Search***
Population Size	20	20
Max Iteration	100	300
Max no Change Iter.	25	100
Max Similarity Value	80	80
No. of Strings to be Replaced	18	18
Selection Type	Proportional	Proportional
Crossover Type	Uniform	Uniform
Mutation and/or Crossover	And	And

Table 2 presents a summary of GABCI's accuracy and its comparison to the other systems' accuracy. The last column indicates the average accuracy after testing the result of GABCI by C4.5 using unseen data. The results are compared with C4.5 and C4.5-Rules [17], which are similarity-based learners, Fringe, Grove and Greedy3 [8], and LFC [9], which are greedy-based CI methods that apply algebraic form for representing features and MRP [10] which is a greedy CI method with a non-algebraic form of representation. In the third column of the table, we show the best results among C4.5-Rules, Fringe, Grove, Greedy3 and LFC, as reported in [10]. Numbers between parentheses indicate standard deviations.

Table 2. The accuracy's result average

Concept	**C4.5**	**The best result**	**MRP**	**GABCI + C4.5**
Parity-4	71.4 (17.7)	**100 (fringe)**	**100 (0.0)**	**100 (0.0)**
Parity-6	49.2 (2.8)	78.7 (fringe)	**99.8 (0.5)**	**99.8 (0.5)**
Parity-8	47.2 (0.2)	49.4 (C4.5-Rules)	88.6 (1.3)	**90.8 (1.9)**
MX-6	98.2 (2.3)	**100 (fringe)**	99.9 (0.4)	99.8 (0.5)
5-of-7	83.2 (1.8)	93.8 (grove)	**98.6 (2.6)**	98.4 (0.8)
6-of-9	78.5 (1.8)	**88.7 (grove)**	87.4 (2.1)	79.4 (2.7)
Maj-4-8	**100 (0.0)**	**100 (grove)**	**100 (0.0)**	**100 (0.0)**
Maj-3-9	88.3 (1.4)	94.0 (grove)	97.3 (1.1)	**98.3 (1.3)**
nm-**4-5**	80.0 (2.7)	90.5 (fringe)	**98.4 (1.1)**	98.1 (2.3)

GABCI's performance is comparable with other CI methods, and sometimes better. Although its accuracy is not always the highest, it is among the best ones. The results show that GABCI has the capacity to be improved to give better accuracy (for instance, by carefully adjusting parameters). The large improvement in accuracy obtained for parity concepts shows that F_{S_i}*-Search* correctly guides the *S-Search* toward a proper subset, and once this subset is found, then F_{S_i}*-Search* constructs a feature defined over this subset that can help the learner to achieve a higher accuracy. In fact, in all experiments GABCI successfully found the subset of interacting attributes, except in one of the 10 experiments with concept *nm*-4-5.

The low performance of GABCI in 6-of-9 concept is due to overfitting. GABCI finds the relevant attributes but generates a function over these attributes, which fits well with training data only, resulting in low accuracy on unseen data; though it is still slightly better than the similarity-based learner C4.5.

4.2 Analysis of Fitness Functions

We also performed experiments to compare the effect of hypothesis-driven and data-driven fitness functions (see Sect. 3.2) on the performance of F_{S_i}*-Search*.

We used C4.5 [17] for hypothesis-driven fitness function. The training data were shuffled three times, each time divided into two parts of equal size: one for training and one for testing. For evaluating each individual, the new feature was added to the original set of attributes and training and test data were updated. Then C4.5 was performed on training data and the accuracy was estimated by test data. The average error rate of three independent runs of C4.5 over three pairs of training and test data determines the fitness of the constructed feature. As a data-driven fitness function, we used (2) of Sect. 3.2.

Since the aim was to observe the effect of fitness function on feature construction, we ran only F_{S_i}*-Search*, once with data-driven fitness and once with hypothesis-driven fitness. A subset of attributes was given to F_{S_i}*-Search* as input to generate new features. We performed these experiments on parity-8 concept. As before, each experiment was performed 10 times, using 10% of data for training and 90% for final evaluation of F_{S_i}*-Search* by C4.5.

Figure 4 shows the average results of 10 independent runs of F_{S_i}*-Search* for each experiment over different subsets of attributes for parity-8. In this figure the horizontal axis shows the number of attributes in the given subset S_i. For each subset S_i, let $S_i = \{x_1, \ldots, x_i\}$ and $|S_i| = i$. The relevant attributes are $x_1, x_2, \ldots, x_8$.

The vertical axis in Fig. 4.a shows the average error rate after running the learner C4.5 on updated data using constructed feature and in Fig. 4.b, represents the size (number of leaves) of the tree generated after pruning. The horizontal lines in these figures show the base-line performance on original set of attributes.

We see that for $i < 8$, C4.5 as fitness function yields lower error than Entropy, but for $i \geq 8$; that is when subset includes all interesting attributes needed for

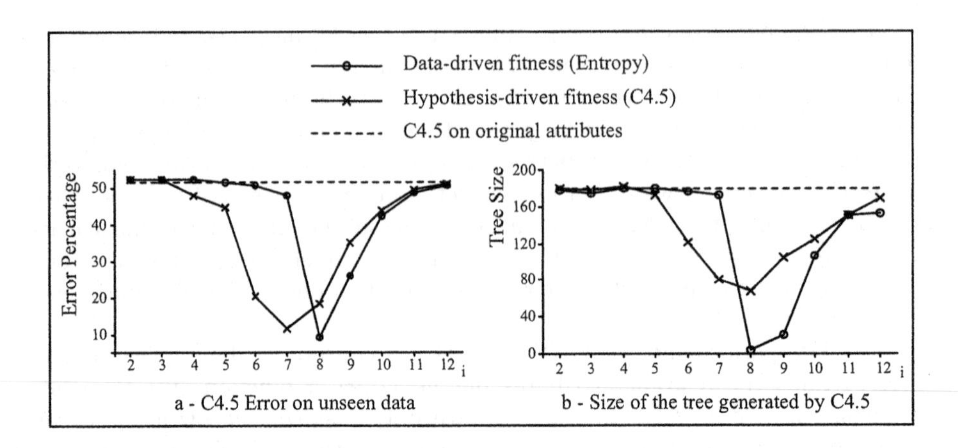

Fig. 4. F_{S_i}*-Search* on Parity-8

constructing feature, Entropy performs better. The tree size in Fig. 4.b indicates that for $i \geq 8$, Entropy generates a feature that abstracts interaction among attributes better than C4.5 fitness function; thus, produces a smaller tree. Focusing on $i = 8$, that is when the best subset (subset of all relevant attributes) is given to F_{S_i}*-Search*, reveals that C4.5 misleads GA to a local optimum and Entropy behaves better.

The reason for these behaviors of Entropy and C4.5 as fitness is that Entropy measures the fitness of the constructed function without involving other attributes, while C4.5 measures the fitness of the constructed feature in relation with original attributes. When the new function is defined over some relevant attributes, C4.5 can combine this function with other relevant attributes (not in the given subset) to incorporate missing attributes. Hence, C4.5 assigns a better fitness to constructed function and produces a lower error rate on unseen data. However, when the function includes all relevant attributes ($i \geq 8$), combining it with relevant attributes cannot help C4.5 and sometimes causes overfitting. By overfitting we mean the program constructs a function with a high accuracy on training data, as it fits well with this data, but when it is tested by unseen data the error rate is increased. These confirm that a hypothesis-driven function may mislead GA to a local optimum.

Furthermore, due to overfitting produced by C4.5 fitness function, the lowest error rate obtained when using this function is higher than the one obtained using entropy. Though the difference is not high, taking into account that computing the fitness using Entropy takes less time than using C4.5, the former is preferable.

5 Conclusion

This paper has presented a new CI method based on genetic (evolutionary) algorithms whose goal is to ease learning problems with complex attribute interaction. The proposed method GABCI intends to outline the attribute interaction by constructing new functions over subsets of related attributes. Our genetic approach makes GABCI more promising than other methods in constructing adequate features when the search space is intractable and the required features are too complex to be found by greedy approaches.

We showed how CI search space could be decomposed into two different search spaces, which can be treated by two different but integrated genetic algorithms. Other CI methods based on genetic algorithms either consider these two spaces as a whole or treat them separately and independently. The former methods often fall in a local optimum by searching over a huge space with many local optima, that is more difficult to be explored; and the latter loses the relation and effects of two search spaces on each other.

Considering such integrated treatment of the decomposed search space, we designed GABCI, using non-algebraic representation of constructed features. The non-algebraic form of representation provides a finite search space of features. Besides, this form of representation assigns an equal degree of complexity to features regardless of the size or complexity of their symbolic representation. This reduces the difficulty of constructing complex features. Empirical results proved that our method is comparable with and in some cases better than other CI methods in terms of accuracy.

References

1. Blake, C.L., Merz, C.J.: UCI Repository of Machine Learning Databases, [http://www.ics.uci.edu/ mlearn/MLRepository.html]. University of California, Department of Information and Computer Science, Irvine, California (1998)
2. Holte, R.C.: Very simple classification rules perform well on most commonly used datasets. Machine Learning, **11**:63–91i (1993)
3. Freitas, A.: Understanding the crucial role of attribute interaction in data mining. Artificial Intelligence Review, **16(3)**:177–199 (November, 2001)
4. Qian, N., Sejnowski, T.J.: Predicting the secondary structure of globular proteins using neural network models. Journal of Molecular Biology, **202**:865–884 (August, 1988)
5. Rendell, L.A., Seshu, R.: Learning hard concepts through constructive induction: framework and rationale. Computational Intelligence, **6**:247–270 (November, 1990)
6. Dietterich, T.G., Michalski, R.S.: Inductive learning of structural description: evaluation criteria and comparative review of selected methods. Artificial Intelligence, **16(3)**:257–294 (July, 1981)
7. Aha, D.W.: Incremental constructive induction: an instance-based approach. In Proceedings of the Eighth International Workshop on Machine Learning, pages 117–121, Evanston, Illinois (1991) Morgan Kaufmann
8. Pagallo G., Haussler, D.: Boolean feature discovery in empirical learning. Machine Learning, **5**:71–99 (1990)
9. Ragavan, H., Rendell, L.A.: Lookahead feature construction for learning hard concepts. In Proceedings of the Tenth International Conference on Machine Learning, pages 252–259, Amherst, Massachusetts (June, 1993) Morgan Kaufmann

10. Pérez, E. Rendell, L.A.: Using multidimensional projection to find relations. In Proceedings of the Twelfth International Conference on Machine Learning, pages 447–455, Tahoe City, California (July 1995) Morgan Kaufmann
11. Holland, J.H.: Adaptation in Natural and Artificial Systems. The University of Michigan Press, Ann Arbor, Michigan (1975)
12. Michalewicz, Z.: Genetic Algorithms + Data Structures = Evolution Programs. Springer-Verlag, Berlin, Heidelberg, New York (1999)
13. Bensusan, H. Kuscu I.: Constructive nduction using genetic programming. In Proceedings of the International Conference on Machine Learning, Workshop on Evolutionary Computing and Machine Learning, Bari, Italy (July, 1996) T. G. Fogarty and G. Venturini (Eds.)
14. Hu, Y.: A genetic programming approach to constructive induction. In Proceedings of the Third Annual Genetic Programming Conference, pages 146–157, Madison, Wisconsin, (July, 1998) Morgan Kauffman
15. Vafaie, H., De Jong, K.: Feature space transformation using genetic algorithms. IEEE Intelligent Systems and Their Applications, **13(2)**:57–65 (March–April, 1998)
16. Koza, J. R.: Genetic Programming: On the Programming of Computers by Means of Natural Selection. MIT Press, Cambridge, Massachusetts (1992)
17. Quinlan, R.J.: C4.5: Programs for Machine Learning. Morgan Kaufmann, San Mateo, California (1993)
18. Quinlan, R.J.: Induction of decision trees. Machine Learning, **1(1)**:81–106 (1986)
19. Levine, D.: Users guide to the PGAPack parallel genetic algorithm library. Technical Report ANL-95/18, Argonne National Laboratory (January 1996)

Active Feature Selection Based on a Very Limited Number of Entities

Yasusi Sinohara and Teruhisa Miura

Central Research Institute of Electric Power Industry,
2-1-11 Iwado-Kita, Komae-shi, 211 Tokyo, Japan
{sinohara, t-miura}@criepi.denken.or.jp

Abstract. In data analysis, the necessary data are not always prepared in a database in advance. If the precision of extracted classification knowledge is not sufficient, gathering additional data is sometimes necessary. Practically, if some critical attributes for the classification are missing from the database, it is very important to identify such missing attributes effectively in order to improve the precision. In this paper, we propose a new method to identify the attributes that will improve the precision of Support Vector Classifiers (SVC) based solely on values of candidate attributes of a very limited number of entities. In experiments, we show the incremental addition of attributes by the proposed method effectively improves the precision of SVC using only a very small number of entities.

1 Introduction

In data analyses such as classifications, all the necessary data is not always prepared in advance. Additional data is sometimes necessary for improving the accuracy of extracted classification knowledge. Notably, if some critical attributes for the classification are missing from the database, it is very important to identify such missing attributes and to add their values to entities in the database in order to improve the classification.

The amount of added data even for one attribute would normally be huge because the number of entities in the database is usually very large and it would be expensive to gather such data. So, most efficient way is to collect values only for effective attributes.

In this paper, we propose a new method of selecting promising attributes effective for improving classification by Support Vector Machines (SVMs), which are powerful tools for data classification. The method selects a promising attribute based solely on the values of the candidate attributes of a very limited number of entities using this method.

There are several related areas. Incremental active learning research[5,7,15] selects unlabeled samples whose labels are useful for improving the precision of knowledge. But here, we would like to select effective attributes. As for the actual selection of attributes, there are many studies on feature selection [3,11, 13,14]. However, the usual feature selection methods are very expensive in this

M.R. Berthold et al. (Eds.): IDA 2003, LNCS 2810, pp. 611–622, 2003.

type of application, because they are only applicable after all the values of all the candidate attributes for all the entities have been prepared. Our goal is to select effective attributes without gathering all such candidate data and with only a small amount of newly collected data.

2 Basic Approach

In our proposed method, a preliminary detailed survey of a limited number of samples will be first conducted and after selection of a promising attribute for improvement of classification, a data collection for all entities in the database (a full-scale survey) will be then conducted.

In this two-phase approach, the problems that should be solved are how to select a small number of samples used in the preliminary survey and how to select the most promising attribute based on the preliminary survey.

We used SVM error estimators as the guide for these selections.

3 Support Vector Machines

SVMs[2] have been successfully applied to classification problems in text mining and other data mining domains [6,10,15]. We use SVMs as our base classification tools. In this section, we briefly describe SVMs and the estimators of classification error rates of SVMs for unknown data (generalization errors).

Let $\{(x_i, y_i)\}_{1 \leq i \leq n}$ be a set of training examples, $x_i \in R^m$, which belongs to a class labeled by $y_i \in \{-1, 1\}$. In SVMs, a datum x is mapped into a (high dimensional) feature space by the function ϕ. The inner product in the feature space $K(x,z) = \phi(x)^\top \phi(z)$ is called the kernel function ($x^\top$ is a transposition of x).

There are two major types of commonly used kernels $K(x,z)$, "*dot-product-based kernels*" and "*distance-based kernels*". Dot-product-based kernels have a form of $K(x,z) = k(x^\top z)$ such as polynomial kernels, and distance based kernels have a form of $K(x,z) = k(||x - z||^2)$ such as RBF kernels.

If the mapped examples $\{\phi(x_i)\}$ are linearly separable, a hard-margin SVM is a hyperplane $w^\top \phi(x) + b$ that separates the set of positive examples from the set of negative examples with maximum inter-class distance, that is, the margin. $f(x) = w^\top \phi(x) + b$ is called the decision function and the sign of $f(x)$ is the predicted class y of an input x.

In the non-separable case, the violation of the separability condition is penalized usually by the L2 norm or the L1 norm. L2-SVMs and L1-SVMs are the solutions of the following optimization problems, respectively.

$$\min_{w,b,\xi} \frac{1}{2} w^\top w + \frac{C}{2} \sum_{i=1}^{l} \xi_i^2, \tag{1}$$

$$\min_{w,b,\xi} \frac{1}{2} w^\top w + C \sum_{i=1}^{l} \xi_i, \tag{2}$$

both subject to $y_i(w^\top \phi(x_i) + b) \geq 1 - \xi_i$ and $\xi_i \geq 0$. $||w||^{-1}$ is the margin.

The duals of (1) and (2) are (3) and (4), respectively.

$$\max_{\alpha} \; e^\top \alpha - \frac{1}{2}\alpha^\top (K^Y + \frac{I}{C})\alpha$$
$$\text{subject to } \; 0 \le \alpha_i, i = 1, \cdots, n, \text{and} y^\top \alpha = 0. \tag{3}$$

$$\max_{\alpha} \; e^\top \alpha - \frac{1}{2}\alpha^\top K^Y \alpha$$
$$\text{subject to } \; 0 \le \alpha_i \le C, i = 1, \cdots, n \text{and } y^\top \alpha = 0. \tag{4}$$

where the component $(K^Y)_{i,j} = y_i K(x_i, x_j) y_j$ and e denotes a vector of ones. An example x_i whose $\alpha_i > 0$ is called a support vector. For L1-SVMs, if $\alpha_i = C$, x_i is called a bounded support vector and if $0 < \alpha_i < C$, x_i is called an inbound support vector. We represent a set of support vectors by SV and a set of inbound vectors IV and a set of bounded vectors BV. We also use the following notations: $X_A = (x_i)_{i \in A}$ for a vector $X = (x_i)_{i=1,\cdots,n}$ and $X_{A,B} = (x_{i,j})_{i \in A, j \in B}$ for a matrix $X = (x_{i,j})$.

The decision function can be expressed using α_i and b as

$$f(x) = \sum_{i=1}^{l} \alpha_i y_i K(x_i, x) + b. \tag{5}$$

The following equality holds for L2-SVMs:

$$\underline{\alpha} \equiv \begin{pmatrix} \alpha \\ b \end{pmatrix} = H_{SV}^{-1} \begin{pmatrix} e_{SV} \\ 0 \end{pmatrix}, H_{SV} = \begin{pmatrix} (K_{SV}^Y + \frac{I}{C}) \; Y_{SV} \\ Y_{SV}^\top \quad\quad 0 \end{pmatrix}, \tag{6}$$

For L1-SVMs:

$$\underline{\gamma} \equiv \begin{pmatrix} \alpha_{IV} \\ \xi_{BV} \\ b \end{pmatrix} = L^{-1} \left(\begin{pmatrix} e_{SV} \\ 0 \end{pmatrix} - \begin{pmatrix} K_{SV,BV}^Y \\ Y_{BV}^\top \end{pmatrix} \alpha_{BV} \right),$$
$$L = \begin{pmatrix} K_{IV,IV}^Y \; 0 \; Y_{IV} \\ K_{BV,IV}^Y \; I \; Y_{BV} \\ Y_{IV}^\top \quad 0 \; 0 \end{pmatrix}. \tag{7}$$

Note that $\alpha_{BV} = C \, e_{BV}$.

Several estimators of generalization errors of SVMs have been proposed. Major estimators are listed in Table 1.

The leave-one-out (LOO) estimator is an almost unbiased estimator of generalization errors but is computationally intensive and time demanding. Other estimators, such as radius margins and span bounds, are analytical approximations of LOO. The empirical estimator introduced in [1] is an approximation of a holdout. It is an estimate of error probability of a holdout set based on a maximum likelihood estimate of probability of SVM output $f(x)$, $\frac{1}{1+\exp(Af(x)+B)}$.

Table 1. Estimators of generalization errors of an SVM classifier

Estimator	Form	SVM type
LOO (leave-one-out)	$E = \frac{1}{l}\sum_{p=1}^{l} \Phi(-y_p f^p(x_p))$	L1,L2
RM (radius margin)	$E = \frac{1}{l}\frac{R^2}{\gamma^2}$	L2,(L1[4])
Span Bound	$E = \frac{1}{l}\sum_{p=1}^{l} \Phi(\alpha_p^2 S_p^2 - 1)$	L2
Empirical Error[1]	$E = \frac{1}{l}\sum_{p=1}^{l}\left(\frac{1-y}{2}(1 - \frac{1}{\exp(Af+b)}) + \frac{1+y}{2}\frac{1}{\exp(Af+b)}\right)$	L2

where $\Phi(x)$ is a step function. That is, 1 if $x > 0$, and otherwise 0, and $f^p(x)$ is a decision function when an entity x_p is not used for learning.

Most analytical estimators are only applicable to hard-margin SVMs and L2-SVMs. The estimator applicable to popular L1-SVMs are LOO and modified radius margin (RM) error estimates, as introduced in [4].

The original form of the radius margin estimate is $4R^2||w||^2 (\geq LOO)$. R^2 is the square of the radius of the smallest hypersphere covering all the training examples in the feature space. R^2 is the object value of the following problem.

$$\max_{\beta} \ \beta^\top \mathrm{diag}(K) - \beta^\top K \beta \tag{8}$$
$$\text{subject to } \ 0 \leq \beta_i, i = 1, \cdots, n, \mathrm{and} e^\top \beta = 1.$$

If we refer to an entity x_i whose $\beta_i > 0$ as a radius vector in this paper and a set of radius vectors as RV, the following equality holds:

$$\underline{\beta} \equiv \begin{pmatrix} \beta \\ c \end{pmatrix} = H_{RV}^{-1} \begin{pmatrix} \mathrm{diag}(K_{RV}) \\ 1 \end{pmatrix}, H_{RV} = \begin{pmatrix} 2K_{RV} & e_{RV} \\ e_{RV}^\top & 0 \end{pmatrix}, \tag{9}$$

where diag(X) is a vector of diagonals of matrix X, and c is a Lagrangian multiplier of the constraint $e^\top \beta = 1$.

LOO and span-bounds are discrete and others are continuous with respect to C and kernel parameters (e.g., σ^2 of RBF), but are usually not differentiable at the points where a set of support vectors SV or a set of inbound support vectors IV or a set of bounded support vectors BV changes. The original radius margin bound is always differentiable to C and kernel parameters for hard-margin and L2- SVMs but not for L1-SVMs. Chung, et. al[4] propose a modified radius margin for L1-SVM using RBF kernels

$$E = \left(R^2 + \frac{1}{C}\right)(||w||^2 + 2Ce^\top \xi)$$

which is always differentiable to C and a scale parameter σ^2 of RBF.

4 Formulation and Algorithm

In this section, we propose a method to select a limited number of entities for a preliminary survey and to select the most promising attribute to improve the precision of an SVM classifier if that attribute is added to the database. Here, we discuss the case of selecting one attribute based on one preliminary survey.

We first show that the reduction of generalization errors E of SVMs by addition of a new attribute can be estimated by a quadratic form involving the attribute values of the support and radius vectors alone.

4.1 Error Reduction Matrix

We treat an introduction of a new attribute a in a database as a change of value of a kernel parameter θ_a in a following kernel K_{θ_a} from zero to some positive number. For a dot-product-based kernel, we assume $K_{\theta_a}(x, z) = k(x^\top z + \theta_a x^a z^a)$. For a distance based kernel, we assume $K_{\theta_a}(x, z) = k(||x-z||^2 + \theta_a(x^a - z^a)^2)$. x^a and y^a is the value of attribute a of entity x and y.

If the change of θ_a is small, the reduction of generalization errors by introducing a new attribute a can be approximated by $-\frac{\partial E}{\partial \theta_a}\theta_a$.

For L2-SVMs, the error estimates shown in Table 1 are functions of support vectors and radius vectors, that is, $\underline{\alpha}$ and $\underline{\beta}$. They are functions of kernel parameter θ_a. Therefore the gradient of error E with respect to θ_a (if it exists) is

$$\frac{\partial E}{\partial \theta_a} = \frac{\partial E}{\partial \underline{\alpha}}\frac{\partial \underline{\alpha}}{\partial \theta_a} + \frac{\partial E}{\partial \underline{\beta}}\frac{\partial \underline{\beta}}{\partial \theta_a}. \tag{10}$$

From (6) and (9), the following equations hold if support vectors and radius vectors do not change by a small change of kernel parameter θ_a.

$$\frac{\partial \underline{\alpha}}{\partial \theta_a} = -H_{SV}^{-1}\frac{\partial H_{SV}}{\partial \theta_a}\underline{\alpha}, \tag{11}$$

$$\frac{\partial \underline{\beta}}{\partial \theta_a} = -H_{RV}^{-1}\frac{\partial H_{RV}}{\partial \theta_a}\underline{\beta}. \tag{12}$$

Therefore, the gradient of the error becomes as follows:

$$\begin{aligned}\frac{\partial E}{\partial \theta_a} &= -\frac{\partial E}{\partial \underline{\alpha}}H_{SV}^{-1}\frac{\partial H_{SV}}{\partial \theta_a}\underline{\alpha}^\top - \frac{\partial E}{\partial \underline{\beta}}H_{RV}^{-1}\frac{\partial H_{RV}}{\partial \theta_a}\underline{\beta}^\top \\ &= -\left(F\mathrm{diag}^{-1}(Y_{SV})\frac{\partial K_{SV}}{\partial \theta_a}\mathrm{diag}^{-1}(Y_{SV})\alpha + F_{RV}\frac{\partial K_{RV}}{\partial \theta_a}\beta\right),\end{aligned} \tag{13}$$

where F is the vector of the first k elements of $(1 \times (k+1))$-matrix $\frac{\partial E}{\partial \underline{\alpha}}H_{SV}^{-1}$ and F_{RV} is the vector of the first l elements of $(1 \times (l+1))$-matrix $\frac{\partial E}{\partial \underline{\beta}}H_{RV}^{-1}$. $\mathrm{diag}^{-1}(x)$ represents a diagonal matrix from a vector x and Y_{SV} is a sub-vector of $Y = (y_i)$ selecting the columns corresponding to the support vectors.

For L1-SVMs, error estimates are functions of $\underline{\gamma}$ and $\underline{\beta}$. The gradient of error E w.r.t θ_a (if it exists) is

$$\frac{\partial E}{\partial \theta_a} = \frac{\partial E}{\partial \underline{\gamma}}\frac{\partial \underline{\gamma}}{\partial \theta_a} + \frac{\partial E}{\partial \underline{\beta}}\frac{\partial \underline{\beta}}{\partial \theta_a}. \tag{14}$$

From (7), the following equation holds if inbound and bounded support vectors do not change by a small change of kernel parameter θ_a,

$$\frac{\partial \underline{\gamma}}{\partial \theta_a} = -L^{-1}\frac{\partial H_{SV}}{\partial \theta_a}\underline{\alpha}. \tag{15}$$

So, (13) also holds for L1-SVMs by setting F to the vector of the first k elements of $(1 \times (k+1))$-matrix $\frac{\partial E}{\partial \underline{\gamma}} L^{-1}$.

We define $\kappa(x,z) \equiv k'(x^\top z)$ for a dot-product-based kernel $K(x,z) = k(x^\top z)$ and $\kappa(x,z) \equiv k'(||x-z||^2)$ for a distance-based kernel $K(x,z) = k(||x-z||^2)$. $k'(x)$ is the first derivative of $k(x)$.

Then, for a dot-product-based kernel,

$$\frac{\partial K_{\theta_a}}{\partial \theta_a}(x,z) = x^a z^a \kappa(x,z).$$

For a distance-based kernel,

$$\frac{\partial K_{\theta_a}}{\partial \theta_a}(x,z) = (x^a - z^a)^2 \kappa(x,z).$$

Because the error reduction is a linear combination of these quadratic forms as shown in (13), it also has a quadratic form

Then we get our main theorems: Theorem 1 and Theorem 2.

In the theorems, $\kappa^Y(x_i, x_i) \equiv y_i y_j \kappa(x_i, x_j)$ and X^a represents the vector of values of the added attribute a of entities (x_i^a).

Theorem 1. *(Error Reduction for dot-product based kernel) For a dot-product based kernel, error reduction $-\frac{dE}{d\theta_a}\theta_a$ can be approximated by a quadratic form.*

$$-\frac{\partial E}{\partial \theta_a} = \left((X_{SV}^a)^\top, (X_{RV}^a)^\top\right) G \begin{pmatrix} X_{SV}^a \\ X_{RV}^a \end{pmatrix}, \tag{16}$$

$$G = \begin{pmatrix} \text{diag}^{-1}(F)\kappa_{SV}^Y \text{diag}^{-1}(\alpha) & 0 \\ 0 & \text{diag}^{-1}(F_{RV})\kappa_{RV}\ \text{diag}^{-1}(\beta) \end{pmatrix}. \tag{17}$$

Theorem 2. *(Error Reduction for distance-based kernel) For a distance-based kernel, the error reduction $-\frac{dE}{d\theta_a}\theta_a$ has a quadratic form.*

$$-\frac{\partial E}{\partial \theta_a} = \left((X_{SV}^a)^\top, (X_{RV}^a)^\top\right)\ G\ \begin{pmatrix} X_{SV}^a \\ X_{RV}^a \end{pmatrix}, \tag{18}$$

$$G = \text{diag}^{-1}\left(e^\top (G_0^\top + G_0)\right) - (G_0^\top + G_0) \tag{19}$$

$$G_0 = \begin{pmatrix} \text{diag}^{-1}(F)\kappa_{SV}^Y \text{diag}^{-1}(\alpha) & 0 \\ 0 & \text{diag}^{-1}(F_{RV})\kappa_{RV}\text{diag}^{-1}(\beta) \end{pmatrix}. \tag{20}$$

Note that $e^\top G = 0$ in Theorem 2. This corresponds to the degree of freedom that a shift of X^a does not change $X_i^a - X_j^a$.

We refer to G as an error reduction matrix.

The error reduction by the introduction of an attribute can be estimated by a quadratic form using G and support vectors and radius vectors, for both dot-product-based kernels and distance-based kernels, and for various kinds of error estimate E.

Algorithm 1.1 SampleSelection and ErrorReductionMatrix (α, SV, β, RV)
1 $V_s := SV \cup RV$
2 $G :=$ ErrorReductionMatrix$(\alpha, \beta, X_{V_s, A_D}, Y_{V_s})$
3 [V, d] := EigenValueDecomposition(G, m)
4 [U,B] := BiQuadraticMaximization(V)
5 $\tilde{U} :=$ SparsifyMatrix(U)
6 $E_s := \{\text{entities } i | \tilde{U_{i,k}} \neq 0 \text{ for some } k\}$ and $\hat{G} := \hat{U}\, B\, \hat{U}^\top$
7 return $(E_s, \hat{G})$

Algorithm 1.2 Bi-quadraticMaximization (V)
1 initialize R such that $R^\top R = I$
2 $R_{old} \leftarrow 0$
3 **while** $1 - \min(|\text{diag}(R^\top R_{old})| > \epsilon$ **do**
4 $\quad R_{old} \leftarrow R$
5 $\quad U \leftarrow VR^\top$
6 $\quad RL \leftarrow U^\top .^3 V$
7 $\quad R \leftarrow RL(RL^\top RL)^{-1/2}$
8 **end while**
9 $B \leftarrow RDR^\top$
10 return(U, B)

4.2 Sparse Approximation of Error Reduction Matrix

In the previous section, we showed that we can estimate the error reduction by introducing an attribute only based on matrix G and the attribute values of support vectors SV and radius vectors RV. Though the size of $SV \cup RV$ is usually much smaller than the number of entities in the database n, it is sometimes still too large to conduct a preliminary survey.

So, we approximate the error reduction matrix G by a sparser matrix $\hat{G}$ and approximate $(X^a)^\top G(X^a)$ by $(X^a_{E_S})^\top \hat{G}(X^a_{E_S})$ using Algorithm 1.1. E_S is a set of entities corresponding to the rows of $\hat{G}$ with non-zero values, which can be much smaller than the size of $SV \cup RV$.

The simplest way to make a matrix sparse is to truncate elements smaller than a given threshold to zero. However, this operation sometimes significantly changes the eigenvectors associated with large magnitude eigenvalues and the value of the quadratic form. We avoid this by making the following approximations.

We assume all the attributes are normalized to be $(X^a)^\top(X^a) = 1$. If we use distance-based kernels such as RBF, we further assume that the mean of each attribute is 0 ($e^\top(X^a) = 0$), corresponding to the degeneration $e^\top G = 0$.

Because the eigenvector v of the most positive eigenvalue of G maximizes $-\frac{dE}{d\theta_a} = (X^a)^\top G(X^a)$ under the constraint $(X^a)^\top X^a = 1$, the attribute whose values of entities are proportional to v reduces the error estimate E most (or, maximizes $-\frac{dE}{d\theta_a}$). On the other hand, the attribute proportional to the eigenvector of the most negative eigenvalues increase the estimated test error E most. The directions of the eigenvectors corresponding to large magnitude eigenvalues have a major influence on the value of the quadratic form. So, we approximate G with the m largest magnitude eigenvalues d_i and the eigenvectors v_i and neglect other directions.

$$G \approx \tilde{G} := \sum_{i=1}^{m} v_i d_i v_i^\top = VDV^\top,$$
$$D = \mathrm{diag}^{-1}(d_i), V = (v_1, \cdots, v_m), V^\top V = I_{mm}.$$

We further approximate matrix $\tilde{G}$ by a sparser matrix $\hat{G}$. To get a sparser matrix, we try to find a decomposition of $\tilde{G} = UBU^\top$ such that $U^\top U = I$ where only a few elements u_{ik} of U are almost equivalent to 1 or -1 and the rest are almost equivalent to 0. When $\sum_i z_i^2 = 1$, $\sum_i z_i^4$ is maximized by z_i such that only one element is 1 or -1 and all the rest are 0. Based on this observation, we solve the following problem to find U having many small elements.

$$\begin{aligned} &\max_R \sum_{i,k} u_{ik}^4 \\ &\text{subject to } U = VR^\top, R^\top R = I_{mm} \end{aligned} \qquad (21)$$

This can be solved by Algorithm 1.2.

Finally, we approximate matrix U by a matrix $\tilde{U}$ and setup the elements with small absolute values to 0. The resulting algorithm is Algorithm 1.1.

Because U has many smaller elements, $\tilde{U}$ becomes sparse and $\hat{G} = \tilde{U}B\tilde{U}^\top$ becomes a sparse approximation of G, generally preserving the large magnitude eigenvectors. The set of entities corresponding to the rows of $\hat{G}$ having non-zero elements E_S can be a very small set. We can use $(X^a_{E_S})^\top \hat{G}(X^a_{E_S})$ as a rough estimate of the reduction of the estimated test errors.

Therefore, in our methods, we conduct a preliminary survey on the values of candidate attributes of entities E_S and select the attribute $\hat{a}$ that has the largest $(X^a_{E_S})^\top \hat{G}(X^a_{E_S})$ as the most promising attribute to improve the precision of the classifier if that attribute was to be added to the database.

Our method can identify the most promising attribute by the attribute values of only a very limited number of entities E_S.

5 Preliminary Experiment and Discussion

In order to evaluate our methods, we applied it to the problem of selecting image pixels important for character recognition. We used the USPS database (ftp://ftp.mpik-tueb.mpg.de/pub/bs/data/) that is a well-known benchmark database of hand written digits. It consists of 7291 training objects and 2007 test objects. The characters are isolated and represented by a 16×16 pixel sized grayscale image. In our experiments, we use each pixel as an attribute, yielding a 256-dimensional input vector.

The task was to select a set of pixels effective to recognize the digit "5." The initial pixels were the best 5 discriminative pixels measured by mutual information criteria. The initial mask pattern is shown in the upper-left in Figure 1. We incrementally applied our attribute selection methods 20 times and finally 25 pixels were selected. The final mask pattern is shown in the lower-left in Figure 1. In each addition of an attribute, the number of selected entities $|E_S|$ was limited to 16 (and the 6 largest eigenvectors were used.). The image data was normalized to zero-means and unit-variance. We use the libSVM package (http://www.csie.ntu.edu.tw/~cjlin/libsvm/) to solve L1-SVM (C=1) with an RBF kernel of the form of $\exp(-||x - x'||^2/(|A_D|c))$. c was adaptively chosen by cross-validation using the training set in each learning step. We used a modified radius margin RM as an error estimate of a L1-SVM.

The dotted line and the solid line in Figure 3 show the reduction of training error rates and test error rates respectively by incremental addition of attributes (pixels) to the mask. The final test error using only selected 25 attributes (pixels) is roughly 3.0%. Because the human error rate of USPS recognition is about 2.5% on the test data and roughly 4.0–5.0% by SVMs using an RBF kernels of the form $\exp(-||x - x'||^2/(256c))$ using the data scaled so that all attributes have an equal range of values [12], the error rate 3.0% is relatively high and the number of entities used in the preliminary surveys is only 320 at most, which is only 4.4% of the training data.

Thus we can conclude that the proposed methods effectively select good attributes based on only a very small number of entities.

When checking the selected images one-by-one, we often found that the set of images as selected in E_S have similar masked images, but some are digit "5" and others are not. Figure 2 is a typical example. The selected attributes based on these are attributes that can discriminate between such images, that is, a pixel whose values are high/low for the images of digit "5" but not for others. It is clear that the addition of such attributes will improve the discrimination of the entities.

The formulation (21) is equivalent to kurtosis maximization, which is used in Independent Component Analysis (ICA) for blind signal separation and Algorithm 1.2 is a special form of the fixed-point ICA algorithm.[9]. Kurtosis maximization is also used in Explanatory Projection Pursuit (EPP) for visualization of clustering structure of the data[8]. In EPP, an independent component (u_i) points to a direction which a cluster (or clusters) of similar data lies. In our formulation, if two data points x_a and x_b are similar in values of x but have opposite class labels y, then they will be placed at opposite positions $(+u_{a,i}, -u_{a,i})$ in some independent component i. So, the proposed algorithms has a tendency to select entities similar in values of known attributes but in different classes.

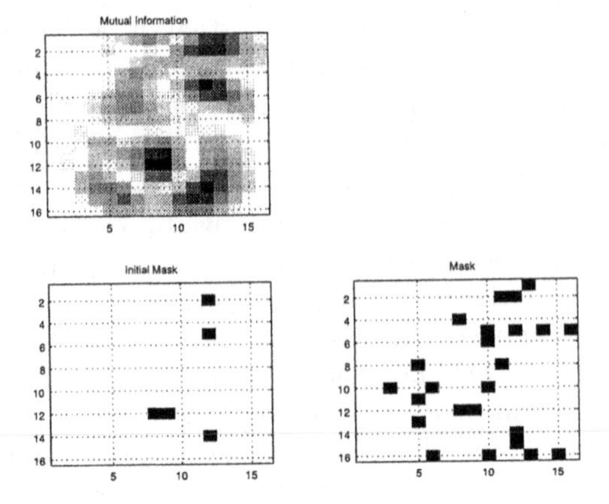

Fig. 1. Mask Patterns

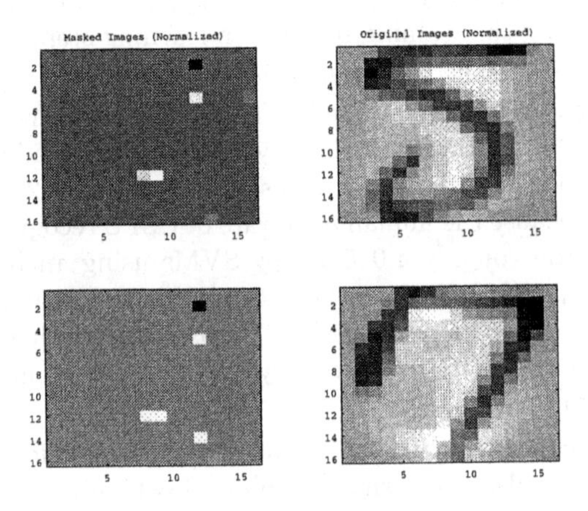

Fig. 2. Selected Images

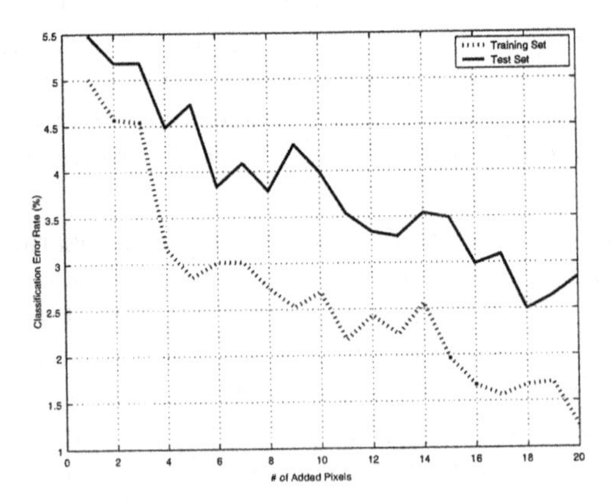

Fig. 3. Error Rates

6 Conclusion

In this paper, we proposed a new method to select attributes to improve the precision of an SVM classifier based on a very limited number of entities. This is based on a sparse approximation of test error estimates of SVM classifiers. The experiment shows that this method can reduce the number of test errors quickly by checking the attribute values of only a small number of entities. We think it is because the proposed method combines the ability of identifying the critical data (support and radius vectors) with SVMs and the ability of selecting representative data by clustering similar data (including data with opposite labels) with ICA in order to identify the entities which is informative for the improvement of classification.

We are now developing a method combining active learning methods for the selection of entities combined with the proposed active feature selection algorithm.

References

1. N. Ayat, M. Cheriet, and C. Y. Suen. Empirical error based on optimization of SVM kernels. In *Proc. of the 8 IWFHR*, 2002.
2. C. Burges. A tutorial on support vector machines for pattern recognition. *Data Mining and Knowledge Discovery*, 2(2):121–167, 1998.
3. S. Chakrabarti, B. Dom, R. Agrawal, and P. Raghavan. Scalable feature selection, classification and signature generation for organizing large text databases into hierarchical topic taxonomies. *VLDB Journal*, 7(3):163–178, 1998.
4. K.-M. Chung et al. Radius margin bounds for support vector machines with the RBF kernel. In *Proc. of ICONIP 2002*, 2002.
5. A. S. Colin Campbell, Nello Cristianini. Query learning with large margin classifiers. In *Proc. 17th International Conf. on Machine Learning (ICML2000)*, pages 111–118, 2000.
6. T. S. Furey, N. Cristianini, N. Duffy, D. W. Bednarski, M. Schummer, and D. Haussler. Support vector machine classification and validation of cancer tissue samples using microarray expression data. *Bioinformatics*, 2001.
7. D. C. G. Harris Drucker, Behzad Shahrary. Relevance feedback using support vector machines. In *Proceeding of the 2001 International Conference on Machine Learning*, 2001.
8. A. Hyvärinen. Beyond independent components. In *Proc. Int. Conf on Articial Neural Networks, Edinburgh, UK*, pages 809–814, 1999.
9. A. Hyvärinen. A fast fixed point algorithm for independent component analysis. *IEEE Transactions on Neural Networks*, 10(3):626–634, 1999.
10. T. Joachims. Text categorization with support vector machines: Learning with many relevant features. In *Proceedings of ECML-98, 10th European Conference on Machine Learning*, 1998.
11. J.Weston, S. Mukherjee, O. Chapelle, M. Pontil, T. Poggio, and V. Vapnik. Feature selection for SVMs. In *Advances in Neural Information Processing Systems*, volume 13, pages 668–674. MIT Press, 2000.

12. Y. LeCun and et. al. Learning algorithms for classification: A comparison on handwritten digit recognition. In J. H. Oh, C. Kwon, and S. Cho, editors, *Neural Networks: The Statistical Mechanics Perspective*, pages 261–276. World Scientific, 1995.
13. H. Liu and H. Motoda, editors. *Feature Extraction, Construction and Selection: A Data Mining Perspective.* Kluwer Academic Publishers, 1998.
14. H. L. M. Dash 1. Feature selection for classification. *Intelligent Data Analysis*, 1:131–156, 1997.
15. D. K. Simon Tong. Support vector machine active learning with applications to text classification. In *Proceedings of ICML-00, 17th International Conference on Machine Learning*, 2000.

Author Index

Lecture Notes in Computer Science

For information about Vols. 1–2698
please contact your bookseller or Springer-Verlag

Vol. 2677: R. de Lemos, C. Gacek, A. Romanovsky (Eds.), Architecting Dependable Systems. XII, 309 pages. 2003.

Vol. 2699: M.G. Hinchey, J.L. Rash, W.F. Truszkowski, C. Rouff, D. Gordon-Spears (Eds.), Formal Approaches to Agent-Based Systems. Proceedings, 2002. IX, 297 pages. 2003. (Subseries LNAI).

Vol. 2700: M.T. Pazienza (Ed.), Information Extraction in the Web Era. XIII, 163 pages. 2003. (Subseries LNAI).

Vol. 2701: M. Hofmann (Ed.), Typed Lambda Calculi and Applications. Proceedings, 2003. VIII, 317 pages. 2003.

Vol. 2702: P. Brusilovsky, A. Corbett, F. de Rosis (Eds.), User Modeling 2003. Proceedings, 2003. XIV, 436 pages. 2003. (Subseries LNAI).

Vol. 2704: S.-T. Huang, T. Herman (Eds.), Self-Stabilizing Systems. Proceedings, 2003. X, 215 pages. 2003.

Vol. 2705: S. Renals, G. Grefenstette (Eds.), Text- and Speech-Triggered Information Access. Proceedings, 2000. VII, 197 pages. 2003. (Subseries LNAI).

Vol. 2706: R. Nieuwenhuis (Ed.), Rewriting Techniques and Applications. Proceedings, 2003. XI, 515 pages. 2003.

Vol. 2707: K. Jeffay, I. Stoica, K. Wehrle (Eds.), Quality of Service – IWQoS 2003. Proceedings, 2003. XI, 517 pages. 2003.

Vol. 2708: R. Reed, J. Reed (Eds.), SDL 2003: System Design. Proceedings, 2003. XI, 405 pages. 2003.

Vol. 2709: T. Windeatt, F. Roli (Eds.), Multiple Classifier Systems. Proceedings, 2003. X, 406 pages. 2003.

Vol. 2710: Z. Ésik, Z, Fülöp (Eds.), Developments in Language Theory. Proceedings, 2003. XI, 437 pages. 2003.

Vol. 2711: T.D. Nielsen, N.L. Zhang (Eds.), Symbolic and Quantitative Approaches to Reasoning with Uncertainty. Proceedings, 2003. XII, 608 pages. 2003. (Subseries LNAI).

Vol. 2712: A. James, B. Lings, M. Younas (Eds.), New Horizons in Information Management. Proceedings, 2003. XII, 281 pages. 2003.

Vol. 2713: C.-W. Chung, C.-K. Kim, W. Kim, T.-W. Ling, K.-H. Song (Eds.), Web and Communication Technologies and Internet-Related Social Issues – HSI 2003. Proceedings, 2003. XXII, 773 pages. 2003.

Vol. 2714: O. Kaynak, E. Alpaydin, E. Oja, L. Xu (Eds.), Artificial Neural Networks and Neural Information Processing – ICANN/ICONIP 2003. Proceedings, 2003. XXII, 1188 pages. 2003.

Vol. 2715: T. Bilgiç, B. De Baets, O. Kaynak (Eds.), Fuzzy Sets and Systems – IFSA 2003. Proceedings, 2003. XV, 735 pages. 2003. (Subseries LNAI).

Vol. 2716: M.J. Voss (Ed.), OpenMP Shared Memory Parallel Programming. Proceedings, 2003. VIII, 271 pages. 2003.

Vol. 2718: P. W. H. Chung, C. Hinde, M. Ali (Eds.), Developments in Applied Artificial Intelligence. Proceedings, 2003. XIV, 817 pages. 2003. (Subseries LNAI).

Vol. 2719: J.C.M. Baeten, J.K. Lenstra, J. Parrow, G.J. Woeginger (Eds.), Automata, Languages and Programming. Proceedings, 2003. XVIII, 1199 pages. 2003.

Vol. 2720: M. Marques Freire, P. Lorenz, M.M.-O. Lee (Eds.), High-Speed Networks and Multimedia Communications. Proceedings, 2003. XIII, 582 pages. 2003.

Vol. 2721: N.J. Mamede, J. Baptista, I. Trancoso, M. das Graças Volpe Nunes (Eds.), Computational Processing of the Portuguese Language. Proceedings, 2003. XIV, 268 pages. 2003. (Subseries LNAI).

Vol. 2722: J.M. Cueva Lovelle, B.M. González Rodríguez, L. Joyanes Aguilar, J.E. Labra Gayo, M. del Puerto Paule Ruiz (Eds.), Web Engineering. Proceedings, 2003. XIX, 554 pages. 2003.

Vol. 2723: E. Cantú-Paz, J.A. Foster, K. Deb, L.D. Davis, R. Roy, U.-M. O'Reilly, H.-G. Beyer, R. Standish, G. Kendall, S. Wilson, M. Harman, J. Wegener, D. Dasgupta, M.A. Potter, A.C. Schultz, K.A. Dowsland, N. Jonoska, J. Miller (Eds.), Genetic and Evolutionary Computation – GECCO 2003. Proceedings, Part I. 2003. XLVII, 1252 pages. 2003.

Vol. 2724: E. Cantú-Paz, J.A. Foster, K. Deb, L.D. Davis, R. Roy, U.-M. O'Reilly, H.-G. Beyer, R. Standish, G. Kendall, S. Wilson, M. Harman, J. Wegener, D. Dasgupta, M.A. Potter, A.C. Schultz, K.A. Dowsland, N. Jonoska, J. Miller (Eds.), Genetic and Evolutionary Computation – GECCO 2003. Proceedings, Part II. 2003. XLVII, 1274 pages. 2003.

Vol. 2725: W.A. Hunt, Jr., F. Somenzi (Eds.), Computer Aided Verification. Proceedings, 2003. XII, 462 pages. 2003.

Vol. 2726: E. Hancock, M. Vento (Eds.), Graph Based Representations in Pattern Recognition. Proceedings, 2003. VIII, 271 pages. 2003.

Vol. 2727: R. Safavi-Naini, J. Seberry (Eds.), Information Security and Privacy. Proceedings, 2003. XII, 534 pages. 2003.

Vol. 2728: E.M. Bakker, T.S. Huang, M.S. Lew, N. Sebe, X.S. Zhou (Eds.), Image and Video Retrieval. Proceedings, 2003. XIII, 512 pages. 2003.

Vol. 2729: D. Boneh (Ed.), Advances in Cryptology – CRYPTO 2003. Proceedings, 2003. XII, 631 pages. 2003.

Vol. 2730: F. Bai, B. Wegner (Eds.), Electronic Information and Communication in Mathematics. Proceedings, 2002. X, 189 pages. 2003.

Vol. 2731: C.S. Calude, M.J. Dinneen, V. Vajnovszki (Eds.), Discrete Mathematics and Theoretical Computer Science. Proceedings, 2003. VIII, 301 pages. 2003.

Vol. 2732: C. Taylor, J.A. Noble (Eds.), Information Processing in Medical Imaging. Proceedings, 2003. XVI, 698 pages. 2003.

Vol. 2733: A. Butz, A. Krüger, P. Olivier (Eds.), Smart Graphics. Proceedings, 2003. XI, 261 pages. 2003.

Vol. 2734: P. Perner, A. Rosenfeld (Eds.), Machine Learning and Data Mining in Pattern Recognition. Proceedings, 2003. XII, 440 pages. 2003. (Subseries LNAI).

Vol. 2735: F. Kaashoek, I. Stoica (Eds.), Peer-to-Peer Systems II. Proceedings, 2003. XI, 316 pages. 2003.

Vol. 2739: R. Traunmüller (Ed.), Electronic Government. Proceedings, 2003. XVIII, 511 pages. 2003.

Vol. 2740: E. Burke, P. De Causmaecker (Eds.), Practice and Theory of Automated Timetabling IV. Proceedings, 2002. XII, 361 pages. 2003.

Vol. 2741: F. Baader (Ed.), Automated Deduction – CADE-19. Proceedings, 2003. XII, 503 pages. 2003. (Subseries LNAI).

Vol. 2742: R. N. Wright (Ed.), Financial Cryptography. Proceedings, 2003. VIII, 321 pages. 2003.

Vol. 2743: L. Cardelli (Ed.), ECOOP 2003 – Object-Oriented Programming. Proceedings, 2003. X, 501 pages. 2003.

Vol. 2744: V. Mařík, D. McFarlane, P. Valckenaers (Eds.), Holonic and Multi-Agent Systems for Manufacturing. Proceedings, 2003. XI, 322 pages. 2003. (Subseries LNAI).

Vol. 2745: M. Guo, L.T. Yang (Eds.), Parallel and Distributed Processing and Applications. Proceedings, 2003. XII, 450 pages. 2003.

Vol. 2746: A. de Moor, W. Lex, B. Ganter (Eds.), Conceptual Structures for Knowledge Creation and Communication. Proceedings, 2003. XI, 405 pages. 2003. (Subseries LNAI).

Vol. 2747: B. Rovan, P. Vojtáš (Eds.), Mathematical Foundations of Computer Science 2003. Proceedings, 2003. XIII, 692 pages. 2003.

Vol. 2748: F. Dehne, J.-R. Sack, M. Smid (Eds.), Algorithms and Data Structures. Proceedings, 2003. XII, 522 pages. 2003.

Vol. 2749: J. Bigun, T. Gustavsson (Eds.), Image Analysis. Proceedings, 2003. XXII, 1174 pages. 2003.

Vol. 2750: T. Hadzilacos, Y. Manolopoulos, J.F. Roddick, Y. Theodoridis (Eds.), Advances in Spatial and Temporal Databases. Proceedings, 2003. XIII, 525 pages. 2003.

Vol. 2751: A. Lingas, B.J. Nilsson (Eds.), Fundamentals of Computation Theory. Proceedings, 2003. XII, 433 pages. 2003.

Vol. 2752: G.A. Kaminka, P.U. Lima, R. Rojas (Eds.), RoboCup 2002: Robot Soccer World Cup VI. XVI, 498 pages. 2003. (Subseries LNAI).

Vol. 2753: F. Maurer, D. Wells (Eds.), Extreme Programming and Agile Methods – XP/Agile Universe 2003. Proceedings, 2003. XI, 215 pages. 2003.

Vol. 2754: M. Schumacher, Security Engineering with Patterns. XIV, 208 pages. 2003.

Vol. 2756: N. Petkov, M.A. Westenberg (Eds.), Computer Analysis of Images and Patterns. Proceedings, 2003. XVIII, 781 pages. 2003.

Vol. 2758: D. Basin, B. Wolff (Eds.), Theorem Proving in Higher Order Logics. Proceedings, 2003. X, 367 pages. 2003.

Vol. 2759: O.H. Ibarra, Z. Dang (Eds.), Implementation and Application of Automata. Proceedings, 2003. XI, 312 pages. 2003.

Vol. 2761: R. Amadio, D. Lugiez (Eds.), CONCUR 2003 - Concurrency Theory. Proceedings, 2003. XI, 524 pages. 2003.

Vol. 2762: G. Dong, C. Tang, W. Wang (Eds.), Advances in Web-Age Information Management. Proceedings, 2003. XIII, 512 pages. 2003.

Vol. 2763: V. Malyshkin (Ed.), Parallel Computing Technologies. Proceedings, 2003. XIII, 570 pages. 2003.

Vol. 2764: S. Arora, K. Jansen, J.D.P. Rolim, A. Sahai (Eds.), Approximation, Randomization, and Combinatorial Optimization. Proceedings, 2003. IX, 409 pages. 2003.

Vol. 2765: R. Conradi, A.I. Wang (Eds.), Empirical Methods and Studies in Software Engineering. VIII, 279 pages. 2003.

Vol. 2766: S. Behnke, Hierarchical Neural Networks for Image Interpretation. XII, 224 pages. 2003.

Vol. 2769: T. Koch, I. T. Sølvberg (Eds.), Research and Advanced Technology for Digital Libraries. Proceedings, 2003. XV, 536 pages. 2003.

Vol. 2776: V. Gorodetsky, L. Popyack, V. Skormin (Eds.), Computer Network Security. Proceedings, 2003. XIV, 470 pages. 2003.

Vol. 2777: B. Schölkopf, M.K. Warmuth (Eds.), Learning Theory and Kernel Machines. Proceedings, 2003. XIV, 746 pages. 2003. (Subseries LNAI).

Vol. 2779: C.D. Walter, Ç.K. Koç, C. Paar (Eds.), Cryptographic Hardware and Embedded Systems – CHES 2003. Proceedings, 2003. XIII, 441 pages. 2003.

Vol. 2782: M. Klusch, A. Omicini, S. Ossowski, H. Laamanen (Eds.), Cooperative Information Agents VII. Proceedings, 2003. XI, 345 pages. 2003. (Subseries LNAI).

Vol. 2783: W. Zhou, P. Nicholson, B. Corbitt, J. Fong (Eds.), Advances in Web-Based Learning – ICWL 2003. Proceedings, 2003. XV, 552 pages. 2003.

Vol. 2786: F. Oquendo (Ed.), Software Process Technology. Proceedings, 2003. X, 173 pages. 2003.

Vol. 2787: J. Timmis, P. Bentley, E. Hart (Eds.), Artificial Immune Systems. Proceedings, 2003. XI, 299 pages. 2003.

Vol. 2789: L. Böszörményi, P. Schojer (Eds.), Modular Programming Languages. Proceedings, 2003. XIII, 271 pages. 2003.

Vol. 2790: H. Kosch, L. Böszörményi, H. Hellwagner (Eds.), Euro-Par 2003 Parallel Processing. Proceedings, 2003. XXXV, 1320 pages. 2003.

Vol. 2794: P. Kemper, W. H. Sanders (Eds.), Computer Performance Evaluation. Proceedings, 2003. X, 309 pages. 2003.

Vol. 2796: M. Cialdea Mayer, F. Pirri (Eds.), Automated Reasoning with Analytic Tableaux and Related Methods. Proceedings, 2003. X, 271 pages. 2003. (Subseries LNAI).

Vol. 2803: M. Baaz, J.A. Makowsky (Eds.), Computer Science Logic. Proceedings, 2003. XII, 589 pages. 2003.

Vol. 2810: M.R. Berthold, H.-J. Lenz, E. Bradley, R. Kruse, C. Borgelt (Eds.), Advances in Intelligent Data Analysis V. Proceedings, 2003. XV, 624 pages. 2003.

Vol. 2817: D. Konstantas, M. Leonard, Y. Pigneur, S. Patel (Eds.), Object-Oriented Information Systems. Proceedings, 2003. XII, 426 pages. 2003.